KAON PHYSICS

Kaon Physics

Edited by
Jonathan L. Rosner and
Bruce D. Winstein

The University of Chicago Press / Chicago and London

Jonathan L. Rosner is professor in the Department of Physics and the Enrico Fermi Institute at the University of Chicago.
Bruce D. Winstein is the Samuel K. Allison Distinguished Service Professor in the Department of Physics and the Enrico Fermi Institute at the University of Chicago.

The University of Chicago Press, Chicago 60637
The University of Chicago Press, Ltd., London

Printed in the United States of America
10 09 08 07 06 05 04 03 02 01 1 2 3 4 5

ISBN: 0-226-90228-5

Library of Congress Cataloging-in-Publication Data

Kaon Physics / edited by Jonathan L. Rosner and Bruce D. Winstein.
p. cm.
Includes bibliographical references.
ISBN 0-226-90228-5 (alk. paper)
1. Kaons. I. Rosner, Jonathan L. II. Winstein, Bruce.

QC793.5.M42 K36 2001
539.7'2162—dc21

00-041166

⊗The paper used in this publication meets the minimum requirements of the American National Standard for Information Sciences—Permanence of Paper for Printed Library Materials, ANSI Z39.48-1992.

ROBERT G. SACHS
1916–1999

A pioneer in the study of fundamental symmetries of nature, our colleague Robert Sachs is sadly not here to enjoy this summing up of the field. Bob passed away in April of 1999 following complications due to surgery. We have dedicated the present volume to his memory.

Contents

Part III Time Reversal Violation and CPT Studies

Part IV Theoretical Topics in Kaon Physics

Part V Rare Kaon Decays

Part VI Hyperon Physics

Part VII Charm: CP Violation and Mixing

Part VIII The Physics of B Mesons

Part IX Future Opportunities in K Physics

Part X Summary and Outlook

Preface and Acknowledgments

Somewhat over 50 years ago, the first of what have come to be called "strange particles" was seen. As the number and variety of these particles proliferated, physicists began to try to make sense of them. It appeared that some had masses around 900 times that of the electron, and existed in both charged and neutral varieties. These are the particles which have come to be known as *kaons*, or K mesons.

During June 21–26, 1999, the University of Chicago played host to a conference on kaon physics (Kaon 99), which has served as the basis for the present collection of articles. Contributors include Richard Dalitz, whose 1953 plot of kinematic variables [1] was invented to analyze the properties of three-body kaon decays; Lincoln Wolfenstein, whose 1964 "superweak" theory [2] was one of the first invented to explain the puzzling phenomenon of CP violation in neutral kaon decays; and theorists and experimentalists providing the latest insights into kaon physics.

It is fitting that this university, which has contributed to the study of kaons—and more generally, strange particles—since the field's inception, should be involved in a review of the field at the turn of the millennium. Dalitz was on the faculty for a decade; Wolfenstein is an alumnus. Murray Gell-Mann elaborated his scheme of "strangeness" [3], classifying the strange particles according to an additive quantum number, in what is now the Enrico Fermi Institute. A question from Fermi during a seminar led Gell-Mann, together with Abraham Pais, to propose the existence of two kinds of neutral kaons [4], a short-lived variety now called K_S and a long-lived variety now called K_L. Riccardo Levi-Setti came to Chicago at Fermi's invitation and pursued a broad program of strange particle studies over twenty years [5], while for many years Roger Hildebrand's experiment [6] was the most sensitive search for the important $K^+ \to \pi^+ \nu \bar{\nu}$ decay about which we heard new results at the conference.

Valentine Telegdi was instrumental in the studies of the weak interactions, including the violation of mirror symmetry in muon decay [7]. In the mid-1960s, he initiated a program of studies of kaon physics that evolved to the present one, which has overthrown Wolfenstein's "superweak" theory and has shed light on the mechanism whereby CP symmetry is violated. This program shows every prospect of continuing in the future. James Cronin, one of the discoverers of CP violation [8], came from Princeton in 1970 to Chicago, where he continued his studies of the phenomenon, winning the Nobel Prize in 1980.

This volume would have not come into existence without the considerable efforts of a number of people. In addition to the contributors, we would like to thank our colleagues on the Local Organizing Committee of Kaon 99: Bill Bardeen, Ed Berger, Ed Blucher, Greg Bock, Peter Cooper, Jim Cronin, Marty Dippel, Paul Mackenzie, Robert Oakes, and Yau Wah; and members of the International Advisory Committee: Doug Bryman, Andrzej Buras, Val Fitch, Paolo Franzini, Belen Gavela, Michael Gronau, Lawrence Hall, Bob Hsiung, Lydia Iconomidou-Fayard, Takao Inagaki, Cecilia Jarlskog, Makoto Kobayashi, Laurie Littenberg, Kam-Biu Luk, Guido Martinelli, Bill Molzon, Tatsuya Nakada, Yosef Nir, Lev Okun, Antonio Pich, Robert Sachs,[1] Walter Schmidt-Parzefall, Evgeny Solodov, Jack Steinberger, Rene Turlay, Heinrich Wahl, Mark Wise, and Lincoln Wolfenstein.

Conference arrangements at the University of Chicago were handled with aplomb by Judith Spurgin and Houston Patterson and their staff at the Office of Special Events, and locally in the High Energy Physics department by Kathy Visak. James Geddes and Theodore Quinn served as our expert assistants in preparing the manuscript for publication. The Vice-Provost for Research at the University of Chicago, Robert Zimmer, graciously welcomed the participants. Music at a concert during the conference was provided by Emilio Colon and Sung Hoon Mo. Elizabeth Pod designed the poster, coordinated the preparation of commemorative University of Chicago gargoyles for participants, and was responsible for much other work behind the scenes. Scientific secretaries were Sami Amasha, Troy André, Colin Bown, Steve Bright, Peter Shawhan, and Scott Slezak. They were supervised in scanning transparencies by Yau Wah and assisted by James Alexander and Rachel Mache. Marty Dippel, our Senior Systems Analyst, coordinated our Web page, arranged for scanning of transparencies, and made sure we had the resources to deal with the resulting flood of information. Van Bistrow, Kate Cleary, and Dennis Gordon of the Physics Department arranged the poster session and provided constant attention to audio-visual needs. Welcome and crucial financial support was provided by Frank Merritt, Chairman of the Physics Department; the U.S. Department of Energy; and the National Science Foundation.

We have arranged this book into parts, each of which contains several chapters preceded by a less specialized introduction. The reader is invited to peruse the introductory material for an overview of this lively field a half-century after its birth.

Jonathan L. Rosner
Bruce D. Winstein
Chicago, February 28, 2000

[1]Deceased April 14, 1999.

References

[1] R.H. Dalitz, *Proc. Conf. Int. Ray. Cosmique (Bagnères de Bigorre)* (Univ. de Toulouse, 1953), p. 236; Phil. Mag. **44**, 1068 (1954); Phys. Rev. **94**, 1046 (1954); E. Fabri, Nuovo Cimento **11**, 479 (1954).

[2] L. Wolfenstein, Phys. Rev. Lett. **13**, 562 (1964).

[3] M. Gell-Mann, Phys. Rev. **92**, 833 (1953); M. Gell-Mann and A. Pais, *Proc. 1954 Glasgow Conf. on Nuclear and Meson Physics*, ed. E.H. Bellamy and R.G. Moorhouse (London, Pergamon, 1955); T. Nakano and K. Nishijima, Prog. Theor. Phys. **10**, 581 (1953); K. Nishijima, Prog. Theor. Phys. **12**, 107 (1954); *ibid.* **13**, 285 (1955).

[4] M. Gell-Mann and A. Pais, Phys. Rev. **97**, 1387 (1955).

[5] R. Levi-Setti and T. Lasinski, *Strongly Interacting Particles* (University of Chicago Press, Chicago, 1973).

[6] J.H. Klems, R.H. Hildebrand, and R. Stiening, Phys. Rev. Lett. **24**, 1086 (1970); Phys. Rev. D **4**, 66 (1971); G.D. Cable, R.H. Hildebrand, C.Y. Pang, and R. Stiening, Phys. Rev. D **8**, 3807 (1973).

[7] J.I. Friedman and V.L. Telegdi, Phys. Rev. **105**, 1681 (1957); **106**, 1290 (1957).

[8] J.H. Christenson, J.W. Cronin, V.L. Fitch, and R. Turlay, Phys. Rev. Lett. **13**, 138 (1964).

[9] See, e.g., R.G. Sachs, *The Physics of Time Reversal* (University of Chicago Press, Chicago, 1987).

Part I

Kaon Physics: History, Progress, Promise

The first kaons were seen in cloud chamber photographs and in nuclear emulsions. With the advent of sufficiently high-energy particle accelerators in the early 1950s, kaons could be studied in sufficient detail to make sense of some initially bewildering properties that they shared with other "strange" particles.

The paradox confronting physicists in the early 1950s was that kaons and certain other heavier particles known as "hyperons" could be very readily produced in the strong interactions but appeared to live for a much longer time than would be expected if the strong interactions also governed their decays. Gell-Mann and Nishijima [1] invented the "strangeness" scheme to account for this peculiar behavior. They assigned an additive quantum number, strangeness or S, to every strongly interacting particle: for example, $S = 0$ for π mesons, protons, and neutrons, $S = 1$ for the neutral and charged kaons K^0 and K^+, and $S = -1$ for their antiparticles $\bar{K}^0$ and K^-. Strangeness was assumed to be conserved in the strong interactions, permitting the production of strange particles in pairs so long as the total strangeness was equal in the initial and final states. Decays via the weak interactions were permitted to violate strangeness.

A puzzle arose regarding the neutral kaons. Both the K^0 and $\bar{K}^0$ were permitted to decay (via the weak interactions) into $\pi^+\pi^-$ or $\pi^0\pi^0$. How, then, was one to distinguish between K^0 and $\bar{K}^0$? This question, asked by E. Fermi, led Gell-Mann and Pais [2] to the conclusion that the neutral kaon states of definite mass and lifetime were

$$K_1 \equiv \frac{K^0 + \bar{K}^0}{\sqrt{2}}, \qquad K_2 \equiv \frac{K^0 - \bar{K}^0}{\sqrt{2}}.$$

The K_1 was permitted to decay to $\pi\pi$, whereas the K_2 was forbidden to decay to $\pi\pi$ but could decay to other, less easily produced states like 3π, $\pi e\nu$, and $\pi\mu\nu$. Thus, the K_2 was expected to live much longer than the K_1. This long-lived neutral kaon was eventually discovered in 1956 [3].

The decays of kaons were a key element in the recognition that the weak interactions did not conserve parity (P), or mirror symmetry. In his contribution to this part, Richard Dalitz recounts that story. The original argument by Gell-Mann and Pais was based on charge-conjugation (C) invariance. With the overthrow of P invariance for the weak interactions in 1957, and the recognition that C invariance was also not valid in such processes, the argument was refashioned on the basis of CP invariance, then thought to be a property of the new theory of the weak interactions. Thus, it was CP invariance that forbade the decay of the K_2 to two pions.

In 1964, Christenson, Cronin, Fitch, and Turlay [4] discovered that in fact the long-lived neutral kaon *did* decay to two pions, leading to the conclusion that even CP invariance is violated by the weak interactions. Shortly after its discovery in neutral kaon decays, CP violation was recognized by Andrei Sakharov as one of the necessary ingredients for the generation of baryon number in the early universe [5].

The present part traces the history of kaon physics (Dalitz), surveys its recent triumphs and prospects (Peccei), puts it in the context of current thinking about the origin of baryons (Worah), and outlines the insights it might provide for an elegant extension, based on *supersymmetry* (Hall; see also Murayama in part II), of our present theoretical framework.

References

[1] M. Gell-Mann, Phys. Rev. **92**, 833 (1953); M. Gell-Mann and A. Pais, *Proc. 1954 Glasgow Conf. on Nuclear and Meson Physics*, ed. E.H. Bellamy and R.G. Moorhouse (London, Pergamon, 1955); T. Nakano and K. Nishijima, Prog. Theor. Phys. **10**, 581 (1953); K. Nishijima, Prog. Theor. Phys. **12**, 107 (1954); *ibid.* **13**, 285 (1955).

[2] M. Gell-Mann and A. Pais, Phys. Rev. **97**, 1387 (1955).

[3] K. Lande, E.T. Booth, J. Impeduglia, and L.M. Lederman, Phys. Rev. **103**, 1901 (1956).

[4] J.H. Christenson, J.W. Cronin, V.L. Fitch, and R. Turlay, Phys. Rev. Lett. **13**, 138 (1964).

[5] A.D. Sakharov, Pis. Zh. Eksp. Teor. Fiz. **5**, 32 (1967) [JETP Lett. **5**, 24 (1967)].

1

Kaon Physics: The First 50+ Years

Richard H. Dalitz

1.1 Historical Introduction

Early in 1947, the nuclear-force meson predicted by H. Yukawa in 1935—known to us today as the pion, with mass about 140 MeV—was discovered by C. F. Powell and his group at Bristol [1], using an early borax-loaded photographic emulsion which been exposed to the cosmic radiation, some of it at altitude 2900m on Pic du Midi (France) and some at altitude 5500m on Mt. Chacaltaya (Bolivia), the latter being prolific in yield. This success stimulated physicists to consider whether there might not exist further mesons, of higher mass, which might also be produced in this way.

In December 1947, G. Rochester and C. Butler [2] reported on two candidate events, a V^0 particle with decay $V^0 \to h^+ + h^-$, where $h^\pm$ denote low-mass particles, most probably pions, observed in October 1946, and a $V^+ \to (h^+ \text{ or } \mu^+) +$ neutral particle(s), where μ denotes the muon, a lepton whose mass is about 106 MeV, observed in May 1947, using a cloud chamber at ground level in Manchester. We have sketched these two events in fig. 1.1. Assuming that the secondaries [a and b in (A), b in (B)] are pions, the masses for the V^0 and V^+ meson were estimated to be about 500 ($\pm$200) MeV. In retrospect, with experience of many later events, the Manchester group now believes that they were examples of $\theta^0 \to \pi^+\pi^-$ and $K^+ \to \mu^+\nu$ decays. These two events were each associated with a penetrating cosmic radiation shower. They came from a batch of 50 such shower events, and these 50 pictures came from 5000 triggered cloud chamber expansions. It is known, and easily understood, that penetrating particles in the cosmic radiation incident on Earth suffer collisions that degrade their energy as they pass down through the atmosphere. Only rather few of them retain much of their energy by the time they reach low altitudes. It was also reasonable to expect the yield of heavy mesons from the collisions of these penetrating particles with the atmosphere to increase with incident energy. Cosmic radiation physicists would have to go to high altitudes for these higher yields, and most of them did so, although first-class work could still be done at ground level.

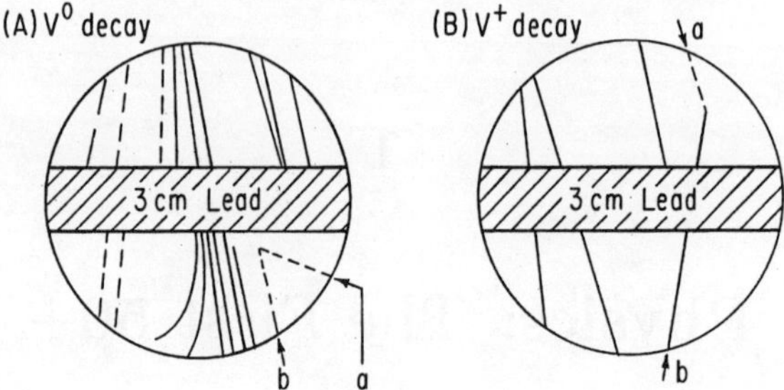

Figure 1.1: Sketch of Rochester and Butler's first V particles. (A) V^0 event (decay products labeled by a and b); (B) V^+ event (incoming V^+ labeled by a, visible decay product labeled by b).

The first heavy meson seen by the Bristol group [3] is shown in fig. 1.2, and was from an exposure in late October 1948 at the Jungfraujoch (Switzerland) using a new electron-sensitive emulsion. The line τ shows the heavy meson coming to rest in the emulsion; grain-counting on this track suggested a mass value of 550 ± 80 MeV. Three tracks a, b, and c emerge. Track a is a π^- meson, coming to rest and then being absorbed by an emulsion nucleus, giving a characteristic π^- star; tracks b and c are due to π^+ mesons, both of which exit the emulsion but slow enough to allow good energy measurements. The three outgoing pions are coplanar; their momenta balance to zero, indicating that the heavy-meson decay is three-body. In retrospect, this was the first example of the τ-decay mode:

$$\tau^+ \rightarrow \pi^+ + \pi^+ + \pi^- \quad \text{(with energy release Q = 75 MeV).} \qquad (1.1)$$

This event was found at the end of October 1948, not long after September 1, when I joined the Bristol Physics Department to work as research assistant to Professor Mott. Thus I became acquainted with the τ meson very early and I followed closely the experimental work being done on τ mesons after that time.

1.2 The *K* Mesons

No further V particles nor τ mesons were reported [4] until about 1950, by which time the cosmic radiation groups were more advantageously equipped and new data on the heavy mesons increased rapidly. Thus "fundamental particle" research developed as a branch of cosmic ray research [5], concerned with the unstable particles generated by the very high-energy penetrating particles incident on the upper atmosphere from outer space. It had been necessary to design and build new chambers, and then to locate them at accessible sites in the high mountains. These chambers needed strong magnetic fields and they were often more complicated than before, to allow the determination of more physical quantities for each event. An

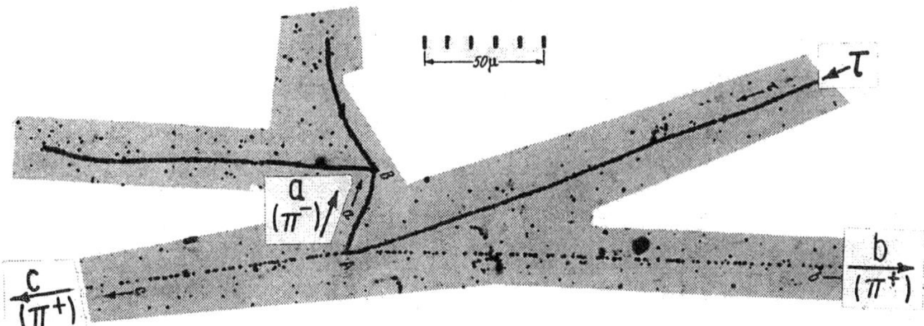

Figure 1.2: First τ^+ event. See text for explanation of tracks.

example [6] was the double cloud chamber, the upper chamber having a massive plate to act as target for the incident penetrating particle, providing information about it and about the particles resulting from this interaction, the lower chamber being multiplate for measuring the energies of these final particles and their decay products.

The emulsion groups expanded in size and many new emulsion groups became established. New techniques were developed to improve the measurements, especially for determining mass values from observations on particle tracks; the mass separation between π^+(mass 140 MeV) and μ^+(mass 106 MeV) was generally difficult. Scattering vs. range and ionization vs. range methods were calibrated to decide between π^+ and μ^+, when a track did not come to rest in the emulsion. Most important ultimately was the development [7] of the "stripped emulsion technique," using a large volume of emulsion, which greatly transformed the emulsion technique to a highly accurate tool. Groups teamed together to combine scanning efforts on sections of the exposed emulsion block, after stripping it into sections, tracks being followed from one section to the next.

Many new particles were reported, seemingly almost one a month. The Bristol group used Greek letters to name new processes as they observed them. Beside the τ^+, there was $\tau'^+ \rightarrow \pi^+\pi^0\pi^0$, a closely related final state. There were also $\kappa^+ \rightarrow \pi^0\mu^+\nu$, and $\kappa^+ \rightarrow \pi^0 e^+\nu$, only the final leptons being different. Neutral mesons were obviously a difficulty for the emulsion technique. The θ^0 was well established from cloud chamber work, through its $\theta^0 \rightarrow \pi^+\pi^-$ mode; its $\pi^0\pi^0$ mode became firm after some considerable time, when θ^0 mesons could be produced plentifully at the Bevatron. Bristol had a χ^+ particle defined by "$\chi^+ \rightarrow \pi^+ + ?$", where ? might be a photon or π^0 or something else neutral.

The landmark conference in this subject was the International Cosmic Ray Conference held at Bagnères-de-Bigorre (France) in July 1953, which was devoted entirely to the new particles. It quickly became clear that all of the cosmic ray workers were observing, measuring, and identifying the same particles. The name K meson was officially adopted as the generic term for these new particles, with masses in the range 500 ± 200 MeV, with the notation $K^{\alpha Q}_{nX}$ (mass), where Q is its charge, n is the number of its decay products, X specifies them (if known), and α could add any other facts about the meson. Thus τ^+ became $K^+_{3\pi}$, while τ'^+ became $K'^+_{3\pi}$. The χ^+ meson became $K^+_{2\pi}$, there being no known K meson with final products $\pi^+\gamma$ or $\pi^+\nu$; there is the decay $K^+ \to \pi^+\gamma\gamma$ now, but this notation is not used for new decay modes today, since we know that there is only one K^+ meson, with many decay modes. To characterize a particle, beyond giving it a name, it was necessary to determine its mass, its decay lifetime, its decay modes and their branching fractions, its spin and parity (if appropriate), its production processes and their dependence on incident energy and its scattering processes, the last two becoming feasible only in the accelerator era.

We were all warned by the senior physicists at the conference, especially by C. F. Powell, *not* to make any simplifying assumption about the relationships between the particles observed. We should use only these "neutral, unbiased" names, the $K^{\alpha Q}_{nX}$ (mass), until we had *firm* evidence of any such relationship, beyond any doubt, since it appeared that we were facing a complicated situation. Although the picture had become clearer at the conference, it was still rather confused, with many pitfalls. To illustrate this, we may quote from a review [8] by Leprince-Ringuet (Ecole Polytechnique, Paris) and Rossi (MIT) late in 1953, of their work on K mesons; they concluded that:

> If all the available experimental data are taken at face value, even the assumption of three different kinds of K-mesons (such as the τ, the χ, and the κ particle, according to the Bristol nomenclature) does not remove all difficulties of interpretation.

Concerning their own data (EP and MIT), they added that:

> it could be accounted for by a single K^+ meson, with mass about 460–510 MeV, if its dominant decay mode were $K_{\mu\nu\gamma}$.

They said that the emulsion data required that this K^+ meson should also have a $K^+_{3\pi}$ mode competing with this $K^+_{\mu\nu\gamma}$ mode at a low rate, although they had no evidence for final pions in their own work. The advice that had been given to us at the Bagnères-de-Bigorre Conference was sound. The year 1953 was far too early for speculations about the relationship between the final states observed.

1.3 Parity Violation

The data on τ^+ decay events increased steadily from 1950. The events were clear and well measured, and usually all three pion energies could be determined. This information was especially valuable since each event configuration requires two variables for its specification, so that there is

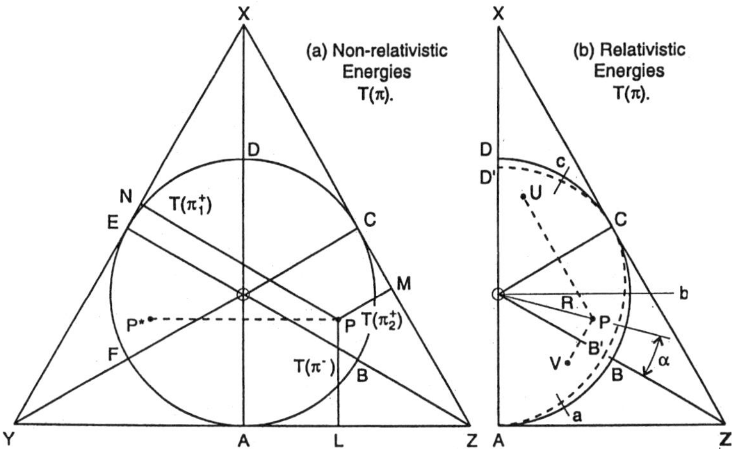

Figure 1.3: Phase space plot depicting kinematic variables in τ^+ decays. P* is the reflection of P in the line of symmetry AODX.

some internal structure involved. These data are usually represented on a phase space plot [9, 10], given as fig. 1.3. Each event is given by a point P; the perpendiculars from P to the three sides of an isosceles triangle are taken in proportion to the kinetic energies of the π^-, π_1^+, and π_2^+ mesons. This is convenient since PL + PM + PN has the same sum Q for all points P lying within this triangle. The coordinates (x, y) for point P are given by

$$x = \sqrt{3}\,(\mathrm{PM} - \mathrm{PN})/\mathrm{Q}, \qquad y = (2\mathrm{PL} - \mathrm{PM} - \mathrm{PN})/\mathrm{Q}, \tag{1.2}$$

for which the axes are Ob on fig. 1.3(b), parallel to AZ, and AODX. However, not all points inside the boundary can represent a physically possible 3π configuration because of the constraints of momentum conservation. With nonrelativistic kinematics, the physical points P must lie within the circle inscribed in the triangle. Although the pion energies are quite low, the use of relativistic kinetic energy $(\sqrt{m^2 + \mathbf{p}^2} - m)$ distorts the boundary of the physical points appreciably, as shown by the dashed boundary in fig. 1.3(b). It turns out that the use of the relativistic kinematics is vital, as we'll make clear below.

An alternate representation for the τ^+ events is given by fig. 1.4. The momentum $\mathbf{p}$ is that of the π^- meson in the rest frame for the τ^+ meson. The momenta for the two π^+ mesons in their rest frame are denoted by $+\mathbf{q}$ and $-\mathbf{q}$, their total energy in that frame being $m_{\pi\pi} = 2\sqrt{m^2 + \mathbf{q}^2}$. In the τ-meson rest frame (laboratory frame), the two π^+ mesons have energy $\sqrt{m_{\pi\pi}^2 + \mathbf{p}^2}$ so that the τ-meson total energy is

$$\sqrt{m^2 + \mathbf{p}^2} + \sqrt{4m^2 + 4\mathbf{q}^2 + \mathbf{p}^2} = M = 3m + Q. \tag{1.3}$$

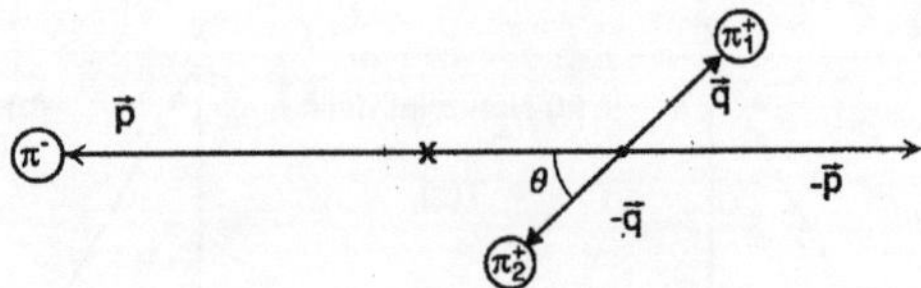

Figure 1.4: Representation of kinematics of τ^+ decay.

This relation connects the two momenta $|\mathbf{p}|$ and $|\mathbf{q}|$, independent of the angle θ between the momenta $\mathbf{p}$ and $\mathbf{q}$ in the two frames. In the τ-rest frame, the π^+ energies are given by

$$m + T(1,2) = \sqrt{m^2 + \mathbf{q}^2 + \mathbf{p}^2/4} \pm |\mathbf{p}||\mathbf{q}| \cos\theta/(2\sqrt{m^2 + \mathbf{q}^2}), \qquad (1.4)$$

where (1,2) refer to the π_1^+ and π_2^+, with $T(\pi_1^+) > T(\pi_2^+)$. As θ decreases from $\pi/2$ (where P is on the axis of symmetry AODX of the triangle) to 0, for fixed $|\mathbf{p}|$, eq. (1.4) shows that the laboratory energy of each π^+ meson varies linearly with $\cos\theta$. For the plus sign in (1.4), P moves to the right, parallel to P*P on fig. 1.3 (the π^--energy PL being fixed) until it reaches the dashed boundary; for the minus sign, P moves to the left until it reaches the left boundary. When θ is zero, the three momenta $\mathbf{p}$, $\mathbf{q}$, and $-\mathbf{q}$ are collinear. In other words, the boundary points are all of the collinear configurations for the three pions, each configuration being characterised by a unit vector (say) $\hat{\mathbf{k}}$ along the direction of the collinearity.

We are aware of the situation at point C on the triangle, where one π^+ is at rest and the other two pions form a two-body system carrying the orbital angular momentum. The π^+ at rest provides only parity -1, and we are accustomed to parity $(-1)^2$ $(-1)^J$ from the orbiting pions, since their wavefunction is $Y_J(\hat{\mathbf{k}})$, as is the case for the θ systems. It is then easy to generalize this remark to all collinear configurations; each pion contributes a parity -1 and there is again only one unit vector $\hat{\mathbf{k}}$ needed to specify the configuration and the wavefunction $Y_J(\hat{\mathbf{k}})$, which provides parity $(-1)^J$. If a three-pion system with angular momentum J had overall parity $P = (-1)^J$, this system must lack any events on the boundary since we have just shown that events on the boundary necessarily have parity $P = -(-1)^J$.

The statistics on $\tau^+ \to \pi^+\pi^+\pi^-$ events from cosmic ray studies increased steadily, until there were 58 of these τ^+ decay events available [see fig. 1.5(a)] at the Pisa Conference in 1955, with an additional 35 events reported from Bevatron experiments. This conference really marked the end of K^+ decay studies based on the cosmic radiation. At the Rochester Conference of April 1956, a phase-space plot of 219 τ^+ (all from Bevatron studies) was presented, shown here on fig. 1.5(b) for comparison. Neither of these plots nor any of the later plots with remarkably high statistics show any sign of the density of events per unit area of phase-space falling to zero at the boundary of the plot.

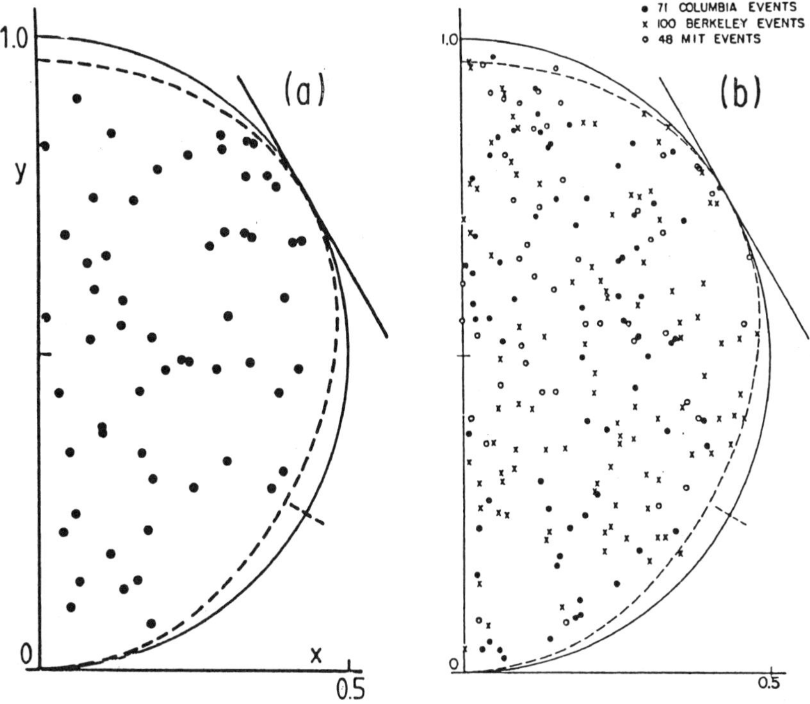

Figure 1.5: Phase space plots for τ^+ decays. (a) 58 cosmic ray events in 1955; (b) 219 Bevatron events in 1956.

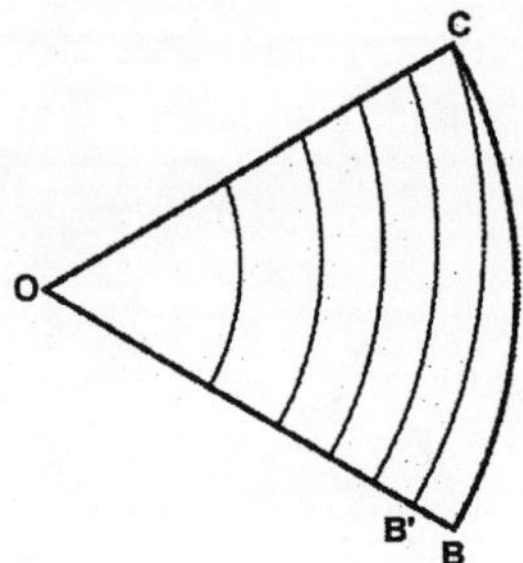

Figure 1.6: Sector phase-space plot COB′ for relativistic kinematics is divided into 5 equal areas, to test boundary behavior. The circular sector COB is that appropriate to nonrelativistic kinematics.

On fig. 1.3(a), the point P*, the reflection of P in AODX, corresponds to interchanging π_1^+ and π_2^+ but this leaves the distribution unchanged since π^+ mesons obey Bose-Einstein statistics. That is why we removed the left half of the phase-space plot in giving fig. 1.3(b). We see that similarly P may be reflected in OZ to give V, or in OC to give U. The points P, U, and V involve the same three meson energies, but with the three different assignments for the meson charges. If we then fold the sector AOB′ over OB, which places A on C, and the sector COD′ over OC, which places D′ on B′, all events are then in sector COB′, and we shall call this the sector plot, shown on fig. 1.6.

It is advantageous to consider the sum $\rho_s(\mathrm{P}) = [\rho(\mathrm{P}) + \rho(\mathrm{U}) + \rho(\mathrm{V})]$ when we are primarily interested in the behavior of the density of events near the boundary of the phase-space plot. We then divide the phase-space shown in fig. 1.6 into n equal areas; we have chosen to divide OC and OB′ being shrunken by factors λ_i for $\lambda_0 = 1$ to $\lambda_n = 0$. The spaces between two successive lines have areas proportional to $\lambda_i^2 - \lambda_{i+1}^2$, the area adjacent to the boundary being $1 - \lambda_1^2$. With such limited data, we chose to use $n = 5$, for which $\{\lambda_i\} = 1,\ 2/\sqrt{5},\ \sqrt{3/5},\ \sqrt{2/5},\ 1/\sqrt{5},\ 0$, and we show on fig. 1.7 how the first 58 cosmic ray events at the Pisa Conference appear on this plot. There is certainly no evidence of a fall in the density of events as λ approaches 1. The same point can be made by using the $(|\mathbf{p}|, \cos\theta)$ distribution, where $\cos\theta$ is defined by fig. 1.4. The phase-space boundary is approached as $\cos\theta \to 1$, for definite value for $|\mathbf{p}|$, and we show this (integrated over $|\mathbf{p}|$) as fig. 1.8, using the larger body of data (892 Bevatron events) available at the April 1957 Rochester Conference.

The θ^+ meson proved relatively difficult to establish. Its secondary π^+ has energy about 248 MeV and this had to be separated from the secondary muon (energy about 259 MeV), a difficult task for the emulsion technique, which was not able to identify the accompanying π^0. With the cloud chamber and its use with electronic techniques, γ-rays could sometimes be detected but not measured accurately enough to identify

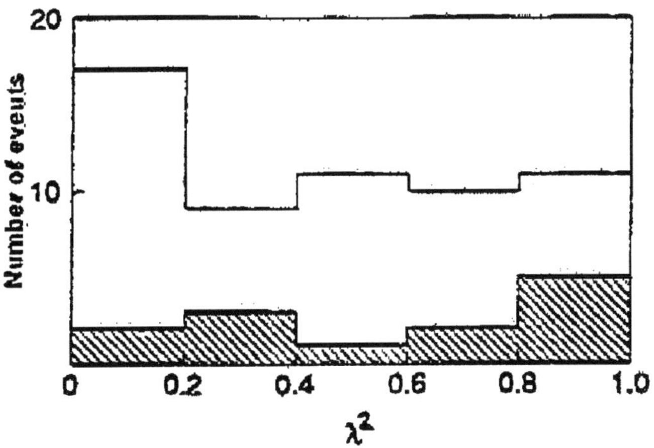

Figure 1.7: Histogram to show the intensity of events in the five chosen areas of fig. 1.6, for all of the 58 cosmic ray events presented at the Pisa Conference. The hatched area corresponds to events reported at Rochester in January 1954.

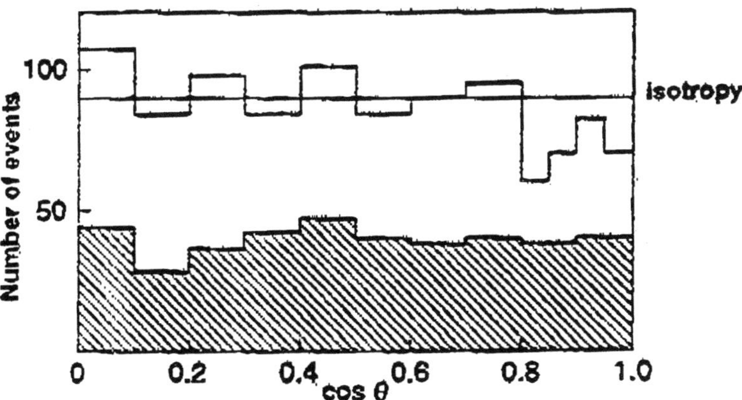

Figure 1.8: Distribution of $\cos\theta$ integrated over $|\mathbf{p}|$, relative to phase-space, for 892 τ^+ decay events from exposures at the Bevatron, for eight emulsion groups [10]. Of these events, 386 (the hatched area) were contributed by the Milan group.

them as due to a $\pi^0 \to \gamma\gamma$ decay. It was still labeled "$\chi^+ \to \pi^+ + ?$" at Pisa in 1955, but it was named $K^+_{\pi 2} \to \pi^+\pi^0$ at the Rochester Conference, completely accepted. With spin J for the θ^+ meson, the parity of the final $\pi^+\pi^0$ system is necessarily $P = (-1)^J$.

The key question was that of the relationship between these two sets of events, the τ^+ and θ^+. By the time of the Rochester Conference of April 1956, we knew that:

- The mass values associated with the $K^+_{\mu\nu}$, θ^+ and the τ^+ decay modes agreed within several MeV;
- The lifetimes associated with the $K^+_{\mu\nu}$, θ^+ and τ^+ decay modes were roughly equal (within ± 30%); and
- The branching fractions $K^+_{\mu\nu} : \theta^+ : \tau^+$ were independent of the origin of the K^+ beam (within ± 20%).

These data were all consistent with the hypothesis that $K^+_{\mu\nu}$, θ^+ and τ^+ were alternative decay modes of one K^+ meson, with a definite mass, but they were certainly no real proof of it. This was *"the τ-θ problem"*:

EITHER θ^+ and τ^+ were from different K^+ mesons, differing in spin or parity or both. They happened to have close mass values and similar lifetimes, for reasons that we did not understand. There was then no reason to question parity conservation.

OR there was only one K^+ meson whose decay gives rise to both the $(\pi\pi\pi)^+$ and the $(\pi\pi)^+$ final states.

Since our arguments above have told us that for any spin J the characteristics of the $\pi^+\pi^+\pi^-$ final state are such that it has opposite parity to the $\pi^+\pi^0$ final state of the same J, we then had to conclude that *"parity conservation fails in these $K \to$ pions electroweak interactions."*

What became clear was that, to make this conclusion convincing by experiment, we had to MEASURE A PSEUDOSCALAR [11], i.e., find an interference between two final states with opposite parity. This could have been done using "virtual kaons". Consider $\Lambda \to p\pi^-$ via a K that couples with pions in either parity, as follows:

$$\Lambda \to p + K^- \overrightarrow{g_2} \;\; p + \pi^-\pi^0 \;\; \to p + \pi^- \qquad \text{(p wave pion)}, \qquad (1.5)$$

$$\Lambda \to p + K^- \overrightarrow{g_3} \; p + \pi^-\pi^-\pi^+ \to p + \pi^- \qquad \text{(s wave pion)}, \qquad (1.6)$$

where g_2 and g_3 denote the weak coupling amplitudes for $K \to \pi\pi$ and $K \to \pi\pi\pi$. As is well-known now, the interference between the s- and p-wave amplitudes for $\Lambda \to p\pi^-$ decay generates a strong polarization term for this decay; the pseudoscalar being measured in this case is $\langle \boldsymbol{\sigma}_\Lambda \cdot \boldsymbol{p} \rangle$.

In fact, as you will all know, the τ-θ puzzle was settled indirectly by an experiment by Wu *et al.* [12] involving quite a different weak interaction, namely the well-known nuclear beta-decay. The pseudoscalar measured to be nonzero was $\langle \mathbf{S} \cdot \mathbf{p} \rangle$, where $\mathbf{S}$ was the spin polarization of ^{60}Co generated by a strong magnetic field and $\mathbf{p}$ the momentum of its decay electron. After this result was obtained, two other experiments quickly confirmed the reality of parity violation but for the decay sequence $\pi^+ \to \mu^+ \to e^+$ rather than for nuclear beta-decay; one [13] used counter techniques with high statistics and the other [14] used the nuclear emulsion technique. After these observations, physicists felt free to accept the possibility of parity violation in any weak interaction process.

1.4 CP Invariance

Since CPT invariance holds for all quantum field theories in use today, CP invariance holds for all field theories for which time reversal T holds. For all these theories, physical phenomena would still satisfy the same laws after a mirror reflection of space in which right-handed and left-handed systems are interchanged, provided that at the same time all particles are replaced by their antiparticles. Philosophically speaking, this matches our intuitive expectation that there should be no handedness preferred in any absolute sense, a point made by more than one physicist, although the name of Landau [15] is usually linked with it. Unfortunately, as we shall see in the next section, CP invariance is not exact and this elegant understanding of the universe does not hold.

With a suitable choice of phases, we can arrange that

$$\text{(a)}\quad \mathrm{CP}|K^0\rangle = |\bar{K}^0\rangle, \quad \text{(b)}\quad \mathrm{CP}|\bar{K}^0\rangle = |K^0\rangle. \tag{1.7}$$

Both K^0 and $\bar{K}^0$ can couple with the $\pi^+\pi^-$ and $\pi^0\pi^0$ systems, through the weak interactions with $\Delta S = \pm 1$, which cause the $|K^0\rangle$ and $|\bar{K}^0\rangle$ states to mix, as pointed out by Gell-Mann and Pais [16]. The mixed states are eigenstates of CP:

$$\text{(a) CP} = +1 : |K_1^0\rangle = \frac{|K^0\rangle + |\bar{K}^0\rangle}{\sqrt{2}}, \text{(b) CP} = -1 : |K_2^0\rangle = \frac{|K^0\rangle - |\bar{K}^0\rangle}{\sqrt{2}}. \tag{1.8}$$

The $(\pi\pi)^0$ states have CP $= +1$ and do not interact with K_2^0. The states K_1^0 and K_2^0 are no longer degenerate in mass, as K^0 and $\bar{K}^0$ were in terms of the strong interactions alone. This corresponds well with observation; the K_S^0 state has decay rate 1.12×10^{10} s^{-1}, dominantly to the $(\pi\pi)^0$ states, while the K_L^0 has a total decay rate of 1.93×10^7 s^{-1}, with comparable rates for $\pi\pi\pi$, $\pi^\pm e^\mp \nu$, and $\pi^\pm \mu^\mp \nu$ final states.

Let us consider a beam of K^0 or $\bar{K}^0$; either is readily made using strong interaction processes, such as

$$\text{(a) } \bar{p}p \to K^-\pi^+K^0, \quad \text{and} \quad \text{(b) } \bar{p}p \to K^+\pi^-\bar{K}^0, \tag{1.9}$$

using the LEAR antiproton storage ring which was at CERN. To be definite, we consider a K^0 particle at $t = 0$, expressing its initial state $|K^0(t = 0)\rangle$

in terms of the eigenstates $|K_1^0\rangle$ and $K_2^0\rangle$ given above:

$$|K^0(t=0)\rangle = [|K_1^0(t=0)\rangle + |K_2^0(t=0)\rangle]/\sqrt{2}, \quad (1.10)$$

leading to the wavefunction at time t,

$$[E_1(t)|K_1^0(t=0)\rangle + E_2(t)|K_2^0(t=0)\rangle]/\sqrt{2}, \quad (1.11)$$

where $E_n(t)$ is $\exp[-i(\omega_n - i\gamma_n/2)t]$, ω_n and γ_n being the energy and decay width of the state K_n^0.

We now express the state at time t in terms of $K^0(t=0)$ and $\bar{K}^0(t=0)$,

$$[\{E_1(t)+E_2(t)\}|K^0(t=0)\rangle + \{E_1(t)-E_2(t)\}|\bar{K}^0(t=0)\rangle]\,/2. \quad (1.12)$$

The E_1 terms die off quickly with increasing t, since γ_1 is three orders of magnitude larger than γ_2. What we notice from (1.12) is that there is a $\bar{K}^0$ component whose intensity rises from zero at $t=0$ to meet the intensity of the K^0 component; as shown in fig. 1.9, there are then a few damped oscillations and the two components finally become equal, ultimately reaching the form (1.8b) for the K_2^0. The $\bar{K}^0$ content of the beam can be detected at any time by the observation of Λ and Σ hyperons as it moves through the medium (*e.g.*, a bubble chamber of hydrogen or propane, or some solid target of light material), as shown by Landé *et al.* [17] and by others. When Pais first mentioned this phenomenon in a seminar at Brookhaven, there was a stampede for the telephones at the end of his seminar, in order to spread the news and to start planning for measurements of these new effects, for the story rang true and was accepted at once when heard. Their study led to a wonderfully accurate measurement of the mass difference, $(m_L - m_S) = 3.49(1) \times 10^{-6}$ eV.

The process of *regeneration* [18] was the next wonder. A beam initially K^0 begins with equal K_1^0 and K_2^0, of which the former rapidly decay, leaving a pure K_2^0 beam. When the beam impinges on an absorptive barrier, the $\bar{K}^0$ component of the K_2^0 is absorbed away and only the K^0 component emerges, leading to a regeneration of the initial beam, but with intensity reduced by a factor 1/4 (see fig. 1.10). This effect has been used with much benefit in today's experiments, which depend on K_1^0 interference with K_2^0.

1.5 CP Violation

Despite the attractive symmetry of the CP-conserved framework, it is natural to search for features of the $(K^0, \bar{K}^0)$ complex that violate CP invariance. To demonstrate their presence was a very difficult task. The most immediate means was to seek the decay mode $K_L^0 \to \pi\pi$. Many experiments attempted to observe these events and to measure their rates, but the experiment of Christenson, Cronin, Fitch, and Turlay [19] was the first to achieve a convincing nonzero result for the ratio

$$\text{Rate }(\mathrm{K_L} \to \pi^+\pi^-)/\text{Rate}(\mathrm{K_L} \to \text{all charged modes}) = 2.0(5) \times 10^{-6}. \quad (1.13)$$

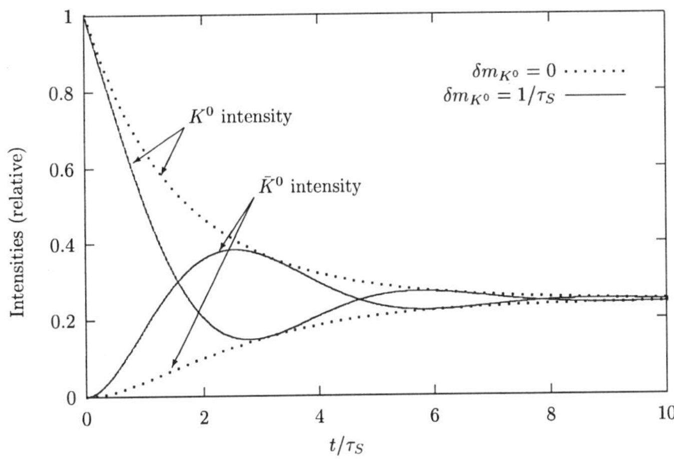

Figure 1.9: K^0 and $\bar{K}^0$ intensities along a beam initially K^0 at $t = 0$, for $\delta m_{K^0} =$ 0 (dotted) and $1/\tau_S$ (solid), its physical value being $\simeq 0.5/\tau_S$.

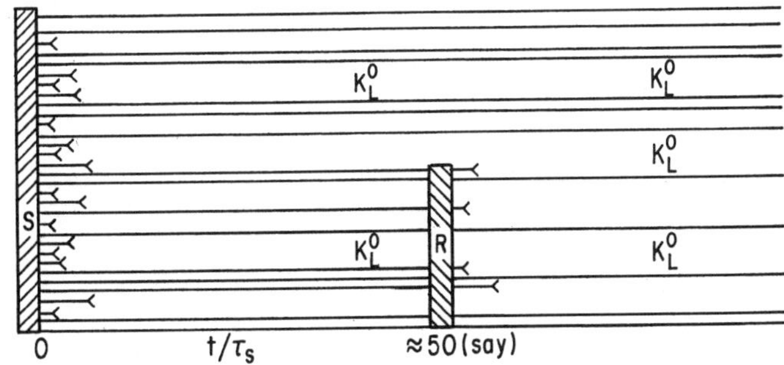

Figure 1.10: Illustration of regeneration. S is a K^0 source. The short lines denote K_S^0 decay. The beam is purely K_L^0 when it reaches the regenerator R. The (idealized) regenerator absorbs completely the $\bar{K}^0$ component of the beam, leaving only K^0 emerging on the downstream side of R, as was the case for S.

This CP violation could occur in the $(K^0, \bar{K}^0)$ mass mixing matrix [20] or in the amplitudes for the K^0 and $\bar{K}^0 \to$ final states, here dominantly the $(\pi\pi)^0$ system. Accepting CPT invariance, the mass matrix may be written in the (K_1, K_2) basis in the form [21]

$$\begin{bmatrix} 0 & M_{12} - i\Gamma_{12}/2 \\ M_{12}^* - i\Gamma_{12}^*/2 & 0 \end{bmatrix} \tag{1.14}$$

and its eigenstates are mixed (K_1^0, K_2^0) states. In a sufficient approximation, since ϵ is small, of order 10^{-3}, they are

$$\text{(a) } K_S^0 = K_1^0 + \epsilon K_2^0, \quad \text{(b) } K_L^0 = \epsilon K_1^0 + K_2^0. \tag{1.15}$$

These two states are not orthogonal although the $K_{1,2}$ states are, $\langle K_S | K_L \rangle$ having the value 2 Re ϵ.

The parameter Re ϵ can be measured directly by the charge asymmetry δ in the $(\pi^{\mp}\ell^{\pm}\nu)$ decay rates along the K_L^0 beam,

$$\delta = \frac{\Gamma(K_L^0 \to \pi^- \ell^+ \nu_\ell) - \Gamma(K_L^0 \to \pi^+ \ell^- \bar{\nu}_\ell)}{\Gamma(K_L^0 \to \pi^- \ell^+ \nu_\ell) + \Gamma(K_L^0 \to \pi^+ \ell^- \bar{\nu}_\ell)} \tag{1.16}$$

$$= 2 \text{ Re } \epsilon. \tag{1.17}$$

Assuming the $\Delta S = \Delta Q$ rule holds for these decays, $\delta = 3.3(1) \times 10^{-3}$ is the present Particle Data Group (PDG) value [22], and we make use of its smallness, which simplifies all analysis of CP-violation effects in the $(K^0, \bar{K}^0)$ system.

The mass matrix (1.14) corresponds to the superweak model [20], the off-diagonal element being its one (complex) parameter. The measured quantities for the decay processes are the amplitudes for the final state f from the eigenstates K_S^0 and K_L^0, their ratio being denoted by η_f:

$$\eta_f = (f|T|K_L^0)/(f|T|K_S^0) = |\eta_f| \exp(i\phi_f). \tag{1.18}$$

For the $(\pi\pi)$ systems, these are now quite accurately known, the values being [22]:

$$\eta_{+-} = 2.285(19) \times 10^{-3} \cdot \exp[i(43.5(6)^0)], \tag{1.19}$$

$$\eta_{00} = 2.275(19) \times 10^{-3} \cdot \exp[i(43.4(10)^0)]. \tag{1.20}$$

In this approximation, the parameters of the analysis are pretty well fixed by the data. The ϵ parameter is well given by

$$\epsilon = \frac{i\text{Im}(M_{12})}{(m_L - m_S) - i(\Gamma_L - \Gamma_S)/2}. \tag{1.21}$$

Going beyond the superweak model, the present analyses are based on the standard model of electroweak interactions. The final states are $(\pi\pi)^I$, where the isospin I can equal 0 or 2, and the $\pi\pi$ system is necessarily s-wave. With $I = 1/2$ and $I_3 = (-1/2, +1/2)$ for the (K^0, K^+) mesons, the $\Delta I = 1/2$ rule allows a weak transition to the $I = 0$ $\pi\pi$ system for the K^0 particle but not for the K^+ particle. This is in accord with their observed transition rates [22]: $\Gamma(K^0 \to \pi\pi) = 1.1193(10) \times 10^{10}\ \mathrm{s}^{-1}$, whilst $\Gamma(K^+ \to \pi^+\pi^0) = 1.708(12) \times 10^7\ \mathrm{s}^{-1}$, a suppression of decay rate by a factor of about 1/655(5), or of decay amplitude by a factor of about 1/25. It has been natural to attribute this suppression to the existence of a $\Delta I = 3/2$ transition for $K \to \pi\pi$ decay, weaker than the $\Delta I = 1/2$ transition, but capable of reaching the $I = 2$ final $\pi\pi$ states. The inclusion of the $I = 2$ terms has been considered by many workers, in many different ways. In a phenomenological spirit, following Wu and Yang [21], we define a parameter ϵ' by

$$\epsilon' = \frac{i}{\sqrt{2}} \cdot \frac{\mathrm{Im}\, A_2}{A_0} \cdot \exp[i(\delta_2 - \delta_0)], \tag{1.22}$$

where δ_0 and δ_2 are s-wave $\pi\pi$ scattering phases for $I = 0$ and 2 at $\pi\pi$ c.m. energy equal to m_K. Taking (1.21) into account, the ratios (1.20) are summarized as

$$\eta_{+-} = \epsilon + \epsilon' \tag{1.23}$$

$$\eta_{00} = \epsilon - 2\epsilon'. \tag{1.24}$$

The estimated values for the phases $\phi(\epsilon)$ and $\phi(\epsilon')$, in the approximations used in this calculation, are $43.49(8)^0$ and $(\delta_2 - \delta_0 + \pi/2) = 48(4)^0$, so that on the Argand plane, ϵ' and ϵ have nearly the same direction. To a good approximation we conclude then that ϵ'/ϵ is almost real, its value being

$$\epsilon'/\epsilon = [1 - |\eta_{00}/\eta_{+-}|^2]/6 \approx 15(8) \times 10^{-4} \tag{1.25}$$

using PDG data. However, you will find in chapter 6 a new, unpublished value for $\mathrm{Re}(\epsilon'/\epsilon)$, standing alone, with magnitude $28.0(4.1) \times 10^{-4}$, and in chapter 8 a value of $\mathrm{Re}(\epsilon'/\epsilon)$ of $18.5(7.3) \times 10^{-4}$.

It is most remarkable how things have come out. The parameter ϵ is quite small, enough to justify approximations that allow us to see in a simple way what the computations depend on. The smallness of ϵ'/ϵ is believed to be linked with the operation of the $\Delta I = 1/2$ rule, which requires the $I = 2$ amplitude to be generated only by a "forbidden" $\Delta I = 3/2$ transition. The earlier estimate of ϵ'/ϵ by Gilman and Wise [23] was quite large, whereas the experimental value was much smaller. Reexamination of the theory showed that smaller estimates were possible, but the experiments indicated still smaller values; the theoreticians then managed to lower their value still more. At present, the theoreticians' values are actually lower than the latest experimental numbers. It is being said that there is a crisis in the present situation; there is an assesment of this in part I, especially chapters 5, 9, and 28, and in part IV, chapter 28 and elsewhere herein.

That there is any CP violation at all is due to the spontaneous symmetry breaking of the SU(2)×U(1) symmetry group of the standard model. As

Kobayashi and Maskawa first pointed out, a model with three quark-lepton families allows a nonzero phase δ in the Cabibbo-Kobayashi-Maskawa (CKM) matrix [24,25], which can be freely adjusted. The contributions to K meson decay rates from this phase necessarily have a factor $s_1 s_2 s_3 \sin\delta$, where $s_i = \sin\theta_i$ and the angles $\{\theta_i\}$ for $i = 1, 2, 3$ parametrize the mixing matrix. The CP-conserving kaon transition amplitudes have only a factor s_1, so that the CP-violating terms always have a factor $s_2 s_3 \sin\delta$ relative to them. Since this factor has a small value, the CP-violation effects in the *uds* sector are necessarily small.

1.6 Time Reversal Invariance T

A failure of T necessarily holds if CP invariance fails, in consequence of the general validity of CPT invariance for all quantum field theories. However, CPT invariance has been challenged, and its validity should be tested whenever possible. In particular, direct tests of T can be made that do not explicitly assume CPT invariance. Kabir [26] has pointed out that, in quantum theory generally, the reciprocity of interaction amplitudes depends on the validity of T so that direct checks on reciprocity for interaction amplitudes do test the validity of T. The simplest amplitudes to consider are those for the processes $K^0 \to \overline{K}^0$ and $\overline{K}^0 \to K^0$ and the test is to determine whether or not the asymmetry

$$A_T = \frac{\text{Rate }(K^0,\ t=0 \to \overline{K}^0, t) - \text{Rate }(\overline{K}^0, t=0 \to K^0, t)}{\text{Rate }(K^0,\ t=0 \to \overline{K}^0, t) + \text{Rate }(\overline{K}^0, t=0 \to K^0, t)} \tag{1.26}$$

is nonzero. This asymmetry has been measured recently by CPLEAR [27], who presented their result at this symposium; it is consistent with the value of $4\mathrm{Re}\epsilon$, the value predicted if CPT holds. Part III of this volume gives this and related results, in particular on the $K_L \to \pi^+\pi^-e^+e^-$ decay, and includes further theoretical discussion.

The question of whether the CPLEAR experiment really measures the failure of T invariance or can be explained by a failure of CPT invariance has been raised by a number of authors since the CPLEAR result was published. Using exact reciprocity and T-invariance, Kabir [28] and Rougé [29] interpret the CPLEAR result in terms of a CPT-violating leptonic charge asymmetry. Rougé claims that the Bell-Steinberger unitarity relation [30] (based on the assumption of a hermitian Hamiltonian) must be used in order to infer T violation from the CPLEAR experiment. Weighing in on the other side of the controversy, a group of theorists at CERN [31] claims that one can conclude that CPLEAR has measured T violation *without* assuming a hermitian Hamiltonian. In the meantime, CPLEAR has published a fresh analysis in which the Bell-Steinberger relation has been used as a constraint [32], leading to a considerable restriction of parameters.

There's still a lot about kaons that we don't know yet!

References

[1] C.M.G. Lattes, G.P.S. Occhialini, and C.F. Powell, Nature **160**, 453 (1947); *ibid.* **160**, 486 (1947).

[2] G.D. Rochester and C.C. Butler, Nature **160**, 855 (1947).

[3] R. Brown, U. Camerini, P.H. Fowler, H. Muirhead, C.F. Powell, and D.M. Ritson, Nature **163**, 82 (1949).

[4] L. Leprince-Ringuet and M. Lhéritier, in J. de Physique et Radium, Series 8, **7**, 66-9 (1946) [see also Phys. Rev. **70**, 791A (1946)] claimed to have observed in 1943 a heavy positive meson colliding elastically with an electron in their cloud chamber, the event being fitted for a meson mass of 506 ± 60 MeV/c^2. However, H.A. Bethe [Phys. Rev. **70**, 821 (1946)] attacked this "new method" of meson mass determination, showing that this event could as well be interpreted as a proton-electron collision, since multiple scattering could modify the apparent change of curvature of the heavy particle track due to this collision. As Peyrou tells us (see pp. C8-(29–30) of ref. [6] below), the triggering of the cloud chamber at that time was such that very few K^+ mesons could be expected to occur in the chamber whereas there were many proton tracks. These were shaky grounds for the claim of a new particle.

[5] This remained so for some five years, when multi-GeV accelerators became available and could produce most of the new particles as beams under well-defined conditions, the changeover being between the years 1955 and 1956.

[6] Ch. Peyrou, in *Proc. Int. Colloq. on the History of Particle Physics*, Paris, 21–23 July, 1982, J. de Physique Coll. **43**, C8, Suppl. au No. 12, pp. 7–66 (1982).

[7] C.F. Powell, P.H. Fowler, and D.H. Perkins, *The Study of Elementary Particles by the Photographic Method* (Pergamon, New York, 1959).

[8] L. Leprince-Ringuet and B. Rossi, Phys. Rev. **92**, 722 (1953).

[9] R.H. Dalitz, in *Proc. Conf. Int. Ray. Cosmiques (Bagnères de Bigorre)*, Toulouse, Univ. of Toulouse, 1953, p. 236; Phil. Mag. **44**, 1068 (1953); Phys. Rev. **94**, 1046 (1954); E. Fabri, Nuovo Cim. **11**, 479 (1954).

[10] R.H. Dalitz, Rep. Prog. Phys. **20**, 163 (1957).

[11] T.D. Lee and C.N. Yang, Phys. Rev. **104**, 254 (1956).

[12] C.S. Wu, E. Ambler, R. Hayward, D. Hopper, and R. Hudson, Phys. Rev. **105**, 1413 (1957); *ibid.* **106**, 1361 (1957).

[13] R.L. Garwin, L.M. Lederman, and M. Weinrich, Phys. Rev. **105**, 1415 (1957).

[14] J.I. Friedman and V.L. Telegdi, Phys. Rev. **105**, 1681, 1957; *ibid.* **106**, 1290 (1957).

[15] L.D. Landau, Nucl. Phys. **3**, 127 (1957).

[16] M. Gell-Mann and A. Pais, Phys. Rev. **97**, 1387 (1955).

[17] K. Lande, E.T. Booth, J. Impeduglia, and L.M. Lederman, Phys. Rev. **103**, 1901 (1956).

[18] A. Pais and O. Piccioni, Phys. Rev. **100**, 1487 (1955).

[19] J.H. Christenson, J.W. Cronin, V.L. Fitch, and R. Turlay, Phys. Rev. Lett. **13**, 138 (1964).

[20] L. Wolfenstein, Phys. Rev. Lett. **13**, 562 (1964).

[21] T.T. Wu and C.N. Yang, Phys. Rev Lett. **13**, 380 (1964). For an up-to-date review, see B. Winstein and L. Wolfenstein, Rev. Mod. Phys. **63**, 1113 (1993).

[22] Particle Data Group, *Review of Particle Physics,* Eur. Phys. J. **3**, 1–794 (1998).

[23] F. Gilman and M.B. Wise, Phys. Lett. **83B**, 83 (1979); Phys. Rev. D **20**, 2392 (1979).

[24] N. Cabibbo, Phys. Rev. Lett. **10**, 531 (1963).

[25] M. Kobayashi and T. Maskawa, Prog. Theor. Phys. **49**, 652 (1973).

[26] P.K. Kabir, Phys. Rev. D **2**, 540 (1970).

[27] CPLEAR collaboration, A. Angelopoulos *et al.*, Phys. Lett. B **444**, 43 (1998).

[28] P.K. Kabir, Phys. Lett. B **459**, 335 (1999).

[29] A. Rougé, LPNHE preprint No. X-LPNHE 99/05, hep-ph/9909205.

[30] J.S. Bell and J. Steinberger, in *Proc. Oxford Int. Conf. on Elementary Particles*, ed. R.G. Moorhouse *et al.*, Rutherford Laboratory, 1966, p. 195.

[31] L. Alvarez-Gaumé, C. Kounnas, S. Lola, and P. Pavlopoulos, Phys. Lett. B **458**, 347 (2000).

[32] CPLEAR collaboration, A. Apostolakis *et al.*, Phys. Lett. B **456**, 297 (1999).

2

Overview of Kaon Physics

R. D. Peccei

Abstract

In this overview, I discuss some of the open issues in kaon physics. After briefly touching on lattice calculations of kaon dynamics and tests of CPT, I focus my attention on ϵ'/ϵ and on constraints on the CKM model. The impact of rare K decays and of experiments with B mesons for addressing the issue of CP violation is also discussed. The importance of looking for signals of flavor-conserving CP-violating phases is emphasized.

2.1 Introductory Remarks

It is difficult to overview a mature field like kaon physics. In thinking about the subject, it seemed natural for me to divide my report into three parts. The first of these parts addresses areas of kaon physics where steady progress continues to be made, but more still needs to be done. Kaon dynamics, as well as tests of conservation laws and of the CKM model [1] properly belong in this category. The second part encompasses what, colloquially, might be called the "hot topic" of kaon physics—the new experimental results on ϵ'/ϵ and their theoretical interpretation. Finally, in the last part, I group together topics in kaon physics that bring new insights into the future, particularly regarding the nature of flavor. Although CP violation in B decays obviously has little to do directly with kaons its study, along with that of rare kaon decays, properly belongs in this last category. So does the hunting for new CP-violating phases.

2.2 Kaon Decay Dynamics

By contrasting the ratio of the charged to neutral kaon lifetimes [2]

$$\tau(K^+)/\tau(K_S^0) = 138.6 \pm 0.4$$

with that for B mesons [2]

$$\tau(B^+)/\tau(B^0) = 1.072 \pm 0.026,$$

it is clear that kaon decays involve strong interaction dynamics. While B decays are essentially reflective of the decays of the b quark, kaon decays are strongly dependent on the underlying QCD dynamics.

A long-standing puzzle of kaon decays has been the so-called $\Delta I = 1/2$ rule, which encodes the dominance of the $I = 0$ $\pi\pi$ final state in kaon decays. Why this should be so, and what fixes the ratio of isospin amplitudes

$$\frac{A_2(K \to \pi\pi)}{A_0(K \to \pi\pi)} \simeq \frac{1}{22} \tag{2.1}$$

is not really known. Although QCD gives a significant short distance enhancement to the $\Delta I = 1/2$ matrix element in the $K \to 2\pi$ amplitudes [3], the dominant effect appears to be nonperturbative in nature. So, to estimate theoretically the ratio (2.1) requires lattice or $1/N_c$ methods.

Recent lattice QCD results for A_2/A_0 [4] show a significant enhancement for this ratio, but do not yet reproduce the experimental result (2.1). There are several reasons one can adduce for the roughly factor-of-two discrepancy between theory and experiment. First, the results obtained are quite sensitive to **chiral corrections**. Technically what is studied are the matrix elements of operators between a kaon and pion state $[\langle\pi|O_i|K\rangle]$, rather than the matrix elements of the relevant operators between a kaon and two pions. These quantities are related exactly only in the soft pion limit, so one must carefully correct for this. In addition, in the recent calculation of Pekurovsky and Kilcup [4], it appears that A_2 itself is strongly dependent on the $m_K^2 \to 0$ extrapolation performed. Finally, for accurate results, fully dynamical quarks must be included. However, present results [4] do not appear to show much difference between quenched and unquenched calculations.

The inclusion of dynamical quarks, perhaps, is more critical to the attempts to extract a reliable value for the strange quark mass—a quantity which is of importance for ϵ'/ϵ. This is nicely illustrated in fig. 2.1, which shows that in the quenched limit the values of m_s are quite sensitive to the discretization used.[1] From this data one can estimate a quenched value [6] for the strange quark mass of

$$m_s(2 \text{ GeV}) = (100 \pm 20) \text{ MeV}. \tag{2.2}$$

My sense is that this value still **overestimates** the true value. However, it is not clear by how much!

2.3 Testing Conservation Laws

Almost from their discovery, kaon decays have been a marvelous test bed for conservation laws and have helped us improve our understanding of

[1]These values are also quite sensitive to the physical input used. For instance, the CP-PACS collaboration [5] reports value for m_s which differs by about 30 MeV, depending on whether they used kaons or ϕ mesons as input.

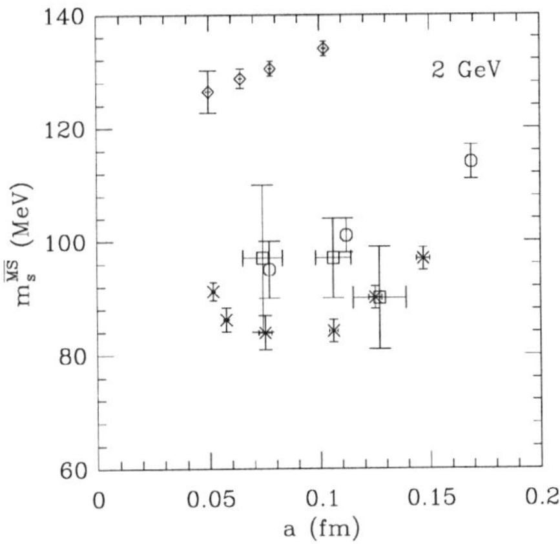

Figure 2.1: Results for m_s in the quenched approximation, for different discretizations, from ref. [6].

particle interactions. The tests done with kaons have ranged from tests of fundamental conservation laws, like CPT, to more mundane tests like those connected to the $\Delta S = \Delta Q$ rule, which basically just reflects the interactions allowed by the standard model. The most famous tests of conservation laws in the kaon system, of course, deal with CP violation. I will address this subject in much more detail in the coming sections. Here, I concentrate on two tests that help probe physics at very large scales—scales much larger than the scale of the weak interactions $v_F \sim (\sqrt{2}\, G_F)^{-1/2} \simeq$ 250 GeV.

2.3.1 Lepton Flavor Violation

The first of these tests involve lepton flavor violation. Here strong bounds exist, in both neutral kaon decays [7]

$$\mathrm{BR}(K_L^0 \to \mu e) < 4.7 \times 10^{-12} \quad (90\%\ \mathrm{CL}) \tag{2.3}$$

and charged kaon decays [8]

$$\mathrm{BR}(K^+ \to \pi^+ \mu e) < 2.1 \times 10^{-10} \quad (90\%\ \mathrm{CL}), \tag{2.4}$$

with more refined values expected at this conference (see chapters 35 and 36). These bounds, typically, can be used to pin down limits on the scale of possible new physics associated with lepton flavor violation. For instance, the existence of massive leptoquarks would violate lepton number. Assuming that the leptoquark coupling to quarks and leptons is of electroweak strength, t-channel leptoquark exchange gives a branching ratio for the process $K_L \to \mu e$:

$$\mathrm{BR}(K_L \to \mu e) \sim \left(\frac{M_W}{M_{LQ}}\right)^4 . \tag{2.5}$$

Thus, branching ratios of order 10^{-12} probe leptoquark masses in the 100 TeV range.

2.3.2 CPT Violation

The CPT theorem [9] is based on rather sacred principles. Any local, Lorentz-invariant, quantum field theory with the normal spin-statistics connection conserves CPT. Thus, the experimental observation of a signal that violates CPT would be a spectacular discovery. Theoretically, a violation of CPT can also occur through a violation of quantum mechanics [10]. Corrections to the Schrödinger equation for the density matrix of the form

$$i\frac{\partial}{\partial t}\rho = H\rho - \rho H^{\dagger} + \delta h \rho \tag{2.6}$$

will produce phenomena that effectively violate CPT, if $\delta h \neq 0$.

In the following, for simplicity, I will assume that quantum mechanics is valid. Even in this case, the phenomenology of CPT violation is quite rich. Essentially, CPT violation causes two modifications to the usual $K^0 - \bar{K}^0$ formalism [11]:

i) The diagonal elements of the effective Hamiltonian for the system, $H = M - \frac{1}{2}\Gamma$, are no longer equal. This introduces into the formalism a CPT-violating parameter

$$\delta = \frac{M_{K^0} - M_{\bar{K}^0} - \frac{i}{2}(\Gamma_{K^0} - \Gamma_{\bar{K}^0})}{2\left[m_S - m_L - \frac{i}{2}(\Gamma_S - \Gamma_L)\right]}. \tag{2.7}$$

Here $m_{S,L}$ and $\Gamma_{S,L}$ are, respectively, the masses and widths of the physical eigenstates of the $K^0 - \bar{K}^0$ complex, the short- and long-lived K mesons, K_S and K_L.

ii) The decay amplitudes for any physical processes are twice as many, since particle and antiparticle decays are no longer simply connected by charge conjugation (modulo strong rescattering phases). Thus, for example, one has [12]

$$\langle 2\pi; I|T|K^0\rangle = (A_I + B_I)e^{i\delta_I}; \quad \langle 2\pi; I|T|\bar{K}^0\rangle = (A_I^* - B_I^*)e^{i\delta_I} \tag{2.8}$$

$$\langle \pi^- \ell^+ \nu_\ell|T|K^0\rangle = a + b; \quad \langle \pi^+ \ell^- \bar{\nu}_\ell|T|\bar{K}^0\rangle = a^* - b^*, \tag{2.9}$$

where the amplitudes B_I and b violate CPT.

The principal test of CPT in the $K^0 - \bar{K}^0$ system at the moment comes from measurements of the parameter ϵ, connected with the amplitude ratio of the K_L and K_S amplitudes into two pions. One can show [11] that ϵ has two components, one CP violating and one CPT violating:

$$\epsilon = \epsilon_{C\not{P}} \exp\left[i\phi_{SW}\right] + \epsilon_{C\not{P}T} \exp\left[i\left(\phi_{SW} + \frac{\pi}{2}\right)\right]. \tag{2.10}$$

Here $\phi_{SW} = \tan^{-1}[2(m_S - m_L)/(\Gamma_S - \Gamma_L)] = (43.64 \pm 0.08)^\circ$ [2], while

$$\epsilon_{C\not{P}T} \simeq \sqrt{2}\mathrm{Im}\delta \simeq \sqrt{2}(\frac{\mathrm{ReB}_0}{\mathrm{ReA}_0} - \mathrm{Re}\delta). \tag{2.11}$$

Because $\phi_{+-} \simeq \phi_{SW}$, it is clear that $\epsilon_{C\not{P}} \gg \epsilon_{C\not{P}T}$. Thus, the decay $K_L \to \pi\pi$ is a sign of CP violation, not CPT violation. Since, ϵ' is small, one can infer a value for $\epsilon_{C\not{P}T}$ from the measurement of ϕ_{+-} [2]. In this way, one arrives at the following values for the CPT-violating parameters:

$$\mathrm{Im}\delta = (\frac{\mathrm{ReB}_0}{\mathrm{ReA}_0} - \mathrm{Re}\delta) = (0.29 \pm 1.58) \times 10^{-5}. \tag{2.12}$$

With only this measurement, it is not possible to bound the $K^0 - \bar{K}^0$ mass difference, since this difference depends on both Reδ and Imδ. However, the CPLEAR collaboration has recently measured Reδ independently, by studying the asymmetry in the time evolution of semileptonic K^0 and $\bar{K}^0$ decays. They find [13]

$$\mathrm{Re}\delta = (3.0 \pm 3.3 \pm 0.6) \times 10^{-4}. \tag{2.13}$$

This result should be improved by KLOE to perhaps the 10^{-4} level. At any rate, one can now bound

$$|M_{K^0} - M_{\bar{K}^0}| = 2(m_S - m_L)(\mathrm{Im}\delta - \mathrm{Re}\delta) = (2.1 \pm 2.8) \times 10^{-18}. \tag{2.14}$$

This is an extremely stringent test of CPT indeed! Further tests are treated in part III of this volume.

2.4 Testing the CKM Paradigm

The CKM model [1] is the simplest example of a model with CP violation. With three generations of quarks, the mixing matrix V_{CKM} in the charged current has **one** CP-violating phase.[2] More importantly, the model gives a consistent and qualitatively understandable description of all the observed phenomena. In particular, the CP-violating parameter ϵ is small not because the CP-violating phase γ in V_{CKM} is small, but as the result of the smallness of the intergenerational mixing.

The consistency of the CKM model with the observed CP-violating phenomena in the kaon system emerges from a careful study of constraints on the CKM mixing matrix. It is useful for these purposes, following Wolfenstein [15], to expand the elements of V_{CKM} in powers of the Cabibbo angle $\lambda = \sin\theta_c = 0.22$:

$$V_{\mathrm{CKM}} \simeq \begin{pmatrix} 1-\lambda^2/2 & \lambda & A\lambda^3(\rho - i\eta) \\ -\lambda & 1-\lambda^2/2 & A\lambda^2 \\ A\lambda^3(1-\rho-i\eta) & -A\lambda^2 & 1 \end{pmatrix} + O(\lambda^4)\ . \tag{2.15}$$

[2]The model has also a CP-violating angle $\bar{\theta}$, which must be extremely tiny to respect the strong bounds one has on the neutron dipole moment. Why this should be so is not really understood [14].

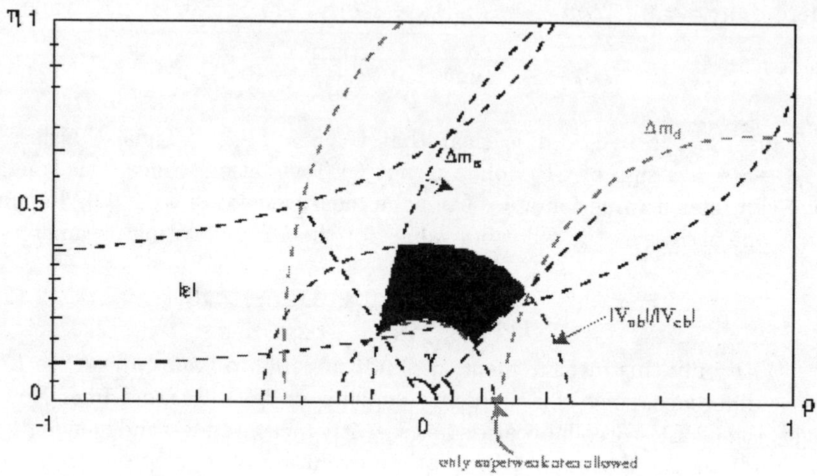

Figure 2.2: Allowed region in the $\rho - \eta$ plane. From ref. [16].

One sees from the above that, to $O(\lambda^4)$, the only complex phases in $V_{\rm CKM}$ enter in the V_{ub} and V_{td} matrix elements:

$$V_{ub} = A\lambda^3(\rho - i\eta) \equiv |V_{ub}|e^{-i\gamma}; \quad V_{td} = A\lambda^3(1-\rho-i\eta) \equiv |V_{td}|e^{-i\beta} \ . \tag{2.16}$$

The unitarity condition $\sum_i V_{ib}^* V_{id} = 0$ on the $V_{\rm CKM}$ matrix elements has a nice geometrical interpretation in terms of a triangle in the $\rho - \eta$ plane with base $0 \leq \rho \leq 1$ and with an apex subtending an angle α, where $\alpha + \beta + \gamma = \pi$.

One can use experimental information on $|\epsilon|$, the $B_d - \bar{B}_d$ mass difference Δm_d, and the ratio of $|V_{ub}|/|V_{cb}|$ inferred from B decays to deduce a 95% CL allowed region in the $\rho - \eta$ plane. If one includes, additionally, information from the recently obtained strong bound on $B_s - \bar{B}_s$ mixing [$\Delta m_s > 12.4$ ps^{-1} (95% CL) [16]], one further restricts the CKM allowed region. Fig. 2.2 shows the result of a recent study for the Babar Physics Book [17]. As one can see, the data is consistent with a rather large CKM phase γ : $45° \leq \gamma \leq 120°$. If one were to imagine that $|\epsilon|$ is due to some other physics, as in the superweak theory [18], then effectively the $\Delta S = 1$ parameter $\eta \simeq \gamma \simeq 0$. In this case one has another allowed region for ρ at the 95% CL: $0.25 \leq \rho \leq 0.27$.

Two experimental results in 1999 have strengthened the case for the validity of the CKM model:

i) The CDF collaboration [19] has announced a first significant result for $\sin 2\beta$ from a study of $B \to \Psi K_S$ decays

$$\sin 2\beta = 0.79^{+0.41}_{-0.44}, \tag{2.17}$$

$\epsilon'/\epsilon \times 10^4$	Reference
$4.6 \pm 3.0 \pm 0.4$	Ciuchini *et al.* [24]
$5.2^{+4.6}_{-3.7}$	Bosch *et al.* [25]
$7.7^{+6.0}_{-3.5}$	Bosch *et al.* [25]
5–22	Paschos *et al.* [26]
17^{+14}_{-10}	Bertolini *et al.* [27]

Table 2.1: Recent theoretical results on ϵ'/ϵ.

whose central value coincides with that emerging from the fit of fig. 2.2.

ii) Recently, the KTeV collaboration announced a new result for ϵ'/ϵ obtained from an analysis of about 20% of the data collected in the last two years [20]. Their result

$$\text{Re } \epsilon'/\epsilon = (28.0 \pm 4.1) \times 10^{-4} \tag{2.18}$$

is much closer to the old CERN result [Re $\epsilon'/\epsilon = (23 \pm 6.5) \times 10^{-4}$] [21] than to the value obtained by KTeV's precursor [Re $\epsilon'/\epsilon = (7.4 \pm 5.9) \times 10^{-4}$] [22]. More significantly, the value obtained is nearly 7σ away from zero, giving strong evidence that the CKM phase γ is indeed nonvanishing.

2.5 Lessons from ϵ'/ϵ

If one were to average the KTeV [20], NA31 [21], and E731 [22] results for ϵ'/ϵ, the resulting number [Re $\epsilon'/\epsilon = (21.8 \pm 3.0) \times 10^{-4}$] is in marvelous agreement with the result announced at Kaon99 by the NA48 Collaboration [23]:

$$\text{Re } \epsilon'/\epsilon = (18.5 \pm 4.5 \pm 5.8) \times 10^{-4}. \tag{2.19}$$

This is very strong evidence that there exists $\Delta S = 1$ CP violation, which is the chief premise of the CKM model.

A nonzero ϵ'/ϵ is good news for the CKM model. However, a value for this ratio of around 20×10^{-4} is unsettling, since this result appears a bit too large. This can be seen from table 2.1, which displays some recent theoretical expectations for ϵ'/ϵ. What is particularly perturbing is that the smaller values in the table correspond, in principle, to more theoretically pristine calculations of the relevant matrix elements, based on lattice methods. However, the calculation of ϵ'/ϵ is challenging, since large cancellations are involved.

As is well known, ϵ'/ϵ depends on both the matrix elements of gluonic and electroweak penguin operators [28]. The former are enhanced over the latter, since they are of $O(\alpha_s)$ rather than of $O(\alpha)$. However, the electroweak penguins are not suppressed by the $\Delta I = 1/2$ rule and are enhanced by the large top mass [29]. As a result, these two contributions are comparable in size. What makes matters worse is that these contributions tend to cancel each other, increasing the uncertainty in the theoretical predictions. One can appreciate the nature of the problem from an approximate formula for ϵ'/ϵ due to Buras and collaborators [25]:

$$\mathrm{Re}\ \epsilon'/\epsilon \simeq 34\eta\,[B_6 - 0.53B_8]\left[\frac{110\,\mathrm{MeV}}{m_s(2\,\mathrm{GeV})}\right]^2 \times 10^{-4}. \tag{2.20}$$

In the vacuum insertion approximation, the contribution of the gluonic penguins, B_6, and that of the electroweak penguins, B_8, are both equal to 1. Since $\eta \simeq 0.3$, in this approximation, one expects $\epsilon'/\epsilon \simeq 5 \times 10^{-4}$.

To get agreement with experiment, one needs an appropriate linear combination of four things to happen: η should be maximized [$\eta \simeq 0.4$?]; the strange quark mass should be smaller [$m_s(2\,\mathrm{GeV}) \simeq 90\,\mathrm{Mev}$?]; B_6 should be bigger than unity [$B_6 \simeq 2$?]; B_8 should be smaller than unity [$B_6 \simeq 0.5$?]. As Bosch *et al.* remark in their recent analysis [25], it is possible to stretch the parameters to get $\epsilon'/\epsilon \simeq 20 \times 10^{-4}$, but it is not easy! What is agreed upon is that the gluonic penguin matrix elements are very uncertain in the lattice. However, whether B_6 can get much bigger than 1, as it appears to be in the chiral quark model [27], remains to be seen. On the other hand, both lattice and 1/N estimates [26] suggest that B_8 can be quite a bit smaller than unity.

The apparent discrepancy between theory and experiment for ϵ'/ϵ has spurred a number of people to invoke new physics explanations. For instance, if somehow the Zds vertex were anomalous [30], then one could get a bigger value for ϵ'/ϵ.[3] However, it is not clear what the physical origin of such an anomaly is. Chanowitz [31] has tried to relate this anomaly to that which seems to affect $Z \to b\bar{b}$. However, it turns out that this connection gives the wrong sign for ϵ'/ϵ! Nevertheless, if such an anomalous Zds vertex existed, it would substantially increase the branching ratio for $K_L \to \pi^0\nu\bar{\nu}$ [32]. Masiero and Murayama [33], on the other hand, noticed that in supersymmetric extensions of the SM it is possible to get contributions to ϵ'/ϵ of order 10^{-3} from rather natural $\tilde{s}_R\tilde{d}_R$ squark mixings. Thus they suggested that, perhaps, the large value of ϵ'/ϵ is the first experimental manifestation of low-energy supersymmetry at work.

I am personally quite skeptical that the ϵ'/ϵ result is a signal of new physics. In my view, the most likely explanation for the discrepancy between theory and experiment is rooted in our inability to accurately calculate K decay matrix elements. Our long and frustrating experience with the $\Delta I = 1/2$ rule should provide an object lesson here! Perhaps the most naive conclusion to draw is that, no matter what else is causing ϵ'/ϵ to be large, the CKM parameter $\eta \sim \sin\gamma$ lies near its maximum. That is,

[3] One needs $ImZ_{ds} \simeq -(2.5\text{–}8)\times 10^{-4}$, rather than the SM value of around $+1\times 10^{-4}$, for this parameter.

perhaps $\gamma \simeq \pi/2$. I do not know a particular reason why this should be so.[4] However, if this is so, there is an interesting phenomenological consequence. It is easy to see that $\gamma \simeq \pi/2$ predicts that

$$\sin 2\alpha \simeq \sin 2\beta \simeq \frac{2\eta}{1+\eta^2}. \tag{2.21}$$

This "prediction" is at the edge of the CKM fits, but suggest that $\sin 2\alpha$, like $\sin 2\beta$, is also large.

2.6 Grappling with the Unitarity Triangle

From the above discussion, it is clear that to make progress one will need further experimental input. Fortunately, help is on the way!

2.6.1 CP Violation in B Decays

With the turn-on of the B factories at SLAC and KEK, and with the upgrade of the Tevatron with the main injector, a new experimental era in the study of CP violation is beginning. Data that will be collected by the new BaBar and Belle detectors, and with CDF and DO, should permit testing the unitarity triangle through separate measurements of α, β, and γ. As I alluded to earlier, one of the most robust predictions of the CKM model is that $\sin 2\beta$ should be large. In contrast to ϵ'/ϵ, this parameter can be extracted in a theoretically clean way by studying B decays to CP self-conjugate states [35]. The decay probability of a state that at $t = 0$ was a B_d into a CP self-conjugate final state, like ΨK_S, has a time evolution that directly isolates $\sin 2\beta$:

$$\Gamma(B_d^{\text{Phys.}}(t) \to \Psi K_S) \sim e^{-\Gamma_B t}\left[1 + \sin 2\beta \sin \Delta m_d t\right]. \tag{2.22}$$

To measure $\sin 2\beta$ experimentally one needs to be able to tag the initial state as a B_d and then to follow its time development. Once this is achieved, as was done recently by CDF [19], then eq. (2.22) yields $\sin 2\beta$ with essentially no theoretical error.

A study done in preparation for the turn-on of the SLAC B factory [17] estimates that, with an integrated luminosity of $30\,\text{fb}^{-1}$, one could measure $\sin 2\beta$ with an error of the order of $\delta \sin 2\beta \leq 0.08$. However, to really test the CKM model, measuring $\sin 2\beta$ is not enough. What one wants really is to measure also the other two angles in the unitarity triangle, to see if the triangle indeed closes. In addition, getting a clean measurement of the Wolfenstein parameter η would be helpful, as this parameter measures the height of the triangle and hence provides redundant information. However, extracting α and γ from B decays to comparable accuracy to that with

[4]Note that this is not the same as the idea of maximal CP violation, suggested by Fritzsch and others [34].

which β will be known is likely to be challenging. To reduce the error on these quantities a variety of processes will need to be studied.

Let me briefly illustrate the nature of the problem by discussing how to obtain α in a manner analogous to β. It is easy to see [35], *mutatis mutandis*, that the time development of the decay $B_d^{\rm phys} \to \pi^+\pi^-$ provides information on α. However, for the decay $B_d \to \pi^+\pi^-$, the quark decay amplitude $b \to u\bar{u}d$ and the penguin amplitude entering in the process depend differently on the weak CP-violating phases. The quark decay is proportional to $e^{-i\gamma}$, while the penguin piece is proportional to $e^{-i\beta}$.[5] Only by neglecting the penguin contributions altogether does one arrive in this case at a formula like (2.22), with $\beta \to \alpha$.

It is possible to estimate and correct for the penguin pollution through an isospin analysis of various channels [37]. This, however, exacts a price in precision. Although only time and real data will tell, it appears difficult for me to imagine measuring $\sin 2\alpha$ to better than $\delta \sin 2\alpha = 0.15$. Coupled with estimates of the error with which one can extract the CP-violating phase γ from B decays [38], my guess is that probably one will not be able to bring down the error on the sum of the angles in the unitarity triangle to better than $\delta[\alpha + \beta + \gamma] \simeq (20 - 30)^\circ$.

2.6.2 $K_L \to \pi^0 \nu\bar{\nu}$

In this respect, the process $K_L \to \pi^0\nu\bar{\nu}$ offers an interesting alternative opportunity to test the CKM model. This decay allows a theoretically very clean extraction of η, and hence of the CKM phase γ. Splitting K_L into its CP-even and CP-odd parts, the amplitude for this process can be written as:

$$A(K_L \to \pi^0\nu\bar{\nu}) = \epsilon A(K_1 \to \pi^0\nu\bar{\nu}) + A(K_2 \to \pi^0\nu\bar{\nu}). \tag{2.23}$$

However, because the semileptonic decay of the CP-even state K_1 is small and this amplitude is suppressed by a factor of ϵ in the above, the decay $K_L \to \pi^0\nu\bar{\nu}$ directly measures $\Delta S = 1$ CP violation [39]. In addition, the amplitude $K_2 \to \pi\nu\bar{\nu}$ is essentially free of hadronic uncertainties since it depends only on a K to π matrix element.

The NLO QCD analysis of Buchalla and Buras [40] gives the following approximate formula, good to 1–2%, for the $K_L \to \pi^0\nu\bar{\nu}$ branching ratio:

$$\mathrm{BR}[K_L \to \pi^0\nu\bar{\nu}] = 4.34 \times 10^{-4} A^4 \eta^2. \tag{2.24}$$

Using the results of the CKM analysis, leads to the expectation [25] $1.6 \times 10^{-11} \leq \mathrm{BR}\left[K_L \to \pi^0\nu\bar{\nu}\right] < 3.9 \times 10^{-11}$. However, this process presents a formidable experimental challenge. Not only is its branching very low, but one is dealing with an all-neutral final state. Nevertheless, it is clear that

[5] There is no penguin pollution [36] for $B_d \to \Psi K_S$, since in this case both the quark decay amplitudes and the penguin amplitudes have the same weak phase.

an experiment that could probe the process $K_L \to \pi^0\nu\bar{\nu}$ to a branching ratio of the order of 3×10^{-11}, with an accuracy better than 25%, would have a significant impact on the CKM model. Of course, as I remarked earlier, if there were indeed an anomalous Zds vertex, then one could expect branching ratios almost an order of magnitude higher [32]

2.7 Looking for Other CP Violating Phases

It is quite possible that the CKM phase γ is the dominant phase connected with **flavor-changing** CP violation. Nevertheless, in my opinion, it is also quite likely that, in addition, there are also some other **flavor-conserving** CP-violating phases. The argument is simple. Given that CP is not conserved, renormalizability requires that any interactions in the Lagrangian of the theory that can be complex should be so. This means that any interactions beyond the SM involving new fields necessarily always will involve new phases. For instance, with two Higgs doublets one has a phase in the mass term connecting these two fields,

$$\mathcal{L}_{H_1H_2} = \mu^2 H_1 H_2^\dagger + \mu^{2^*} H_1^\dagger H_2, \tag{2.25}$$

which just cannot be avoided if CP is violated.[6]

Kaon decays afford a wonderful opportunity to search for the presence of these flavor-conserving CP-violating phases. The nicest example, perhaps, is provided by quantities where CKM effects either vanish or are negligible. A good case in point is the triple correlation in $K_{\mu 3}$ decays that measures the polarization of the outgoing muon perpendicular to the plane of production, $\langle P_\perp^\mu \rangle$—a quantity that vanishes in the CKM model [42].

Since $\langle P_\perp^\mu \rangle \sim \langle \vec{s}_\mu \cdot (\vec{p}_\mu \times \vec{p}_\pi)\rangle$, the transverse muon polarization is T-odd, and therefore is sensitive to CP-violating contributions. However, as Sakurai [43] pointed out long ago, $\langle P_\perp^\mu \rangle$ can also arise from final-state rescattering effects. Fortunately, although $\langle P_\perp^\mu \rangle$ is affected by final-state interactions (FSI), for the process $K^+ \to \pi^0\mu^+\nu_\mu$, because the final hadron is neutral, these FSI are very small [$\langle P_\perp \rangle^\mu_{\rm FSI} \sim 10^{-6}$] [44].

The transverse muon polarization is particularly sensitive to any scalar interactions present in the decay amplitude. Writing the effective amplitude for the process $K^+ \to \pi^0\mu^+\nu_\mu$ as

$$\begin{aligned} A(K^+ \to \pi^0\mu^+\nu_\mu) &= G_F\lambda f_+(q^2)[p_\alpha \bar{u}_\mu \gamma^\alpha (1-\gamma_5)u_{\nu_\mu} \\ &\quad + f_o(q^2) m_\mu \bar{u}_\mu (1-\gamma_5) u_{\nu_\mu}], \end{aligned} \tag{2.26}$$

one finds that the transverse muon polarization $\langle P_\perp^\mu \rangle$ is determined by the imaginary part of the scalar form factor:

$$\langle P_\perp^\mu \rangle \simeq 0.2 \ \mathrm{Im} f_s, \tag{2.27}$$

[6] There are other arguments that point towards the existence of other CP violating phases, besides γ. For instance, one needs to have some non-GIM-suppressed CP-violating phase in the theory to generate the matter-antimatter asymmetry in the universe at the electroweak phase transition [41].

with the numerical constant being essentially a kinematical factor.

The simplest models which have a nontrivial Im f_s are multi-Higgs models [45]. In these models one can, in fact, obtain values for Im f_s that are at the verge of observability [46]. Bounds on Im f_s coming from other processes allow $\langle P_\perp^\mu \rangle \leq 10^{-2}$ [47]. This is precisely the level of sensitivity of a measurement of $\langle P_\perp^\mu \rangle$ done at Brookhaven over 15 years ago [48]:

$$\langle P_\perp^\mu \rangle = [-3.1 \pm 5.3] \times 10^{-3}. \tag{2.28}$$

An ongoing experiment at KEK, KEK 246, should be able to improve this measurement slightly. However, it would be very interesting if one were able to mount an experiment to get to $\delta\langle P_\perp^\mu \rangle \sim 10^{-4}$.

2.8 Concluding Remarks

We are at a very exciting juncture in kaon physics and in the study of CP violation. At long last, we can now say that ϵ'/ϵ is at hand. Even though we are now entering an era where crucial information on CP violation will be learned from B decays, experiments with kaons will continue to play a central role. Experiments at the Frascati ϕ factory should further refine our knowledge of fundamental conservation laws, like CPT, and rare kaon decays will provide precious windows into new phenomena. It is particularly important that, experimentally, we not be afraid to push into obscure corners—like flavor-conserving CP violation.

On the theoretical front, dynamical calculations of weak decay matrix elements are approaching the level of accuracy needed to really test the theory. However, it is crucial to properly estimate theoretical uncertainties, so that one can better gauge if discrepancies with experiment really signal new physics effects. Indeed, the real issue with ϵ'/ϵ is whether the large value seen represents new physics or old matrix elements!

Acknowledgments

This work was supported in part by the Department of Energy under contract No. DE-FG03-91ER40662, Task C.

References

[1] N. Cabibbo, Phys. Rev. Lett. **12**, 531 (1963); M. Kobayashi and T. Maskawa, Prog. Theor. Phys. **49**, 652 (1973).

[2] Particle Data Group, C. Caso *et al.*, Europ. Phys. J. C, **3** 1 (1998).

[3] M.K. Gaillard and B.W. Lee, Phys. Rev. Lett. **33**, 108 (1974); G. Altarelli and L. Maiani, Phys. Lett. B **52**, 351 (1974).

[4] For a recent discussion, see, for example, D. Pekurovsky and G. Kilcup, hep-lat/9903025, in Proceedings of the DPF99 Conference, Los Angeles, ed. K. Arisaka and Z. Bern, http://www.dpf99.library.ucla.edu.

[5] CP-PACS Collaboration, R. Burkhalter, in Proceedings of Lattice 98, Nucl. Phys. B. (Proc. Suppl.) **73**, 3 (1999).

[6] T. Blum, A. Soni, and M. Wingate, hep-lat/9902016.

[7] BNL 871 Collaboration, D. Ambrose *et al.* Phys. Rev. Lett. **81**, 5734 (1998).

[8] BNL 777 Collaboration, A.M. Lee *et al.*, Phys. Rev. Lett **64**, 165 (1990).

[9] W. Pauli, in *Niels Bohr and the Development of Physics*, ed. W. Pauli (Pergamon Press, New York 1955); J. Schwinger, Phys. Rev. **82**, 914 (1951); G. Lüders, Dansk Mat. Fys. Medd **28**, 5 (1954); G. Lüders and B. Zumino, Phys. Rev. **110**, 1450 (1958).

[10] J. Ellis, J.S. Hagelin, D.V. Nanopoulos, and M. Srednicki, Nucl. Phys. B **241**, 381 (1984).

[11] C.D. Buchanan, R. Cousins, C.O. Dib, R.D. Peccei, and J. Quackenbush, Phys. Rev. D **45** 4088 (1992); C.O. Dib and R.D. Peccei, Phys. Rev. D **46**, 2265 (1992).

[12] V.V. Barmin *et al.*, Nucl. Phys. B **247**, 293 (1984): Erratum B **254**, 747 (1985); see also N.W. Tanner and R.H. Dalitz, Ann. Phys **171**, 463 (1986).

[13] CPLEAR Collaboration, A. Angelopoulos *et al.*, Phys. Lett. B **444**, 52 (1998).

[14] For a discussion see, for example, R.D. Peccei, Nucl. Phys. B (Proc. Suppl.), **72**, 3 (1999).

[15] L. Wolfenstein, Phys. Rev. Lett. **51**, 1945 (1983).

[16] F. Parodi, in the Proceedings of the 29th Conference on High Energy Physics, ICHEP98, Vancouver, Canada, July 1998, eds. A. Astbury, D. Axen, and J. Robinson (World Scientific, Singapore, 1999) p. 1148.

[17] *The Babar Physics Book*, ed. P.F. Harrison and H.R. Quinn, SLAC-R 504, October 1998.

[18] L. Wolfenstein, Phys. Rev. Lett. **13**, 562 (1964).

[19] I.J. Kroll, this volume, chapter 51; see also CDF pub. CDF/PUB/Bottom/CDF/4855.

[20] KTeV Collaboration, A. Alavi-Harati *et al.*, Phys. Rev. Lett. **83**, 22 (1999).

[21] NA31 Collaboration, G.D. Barr *et al.*, Phys. Lett. B **317**, 1233 (1993).

[22] E731 Collaboration, L.K. Gibbons *et al.*, Phys. Rev. Lett. **70**, 1203 (1993).

[23] M.S. Sozzi, this volume, chapter 8.

[24] M. Ciuchini, E. Franco, G. Martinelli and L. Reina, Phys. Lett. B **301**, 263 (1993); M. Ciuchini, E. Franco, G. Martinelli, L. Reina, and L. Silvestrini, Z. Phys. C**68**, 239 (1995).

[25] S. Bosch, A.J. Buras, M. Gorbahn, S. Jäger, M. Jamin, M.E. Lautenbacher, and L. Silvestrini, hep-ph/9904408.

[26] E.A. Paschos, DO-TH 99/04, to be published in the Proceedings of the 17th International Workshop on Weak Interactions and Neutrinos (WIN99), Cape Town, South Africa, 1999; T. Hambye, G.O. Koehler, E.A. Paschos, and P.H. Soldan, hep-ph/9906434.

[27] S. Bertolini, J.O. Eeg, and M. Fabbrichesi, hep-ph/9802405, to appear in Rev. Mod. Phys.; S. Bertolini, J.O. Eeg, M. Fabbrichesi, and E. I Lashin, Nucl. Phys. B **514** 93 (1998).

[28] G. Buchalla, A.J. Buras and M.E. Lautenbacher, Rev. Mod. Phys. **68** 1125 (1996).

[29] J. Flynn and L. Randall, Phys. Lett. B **216**, 221 (1989); *ibid.* B **224**, 221 (1989); Nucl. Phys. B **326**, 3 (1989).

[30] Y.-Y. Keum, U. Nierste, and A.I. Sanda, Phys. Lett. **B457** 157 (1999); L. Silvestrini, hep-ph/9906202, to appear in the Proceedings of the XXXIV Rencontres de Moriond "Electroweak Interactions and Unified Theories," Les Arcs, France, 1999; see also A.J. Buras and L. Silvestrini, Nucl. Phys. B **546**, 299 (1999).

[31] M. Chanowitz, hep-ph/9905478.

[32] G. Colangelo and G. Isidori, JHEP **09**, 9 (1998).

[33] A. Masiero and H. Murayama, Phys. Rev. Lett. **83**, 907 (1999).

[34] For a review see, for example, H. Fritzsch, in *Broken Symmetries*, ed. L. Mathelitsch and W. Plessas, Lecture Notes in Physics 521 (Springer Verlag, Berlin, 1999); see also A. Mondragon and E. Rodriguez-Jauregui, hep-ph/9906429.

[35] See, for example, I.I. Bigi, V.A. Khoze, N.G. Uraltsev, and A. Sanda, in *CP Violation* ed. C. Jarlskog (World Scientific, Singapore, 1989), p. 175.

[36] M. Gronau, Phys. Rev. Lett. **63**, 1451 (1989); D. London and R.D. Peccei, Phys. Lett. B **223**, 257 (1989).

[37] M. Gronau and D. London, Phys. Rev. Lett. **65**, 3381 (1990).

[38] R. Fleischer, Phys. Lett. B **459**, 306 (1999).

[39] L. Littenberg, Phys. Rev. D **39**, 3322 (1989).

[40] G. Buchalla and A.J. Buras, Nucl. Phys. B **400** 225 (1993); B **548**, 309 (1999).

[41] M.E. Shaposhnikov, Nucl. Phys. B **287**, 757 (1987).

[42] M. Leurer, Phys. Rev. Lett. **62**, 1967 (1989).

[43] J.J. Sakurai, Phys. Rev. **109**, 980 (1958).

[44] A.R. Zhitnitski, Sov. J. Nucl. Phys. **31**, 529 (1980).

[45] S. Weinberg, Phys. Rev. Lett. **37** 657 (1976).

[46] R. Garisto, and G. Kane, Phys. Rev. D **44**, 2038 (1991); G. Belanger, and C.O. Geng, Phys. Rev. D **44**, 2789 (1991).

[47] Y. Grossman, Nucl. Phys. B **426**, 355 (1994).

[48] S.R. Blatt *et al.*, Phys. Rev. D **27**, 1056 (1983).

3

Baryogenesis and Low-Energy CP Violation

Mihir P. Worah

Abstract

CP violation is a crucial component in the creation of the matter-antimatter asymmetry of the universe. An important open question is whether the CP-violating phenomena observable in terrestrial experiments have any relation with those responsible for baryogenesis. We discuss two mechanisms of baryogenesis where this question can be meaningfully posed: "electroweak baryogenesis" and "baryogenesis via leptogenesis." We show how these scenarios can be constrained by existing and forthcoming experimental data. We present a specific example of both these scenarios where the CP-violating phase in the Cabibbo Kobayashi Maskawa matrix is related in a calculable way to the CP-violating phase responsible for baryogenesis.

The world that we observe is manifestly baryon asymmetric. All the stable matter we see is made up of baryons, with antibaryons being created only in high-energy collisions (either in the laboratories or out in the cosmos). There is evidence that this asymmetry persists even at much larger scales. Matter and antimatter galaxies within the same galactic cluster would result in strong γ ray emission due to annihilations. The absence of these confirms a baryon asymmetric region on the 20 Mpc scale [1]. More recently, a bound on the scale of the observable universe has been obtained by ruling out a contribution to the diffuse γ ray spectrum from particle-antiparticle annihilation [2]. The observed nuclear abundances in the stars then allow us to estimate that the current baryon-to-photon ratio, $n_B/n_\gamma = (4-7) \times 10^{-10}$. This corresponds to a baryon-antibaryon asymmetry of 1 part in 10^8 in the early universe.

One possible explanation for the asymmetry is that it is an initial condition that we cannot hope to understand. The other, more appealing, possibility is that although the universe initially had no net baryon number, microphysical processes that we can hope to understand led it to develop one during its evolution from the big bang to the present epoch. There are three requirements for such a baryon asymmetry to develop [3]:

1. There must be a departure from thermal equilibrium. CPT invariance gaurantees the equality of particle and antiparticle masses. Hence in thermal equilibrium both will have the same number density as dictated by Boltzmann statistics.

2. There must be baryon number violation. This requirement is self-explanatory.

3. There must be C and CP violation. This is required for the above baryon number–violating interactions to preferentially produce baryons rather than antibaryons.

The discovery of CP violation in the neutral K mesons thus made possible a meaningful discussion, in terms of physical processes, of why the universe consists of only matter and no antimatter.

It was later realized that the standard model, in fact, contains all of the three ingredients that are required for baryogenesis [4]. At a temperature $T \sim 100$ GeV in the early universe, the electroweak symmetry was broken due to the Higgs field acquiring a vacuum expectation value. This resulted in a phase transition that, if strong enough, could provide the departure from thermal equilibrium needed for baryogenesis. Although baryon number is conserved in the standard model at the classical level, it is broken at the quantum level due to the anomalous coupling of the $B + L$ (baryon number plus lepton number) current to two W bosons. This baryon number violation is unobservably small at zero temperature, but it is enhanced at high temperatures, and could be a viable source for the asymmetric creation of baryons over antibaryons. Finally, CP violation has been observed in the neutral K's, and is explained by a complex phase in the Cabibbo-Kobayashi-Maskawa (CKM) matrix.

Unfortunately, in the standard model, neither is the phase transition strong enough, nor is the CP violation efficient enough, to explain the observed baryon asymmetry. The requirement on the strength of the phase transition in order to be able to generate and mantain a baryon asymmetry is given by [5]

$$\frac{H(T_0)}{T_0} \geq 1. \tag{3.1}$$

Here $T_0 \sim 100$ GeV is the critical temperature for the phase transition, and $H(T_0)$ is the Higgs vacuum expectation value at this temperature. This strength is governed by the ratios of boson masses that are generated by the spontaneous symmetry breaking (SSB) to the mass of the Higgs boson. In the standard model, the W and Z bosons get their masses by SSB, and one obtains the approximate relationship

$$\frac{H(T_0)}{T_0} \sim \frac{2M_W^3 + M_Z^3}{2m_H^2 v} \sim \frac{1}{2}, \tag{3.2}$$

where we have used $M_W = 80$ GeV, $M_Z = 90$ GeV, $m_H = 95$ GeV (which is the current LEP lower bound), and $v = 246$ GeV is the zero temperature Higgs vacuum expectation value.

Assuming a strong enough phase transition and perfectly efficient baryon number violation one can obtain the estimate $n_B/n_\gamma \sim 10^{-2}\delta$ where δ is a dimensionless measure of CP violation [6]. However, the CKM mecahnism of CP violation in the standard model requires the participation of all three fermion families, and δ will be proportional to $Det\ C/T_0^{12} \sim 10^{-21}$, where $Det\ C$ is the Jarlskog determinant [7], and we have used $T_0 = 100$ GeV. There is a further suppression since the time scale needed for such interactions is so large that finite temperature plasma effects cause the participating particle wave functions to decohere before they can interfere enough to generate a significant CP asymmetry [8]. Thus, it is clear that one needs to invoke physics beyond the standard model in order to explain the baryon asymmetry of the universe.

In this talk we give an overview of baryogenesis in two extensions of the standard model. These models are motivated by the fact that they offer explanations for observed phenomena other than the baryon asymmetry that cannot be explained by the standard model and, most importantly, have low-energy experimental consequences. We will demonstrate that in these models it is possible to relate the CP violation responsible for baryogenesis with the CP violation observed in the neutral kaons.

One obvious possibility is to augment the standard model with new particles in the 100 GeV mass range that would remedy the deficiencies pointed out above [9]. Additional bosons that get their masses by the Higgs mechanism could enhance the strength of the electroweak phase transition. Moreover, the richer particle content could make CP violation more efficient. The most attractive such extension is the minimal supersymmetric standard model (MSSM), which we consider here. This model has its primary motivations in the facts that it stabilizes the hierarchy between the electroweak scale and the Planck scale, and that it provides a natural explanation of electroweak symmetry breaking.

The other distinct possibility is to use the baryon and/or lepton number– and CP-violating decays of some superheavy particle. The departure from thermal equilibrium typically occurs because the decay rate of the particle is slower than the expansion rate of the universe [10]. These processes must occur in the very early history of the universe because it is only then that the expansion rate was rapid enough to provide the out-of-equilibrium conditions needed for baryogenesis. The situation we will consider is where the standard model is augmented with massive ($\sim 10^{10}$ GeV) right-handed Majorana neutrinos that have lepton number– and CP-violating mass matrices. Their out-of-equilibrium decays generate a net lepton number, which is then processed by the anomolous $B + L$ violation in the standard model into a net baryon number. The primary motivation for this extension lies in the fact that it provides, via the see-saw mechanism, a framework for understanding the smallness of the left-handed neutrino masses suggested by the atmospheric and solar neutrino data.

3.1 Baryogenesis in the MSSM

The squarks (scalar partners of the quarks) present in the MSSM get contributions to their masses from supersymmetry breaking, as well as from electroweak symmetry breaking via the Higgs mechanism. In particular, $\tilde{t}_L$ and $\tilde{t}_R$, the scalar partners of the top quark, get a large contribution from the Higgs mechanism due to the size of the top quark Yukawa coupling to the Higgs boson. If the supersymmetry breaking mass of the $\tilde{t}_R$ is negligible, and there is no $\tilde{t}_L$- $\tilde{t}_R$ mixing (the $\tilde{t}_R$ is chosen to be light in order to avoid conflicts with the ρ parameter if the $\tilde{t}_L$ were light), eq. (3.2) gets modified to

$$\frac{H(T_0)}{T_0} \sim \frac{2M_W^3 + M_Z^3 + 2m_t^3}{2m_H^2 v} \sim 3 \tag{3.3}$$

for $m_t = 175$ GeV. Thus, we see the condition of eq. (3.1) can be satisfied and we have a strongly first-order phase transition. This simple relation is modified by the presence of supersymmetry breaking masses, $\tilde{t}_L - \tilde{t}_R$ mixing, and finite temperature effects. A detailed analysis [11] shows that an electroweak phase transition strong enough to allow baryogenesis is possible if $m_{\tilde{t}_R} \leq 175$ GeV and $m_H \leq 115$ GeV. Moreover, efficient baryogenesis requires rapid intraconversion between the particles and their supersymmetric partners. This means that most of the supersymmetric particles and especially the gauginos (fermionic partners of the gauge bosons) must also have masses of order $T_0 \sim 100$ GeV, where T_0 is the critical temperature for the electroweak phase transition. Besides the obvious direct search implications of these light sparticles and Higgs boson, the light $\tilde{t}_R$ and charginos also result in large contributions to $B - \bar{B}$ mixing. This is because the $b_L - \tilde{t}_R - \tilde{h}$ coupling, proportional to the top quark mass, removes the possibility of any GIM cancellation of its contribution [12].

The most effective way to generate a particle number asymmetry for some species is to arrange that, during the electroweak phase transition, a CP-violating space-time-dependent phase appears in the mass matrix for that species. If this phase cannot be rotated away at subsequent points by the same unitary transformation, it leads to different propagation probabilities for particles and antiparticles, thus resulting in a particle number asymmetry. The existence of two Higgs fields in the MSSM makes this possible. If $\tan\beta$ (the ratio of the expectation values of the two Higgs fields) changes as one traverses the bubble wall separating the symmetric phase from the broken one, particle number asymmetries can be generated, which will be proportional to $\Delta\beta$, the change in β across the bubble wall [13].

It has been estimated that $\Delta\beta \propto m_h^2/m_A^2 \sim 0.01$ for the pseudoscalar Higgs boson mass $m_A = 200 - 300$ GeV [14]. This can actually be turned into an upper bound for $\Delta\beta$ using the relation $m_{h_+}^2 = m_A^2 + m_W^2$, where m_{h_+} is the charged Higgs boson mass. Charged Higgs bosons make large positive contributions to the $b \to s\gamma$ decay rate. The current experimental value for $Br(b \to s\gamma)$ already sets the limit $m_{h_+} \gtrsim 300$ GeV at the 2σ level [16]. This then implies $\Delta\beta \lesssim 0.01$ through the relations above.

Baryogenesis in the MSSM proceeds most efficiently through the generation of higgsino number or axial squark number in the bubble wall, which then diffuses to the symmetric phase. Here, they bias the standard model $B + L$ violation to produce a net baryon number [13–15]. In this paper we present the special case of baryogenesis through the production of axial squark number, where the CKM phase responsible for kaon CP violation is also directly responsible for baryogenesis [15].

Consider the mass squared matrix for the up-type squarks:

$$M^2_{\tilde{u}} = \begin{pmatrix} M^2_{\tilde{u}_{LL}} & M^2_{\tilde{u}_{LR}} \\ M^{2\dagger}_{\tilde{u}_{LR}} & M^2_{\tilde{u}_{RR}} \end{pmatrix} \tag{3.4}$$

where

$$\begin{aligned} M^2_{\tilde{u}_{LL}} &= m_Q^2 A_{U_{LL}} + (F,\ D) \text{ terms}, \\ M^2_{\tilde{u}_{RR}} &= m_U^2 A_{U_{RR}} + (F,\ D) \text{ terms}, \\ M^2_{\tilde{u}_{LR}} &= m_A v_2 \lambda_U A_{U_{LR}} + \mu v_1 \lambda_U, \end{aligned} \tag{3.5}$$

where M_Q, M_U, and M_A are supersymmetry breaking masses, λ_U is the Yukawa coupling matrix for up-type quarks, and the A_Us are dimensionless matrices. Concentrating only on the production of $\tilde{t}_R$, and using $m_{\tilde{t}_R} = 175$ GeV, $m_{\tilde{t}_L} = 300$ GeV, and $\tan\beta \sim 1$, we obtain the result

$$\frac{n_B}{s} \simeq 10^{-8} \frac{\kappa \Delta\beta}{v_w} \frac{m_A}{T_0} \frac{|\mu|}{T_0} Im[e^{i\phi_B} A^\dagger_{U_{LR}} \lambda^\dagger_U \lambda_U]_{(3,3)}, \tag{3.6}$$

where κ is related to the rate of anomalous $B+L$ violation, $\Gamma_{B+L} = \kappa\alpha_w^4 T$. There is a large uncertainty in its precise value, with current estimates giving $\kappa = 1 - 0.03$ [17]. $v_w \simeq 0.1$ is the velocity of the wall separating the phase where electroweak symmetry is broken (the Higgs field has an expectation value) from where it is unbroken (the Higgs field has no expectation value). $\Delta\beta \lesssim 0.01$, and $T_0 \sim m_A \sim |\mu| \sim 100$ GeV. The approximations made in deriving eq. (3.6) and their validity are outlined in [13]. If $\tilde{t}_L$ and $\tilde{t}_R$ have very different masses there is a suppression of the baryon asymmetry by $m^2_{\tilde{t}_R}/m^2_{\tilde{t}_L}$ that is not explicit in their work. Thus the estimate of eq. (3.6) would be modified if $m_{\tilde{t}_L} \gg 300$ GeV.

Consider the possibility that the supersymmetric parameters $A_{U_{LR}}$ and μ are real, with all the CP violation being in the quark mass matrix [15]. Notice that $\lambda^\dagger_U \lambda_U$ in eq. (3.6) is Hermitian, hence the phase is on one of the off-diagonal terms. One then requires $A_{U_{LR}}$ to have off-diagonal entries in order to move this phase to the (3,3) element of the product $A^\dagger_{U_{LR}} \lambda^\dagger_U \lambda_U$. These large off-diagonal terms in $A_{U_{LR}}$ always lead to large $D - \bar{D}$ mixing due to gluino-mediated box diagrams. The magnitude of the mixing is generically about an order of magnitude lower than the current experimental bound $\Delta(m_D) < 1.3 \times 10^{-13}$ GeV. Further, given the hierarchical structure of the quark masses and mixings, one expects the largest off-diagonal entry in $\lambda_U^\dagger \lambda_U$ to be $\sim \theta_C^2 \sim 0.04$. For example the ansatz $\lambda_U = V_{CKM}^\dagger \hat{\lambda}_U V_{CKM}$ where V_{CKM} is the CKM matrix, and $\hat{\lambda}_U$ is the diagonal matrix of up-type Yukawa couplings, can lead to

$$Im[A_{U_{LR}}^\dagger \lambda_U^\dagger \lambda_U]_{(3,3)} = \lambda_t^2 |V_{cb}| \sin\gamma \tag{3.7}$$

for

$$A_{U_{LR}} = \begin{pmatrix} 1 & 0 & 0 \\ 0 & 1 & 1 \\ 0 & 1 & 1 \end{pmatrix}, \tag{3.8}$$

where $\gamma \sim 1$ is the phase in the CKM matrix. Thus, we see that the baryon asymmetry is directly related to the phase responsible for CP violation in $K - \bar{K}$ mixing. We can obtain a large enough baryon asymmetry [cf. eq. (3.6)] for $\kappa = 1,\ \Delta\beta = 0.01$.

3.2 Baryogenesis via Leptogenesis

The idea that one can obtain a baryon asymmetry by first generating a lepton asymmetry was first proposed in [18], and subsequently explored in several papers [19]. As mentioned earlier, $B + L$ is anomalously violated in the standard model, and the rate for this process is large at high temperatures. However, $B - L$ is conserved. Thus given enough time for the $B + L$–violating processes to act, we obtain the relations

$$\begin{aligned} (B-L)_f &= (B-L)_i \\ (B+L)_f &= 0, \end{aligned} \tag{3.9}$$

where the subscripts f and i stand for final and initial respectively. Thus if one started with zero initial baryon number, but nonzero initial lepton number, one would obtain the final condition $B_f = -L_i$ (this relationship is slightly modified by a careful consideration of all the standard model interactions [20]). The initial lepton number asymmetry is obtained by the CP- and lepton number–violating decay of heavy right-handed Majorana neutrinos.

Consider a model with right-handed Majorana neutrinos N_R. By definition these fields are self-conjugate, $N_R^c = N_R$ where the superscript c denotes the charge-conjugated field. Thus given the Yukawa interaction

$$\mathcal{L}_Y = -h_{ij} \bar{l}_L^i N_R^j H + h.c. \tag{3.10}$$

where h_{ij} is the matrix of Yukawa couplings, the l_L are left-handed standard model leptons, and H is the Higgs field one finds that N_R can decay into both light leptons and antileptons. If these decays are CP violating they will generate an excess of one over the other. Let us define an asymmetry

$$\delta = \frac{\Gamma - \Gamma^{CP}}{\Gamma + \Gamma^{CP}} \tag{3.11}$$

where Γ is the decay rate into leptons and Γ^{CP} into antileptons. In the case that the heavy neutrinos are not degenerate in mass, which is the case we study here, it is sufficient to consider only CP violation in the decays of the heavy neutrinos (direct CP violation). One then obtains the result [18]

$$\delta = \frac{1}{2\pi(h^\dagger h)_{11}} \sum_{j=1}^{6} \mathrm{Im}[(h^\dagger h)_{1j}]^2 f(m_j^2/m_1^2), \tag{3.12}$$

where $f(x)$ is a kinematic function of order one for reasonable choices of the masses [18]. The subscript 1 in the terms above is due to the fact that the lepton asymmetry is generated by the decay of the lightest of the right-handed neutrinos (any asymmetry generated by the heavier right-handed neutrinos will be washed out by the decays of the lightest).

The first constraint on the mass scale of the N_R is obtained by insisting that it be out of thermal equilibrium with the rest of the universe when it decays. This will hold if it lives till the universe has cooled to a temperature below the mass of the particle. This condition is encoded in the requirement that

$$\Gamma_R \leq H \qquad (T = m_R), \tag{3.13}$$

where Γ_R is the decay rate of the right-handed neutrino with mass m_R, and H is the Hubble constant. This translates to

$$\frac{(h^\dagger h)_{11} m_R}{8\pi} \lesssim \frac{20 m_R^2}{M_P} \Rightarrow \frac{(h^\dagger h)_{11}}{m_R} \lesssim 10^{-16}\ \mathrm{GeV}^{-1} \tag{3.14}$$

if the dominant decay is via the Yukawa coupling of eq. (3.10). The second constraint is obtained by insisting that the heavy Majorana mass scale explain the solar and atmospheric neutrino data. If we assume that the observed deficit in ν_es from the sun is due to $\nu_e - \nu_\mu$ mixing, then the mass squared difference $\Delta m^2 \sim 10^{-6}$ eV^2, preferred by the data, implies $m_{\nu_\mu} \sim 10^{-3}$ eV. Similarly, assuming the deficit in atmospheric ν_μs is due to $\nu_\mu - \nu_\tau$ mixing, then the preferred mass squared difference, $\Delta m^2 \sim 10^{-3}$ eV^2, implies $m_{\nu_\tau} \sim 3 \times 10^{-2}$ eV. The see-saw mass relations

$$m_{\nu_\mu} \sim \frac{m_\mu^2}{m_R}; \quad m_{\nu_\tau} \sim \frac{m_\tau^2}{m_R} \tag{3.15}$$

then imply that $m_R \sim 10^{10} - 10^{11}$ GeV. Eq. (3.14) then tells us that $h_{11} \sim 10^{-2} - 10^{-3}$ (assuming a hierarchical matrix of Yukawa couplings). Note that if the electron gets its mass at tree level from the Yukawa coupling h as in the standard model, one would obtain $m_e = h_{11} v = 1$ GeV, for

$v = 246$ GeV, which is too large by several orders of magnitude. In order for this model to work, one has to impose symmetries such that the standard model fermions get their masses only at the loop level. In such a case the fermion masses would be proportional to the squares of the Yukawa coupling constants, and one obtains $m_e = h_{11}^2 v = 0.2$ MeV for $h_{11} = 10^{-3}$, which is the correct order of magnitude. It is indeed possible to construct a model that incorporates all these requirements [21]. Moreover, in this model, the CP violation responsible for baryogenesis is related in a calculable way to the CP violation present in the CKM matrix.

The model is based on the $SU(4) \times SU(2)_L \times SU(2)_R$ group. The standard model fermions transform in the usual representations:

$$\Psi_L^i \sim (4,2,1)^i \equiv \begin{pmatrix} u_1 & u_2 & u_3 & \nu \\ d_1 & d_2 & d_3 & e^- \end{pmatrix}_L^i \tag{3.16}$$

$$\Psi_R^i \sim (4,1,2)^i \equiv \begin{pmatrix} u_1 & u_2 & u_3 & N \\ d_1 & d_2 & d_3 & e^- \end{pmatrix}_R^i \tag{3.17}$$

where $i = 1,2,3$ is a generation index, and we have included a right-handed neutrino N. We add to this three generations of (right-handed) sterile neutrinos

$$s^i \sim (1,1,1)^i. \tag{3.18}$$

The matter spectrum is supersymmetric, so the scalars $\tilde{\Psi}_L^i$, $\tilde{\Psi}_R^i$, and $\tilde{s}^i$ in the model transform in exactly the same way. We will impose a discrete Z_3 symmetry on the gauge singlets (broken by the interactions of the standard model particles) under which $s^j \to e^{-i(j\pi)/3} s^j$ and $\tilde{s}^j \to e^{i(2j\pi)/3} \tilde{s}^j$. This permits us to make the Lagrangian CP invariant, with the vacuum expectation values of the $\tilde{s}^j$ breaking CP spontaneously. We can choose parameters for the scalar potential such that it is minimized when

$$\langle \tilde{s} \rangle_j = \frac{v_0}{\sqrt{2}} e^{i\alpha_j}; \quad \langle \tilde{N} \rangle_j = \frac{v_R}{\sqrt{2}} \delta_{1j}; \quad \langle \tilde{\nu} \rangle_j = \frac{v_L}{\sqrt{2}} \delta_{1j} \tag{3.19}$$

with $|v_0| > |v_R| \gg |v_L|$. This provides the correct pattern of symmetry breaking.

The Yukawa interactions are given by

$$\mathcal{L}_Y = -y_i (\bar{s}^c)^i s^i \tilde{s}^i - (\kappa_L^a)_{ij} \bar{\Psi}_L^i s^j \tilde{\Psi}_L^a - (\kappa_R^a)_{ij}^T \bar{\Psi}_R^i (s^c)^j \tilde{\Psi}_R^a + \text{h.c.}, \tag{3.20}$$

with all of the coupling constants real. However, the mass matrix of the s_i will contain the phases α due to the spontaneous breaking of CP invariance when the $\tilde{s}_i$ obtain a vacuum expectation value. Note that since $\bar{\Psi}_L \Psi_R$ transforms as $(1,2,2)$ and there are no scalars in this representation, none of the standard model fermions get masses at tree level. Their masses are generated at one loop by diagrams involving the (s_i) on the internal lines. The CP-violating phases in the quark mass matrices and hence in the CKM matrix are a function of the phases α in the masses of the s_i.

The out-of-equilibrium decays of the s_i generate the lepton (and hence baryon) asymmetry. It is this same phase α that is responsible for the CP violation in these decays. Thus one obtains a relationship between the CKM phase and the phase responsible for the baryon asymmetry.

3.3 Conclusions

We have presented an overview of two models of baryogenesis that also have other low-energy experimental consequences. Baryogenesis in the MSSM is possible if the Higgs and $\tilde{t}_R$ are light. Moreover, one expects large contributions to the $b \to s\gamma$ rate, and the $B-\bar{B}$ mixing amplitude. Baryogenesis via the decay of heavy neutrinos can be constrained by insisting that they be at the see-saw scale implied by the solar and atmospheric neutrino data. We have presented specific implementations of these models where the CKM phase responsible for CP violation in the neutral kaons is related to the phase responsible for the baryogenesis.

References

[1] G. Steigman, Ann. Rev. Astr. Ap. **14**, 339 (1976).

[2] A. Cohen, A. De Rújula, and S. Glashow, Astrophys. J. **495**, 539 (1988).

[3] A. Sakharov, JETP Lett. **5**, 24 (1967).

[4] V. Kuzmin, V. Rubakov, and M. Shaposhnikov, Phys. Lett. B **155**, 36 (1985).

[5] M. Shaposhnikov, JETP Lett. **44**, 465 (1988).

[6] M. Dine *et al.*, Phys. Lett. B **283**, 319 (1992).

[7] C. Jarlskog, Phys. Rev. Lett. **55**, 1039 (1985).

[8] M. Gavela *et al.*, Mod. Phys. Lett. A **9**, 795 (1994); P. Huet and E. Sather, Phys. Rev. D **51**, 379 (1995).

[9] For a review, see A. Cohen, D. Kaplan, and A. Nelson, Ann. Rev. Nucl. Part. Sci. **43**, 27 (1993).

[10] For a review see K. Olive, Lectures given at 33rd Internationale Universitaetswochen fuer Kern- und Teilchenphysik. UMN-TH-1249-94

[11] M. Carena, M. Quiros and C. Wagner, Phys. Lett. B **380**, 81 (1996); J. Espinosa, Nucl. Phys. B **475**, 273 (1996).

[12] M. Worah, Phys. Rev. D **54**, 2198 (1996).

[13] P. Huet and A. Nelson, Phys. Rev. D **53**, 4578 (1996).

[14] M. Carena *et al.*, Nucl. Phys. B **503**, 387 (1997).

[15] M. Worah, Phys. Rev. D **56**, 2010, (1997); M. Worah, Phys. Rev. Lett. **79**, 3810 (1997).

[16] M. Misiak, S. Pokorski, and J. Rosiek in *Heavy Flavors II*, ed. A. Buras and M. Lindner (World Scientific, Singapore, 1997), p. 795.

[17] P. Arnold, D. Son, and L. Yaffe, Phys. Rev. D **55**, 6264 (1997); J. Ambjorn and A. Krasnitz, Nucl. Phys. B **506**, 387 (1997).

[18] M. Fukugita and T. Yanagida, Phys. Rev. D **42**, 1285 (1990).

[19] For a review, see A. Pilaftsis, Int. J. Mod. Phys. A **14**, 1811 (1999).

[20] J. Harvey and M. Turner, Phys. Rev. D **42**, 3344 (1990).

[21] M. Worah, Phys. Rev. D **53**, 3902 (1996).

4

Kaon Physics in Supersymmetric Theories

Lawrence Hall

4.1 Introduction

This brief review of flavor in supersymmetric theories aims to stress a few simple but important highlights, and is certainly far from a comprehensive review. In section 4.2 the role of symmetries in nature is discussed, with the aim of motivating why supersymmetry might occur in nature at the weak scale. A discussion of superpartner masses and flavor symmetries is also given.

In section 4.3 we stress that supersymmetric theories contain more flavor symmetry–breaking interactions than the standard model, and that this leads to new flavor-mixing matrices. The need for a super-GIM mechanism is discussed, and we focus on the role of flavor symmetries as the origin of the small flavor parameters in both the fermion and scalar sectors. Two broad classes of supersymmetric contributions to flavor- and CP-violating processes are introduced and discussed in a model-independent way. The role of the rare decay $K \to \pi\nu\bar{\nu}$ is also discussed, and conclusions are drawn.

Despite the title, kaon physics is treated as just one piece of the flavor/CP puzzle in this review. It is the totality of the rare process signals that may allow us to uncover simplicity behind the apparent complexity.

4.2 Symmetry and Symmetry Breaking

4.2.1 Vertical and Horizontal Symmetries

Symmetries play a central role in our understanding of the elementary particles and their interactions. The symmetry structure of the known matter is shown in fig. 4.1.

As we move in the vertical direction we move amongst particles with different charges and gauge interactions—this is also known as the gauge direction. The non-Abelian gauge bosons act in this direction—for example

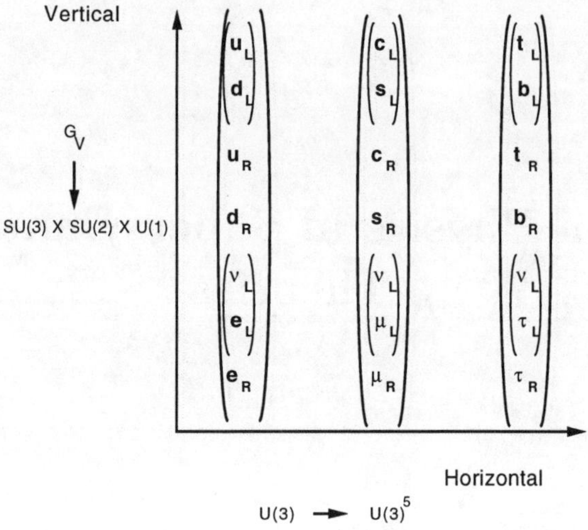

Figure 4.1: Vertical and horizontal symmetries of matter. In a unified picture, the vertical and horizontal groups are G_V and $U(3)$, whereas in the standard model they are $SU(3) \times SU(2) \times U(1)$ and $U(3)^5$, both of which are broken, as shown in equations (4.1) and (4.2).

the W^+ connects u_L to d_L, and the gluons change the quark colors, which are understood to be part of this direction. There are 15 states in the vertical direction, forming a generation. Not all of them are connected by the known gauge interactions—although this could occur if the $SU(3) \times SU(2) \times U(1)$ gauge group of the standard model was embedded in a larger vertical group. Such a grand unified theory could provide an understanding for the gauge quantum numbers of the quarks and leptons.

The horizontal direction of fig. 4.1 shows a threefold repetition of the basic vertical structure, corresponding to the three generations. The horizontal symmetry group acting on these three generations is $U(3)$. In fact, the horizontal symmetry group of the known gauge interactions is much larger: there is a $U(3)$ factor for each of the 5 multiplets within each generation. The horizontal, or flavor, symmetry group of the standard model gauge interactions is therefore $U(3)^5$, with one $U(3)$ factor for each of $q_L = (u, d)_L, u_R, d_R, l_L = (\nu, e)_L$ and e_R.

The vertical and horizontal symmetry groups of the quarks and leptons are both broken. The electroweak symmetry is broken at the weak scale, which I label by M_Z, to the $U(1)$ of electromagnetism:

$$SU(3) \times SU(2) \times U(1)_Y \xrightarrow{M_Z} SU(3) \times U(1)_{EM}. \tag{4.1}$$

Experiments give very tight constraints on any breaking of color or electromagnetism. The flavor group is broken by the quark and lepton masses and by the CKM mixing to baryon number, B, and the three individual lepton numbers $L_{e,\mu,\tau}$:

$$U(3)^5 \xrightarrow{M_F,\epsilon} B \times L_{e,\mu,\tau} \tag{4.2}$$

While the experimental constraints on B and $L_{e,\mu,\tau}$ breaking are strong, it is far from clear that they are exact symmetries of nature.

There is a crucial difference between the flavor and vertical symmetry directions: while we know that at least some of the vertical symmetry is gauged, we do not know that any of the horizontal direction is gauged. This means that the breaking of eq. (4.1) necessarily occured spontaneously—the underlying theory must have interactions that possess the $SU(3) \times SU(2) \times U(1)$ symmetry. On the other hand, we cannot be sure that there is an underlying theory with an exact flavor symmetry. It is logically possible that, at the most fundamental level, the theory possesses interactions that contain small dimensionless parameters that explicitly break the flavor group. This would mean that, even in principle, it is not possible to understand the origin of the quark and lepton masses. This seems unreasonable to me. Instead, I will assume that there is an underlying theory that possesses an exact flavor symmetry group, G_F. This theory must contain new interactions that lead to a breaking of G_F at a scale M_F—the fundamental scale of flavor physics. Furthermore, these interactions must break G_F in a way that generates a set of small dimensionless G_F-breaking parameters, ϵ, which are the origin of the small quark and lepton mass ratios and mixing angles. The flavor group G_F may be smaller than $U(3)^5$. The maximal flavor group gets smaller as the vertical gauge group is extended, and we cannot be sure that nature has chosen the maximal possible flavor group.

4.2.2 Mass Scales

There is a mass scale associated with each of the known forces: m_γ, Λ_{QCD}, M_Z, and M_{Pl} for the electromagnetic, strong, weak, and gravitational forces respectively. The masses of the quarks and charged leptons are all proportional to the weak scale, M_Z, while the large hierarchies in these masses is due to the set of small G_F-breaking parameters, ϵ, discussed above. Since short-distance physics is the most fundamental, we can use the Planck scale, M_{Pl}, to define our units of mass. It is convenient to set $M_{Pl} = 1$, which is analogous to defining units by setting $\hbar$ and c equal to unity. Having defined the unit of mass, our experience with quantum field theory tells us that all further mass scales should have a symmetry description. This is not a mathematical requirement—it is simply a requirement that we have a physical understanding of the new mass scales. The massless photon is understood because the $U(1)_{EM}$ symmetry is unbroken, while the scale of strong interactions arises from the dynamics of QCD, which is based on the symmetry group $SU(3)$. However, *the standard model provides no symmetry description for the weak scale.* To date, the standard model provides a successful mathematical description of particle phenomena, but I believe that it is not tenable as the ultimate theory of the weak interactions. These thoughts are summarized in table 4.1.

Mass	Symmetry description
$m_\gamma = 0$	$U(1)_{EM}$ unbroken
$\Lambda_{QCD} = 10^{-20}$	$SU(3)$ dynamics
$M_Z = 10^{-17}$	????

Table 4.1: Symmetry descriptions of mass scales (Planckian units).

There are several possible extensions of the standard model that allow a symmetry description of the weak scale. The standard model, with an elementary Higgs boson, certainly provides an economical description of weak symmetry breaking. If we require that our theory include this picture, then the *only* known symmetry description of the weak scale is provided by supersymmetry.

4.2.3 Supersymmetry

Supersymmetry is an extension of the continuous symmetries of space-time. The familiar generators of translations (P), rotations (J), and Lorentz boosts (K) are augmented by those of supersymmetry (Q), and the Poincaré algebra of commutation relations amongst (P, J, K) is augmented by the anticommutation relation $\{Q, Q\} \sim P$. It is this relation that implies that the structure of space-time has changed.

In fig. 4.1 the vertical and horizontal symmetries of matter were shown, but the space-time symmetry axis was ignored. This is rectified in fig. 4.2, where the third axis represents increasing space-time symmetry. The Lorentz symmetry implies that there is a repetition of the entire sheet of three generations of matter, giving particles of the same spin but opposite gauge charges—this is the antimatter. The particles of this replicated sheet could aptly be refered to as "Lorentz partners" of the original matter particles. Supersymmetry represents a further step along the space-time axis and generates an entirely new sheet, once again replicating the three generations. This time the particles have the same gauge quantum numbers as the original matter, but have spin 0 instead of spin 1/2. They are known as superpartners, and are denoted by a tilde above the particle symbol.

The space-time structure of the electron is shown in table 4.2. When first discovered a century ago, it was believed to be a particle with just two attributes: mass and electric charge. However, the electron has nontrivial space-time properties also; the discovery of its nature due to rotations and Lorentz boosts each led to a doubling of states. It is not surprising that a further increase in space-time would result in one more such doubling, leading to the selectron $\widetilde{e}$.

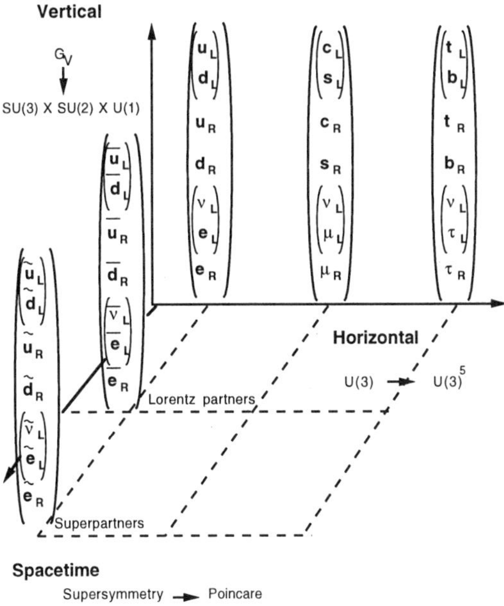

Figure 4.2: The three symmetry axes. For clarity, the Lorentz partners and the superpartners of the second and third generations are not shown.

Symmetry Generator	Electron states	Discovered
Momentum	$e(m)$	1897 J. J. Thompson
Angular Momentum	$(e^{\uparrow}, e^{\downarrow})$	1922 Stern-Gerlach
Lorentz Boost	$(e, \bar{e})$	1932 Anderson
Supersymmetry	$(e, \widetilde{e})$	??

Table 4.2: Space-Time structure of the electron.

In fig. 4.2 we know that the underlying fundamental symmetries of the world in both the vertical and horizontal directions are broken. Understanding these breakings, shown in eqs. (4.1) and (4.2), is perhaps the most important task of particle physics. If supersymmetry is also a fundamental symmetry of nature, then there is also symmetry breaking along the third axis of fig. 4.2:

$$\text{Supersymmetry} \xrightarrow{\widetilde{m}} \text{Poincaré}. \tag{4.3}$$

In this case the Lorentz and translation symmetries are simply unbroken remnants of a larger fundamental symmetry, in analogy with electromagnetism, QCD, baryon number, and lepton numbers.

The discovery of the Lorentz partners led to a profound puzzle: why does the universe predominantly contain baryons and leptons rather than the corresponding antimatter? If supersymmetry is correct there is no such analogous puzzle: the superleptons and superbaryons are not present in the universe because the supersymmetry breaking of eq. (4.3) gives them a mass of order $\widetilde{m}$, so that they are unstable and decay into ordinary matter. In fact, the lightest superpartner may be stable—this could be the dark matter in the universe, thus solving a puzzle rather than creating one.

I have argued that a space-time with broken supersymmetry allows a symmetry understanding of the weak scale. For this to occur, it is perhaps not surprising that the scale of the superpartner masses, $\widetilde{m}$, should be of order the weak scale, M_Z. Three questions come to mind:

• *Has any progress really been made? What is the advantage of replacing electroweak symmetry breaking at scale M_Z by supersymmetry breaking at scale $\widetilde{m}$?*

Progress has been made, because many quantum field theories are known that allow an understanding of supersymmetry breaking, including the generation of $\widetilde{m}$ at a scale much beneath M_{Pl}. This allows a dynamical understanding of the ratio M_Z/M_{Pl}, and hence a solution to the hierarchy problem.

A very significant result of weak scale supersymmetry is the prediction of the weak mixing angle from coupling constant unification [1], $\sin^2\theta_{susy} = 0.233 \pm 0.003$, which is to be compared with the result from Z factories $\sin^2\theta_{exp} = 0.2312 \pm 0.0003$ [2]. In 25 years of theoretical physics beyond the standard model, this is the only case in which a free parameter of the standard model was correctly predicted to the 1% level of accuracy.

• *What is the mechanism that connects electroweak symmetry breaking to supersymmetry breaking?*

The dynamics of the theory provides the connection in a very convincing way [3], explaining why the Higgs mass squared becomes negative while the mass squared for all squarks and sleptons is positive. Indeed, electroweak symmetry breaking is hard to avoid in supersymmetric theories, as it is triggered by the same interaction that generates the large top quark mass.

• *Can the superpartner masses be predicted quantitatively in terms of M_Z?*

Unfortunately not; one can only make naturalness arguments, which are not precise. In any supersymmetric theory in which electroweak symmetry breaking is triggered as a heavy top quark effect, as shown in eq. (4.4), it is possible to derive a formula for the weak scale that has the form

$$M_Z^2 = \sum_i c_i \widetilde{m}_i^2, \tag{4.4}$$

where $\widetilde{m}_i$ represent the several possible fundamental parameters of order the scale $\widetilde{m}$ that control the superpartner masses, and the c_i are constants. Some c_i are of order unity, while others may be small, but none are large.

One sees that the $\widetilde{m}_i$ that have c_i of order unity cannot become much larger than M_Z, unless several terms in the sum conspire to cancel against each other. As these $\widetilde{m}_i$ become larger, such a fine tuning renders the theory artificial and physically unacceptable. Hence, there are always some $\widetilde{m}_i$, and some superpartner masses, that cannot be made much larger than M_Z. In the minimal supersymmetric standard model, all the $\widetilde{m}_i$ parameters have c_i of order unity, and hence all of the superpartners must be close in mass to the weak scale. How "close" depends on the particle and the degree of cancellation allowed in eq. (4.4) and is illustrated in fig. 4.3.

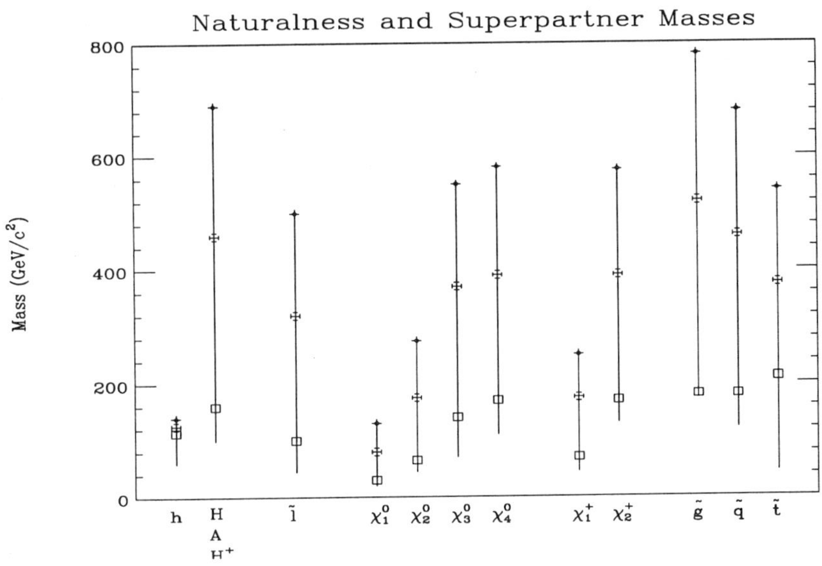

Figure 4.3: Expectations for superpartner masses in the minimal supersymmetric extension of the standard model. For each particle, the line extends up to a point where the fine tuning of parameters is 1 part in 10. (Figure supplied by G. Anderson.)

While not predictions, the expectations of fig. 4.3 are certainly exciting. The particles of the standard model were found one at a time—some new facilities discovered one, some none. With weak scale supersymmetry, we expect that all the superpartners will show up at 1 or 2 new facilities. Finding 21 supermatter particles and 7 supergauge/superHiggs particles at once will present unusual experimental challenges! The expectations of fig. 4.3 are broadly applicable to many more models than the so-called minimal one. On the other hand, it is possible that many superpartners are much heavier, with multi-TeV masses that do not substantially contribute to the right-hand side of eq. (4.4) due to certain c_i being small. Even in this case, there are always some colored superpartners, and some noncolored superpartners, for which the expectations of fig. 4.3 are still broadly correct.

4.3 Flavor in Supersymmetric Theories

4.3.1 A New Probe of Flavor Physics

People sometimes say that supersymmetry doesn't have anything to say about the question of how quarks and leptons get their mass. They are wrong. It is true that we don't know the origin of the fermion masses, or even the fundamental flavor physics scale, M_F, at which the $U(3)^5$ flavor symmetry is broken. It is possible that we may never know these things. However, a question that we must address is: ***what interactions describe flavor breaking at the weak scale?*** Supersymmetry changes the answer to this crucial queston.

In the standard model only the Yukawa interactions, λ, violate flavor. Hence the only window to the fundamental scale M_F is the quark and lepton masses and the CKM mixing matrix. Even though these parameters are now all known, with varying levels of accuracy, it has proven extraordinarily difficult to use them to construct a convincing theory of flavor. Going beyond the standard model, there may be other interactions involving higher-dimension operators with coefficients suppressed by inverse powers of M_F. Whether such effects will ever be observed is very sensitive to the value of M_F.

With weak scale supersymmetry, flavor symmetries are also broken by the soft mass-squared matrices of the superpartners, m^2, and by the trilinear scalar interactions, A. Thus at the weak scale there are necessarily more experiment probes of flavor:

$$\lambda \longrightarrow \lambda, m^2, A. \tag{4.5}$$

These provide new windows to the flavor physics at scale M_F [4] and offer the hope that we may learn enough to construct convincing theories of flavor.

4.3.2 The New Flavor Mixing Matrices

Flavor and CP violation arises in the standard model when the quark mass matrices are diagonalized—the CKM matrix, V, results as a relative rotation between u_L and d_L type quarks in the 3-dimensional generation space. The W gauge boson couples to $\bar{u}_{Li} V_{ij} d_{Lj}$. If either the three up quarks or the three down quarks were degenerate, V could be rotated to the unit matrix by rotating the basis for the u_L or d_L quarks. Thus the standard model CP and flavor violation is a consequence of the nontrivial flavor structure of the quark mass matrices.

In any supersymmetric extension of the standard model, there are 3×3 mass matrices for the squarks and for the sleptons, in addition to those for the quarks and leptons. The m^2 parameters discussed above actually form 3×3 matrices in each charge sector. If these scalar mass matrices are nontrivial, then the supersymmetric gauge interactions will involve new flavor-mixing matrices, W, analogous to the CKM matrix. For example, a relative rotation between the up quarks and the up squarks leads to a matrix, W^u, appearing at the gluino vertex: $\tilde{u}_i^\dagger W_{ij}^u u_j \tilde{g}$. Similarly if there is a relative rotation between the charged leptons and charged sleptons necessary to diagonalize the mass matrices, then the photino vertex allows CP and flavor violation via a matrix W^e: $\tilde{e}_i^\dagger W_{ij}^e e_j \tilde{\gamma}$. These W mixing matrices, one each for $u_{L,R} d_{L,R}$ and $e_{L,R}$ sectors, are a generic feature of supersymmetric theories.

4.3.3 The Super-GIM Mechanism

In a superfield basis where the fermion masses are diagonal, it is the structure of the squark and slepton mass matrices that determines the flavor and CP violation from the W matrices. Scalar mass matrices in field theory are quite unlike fermion mass matrices, since they have different symmetry properties. If all entries of the scalar mass matrices are of order the weak scale, and there are no precise degeneracies amongst the entries, Feynman diagrams with virtual superpartners and vertices involving the W matrices give contributions to processes that violate flavor (e.g., $\mu \to e\gamma$) and CP (e.g., ϵ_K) that are several orders of magnitude larger than allowed by experiment.

To be concrete, consider $K_L - K_S$ mixing. In the standard model the GIM mechanism [5] explains why the mixing is so small. The box diagram gives a $\Delta S = 2$ amplitude

$$A_{SM} \propto \frac{1}{16\pi^2 M_W^4} \left[V_{us}^2 (m_c^2 - m_u^2) + (V_{td}^* V_{ts})^2 (m_t^2 - m_u^2) \right]. \qquad (4.6)$$

The first term, which arises from the exchange of only the two lightest generations, I call the "12" contribution. It is small because the quark masses are small compared to the W mass. The contribution from the third generation, the "3" contribution, is small because the CKM mixings between the third and lighter two generations, V_{td} and V_{ts}, are small. Of course, we need a theory of flavor to understand why these parameters are small, but given the experimental fact, we can explain the small mixing of the neutral K mesons.

In supersymmetric theories there are additional box diagrams. The largest presumably comes from supersymmetric QCD with internal squarks and gluinos. The flavor violation now comes from the mixing matrix W at the gluino vertices. One finds a $\Delta S = 2$ amplitude

$$A_{susy} \propto \frac{1}{16\pi^2 \tilde{m}^4} \left[W_{ds}^2 (m_{\tilde{s}}^2 - m_{\tilde{d}}^2) + (W_{bd}^* W_{bs})^2 (m_{\tilde{b}}^2 - m_{\tilde{d}}^2) \right]. \qquad (4.7)$$

For superpartner masses of order the weak scale, such mixing looks to be enormous. This is frequently called the supersymmetric flavor problem. There is a problem for both terms in the square bracket of eq. (4.7), the

"12" and "3" contributions. Clearly it is not an option to assume that these contributions are small because the squarks are much lighter than the W boson!

This is no longer viewed as a problem for supersymmetry; rather, it provides information about how supersymmetry should be broken, and, broadly speaking, has led to two classes of theories. In the first, the scalar masses are generated by some new generation independent dynamics, so that the mass matrices are proportional to the unit matrix. In this case a mass basis for the scalars can be found where the W matrices are also the unit matrix and hence do not violate flavor or CP. I will not discuss this case further.

The second class of theories has a nontrivial structure for the scalar mass matrices, but like the quark mass matrices, this structure involves a set of small parameters. For example the "12" contribution may be small because the d and s squarks are nearly degenerate, while the "3" contribution may be small either because of a near degeneracy between b and d squarks or because the W mixing matrix has only small mixings between the third and lighter generations. This latter option appears plausible: perhaps these small mixings in the V and W matrices are related in a deeper theory. This leads to an important conclusion: if W and V matrices have similar structures, and if the third-generation scalars have no special degeneracies, then the supersymmetric "3" contribution is expected to be broadly comparable to the standard model contribution, which is also dominated by the "3" effect.

4.3.4 Approximate Flavor Symmetries

In both the fermion and scalar sector there is a need for small parameters. What are these small parameters? In physics we are used to symmetries that are not exact but are broken only by small amounts. In nuclear and particle physics isospin provides a familiar example of an approximately broken symmetry. Even if we do not understand the ultmate origin and size of the symmetry breaking, great progress results from introducing small symmetry-breaking parameters. Flavor $SU(3)$ provides another example: if the small symmetry-breaking parameters are assumed to transform as octets, one obtains the Gell-Mann–Okubo mass formula amongst the pseudoscalar meson masses $m_\pi^2 + 3m_\eta^2 = 4m_K^2$. *Could it be that some simple symmetry, with a few small symmetry-breaking parameters, could provide a compelling description of flavor in supersymmetric theories?*

An extremely attractive hypothesis is to assume that the two sets of small parameters, those in the fermion mass matrices and those in the scalar mass matrices, have a common origin: they are the small symmetry-breaking parameters of an approximate flavor symmetry group. This provides a link between the fermion mass and flavor-changing problems; both are addressed by the same symmetry. There has been a great deal of supersymmetric flavor model building based on this idea over the last decade.

For the structure and tests of the $U(2)$ model, as well as references to early literature on such models, see [6]. Rather than discuss models, I will present predictions for flavor-changing and CP-violating phenomena framed in a model-independent way. Let me stress again that there are other ways supersymmetric theories can be constructed: Gauge mediation of supersymmetry breaking removes the signals from these interesting rare processes, as does the possibility that the squarks are much heavier than generally expected. Even with approximate flavor symmetries, it is possible to obtain the Cabibbo angle from the up-type quark sector, so that the signal is in $D-\bar{D}$ mixing rather than in $K-\bar{K}$ mixing [7].

4.3.5 Predictions from the "3" Signal

The "3 signal" occurs when the third-generation scalars have masses very different from those of the first two generations, and when the W matrices are CKM-like. There are several interesting cases depending on which flavor $U(3)$ factors are strongly broken.

The minimal case is that the only large flavor symmetry breaking is $U(3)_{q_L} \times U(3)_{u_R} \to U(2)_{q_L} \times U(2)_{u_R}$. This breaking is necessary to produce the heavy top quark mass, and it results in $m^2_{q_L,3}$ and $m^2_{u_R,3}$ being considerably different from the corresponding masses for the first two generations. This is the case of the minimal supersymmetric standard model (with universal boundary conditions). All 1-loop diagrams involving superpartners give only small corrections to the standard model predictions for flavor-changing and CP-violating processes.[1]

Another case occurs when there is the additional strong breaking $U(3)_{e_R} \to U(2)_{e_R}$. This occurs in $SU(5)$ grand unified theories because q_L and u_R occur in the same multiplet as e_R. In these theories there is an interesting one-loop contribution to the dipole operator for $\mu \to e\gamma$ involving the exchange of $\tilde{\tau}_R$ [8].

The most interesting case occurs when all components of the third generation feel the large flavor symmetry breaking: $U(3)^5 \to U(2)^5$. This occurs in $SO(10)$ grand unified theories, where an entire generation is a single multiplet, and also whenever the flavor group acts purely in the vertical direction with no dependence on the horizontal component, as in the case of $U(3)$. In this case there are many interesting "3 signals" in both the quark and lepton sectors. A close inspection shows that the most important effects occur in ϵ_K, $B-\bar{B}$ mixing, electric dipole moments of the electron (d_e) and the up and down quarks (d_u, d_d) and, finally, in $\mu \to e\gamma$ and $\mu \to e$ conversion in atoms [6]. In the case of ϵ_K and $B-\bar{B}$ mixing one obtains effects comparable to those present in the standard model. On the other hand, the effects in the dipole moments and in the lepton flavor-violating processes are at the level of the present experimental limits. The

[1]However, there is a very interesting contribution to radiative decays, such as $b \to s\gamma$ from a charged Higgs exchange diagram.

calculation of some typical, although partial, supersymmetric contributions to these observables gives[2]

$$\epsilon_K \approx 2 \cdot 10^{-3} \left(\frac{500\,\mathrm{GeV}}{m_{\tilde{q}}} \right)^2 \tag{4.8}$$

$$|\Delta m_{B_d}| \approx 0.1\,\mathrm{ps}^{-1} \left(\frac{500\,\mathrm{GeV}}{m_{\tilde{q}}} \right)^2 \tag{4.9}$$

$$\mathrm{BR}(\mu \to e\gamma) \approx 2 \cdot 10^{-11} \left(\frac{100\,\mathrm{GeV}}{m_{\tilde{l}}} \right)^4 \tan\beta^2 \tag{4.10}$$

$$d_e \approx 6 \cdot 10^{-27} \mathrm{ecm} \left(\frac{100\,\mathrm{GeV}}{m_{\tilde{l}}} \right)^2 \tan\beta \tag{4.11}$$

$$d_d \approx 1 \cdot 10^{-26} \mathrm{ecm} \left(\frac{500\,\mathrm{GeV}}{m_{\tilde{q}}} \right)^2 \tan\beta^2, \tag{4.12}$$

where $\tan\beta$ is the ratio of the vacuum expectation values of the Higgs doublets. We have set $|W_{ij}|$ equal to $|V_{\mathrm{CKM}ij}|$, and taken unknown phases to be unity, so that there are considerable uncertainties of order unity. For these simple estimates, all (non)colored superpartners have been put at (100 GeV) 500 GeV.

The "3" contribution to B meson mixing leads to a modification of the phase in the neutral B meson eigenstates: $(q/p)_{B_{d,s}}$ acquire an extra phase $e^{-2i\phi_{B_{d,s}}}$. This modifies the CP asymmetries in neutral B meson decays: for example, the asymmetry in B_d decay to ψK_S $(\pi\pi)$ becomes $-\sin 2(\beta + \phi_{Bd})$ $(\sin 2(\alpha - \phi_{Bd}))$ [9]. A careful study shows that a large "3" effect in B mixing is less likely than for ϵ_K [10].

4.3.6 Predictions from the "LR" Signal

Flavor symmetry is also violated in scalar trilinear interactions, as shown in eq. (4.5). At the 1-loop level these interactions can lead to interesting dipole moment operators. A large contribution to ϵ'/ϵ is possible from the operator $\tilde{s}_L^\dagger \tilde{d}_R h$ [11]. We call such contributions "12_{LR}" because there is one squark from each of the first and second generations, and one is a partner of a left-handed quark and the other of a right-handed one. Recently it has been argued that the large observed value for ϵ'/ϵ can be explained by such a "12_{LR}" effect, with the magnitude agreeing very well with expectations from approximate flavor symmetry arguments [12]. If this is the correct origin for ϵ'/ϵ, there is also an important "11_{LR}" contribution to the neutron electric dipole moment, by a Cabibbo rotation of the "12" operator, which should be seen soon. However, there is a puzzle: if there is a similar approximate flavor symmetry in the lepton sector, then

[2] In eq. (4.10) and (4.11) we consider the photino contribution. For the EDM of the u-quark one has $d_u \approx 8 d_d (v_1/v_2)^2$.

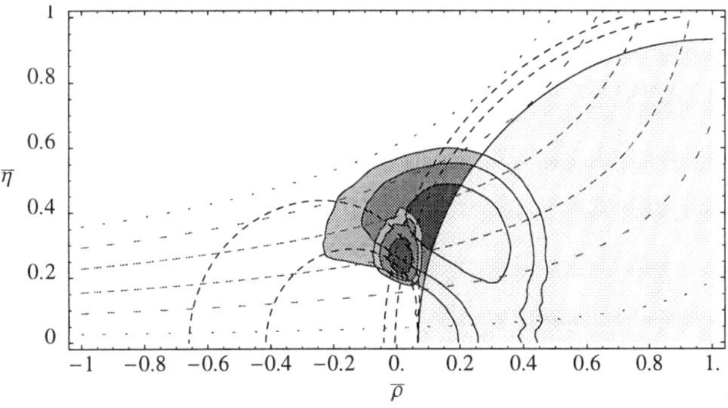

Figure 4.4: The ρ/η plane: expermental fit and $U(2)$ prediction. The smaller shaded regions show the predictions of the $U(2)$ theory. The three larger areas show the result of a fit in the standard model using experimental values of $|V_{ub}/V_{cb}|, \Delta m_{Bd}, \Delta m_{Bs}$ but not ϵ_K, whose constraint is shown independently. Also shown is a solid quarter circle line which excludes much of the standard model region by the requirement that $m_u/m_d > 0$. For both fits, to the $U(2)$ theory and to the standard model, the contours are at 68, 95, and 99% CL. This figure is taken from [14], where further details of the fit are given, as well as the meaning of the other dashed curves.

simple estimates show that the slepton masses must be heavier than about 1 TeV to avoid the recent MEGA bound on $\mu \to e\gamma$.

4.3.7 $K \to \pi\nu\bar{\nu}$

In supersymmetric theories with approximate flavor symmetries, the 1 loop supersymmetric contributions to $K^+ \to \pi^+\nu\bar{\nu}$ and $K^0 \to \pi^0\nu\bar{\nu}$ are small [13], largely because there is no loop diagram involving a gluino. This means that these processes provide an important opportunity to measure the CKM parameters, ρ and η. These parameters cannot be constrained from ϵ_K, because a significant supersymmetric "3 effect" is expected, and possibly also in B mixing.

Theories with approximate flavor symmetry constrain the structure of the fermion masses, and in certain cases predict ρ and η—the prediction for the $U(2)$ theory is shown in fig. 4.4. Clearly the $U(2)$ theory could also be tested soon by measurements of B_s mixing and CP violation in the B mesons, in the case that B mixing is not infected with supersymmetric contributons.

4.4 Conclusions

The idea of weak scale supersymmetry is closely linked to understanding the origin and scale of electroweak symmetry breaking. However, it may also provide a critical new window to flavor physics through the masses and interactions of the scalar quarks and leptons. If flavor symmetry breaking is imprinted on these scalar masses and interactions, then continued searches for rare processes involving $\mu \to e$ transitions, and for electric dipole moments of the neutron and electron, take on a new importance, as do the measurements of $K \to \pi\nu\bar{\nu}$ and B_s mixing. Solid arguments, based on approximate flavor symmetries, show that the supersymmetric contributions to these processes are measurable, except in the case of $K \to \pi\nu\bar{\nu}$, where the measurements will serve to pin down the CKM matrix parameters. It may then become apparent that the first observation of a supersymmetric effect occured in the mixing and CP violation in the neutral K mesons. The flavor puzzle is not made easier in supersymmetric theories, even though it presents more ways of probing flavor violation. It will take many years to uncover the supersymmetric contributions to flavor- and CP-violating observables, and it may well not conform to the ideas presented in this short review.

Acknowledgments

I thank the conference organisers for a superb and stimulating meeting. This work was supported in part by the U.S. Department of Energy under Contracts DE-AC03-76SF00098, in part by the National Science Foundation under grant PHY-95-14797.

References

[1] H. Georgi, H. Quinn, and S. Weinberg, Phys. Rev. Lett. **33**, 451 (1974); S. Dimopoulos, S. Raby, and F. Wilczek, Phys. Rev. D **24**, 1681 (1981); S. Dimopoulos and H. Georgi, Nucl. Phys. B **193**, 150 (1981); L. Ibanez and G. G. Ross, Phys. Lett. B **105**, 439 (1981).

[2] Particle Data Group, *Review of Particle Physics,* Eur. Phys. J C **3**, 1 (1998).

[3] L. Ibanez and G. Ross, Phys. Lett. B **110**, 215 (1982); L. Alvarez-Gaume, M. Claudson and M. Wise, Nucl. Phys. B **207**, 96 (1982).

[4] L. J. Hall, A. Kostelecky, and S. Raby, Nucl. Phys. B **267**, 415 (1986).

[5] S. Glashow, J. Iliopoulos, and L. Maiani, Phys. Rev. D **2**, 1285 (1970).

[6] R. Barbieri, G. Dvali, and L.J. Hall, hep-ph/9512388, Phys. Lett. B **377**, 76 (1996); R. Barbieri, L.J. Hall, and A. Romanino, Phys. Lett. B **401**, 47 (1997); R. Barbieri, L.J. Hall, and A. Romanino, Nucl. Phys. B **551**, 93 (1999).

[7] Y. Nir and N. Seiberg, Phys. Lett. B **309**, 337 (1993), M. Leurer, Y. Nir, and N. Seiberg, Nucl. Phys. B **420**, 468 (1994); Nucl. Phys. B **398**, 319 (1993).

[8] R. Barbieri and L.J. Hall, Phys. Lett. B **338**, 212 (1994).

[9] I. Bigi, *Perspectives in Particle Physics*, La Thuile, March 1994, p. 137.

[10] R. Barbieri, and A. Strumia, hep-ph/9704402, Nucl. Phys. B **508**, 3 (1997).

[11] M. Dugan, B. Grinstein, and L. Hall, Nucl. Phys. B **255**, 413 (1985).

[12] A. Masiero and H. Murayama, hep-ph/9903363; H. Murayama, this volume, chapter 10.

[13] R. Barbieri, L.J. Hall, and A. Romanino, Phys. Lett. B **401**, 47 (1997); Y. Nir and M. Worah, Phys. Lett. B **423**, 326 (1998); A. Buras, A. Romanino, and L. Silvestrini, Nucl. Phys. B **520**, 3 (1998); G. Colangelo and G. Isidori, JHEP 9809, 009 (1998).

[14] R. Barbieri, L.J. Hall, and A. Romanino, hep-ph/9812384, Nucl. Phys. B **551**, 93 (1999).

Part II

Direct CP Violation in Kaon Decays

With the discovery of CP violation in 1964, the neutral kaon states of definite mass and lifetime could no longer be expressed as CP eigenstates (as written in the introduction to Part I), but could be written as

$$|K_{S,L}\rangle = \frac{1}{\sqrt{(1+|\epsilon|^2)}}[(1+\epsilon)|K^0\rangle \pm (1-\epsilon)|\bar{K}^0\rangle],$$

where ϵ is a complex parameter. For many years it was the sole expression of CP violation. Such CP violation is said to be "indirect"; it can originate in various ways. One of these is a "superweak" interaction first suggested by Lincoln Wolfenstein [1], which has no other observable consequence except to generate ϵ through the mixing of K^0 and $\bar{K}^0$.

The theory of CP violation based on phases in weak couplings of quarks with one another, as expressed through the Cabibbo-Kobayashi-Maskawa (CKM) matrix [2], implies as well the presence of "direct" CP violation, as expressed in terms of a parameter ϵ' describing the difference between CP-violating $K_L \to \pi^+\pi^-$ and $K_L \to \pi^0\pi^0$ decays. Specifically, if one defines

$$\eta_{+-} \equiv \frac{\mathcal{A}(K_L \to \pi^+\pi^-)}{\mathcal{A}(K_S \to \pi^+\pi^-)}, \qquad \eta_{00} \equiv \frac{\mathcal{A}(K_L \to \pi^0\pi^0)}{\mathcal{A}(K_S \to \pi^0\pi^0)},$$

then $\eta_{+-} = \epsilon + \epsilon'$ while $\eta_{00} = \epsilon - 2\epsilon'$, and a double ratio of rates then measures $\mathrm{Re}(\epsilon'/\epsilon)$:

$$\frac{\Gamma(K_L \to \pi^+\pi^-)}{\Gamma(K_S \to \pi^+\pi^-)} / \frac{\Gamma(K_L \to \pi^0\pi^0)}{\Gamma(K_S \to \pi^0\pi^0)} = 1 + 6\ \mathrm{Re}\left(\frac{\epsilon'}{\epsilon}\right).$$

In this Part the predictions for ϵ'/ϵ are reviewed and compared with recent measurements at Fermilab and CERN. Prospects for improved measurements and calculations are also discussed. Since the measured value seems larger than at least some theoretical estimates, the question arises as to whether there is a signal for new physics beyond that of the Cabibbo-Kobayashi-Maskawa matrix. This question is hotly debated in the following pages!

References

[1] L. Wolfenstein, Phys. Rev. Lett. **13**, 562 (1964).

[2] N. Cabibbo, Phys. Rev. Lett. **10**, 531 (1963); M. Kobayashi and T. Maskawa, Prog. Theor. Phys. **49**, 652 (1973).

5

Theoretical Status of ε'/ε

Andrzej J. Buras

Abstract

We review the present theoretical status of the CP-violating ratio ε'/ε in the standard model and confront its estimates with the most recent data. In particular we review the present status of the most important parameters m_s, $B_6^{(1/2)}$, $B_8^{(3/2)}$, $\Lambda_{\overline{MS}}^{(4)}$, and $\Omega_{\eta+\eta'}$. While the sign and the order of magnitude of standard model estimates for ε'/ε agree with the data, for central values of these parameters most estimates of ε'/ε in the standard model are substantially below the grand experimental average $(21.2 \pm 4.6) \cdot 10^{-4}$. Only in a small corner of the parameter space can ε'/ε in the standard model be made consistent with experimental results. In view of very large theoretical uncertainties, it is impossible to conclude at present that the data on ε'/ε indicate new physics. A brief discussion of ε'/ε beyond the standard model is presented.

5.1 Introduction

The purpose of this talk is to summarize the present theoretical status of the CP-violating ratio ε'/ε and to confront it with the recent experimental findings. The parameters ε and ε' describe two types of CP violation in the decays $K_L \to \pi\pi$, which could not take place if CP was conserved. In the standard model a nonvanishing value of ε originates in the fact that the mass eigenstate K_L is not a CP eigenstate due to the complex CKM couplings [1] in the box diagrams responsible for the $K^0 - \bar{K}^0$ mixing. Indeed, K_L is a linear combination of CP eigenstates K_2 (CP=$-$) and K_1 (CP=+): $K_L = K_2 + \bar{\varepsilon} K_1$, where $\bar{\varepsilon}$ is a small parameter. The decay of K_L via K_1 into CP=+ state $\pi\pi$ is termed *indirectly* CP violating as it proceeds not via explicit breaking of the CP symmetry in the decay itself but via the admixture of the CP state with opposite CP parity to the dominant one. The measure for this indirect CP violation is defined as

$$\varepsilon = \frac{A(K_L \to (\pi\pi)_{I=0})}{A(K_S \to (\pi\pi)_{I=0})}, \tag{5.1}$$

which can be rewritten as

$$\varepsilon = \frac{\exp(i\pi/4)}{\sqrt{2}\Delta M_K}\left(\mathrm{Im}M_{12} + 2\xi\mathrm{Re}M_{12}\right), \qquad \xi = \frac{\mathrm{Im}A_0}{\mathrm{Re}A_0}, \tag{5.2}$$

where M_{12} represents $K^0 - \bar{K}^0$ mixing, ΔM_K is the $K_L - K_S$ mass difference, and A_0 is the isospin amplitude for $\pi\pi$ in the $I = 0$ state. The second term involving ξ cancels the phase convention dependence of the first term but in the usual CKM convention can be safely neglected. In this limit $\varepsilon = \bar{\varepsilon}$.

On the other hand, so-called *direct* CP violation is realized via a direct transition of a CP odd to a CP even state or vice versa. $K_2 \to \pi\pi$ in the case of $K_L \to \pi\pi$. A measure of such a direct CP violation in $K_L \to \pi\pi$ is characterized by a complex parameter ε' defined as

$$\varepsilon' = \frac{1}{\sqrt{2}}\mathrm{Im}\left(\frac{A_2}{A_0}\right)\exp(i\Phi_{\varepsilon'}), \qquad \Phi_{\varepsilon'} = \frac{\pi}{2} + \delta_2 - \delta_0 \,. \tag{5.3}$$

Here the subscript $I = 0, 2$ denotes $\pi\pi$ states with isospin $0, 2$ equivalent to $\Delta I = 1/2$ and $\Delta I = 3/2$ transitions, respectively, and $\delta_{0,2}$ are the corresponding strong phases. The weak CKM phases are contained in A_0 and A_2. The isospin amplitudes A_I are complex quantities that depend on phase conventions. On the other hand, ε' measures the difference between the phases of A_2 and A_0 and is a physical quantity. The strong phases $\delta_{0,2}$ can be extracted from $\pi\pi$ scattering. Then $\Phi_{\varepsilon'} \approx \pi/4$.

Experimentally we have [2, 3]

$$\varepsilon_{\mathrm{exp}} = (2.280 \pm 0.013) \cdot 10^{-3}\ \exp i\Phi_\varepsilon, \qquad \Phi_\varepsilon \approx \frac{\pi}{4}. \tag{5.4}$$

Until recently the experimental situation on ε'/ε was rather unclear:

$$\mathrm{Re}(\varepsilon'/\varepsilon) = \begin{cases} (23 \pm 7) \cdot 10^{-4} & \text{(NA31)}\ [4] \\ (7.4 \pm 5.9) \cdot 10^{-4} & \text{(E731)}\ [5]. \end{cases} \tag{5.5}$$

While the result of the NA31 collaboration at CERN [4] clearly indicated direct CP violation, the value of E731 at Fermilab [5] was compatible with superweak theories [6] in which $\varepsilon'/\varepsilon = 0$. After heroic efforts on both sides of the Atlantic, this controversy is now settled with the two new measurements by KTeV at Fermilab [7] and NA48 at CERN [8]

$$\mathrm{Re}(\varepsilon'/\varepsilon) = \begin{cases} (28.0 \pm 4.1) \cdot 10^{-4} & \text{(KTeV)}\ [7] \\ (18.5 \pm 7.3) \cdot 10^{-4} & \text{(NA48)}\ [8], \end{cases} \tag{5.6}$$

which together with the NA31 result confidently establish direct CP violation in nature ruling out superweak models [6]. The grand average including NA31, E731, KTeV, and NA48 results reads [8]

$$\mathrm{Re}(\varepsilon'/\varepsilon) = (21.2 \pm 4.6) \cdot 10^{-4}, \tag{5.7}$$

very close to the NA31 result but with a smaller error. The error should be further reduced once complete data from both collaborations have been analyzed. It is also of great interest to see what value for ε'/ε will be measured by KLOE at Frascati, which uses a different experimental technique than KTeV and NA48.

5.2 ε in the Standard Model

In the standard model ε is found by calculating the box diagrams with internal u, c, t, $W^{\pm}$ exchanges and including short distance QCD corrections. The final result can be written as

$$\begin{aligned}\varepsilon &= C_{\varepsilon}\hat{B}_K \mathrm{Im}\lambda_t \left\{\mathrm{Re}\lambda_c \left[\eta_1 S_0(x_c) - \eta_3 S_0(x_c, x_t)\right]\right.\\ &\quad \left. - \mathrm{Re}\lambda_t \eta_2 S_0(x_t)\right\} \exp(i\pi/4), \qquad (5.8)\end{aligned}$$

where C_{ε} is a numerical factor, $\lambda_i = V_{is}^* V_{id}$, and $S_0(x_i)$ with $x_i = m_i^2/M_{\mathrm{W}}^2$ are Inami-Lim functions [9]. Explicit expressions can be found in [10]. The NLO values of the QCD factors η_i are given as follows [11]:

$$\eta_1 = 1.38 \pm 0.20, \qquad \eta_2 = 0.57 \pm 0.01, \qquad \eta_3 = 0.47 \pm 0.04\,. \qquad (5.9)$$

The main theoretical uncertainty in (5.8) resides in the nonperturbative parameter $\hat{B}_K$, which parametrizes the relevant hadronic matrix element $\langle \bar{K}^0 | (\bar{s}d)_{V-A} (\bar{s}d)_{V-A} | K^0 \rangle$. Recent reviews of $\hat{B}_K$ are given in [10, 12, 13]. In our numerical analysis presented below we will use $\hat{B}_K = 0.80 \pm 0.15$, which is in the ballpark of various lattice and large-N estimates.

It is well known that the experimental value of ε in (5.4) can be accommodated in the standard model. We know this from numerous analyses of the unitarity triangle that in addition to (5.4) take into account data on $B^0_{d,s} - \bar{B}^0_{d,s}$ mixings and the values of the CKM elements $|V_{us}|$, $|V_{ub}|$, and $|V_{cb}|$. From this analysis one extracts in particular [13]

$$\mathrm{Im}\lambda_t = \begin{cases} (1.33 \pm 0.14) \cdot 10^{-4} & \text{(Monte Carlo)} \\ (1.33 \pm 0.30) \cdot 10^{-4} & \text{(Scanning).} \end{cases} \qquad (5.10)$$

The "Monte Carlo" error stands for an analysis in which the experimentally measured parameters, like $|V_{us}|$, $|V_{ub}|$, and $|V_{cb}|$, are used with Gaussian errors and the theoretical input parameters, like $\hat{B}_K$, $F_B\sqrt{\hat{B}_B}$ are taken with flat distributions. In the "scanning" estimate all input parameters are scanned independently in the appropriate ranges. Details can be found in [13]. This analysis gives also $\sin 2\beta = 0.73 \pm 0.09$ (0.71 ± 0.13), where β is one of the angles in the unitarity triangle. The recent study of CP violation in $B^0 \to J/\psi K_S$ by CDF [14] gives $\sin 2\beta = 0.79 \pm 0.44$ in good agreement with the value above, although the large experimental error precludes any definite conclusion.

5.3 ε'/ε in the Standard Model

5.3.1 Basic Formulae

In the standard model the ratio ε'/ε is governed by QCD penguins and electroweak penguins (γ and Z^0 penguins). With increasing value of m_t the electroweak penguins become increasingly important [15, 16] and entering ε'/ε with the opposite sign to QCD penguins, suppress this ratio for large m_t. The size of ε'/ε is also strongly affected by QCD renormalization group effects. Without these effects ε'/ε would be at most $\mathcal{O}(10^{-5})$ independently of m_t [16].

The parameter ε' is given in terms of the isospin amplitudes A_I in (5.3). Applying the operator product expansion to these amplitudes one finds

$$\frac{\varepsilon'}{\varepsilon} = \mathrm{Im}\lambda_t \cdot F_{\varepsilon'}(m_t, \Lambda_{\overline{\mathrm{MS}}}^{(4)}, m_s, B_6^{(1/2)}, B_8^{(3/2)}, \Omega_{\eta+\eta'}), \tag{5.11}$$

where

$$F_{\varepsilon'} = \left[P^{(1/2)} - P^{(3/2)}\right] \exp(i\Phi), \qquad \Phi = \Phi_{\varepsilon'} - \Phi_\varepsilon \tag{5.12}$$

with

$$P^{(1/2)} = r \sum y_i(\mu) \langle Q_i(\mu)\rangle_0 (1 - \Omega_{\eta+\eta'}), \tag{5.13}$$

$$P^{(3/2)} = \frac{r}{\omega} \sum y_i(\mu) \langle Q_i(\mu)\rangle_2 . \tag{5.14}$$

Here

$$r = \frac{G_F \omega}{2|\varepsilon| \mathrm{Re}A_0}, \qquad \langle Q_i \rangle_I \equiv \langle (\pi\pi)_I | Q_i | K \rangle, \qquad \omega = \frac{\mathrm{Re}A_2}{\mathrm{Re}A_0}, \tag{5.15}$$

and μ is the renormalization scale, which is $\mathcal{O}(1\,\mathrm{GeV})$. Since $\Phi \approx 0$, $F_{\varepsilon'}$ and ε'/ε are real to an excellent approximation. The arguments of $F_{\varepsilon'}$ will be discussed shortly. Explicit expressions for the operators $Q_{1,2}$ (current-current), Q_{3-6} (QCD penguins), and Q_{7-10} (electroweak penguins) are given in [10]. The dominant are these two:

$$Q_6 = (\bar{s}_\alpha d_\beta)_{V-A} \sum_{q=u,d,s} (\bar{q}_\beta q_\alpha)_{V+A},$$

$$Q_8 = \frac{3}{2} (\bar{s}_\alpha d_\beta)_{V-A} \sum_{q=u,d,s} e_q (\bar{q}_\beta q_\alpha)_{V+A}, \tag{5.16}$$

where e_q denotes the electric quark charges.

The Wilson coefficient functions $y_i(\mu)$ were calculated including the complete next-to-leading order (NLO) corrections in [17–19]. The details of these calculations can be found there and in the review [20]. Their numerical values for $\Lambda_{\overline{\mathrm{MS}}}^{(4)}$ corresponding to $\alpha_{\overline{MS}}^{(5)}(M_Z) = 0.119 \pm 0.003$ and two renormalization schemes (Naive Dimensional Regularization, NDR, and 't Hooft-Veltman, HV) can be found in [13].

It is customary in phenomenological applications to take $\mathrm{Re}A_0$ and ω from experiment, i.e.,

$$\mathrm{Re}A_0 = 3.33 \cdot 10^{-7}\,\mathrm{GeV}, \qquad \omega = 0.045, \tag{5.17}$$

where the last relation reflects the so-called $\Delta I = 1/2$ rule. This strategy avoids to a large extent the hadronic uncertainties in the real parts of the isospin amplitudes A_I.

The sum in (5.13) and (5.14) runs over all contributing operators. $P^{(3/2)}$ is fully dominated by electroweak penguin contributions. $P^{(1/2)}$ on the other hand is governed by QCD penguin contributions which are suppressed by isospin breaking in the quark masses ($m_u \neq m_d$). The latter effect is described by [21–23]:

$$\Omega_{\eta+\eta'} = \frac{1}{\omega}\frac{(\mathrm{Im}A_2)_{\mathrm{I.B.}}}{\mathrm{Im}A_0} = 0.25 \pm 0.08. \tag{5.18}$$

5.3.2 History of ε'/ε

There is a long history of calculations of ε'/ε in the standard model. As it has been already described in [10, 13], I will be very brief here. The first calculation of ε'/ε for $m_\mathrm{t} \ll M_\mathrm{W}$ without the inclusion of renormalization group effects can be found in [24]. Renormalization group effects in the leading logarithmic approximation have been first presented in [25]. For $m_\mathrm{t} \ll M_\mathrm{W}$ only QCD penguins play a substantial role. With increasing m_t the role of electroweak penguins becomes important. The first leading log analyses for arbitrary m_t can be found in [15, 16], where a strong cancellation between QCD penguins and electroweak penguin contributions to ε'/ε for $m_t > 150$ GeV has been found. Finally, during the 1990s considerable progress has been made by calculating complete NLO corrections to the Wilson coefficients relevant for ε [11] and ε' [17–19]. The progress in calculating the corresponding hadronic matrix elements was substantially slower. It will be summarized below.

Now, the function $F_{\varepsilon'}$ in (5.11) can be written in a crude approximation (not to be used for any serious analysis) as follows .

$$\begin{aligned} F_{\varepsilon'} &\approx 13 \cdot \left[\frac{110\,\mathrm{MeV}}{m_\mathrm{s}(2\ \mathrm{GeV})}\right]^2 \left(\frac{\Lambda^{(4)}_{\overline{\mathrm{MS}}}}{340\ \mathrm{MeV}}\right) \\ &\cdot \left[B_6^{(1/2)}(1-\Omega_{\eta+\eta'}) - 0.4 \cdot B_8^{(3/2)}\left(\frac{m_\mathrm{t}}{165\,\mathrm{GeV}}\right)^{2.5}\right]. \end{aligned} \tag{5.19}$$

Here B_i are hadronic parameters defined through

$$\langle Q_6\rangle_0 \equiv B_6^{(1/2)}\,\langle Q_6\rangle_0^{(\mathrm{vac})}, \qquad \langle Q_8\rangle_2 \equiv B_8^{(3/2)}\,\langle Q_8\rangle_2^{(\mathrm{vac})}. \tag{5.20}$$

The label "vac" stands for the vacuum insertion estimate of the hadronic matrix elements in question for which $B_6^{(1/2)} = B_8^{(3/2)} = 1$. The dependence on m_s in (5.19) originates in the m_s dependence of $\langle Q_6\rangle_0^{(\mathrm{vac})}$ and

$\langle Q_8\rangle_2^{(\text{vac})}$, so that $\langle Q_6\rangle_0$ and $\langle Q_8\rangle_2$ are roughly proportional to

$$R_6 \equiv B_6^{(1/2)}\left[\frac{137\,\text{MeV}}{m_\text{s}(m_\text{c})+m_\text{d}(m_\text{c})}\right]^2$$
$$R_8 \equiv B_8^{(3/2)}\left[\frac{137\,\text{MeV}}{m_\text{s}(m_\text{c})+m_\text{d}(m_\text{c})}\right]^2 \tag{5.21}$$

respectively. The scale $m_\text{c} = 1.3\,\text{GeV}$ turns out to be convenient. $m_\text{s}(m_\text{c}) \approx 1.17 \cdot m_\text{s}(2\,\text{GeV})$. The formula (5.19) exhibits very clearly the dominant uncertainties in $F_{\varepsilon'}$ that reside in the values of m_s, $B_6^{(1/2)}$, $B_8^{(3/2)}$, $\Lambda_{\overline{\text{MS}}}^{(4)}$, and $\Omega_{\eta+\eta'}$. Moreover, the partial cancellation between QCD penguin ($B_6^{(1/2)}$) and electroweak penguin ($B_8^{(3/2)}$) contributions requires accurate values of $B_6^{(1/2)}$ and $B_8^{(3/2)}$ for an acceptable estimate of ε'/ε. Because of the accurate value $m_\text{t}(m_\text{t}) = 165 \pm 5$ GeV, the uncertainty in ε'/ε due to the top quark mass amounts only to a few percent. A very accurate analytic formula for $F_{\varepsilon'}$ has been derived in [26] and its update can be found in [13].

Now, it has been known for some time that for central values of the input parameters the size of ε'/ε in the standard model is well below the NA31 value of $(23.0 \pm 6.5) \cdot 10^{-4}$. Indeed, extensive NLO analyses with lattice and large-N estimates of $B_6^{(1/2)} \approx 1$ and $B_8^{(3/2)} \approx 1$ performed first in [18,19] and after the top discovery in [27–29] have found ε'/ε in the ball park of $(3-7)\cdot 10^{-4}$ for $m_\text{s}(2\ \text{GeV}) \approx 130$ MeV. On the other hand it has been stressed repeatedly in [10,28] that for extreme values of $B_6^{(1/2)}$, $B_8^{(3/2)}$ and m_s still consistent with lattice, QCD sum rules and large-N estimates as well as sufficiently high values of $\text{Im}\lambda_t$ and $\Lambda_{\overline{\text{MS}}}^{(4)}$, a ratio ε'/ε as high as $(2-3)\cdot 10^{-3}$ could be obtained within the standard model. Yet, it has also been admitted that such simultaneously extreme values of all input parameters and consequently values of ε'/ε close to the NA31 result are rather improbable in the standard model. Different conclusions have been reached in [30], where values $(1-2)\cdot 10^{-3}$ for ε'/ε can be found. Also the Trieste group [31], which calculated the parameters $B_6^{(1/2)}$ and $B_8^{(3/2)}$ in the chiral quark model, found $\varepsilon'/\varepsilon = (1.7^{+1.4}_{-1.0})\cdot 10^{-3}$. On the other hand using an effective chiral Lagrangian approach, the authors in [32] found ε'/ε consistent with zero.

5.3.3 Hadronic Matrix Elements

The main sources of uncertainty in the calculation of ε'/ε are the hadronic matrix elements $\langle Q_i(\mu)\rangle_I$. They generally depend on the renormalization scale μ and on the scheme used to renormalize the operators Q_i. These two dependences are canceled by those present in the Wilson coefficients $y_i(\mu)$ so that the resulting physical ε'/ε does not (in principle) depend on μ and on the renormalization scheme of the operators. Unfortunately, the accuracy of the present nonperturbative methods used to evalutate $\langle Q_i \rangle_I$ is not sufficient to have the μ and scheme dependences of $\langle Q_i \rangle_I$ fully under control. We believe that this situation will change once the nonperturbative calculations, in particular lattice calculations improve.

As pointed out in [18] the contributions of $(V-A)\otimes(V-A)$ operators (Q_i with i=1,2,3,4,9,10) to ε'/ε can be determined from the leading CP-conserving $K \to \pi\pi$ decays, for which the experimental data is summarized in (5.17). The details of this approach will not be discussed here. For the central value of $\text{Im}\lambda_t$ these operators give a negative contribution to ε'/ε of about $-2.5 \cdot 10^{-4}$. This shows that these operators are relevant only if ε'/ε is below $1 \cdot 10^{-3}$. Unfortunately the matrix elements of the dominant $(V-A)\otimes(V+A)$ operators (Q_6 and Q_8) cannot be determined by the CP-conserving data and one has to use nonperturbative methods to estimate them. Let us then briefly review the present status of the corresponding nonperturbative parameters $B_6^{(1/2)}$ and $B_8^{(3/2)}$ as well as of m_s, $\Omega_{\eta+\eta'}$, and $\Lambda^{(4)}_{\overline{\text{MS}}}$, which all enter the evaluation of ε'/ε as seen in (5.19).

5.3.4 The Status of m_s, $B_6^{(1/2)}$, $B_8^{(3/2)}$, $\Omega_{\eta+\eta'}$, and $\Lambda^{(4)}_{\overline{\text{MS}}}$

m_s

The values for $m_s(2\,\text{GeV})$ extracted from quenched lattice calculations and QCD sum rules before summer 1999 were

$$m_s(2\,\text{GeV}) = \begin{cases} (110 \pm 20)\ \text{MeV} & \text{(Lattice)}\ [12,33] \\ (124 \pm 22)\ \text{MeV} & \text{(QCDS)}\ [34] \end{cases} \tag{5.22}$$

The value for QCD sum rules is an average [13] over the results given in [34]. QCD sum rules also allow one to derive lower bounds on the strange quark mass. It was found that generally $m_s(2\,\text{GeV}) \gtrsim 100$ MeV [35]. The most recent quenched lattice results: $m_s(2\,\text{GeV}) = 106 \pm 7$ MeV [36], $m_s(2\,\text{GeV}) = 97 \pm 4$ MeV [37], $m_s(2\,\text{GeV}) = 105 \pm 5$ MeV [38] are consistent with these bounds as well as (5.22) but have a smaller error. The unquenching seems to lower these values down to $m_s(2\,\text{GeV}) = 84 \pm 7$ MeV [39]. We refer to the talks of Ryan and Martinelli [40] for more details. The most recent determination of m_s from the hadronic τ-spectral function [41] reads $m_s(2\,\text{GeV}) = (114 \pm 23)$ MeV [42] in a very good agreement with (5.22) and lower than the value $m_s(2\,\text{GeV}) = (170^{+44}_{-55})$ MeV obtained by ALEPH [43] using this method early this year. We conclude that the error on m_s decreased considerably during the last two years and the central value is in the ballpark of $m_s(2\,\text{GeV}) = 110$ MeV with smaller values coming from unquenched lattice QCD.

$B_6^{(1/2)}$ and $B_8^{(3/2)}$

We recall that in the large-N limit $B_6^{(1/2)} = B_8^{(3/2)} = 1$ [44, 45]. The values for $B_6^{(1/2)}$ and $B_8^{(3/2)}$ obtained in various approaches are collected in table 5.1. The lattice results have been obtained at $\mu = 2\,\mathrm{GeV}$. The results in the large-N approach [44,45] and the chiral quark model correspond to scales below 1 GeV. However, as a detailed numerical analysis in [18] showed, $B_6^{(1/2)}$ and $B_8^{(3/2)}$ are only weakly dependent on μ. Consequently the comparison of these parameters obtained in different approaches at different μ is meaningful. Next, the values coming from lattice and chiral quark model are given in the NDR renormalization scheme. The corresponding values in the HV scheme can be found using approximate relations [13]

$$(B_6^{(1/2)})_{\mathrm{HV}} \approx 1.2 (B_6^{(1/2)})_{\mathrm{NDR}}, \qquad (B_8^{(3/2)})_{\mathrm{HV}} \approx 1.2 (B_8^{(3/2)})_{\mathrm{NDR}}. \tag{5.23}$$

The present results in the large-N approach are unfortunately not sensitive to the renormalization scheme but this can be improved [45,46].

Concerning the lattice results for $B_6^{(1/2)}$, the old results were in the range 0.6–1.2 [47]. However, a recent work [48] shows that lattice calculations of $B_6^{(1/2)}$ are very uncertain and one has to conclude that there are no solid predictions for $B_6^{(1/2)}$ from the lattice at present. The average value of $B_6^{(1/2)}$ in the large-N approach including full p^2 and p^0/N contributions and given in table 5.1 is close to 1.0 where the uncertainty comes from the variation of the cut-off Λ_c in the effective theory. On the other hand, it has been found [54,55] that a higher order term $\mathcal{O}(p^2/N)$ enhances $B_6^{(1/2)}$ to 1.5. This result is clearly interesting. Yet, in view of the fact that other p^2/N terms as well as p^4 and p^0/N^2 terms have not been calculated, it is premature to take this enhancement seriously. Finally, the chiral quark model gives in the NDR scheme the value for $B_6^{(1/2)}$ as high as 1.33 ± 0.25.

Method	$B_6^{(1/2)}$	$B_8^{(3/2)}$
Lattice [49–51]	–	0.69 – 1.06
Large-N [52,53]	0.72 – 1.10	0.42 – 0.64
ChQM [31]	1.07 – 1.58	0.75 – 0.79

Table 5.1: Results for $B_6^{(1/2)}$ and $B_8^{(3/2)}$ obtained in various approaches.

The status of $B_8^{(3/2)}$ looks better. Most nonperturbative approaches find $B_8^{(3/2)}$ below unity. The suppression of $B_8^{(3/2)}$ below unity is rather modest (at most 20%) in the lattice approaches and in the chiral quark model. It is stronger in the large-N approach. Interestingly in the latter approach the ratio $B_6^{(1/2)}/B_8^{(3/2)} \approx 1.72$ independently of the cut-off Λ_c.

Guided by the results presented above and biased to some extent by the results from the large-N approach and lattice calculations, we will use in our numerical analysis below $B_6^{(1/2)}$ and $B_8^{(3/2)}$ in the ranges:

$$B_6^{(1/2)} = 1.0 \pm 0.3, \qquad B_8^{(3/2)} = 0.8 \pm 0.2 \tag{5.24}$$

keeping always $B_6^{(1/2)} \geq B_8^{(3/2)}$.

$\Omega_{\eta+\eta'}$ and $\Lambda_{\overline{\rm MS}}^{(4)}$

The last estimates of $\Omega_{\eta+\eta'}$ have been done more than ten years ago [21–23] and it is desirable to update these analyses, which can be summarized by $\Omega_{\eta+\eta'} = 0.25 \pm 0.08$. In the numerical analysis presented below we have incorporated the uncertainty in $\Omega_{\eta+\eta'}$ by increasing the error in $B_6^{(1/2)}$ from ± 0.2 to ± 0.3. Finally from $\alpha_s(M_{\rm Z}) = 0.119 \pm 0.003$ one deduces $\Lambda_{\overline{\rm MS}}^{(4)} = (340 \pm 50)\,{\rm MeV}$.

5.3.5 *Comments*

General Remark

We would like to emphasize that it would not be appropriate to fit $B_6^{(1/2)}$, $B_8^{(3/2)}$, $m_{\rm s}$, $\Lambda_{\overline{\rm MS}}^{(4)}$, $\Omega_{\eta+\eta'}$, and $\hat{B}_K$ in order to make the standard model compatible simultaneously with experimental values on ε'/ε, ε and the analysis of the unitarity triangle. Such an approach would be against the whole philosophy of searching for new physics with the help of loop-induced transitions as represented by ε'/ε and ε. Moreover, it would not give us any clue whether the standard model is consistent with the data on ε'/ε. Indeed, it should be kept in mind that:

- $B_6^{(1/2)}$, $B_8^{(3/2)}$, $\hat{B}_K$, and $\Omega_{\eta+\eta'}$, in spite of carrying the names of nonperturbative parameters, are really not parameters of the standard model as they can be calculated by means of nonperturbative methods in QCD.
- $m_{\rm s}$, $\Lambda_{\overline{\rm MS}}^{(4)}$, $m_{\rm t}$, $|V_{cb}|$, and $|V_{ub}|$ are parameters of the standard model but there are better places than ε'/ε to determine them. In particular the usual determinations of these parameters can only marginally be affected by physics beyond the standard model, which is not necessarily the case for ε and ε'/ε.

Consequently, the only parameter to be fitted by direct CP violation is ${\rm Im}\lambda_t$. The numerical analysis of ε'/ε as a function of $B_6^{(1/2)}$, $B_8^{(3/2)}$, $m_{\rm s}$, and $\Lambda_{\overline{\rm MS}}^{(4)}$ presented below should only give a global picture for which ranges of parameters the presence of new physics in ε'/ε should be expected.

The Issue of Final State Interactions

In (5.3) and (5.12) the strong phases $\delta_0 \approx 37°$ and $\delta_2 \approx -7°$ are taken from experiment. They can also be calculated from NLO chiral perturbation for $\pi\pi$ scattering [56]. However, generally nonperturbative approaches to hadronic matrix elements are unable to reproduce them at present. As δ_I

are factored out in (5.3), in nonperturbative calculations in which some final state interactions are present in $\langle Q_i \rangle_I$ one should make the following replacements in (5.13) and (5.14)—

$$\langle Q_i \rangle_I \to \frac{\mathrm{Re}\langle Q_i \rangle_I}{(\cos\delta_I)_{\mathrm{th}}} \tag{5.25}$$

—in order to avoid double counting of final-state interaction phases. Here $(\cos\delta_I)_{\mathrm{th}}$ is obtained in a given nonperturbative calculation. Yet, in most calculations the phases are substantially smaller than found in experiment [31, 44, 52, 53] and $\langle Q_i \rangle_I \approx \mathrm{Re}\langle Q_i \rangle_I$.

The above point has been first discussed by the Trieste group [31], who suggested that in models in which at least the real part of $\langle Q_i \rangle_I$ can be calculated reliably, one should use $(\cos\delta_I)_{\mathrm{exp}}$ in (5.25). As $(\cos\delta_0)_{\mathrm{exp}} \approx 0.8$ and $(\cos\delta_2)_{\mathrm{exp}} \approx 1$ this modification enhances $P^{(1/2)}$ by 25% leaving $P^{(3/2)}$ unchanged. To our knowledge there is no method for hadronic matrix elements that can provide $\delta_0 \approx 37^\circ$, and consequently this suggestion may lead to an overestimate of the matrix elements and of ε'/ε.

ε'/ε and the $\Delta I = 1/2$ Rule

In one of the first estimates of ε'/ε, Gilman and Wise [25] used the suggestion of Vainshtein, Zakharov, and Shifman [57] that the amplitude ReA_0 is dominated by the QCD penguin operator Q_6. Estimating $\langle Q_6 \rangle_0$ in this manner they predicted a large value of ε'/ε. Since then it has been understood [48, 52, 58] that as long as the scale μ is not much lower than 1 GeV the amplitude ReA_0 is dominated by the current-current operators Q_1 and Q_2, rather than by Q_6. Indeed, at least in the HV scheme the operator Q_6 does not contribute to ReA_0 for $\mu = m_{\mathrm{c}}$ at all, as its coefficient $z_6(m_{\mathrm{c}})$ relevant for this amplitude vanishes. Also in the NDR scheme $z_6(m_{\mathrm{c}})$ is negligible.

For decreasing μ the coefficient $z_6(\mu)$ increases and the Q_6 contribution to ReA_0 is larger. However, if the analyses in [48, 52, 58] are taken into account, the operators Q_1 and Q_2 are responsible for at least 90% of ReA_0 if the scale $\mu = 1$ GeV is considered. Moreover, it should be stressed that whereas the operator Q_8 is irrelevant for the $\Delta I = 1/2$ rule, it is important for ε'/ε. Similarly, whereas the Wilson coefficients $y_6(\mu)$ and $y_8(\mu)$ entering ε'/ε can be sensitive to new physics as they receive contributions from very short-distance scales, this is not the case for $z_6(\mu)$, which due to GIM mechanism receives contribution only from $\mu \leq m_{\mathrm{c}}$. Therefore, even if $\langle Q_6 \rangle_0$ enters both ε'/ε and ReA_0, there is no strict relation between the large value of ε'/ε and the $\Delta I = 1/2$ rule as sometimes stated in the literature. On the other hand if the long-distance dynamics responsible for the enhancement of $\langle Q_{1,2} \rangle_0$ also enhances $\langle Q_6 \rangle_0$, then some connection between ε'/ε and the $\Delta I = 1/2$ rule is possible. There are some indications that this indeed could be the case [31, 54, 55].

ε'/ε, $B_6^{(1/2)}$, $B_8^{(3/2)}$, and m_s

At this symposium there has been a vigorous discussion whether the value of m_s is really relevant for the estimate of ε'/ε. In nonperturbative approaches in which hadronic matrix elements can only be calculated in terms of m_s, F_π, F_K, etc., it is obvious that the value of m_s enters the estimate of ε'/ε. This is in particular the case of the large-N approach. Moreover in this approach the values of $B_6^{(1/2)}$ and $B_8^{(3/2)}$ are independent of m_s.

On the other hand, as seen in (5.19), ε'/ε depends approximately on $B_6^{(1/2)}$, $B_8^{(3/2)}$, and m_s only through R_6 and R_8 defined in (5.21). If a given nonperturbative approach is able to calculate directly the relevant hadronic matrix elements and consequently R_6 and R_8, then in principle the value of m_s may not matter. If indeed R_6 and R_8 are independent of m_s, then through (5.21) there must be a quadratic dependence of $B_6^{(1/2)}$ and $B_8^{(3/2)}$ on m_s. As stated above this dependence is not observed in the large-N approach [44, 52, 53]. As discussed by Martinelli at this symposium [59], this question is being investigated in the lattice approach at present. On the other hand there are results in the literature showing a strong m_s dependence of the B_i parameters. This is the case for $B_6^{(1/2)}$ in the chiral quark model, where $B_6^{(1/2)}$ scales like m_s [31]. Similarly values for $B_7^{(3/2)}$ calculated in [60] show a strong m_s dependence. However, $B_8^{(3/2)}$ calculated in the chiral quark model shows a very weak dependence on m_s. This discussion shows that comparision of $(B_6^{(1/2)}, B_8^{(3/2)})$ obtained in various approaches has to be done with care.

Personally, I think that the value of m_s is relevant for the matrix elements $\langle Q_6\rangle_0$ and $\langle Q_8\rangle_2$ and that $B_6^{(1/2)}$ and $B_8^{(3/2)}$ are nearly m_s independent. As both operators (after Fierz transformation) have the density-density structure, their matrix elements are proportional to the square of the quark condensate and hence proportional to $1/m_s^2$. In this context it should be recalled that the anomalous dimensions of Q_6 and Q_8 are in a good approximation equal to twice the anomalous dimension of the mass operator. As a result of this, the products $y_6(\mu)\langle Q_6(\mu)\rangle_0$ and $y_8(\mu)\langle Q_8(\mu)\rangle_2$ and the corresponding contributions to ε'/ε are only very weakly μ dependent.

5.3.6 *Numerical Estimates of ε'/ε*

Munich Analysis

We will begin by presenting the analysis in [13]. In table 5.2 we summarize the input parameters used there. The value of $m_s(m_c)$ corresponds roughly to $m_s(2\text{ GeV}) = (110 \pm 20)$ MeV as obtained in lattice simulations. Imλ_t is given in (5.10) except that in evaluating ε'/ε the correlation in m_t between Imλ_t and $F_{\varepsilon'}$ has to be taken into account. In what follows we will present the results of two types of numerical analyses of ε'/ε that use Monte Carlo and scanning methods discussed already in section 5.2.

Quantity	Central	Error	Reference
$\Lambda^{(4)}_{\overline{\rm MS}}$	340 MeV	± 50 MeV	[3,61]
$m_{\rm s}(m_{\rm c})$	130 MeV	± 25 MeV	see text
$B_6^{(1/2)}$	1.0	± 0.3	see text
$B_8^{(3/2)}$	0.8	± 0.2	see text

Table 5.2: Collection of input parameters. $m_{\rm s}(m_{\rm c}) = 1.17 m_{\rm s}(2\,{\rm GeV})$.

Using the first method we find the probability density distributions for ε'/ε in fig. 5.1. From this distribution we deduce the following results:

$$\varepsilon'/\varepsilon = \left\{ \begin{array}{ll} (7.7\,{}^{+6.0}_{-3.5}) \cdot 10^{-4} & ({\rm NDR}) \\ (5.2\,{}^{+4.6}_{-2.7}) \cdot 10^{-4} & ({\rm HV}). \end{array} \right. \tag{5.26}$$

The difference between these two results indicates the leftover renormalization scheme dependence. Since the resulting probability density distributions for ϵ'/ϵ are very asymmetric with very long tails towards large values we quote the medians and the 68% (95%) confidence level intervals. Using the second method we find

$$1.05 \cdot 10^{-4} \le \varepsilon'/\varepsilon \le 28.8 \cdot 10^{-4} \qquad ({\rm NDR}) \tag{5.27}$$

and

$$0.26 \cdot 10^{-4} \le \varepsilon'/\varepsilon \le 22.0 \cdot 10^{-4} \qquad ({\rm HV}). \tag{5.28}$$

We observe that ε'/ε is generally lower in the HV scheme if the same values for $B_6^{(1/2)}$ and $B_8^{(3/2)}$ are used in both schemes. Since the present nonperturbative methods do not have renormalization scheme dependence fully under control we think that such a treatment of $B_6^{(1/2)}$ and $B_8^{(3/2)}$ is the proper way of estimating scheme dependences at present. Assuming, on the other hand, that the values in (5.24) correspond to the NDR scheme and using the relation (5.23), we find for the HV scheme the range $0.58 \cdot 10^{-4} \le \varepsilon'/\varepsilon \le 26.9 \cdot 10^{-4}$, which is much closer to the NDR result in (5.27). This exercise shows that it is very desirable to have the scheme dependence under control.

We observe that the most probable values for ε'/ε in the NDR scheme are in the ballpark of $1 \cdot 10^{-3}$. They are lower by roughly 30% in the HV scheme if the same values for $(B_6^{(1/2)}, B_8^{(3/2)})$ are used. On the other hand the ranges in (5.27) and (5.28) show that for particular choices of the input parameters, values for ε'/ε as high as $(2-3) \cdot 10^{-3}$ cannot be excluded at present. Let us study this in more detail.

In table 5.3 we show the values of ε'/ε in units of 10^{-4} for specific values of $B_6^{(1/2)}$, $B_8^{(3/2)}$, and $m_{\rm s}(m_{\rm c})$ as calculated in the NDR scheme. The

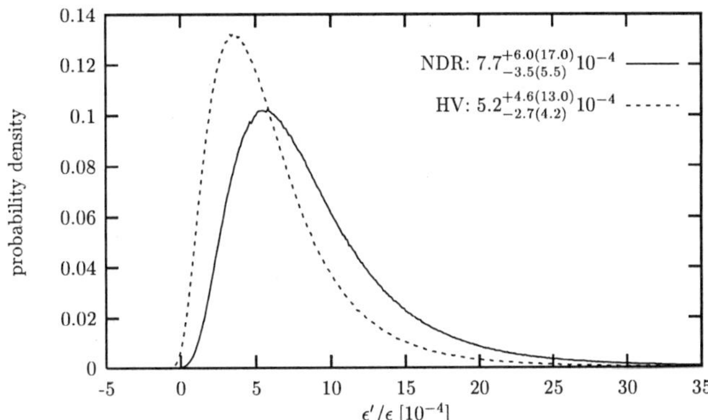

Figure 5.1: Probability density distributions for ε'/ε in NDR and HV schemes [13].

corresponding values in the HV scheme are lower as discussed above. The fourth column shows the results for central values of all remaining parameters. The comparison of the the fourth and the fifth columns demonstrates how ε'/ε is increased when $\Lambda^{(4)}_{\overline{\rm MS}}$ is raised from 340 MeV to 390 MeV. As stated in (5.19) ε'/ε is roughly proportional to $\Lambda^{(4)}_{\overline{\rm MS}}$. Finally, in the last column maximal values of ε'/ε are given. To this end we have scanned all parameters relevant for the analysis of $\mathrm{Im}\lambda_t$ within one standard deviation and have chosen the highest value of $\Lambda^{(4)}_{\overline{\rm MS}} = 390\,\mathrm{MeV}$. Comparison of the last two columns demonstrates the impact of the increase of $\mathrm{Im}\lambda_t$ from its central to its maximal value and of the variation of $m_{\rm t}$.

Table 5.3 gives a good insight in the dependence of ε'/ε on various parameters, which is roughly described by (5.19). We observe the following hierarchies:

- The largest uncertainties reside in $m_{\rm s}$, $B_6^{(1/2)}$ and $B_8^{(3/2)}$.
- The combined uncertainty due to $\mathrm{Im}\lambda_t$ and $m_{\rm t}$, present in both $\mathrm{Im}\lambda_t$ and $F_{\varepsilon'}$, is approximately $\pm 25\%$. The uncertainty due to $m_{\rm t}$ alone is only $\pm 5\%$.
- The uncertainty due to $\Lambda^{(4)}_{\overline{\rm MS}}$ is approximately $\pm 16\%$.
- The uncertainty due to $\Omega_{\eta+\eta'}$ is approximately $\pm 12\%$.

The large sensitivity of ε'/ε to $m_{\rm s}$ has been known since the analyses in the 1980s. In the context of the KTeV result this issue has been first analyzed in [62]. It has been found that provided $2B_6^{(1/2)} - B_8^{(3/2)} \leq 2$ the consistency of the standard model with the KTeV result requires the 2σ bound $m_{\rm s}(2\,\mathrm{GeV}) \leq 110\,\mathrm{MeV}$. Our analysis is compatible with these findings.

$B_6^{(1/2)}$	$B_8^{(3/2)}$	$m_s(m_c)$[MeV]	Central	$\Lambda_{\overline{MS}}^{(4)} = 390\,\text{MeV}$	Maxim.
		105	20.2	23.3	28.8
1.3	0.6	130	12.8	14.8	18.3
		155	8.5	9.9	12.3
		105	18.1	20.8	26.0
1.3	0.8	130	11.3	13.1	16.4
		155	7.5	8.7	10.9
		105	15.9	18.3	23.2
1.3	1.0	130	9.9	11.5	14.5
		155	6.5	7.6	9.6
		105	13.7	15.8	19.7
1.0	0.6	130	8.4	9.8	12.2
		155	5.4	6.4	7.9
		105	11.5	13.3	16.9
1.0	0.8	130	7.0	8.1	10.4
		155	4.4	5.2	6.6
		105	9.4	10.9	14.1
1.0	1.0	130	5.5	6.5	8.5
		155	3.3	4.0	5.2

Table 5.3: Values of ε'/ε in units of 10^{-4} for specific values of $B_6^{(1/2)}$, $B_8^{(3/2)}$ and $m_s(m_c)$ and other parameters as explained in the text [13].

Comparision with Other Analyses

In table 5.4 we compare the results discussed above with the most recent results obtained by other groups. Details of these analyses can be found in other contributions to this symposium and in references quoted in the table. There exists no recent phenomenological analysis from the Trieste group [31] and we quote their 1998 result. The values for $B_6^{(1/2)}$ and $B_8^{(3/2)}$ are given in the NDR scheme. The corresponding values in the HV scheme can be found by using (5.23). All groups use the Wilson coefficients calculated in [17–19] and the differences in ε'/ε result dominantly from different values of $B_6^{(1/2)}$, $B_8^{(3/2)}$, and m_s and to some extent in different values in input parameters (like B_K) needed for the determination of $\text{Im}\lambda_t$. The results for ε'/ε are given in the NDR scheme except for the result from Trieste, which corresponds to the HV scheme. The values of $m_s(m_c)$ used by all groups are very close to the ones given in table 5.2.

We observe that the results from Munich, Rome, and Dortmund are compatible with each other and generally well below the experimental data. The values of ε'/ε for central values of parameters obtained by the

Dortmund group are in good agreement with those of the Munich group. On the other hand as discussed above and also found by the Dortmund group for extreme choices of input parameters, values for ε'/ε consistent with experiment can be obtained. In the framework of an effective chiral Lagrangian approach [63] $B_6^{(1/2)}$ and $B_8^{(3/2)}$ cannot be calculated. For $B_6^{(1/2)} = B_8^{(3/2)} = 1$ values for ε'/ε consistent with zero are obtained.

The Trieste group finds generally higher values of ε'/ε, with the central value around $17 \cdot 10^{-4}$ and consequently consistent with the experimental findings. Similarly the Dortmund group finds the central values for ε'/ε in the ballpark of $15 \cdot 10^{-4}$ if they use $B_6^{(1/2)} \approx 1.5$ as obtained by including one p^2/N term.

Reference	$B_6^{(1/2)}$	$B_8^{(3/2)}$	$\varepsilon'/\varepsilon[10^{-4}]$
Munich [13]	1.0 ± 0.3	0.8 ± 0.2	$7.7^{+6.0}_{-3.5}$ (MC)
Munich [13]	1.0 ± 0.3	0.8 ± 0.2	$1.1 \rightarrow 28.8$ (S)
Rome [59]	0.83 ± 0.83	0.71 ± 0.13	$4.7^{+6.7}_{-5.9}$ (MC)
Dortmund [54]	$0.72 - 1.10$	$B_6^{(1/2)}/1.72$	$2.1 \rightarrow 26.4$ (S)
Trieste [31]	1.33 ± 0.25	0.77 ± 0.02	$7 \rightarrow 31$ (S)
Dubna-DESY [63]	1.0	1.0	$-3.2 \rightarrow 3.3$ (S)

Table 5.4: Results for ε'/ε in units of 10^{-4} obtained by various groups. The labels (MC) and (S) in the last column stand for "Monte Carlo" and "Scanning" respectively, as discussed in the text.

5.4 ε'/ε beyond the Standard Model

As we have seen the estimates of ε'/ε within the standard model are generally below the experimental data, but in view of large theoretical uncertainties stemming from hadronic matrix elements one cannot firmly conclude that the data on ε'/ε indicate new physics. However, there is still a lot of room for nonstandard contributions to ε'/ε and the apparent discrepancy between the standard model estimates and the data invites speculations. Indeed, results from NA31, KTeV, and NA48 prompted several analyses of ε'/ε within various extensions of the standard model like general supersymmetric models [62,64,65], models with anomalous gauge couplings [66], and models with additional fermions and gauge bosons [67]. Unfortunately several of these extensions have many free parameters and are not very conclusive.

On the other hand it is clear that the ε'/ε data puts models in which there are new positive contributions to ε and negative contibutions to ε' in serious difficulties. In particular as analyzed in [13,68] either the two Higgs doublet model II can be ruled out with improved hadronic matrix elements or a powerful lower bound on $\tan\beta$ can be obtained from ε'/ε. In the

Minimal Supersymmetric Standard Model, in addition to charged Higgs exchanges in loop diagrams, also charginos contribute. For suitable choice of the supersymmetric parameters, the chargino contribution can enhance ε'/ε with respect to the standard model expectations [69]. Yet, generally the most conspicuous effect of minimal supersymmetry is a depletion of ε'/ε. The situation can be different in more general models in which there are more parameters than in the two Higgs doublet model II and in the MSSM, in particular new CP-violating phases. As an example, in general supersymmetric models ε'/ε can be made consistent with experimental findings through the contributions of the chromomagnetic penguins [62,64, 65,70] and enhanced Z^0 penguins with the opposite sign to the one in the standard model [65,71,72].

While substantial new physics contributions to ε'/ε from chromomagnetic penguins and modified Z penguins appear rather plausible, one can give rather solid arguments that new physics should have only a minor impact on the QCD penguins represented by the operator Q_6. The point is that the contribution of Q_6 to ε'/ε can generally be written as

$$(\varepsilon'/\varepsilon)_{Q_6} \approx \tilde{R}_6\,[11 - 1.3E(m_{\rm t}, \ldots)]\,, \tag{5.29}$$

where

$$\tilde{R}_6 = {\rm Im}\lambda_t \left[\frac{110\,{\rm MeV}}{m_{\rm s}(2\ {\rm GeV})}\right]^2 \left(\frac{\Lambda^{(4)}_{\overline{\rm MS}}}{340\ {\rm MeV}}\right) B_6^{(1/2)} \tag{5.30}$$

and $E(m_{\rm t}, \ldots)$ results from QCD penguin diagrams in the full theory, which in addition to W boson and top-quark exchanges may receive contributions from new particles. As in the standard model $E(m_{\rm t}) \approx 0.3$, the contribution of Q_6 to ε'/ε is dominated by "11," which results from the operator mixing between Q_6 and other operators, in particular the current-current operator Q_2. The latter mixing taking place at scales below $M_{\rm W}$ is unaffected by new physics contributions that can enter (5.29) only through the function E. One would need an order of magnitude enhancement of E through new physics contributions in order to see a 40% effect in the contribution of Q_6 to ε'/ε, and unless the sign of E is reversed one would find rather a suppression of ε'/ε with respect to the standard model expectations than a required enhancement. This discussion shows that a substantial enhancement of QCD penguins through new physics contribution as suggested recently in [73] is rather implausible. On the other hand, in the case of the operator Q_8 one has instead of (5.29)

$$(\varepsilon'/\varepsilon)_{Q_8} \approx \tilde{R}_8\,[-10]\,Z(m_{\rm t}, \ldots), \tag{5.31}$$

where $\tilde{R}_8$ is given by (5.30) with $B_6^{(1/2)}$ replaced by $B_8^{(3/2)}$ and $Z(m_{\rm t}, \ldots)$ results from electroweak penguin diagrams, in particular Z penguin diagrams. Again "-10" comes from operator mixing under renormalization and new physics can enter (5.31) only through the function Z. But this time any sizable impact of new physics on the function Z translates directly into a sizable impact on ε'/ε.

5.5 Summary

As we have seen, the estimates of ε'/ε in the standard model are typically below the experimental data. However, as the scanning analyses show, for suitably chosen parameters, ε'/ε in the standard model can be made consistent with data. Yet, this happens only if all relevant parameters are simultaneously close to their extreme values. This is clearly seen in table 5.3. Moreover, the probability density distributions for ε'/ε in fig. 5.1 indicate that values of ε'/ε in the ballpark of the experimental grand average $21.2 \cdot 10^{-4}$ are rather improbable.

In spite of a possible "disagreement" of the standard model with the data, one should realize that certain features present in the standard model are confirmed by the experimental results. Indeed, the sign and the order of magnitude of ε'/ε predicted by the standard model turn out to agree with the data. In obtaining these results renormalization group evolution between scales $\mathcal{O}(M_{\rm W})$ and $\mathcal{O}(1\,\text{GeV})$, an important ingredient in the evaluation of the Wilson coefficients $y_i(\mu)$, plays a crucial role. As analyzed in [16] without these renormalization group effects y_6 and y_8 would be tiny and ε'/ε at most $\mathcal{O}(10^{-5})$ in vast disagreement with the data. Since these effects are present in all extensions of the standard model, we conclude that we are probably on the right track.

Unfortunately, in view of very large hadronic and substantial parametric uncertainties, it is impossible to conclude at present whether new physics contributions are indeed required to fit the data. Yet as we stressed above, there is still a lot of room for nonstandard contributions to ε'/ε. The most plausible sizable new contributions could come from chromomagnetic penguins in general supersymmetric models and modified Z penguins. On the other hand substantial modification of QCD penguins through new physics are rather implausible.

In view of large hadronic uncertainties it is difficult to conclude what is precisely the impact of the ε'/ε-data on the CKM matrix. However, as analyzed in [13] there are indications that the lower limit on $\text{Im}\lambda_t$ is improved. The same applies to the lower limits for the branching ratios for $K_L \to \pi^0\nu\bar{\nu}$ and $K_L \to \pi^0 e^+e^-$ decays.

The future of ε'/ε in the standard model and in its extensions depends on the progress in the reduction of parametric and in particular hadronic uncertainties. In this context it is essential to get full control over renormalization scheme and renormalization scale dependences of hadronic matrix elements. Personally I believe that the only hope for making the standard model naturally consistent with the data, without stretching all parameters to their extreme values, is the value of $m_{\rm s}(2\ \text{GeV})$ in the ballpark of 90 to 100 MeV, $B_6^{(1/2)}$ in the ballpark of 1.5 to 2.0 and $B_8^{(3/2)} < 1.0$ (both in the NDR scheme). The required values for $m_{\rm s}$ seem to be found in lattice calculations but the story is not finished yet. Values of $B_6^{(1/2)}$ in the ballpark of 1.5 to 2.0 are suggested by the chiral quark model calculations [31] and a higher-order term $\mathcal{O}(p^2/N)$ in the large-N approach [54]. Yet, from my point of view, it is not clear how well the chiral quark model approximates QCD and whether other higher-order terms in the large-N approach will weaken the indicated enhancement of $B_6^{(1/2)}$. We should hope that the new efforts by the lattice community will help in clarifying the situation. An interesting work in this direction in the framework of the large–N approach can also be found in [46]. In any case ε'/ε already played a decisive role in establishing direct CP violation in nature and its rather large value gives additional strong motivation for searching for this phenomenon in other decays.

Acknowledgments

I would like to thank Jonathan Rosner and Bruce Winstein for inviting me to such an exciting symposium. I would also like to thank all my collaborators for the most enyojable time we had together and the authors of [31, 54, 59, 63] for informative discussions.

This work has been supported by the German Bundesministerium für Bildung und Forschung under contract 06 TM 874. Travel support from Max-Planck Institute for Physics in Munich is gratefully acknowledged.

References

[1] M. Kobayashi and K. Maskawa, Prog. Theor. Phys. **49**, 652 (1973); N. Cabibbo, Phys. Rev. Lett. **10**, 531 (1963).

[2] J.H. Christenson, J.W. Cronin, V.L. Fitch, and R. Turlay, Phys. Rev. Lett. **13**, 128 (1964).

[3] Particle Data Group, Euro. Phys. J. C **3**, 1 (1998).

[4] H. Burkhardt *et al.*, Phys. Lett. B **206**, 169 (1988); G.D. Barr *et al.*, Phys. Lett. B **317**, 233 (1993).

[5] L.K. Gibbons *et al.*, Phys. Rev. Lett. **70**, 1203 (1993).

[6] L. Wolfenstein, Phys. Rev. Lett. **13**, 562 (1964).

[7] A. Alavi-Harati *et al.*, Phys. Rev. Lett. **83**, 22 (1999).

[8] Seminar presented by P. Debu for NA48 collaboration, CERN, June 18, 1999; http://www.cern.ch/NA48/First Result/slides.html.

[9] T. Inami and C.S. Lim, Progr. Theor. Phys. **65**, 297 (1981).

[10] A.J. Buras and R. Fleischer, hep-ph/9704376, in *Heavy Flavours II*, ed. A.J. Buras and M. Lindner (World Scientific, Singapore, 1998), p. 65. A.J. Buras, hep-ph/9806471, in *Probing the Standard Model of Particle Interactions*, ed. R. Gupta, A. Morel, E. de Rafael, and F. David (Elsevier Science B.V., Amsterdam, 1998), p. 281; A.J. Buras, hep-ph/9905437, Lectures given at the 14th Lake Louise Winter Institute.

[11] A.J. Buras, M. Jamin, and P.H. Weisz, Nucl. Phys. B **347**, 491 (1990); S. Herrlich and U. Nierste, Nucl. Phys. B **419**, 292 (1994), Phys. Rev. D **52**, 6505 (1995), Nucl. Phys. B **476**, 27 (1996); J. Urban, F. Krauss, U. Jentschura, and G. Soff, Nucl. Phys. B **523**, 40 (1998).

[12] R. Gupta, Nucl. Phys. (Proc. Suppl.) **63**, 278 (1998) and hep-ph/9801412; L. Lellouch, hep-ph/9906497.

[13] S. Bosch, A.J. Buras, M. Gorbahn, S. Jäger, M. Jamin, M.E. Lautenbacher, and L. Silvestrini, TUM-HEP-347/99 [hep-ph/9904408].

[14] CDF Collaboration, CDF/PUB/BOTTOM/CDF/4855,1999.

[15] J.M. Flynn and L. Randall, Phys. Lett. B **224**, 221 (1989); erratum Phys. Lett. B **235**, 412 (1990).

[16] G. Buchalla, A.J. Buras, and M.K. Harlander, Nucl. Phys. B **337**, 313 (1990).

[17] A.J. Buras, M. Jamin, M.E. Lautenbacher, and P.H. Weisz, Nucl. Phys. B **370**, 69 (1992); Nucl. Phys. B **400**, 37 (1993); A.J. Buras, M. Jamin, and M.E. Lautenbacher, Nucl. Phys. B **400**, 75 (1993); M. Ciuchini, E. Franco, G. Martinelli, and L. Reina, Nucl. Phys. B **415**, 403 (1994).

[18] A.J. Buras, M. Jamin, and M.E. Lautenbacher, Nucl. Phys. B **408**, 209 (1993).

[19] M. Ciuchini, E. Franco, G. Martinelli, and L. Reina, Phys. Lett. B **301**, 263 (1993).

[20] G. Buchalla, A.J. Buras, and M. Lautenbacher, Rev. Mod. Phys **68**, 1125 (1996).

[21] J.F. Donoghue, E. Golowich, B.R. Holstein, and J. Trampetic, Phys. Lett. B **179**, 361 (1986).

[22] A.J. Buras and J.-M. Gérard, Phys. Lett. B **192**, 156 (1987);

[23] H.-Y. Cheng, Phys. Lett. B **201**, 155 (1988); M. Lusignoli, Nucl. Phys. B **325**, 33 (1989).

[24] J. Ellis, M.K. Gaillard, and D.V. Nanopoulos, Nucl. Phys. B **109**, 213 (1976).

[25] F.J. Gilman and M.B. Wise, Phys. Lett. B **83**, 83 (1979); B. Guberina and R.D. Peccei, Nucl. Phys. B **163**, 289 (1980).

[26] A.J. Buras and M.E. Lautenbacher, Phys. Lett. B **318**, 212 (1993).

[27] M. Ciuchini, E. Franco, G. Martinelli, L. Reina, and L. Silvestrini, Z. Phys. C **68**, 239 (1995).

[28] A.J. Buras, M. Jamin, and M.E. Lautenbacher, Phys. Lett. B **389**, 749 (1996).

[29] M. Ciuchini, Nucl. Phys. B. Proc. Suppl. **59**, 149 (1997).

[30] J. Heinrich, E.A. Paschos, J.-M. Schwarz, and Y.L. Wu, Phys. Lett. B **279**, 140 (1992); E.A. Paschos, review presented at the 27th Lepton-Photon Symposium, Beijing, China (August 1995).

[31] S. Bertolini, M. Fabbrichesi, and J.O. Eeg, hep-ph/9802405, to appear in Reviews of Modern Physics; S. Bertolini, hep-ph/9908268.

[32] A.A. Belkov, G. Bohm, A.V. Lanyov, and A.A. Moshkin, hep-ph/9704354.

[33] R.D. Kenway, Plenary talk at LATTICE 98, hep-ph/9810054.

[34] M. Jamin and M. Münz, Z. Phys. C **66**, 633 (1995); S. Narison, Phys. Lett. B **358**, 113 (1995); K. G. Chetyrkin, D. Pirjol, and K. Schilcher, Phys. Lett. B **404**, 337 (1997); P. Colangelo, F. De Fazio, G. Nardulli, and N. Paver, Phys. Lett. B **408**, 340 (1997); M. Jamin, Nucl. Phys. B. Proc. Suppl. **64**, 250 (1998); S. Narison, hep-ph/9905264.

[35] L. Lellouch, E. de Rafael, and J. Taron, Phys. Lett. B **414**, 195 (1997); F.J. Yndurain, Nucl. Phys. B **517**, 324 (1998). H.G. Dosch and S. Narison, Phys. Lett. B **417**, 173 (1998).

[36] S. Aoki *et al.* (JLQCD), Phys. Rev. Lett. **82**, 4392 (1999).

[37] J. Garden *et al.* (ALPHA/UKQCD), hep-lat/9906013.

[38] M. Göckeler *et al.* (QSDSF), hep-lat/9908005.

[39] R. Burkhalter (CP-PACS), talk given at Lattice 99.

[40] S. Ryan, hep-ph/9908386, this volume, chapter 29; M. Ciuchini *et al.*, this volume, chapter 28.

[41] J. Prades and A. Pich, JHEP06 013 (1998); Nucl. Phys. B. Proc. Suppl. **74**, 309 (1999); J. Prades, Nucl. Phys. B. Proc. Suppl. **76**, 341 (1999).

[42] J. Prades and A. Pich, private communication.

[43] ALEPH collaboration, CERN-EP/99-026, hep-ex/9903015.

[44] W.A. Bardeen, A.J. Buras, and J.-M. Gérard, Phys. Lett. B **180**, 133 (1986), Nucl. Phys. B **293**, 787 (1987), Phys. Lett. B **192**, 138 (1987).

[45] W.A. Bardeen, Nucl. Phys. Proc. **7 A**, 149 (1989); W.A. Bardeen, this volume, chapter 14.

[46] J. Bijnens and J. Prades, JHEP **01**, 023 (1999) [hep-ph/9811472]; J. Bijnens, hep-ph/9907307, hep-ph/9907514.

[47] G.W. Kilcup, Nucl. Phys. (Proc. Suppl.) B **20**, 417 (1991); S.R. Sharpe, Nucl. Phys. (Proc. Suppl.) B **20**, 429 (1991); D. Pekurovsky and G. Kilcup, hep-lat/9709146.

[48] D. Pekurovsky and G. Kilcup, hep-lat/9812019.

[49] G. Kilcup, R. Gupta, and S.R. Sharpe, Phys. Rev. D **57**, 1654 (1998).

[50] L. Conti, A. Donini, V. Gimenez, G. Martinelli, M. Talevi, and A. Vladikas, Phys. Lett. B **421**, 273 (1998).

[51] R. Gupta, T. Bhattacharaya, and S.R. Sharpe, Phys. Rev. D **55**, 4036 (1997).

[52] T. Hambye, G.O. Köhler, and P.H. Soldan, hep-ph/9902334.

[53] T. Hambye, G.O. Köhler, E.A. Paschos, P.H. Soldan, and W.A. Bardeen, Phys. Rev. D **58**, 014017 (1998).

[54] T. Hambye, G.O. Köhler, E.A. Paschos, and P. H. Soldan, hep-ph/9906434.

[55] P.H. Soldan, hep-ph/9608281; T. Hambye and P.H. Soldan, hep-ph/9908232.

[56] J. Gasser and U.G. Meissner, Phys. Lett. B **258**, 219 (1991).

[57] A.I. Vainshtein, V.I. Zakharov, and M.A. Shifman, JEPT **45**, 670 (1977).

[58] W.A. Bardeen, A.J. Buras, and J.-M. Gérard, Phys. Lett. B **192**, 138 (1987); A. Pich and E. de Rafael, Nucl. Phys. B **358**, 311 (1991); M. Neubert and B. Stech, Phys. Rev. D **44**, 775 (1991); M. Jamin and A. Pich, Nucl. Phys. B **425**, 15 (1994); J. Kambor, J. Missimer, and D. Wyler, Nucl. Phys. B **346**, 17 (1990); Phys. Lett. B **261**, 496 (1991); V. Antonelli, S. Bertolini, M. Fabbrichesi, and E.I. Lashin, Nucl. Phys. B **469**, 181 (1996).

[59] M. Ciuchini *et al.*, this volume, chapter 28; M. Ciuchini, private communication.

[60] M. Knecht, S. Peris, and E. de Rafael, hep-ph/9812471.

[61] S. Bethke, hep-ex/9812026.

[62] Y.-Y. Keum, U. Nierste, and A.I. Sanda, hep-ph/9903230.

[63] A.A. Belkov, G. Bohm, A.V. Lanyov, and A.A. Moshkin, hep-ph/9907335.

[64] A. Masiero and H. Murayama, hep-ph/9903363; K.S. Babu, B. Dutta, and R. N. Mohapatra, hep-ph/9905464; S. Khalil and T. Kobayashi, hep-ph/9906374; E. Accomondo, R. Arnowitt, and B. Dutta, hep-ph/9907446; S. Baek, J.-H. Jang, P. Ko, and J.H. Park, hep-ph/9907572; R. Barbieri, R. Contino, and A. Strumia, hep-ph/9908255; G. Eyal, A. Masiero, Y. Nir, and L. Silvestrini, hep-ph/9908382.

[65] A.J. Buras, G. Colangelo, G. Isidori, A. Romanino, and L. Silvestrini, hep-ph/9908371.

[66] X-G. He, hep-ph/9903242.

[67] J. Agrawal and P. Frampton, Nucl. Phys. B **419**, 254 (1994).

[68] G. Buchalla, A.J. Buras, M.K. Harlander, M.E. Lautenbacher, and C. Salazar, Nucl. Phys. B **355**, 305 (1991).

[69] E. Gabrielli and G.F. Giudice, Nucl. Phys. B **433**, 3 (1995).

[70] E. Gabrielli, A. Masiero, and L. Silvestrini, Phys. Lett. B **374**, 80 (1996); F. Gabbiani, E. Gabrielli, A. Masiero, and L. Silvestrini, Nucl. Phys. B **477**, 321 (1996).

[71] G. Colangelo and G. Isidori, JHEP **09** 009 (1998).

[72] A.J. Buras and L. Silvestrini, Nucl. Phys. B **546**, 299 (1999); hep-ph/9811471.

[73] M.S. Chanowitz, hep-ph/9905478.

6

Re(ϵ'/ϵ) Result from KTeV: An Observation of Direct CP Violation

Yee Bob Hsiung[1]

Abstract

Using a subset of data collected in the 1996–97 fixed target run at Fermilab, we report the first KTeV measurement for the search of direct CP violation. The result is Re$(\epsilon'/\epsilon) = (28.0 \pm 4.1) \times 10^{-4}$, nearly 7σ above zero obtained by a blind analysis. This establishes the long-sought "direct CP violation" effect in the two-pion system of neutral kaon decays. Other new measurements of Δm, τ_S, $\Delta\phi$ and a limit on the diurnal variation of ϕ_{+-} are also presented.

One of the long-lasting puzzles in particle physics is the origin of the CP violation, where C stands for charge conjugation (exchange particle and antiparticle) and P stands for parity (space inversion). Thirty-five years after the first unexpected discovery [1] of CP violation in $K_L \to \pi\pi$ decays in 1964, we can only explain the dominant effect as due to a small asymmetry of the $K^0 - \bar{K}^0$ mixing or admixture of wrong CP states in the K_S and K_L neutral kaons, parametrized by ϵ (about 0.0023). The question is "Does CP violation also occur in the $K \to \pi\pi$ decay process itself?" An effect referred as "direct" CP violation [2], parametrized by ϵ', contributes differently to the decay rates of $K_L \to \pi^+\pi^-$ and $K_L \to \pi^0\pi^0$ (relative to the corresponding K_S decays), and would be observed as a nonzero value in the ratio of Re(ϵ'/ϵ).

Experimentally we measure the double ratio R,

$$R = \frac{\Gamma(K_L \to \pi^+\pi^-)/\Gamma(K_S \to \pi^+\pi^-)}{\Gamma(K_L \to \pi^0\pi^0)/\Gamma(K_S \to \pi^0\pi^0)} = \frac{|\eta_{+-}|^2}{|\eta_{00}|^2} \approx 1 + 6\mathrm{Re}(\epsilon'/\epsilon). \quad (6.1)$$

The standard Cabibbo-Kobayashi-Maskawa (CKM) model [3] accomodates CP violation with a complex phase in the quark-mixing matrix, but the calculations of Re(ϵ'/ϵ) are still uncertain depending on several input

[1]For KTeV Collaboration: Arizona, Chicago, Colorado, Elmhurst, Fermilab, Osaka, Rice, Rutgers, UCLA, UCSD, Virginia, and Wisconsin

parameters and on the method used to estimate the hadronic matrix elements. Most recent estimates [4,5] had given nonzero values slightly below 10^{-3}; however, another group [6] gave somewhat larger estimates. Alternatively, a "superweak" interaction [7] could also produce the observed CP-violating mixing effect (ϵ) but would give $\text{Re}(\epsilon'/\epsilon) = 0$. Therefore, a nonzero measurement of $\text{Re}(\epsilon'/\epsilon)$ would rule out the possibility that a superweak interaction is the sole source of CP violation and would establish the "direct" CP violation from the decay process itself.

The earlier two measurements of $\text{Re}(\epsilon'/\epsilon)$ from Fermilab-E731 [8] and CERN-NA31 [9] were

$$\text{Re}(\epsilon'/\epsilon) = (7.4 \pm 5.9) \times 10^{-4}, \qquad \text{(E731)} \tag{6.2}$$

$$\text{Re}(\epsilon'/\epsilon) = (23.0 \pm 6.5) \times 10^{-4}, \qquad \text{(NA31)} \tag{6.3}$$

and the PDG average [10] was $(15 \pm 8) \times 10^{-4}$, which gave inconclusive interpretations between standard model and superweak CP violation. New experiments have been constructed at Fermilab (KTeV), CERN (NA48), and Frascati (KLOE) [11] to measure $\text{Re}(\epsilon'/\epsilon)$ with a precision of $(1 \sim 2) \times 10^{-4}$ for the search of "direct" CP violation and determining its magnitude.

We report here a new measurement of $\text{Re}(\epsilon'/\epsilon)$ from 23% of the data collected by the KTeV experiment (E832) during the 1996–97 run at Fermilab. This result has recently been published in [12] after a preliminary announcement in February 1999.

The KTeV experiment was designed to improve on the previous experiments and ultimately to have the sensitivity to establish direct CP violation if $\text{Re}(\epsilon'/\epsilon)$ is on the order of 10^{-3}. The experimental technique is the same as in E731 [13] with many improvements in beam and detector performance. Double kaon beams from a single target are used to enable the simultaneous collection of K_L and K_S decays to minimize the systematics due to time variation of beam flux and detector inefficiencies. A precision magnetic spectrometer is used to minimize backgrounds in the $\pi^+\pi^-$ samples and to allow in situ calibration of the calorimeter with electrons. A high-precision electromagnetic calorimeter, a cesium iodide (CsI) array, is used for $\pi^0\pi^0$ reconstruction and better background suppression. Nearly hermetic photon vetoes are used for further background reduction for the $\pi^0\pi^0$ mode. A new beamline was constructed for KTeV with cleaner beam collimation and improved muon sweeping. While the method of producing the K_S beam (by passing a K_L beam through a "regenerator") is also the same as E731, the KTeV regenerator is made of scintillator and is fully active to reduce the scattered background to the coherently regenerated K_S.

The KTeV detector is shown in fig. 6.1. Two beams (called "regenerator" and "vacuum") enter the evacuated decay region, with the main detector elements located downstream. The regenerator switches sides once every accelerator cycle to minimize the effect of any left-right asymmetry of beam or detectors. A movable "shadow absorber" far upstream atten-

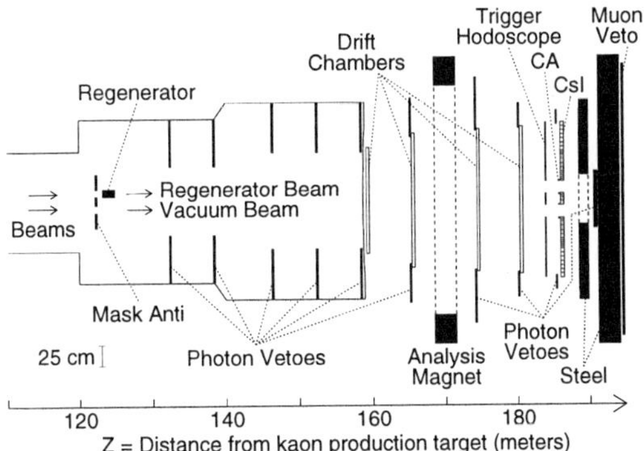

Figure 6.1: Plan view of the KTeV apparatus with a double kaon beam configured to measure $\mathrm{Re}(\epsilon'/\epsilon)$. The evacuated decay volume ends with a thin vacuum window at $Z = 159$ m. The label "CsI" indicates the electromagnetic calorimeter.

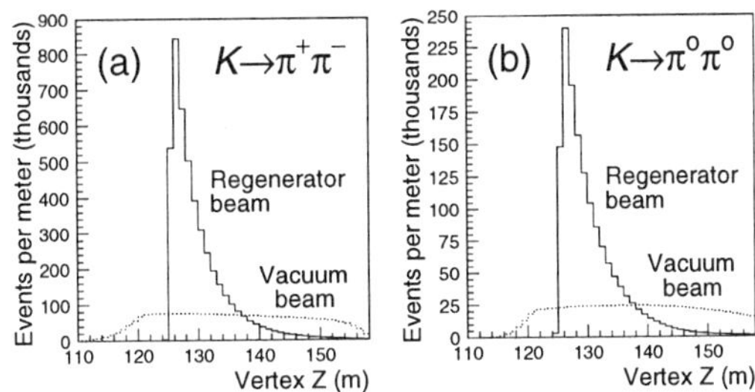

Figure 6.2: Decay vertex distributions for the (a) $K \to \pi^+\pi^-$ and (b) $K \to \pi^0\pi^0$ decay modes, showing the difference between the "regenerator" (K_S) and "vacuum" (K_L) beams.

uates the kaon beam onto the regenerator. To measure the double ratio of decay rates for $\mathrm{Re}(\epsilon'/\epsilon)$, we need to understand the difference between the acceptances for K_S versus K_L decays to each $\pi\pi$ final state. Event reconstruction and selection are done with identical criteria for decays in either beam, so the principal difference between the K_S and K_L data samples is in the decay vertex distributions, shown in fig. 6.2 as a function of Z, the decay distance from the target. Therefore, the most crucial requirement of measuring $\mathrm{Re}(\epsilon'/\epsilon)$ with this technique is a precise understanding of the Z dependence of the detector acceptance. The acceptance for decays upstream of $Z = 122$ m in the vacuum beam is defined by a lead scintillator counter, "mask-anti" (MA), with two square holes 50% larger than the beams. In the regenerator beam, the beginning of the decay region is sharply defined by a thin lead scintillator module at the end of the 1.7 m long regenerator.

The momenta of charged particles from the decay are measured by the spectrometer, consisting of four drift chambers, each with two horizontal and two vertical planes of sense wires, and a large dipole magnet with 412 MeV/c transverse kick. The position resolution is typically 110 μm for the chamber and the momentum resolution is 0.4% at the mean pion momentum of 36 GeV/c.

The electromagnetic calorimeter consists of 3100 blocks of pure CsI in a square array 1.9 m on a side and 50 cm deep. Two 15 cm square holes allow passage of the neutral beams through the calorimeter. The calorimeter was calibrated using 190 million momentum-analyzed electrons from $K_L \to \pi e \nu$ decays collected during the normal running. The average energy resolution for photons from $\pi^0\pi^0$ is 0.7% with a mean photon energy of 19 GeV.

The inner aperture at the CsI is defined by a tungsten scintillator counter (CA) around each beam hole. In addition, there are ten lead scintillator "photon veto" counters to detect particles escaping from the decay volume or missing the CsI, in order to suppress the major $K_L \to 3\pi^0$ background to the CP-violating $\pi^0\pi^0$ signal mode.

The charged mode (for $\pi^+\pi^-$, $Ke3$) trigger is based on a scintillator hodoscope located just upstream of the CsI calorimeter and requirements on the number and pattern of hits in the drift chambers. The neutral mode trigger is based on a fast energy sum from the calorimeter and a hardware cluster-finding processor (e.g., for $\pi^0\pi^0$ it must find 4 or 5 clusters of energy in CsI). Additional fast veto signals from the regenerator, photon vetoes, and a downstream muon veto hodoscope located behind 4 m of steel are also used in the trigger to keep the trigger rate at a manageable level. A CPU-based "Level 3 filter" reconstructs events on-line and applies loose kinematic cuts to select $\pi^+\pi^-$ and $\pi^0\pi^0$ candidates. Large samples of $K_L \to \pi e \nu$ and $K_L \to 3\pi^0$ decays are also recorded for detector calibration and acceptance studies. In addition, an "accidental" trigger is formed by using small scintillation counters near 90° to the target to randomly record the underlying activity in the KTeV detector with the same instantaneous rate as the physics data for study.

The $\pi^0\pi^0$ samples in this analysis were collected in 1996, while $\pi^+\pi^-$ samples were from the first 18 days of data in 1997. The 1996 $\pi^+\pi^-$ samples were analyzed but not used for this result because of a large Level 3 tracking inefficiency (about 22% loss) from an unanticipated drift chamber effect that could sometimes delay a hit by 20 ns or more (due to lower gas gain and higher threshold relative to the first avalanche pulse in the chamber). The inefficiency was nearly the same for both beams but still has led to a larger systematic error. The Level 3 software was modified for the 1997 run to allow for this effect, resulting in an inefficiency of less than 0.1% and reduced systematics. The double ratio [eq. (6.1)] allows us to use data from two different time periods with small systematics.

Event reconstruction and selection are done with identical cuts and criteria for decays in either beam to minimize the systematics. The $\pi^+\pi^-$ candidates are selected, by requiring each pion to have a momentum $p_\pi >$ 8 GeV/c and to deposit less than 85% of its energy in the CsI calorimeter.

Cuts are also made near the edges of the detectors, and on the separation between the two pions at the chambers and calorimeter to avoid poorly reconstructed event topology and to cleanly define the acceptance. The $\pi^+\pi^-$ invariant mass is required to be between 488 and 508 MeV/c^2 (where the mass resolution is about 1.6 MeV/c^2) and the square of the transverse momentum of the $\pi^+\pi^-$ system relative to the initial kaon trajectory from target, p_T^2, is required to be less than 250 MeV^2/c^2.

The $\pi^0\pi^0$ candidates are reconstructed from four-photon events by choosing the photon-pairing combination consistent with two π^0 decays at a common kaon decay vertex. Each photon is required to have an energy $E_\gamma > 3$ GeV and to be at least 5 cm away from the outer edge of the CsI and 7.5 cm from any other photon. The $\pi^0\pi^0$ invariant mass is required to be between 490 and 505 MeV/c^2 (where the mass resolution is about 1.5 MeV/c^2). The initial kaon trajectory is not well known, so the only available indicator of kaon scattering is the position of the energy centroid of the four photons at the CsI. This is used to calculate a box "ring number," defined as four times the square of the larger normal distance (either horizontal or vertical), in units of cm, from the centroid to the center of the closer beam. Its value is required to be less than 110, which selects events with energy centroid lying within a square region of area 110 cm^2 centered on each beam.

In both $\pi^+\pi^-$ and $\pi^0\pi^0$ analyses, cuts are made on energy deposits in the MA, photon veto counters, and regenerator. The final samples consist of events with $110 < Z < 158$ m and $40 < E_K < 160$ GeV.

Detailed Monte Carlo (MC) simulation is used to determine the detector acceptance and to evaluate backgrounds. The simulation models kaon production and regeneration to generate decays with the same energy E_K and decay vertex Z distributions as the data. The decay products are traced through the KTeV detector, allowing for electromagnetic interactions with detector material and pion decay. The acceptance is largely determined by the geometry of the detector and by geometric analysis cuts. Simulation of the detector response is also included to understand the reconstruction biases. High statistics $\pi e\nu$ and $3\pi^0$ data samples are used to check or tune various aspects of the detector geometry and simulation. To reproduce the biases due to underlying activity in the detector, an event from the "accidental" trigger is overlaid on top of each simulated decay; the net effect on Re(ϵ'/ϵ) is of order 10^{-4}.

Backgrounds to the $\pi^+\pi^-$ samples are determined by using sidebands in the mass and p_T^2 distributions to normalize MC predictions from various background processes. Figs. 6.3(a) and (b) show that the p_T^2 distributions for data are well described by the sum of coherent $\pi\pi$ MC and total background MC. Semileptonic $K_L \to \pi e\nu$ and $K_L \to \pi\mu\nu$ decays, with the electron or muon misidentified as a pion, contribute 0.069% mainly to the vacuum beam. The dominant regenerator beam background (0.072%) is from kaons that scatter in the regenerator before decaying to $\pi^+\pi^-$. Kaons that scatter in the final beam-defining collimator contribute an additional

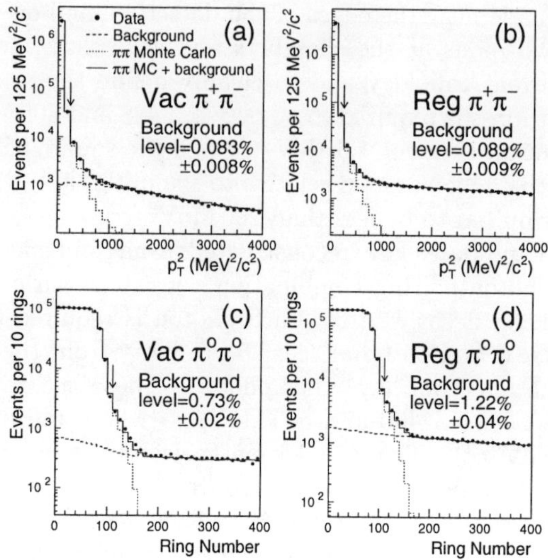

Figure 6.3: Distributions of p_T^2 for the $\pi^+\pi^-$ samples and ring number for the $\pi^0\pi^0$ samples. Total background levels and uncertainties (dominated by systematics) are given for the samples passing the analysis cuts (arrows).

0.014% to each beam. Each type of scattering is parametrized using the $\pi^+\pi^-$ data of the same running period and incorporated into the MC simulation.

Background levels are larger for the $\pi^0\pi^0$ samples since the ring-number variable is not as effective as p_T^2 at identifying scattered kaons and detecting "crossover" scattering from the regenerator into the vacuum beam. Ring-number distributions are shown in figs. 6.3(c) and (d). The upturn under the peak in (c) is due to $K_L \to 3\pi^0$ decays with lost and/or overlapping photons; it is determined, using mass sidebands, to contribute a background of 0.27% mainly to the vacuum beam. Ring-number sidebands are used to normalize MC distributions from kaons that scatter in the regenerator or collimator before decaying to $\pi^0\pi^0$. The vacuum (regenerator) beam background includes 0.30% (1.07%) from regenerator scattering and 0.16% (0.14%) from collimator scattering.

After background subtraction, the net yields are 2.607M $\pi^+\pi^-$ events in the vacuum beam, 4.516M $\pi^+\pi^-$ in the regenerator beam, 862K $\pi^0\pi^0$ in the vacuum beam, and 1.434M $\pi^0\pi^0$ in the regenerator beam.

Re(ϵ'/ϵ) is extracted from the background-subtracted data using a fitting program which analytically calculates regeneration and decay distributions accounting for $K_S - K_L$ interference. After the acceptance correction, the resulting prediction for each decay mode is integrated over Z and compared to data in 10 GeV bins of kaon energy. CPT symmetry is assumed, and the values of $K_S - K_L$ mass difference (Δm) and K_S lifetime (τ_S) are fixed to PDG values [10]. The regeneration amplitude is allowed to float in the fit, but constrained to have a power law dependence on kaon energy, with the phase determined by analyticity [13,14]. The kaon energy distributions are also allowed to float for $\pi^+\pi^-$ and $\pi^0\pi^0$ modes in each energy bin (24 fit parameters in all).

Fitting was done "blind," by hiding the value of Re(ϵ'/ϵ) with an unknown offset between η_{+-} and η_{00}, until after the analysis and systematic error evaluation were finalized. The final fit result is Re(ϵ'/ϵ) = $(28.0 \pm 3.0) \times 10^{-4}$, where the error is statistical only with a χ^2 equal to 30 for 21 degrees of freedom.

Only biases that affect the K_L and K_S samples differently will lead to systematic errors on Re(ϵ'/ϵ), a virtue of double-ratio and double-beam method. Possible sources are divided into four classes: (1) data collection inefficiencies; (2) biases in event reconstruction, sample selection, and background subtraction; (3) misunderstanding of the detector acceptance; and (4) uncertainties in kaon flux and physics parameters. Table 6.1 summarizes all of the estimated contributions, and the detailed discussion on systematics can be found in [12].

A systematic shift in measured energy scales can shift the reconstructed Z vertex and E_K distributions for the $\pi^0\pi^0$ sample and thus can bias Re(ϵ'/ϵ), mainly by moving K_L events past the fiducial Z cut at 158 m. After calibrating the calorimeter with electrons (and allowing for a small expected electron-photon difference), a final energy scale correction for photons of -0.125% is determined by matching the sharp turn-on of the $\pi^0\pi^0$ Z distribution at the regenerator edge between data and MC. After making this correction, a check using π^0 pairs produced by hadronic interactions in the vacuum window reveals a Z mismatch of 2 cm at the downstream end of the decay region, leading to a systematic error of 0.7×10^{-4} on Re(ϵ'/ϵ). Residual nonlinearities in the calorimeter response contribute an additional error of 0.6×10^{-4}.

The accuracy of the background determination for the $\pi^0\pi^0$ samples depends on our understanding of kaon scattering in the regenerator and collimator. We consider several variations in the procedure for determining the scattering distributions from the $K \to \pi^+\pi^-$ data; these affect the shapes of the background MC ring-number distributions, but the sideband normalization procedure limits the impact on Re(ϵ'/ϵ) ($\leq 0.8 \times 10^{-4}$).

The largest systematics in table 6.1 comes from the detector acceptance. Many potential detector modeling problems would affect the acceptance as a function of Z, so a crucial check of our understanding of the acceptance is to compare the Z distribution for the data against the MC simulation. Fig. 6.4 shows the vacuum-beam comparisons for the $\pi^+\pi^-$ and $\pi^0\pi^0$ signal modes as well as for the two high statistics $\pi e\nu$ and $3\pi^0$ samples. The overall agreement is fairly good, but since the mean Z positions for K_L and K_S decays differ by about 6 m, a relative slope of 10^{-4} per meter in the data/MC ratio would cause an error of 10^{-4} on Re(ϵ'/ϵ). As shown in fig. 6.4(b), the $\pi e\nu$ comparison agrees to better than this level; however, the $\pi^+\pi^-$ comparison has a slope of $(-1.60 \pm 0.63) \times 10^{-4}$ per meter.

Source of Uncertainty	Uncertainty ($\times 10^{-4}$) $\pi^+\pi^-$	$\pi^0\pi^0$
1. Data Collection		
Trigger and Level 3 filter	0.5	0.3
2. Reconstruction, Selection, Backgrounds		
Energy scale	0.1	0.7
Calorimeter nonlinearity	...	0.6
Detector calibration, alignment	0.3	0.4
Analysis cut variations	0.6	0.8
Background subtraction	0.2	0.8
3. Detector Acceptance		
Limiting apertures	0.3	0.5
Detector resolution	0.4	<0.1
Drift chamber simulation	0.6	...
Z dependence of acceptance	1.6	0.7
Monte Carlo statistics	0.5	0.9
4. Kaon Flux and Physics Parameters		
Regenerator-beam attenuation:		
1996 versus 1997	0.2	
Energy dependence	0.2	
Δm, τ_S, regeneration phase	0.2	
TOTAL	2.8	

Table 6.1: Systematic uncertainties on Re(ϵ'/ϵ).

Although the significance of the $\pi^+\pi^-$ slope is marginal ($\sim 2.5\sigma$), we assign a systematic error on Re(ϵ'/ϵ) based on the full size of the slope, 1.6×10^{-4}. The $3\pi^0$ and $\pi^0\pi^0$ Z distributions agree well, and we place a limit of 0.7×10^{-4} for the possible bias from the neutral-mode acceptance.

Other checks on the acceptance include data/MC comparisons of track illuminations at the drift chambers and CsI, photon illumination at the CsI, and minimum photon separation distance. These all agree well and indicate no other sources of acceptance misunderstanding.

The final class of systematics includes possible differences in the K_S/K_L flux ratio between the $\pi^+\pi^-$ and $\pi^0\pi^0$ samples. The fact that using $\pi^+\pi^-$ and $\pi^0\pi^0$ data from different running periods is of little concern because the regenerator and movable absorber were the same for both periods; however, we still assign a small uncertainty due to a possible temperature difference which might change their densities and thus the regenerator beam attenuation. In addition, a small difference in the energy dependence of the attenuation (measured using $\pi^+\pi^-\pi^0$ and $3\pi^0$ data) leads to a small

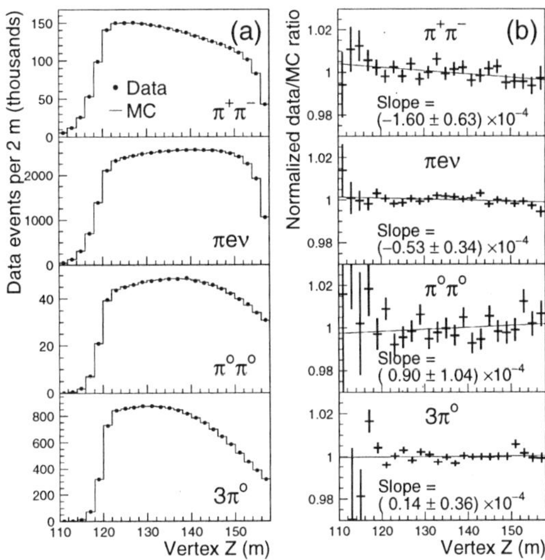

Figure 6.4: (a) Data versus Monte Carlo comparisons of vacuum beam decay vertex Z distributions for $\pi^+\pi^-$, $\pi e\nu$, $\pi^0\pi^0$, and $3\pi^0$ decays. (b) Linear fits to the data/MC ratio of Z distributions for each of the four decay modes.

uncertainty on Re(ϵ'/ϵ).

Finally, we assign uncertainties corresponding to one-sigma variations of Δm and τ_S from the PDG averages [10], and from a deviation of the phase of the regeneration amplitude by $\pm 0.5°$ from the value given by analyticity [14]. Adding all contributions in quadrature, the total systematic uncertainty on Re(ϵ'/ϵ) is 2.8×10^{-4}.

We have performed several cross-checks on the Re(ϵ'/ϵ) result. Consistent values are obtained at all kaon energies (see fig. 6.5), and there is no significant variation as a function of time or beam intensity. Relaxing the power-law constraint on the regeneration amplitude yields a consistent value with the same precision.

We have also extracted Re(ϵ'/ϵ) using an alternative fitting technique which compares the vacuum and regenerator beam Z distributions directly, eliminating the need for a MC simulation to determine the acceptance. While statistically less powerful, this technique yields a value of Re(ϵ'/ϵ) that is consistent with the standard analysis based on the uncorrelated parts of the statistical and systematic errors. In the end, using $\pi^+\pi^-$ data from 1996 (collected simultaneously with the $\pi^0\pi^0$ data) instead of from 1997 yields a consistent value of Re(ϵ'/ϵ), 25×10^{-4}, allowing for a larger systematic error of 4×10^{-4} due to the 1996 Level 3 inefficiency.

In conclusion, we measured Re(ϵ'/ϵ) $= (28.0 \pm 3.0\ (\text{stat}) \pm 2.8\ (\text{syst})) \times 10^{-4}$; combining errors in quadrature, Re(ϵ'/ϵ) $= (28.0 \pm 4.1) \times 10^{-4}$. This result [12], nearly 7σ above zero, firmly establishes the existence of CP violation in a "decay process," agreeing better with the earlier measurement

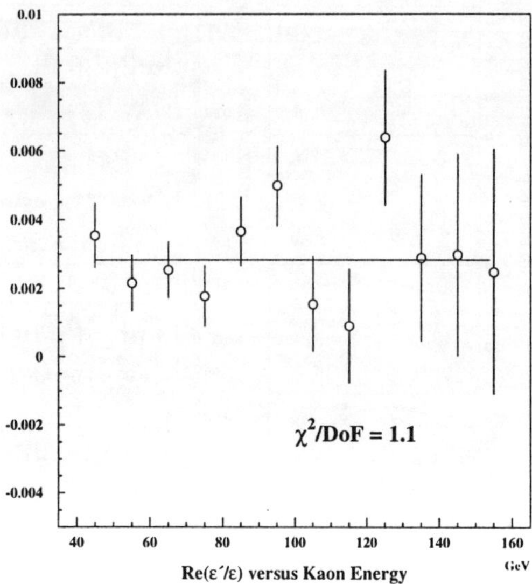

Figure 6.5: A systematic check of Re(ϵ'/ϵ) vs kaon energy.

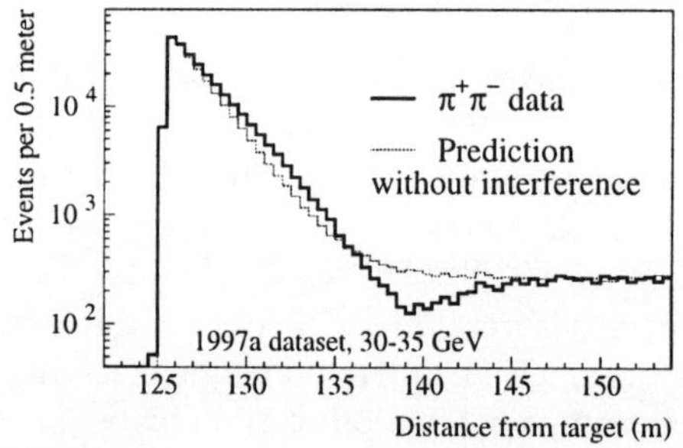

Figure 6.6: The interference between K_S and K_L behind the regenerator for $\pi^+\pi^-$ mode.

from NA31 than with E731 [15] and shows that a superweak interaction cannot be the sole source of CP violation in the K meson system [16]. The average of the three measurements (KTeV, NA31, and E731), $(21.7\pm3.0)\times10^{-4}$, while at the high end of standard model predictions, supports the notion of a nonzero phase in the CKM matrix. Further theoretical and experimental advances are needed before one can say whether or not there are other sources of CP violation beyond the standard model.

With the $K_S - K_L$ interference downstream of the regenerator beam (see fig. 6.6) we have also performed the fits to extract Δm, τ_S, ϕ_{+-}, and $\Delta\phi$ as a test of CPT symmetry without assuming CPT invariance. The

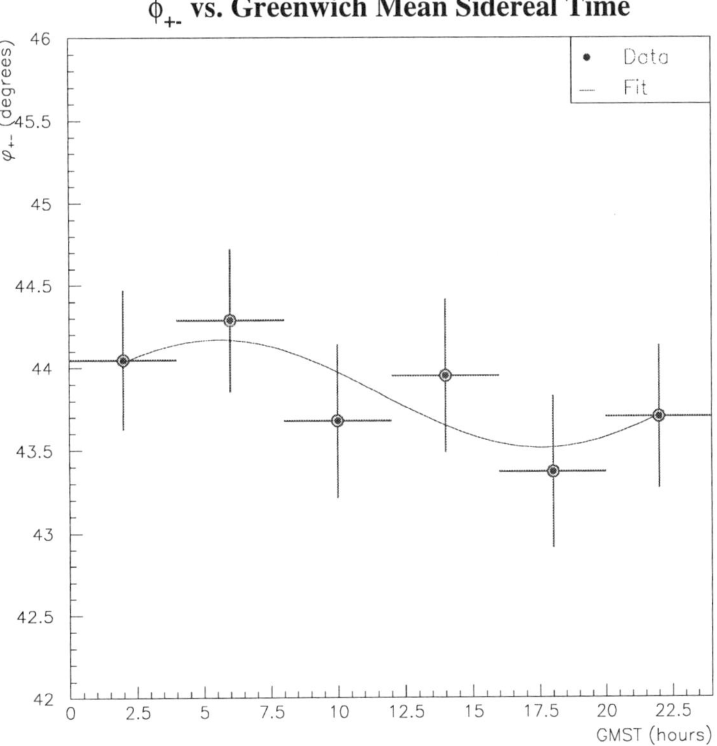

Figure 6.7: Diurnal fit for the ϕ_{+-} vs Greenwich mean sidereal time.

result is $\Delta m = (0.5286 \pm 0.0023) \times 10^{-10}\ \hbar s^{-1}$ and for ϕ_{+-} is

$$\phi_{+-} = (43.66 \pm 0.30)^\circ + 0.23^\circ \times \frac{\Delta m - 0.5286}{0.0010} - 0.26^\circ \times \frac{\tau_s - 0.8967}{0.0010}. \quad (6.4)$$

The ϕ_{+-} is in excellent agreement with $\phi_{SW} = (43.5 \pm 0.08)^\circ$ [10] in the limit of CPT invariance, and the Δm is also in good agreement with the fit assuming CPT ($\phi_{+-} = \phi_{SW}$) and the τ_S is $(0.8967 \pm 0.0007) \times 10^{-10}$ s. The fit χ^2 is 234 for 214 degrees of freedom. The Δm result agrees well with all the measurements since 1992, but about 2σ lower than earlier measurements. The τ_S result is somewhat higher than PDG average, which is affected by the correlation with Δm.

Another test is to measure the CPT-violating phase difference $\Delta\phi = \phi_{00} - \phi_{+-}$, and we get

$$\Delta\phi = (0.09 \pm 0.43(\text{stat}) + 0.15(\text{syst}))^\circ = (0.09 \pm 0.46)^\circ, \quad (6.5)$$

consistent with zero. The grand average is then $\Delta\phi = (-0.01 \pm 0.40)^\circ$. Assuming $\Gamma_{K^0} = \Gamma_{\overline{K}^0}$, the above results would give the bound for $K^0 - \bar{K}^0$

mass difference,

$$\frac{m_{K^0} - m_{\overline{K}^0}}{m_K} = (4.5 \pm 3) \times 10^{-19}. \qquad (6.6)$$

Finally, we have also tested the CPT-violating hypothesis, suggested by [17], that there might be a diurnal variation of ϕ_{+-} with respect to sidereal time. (See the contribution of A. Kostelecký, chapter 28, this volume.) Fig. 6.7 shows our ϕ_{+-} versus Greenwich mean sidereal time and the fit of sine variation gives $(0.34 \pm 0.27)^\circ$ over a 24-hour period, consistent with zero. Therefore, the CPT invariance in $K^0 \to \pi\pi$ has been tested very precisely and holds well.

More data from KTeV are currently being processed through calibration and analysis to reduce both statistical and systematic uncertainties. At the same time, we are taking more data again in 1999 with the aim of doubling the statistics with much improved detector performance and additional systematics checks. We expect to reduce the $\mathrm{Re}(\epsilon'/\epsilon)$ statistical uncertainty to $\sim 1 \times 10^{-4}$ and lower the systematics to a similar level.

In next few years we expect $\mathrm{Re}(\epsilon'/\epsilon)$ can be precisely measured by experiments to 5–10% of itself, which would challenge the theorists to refine their calculations for the origin of direct CP violation.

References

[1] J.H. Christenson, J.W. Cronin, V.L. Fitch, and R. Turlay, Phys. Rev. Lett. **13**, 138 (1964).

[2] B. Winstein and L. Wolfenstein, Rev. Mod. Phys. **65**, 1113 (1993).

[3] M. Kobayashi and T. Maskawa, Prog. Theo. Phys. **49**, 652 (1973).

[4] M. Ciuchini, Nucl. Phys. Proc. Suppl. **59**, 149 (1997).

[5] A.J. Buras in *Probing the Standard Model of Particle Interactions*, ed. R. Gupta *et al.* (Elsevier Science, Amsterdam 1999); hep-ph/9806471.

[6] S. Bertolini *et al.*, Nucl. Phys. B **514**, 93 (1998).

[7] L. Wolfenstein, Phys. Rev. Lett. **13**, 569 (1964).

[8] L.K. Gibbons *et al.*, Phys. Rev. Lett. **70**, 1203 (1993).

[9] G.D. Barr *et al.*, Phys. Lett. B **317**, 233 (1993).

[10] Particle Data Group, C. Caso *et al.*, Eur. Phys. J. C **3**, 1 (1998).

[11] See A. Antonelli, this volume, chapter 7; M.S. Sozzi, this volume, chapter 8.

[12] A. Alavi-Harati *et al.*, Phys. Rev. Lett. **83**, 22 (1999).

[13] L.K. Gibbons *et al.*, Phys. Rev. D **55**, 6625 (1997).

[14] See R.A. Briere and B. Winstein, Phys. Rev. Lett. **75**, 402 (1995).

[15] Scrutiny of the E731 analysis has not revealed any explanation for its 2.9σ lower value other than a possible, if improbable, fluctuation.

[16] See L. Wolfenstein, this volume, chapter 9.

[17] V.A. Kostelecký, Phys. Rev. Lett. **80**, 1818 (1998).

7

KLOE at DAΦNE: Present Status and Progress

A. Antonelli[1]

Abstract

The KLOE experiment at the Frascati ϕ factory DAΦNE is fully operational and has recently started to collect its first data. During a short test run period 30 nb^{-1} of integrated luminosity has been accumulated and has been used to study the detector performance and to fully test the reconstruction and analysis program. In the near future DAΦNE should provide KLOE with 100 pb^{-1} of data, which should allow the measurement of $\Re(\epsilon'/\epsilon)$ to a statistical accuracy 10^{-3}.

7.1 Introduction

In a ϕ factory kaon pairs are produced in a very clean environment; they are almost[2] back-to-back with a momentum ~110 MeV/c, and the initial state has well defined quantum numbers, those of the photon. These unique

---

[1]For the KLOE collaboration: M. Adinolfi, A. Aloisio, F. Ambrosino, A. Andryakov, A. Antonelli, C. Bacci, A. Bankamp, G. Barbiellini, G. Bencivenni, S. Bertolucci, C. Bini, C. Bloise, V. Bocci, F. Bossi, P. Branchini, G. Cabibbo, R. Caloi, P. Campana, G. Capon, G. Carboni, A. Cardini, G. Cataldi, F. Ceradini, F. Cervelli, F. Cevenini, G. Chiefari, P. Ciambrone, S. Conticelli, E. De Lucia, G. De Robertis, P. De Simone, G. De Zorzi, S. Dell'Agnello, A. Denig, A. Di Domenico, S. Di Falco, A. Doria, E. Drago, O. Erriquez, A. Farilla, G. Felici, A. Ferrari, M. L. Ferrer, G. Finocchiaro, C. Forti, G. Foti, A. Franceschi, P. Franzini, M. L. Gao, P. Gauzzi, S. Giovannella, V. Golovatyuk, E. Gorini, F. Grancagnolo, E. Graziani, P. Guarnaccia, X. Huang, M. Incagli, L. Ingrosso, Y. Y. Jiang, W. Kim, W. Kluge, V. Kulikov, F. Lacava, G. Lanfranchi, J. Lee-Franzini, T. Lomtadze, C. Luisi, C. S. Mao, A. Martini, W. Mei, L. Merola, R. Messi, S. Miscetti, S. Moccia, M. Moulson, S. Mueller, F. Murtas, M. Napolitano, A. Nedosekin, L. Pacciani, P. Pagès, M. Palutan, L. Paoluzi, E. Pasqualucci, L. Passalacqua, A. Passeri, V. Patera, E. Petrolo, D. Picca, G. Pirozzi, L. Pontecorvo, M. Primavera, F. Ruggieri, P. Santangelo, E. Santovetti, G. Saracino, R. D. Schamberger, B. Sciascia, A. Sciubba, F. Scuri, I. Sfiligoi, T. Spadaro, E. Spiriti, C. Stanescu, L. Tortora, P. Valente, G. Venanzoni, S. Veneziano, Y. Wu.

[2]Due to the small crossing angle, 12.5 mr, the ϕ mesons are produced with a momentum of ~13 MeV/c.

features of the ϕ factory allow both the efficient identification of the nature of a kaon by looking at the decay of the companion in the opposite direction (tagging) and the performance of quantum interferometry measurements.

At full luminosity, 5×10^{32} cm^{-2}s^{-1}, in one year[3] of data taking we can collect $\sim$2 $\times10^{10}$ ϕ meson, which corresponds, according to the ϕ branching ratios, to a large amount of kaon pairs ($\sim$7$\times10^{9}$ K_LK_S and $\sim10^{10}$ K^+K^-). This will allow a precise determination of all the relevant CP and CPT violation parameters from the "double ratio" measurement and from interferometry [1], a measurement of the kaon form factor [2] and for the first time a measurement of the K_S semileptonic asymmetry [1]. In particular, from the "double ratio," $\Re(\epsilon'/\epsilon)$ can be obtained with a statistical precision of $\sim10^{-4}$ in two years of data taking while keeping the systematic error down to the same level. Moreover η, η', f_0, and a_0 mesons are copiously produced in ϕ radiative decays. This will allow coverage of a rich wealth of physics topics [3] beside CP and CPT studies.

7.2 The DAΦNE Machine

DAΦNE [4] is an e^+e^- collider optimized to work at the ϕ mass with a target luminosity L= 5×10^{32} cm^{-2}s^{-1}. The strategy used to reach the target luminosity is to obtain the same single-bunch luminosity achieved at VEPP-2M and to increase the number of bunches (up to 120). To suppress multibunch instabilities the electrons and the positrons circulate in two separate storage rings and collide at two interaction points with a horizontal crossing angle of 12.5 mrad to minimize the effect of parasitic collisions. In one of the two interaction regions, equipped with two triplets of low-β quadrupoles, KLOE is installed. Some machine parameters are given in table 7.1 and compared with those of KEK-B and PEP-II.

	DAΦNE	KEK-B	PEP-II
Maximum beam energy (GeV)	0.55	3.5/8.0	3.1/9.0
Single bunch luminosity (cm^{-2} s^{-1} 10^{30})	4.4	1.95	1.8
Bunches per ring per beam	120	5120	2200
Particles/bunch (10^{10})	8.9	3.3/1.4	5.9/2.7
Energy spread (units 10^{-3})	0.4	0.7	0.6
Single ring circumference (m)	97.69	3016.26	2199.32
Minimum bunch separation (cm)	81.4	58.8	126
Crossing half-angle (mrad)	12.5	11	0
Beam-beam tune shift H/V	0.04/0.04	0.04/0.05	0.03/0.03
Beam r.m.s. size at IP (mm) H/V	2/0.02	0.08/0.002	0.016/0.006
Energy loss/turn (KeV)	9.3	1500/3500	700 /3570
r.m.s. bunch length (cm)	3.0	0.4	1.0
Nat. emittance (mm $\times$ mrad) H/V	1.0/0.01	0.018/0.00036	0.064/0.0026

Table 7.1: DAΦNE, KEK-B, and PEP-II machine design parameters.

[3]Assuming 1 year = 10^7 sec and a ϕ cross section of $\sim$4.4 μbarn

The machine commissioning without the experiment was successfully completed during 1998. In the single-bunch mode, a luminosity of 1.5×10^{30} $\text{cm}^{-2}\text{s}^{-1}$ was achieved with a current of 20 mA/beam and in the multibunch mode, a luminosity of $\sim10^{31}$ $\text{cm}^{-2}\text{s}^{-1}$ was reached with 13 bunches and 200 mA/beam. This is about 33%, 20% of the design luminosity obtained with about 50%, 40% of the design current.

In March 1999, the first test run with the KLOE experiment began. The machine has been operated in the single- and multibunch mode and a luminosity of $\sim10^{30}$ $\text{cm}^{-2}\text{s}^{-1}$ has been reached with a lifetime $\sim1/2$ hour. In fig. 7.1 the beam current and the luminosity for a typical run are shown as a function of time. The energy scan has also been performed to find the ϕ resonance peak, as shown in fig. 7.2.

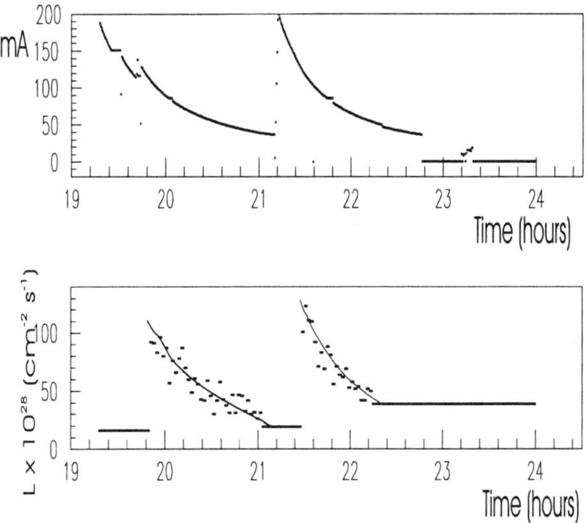

Figure 7.1: Electron current and luminosity vs. time.

At present, machine commissioning with the KLOE detector is in progress and a luminosity $\sim5\ \times10^{31}$ $\text{cm}^{-2}\text{s}^{-1}$ is foreseen for the summer run period.

7.3 The KLOE Detector

The KLOE [5, 6] detector, fig. 7.3, has been designed primarily with the goal of detecting CP violation in K^0 decays. It is a typical e^+e^- detector, of $\sim$5 m diameter and $\sim$4 m length, composed of a superconducting solenoid that provides a magnetic field of 6 kG, a central drift chamber, and an electromagnetic calorimeter installed inside the coil. The beam pipe at the interaction zone is a 10 cm ($\sim16\ \lambda_S$) beryllium sphere 0.5 mm thick. This allows us to define a fiducial volume for K_S without complications

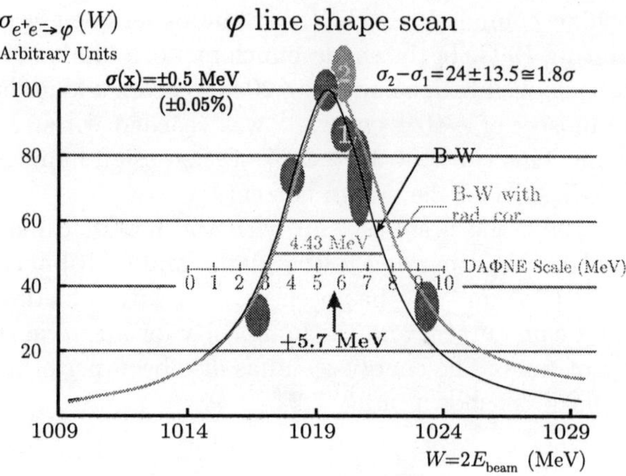

Figure 7.2: ϕ resonance energy scan.

from regeneration and to minimize multiple scattering and energy loss for charged particles. The two low-beta quadrupole triplets in the interaction region are also instrumented to improve the photon acceptance. This helps in the rejection of $K_L \to \pi^0\pi^0\pi^0$.

As demonstrated below, the performance of the KLOE detector, the final trigger configuration and the data acquisition reliability have been studied using the first 30 nb^{-1} of data. The results obtained are well in agreement with the design specifications.

7.3.1 Electromagnetic Calorimeter Performance

To discriminate between $K_L \to \pi^0\pi^0 / K_L \to \pi^0\pi^0\pi^0$ the calorimeter should have above all full efficiency for γs in the energy range 20–280 MeV. Of course, hermeticity and good energy resolution are also required. The calorimeter should also provide a fast trigger, should have some capability for particle identification, and should have a reasonable cost. Good timing performance is also very useful for reconstruction of the K_L neutral vertex with high resolution, thanks to the method shown in fig. 7.4.

We have chosen to use a lead scintillator sampling calorimeter consisting of 0.5 mm lead layer in which 1 mm diameter scintillating fibers are embedded. The energy-sampling fraction is 13%. The calorimeter is composed of a central part (barrel), with an inner radius of 2 m, a thickness of 23 cm ($\sim 15X_0$). and a length of 4.3 m, and two end-caps that hermetically close the calorimeter. The total acceptance is 98% of the full solid angle. The 24 barrel modules and the 68 end-cap modules are read out at both ends by photomultipliers for a grand total of about 5000 tubes. The readout granularity is 4.4×4.4 cm^2. The spatial resolution in reconstructing the shower apex is $\sigma_x \sim \sigma_y \sim 1.2$ cm and $\sigma_z \sim 2.5$ cm as obtained by the time difference.

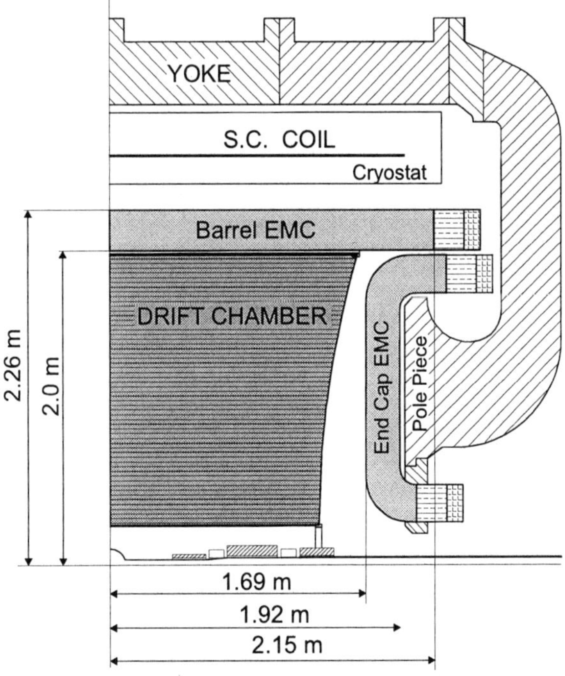

Figure 7.3: Schematic view of the KLOE detector.

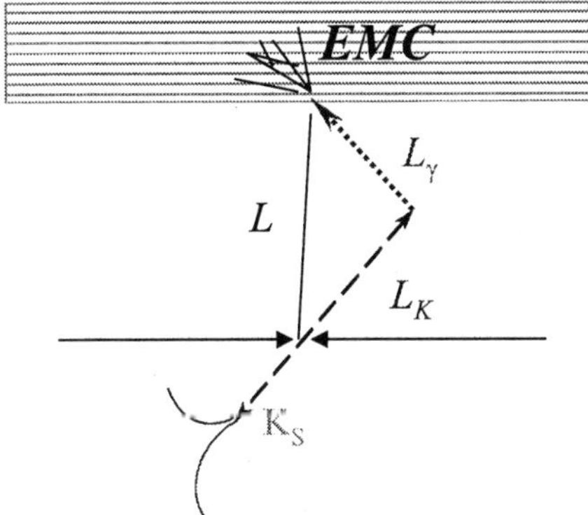

Figure 7.4: Neutral vertex position measurement: L is deduced from the photon conversion point, L_K is deduced from K_S direction. In addition the photon arrival time is measured, and the K_L speed is known.

The calorimeter has been fully calibrated in energy and in time using cosmic rays and Bhabha events. The energy calibration is performed by equalizing the response of the photomultipliers to minimum ionizing particles and the energy scale is deduced from Bhabha and $\gamma\gamma$ events. The energy measured for Bhabha and $\gamma\gamma$ events is shown in figs. 7.5 and 7.6. The energy resolution is $\sigma(E)/E$ ~8% at 510 MeV, which corresponds to $\sigma(E)/E$ ~$5.7\%/\sqrt{E\text{ (GeV)}}$ in fine agreement with test beam results and the design performance. The difference between the measured and the

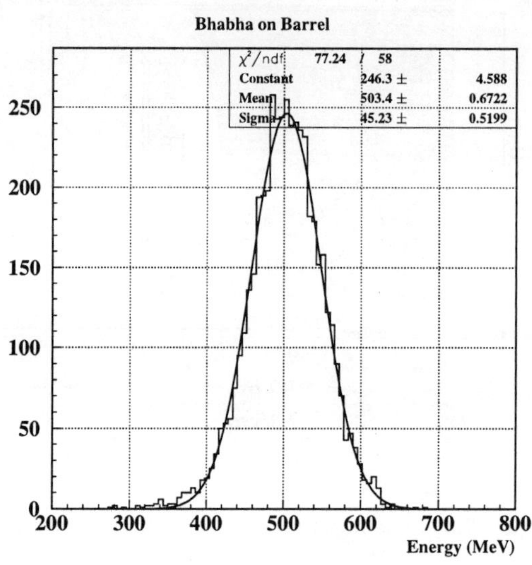

Figure 7.5: Calorimeter energy for Bhabha events.

expected times for $\gamma\gamma$ events is shown in fig. 7.7; the time resolution is σ_T~$80\text{ ps}/\sqrt{E\text{ (GeV)}}$.

The excellent time resolution is also used to recognize K_Ls that interact in the calorimeter by measuring their speed and to reject cosmic rays. In fig. 7.8 the speed of the cluster found in a cone of 20 degrees around the K_L direction (obtained from the $K_S \to \pi^+\pi^-$) is shown. A clear K_L signal is seen with β ~0.21. In fig. 7.9 the time difference between last and first calorimeter planes is plotted; the ingoing and outgoing cosmics are nicely recognized. Particle masses can also be deduced using the time information and the momenta measured by the chamber. An example can be seen in fig. 7.10, where the speed of cosmic muons is plotted as a function of momentum. As mentioned above the time information is used in KLOE to measure the neutral decay vertex position. The resolution on this measurement should be ~1 cm in order to minimize the effect on the "double ratio" of event miscounting in the vicinity of the fiducial volume boundary for K_L. In particular, the relative scale for charged and neutral decays must coincide within a fraction of a millimeter. The

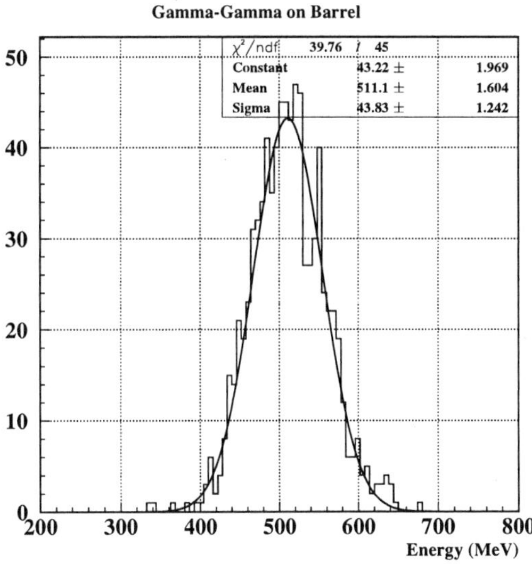

Figure 7.6: Calorimeter energy for $\gamma\gamma$ events.

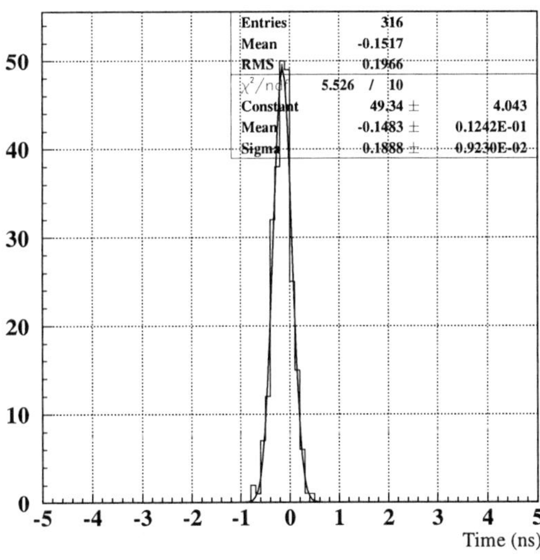

Figure 7.7: Time resolution for $\gamma\gamma$ events.

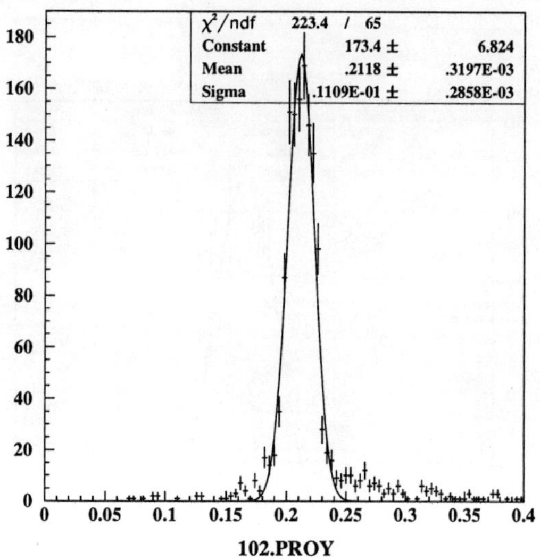

Figure 7.8: β for K_L interacting in the calorimeter.

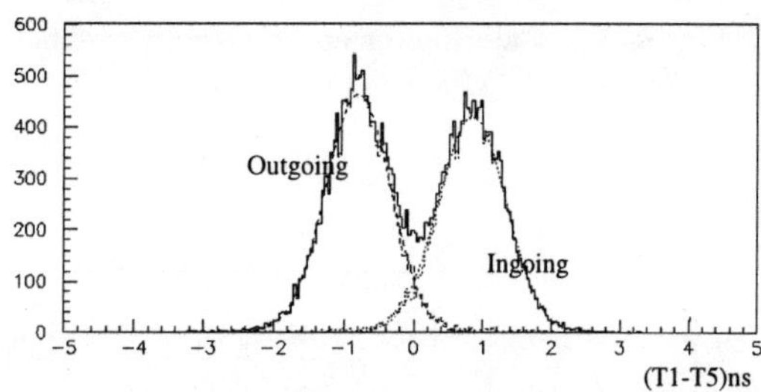

Figure 7.9: First and last calorimeter plain time difference.

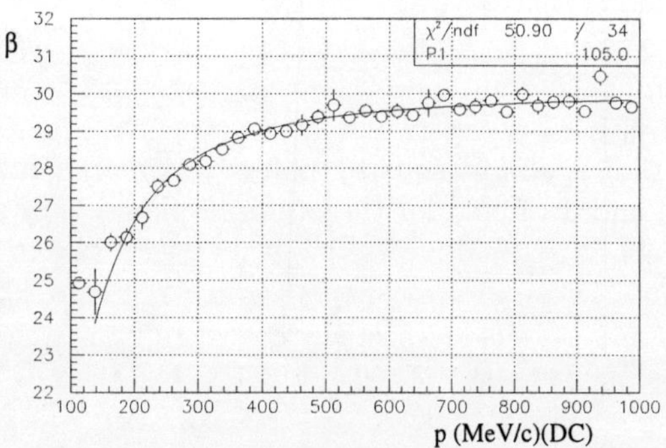

Figure 7.10: Muon mass measured from time and momentum information. The mass value from the fit is $m_\mu = 105$ MeV/c^2.

intercalibration of the scales can be obtained using events where the neutral and charged vertices coincide such as $K_L \to \pi^+\pi^-\pi^0$ or $K^+ \to \pi^+\pi^0$. A plot of the difference between neutral and charged decay vertex positions is shown in fig. 7.11 from a preliminary analysis of the $K_L \to \pi^+\pi^-\pi^0$ channel. The resolution is still poor, ~3 cm, but the statistics are low.

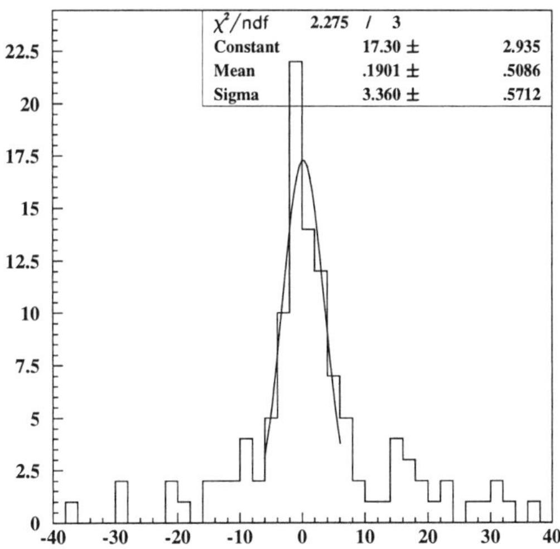

Figure 7.11: Neutral and charged decay vertex position difference (cm) for $K_L \to \pi^+\pi^-\pi^0$.

7.3.2 Tracking Chamber Performance

The main characteristics of the tracking system are:

• Large radius and uniform sampling to collect as many K_L decays as possible, taking into account that the K_L mean decay length is 3.5 m.

• Optimized resolution for low momentum tracks by using low Z gas and thin walls to minimize multiple scattering, regeneration, and γ conversion.

We have chosen a 2 m radius and 3.4 m long drift chamber [7] with a cell configuration that is almost square, with effective area of 2×2 cm^2 for the first 12 planes and 3×3 cm^2 for the outer 46 planes. For a uniform filling of the sensitive volume only stereo wires are used. The stereo angle varies with radius from 50 mrad to 120 mrad going outward. The gas mixture is 90% helium and 10% isobutane for a total radiation length of ~900 m, including the ~52,000 wires. The mechanical structure is done entirely of carbon-fiber/epoxy and adds up to ~0.1 X_0.

The chamber is fully operational (see fig. 7.12) with very few dead channels (< 0.1%). It is very clean even down to low thresholds (4 mV) and the cell efficiency is ~99%. Chamber calibration is performed using cosmic and Bhabha events. With 4 hours of data taking, some 230 space-to-

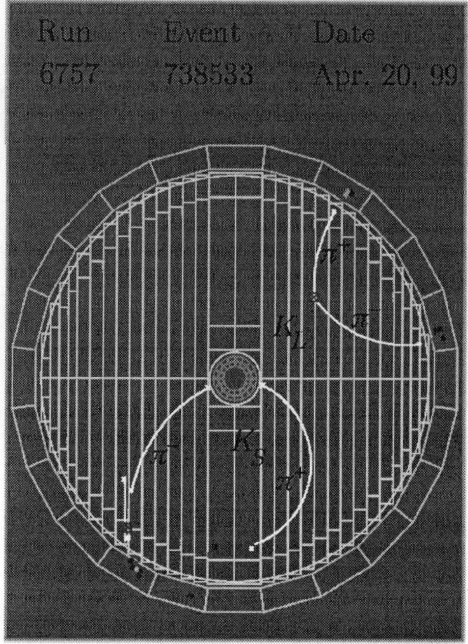

Figure 7.12: First CP violating event $K_S \rightarrow \pi^+\pi^-$, $K_L \rightarrow \pi^+\pi^-$.

time relations are computed for the different cell configurations. The track fit residuals and the fit resolution, after chamber calibration, are shown in fig. 7.13 for Bhabha events. The spatial resolution is $\sim$150 μm over the entire area of the effective cell with an expected deterioration very close to the sense wire. The vertex resolution for Bhabha events is $\sigma_x \sim \sigma_y \sim$2mm, $\sigma_z \sim$5 mm (see fig. 7.14) in fine agreement with expectations. The same resolutions are found for lower momentum tracks such the one produced in the $\phi \rightarrow \rho\pi$ decay. The expected momentum resolution, entirely dominated by multiple scattering in our momentum range, is σ_P /p $\sim$0.4%, and agrees with the value we find for Bhabha events. In fig. 7.15 the momentum resolution is plotted against the polar angle. The usual deterioration for low angle is present. The invariant mass resolution for $K_S \rightarrow \pi^+\pi^-$ is shown in fig. 7.16. The resolution is very good, of the order of 1.4 MeV/c^2.

7.4 Trigger and DAQ

The KLOE trigger [8] has been designed to have high efficiency for ϕ decays ($\sim$99% for the CP violation decays) and to reject or downscale to a few kHz the huge amount of Bhabha scattering ($\sim$20 kHz), cosmic rays, and machine background ($\sim$3 kHz). A two-level trigger has been developed based on the energy deposited in the calorimeter and on the hit multiplicity in the chamber. The level-1 trigger occurs within 150 ns after the beam

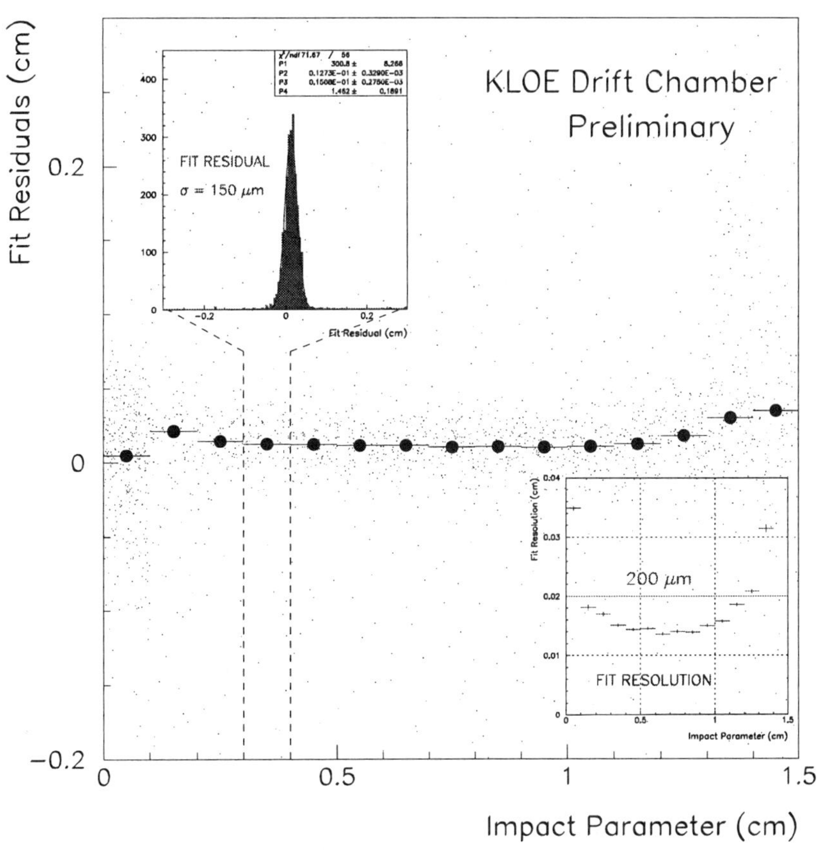

Figure 7.13: Track fit residuals and resolution.

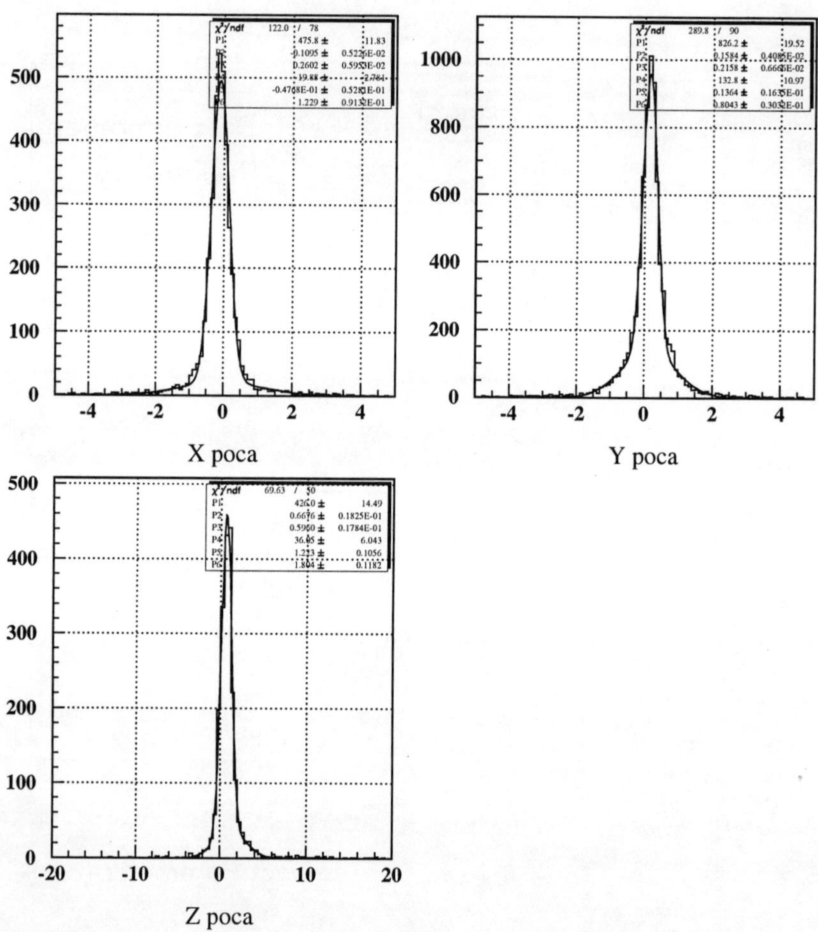

Figure 7.14: Vertex fit resolution for Bhabha events.

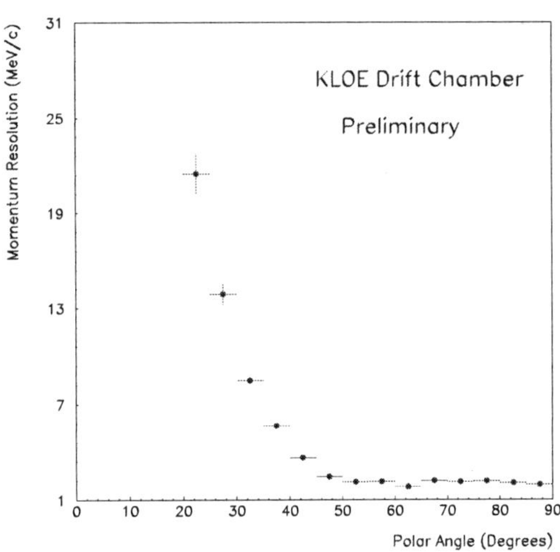

Figure 7.15: Drift chamber momentum resolution.

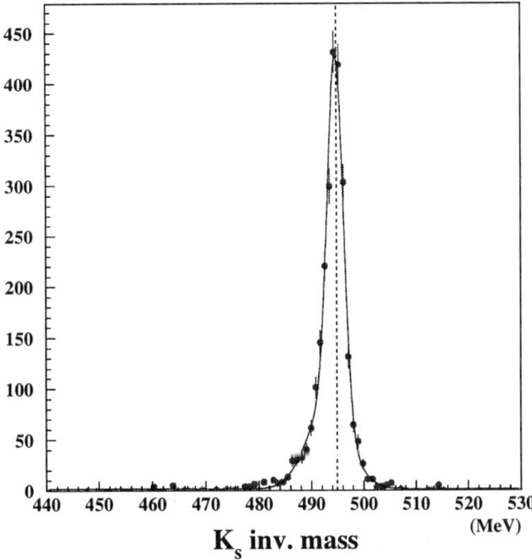

Figure 7.16: $K_S \to \pi^+\pi^-$ invariant mass.

crossing and is mainly generated by the calorimeter; the level-2 occurs 850 ns after the level-1 and is mainly generated by the chamber. The trigger rate after level-2 will be ~10 kHz at full luminosity. The final trigger configuration has been used in the recent data taking and the study of its performance is in progress.

The DAQ system [9] has to handle about 23000 FEE channels, organized in 10 parallel VME crate chains, and has to handle up to 50 Mbytes/s. The storage capability should be 500 Tbyte/year. The final DAQ system has been fully tested up to more than 24 hours of continuous running with peak rates of 10 kHz in multibunch mode. The performances largely fulfill the design requirements.

7.5 Conclusions

The KLOE detector performance, the DAQ and the trigger reliability, the online monitoring, and the reconstruction and filtering procedures have been tested successfully with the first data collected. In the near future DAΦNE should provide KLOE with 100 pb^{-1} of data, which will allow the measurement of $\Re(\epsilon'/\epsilon)$ to a statistical accuracy of 10^{-3}, the measurement of the K_{l3} form factor, and the first measurement of BR($\phi \rightarrow f_0\gamma$).

References

[1] S. Di Falco and M. Incagli, this volume, chapter 24.

[2] G. Saracino, for the KLOE Collaboration, "*K* Meson Semileptonic Decay Form Factors with KLOE," *Proceedings of Workshop on Hadron Spectroscopy*, Frascati Physics Series, vol. 15 (1999), 555–600.

[3] S. Giovannella, for the KLOE Collaboration, "Detection of Scalar Mesons with KLOE," *Proceedings of Workshop on Hadron Spectroscopy*, Frascati Physics Series, vol. 15 (1999), 473–78.

[4] G. Vignola, *Workshop on Physics and Detector for DAΦNE 95*, ed. R. Baldini *et al.*, SIS-Pubblicazioni, Frascati, 1995.

[5] The KLOE Collaboration, *A General Purpose Detector for DAΦNE*, LNF-92/019 (1992).

[6] The KLOE Collaboration, *The KLOE Detector, Technical Proposal*, LNF-93/002 (1993).

[7] The KLOE Collaboration, *The KLOE Central Drift Chamber*, LNF-94/028 (1994).

[8] The KLOE Collaboration, *The KLOE Trigger System*, LNF-96/043 (1996).

[9] The KLOE Collaboration, *The KLOE Data Acquisition System*, LNF-95/014 (1995).

8

First Result on Re(ε'/ε) Measurement from the NA48 Experiment at CERN

M. S. Sozzi[1]

8.1 Introduction

The breakdown of CP symmetry and time reversal symmetry [1] has been observed so far only using decays of the neutral kaons [2], and is mainly manifested in the mixing of the two CP eigenstates, parametrized by ϵ, and caused by an effective $\Delta S = 2$ interaction [3]. In the context of the standard model, CP violation arises in the Higgs sector [4] as a universal feature of weak interactions, and direct CP violation in the decay amplitudes $K_L \to 2\pi$, parametrized by ϵ', is also predicted.

Computations of ϵ'/ϵ within the standard model [5] mostly fall in the range $(0 \div 10) \cdot 10^{-4}$ [6].

To first order in the small parameter $|\epsilon'/\epsilon|$, the measurable double ratio of decay widths is

$$R \equiv \frac{\Gamma(K_L \to \pi^0\pi^0)/\Gamma(K_S \to \pi^0\pi^0)}{\Gamma(K_L \to \pi^+\pi^-)/\Gamma(K_S \to \pi^+\pi^-)} \simeq 1 - 6\,\mathrm{Re}(\epsilon'/\epsilon).$$

Two previous measurements of Re(ϵ'/ϵ) were only marginally consistent: experiment NA31 at CERN [7] measured $(23 \pm 6.5) \cdot 10^{-4}$, showing evidence of direct CP violation, while experiment E731 at FNAL [8] measured $(7.4 \pm 5.9) \cdot 10^{-4}$, consistent with no effect. A recent result from Fermilab experiment E832 [9] supports the presence of direct CP violation: $(28.0 \pm 4.1) \cdot 10^{-4}$.

The NA48 collaboration at CERN has recently obtained a new measurement of $\mathrm{Re}(\epsilon'/\epsilon) = (18.5 \pm 7.3) \cdot 10^{-4}$ from the above double ratio, based on the first data collected by the experiment. The event statistics on which this result is based represents $\approx 10\%$ of the total amount expected at the end of the experiment.

[1]On behalf of the NA48 collaboration.

8.2 The NA48 Technique

The double ratio measurement exploits the cancellation of several effects, thus allowing a good control of systematic uncertainties. In NA48 the four decay modes are collected simultaneously in the same decay region, thereby minimizing the sensitivity to accidental activity and variations of beam intensity and detection efficiencies. K_L and K_S decays are provided by two quasi-collinear neutral beams, produced by protons hitting two targets at different distances from the detector.

The K_L and K_S decay momentum spectra are made similar by the choice of targeting angles, and residual differences are accounted for by performing the analysis in small kaon energy bins.

The difference in longitudinal decay position (lifetime) distributions leads to acceptance differences between K_S and K_L that do not cancel in the double ratio. This issue is addressed by collecting K_L and K_S decays only in the region where both are present, and by weighting the K_L events in the analysis with a function of their proper time of decay. In such a way, at the loss of some statistical power, the longitudinal decay position distributions are made equal for 2π decays originating from either beam, achieving a good cancellation of acceptance differences, and thus avoiding large corrections—which would require relying on very accurate modeling and simulations of the apparatus.

The above minimization of biases and corrections represents the general approach of the NA48 experiment, and is consistently applied wherever it is relevant.

8.2.1 Beams and K_S Tagging

The K_L beam is produced by $1.1 \cdot 10^{12}$ protons (450 GeV/c momentum) slow-extracted from the SPS during each 2.4 s spill impinging on a beryllium target located 126 m upstream of the beginning of the fiducial decay region; the effect of the small K_S component in this beam is significant only at energies above 140 GeV.

A small fraction of the noninteracting protons ($3 \cdot 10^7$ per spill) is deflected towards a second target, similar to the first and located 6 m upstream of the beginning of the fiducial region, by exploiting *channeling* on a single bent crystal [10]. On its way to this K_S production target, the proton beam crosses an array of scintillator counters. This device [11] measures the time of passage of individual protons with 120 ps resolution and $\simeq 4$ ns double-pulse separation, and is used to *tag* the K_L or K_S character of a decay, by comparing the event time, as reconstructed by the detectors, with the time of the closest proton on this beam line.

The neutral beam emerging from the second target, producing mostly K_S (and hyperon) decays, crosses a scintillator veto counter (AKS) preceded by an iridium crystal photon converter, which defines the upstream end of the decay region.

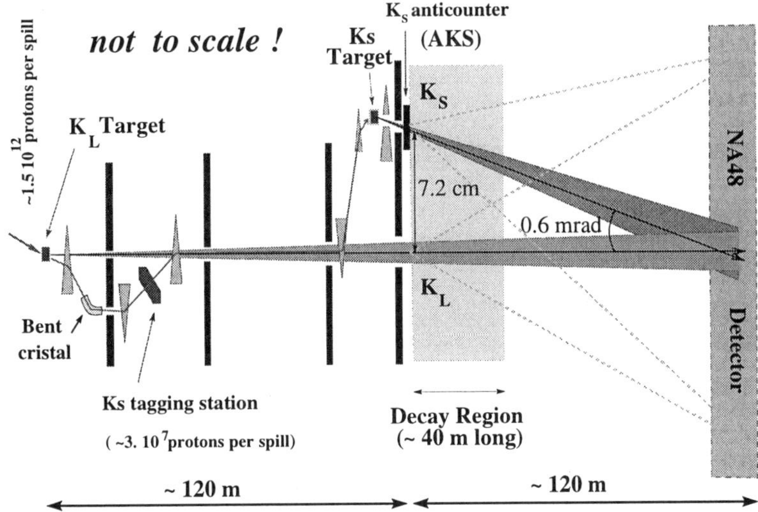

Figure 8.1: Schematic drawing of the NA48 beam setup.

The two neutral beams converge in the center of the detector at a 0.6 mrad angle, traveling in vacuum in order to reduce hit rates and accidental effects. The beam setup is shown schematically in fig. 8.1.

8.2.2 Detector and Performance

To collect the large amount of events required, while minimizing the amount of background events, fast and high-performance detectors and trigger systems are used, with fully pipelined and dead-time-less readout.

$K_{L,S} \to \pi^0\pi^0$ decays are detected in a fine-grained quasi-homogeneous liquid krypton calorimeter, working as an ionization chamber with longitudinal tower structure [12]. By exploiting projective geometry, the calorimeter allows the measurement of angles between photons independently of their initial conversion depth.

The initial current induced by the ionization drift on copper ribbon electrodes in the $\approx$ 13000 2×2 cm^2 cells is read and continuously flash-digitized at 40 MHz, allowing for a fast response and an intrinsic time resolution better than 300 ps for photon energies above 20 GeV.

During 1997 the calorimeter had to be operated at reduced electrode voltage, resulting in a slightly worse signal-to-noise ratio and a small ($\approx$ 0.5%) correction as a function of event time during the spill for space charge accumulation; moreover, a 4 cm wide column of cells was not connected to the voltage supply, resulting in a $\approx$ 20% acceptance loss.

The energy resolution measured in 1997 was

$$\sigma(E)/E = 3.2\%/\sqrt{E(\text{GeV})} \oplus 125\,\text{MeV}/E \oplus 0.5\%$$

and the spatial resolution better than 1.3 mm above 20 GeV, resulting in a π^0 mass resolution of 1.1 MeV/c^2. The calorimeter response was stable during the run within 0.1%.

The main detector for $K_{L,S} \to \pi^+\pi^-$ decays is a magnetic spectrometer, consisting of four large drift chambers and a dipole magnet providing a 265 MeV/c transverse momentum kick. Each chamber [13] has four double planes of staggered wires strung at 45° with respect to each other; their spatial resolution is $\simeq 90\ \mu$m and their efficiency $\simeq 99.5\%$. The momentum resolution was measured in 1997 to be $\sigma(p)/p = 0.5\% \oplus 0.009\% \cdot p$ (GeV/c), resulting in a K^0 mass resolution of 2.5 MeV/c^2.

A two-plane plastic scintillator hodoscope in front of the calorimeter is used to measure the arrival time of charged pions with a resolution of $\simeq$ 200 ps per track.

Other detectors comprise a scintillating fiber hodoscope inside the liquid krypton volume, giving redundant time information for neutral decays, an iron scintillator hadron calorimeter for triggering purposes, and a plastic scintillator muon veto counter system.

8.2.3 Triggers

The trigger system for neutral decays [14] is a fully pipelined, dead-timeless, programmable system, based on the information provided by 64 + 64 horizontal and vertical electromagnetic calorimeter projections. This system continuously computes, for each 25 ns time bin, the deposited energy momenta, the longitudinal decay vertex position, and the number of clusters close in time, allowing a massive reduction of the $K_L \to 3\pi^0$ background, resulting in an output rate of $\simeq$ 1.9 kHz.

The trigger for charged decays [15] is a two-level system, the first one being provided by hardware hodoscope cross quadrants coincidences and an energy threshold from the combination of the two calorimeters' outputs. The second level, based on drift chambers' information, consists of hardware coordinate builders and a farm of fast processors that compute the invariant mass and the longitudinal decay vertex position of the kaon candidates, allowing a large reduction of the background from three-body K_L decays, with $\simeq$ 1.2 kHz output rate. The charged trigger system introduces 0.3% dead time, which is continuously monitored.

8.3 Data Analysis

The double ratio is insensitive in first order to all sources of biases that are common to the two beams or to the two 2π decay charge modes, being biased only by the small effects that can be ascribed to only one of the four decay modes, or by the differential component of beam- or detector-related effects.

A first source of possible biases is due to the K_S tagging technique: K decays are labeled as K_S whenever a proton is detected by the tagger

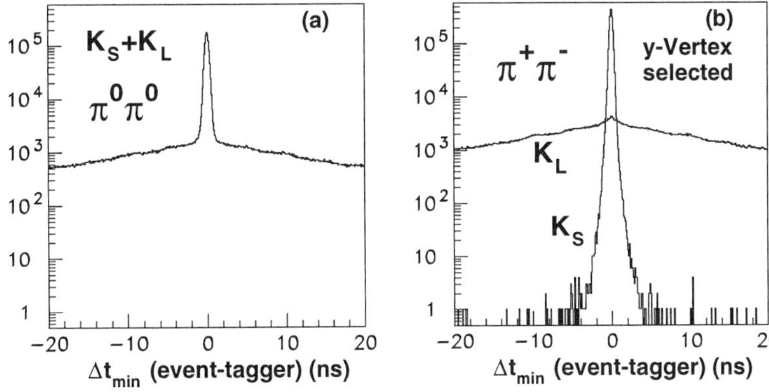

Figure 8.2: Minimum event-tagger time difference: (a): $\pi^0\pi^0$, (b) $\pi^+\pi^-$.

system on the beam line towards the K_S production target, within a $\pm$ 2 ns window around the event time, as reconstructed by the charged hodoscope or the electromagnetic calorimeter.

For charged decays, the decay vertex transverse position resolution is good enough to identify the beam to which a decaying K belongs without ambiguities. Not being available for neutral decays, this independent tagging information is not used for the final analysis, but it is used to study the tagging systematics (see fig. 8.2).

The presence in the beam line of an accidental proton close in time to a K_L decay can lead to the mistagging of this decay as a K_S: the probability of this happening is measured for charged decays to be $(11.19 \pm 0.03)\%$, and its effect on R is negligible in first order, being a priori the same for charged and neutral decays.[2] A very small effect is actually induced by the rate dependence of the charged trigger inefficiency, which leads to a slightly higher effective beam rate for detected neutral decays. The K_L-to-K_S mistagging probability for neutral decays is measured by studying the number of accidental proton coincidences in several 4 ns wide time windows, offset from the central one used for tagging; the effect on the double ratio due to the charged/neutral difference in these mistagging probabilities is $\Delta R = (-18 \pm 9) \cdot 10^{-4}$.

A second potential tagging-related bias is due to tagging inefficiencies or time reconstruction tails, resulting in a K_S event being mistagged as K_L. This effect could be in principle quite different for charged and neutral decays; however, redundant time information shows that most of the effect is due to inefficiencies in the tagger itself, and therefore charged/neutral symmetric. The K_S-to-K_L mistagging probability for charged decays is directly measured to be $(1.5 \pm 0.1) \cdot 10^{-4}$ by using the decay vertex transverse position information. For neutral decays, this probability was measured

[2]This effect, if not corrected for, would only dilute a non-zero value of Re(ϵ'/ϵ), but it could not create a fake signal of direct CP violation.

using K_S-only special runs, $K_S \to 2\pi^0$ decays with either a Dalitz pair or a photon conversion in the material in front of the detector; in these cases the beam to which the K belongs either is known or can be identified by the transverse vertex position. The resulting effect on the double ratio is $\Delta R = (0 \pm 6) \cdot 10^{-4}$.

Trigger efficiencies were measured by using data continuously collected with less restrictive independent trigger conditions. The neutral trigger efficiency was measured to be (0.9988 ± 0.0004) and equal for K_S and K_L decays. The overall charged trigger efficiency was measured to be (0.9168 ± 0.0009), the inefficiency being mostly due to an accidental time misalignment in the hardware. The net effect on R due to the K_S-K_L asymmetry of this inefficiency is measured to be $\Delta R = (-9 \pm 23) \cdot 10^{-4}$, limited in its precision by the size of the control sample.

To reconstruct $2\pi^0$ events, the longitudinal vertex position is computed from the measured energies E_i and positions (x_i, y_i) of the four clusters in the calorimeter:

$$z_{\text{vertex}} = z_{\text{cal}} - \sqrt{\textstyle\sum_{i,j>i} E_i E_j[(x_i - x_j)^2 + (y_i - y_j)^2]}/m_K$$

assuming the invariant mass of the 4 detected photons to be the neutral kaon mass m_K. A χ^2 variable is defined, describing the match of each photon pair's invariant mass with the π^0 mass. The residual $K_L \to 3\pi^0$ background after cutting on this variable is measured to be less than 1 per mille, resulting in a shift $\Delta R = (8 \pm 2) \cdot 10^{-4}$.

The $\pi^+\pi^-$ reconstruction exploits the detailed knowledge of the measured magnetic field, although the K energy is defined using the tracks' opening angle and ratio of momenta, therefore being largely insensitive to it. The background from K_L semileptonic decays is reduced by comparing the measured track momenta and deposited energy in the electromagnetic calorimeter (for K_{e3}), and by using the muon veto counters (for $K_{\mu 3}$). Further cuts are made on the reconstructed $\pi^+\pi^-$ invariant mass and on a rescaled K transverse momentum variable, in which K_S and K_L have similar resolution despite the large difference in target to average decay vertex distance.

The residual background in the charged mode is measured to be around 2 per mille, and the resulting effect on R is $\Delta R = (-23 \pm 2 \pm 4) \cdot 10^{-4}$, where the first error is statistical and the second systematic. Fig. 8.3 shows the signal and background distributions of two relevant variables.

A fraction $\approx 20\%$ of $\pi^+\pi^-$ events were lost in 1997 due to the loss of drift chambers' data in case of high hit multiplicity due to e.m. showering in the beam tube in the spectrometer region; the occurrence of this condition was monitored, and an equivalent requirement was imposed on $\pi^0\pi^0$ events, resulting in a negligible measured effect on R.

A small fraction ($\approx 0.5 \cdot 10^{-3}$) of $K \to \pi^+\pi^-$ with high transverse momentum are due to multiple K_L scattering and regeneration on collimators. Since this kind of events are not removed in the $\pi^0\pi^0$ sample, they cause a shift $\Delta R = (+12 \pm 3) \cdot 10^{-4}$.

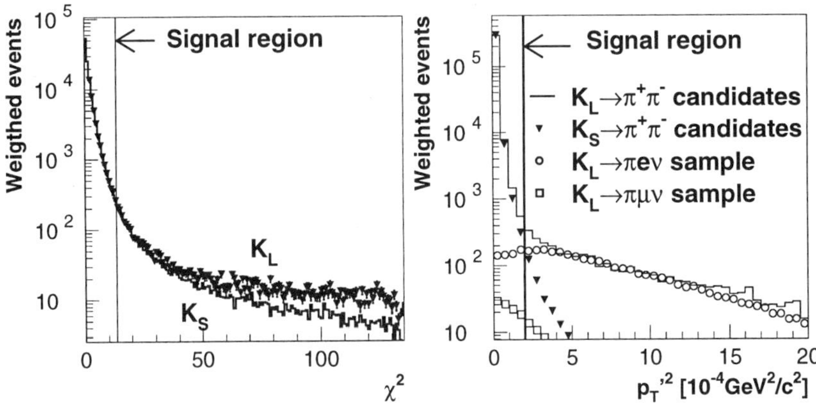

Figure 8.3: Backgrounds to neutral and charged K_L $\pi\pi$ decays.

Using simultaneous beams, R is largely insensitive to accidental effects caused by the high-rate environment, and moreover the beam intensities are highly correlated, resulting in the same measured accidental activity for K_S and K_L events. The effect of accidentals is studied by software overlaying randomly triggered events (proportional to beam intensities) to normal events: the observed $\approx 2\%$ net variation in the number of accepted events is highly K_S- K_L symmetric and corresponds to an effect $\Delta R = (2\pm14)\cdot10^{-4}$, determined with limited precision due to the amount of overlaid events.

The effect of extra activity in the detectors correlated in time with K_S production was measured to affect R by less than $3 \cdot 10^{-4}$.

The absolute energy scale, which is directly linked to the longitudinal decay position for $\pi^0\pi^0$, is fixed by fitting the position of the edge of the AKS veto counter in the data to a precision of $5 \cdot 10^{-4}$. The scale factor is cross-checked using neutral decays with charged tracks and special runs in which a π^- beam (replacing the K_L one) impinges on a thin target producing π^0 and η decays with a well-known vertex position. The scale is stable in time within $5 \cdot 10^{-4}$, and the overall uncertainty on R due to it is $\pm 5 \cdot 10^{-4}$.

Other effects linked to the 0.3% maximum calorimeter nonlinearity and to residual nonuniformities give an overall uncertainty of $\pm 11 \cdot 10^{-4}$ on R.

The accepted range of kaon energies is between 70 and 170 GeV; the analysis is performed in 20 energy bins, to be insensitive to the different K_S and K_L decay energy spectra, and the results are averaged using an unbiased estimator.

The beginning of the fiducial region for K_S, which have a steep longitudinal decay position distribution, is defined in hardware by the absence of signals from the AKS veto counter, located ≈ 6 m downstream of the K_S production target, and its length is defined to be 3.5 proper K_S lifetime units: in this way the sensitivity of R to the knowledge of the absolute neutral energy scale is reduced.

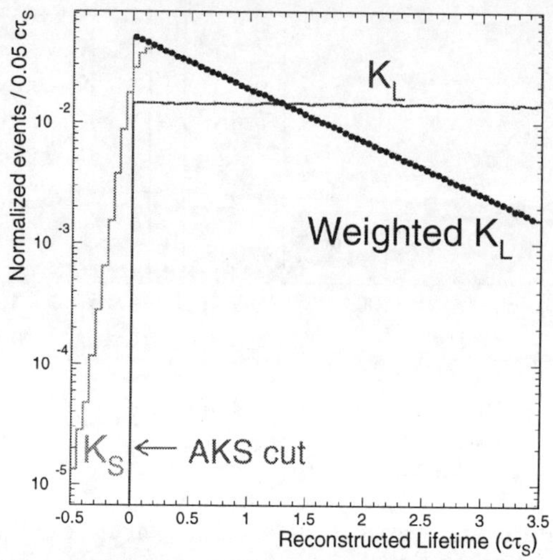

Figure 8.4: K_S and K_L longitudinal decay vertex distributions.

In order to minimize the K_S-K_L acceptance differences due to the different decay vertex distributions, the K_L decays are weighted according to the ratio of their expected 2π decay rate to the one from the K_S beam, thus effectively equalizing the two distributions as shown in fig. 8.4. Small residual acceptance differences, due to the acollinearity and different divergence of the two beams, have been modeled by a detailed Monte Carlo simulation of the apparatus. Their overall effect is $\Delta R = (-29 \pm 11 \pm 5) \cdot 10^{-4}$ (MC statistical and systematic error). Table 8.1 summarizes the corrections applied on R and their uncertainties. Table 8.2 shows the number of events.

8.4 Result and Conclusions

The final result after corrections is

$$\mathrm{Re}(\epsilon'/\epsilon) = (18.5 \pm 4.5 \pm 5.8) \cdot 10^{-4},$$

where the first error is statistical and the second is systematic although, as mentioned above, its main component is statistical in nature, being given by the size of the auxiliary samples used to measure the various effects.

Combining the errors in quadrature the result is $\mathrm{Re}(\epsilon'/\epsilon) = (18.5 \pm 7.3) \cdot 10^{-4}$. This measurement confirms the existence of direct CP violation in $K_L \to \pi\pi$ decays.

	Correction (10^{-4})	Uncertainty (10^{-4})	
Tagging dilution	+18	± 9	(stat.)
Tagging inefficiency	0	± 6	(stat.)
Charged trigger eff.	+9	± 23	(stat.)
Reconstruction eff.	0	± 3	
Neutral background	-8	± 2	
Charged background	+23	± 4	
Beam scattering	-12	± 3	
Accidental activity	-2	± 14	(stat.)
Energy scale/linearity	0	± 12	
Charged vertex	0	± 5	
Acceptance	+29	± 12	(MC stat.)
Total	+57	± 35	

Table 8.1: Corrections applied on R and their systematic uncertainties (divide by 6 to get the corresponding values for $\mathrm{Re}(\epsilon'/\epsilon)$).

$K_L \to \pi^0\pi^0$	$K_S \to \pi^0\pi^0$	$K_L \to \pi^+\pi^-$	$K_S \to \pi^+\pi^-$
$0.49 \cdot 10^6$	$0.98 \cdot 10^6$	$1.07 \cdot 10^6$	$2.09 \cdot 10^6$

Table 8.2: Sample sizes after cuts, background and tagging corrections (without lifetime weighting).

The world average (rescaling errors as suggested by the PDG [16]) was $\mathrm{Re}(\epsilon'/\epsilon) = (21.7 \pm 6.1) \cdot 10^{-4}$ before this measurement; including it, the average becomes $\mathrm{Re}(\epsilon'/\epsilon) = (21.3 \pm 4.6) \cdot 10^{-4}$.

It should be noted that NA48 has collected data in 1998 and is currently running with significant improvements in detector performance and rate capability, corresponding to several times the statistics on which the present result is based.

Acknowledgments

The author wishes to thank the organizing committee and the participants in the KAON '99 workshop for the scientifically intense and interesting experience. The contribution of all the colleagues of the NA48 collaboration is gratefully acknowledged.

References

[1] R.G. Sachs, *The Physics of Time Reversal* (University of Chicago Press, Chicago 1987).

[2] J.H. Christenson *et al.*, Phys. Rev. Lett. **13**, 138 (1964).

[3] L. Wolfenstein, Phys. Rev. Lett. **13**, 562 (1964).

[4] M. Kobayashi and T. Maskawa, Prog. Theor. Phys. **49**, 652 (1973).

[5] For recent reviews, see *e.g.*, S. Bertolini *et al.*, hep-ph/9802405, and A. Buras, this volume, chapter 5.

[6] For an exception, see J. Heinrich *et al.*, Phys. Lett. B **279**, 140 (1992).

[7] G.D. Barr *et al.*, Phys. Lett. B **317**, 233 (1993).

[8] L.K. Gibbons *et al.*, Phys. Rev. Lett. **70**, 1203 (1993).

[9] A. Alavi-Harati *et al.*, Phys. Rev. Lett. **83**, 22 (1999); Y.B. Hsiung, this volume, chapter 6.

[10] N. Doble *et al.*, Nucl. Instr. Meth. B **119**, 181 (1996).

[11] H. Bergauer *et al.*, Nucl. Instr. Meth. A **419**, 623 (1998).

[12] G.D. Barr *et al.*, Nucl. Instr. Meth. A **370**, 413 (1996).

[13] D. Bédérède *et al.*, Nucl. Instr. Meth. A **367**, 88 (1995).

[14] B. Gorini *et al.*, IEEE Trans. Nucl. Sci. **45**, 1771 (1998).

[15] S. Anvar *et al.*, Nucl. Instr. Meth. A **419**, 75 (1999).

[16] Particle Data Group, Eur. Phys. J. C **3**, 11 (1998).

9

The Superweak Theory 35 Years Later

L. Wolfenstein

Abstract

The origins, possible rationale, and definition of the superweak theory are reviewed. The observation of direct CP violation in K_L decay provides the first significant evidence against the theory. The much larger direct CP violation expected in B decays should definitively kill it.

When CP violation was discovered in 1964, there existed a very successful theory of weak interactions called the V–A theory. This closely resembled the original Fermi theory, but there were added the helicity projection operators that led to maximal parity (P) violation and particle-antiparticle symmetry (C) violation. However, this theory was CP invariant.

Thus, two possibilities emerged. Either one modified the V–A theory in some way or one assumed there existed a much weaker interaction that led to CP violation. The second possibility existed because the observed effect could be limited to CP violation in $K^0 - \bar{K}^0$ mixing causing a small admixture of the CP-even state K_1 into K_L

$$|K_L\rangle = (|K_2\rangle + \epsilon|K_1\rangle)(1 + |\epsilon|^2)^{-1}, \tag{9.1}$$

where K_2 is the CP-odd state. Because $K^0 - \bar{K}^0$ mixing is a second-order weak effect the CP-violating effect could be limited to a new $\Delta S = 2$ interaction of the form

$$G_{sdsd}(\bar{s}\mathcal{O}d)(\bar{s}\mathcal{O}d) + \text{h.c.}, \tag{9.2}$$

with G_{sdsd} having a value of order 10^{-10} to 10^{-11} times the standard weak interaction coupling G_F. This came to be called the *superweak theory.*

There are several consequences of the superweak theory:

1. The CP violation would be the same for different decay modes such as $\pi^+\pi^-$ and $\pi^0\pi^0$ since it was determined entirely by the mixing ϵ. In a standard notation, this says ϵ' vanishes.

2. Assuming CPT invariance the phase of ϵ is given exactly by

$$\theta_\epsilon = \tan^{-1}(2\Delta M/\Delta\Gamma) \approx 43.5°,$$

where $\Delta M = M_L - M_S$ and $\Delta\Gamma = \Gamma_S - \Gamma_L \approx \Gamma_S$. This is referred to as the superweak phase.

3. It would be very unlikely that CP violation would be discovered outside the K^0 system.

All these predictions appeared to be true for the next 25 years.

The superweak theory is really a large class of theories with the common feature that their effect at low energies contains the term (9.2). The original form I wrote in 1964 [1] had the standard $\Delta Q = \Delta S$ current coupled very weakly to a postulated $\Delta Q = -\Delta S$ current. With the advent of the quark picture, this is no longer a possibility. It is possible that the superweak character arises from the exchange of a superheavy particle, from very weak coupling constants, or from higher-order effects.

In 1973, with the discovery of neutral currents, a new theory of weak interactions came to be accepted, the Weinberg-Salam model. As shown by Glashow, Iliopoulos, and Maiani, this required a fourth quark. This theory was more elegant that the previous theory because it was based on a gauge principle and was renormalizable. We all loved it, but it was CP invariant, not by design, but because of its basic form. I'm not sure why people didn't worry more about that, but perhaps it was because the one small effect could always be explained by superweak.

However, in 1973, a paper was published in the *Progress of Theoretical Physics* [2], a paper that, at first, few people noticed. This paper pointed out in one paragraph that if there were six quarks instead of four (at the time, in fact, only three were known), then CP violation could be accommodated in the Weinberg-Salam model. With the discovery of the b quark system, this became the standard model of CP violation. In this model, the up quarks (u_j) are connected to the down quarks (d_i) in a weak decay by

$$d_i \to V_{ji} u_j,$$

where V_{ji} is the 3×3 CKM matrix. The one significant phase in the matrix is the origin of CP violation.

At first we thought that the small value of ϵ meant that the phase in the CKM matrix was small. The true picture came with the determination of the relatively long B lifetime. It then became clear that the matrix had a hierarchical structure. It could be expanded in powers of the sine of the Cabibbo angle ($\lambda = \sin\theta_c = .22$) and the phase entered only at order λ^3 in the elements connecting the first and third generations:

$$\begin{aligned} V_{ub} &= A\lambda^3(\rho - i\eta) \\ V_{td} &= A\lambda^3(1 - \rho - i\eta). \end{aligned} \tag{9.3}$$

The rate of $b \to u$ decays gives roughly

$$(\rho^2 + \eta^2)^{1/2} \approx 0.4 \pm 0.15,$$

while the value of ϵ requires η greater than 0.2. As a result, the phases of eq. (9.3) are not small.

Given the standard model, is there any motivation for thinking about superweak? The most compelling arguments are those given by Steve Barr [3]. One would like a more elegant way to introduce CP violation than by putting a whole set of arbitrary phases in the Yukawa matrices. Also, if one assumes CP violation is allowed everywhere you can put it, then the standard model contains an arbitrary parameter called θ_{QCD}, which is restricted to values less than 10^{-9} by the neutron electric dipole moment. A possible solution to these problems is to assume CP invariance is valid at some higher mass scale, but it is broken softly, either spontaneously or in terms of dimensions less than 4. Then θ_{QCD} is calculable and in suitable theories is small enough. Similarly, the parameter η in the CKM matrix must be calculated from the same origin and it may turn out [4] to be much too small to explain ϵ. In that case, the observed CP violation must be explained by other effective interactions caused by the new physics at the higher mass scale, which may be effectively superweak.

The major feature of the weak Hamiltonian of the standard CKM theory that distinguishes it from superweak is the presence of the CP-violating terms that change flavor by one unit that are of the order $G_F A \lambda^3$. Thus, besides the CP violation in the mixing, we expect *direct CP violation*; that is, CP violation in the decay amplitude. However, direct CP violation for the K^0 system given by the parameter ϵ' is very small and, as discussed in these pages, the value of ϵ' in the CKM model is very uncertain.

A question arises as to the definition of a superweak model since there exist many examples in the literature. In general, a superweak model is defined by the effective Lagrangian at the electroweak scale or below:

$$L_{\text{eff}} = L_0 + L_{\text{SW}},$$

where L_0 is the standard model with η much less than that required to explain ϵ and L_{SW} contains various four-quark interaction with coefficients G_{ijkl} including the term (9.2) needed to explain ϵ.

A theory is superweak if all the G_{ijkl} are much smaller than the usual weak interactions in L_0. An extreme superweak theory would be one in which all G_{ijkl} were as small as G_{sdsd} and involved a change of flavor ΔF of 2 units. However, I would consider theories with larger values of G_{ijkl} as long as they were much smaller than those of the standard model and allow for $\Delta F = 1$ as well as 2. This raises the interesting point of whether the observed value of ϵ'/ϵ could possibly be consistent with some theories that are essentially superweak. The simplest superweak models involve the tree-level exchange of a heavy boson. The $\Delta F = 1$ CP-violating term in L_{SW} then might have G_{sddd} of order G_{sdsd}, in which case ϵ' would be of order 10^{-9} to 10^{-10} or $\epsilon'/\epsilon \sim 10^{-6}$ to 10^{-7}. However, it is also possible that the exchange might involve a relatively light Higgs boson in a multi-Higgs scenario with very small couplings. In general, the contribution of the neutral Higgs exchange to ϵ'/ϵ in such models is expected to be small, but it is possible that charged Higgs exchange could give ϵ'/ϵ as large as 10^{-3} [5].

There exists a large class of models that fit our definition of superweak although they are called "milliweak" by Barr [3]. These are models in which ϵ occurs only in a box diagram involving a new boson H and either a quark, a lepton, or a new fermion. A large catalogue of such models is given by Barr. If H is a charged diquark, ϵ'/ϵ can actually arise at tree level and be as large as 3×10^{-3}; in contrast, if H is a leptoquark, ϵ'/ϵ occurs only via a loop (penguin-type) graph and is no larger than 10^{-7} [6]. In another model [7], H is a charged scalar and the fermion is a new "quark" with charge 4/3; here again ϵ'/ϵ occurs at loop level and is found to be of order 10^{-4}. It has been suggested [8] that models that are effectively superweak can be constructed with SUSY particles. The possibility that ϵ'/ϵ could be as large as 2×10^{-3} from loops with SUSY particles with particle mixing has recently been noted [9].

We conclude that while it is reasonable to expect ϵ'/ϵ to be small in superweak models, one cannot rule out models that are essentially superweak and still have ϵ'/ϵ of the order 2×10^{-3} as observed. The reason is that ϵ' is only 5×10^{-6}, which corresponds to G_{sdqq} of order 10^{-5} to $10^{-6} G_F$.

In fact, one might say that if L_{eff} is defined below the b-mass scale, then the standard model becomes practically superweak: L_0 is then the CP-conserving four-quark model and L_{SW} represents the effective four-fermion interactions due to virtual effects of t quarks.

The most definitive way to distinguish the standard model from superweak is in B physics. The first observation of CP violation should come from the asymmetry in the decay $B^0(\bar{B}^0) \to J/\psi K_S$. Within the standard model this measures $\sin 2\beta$, where β is the phase of $(1 - \rho + i\eta)$. The standard model predicts a large positive value for $\sin 2\beta$ between 0.3 and 0.9. This could be blamed entirely on CP violation in $B - \bar{B}$ mixing and is thus analagous to the ϵ parameter of the K system. It could be explained by a superweak interaction with the coupling G_{bdbd} of order $10^{-7} G_F$. The final, completely definitive death of any superweak theory will come from the observation of direct CP violation in the B system.

This arises from the phase γ in $b \to u$ decays (where γ is the phase of $\rho + i\eta$). To be more accurate, one is comparing the phase of $b \to u$ decays to that of $b \to c$ (such as $B \to J/\psi K_S$). Evidence for such direct CP violation would be given by the difference between the asymmetry parameter in a decay such as $B^0(\bar{B}^0) \to \pi^+\pi^-$ from that of $J/\psi K_S$. This can be considered [10] the ϵ' experiment for the B system. In the tree approximation, this is given by $\sin 2(\beta + \gamma)$, which is expected to be very different from $\sin 2\beta$. Thus, ϵ' for the B system is of order unity in contrast to the K system, where it is 5×10^{-6}. It should be emphasized that "penguin pollution" does not affect the yes-no question as to the existence of direct CP violation; all that is required is a significant difference in the two asymmetries.

It may well be that $B \to \pi\pi$ will not be the first place to find direct CP violation. It may be found in the difference in partial rates of B^+ and B^- decays if final state interactions are large enough. In thinking about B physics, it is important to recognize the importance of finding evidence for direct CP violation even if the extraction of (β, γ) or (ρ, η) is not directly possible from the data.

In retrospect, it is very surprising that it has taken over 30 years to kill the superweak theory as the origin of CP violation. It will still take a long time to establish whether the CKM phase is the only source of CP violation or whether there are important effects from physics beyond the standard model.

Acknowledgments

I am indebted to discussions with Darwin Charg and Steve Barr. This research was supported, in part, by the Department of Energy under Contract DE-FG02-91-ER-40682.

References

[1] L. Wolfenstein, Phys. Rev. Lett. **13**, 562 (1964).

[2] M. Kobayshi and T. Maskawa, Prog. Theor. Phys. **49**, 652 (1973).

[3] S.M. Barr, Phys. Rev. D **34**, 1567 (1986).

[4] See, for example, S.L. Glashow and H. Georgi, Phys. Lett. B **451**, 372 (1999).

[5] Y.L. Wu and L. Wolfenstein, Phys. Rev. Lett. **73**, 1762 (1994).

[6] S.M. Barr and E. Frère, Phys. Rev. D **41**, 2129 (1990).

[7] D. Bowser-Chao, D. Chang, and W.Y. Keung, Phys. Rev. Lett. **81**, 2028 (1998).

[8] R. Barbieri *et al.*, Phys. Lett. B **425**, 119 (1998).

[9] A. Masiero and H. Murayama, Berkeley preprint hep-ph/9903363.

[10] L. Wolfenstein, Nucl. Phys. B **246**, 45 (1984).

10

Can ε'/ε Be Supersymmetric?

Hitoshi Murayama

Abstract

I first motivate why we may want to look at possible new physics contributions to ε' given relatively clear experimental but unclear theoretical situations. I reexamine the supersymmetric contribution to ε' and find an important one generally missed in the literature. Based on rather model-independent arguments based on flavor symmetries, an estimate of the possible supersymmetric ε' is given, which interestingly come around the reported values without fine-tuning. If the observed values are dominated by supersymmetry, it is likely to give interesting consequences on hyperon CP violation, $\mu \to e\gamma$, and neutron and electron electric dipole moments.

10.1 Motivation

The year 1999 has already seen impressive progress in flavor physics. CDF reported the first measurement of $\sin 2\beta$ from B decay [1] which strongly hints to CP violation in a system other than the neutral kaon system. The situation of direct CP violation ε' in the neutral kaon system used to be somewhat unclear, but the KTeV result [2] and the NA48 result reported at this meeting [3] basically settled the experimental situation. The numbers on ε'/ε reported are[1]

E731	$(7.4 \pm 5.9) \times 10^{-4}$
NA31	$(23.0 \pm 6.5) \times 10^{-4}$
KTeV	$(28.0 \pm 4.1) \times 10^{-4}$
NA48	$(18.5 \pm 7.3) \times 10^{-4}$
W.A.	$(21.2 \pm 4.6) \times 10^{-4}$

On the other hand, the theoretical situation is rather unclear. The calculation of ε' in the standard model is difficult partly because of a cancellation between gluon and electroweak penguins that makes the result

[1] The world average (W.A.) value includes the rescaling of errors due to the PDG prescription to account for still-not-too-good $\chi^2/\text{d.o.f.} = 2.8$.

Reference	ε'/ε
Ciuchini [4]	$(4.6 \pm 3.0 \pm 0.4) \cdot 10^{-4}$
Bosch (NDR) [5]	$(7.7^{+6.0}_{-3.5}) \cdot 10^{-4}$
Bosch (HV) [5]	$(5.2^{+4.6}_{-2.7}) \cdot 10^{-4}$
Bertolini [6]	$(17^{+14}_{-10}) \cdot 10^{-4}$

Table 10.1: Various standard model estimates of ε'/ε in the literature. The two estimates by Bosch et al. use different renormalization schemes (NDR and HV).

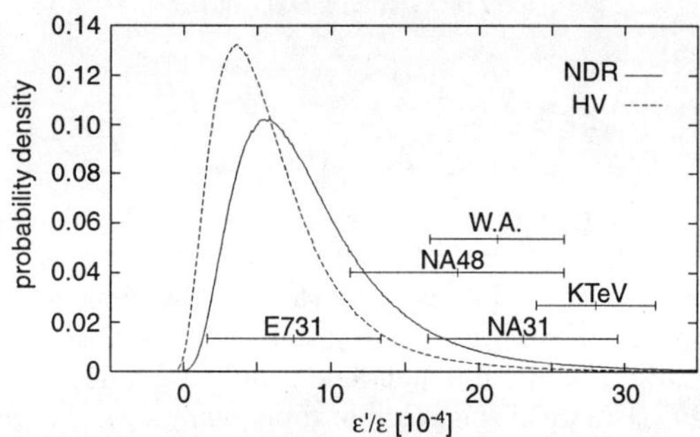

Figure 10.1: Probability density distributions for the standard model calculations of ε'/ε in NDR and HV schemes from [5] compared to experimental values.

sensitive to the precise values of the hadronic matrix elements. A (not complete) list of theoretical calculations is given in table 10.1. See [7] and [8] for more details on this issue. The experimental values are compared to the probability density distributions for ε'/ε [5] in fig. 10.1. There is a feeling in the community that the data came out rather high, even though one cannot draw a definite conclusion if the standard model accounts for the observed high value because of theoretical uncertainties.

Of course the correct strategy to resolve this issue is to improve theoretical calculations, probably relying on the progress in lattice calculations. Unfortunately, this is a challenging program and we cannot expect an immediate resolution. What I instead attempt in this talk is the alternative approach: to think about new physics candidates that are "reasonable" and at the same time account for the observed value of ε'/ε. Since any new physics explanation would probably give rise to other consequences, the cross-check can start eliminating such possibilities. If all "reasonable" candidates get excluded, one can draw a conclusion that the standard model should account for the observed high value of ε'. Or such a cross-check might confirm other consequences that would be truly exciting.

Naturally, I turned my attention to supersymmetry, the most widely discussed candidate of physics beyond the standard model, and asked the question if the supersymmetric contribution to ε' can be interesting at all.

10.2 Lore

There are many studies of ε' in supersymmetric models, most notably that of Gabrielli and Giudice [9]. Their detailed study found the possible range of supersymmetric contribution, $(\varepsilon'/\varepsilon)_{\rm SUSY} = (-0.4\text{–}7.4) \times 10^{-4}$. Then clearly supersymmetry cannot account for the observed high value, and indeed I do not have anything new to add to their beautiful analysis within the framework they used: the minimal supergravity model.

The reason why the supersymmetric contribution to ε' is small can be quite easily understood. Within the minimal supergravity framework, the main contribution to flavor-changing effects between the first and second generations originate in the left-handed squark mass-squared matrix. In general, the superpartners of the left-handed quarks have a mass-squared matrix

$$\mathcal{L}_{LL} = - \begin{pmatrix} \tilde{d}_L^* & \tilde{s}_L^* \end{pmatrix} \begin{pmatrix} m_{11}^2 & m_{12}^2 \\ m_{12}^{2*} & m_{22}^2 \end{pmatrix} \begin{pmatrix} \tilde{d}_L \\ \tilde{s}_L \end{pmatrix}. \tag{10.1}$$

If there is a nonvanishing off-diagonal element m_{12}^2 in the above mass-squared matrix, it would contribute to flavor-changing processes from loop diagrams. The easiest way to study such contributions is to use the mass insertion formalism where one treats the off-diagonal element perturbatively as an "interaction" in the squark propagator, because the existent constraints require the off-diagonal element to be small anyway. The size of such a perturbation can be nicely parameterized by $(\delta_{12}^d)_{LL} \equiv m_{12}^2/m^2$, where m^2 is the average of the two diagonal elements.

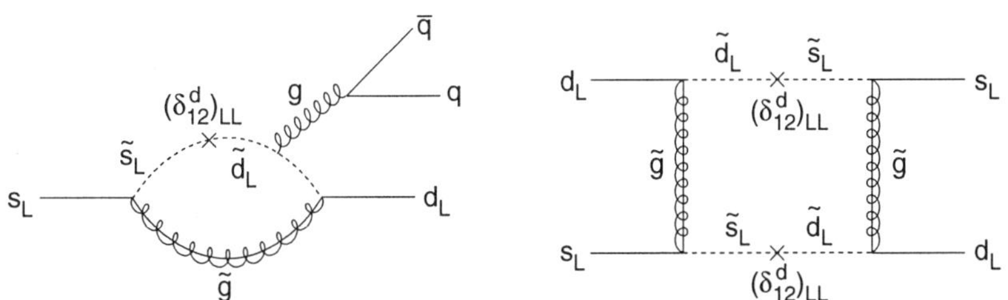

Figure 10.2: Representative Feynman diagrams for LL contributions to ε' (left) and Δm_K, ε (right).

The contribution of the mass insertion to ε' is given by[2] [10]:

$$\left(\frac{\varepsilon'}{\varepsilon}\right)_{\rm SUSY} \simeq 27 \times 10^{-4} \left(\frac{500\,\text{GeV}}{m}\right)^2 \frac{\text{Im}(\delta_{12}^d)_{LL}}{0.50}. \tag{10.2}$$

A representative Feynman diagram is shown in fig. 10.2. On the other hand, the insertion of the same parameter $(\delta_{12}^d)_{LL}$ induces kaon mixing

[2]Here and hereafter, I take $M_{\tilde{g}} = m$ for simplicity. See [10] for more details.

parameters

$$(\Delta m_K)_{\rm SUSY} \simeq (\Delta m_K)_{\rm experiment} \left(\frac{500\,{\rm GeV}}{m}\right)^2 \frac{{\rm Re}(\delta^d_{12})^2_{LL}}{(0.04)^2} \quad (10.3)$$

$$\varepsilon_{\rm SUSY} \simeq \varepsilon_{\rm experiment} \left(\frac{500\,{\rm GeV}}{m}\right)^2 \frac{{\rm Im}(\delta^d_{12})^2_{LL}}{(0.003)^2}. \quad (10.4)$$

Therefore, once the constraints from Δm_K and ε are satisfied, a large contribution to ε' is not allowed unless some fine-tuning is done. This simple comparison already gives a typical order of magnitude $(\varepsilon'/\varepsilon)_{\rm SUSY} \lesssim 2 \times 10^{-4}$, with some wiggle room by varying $M_{\tilde{g}}$, allowing chargino contributions, etc.[3] This leads to the basic conclusion in [9] that the supersymmetric contribution to ε' is never important.[4]

10.3 New Contribution

Antonio Masiero and I found that there is a loophole in previous discussions of why the supersymmetric contribution to ε' is small [12].[5] We pointed out that a broad class of supersymmetric models actually gives an important contribution,[6]

$$\left(\frac{\varepsilon'}{\varepsilon}\right)_{\rm SUSY} \simeq 30 \times 10^{-4} \left(\frac{500\,{\rm GeV}}{m}\right)^2 \frac{\sin\phi}{0.5} \times (0.5\text{–}2) \times (0.5\text{–}2), \quad (10.5)$$

where the uncertainties come from model dependence and hadronic matrix elements. I discuss the angle ϕ and the origin of this new contribution below.

The masses of quarks arise from the coupling of left-handed and right-handed quarks to the Higgs expectation value,

$$\mathcal{L}_{\rm quark\ mass} = -\begin{pmatrix} \overline{d_R} & \overline{s_R} \end{pmatrix} \begin{pmatrix} y_{11} & y_{12} \\ y_{21} & y_{22} \end{pmatrix} \begin{pmatrix} d_L \\ s_L \end{pmatrix} \langle H_d \rangle + c.c. \quad (10.6)$$

Here, H_d is the Higgs field that couples to the down-type quarks in the Minimal Supersymmetric Standard Model. Similarly, in addition to the

[3] One can maximize $\varepsilon' \propto {\rm Im}(\delta^d_{12})_{LL}$ beyond this estimate by assuming that $(\delta^d_{12})_{LL}$ is nearly pure imaginary because $(\delta^d_{12})^2_{LL}$ is then nearly pure real and hence the ε constraint is satisfied.

[4] This conclusion, however, relied only on the $\Delta I = 1/2$ amplitude from supersymmetry. A generally missed contribution to $\Delta I = 3/2$ amplitude due to the isospin breaking $m^2_{\tilde{d}_R} \neq m^2_{\tilde{u}_R}$ leads to a too-large ε' if the splitting is $O(1)$. A more natural size of splitting from the renormalization group point of view is of order 10%, and with a quarter of the maximum possible value of ${\rm Im}(\delta^d_{12})_{LL}$, the resulting size of ε' is in the ballpark of the observed value [11].

[5] Another possibility of generating ε' from an enhanced $\overline{d_L}\gamma^\mu s_L Z_\mu$ vertex was suggested [13] that is subject to tighter constraints phenomenologically [14].

[6] This estimate is larger by a factor of two than that in [12], which ignored the running effect of the dipole moment operator.

mass-squared mass matrix (10.1) for the left-handed squarks and the analogous one for the right-handed squarks, there is another contribution to the masses of squarks that mix left-handed and right-handed particles,

$$\mathcal{L}_{LR} = -\begin{pmatrix} \tilde{d}_R^* & \tilde{s}_R^* \end{pmatrix} \begin{pmatrix} A_{11} & A_{12} \\ A_{21} & A_{22} \end{pmatrix} \begin{pmatrix} \tilde{d}_L \\ \tilde{s}_L \end{pmatrix} \langle H_d \rangle + c.c. \tag{10.7}$$

The important point here, without going into detailed discussions of models, is that we expect the matrix A_{ij} to have similar structure to the Yukawa matrix y_{ij} because they have exactly the same property under the flavor symmetries. I will describe this point more clearly in the context of models later on, but based on this simple fact, I expect

$$A_{ij} = m_{\rm SUSY} Y_{ij} \times O(1), \tag{10.8}$$

where $m_{\rm SUSY}$ is the typical mass scale of supersymmetry breaking (*i.e.*, superparticle masses). It is possible that A_{ij} and Y_{ij} are exactly proportional to each other as matrices, which was assumed in the minimal supergravity and hence in previous analyses, but not necessarily. Generically, one expects the following patterns for Yukawa and A matrices,

$$\begin{pmatrix} y_{11} & y_{12} \\ y_{21} & y_{22} \end{pmatrix} \langle H_d \rangle = \begin{pmatrix} m_d & m_s\lambda \\ & m_s \end{pmatrix}, \tag{10.9}$$

$$\begin{pmatrix} A_{11} & A_{12} \\ A_{21} & A_{22} \end{pmatrix} \langle H_d \rangle = m_{\rm SUSY} \begin{pmatrix} cm_d & bm_s\lambda \\ & am_s \end{pmatrix} + c.c., \tag{10.10}$$

where a, b, c are $O(1)$ unknown constants and $\lambda = \sin\theta_C$. The (2,1) elements are intentionally left blank because we do not know the mixing angles of right-handed quarks (read: model-dependent).

In order to discuss flavor-changing effects from squarks, one first needs to go in to the mass basis for quarks, which is achieved by rotating the Yukawa matrix by the Cabibbo angle θ_C. The same rotation acts on the A matrix, but unless $a = b$, it still leaves the off-diagonal element after rotation $A'_{12}\langle H_d\rangle = (b-a)m_{\rm SUSY}m_s\lambda$. We can again describe its effect in terms of a mass insertion parameter

$$\begin{aligned} (\delta_{12}^d)_{LR} &= \frac{A'_{12}\langle H_d\rangle}{m^2} \\ &= 4\times 10^{-5}\left(\frac{m_s}{100\ \text{MeV}}\right)\left(\frac{m_{\rm SUSY}}{m}\right)\left(\frac{500\ \text{GeV}}{m}\right)\times(b-a). \end{aligned} \tag{10.11}$$

It contributes to the flavor-changing chromo-electric dipole moment $\frac{g_s}{8\pi^2}m_s$ $\overline{d_L}\sigma^{\mu\nu}G_{\mu\nu}s_R$ (fig. 10.3), where $G_{\mu\nu}$ is the gluon field strength. This operator generates a supersymmetric contribution to ε' as $(\varepsilon'/\varepsilon)_{\rm SUSY} = 30 \times 10^{-4}({\rm Im}(\delta_{12}^d)_{LR}/2\times 10^{-5})$, which gives eq. (10.5). The model uncertainty (0.5–2) comes from the lack of knowledge on the $O(1)$ constant a, b, and also that we do not know the ratio of $m_{\rm SUSY}$ in the LR mass matrix to the

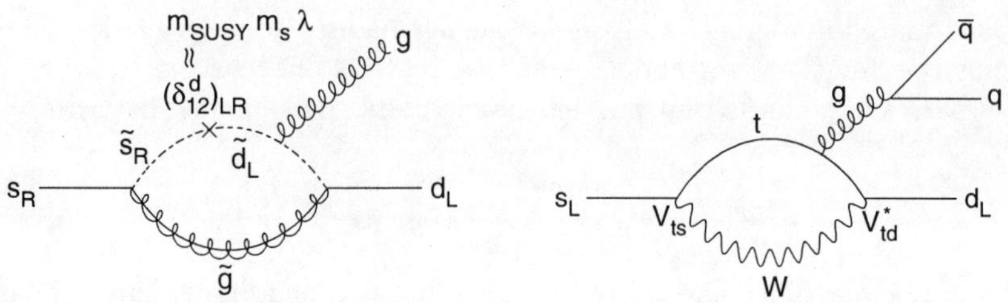

Figure 10.3: Representative Feynman diagrams for the supersymmetric contribution from LR mass insertion (left), and the standard model contribution (right) to ε'.

average squark mass. The phase factor is $\phi = \arg(b - a)$. Finally the matrix element of this operator between kaon and two-pion states vanishes at the lowest order in chiral perturbation theory, and it was estimated in [15] using the chiral quark model. I have arbitrarily assigned a factor of two uncertainty (0.5–2) as a guess.[7]

Note that the exact proportionality between Yukawa and A matrices assumed in the minimal supergravity automatically makes the off-diagonal element vanish after diagonalizing the Yukawa matrix and hence no flavor-changing effects. This is one of the reasons why the LR mass insertion has been regarded as unimportant in the literature. But such an exact proportionality is a strong assumption with no justification. The constraint from ε on the LR mass insertion is an order of magnitude weaker: $\text{Im}(\delta^d_{12})^2_{LR} \lesssim (3.5 \times 10^{-4})^2 (m^2/500 \text{ GeV})^2$ [10]. This phenomenological fact that the constraint from ε still allows an interesting LR contribution to ε' has been known (see, *e.g.*, [10]). But the natural size of $(\delta^d_{12})_{LR}$ has not been discussed and it was always assumed to be smaller than the size estimated here due to various prejudices partly because of the constraint from the neutron electric dipole moment (see discussions later).

The reason why this supersymmetric contribution is competitive (or larger) than the standard model contribution is basically because of the larger KM factor as seen in the following back-of-the-envelope estimate. The estimate of the supersymmetry contribution is given by

$$\frac{g_s^3}{8\pi^2} \frac{m_s \lambda}{m_{\text{SUSY}}^2} \overline{d_L} \sigma^{\mu\nu} G_{\mu\nu} s_R$$

and is suppressed only by the Cabibbo angle in the LR mass insertion. On the other hand the standard model QCD penguin is suppressed because all three generations should participate for CP violation:

$$\frac{g_s^2 g^2}{8\pi^2} \frac{V_{ts} V_{td}^*}{m_t^2} \overline{d_L} \gamma^\mu T^a s_L \overline{q} \gamma_\mu T^a q.$$

[7]The authors of [14] instead assign an uncertainty of $B_G = (1\text{–}4)$ in their notation and have emphasized that the sign of the matrix element is model dependent.

To compare them, m_s in the supersymmetric contribution can be regarded as an $O(1)$ parameter in kaon decays. The mass scale is higher $m_{\rm SUSY} > m_t$, but the KM factor is larger by 100. However, the matrix element of the chromo-electric dipole moment operator vanishes at the leading order in chiral perturbation theory [15], resulting in about a factor of ten suppression, which brings the supersymmetry contribution roughly on par with the QCD penguin. Finally, the QCD model penguin is partially canceled by the electroweak penguin. This leaves the supersymmetry contribution important in ε'.

10.4 Models

Now that we have found the new supersymmetric contribution to ε' based on general grounds, it is important to ask what class of models actually gives such a contribution. Because we are talking about flavor physics issues in supersymmetry here (see [16] for a good summary in this context), we need to look at models that naturally explain the hierarchy in fermion masses.

The easiest example to explain is the model based on U(2) flavor symmetry [17]. In this model, the first and second generations form a doublet under the U(2) flavor symmetry, and the third generation a singlet. This makes the LL squark mass-squared matrix (10.1) an identity matrix and helps to keep the first- and second-generation scalars from inducing too-large flavor-changing effects in Δm_K and ε. The flavor symmetry allows the top Yukawa coupling, but forbids down and strange Yukawa couplings and hence needs to be broken by some vacuum expectation values (VEVs). In [17] two such VEVs are introduced, one in the symmetric tensor of U(2) $\sigma_{ij} = \sigma_{ji}$ and the other in the antisymmetric tensor $\alpha_{ij} = -\alpha_{ji}$, which can generate the Yukawa matrix

$$\frac{1}{M}\begin{pmatrix} \overline{d_R} & \overline{s_R} \end{pmatrix}\begin{pmatrix} 0 & \langle\alpha_{12}\rangle \\ -\langle\alpha_{12}\rangle & \langle\sigma_{22}\rangle \end{pmatrix}\begin{pmatrix} d_L \\ s_L \end{pmatrix} H_d, \tag{10.12}$$

where M is the energy scale of the flavor physics. The point here is that (2,2) and (1,2) elements originate from couplings to different fields σ and α, and hence their supersymmetry breaking counterparts can have different coefficients, giving in general $a \neq b$ in the LR mass matrix (10.10).

A similar situation occurs in string theory, where the Yukawa couplings y_{ij} are in general functions of so-called moduli fields T: $y_{ij}(T)$. The moduli fields also acquire supersymmetry breaking vacuum expectation value (F_T), and the LR mass matrix is given by $A_{ij} = (\partial y_{ij}(T)/\partial T)F_T$. Generally, these two matrices are not proportional to each other, even though their structures are likely to be similar if the Yukawa hierarchy is natural, *i.e.*, stable under small changes of the moduli. A particular realization of this idea is an attempt to generate Yukawa hierarchy as a consequence of a somewhat large compactification radius of an orbifold and fields of different generations belonging to twisted sectors of different fixed points. Then the modular invariance requires Yukawa couplings to be proportional to powers of Dedekind eta function $\eta(T)$: $Y_{ij} \simeq \eta(T)^{3+n_i+n_j+n_H}$ (n_i is the modular weight) [18]. This mechanism can generate a Yukawa hierarchy because $\eta(T) \sim e^{-T}$ for a large T. Then the derivative picks the power $3+n_i+n_j+n_H$, which is generation dependent, and hence $b \neq a$ again. The flavor-changing effects from the LL mass insertion, however, are a problem in general string models.

I have to emphasize that not all supersymmetric models give this type of contributions to ε'. For instance, the models with gauge-mediated supersymmetry breaking (see [19] for a review) give either a vanishing A matrix or one proportional to the Yukawa matrix and hence does not give interesting ε'. If the Cabibbo angle originates in the up sector, the naive analysis here also fails. Also, the A matrix may be real. All I can argue here is that we generically expect a supersymmetric contribution to ε' as in (10.5), but there are exceptions. See [20] for more discussions on model dependence.

10.5 Other Observables

Since supersymmetry can give a contribution to ε' of an interesting size, we should now ask if there are other consequences of this. The immediate question is the neutron electric dipole moment. Again using the mass insertion technique, the current constraint is $|\text{Im}(\delta^d_{11})_{LR}| \lesssim 3 \times 10^{-6}$ [10], while my estimate is $(\delta^d_{11})_{LR} \equiv A_{11}\langle H_d\rangle/m^2 \simeq m_{\text{SUSY}} m_d/m^2 \simeq 1 \times 10^{-5}(500 \text{ GeV}/m_{\text{SUSY}})$. If we are interested in a large supersymmetric contribution to ε', it is natural to expect a phase of order unity in A_{11} as well. This basically saturates the current upper limit for a phase of order 1/3, and an improved experimental bound would be extremely interesting. The phase in A_{11}, however, is in principle unrelated to that in A_{12}, and in fact some models give a real A_{11} despite a complex A_{12} [21].

A more exciting aspect is that hyperon CP violation arises from the same chromo-electric dipole moment operator that generates supersymmetric ε', and in principle there is a perfect correlation. Unfortunately the matrix element uncertainties obscure the correlation, but Pakvasa argued that the hyperon CP violation should appear within the sensitivity of the current HyperCP experiment if ε' is saturated by supersymmetry in the way I described [22]. It would the most direct test of this possibility.

Another important question is what the corresponding flavor-changing effects in the lepton sector would be. In fact, we found a new contribution to $\mu \to e\gamma$ that was also generally missed in the literature. Unfortunately the estimate is uncertain because we do not know the mixing angle in the lepton sector, but a reasonable guess is $V_{\nu_e\mu} \simeq 0.016$–$0.05$ suggested by the small angle MSW solution. The new limit from MEGA collaboration [23] $\text{BR}(\mu \to e\gamma) < 1.2 \times 10^{-11}$ (90% CL) leads to the bound on the mass insertion parameter

$$(\delta^l_{12})_{LR} < 4.2 \times 10^{-6}(m_{\tilde{l}}/500\,\text{GeV}).$$

This is to be compared to the estimate

$$\begin{aligned} |(\delta^l_{12})_{LR}| &\simeq m_{\text{SUSY}} m_\mu V_{\nu_e \mu}/m^2_{\tilde{l}} \\ &\simeq 6.6 \times 10^{-6}(500\,\text{GeV}/m_{\tilde{l}})(V_{\nu_e\mu}/0.032)(m_{\text{SUSY}}/m_{\tilde{l}}). \end{aligned}$$

Note that this bound is on the *magnitude* of the parameter, not its imaginary part, and is hence more difficult to avoid. One can regard this estimate as a lower bound on $m_{\tilde{l}}$ of about 500 GeV, which is much stronger than previously discussed limits. The electron electric dipole moment

$$d_e = (0.18 \pm 0.16) \times 10^{26} < 0.38 \times 10^{-26} e \cdot \text{cm} \quad (90\%\ \text{CL})$$

translates to the limit

$$|\text{Im}(\delta^l_{11})_{LR}| < 1.0 \times 10^{-6}(m_{\tilde{l}}/500\,\text{GeV}),$$

compared to the estimate

$$(\delta^l_{11})_{LR} \simeq 1.0 \times 10^{-6}(m_{\tilde{l}}/500\,\text{GeV}).$$

The supersymmetric contribution again nearly saturates the constraint and a future improvement of the bound would be useful.

More recently, it has been suggested that there may also be enhancements in $K_L \to \pi^0 e^+ e^-$ [14] and $K \to \pi\pi\gamma$ [24]. See also discussions in [25].

Overall, the naive estimate of the LR mass insertion I've described gives interesting and important contributions to ε', $\mu \to e\gamma$, d_n, and d_e, and those for the first two had been generally missed in the literature.

10.6 Conclusions

We reexamined the supersymmetric contribution to ε' and found that an important contribution had been missed in the literature. The assumptions that went into the estimate are: (a) the hierarchical Yukawa matrix is probably controlled by certain flavor symmetries, (b) the A matrix is subject to the same control that makes its structure similar to the Yukawa matrix but not necessarily exactly proportional to it, (c) the Cabibbo rotation arises in the down sector, and (d) a phase of order 1 in the A matrix. Given these assumptions, we came up with the estimate eq. (10.5), which surprisingly is in the ballpark of reported experimental values without any fine-tuning.

If ε' is dominated by the supersymmetric contribution described here, it should give an important contribution to the hyperon CP violation and possibly also to the neutron electric dipole moment. Correspondingly, there is an important contribution to $\mu \to e\gamma$ and possibly also to the electron electric dipole moment.

I thank Antonio Masiero for collaboration. This work was supported in part by the U.S. Department of Energy under Contracts DE-AC03-76SF00098, in part by the National Science Foundation under grant PHY-95-14797, and in part by Alfred P. Sloan Foundation.

References

[1] M. Paulini, Int. J. Mod. Phys. A **14**, 2791 (1999), hep-ex/9903002.

[2] KTeV Collaboration (A. Alavi-Harati et al.), Phys. Rev. Lett. **83**, 22 (1999), hep-ex/9905060.

[3] M.S. Sozzi, this volume, chapter 8.

[4] M. Ciuchini, invited talk at 4th KEK Topical Conference on Flavor Physics, Tsukuba, Japan, 29–31 Oct. 1996, Nucl. Phys. Proc. Suppl. **59**, 149 (1997).

[5] S. Bosch, A.J. Buras, M. Gorbahn, S. Jager, M. Jamin, M.E. Lautenbacher, and L. Silvestrini, hep-ph/990440.

[6] S. Bertolini, J.O. Eeg, M. Fabbrichesi, and E.I. Lashin, Nucl. Phys. B **514**, 93 (1998).

[7] A.J. Buras, this volume, chapter 5.

[8] S. Bertolini, this volume, chapter 12; R. Gupta, this volume, chapter 13; M. Ciuchini *et al.*, this volume, chapter 31; B. Bardeen, this volume, chapter 14; A.J. Buras, this volume, chapter 5.

[9] E. Gabrielli and G. F. Giudice, Nucl. Phys. B **433** 3 (1995), hep-lat/9407029.

[10] F. Gabbiani, E. Gabrielli, A. Masiero, and L. Silvestrini, Nucl. Phys. B **477**, 321 (1996), hep-ph/9604387.

[11] A.L. Kagan and M. Neubert, hep-ph/9908404.

[12] A. Masiero and H. Murayama, Phys. Rev. Lett. **83**, 907 (1999), hep-ph/9903363.

[13] G. Colangelo and G. Isidori, JHEP **9809**, 009 (1998), hep-ph/9808487.

[14] A.J. Buras, G. Colangelo, G. Isidori, A. Romanino, and L. Silvestrini, hep-ph/9908371.

[15] S. Bertolini, J.O. Eeg, and M. Fabbrichesi, Nucl. Phys. B **449** 197 (1995), hep-ph/9409437.

[16] L. Hall, this volume, chapter 4.

[17] R. Barbieri, G. Dvali, and L.J. Hall, Phys. Lett. B **377**, 76 (1996), hep-ph/9512388.

[18] L.E. Ibañez, Phys. Lett. B **181** 269 (1986).

[19] G.F. Giudice and R. Rattazzi, hep-ph/9801271.

[20] R. Barbieri, R. Contino, and A. Strumia, hep-ph/9908255; G. Eyal, A. Masiero, Y. Nir, and L. Silvestrini, hep-ph/9908382.

[21] K.S. Babu, B. Dutta, and R.N. Mohapatra, hep-ph/9905464. See also S. Baek, J.-H. Jang, P. Ko, J.H. Park, hep-ph/9907572.

[22] S. Pakvasa, this volume, chapter 42.

[23] MEGA Collaboration, hep-ex/9905013.

[24] G. Colangelo, G. Isidori, and J. Portoles, hep-ph/9908415.

[25] K.S. Babu and R.N. Mohapatra, hep-ph/9906271; J.L. Diaz-Cruz, hep-ph/9906330; S. Khalil and T. Kobayashi, hep-ph/9906374; S. Baek, J.-H. Jang, P. Ko, and J.H. Park, hep-ph/9907572.

11

ε'/ε in the $1/N_c$ Expansion

T. Hambye and P. H. Soldan

Abstract

We present a new analysis of the ratio ε'/ε, which measures the direct CP violation in $K \to \pi\pi$ decays. We use the $1/N_c$ expansion within the framework of the effective chiral Lagrangian for pseudoscalar mesons. From general counting arguments we show that the matrix element of the operator Q_6 is not protected from possible large $1/N_c$ corrections beyond its large-N_c or VSA value. Calculating the $1/N_c$ corrections, we explicitly find that they are large and positive. Our result indicates that a $\Delta I = 1/2$ enhancement is operative for Q_6 similar to the one of Q_1 and Q_2 which dominate the CP conserving amplitude. This enhances ε'/ε and can bring the standard model prediction close to the measured value for central values of the parameters.

11.1 Introduction

Recently, direct CP violation in $K \to \pi\pi$ decays was observed by the KTeV [1] and NA48 [2] collaborations. The new measurements are in agreement with the results of the NA31 experiment [3]. The present world average for the parameter ε'/ε is [2]

$$\mathrm{Re}\,\varepsilon'/\varepsilon = (21.2 \pm 4.6) \cdot 10^{-4}, \tag{11.1}$$

which differs from zero by 4.6 standard deviations. In the standard model CP violation originates in the CKM phase, and direct CP violation is governed by loop diagrams of the penguin type, in which the three quark generations are present. The value of ε'/ε is determined by a nontrivial interplay of the strong, electromagnetic, and weak interactions and depends on almost all standard model parameters. This makes its calculation quite complex. The main source of uncertainty in the calculation of the CP ratio is the QCD nonperturbative contribution related to the hadronic nature of the $K \to \pi\pi$ decays. Using the $\Delta S = 1$ effective hamiltonian,

$$\mathcal{H}_{\mathrm{eff}}^{\Delta S=1} = \frac{G_F}{\sqrt{2}}\,\lambda_u \sum_{i=1}^{8} c_i(\mu)\,Q_i(\mu) \qquad (\mu < m_c)\,, \tag{11.2}$$

the nonperturbative contribution, contained in the hadronic matrix elements of the four-quark operators Q_i, can be separated, at the renormalization scale $\mu \simeq 1\,\text{GeV}$, from the perturbative Wilson coefficients $c_i(\mu) = z_i(\mu) + \tau y_i(\mu)$ (with $\tau = -\lambda_t/\lambda_u$ and $\lambda_q = V_{qs}^* V_{qd}$). Introducing the matrix elements $\langle Q_i \rangle_I \equiv \langle (\pi\pi)_I | Q_i | K \rangle$, ε'/ε can be written as

$$\frac{\varepsilon'}{\varepsilon} = \frac{G_F}{2} \frac{\omega \, \text{Im}\lambda_t}{|\varepsilon| \, \text{Re}A_0} \left[\left| \sum_i y_i \langle Q_i \rangle_0 \right| \left(1 - \Omega_{\eta+\eta'}\right) - \frac{1}{\omega} \left| \sum_i y_i \langle Q_i \rangle_2 \right| \right] . \tag{11.3}$$

$\omega = \text{Re}A_0/\text{Re}A_2 = 22.2$ is the ratio of the CP-conserving $K \to \pi\pi$ isospin amplitudes; $\Omega_{\eta+\eta'}$ takes into account the effect of the isospin breaking in the quark masses. In an exact realization of nonperturbative QCD, the scale dependences of the y_i and $\langle Q_i \rangle_I$ in eq. (11.3) must cancel. ε'/ε is dominated by $\langle Q_6 \rangle_0$ and $\langle Q_8 \rangle_2$, which cannot be fixed from the CP-conserving data, different from most of the matrix elements of the current-current operators [4, 5]. Beside the theoretical uncertainties coming from the calculation of the $\langle Q_i \rangle_I$ and of $\Omega_{\eta+\eta'}$, the analysis of the CP ratio suffers from the uncertainties on the values of various input parameters, in particular of the CKM phase in $\text{Im}\lambda_t$, of $\Lambda_{\text{QCD}} \equiv \Lambda^{(4)}_{\overline{\text{MS}}}$, and of the strange quark mass. The uncertainty on m_s is especially important due to the density-density structure of Q_6 and Q_8, which implies that their matrix elements are proportional to the square of the quark condensate and are hence proportional to $1/m_s^2$. In all methods the calculation of the hadronic matrix element of Q_6, which is a peculiar operator, appears to be the most difficult one.

In this talk we present the results we obtained for ε'/ε together with G.O. Köhler and E.A. Paschos [6]. We briefly explain the method used to compute the hadronic matrix elements and discuss in detail the peculiarities in the calculation of the matrix element of Q_6.

11.2 General Framework

To calculate the hadronic matrix elements we start from the effective chiral Lagrangian for pseudoscalar mesons, which involves an expansion in momenta where terms up to $\mathcal{O}(p^4)$ are included [7]. Keeping only terms of $\mathcal{O}(p^4)$ that contribute, at the order we calculate, to the $K \to \pi\pi$ amplitudes, for the Lagrangian we obtain

$$\begin{aligned} \mathcal{L}_{\text{eff}} &= \frac{f^2}{4} \Big(\langle D_\mu U^\dagger D^\mu U \rangle + \frac{\alpha}{4N_c} \langle \ln U^\dagger - \ln U \rangle^2 + \langle \chi U^\dagger + U \chi^\dagger \rangle \Big) \\ &\quad + L_5 \langle D_\mu U^\dagger D^\mu U (\chi^\dagger U + U^\dagger \chi) \rangle \\ &\quad + L_8 \langle \chi^\dagger U \chi^\dagger U + \chi U^\dagger \chi U^\dagger \rangle , \end{aligned} \tag{11.4}$$

with $\langle A \rangle$ denoting the trace of A, $\alpha = m_\eta^2 + m_{\eta'}^2 - 2m_K^2$, $\chi = r\mathcal{M}$, and $\mathcal{M} = \text{diag}(m_u, m_d, m_s)$. f and r are parameters related to the pion decay constant F_π and to the quark condensate, with $r = -2\langle \bar{q}q \rangle / f^2$. The complex matrix U is a nonlinear representation of the pseudoscalar meson nonet. The conventions and definitions we use are the same as those in [6, 8, 9].

The method we use is the $1/N_c$ expansion introduced in [10]. In this approach, we expand the matrix elements in powers of the external momenta and of $1/N_c$. From the Lagrangian the mesonic representations of the quark currents and densities can be obtained by usual bosonization techniques. It is then straightforward to calculate the tree level (leading-N_c) matrix elements. For the $1/N_c$ corrections to the matrix elements $\langle Q_i \rangle_I$ we calculated chiral loops as described in [8, 9]. Especially important to this analysis are the nonfactorizable $1/N_c$ corrections, which are UV divergent and must be matched to the short-distance part. They are regularized by a finite cutoff that is identified with the short-distance renormalization scale [8–11]. The definition of the momenta in the loop diagrams, which are not momentum translation invariant, is discussed in detail in [8].

For the Wilson coefficients we use both the leading logarithmic (LO) and the next-to-leading logarithmic (NLO) values [4]. In the pseudoscalar approximation, the matching has to be done below 1 GeV. Values of the y_i at scales $0.6\,\text{GeV} \leq \mu \leq 0.9\,\text{GeV}$ were communicated to us by M. Jamin. The NLO values are scheme dependent and are calculated within naive dimensional regularization (NDR) and in the 't Hooft-Veltman scheme (HV). The absence of any reference to the renormalization scheme in the low-energy calculation, at this stage, prevents a complete matching at the next-to-leading order [12]. Nevertheless, a comparison of the numerical results obtained from the LO and NLO coefficients is useful as regards estimating the uncertainties and testing the validity of perturbation theory.

11.3 Analysis of ε'/ε

At next-to-leading order, in the twofold expansion in powers of external momenta and of $1/N_c$, we investigate the tree-level contributions from the $\mathcal{O}(p^2)$ and the $\mathcal{O}(p^4)$ Lagrangian as well as the one-loop contribution from the $\mathcal{O}(p^2)$ Lagrangian. For the matrix elements of the current-current operators, the corresponding terms are $\mathcal{O}(p^2)$, $\mathcal{O}(p^4)$, and $\mathcal{O}(p^2/N_c)$, respectively; for density-density operators they are $\mathcal{O}(p^0)$, $\mathcal{O}(p^2)$, and $\mathcal{O}(p^0/N_c)$.

Analytical formulas for all matrix elements at these orders are given in [8, 9]. Among them four are particularly interesting and important:

$$\begin{aligned}
\langle Q_1 \rangle_0 &= -\frac{1}{\sqrt{3}} F_\pi \left(m_K^2 - m_\pi^2\right) \Bigg[1 + \frac{4\hat{L}_5^r}{F_\pi^2} m_\pi^2 + \frac{1}{(4\pi)^2 F_\pi^2} \\
&\quad \times \left(6\Lambda_c^2 - \left(\frac{1}{2} m_K^2 + 6 m_\pi^2\right) \log \Lambda_c^2 + \cdots \right) \Bigg], \qquad (11.5) \\
\langle Q_2 \rangle_0 &= \frac{2}{\sqrt{3}} F_\pi \left(m_K^2 - m_\pi^2\right) \Bigg[1 + \frac{4\hat{L}_5^r}{F_\pi^2} m_\pi^2 + \frac{1}{(4\pi)^2 F_\pi^2}
\end{aligned}$$

$$\times\left(\frac{15}{4}\Lambda_c^2+\left(\frac{11}{8}m_K^2-\frac{15}{4}m_\pi^2\right)\log\Lambda_c^2+\cdots\right)\Bigg], \tag{11.6}$$

$$\langle Q_6\rangle_0 = -\frac{4\sqrt{3}}{F_\pi}R^2(m_K^2-m_\pi^2)\left[\hat{L}_5^r-\frac{3}{16\,(4\pi)^2}\log\Lambda_c^2+\cdots\right], \tag{11.7}$$

$$\begin{aligned}\langle Q_8\rangle_2 &= \frac{\sqrt{3}}{2\sqrt{2}}F_\pi R^2\Bigg[1+\frac{8m_K^2}{F_\pi^2}\left(\hat{L}_5^r-2\hat{L}_8^r\right)-\frac{4m_\pi^2}{F_\pi^2}\left(3\hat{L}_5^r-8\hat{L}_8^r\right)\\ &\quad -\frac{1}{(4\pi)^2F_\pi^2}\left(m_K^2-m_\pi^2+\frac{2}{3}\alpha\right)\log\Lambda_c^2+\cdots\Bigg],\end{aligned} \tag{11.8}$$

with $R\equiv 2m_K^2/(m_s+m_d)$. The ellipses denote finite terms that are not written explicitly here. The constants $\hat{L}_5^r$ and $\hat{L}_8^r$ are renormalized couplings whose values are $\hat{L}_5^r=2.07\cdot10^{-3}$ and $\hat{L}_8^r=1.09\cdot10^{-3}$. Eqs. (11.5–11.8) have several interesting properties [8,9]. First, the VSA values for $\langle Q_1\rangle_0$ and $\langle Q_2\rangle_0$ are far too small to account for the large $\Delta I=1/2$ enhancement observed in the CP-conserving amplitudes. Using the large-N_c limit $[B_1^{(1/2)}=3.05,\ B_2^{(1/2)}=1.22;\ B_i^{(\Delta I)}\equiv \mathrm{Re}\langle Q_i\rangle_I/\langle Q_i\rangle_I^{\mathrm{VSA}}]$ improves the agreement between theory and experiment, but it still provides a gross underestimate. However, the nonfactorizable $1/N_c$ corrections in eqs. (11.5) and (11.6) contain quadratically divergent terms that are not suppressed with respect to the tree-level contribution, since they bring about a factor of $\Delta\equiv\Lambda_c^2/(4\pi F_\pi)^2$ and have large prefactors. Varying Λ_c between 600 and 900 MeV, $B_1^{(1/2)}$ and $B_2^{(1/2)}$ take the values 8.2 – 14.2 and 2.9 – 4.6, respectively. Quadratic terms in $\langle Q_1\rangle_0$ and $\langle Q_2\rangle_0$ produce a large enhancement that brings the $\Delta I=1/2$ amplitude in agreement with the data [9]. Corrections beyond the chiral limit ($m_q=0$) in eqs. (11.5) and (11.6) are suppressed by a factor of $\delta=m_{K,\pi}^2/(4\pi F_\pi)^2\simeq 20\,\%$ and were found to be small.

For the operators Q_6 and Q_8, values rather close to the VSA $[B_6^{(1/2)}=B_8^{(3/2)}=1]$ are used in the literature. As a result the experimental range for ε'/ε can be accommodated in the standard model only if there is a conspiracy of the input parameters m_s, $\Omega_{\eta+\eta'}$, $\mathrm{Im}\lambda_t$, and Λ_{QCD} (see e.g. [5]). The fact that the VSA fails completely in explaining the $\Delta I=1/2$ rule therefore raises the question whether it can be used for Q_6 and Q_8. In fact, at the present stage of the calculation, the case of $\langle Q_6\rangle_0$ and $\langle Q_8\rangle_2$ is different from that of $\langle Q_{1,2}\rangle_0$. The leading-N_c values are very close to the corresponding VSA values. Moreover, the nonfactorizable loop corrections in eqs. (11.7) and (11.8), which are of $\mathcal{O}(p^0/N_c)$, are found to be only logarithmically divergent [8]. Consequently, in the case of $\langle Q_8\rangle_2$ they are suppressed by a factor of δ compared to the leading $\mathcal{O}(p^0)$ term and are expected to be of the order of 20 % to 50 % depending on the prefactors. We note that eq. (11.8) is a full leading plus next-to-leading order analysis of the Q_8 matrix element. The case of $B_6^{(1/2)}$ is more complicated since the $\mathcal{O}(p^0)$ term vanishes for Q_6. Nevertheless, the nonfactorizable loop corrections to this term remain and have to be matched to the short-distance part of the amplitudes [8]. These $\mathcal{O}(p^0/N_c)$ nonfactorizable corrections must be considered at the same level, in the twofold expansion, as the $\mathcal{O}(p^2)$ tree contribution. Consequently, a value of $B_6^{(1/2)}$ around one [which corresponds to the $\mathcal{O}(p^2)$ term alone] is not a priori expected. However, numerically it turns out that the $\mathcal{O}(p^0/N_c)$ contribution is only moderate. This property can be understood from the $(U^\dagger)_{dq}(U)_{qs}$ structure of the Q_6 operator, which vanishes to $\mathcal{O}(p^0)$ implying that the factorizable and nonfactorizable $\mathcal{O}(p^0/N_c)$ contributions cancel to a large extent [8]. This explains why the deviations we observe with respect to the VSA values are smaller than for Q_1 and Q_2 and why in particular for Q_6 to $\mathcal{O}(p^0/N_c)$ we do not observe a $\Delta I = 1/2$ enhancement. Varying Λ_c between 600 and 900 MeV, $B_6^{(1/2)}$ and $B_8^{(3/2)}$ take the values $1.10 - 0.72$ and $0.64 - 0.42$, respectively. $B_6^{(1/2)}$ and $B_8^{(3/2)}$ are therefore more efficiently protected from possible large $1/N_c$ corrections of the $\mathcal{O}(p^2)$ Lagrangian than $B_{1,2}^{(1/2)}$. The effect of the $\mathcal{O}(p^0/N_c)$ term is, however, important for $B_6^{(1/2)}$ as for $B_8^{(3/2)}$ because it gives rise to a noticeable dependence on the cutoff scale [8]. We note that $B_8^{(3/2)}$ shows a scale dependence that is very similar to the one of $B_6^{(1/2)}$ leading to a stable ratio $B_6^{(1/2)}/B_8^{(3/2)} \simeq 1.72$ for Λ_c between 600 and 900 MeV. The $\mathcal{O}(p^0/N_c)$ corrections consequently make the cancellation of Q_6 and Q_8 in ε'/ε less effective, but the values of the matrix elements are reduced. Hence we obtain values for ε'/ε [6] close to the ones found with the VSA values of $B_6^{(1/2)}$ and $B_8^{(3/2)}$; in particular, for central values of the input parameters [$m_s(1\text{ GeV})$=150 MeV, $\Omega_{\eta+\eta'} = 0.25$, $\text{Im}\lambda_t = 1.33 \cdot 10^{-4}$, and $\Lambda_{\text{QCD}} = 325$ MeV, see [6] and references therein] the values for the CP ratio are significantly smaller than the data.

However, since the leading $\mathcal{O}(p^0)$ contribution vanishes for Q_6, corrections from higher-order terms beyond the $\mathcal{O}(p^2)$ and $\mathcal{O}(p^0/N_c)$ are expected to be large [6]. The terms of $\mathcal{O}(p^2)$ and $\mathcal{O}(p^0/N_c)$ correspond to the lowest (nonvanishing) order, and the calculation of the next order terms is very desirable. In the twofold expansion, the higher-order corrections to the matrix element of Q_6 are of orders $\mathcal{O}(p^4)$, $\mathcal{O}(p^0/N_c^2)$, and $\mathcal{O}(p^2/N_c)$. A full calculation of these terms is beyond the scope of our study. In particular, higher-order terms in the p^2 expansion, which are chirally suppressed, cannot be calculated because the low-energy couplings in the $\mathcal{O}(p^6)$ Lagrangian are very uncertain or even unknown. In [6] we investigated the $\mathcal{O}(p^2/N_c)$ contribution, i.e., the $1/N_c$ correction at the next order in the chiral expansion, because it brings about, for the first time, quadratic corrections on the cutoff. We remind our readers that for the CP-conserving amplitude it is mainly the (quadratic) $\mathcal{O}(p^2/N_c)$ corrections that bring to $\langle Q_{1,2}\rangle_0$ a large enhancement relative to the (leading-N_c) $\mathcal{O}(p^2)$ values. As the leading-N_c value for Q_6 is also $\mathcal{O}(p^2)$ we cannot a priori exclude that the value of $\langle Q_6\rangle_0$ is largely affected by $\mathcal{O}(p^2/N_c)$ corrections too, removing from Q_6 the property observed to $\mathcal{O}(p^0/N_c)$, to be protected from large $1/N_c$ corrections. As explained above, quadratic $\mathcal{O}(p^2/N_c)$ corrections are proportional to the factor $\Delta \equiv \Lambda_c^2/(4\pi F_\pi)^2$ relative to the $\mathcal{O}(p^2)$ tree-level contribution. Different is the case of Q_8 since its leading-N_c value is $\mathcal{O}(p^0)$ at lowest order in the chiral expansion. Quadratic terms for Q_8 are consequently chirally suppressed with respect to the leading-N_c value.

Calculating the quadratic term of $\mathcal{O}(p^2/N_c)$ for matrix element of Q_6 and adding it to the $\mathcal{O}(p^2)$ and $\mathcal{O}(p^0/N_c)$ result in eq. (11.7) we obtain:

$$\langle Q_6\rangle_0 = -\frac{4\sqrt{3}}{F_\pi} R^2 (m_K^2 - m_\pi^2)\left[\hat{L}_5^r\left(1 + \frac{3}{2}\frac{\Lambda_c^2}{(4\pi)^2 F_\pi^2}\right) - \frac{3}{16}\frac{\log \Lambda_c^2}{(4\pi)^2} + \cdots\right]. \tag{11.9}$$

The result for the $\mathcal{O}(p^2/N_c)$ term we already presented in [13]. Numerically, we observe a large positive correction from the quadratic term in eq. (11.9). The slope of this correction is qualitatively consistent and welcome since it compensates for the logarithmic decrease at $\mathcal{O}(p^0/N_c)$. Varying Λ_c between 600 and 900 MeV, the $B_6^{(1/2)}$ factor takes the values $1.50 - 1.62$. The approximate stability of $B_6^{(1/2)}$ is in accordance with the perturbative evolution, since the nondiagonal dependence of y_6 on the renormalization scale, i.e., the one beyond the leading-N_c scale dependence of R^2 in eq. (11.9), was found to be small. The quadratic term of $\mathcal{O}(p^2/N_c)$ is of the same magnitude as the $\mathcal{O}(p^2)$ tree term. Q_6 is a $\Delta I = 1/2$ operator, and the enhancement of $\langle Q_6\rangle_0$ indicates that at the level of the $1/N_c$ corrections the dynamics of the $\Delta I = 1/2$ rule applies to Q_6 as to $Q_{1,2}$. One might, however, note that the enhancement observed for Q_6 is smaller than for Q_1 and Q_2.

	Case 1	Case 2
LO	$19.5 \leq \varepsilon'/\varepsilon \leq 24.7$	$14.8 \leq \varepsilon'/\varepsilon \leq 19.4$
NDR	$16.1 \leq \varepsilon'/\varepsilon \leq 23.4$	$12.5 \leq \varepsilon'/\varepsilon \leq 18.3$
HV	$9.3 \leq \varepsilon'/\varepsilon \leq 19.3$	$7.0 \leq \varepsilon'/\varepsilon \leq 14.9$
LO+NDR+HV	$9.3 \leq \varepsilon'/\varepsilon \leq 24.7$	$7.0 \leq \varepsilon'/\varepsilon \leq 19.4$

Table 11.1: Central ranges for ε'/ε (in units of 10^{-4}).

Using the quoted values for $B_6^{(1/2)}$ together with the full leading plus next-to-leading order B factors for the remaining operators [6] we calculated ε'/ε for central values of m_s, $\Omega_{\eta+\eta'}$, $\mathrm{Im}\lambda_t$, and Λ_{QCD}. The results for the three sets of Wilson coefficients LO, NDR, and HV and for Λ_c between 600 and 900 MeV are given in table 11.1. The numbers are obtained with two different methods for analyzing the sensitivity on the imaginary part coming from the final states interactions. In the first case, we use the real part of our calculation and the phenomenological phases $\delta_0 = (34.2 \pm 2.2)^\circ$ and $\delta_2 = (-6.9 \pm 0.2)^\circ$ [14], and replace $|\sum_i y_i \langle Q_i\rangle_I|$ in eq. (11.3) by $\sum_i y_i \mathrm{Re}\langle Q_i\rangle_I / \cos\delta_I$ [15]. In the second case, we use only the real part assuming zero phases. The latter case is very close to the results we would get if we used the small imaginary part obtained at the one-loop level [6]. Collecting together the LO, NDR, and HV results for the two cases and for central values of the parameters, we find the following (conservative) range:

$$7.0 \cdot 10^{-4} \;\leq\; \varepsilon'/\varepsilon\,(\text{central}) \;\leq\; 24.7 \cdot 10^{-4}\,,$$

which is in the ball park of the experimental result in eq. (11.1). Performing a complete scanning of the parameters [$125\,\text{MeV} \leq m_s(1\,\text{GeV}) \leq 175\,\text{MeV}$, $0.15 \leq \Omega_{\eta+\eta'} \leq 0.35$, $1.04 \cdot 10^{-4} \leq \text{Im}\lambda_t \leq 1.63 \cdot 10^{-4}$, and $245\,\text{MeV} \leq \Lambda_{\text{QCD}} \leq 405\,\text{MeV}$] we obtain $2.2 \cdot 10^{-4} \leq \varepsilon'/\varepsilon\,(\text{scanned}) \leq 63.2 \cdot 10^{-4}$ (see table 11.2). The numerical values in the tables can be compared with the results of the Munich, Trieste, and Rome groups [5, 15, 16]. The values for $B_6^{(1/2)}$ can also be compared with [17]. The large ranges reported in table 11.2 can be traced back to the large ranges of the input parameters. The parameters, to a large extent, act multiplicatively, and the larger range for ε'/ε is due to the fact that the central value(s) for the ratio are enhanced roughly by a factor of two compared to the results obtained with B factors for Q_6 and Q_8 close to the VSA. More accurate information on the parameters, from theory and experiment, will restrict the values for the CP ratio.

	Case 1	Case 2
LO	$8.0 \leq \varepsilon'/\varepsilon \leq 62.1$	$6.1 \leq \varepsilon'/\varepsilon \leq 48.5$
NDR	$6.8 \leq \varepsilon'/\varepsilon \leq 63.9$	$5.2 \leq \varepsilon'/\varepsilon \leq 49.8$
HV	$2.8 \leq \varepsilon'/\varepsilon \leq 49.8$	$2.2 \leq \varepsilon'/\varepsilon \leq 38.5$
LO+NDR+HV	$2.8 \leq \varepsilon'/\varepsilon \leq 63.9$	$2.2 \leq \varepsilon'/\varepsilon \leq 49.8$

Table 11.2: Same as in table 11.1, but for the complete scanning of the parameters.

11.4 Summary

We have shown that the operator Q_6, similar to Q_1 and Q_2, is not protected from large $1/N_c$ corrections coming from quadratic terms of $\mathcal{O}(p^2/N_c)$. From general counting arguments we have good indications that among the various next-to-leading order terms in the p^2 and $1/N_c$ expansions they are the dominant ones. Calculating those terms we find that they enhance $B_6^{(1/2)}$ and bring ε'/ε much closer to the data for central values of the parameters. We obtain a quadratic evolution for Q_6 that indicates that a $\Delta I = 1/2$ enhancement is operative for Q_6 similar to the one of Q_1 and Q_2. $B_8^{(3/2)}$ is expected to be affected much less by terms of $\mathcal{O}(p^2/N_c)$ due to an extra p^2 suppression factor relative to the leading $\mathcal{O}(p^0)$ tree term.

One should recall that our analysis of the $\mathcal{O}(p^2/N_c)$ terms for Q_6 is performed in the chiral limit. It would be desirable to calculate the corrections beyond the chiral limit, from logarithms and finite terms. It would also be interesting to investigate the effect of higher resonances. Each of the additional effects separately is not expected to counteract largely the enhancement found for $B_6^{(1/2)}$. Nevertheless, in the extreme (and unlikely) case where all these effects would come with the same sign a significant modification of the result cannot be excluded formally. In order to reduce the scheme dependence in the result for ε'/ε, appropriate subtractions would be necessary (see [17,18]).

Acknowledgments

The talk is based on the work carried through with G.O. Köhler and E.A. Paschos [6]. This work was supported in part by BMBF, 057D093P(7), Bonn, FRG, and DFG Antrag PA-10-1. T.H. acknowledges partial support from EEC, TMR-CT980169.

References

[1] A. Alavi-Harati *et al.*, Phys. Rev. Lett. **83**, 22 (1999); Y.B. Hsiung, this volume, chapter 6.

[2] M.S. Sozzi (NA48 Collaboration), this volume, chapter 8.

[3] G.D. Barr *et al.*, Phys. Lett. B **317**, 233 (1993).

[4] A.J. Buras, M. Jamin, M.E. Lautenbacher, and P.H. Weisz, Nucl. Phys. B **370**, 69 (1992), *ibid.* B **400**, 37 (1993); A.J. Buras, M. Jamin, and M.E. Lautenbacher, Nucl. Phys. B **400**, 75 (1993); *ibid.* B **408**, 209 (1993). See also: M. Ciuchini, E. Franco, G. Martinelli, and L. Reina, Phys. Lett. B **301**, 263 (1993); Nucl. Phys. B **415**, 403 (1994).

[5] S. Bosch, A.J. Buras, M. Gorbahn, S. Jäger, M. Jamin, M.E. Lautenbacher, and L. Silvestrini, Nucl. Phys. B **565**, 3 (2000); A.J. Buras, this volume, chapter 5.

[6] T. Hambye, G.O. Köhler, E.A. Paschos, and P.H. Soldan, Nucl. Phys. B **564**, 391 (2000).

[7] J. Gasser and H. Leutwyler, Nucl. Phys. B **250**, 465 (1985).

[8] T. Hambye, G.O. Köhler, E.A. Paschos, P.H. Soldan, and W.A. Bardeen, Phys. Rev. D **58**, 014017 (1998); T. Hambye and P. Soldan, hep-ph/9806203, talk at CPMASS 97, Lisbon, Portugal, 6–15 Oct 1997.

[9] T. Hambye, G.O. Köhler, and P. Soldan, hep-ph/9902334 (to appear in Eur. Phys. J. C; T. Hambye, talk at USTRON97, Ustron, Poland, 19–24 Sep. 1997, Acta Phys. Polon. B **28**, 2479 (1997).

[10] W.A. Bardeen, A.J. Buras, and J.-M. Gérard, Phys. Lett. B **180**, 133 (1986); Nucl. Phys. B **293**, 787 (1987); Phys. Lett. B **192**, 138 (1987).

[11] J. Heinrich, E.A. Paschos, J.-M. Schwarz, and Y.L. Wu, Phys. Lett. B **279**, 140 (1992); E.A. Paschos, talk at the 17th Lepton-Photon Symposium, Beijing, China, Aug. 1995.

[12] A.J. Buras, in *Probing the Standard Model of Particle Interactions*, ed. F. David and R. Gupta (Elsevier Science, Amsterdam, 1998).

[13] P. Soldan, hep-ph/9608281, talk at the Workshop on K Physics, Orsay, France, Jun. 1996.

[14] E. Chell and M.G. Olsson, Phys. Rev. D **48**, 4076 (1993).

[15] S. Bertolini, J.O. Eeg, and M. Fabbrichesi, hep-ph/9802405; S. Bertolini, J.O. Eeg, M. Fabbrichesi, and E.I. Lashin, Nucl. Phys. B **514**, 93 (1998).

[16] M. Ciuchini *et al.*, this volume, chapters 28 and 31.

[17] J. Bijnens and J. Prades, JHEP 9901:023 (1999); J. Bijnens, this volume, chapter 37 (hep-ph/9907514).

[18] W.A. Bardeen, Nucl. Phys. Proc. Suppl. A **7**, 149 (1989).

12

Estimating ε'/ε in the Standard Model

S. Bertolini

Abstract

I discuss the comparison of the current theoretical calculations of ε'/ε with the experimental data. Lacking reliable "first principle" calculations, phenomenological approaches may help in understanding correlations among different contributions and available experimental data. In particular, in the chiral quark model approach the same dynamics that underlies the $\Delta I = 1/2$ selection rule in kaon decays appears to enhance the $K \to \pi\pi$ matrix element of the Q_6 gluonic penguin, thus driving ε'/ε in the range of the recent experimental measurements.

The results announced by the KTeV Collaboration last February and by the NA48 Collaboration at this conference [1] (albeit preliminary) have marked a great experimental achievement, establishing 35 years after the discovery of CP violation in the neutral kaon system the existence of a much smaller violation acting directly in the decays.

While the standard model of strong and electroweak interactions provides an economical and elegant understanding of the presence of indirect (ε) and direct (ε') CP violation in terms of a single phase, the detailed calculation of the size of these effects implies mastering strong interactions at a scale where perturbative methods break down. In addition, CP violation in $K \to \pi\pi$ decays is the result of a destructive interference between two sets of contributions (for a suggestive picture of the gluonic and electroweak penguin diagrams see the talk by Buras at this conference [2]), thus potentially inflating up to an order of magnitude the uncertainties on the individual hadronic matrix elements of the effective four-quark operators.

In fig. 12.1, taken from ref. [3], the comparison of the theoretical predictions and the experimental results available before the Kaon 99 conference is summarized. The gray horizontal band shows the two-sigma experimental range obtained averaging the recent KTeV result with the older NA31 and E731 data, corresponding to $\varepsilon'/\varepsilon = (21.8 \pm 3) \times 10^{-4}$. The vertical

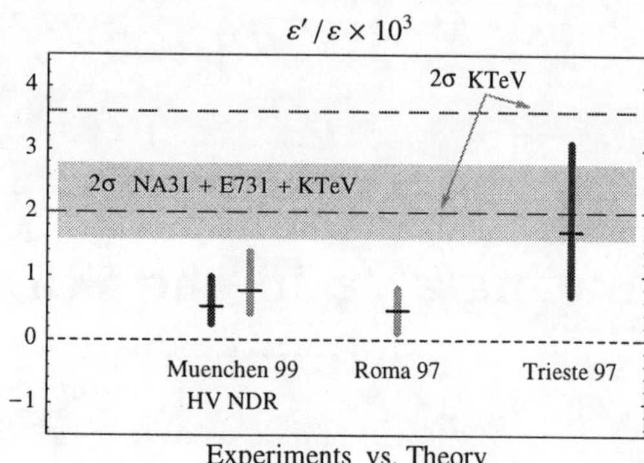

Figure 12.1: The recent KTeV result (2σ) is shown by the area enclosed by the long-dashed lines. The combined 2σ average of the KTeV, NA31 and E731 results is shown by the gray band. The predicted München, Roma and Trieste theoretical ranges for ε'/ε are shown by the vertical bars with their central values.

lines show the ranges of the most recent published theoretical predictions, identified with the cities where most of the group members reside. The figure does not include two new results announced at this conference: on the experimental side, the first NA48 measurement [1] and, on the theoretical side, the new prediction based on the $1/N$ expansion [4], which I will refer to in the following as the Dortmund group estimate. The inclusion of the NA48 result $\varepsilon'/\varepsilon = (18.5 \pm 7.3) \times 10^{-4}$ lowers the experimental average shown in fig. 12.1 by about 4%.

Looking at fig. 12.1 two comments are in order. On the one hand, we should appreciate the fact that within the uncertainties of the theoretical calculations, there is indeed an overall agreement among the different predictions. All of them agree on the presence of a nonvanishing positive effect in the standard model. On the other hand, the central values of the Munich (phenomenological $1/N$) and Rome (lattice) calculations are lower by a factor 3 to 5 than the averaged experimental central value.

In spite of the complexity of the calculations, I would like to emphasize that the difference between the predictions of the two estimates above and that of the Trieste group, based on the Chiral Quark Model (χQM) [5], is mainly due to the different size of the hadronic matrix element of the gluonic penguin Q_6. In addition, I will show that the enhancement of the Q_6 matrix element in thc χQM approach can be simply understood in terms of chiral dynamics and, in this respect, it is related to the phenomenological embedding of the $\Delta I = 1/2$ selection rule.

The $\Delta I = 1/2$ selection rule in $K \to \pi\pi$ decays has been known for some 40 years [6] and it states the fact that kaons are 400 times more likely to decay in the isospin zero two-pion state than in the isospin two component. This rule is not justified by any symmetry consideration and,

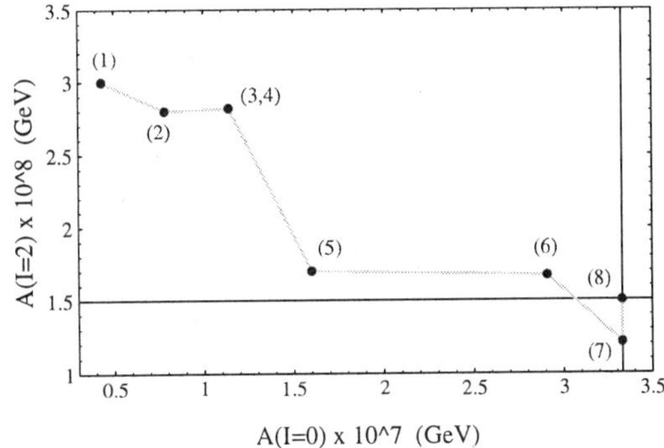

Figure 12.2: Anatomy of the $\Delta I = 1/2$ rule in the χQM [8]. See the text for explanations. The cross-hairs indicate the experimental point.

although it is common understanding that its explanation must be rooted in the dynamics of strong interactions, there is no up-to-date derivation of this effect from first-principle QCD.

As summarized by Martinelli at this conference [7] lattice gauge theory cannot provide us at present with reliable calculations of the $I = 0$ penguin operators relevant to ε'/ε, as well as of the $I = 0$ components of the hadronic matrix elements of the tree-level current-current operators (penguin contractions), which are relevant for the $\Delta I = 1/2$ selection rule.

In the Munich approach the $\Delta I = 1/2$ rule is used in order to determine phenomenologically the matrix elements of $Q_{1,2}$ and, via operatorial relations, some of the matrix elements of the left-handed penguins. Unfortunately, the approach does not allow for a phenomenological determination of the matrix elements of the penguin operators that are most relevant for ε'/ε, namely the gluonic penguin Q_6 and the electroweak penguin Q_8. Values in the ballpark of the leading $1/N$ estimate are assumed for these matrix elements, taking also into account that all present approaches show a suppression of $\langle Q_8 \rangle$ with respect to its vacuum saturation approximation (VSA).

In the χQM approach, the hadronic matrix elements can be computed as an expansion in momenta in terms of three parameters: the constituent quark mass, the quark condensate, and the gluon condensate. The Trieste group has computed the $K \to \pi\pi$ matrix elements of the $\Delta S = 1, 2$ effective Lagrangian up to $O(p^4/N)$ in the chiral and $1/N$ expansions [8].

Hadronic matrix elements and short-distance Wilson coefficients are matched at a scale of 0.8 GeV as a reasonable compromise between the ranges of validity of perturbation theory and chiral Lagrangian. By requiring the $\Delta I = 1/2$ rule to be reproduced within a 20% uncertainty one obtains a phenomenological determination of the three basic parameters of the model. This step is needed in order to make the model predictive, since

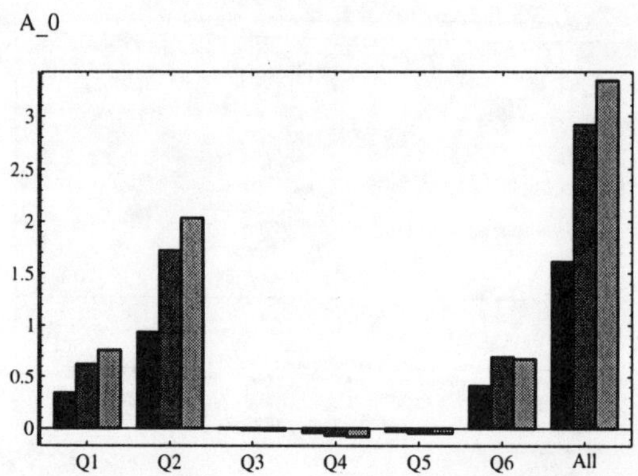

Figure 12.3: Anatomy of the $A(K^0 \to \pi\pi)_{I=0}$ amplitude (A_0) in units of 10^{-7} GeV for central values of the χQM input parameters [8]: $O(p^2)$ calculation (black), with minimally subtracted chiral loops (half-tone), complete $O(p^4)$ result (light gray).

there is no a priori argument for the consistency of the matching procedure. As a matter of fact, all computed observables turn out to be very weakly scale dependent in a few hundred MeV range around the matching scale.

Fig. 12.2 shows an anatomy of the (model-dependent) contributions that lead in the Trieste approach to reproducing the $\Delta I = 1/2$ selection rule.

Point (1) represents the result obtained by neglecting QCD and taking the factorized matrix element for the tree-level operator Q_2, which is the only one present. The ratio A_0/A_2 is found equal to $\sqrt{2}$: a long way from the experimental point (8). Step (2) includes the effects of perturbative QCD renormalization on the operators $Q_{1,2}$ [9]. Step (3) shows the effect of including the gluonic penguin operators [10]. Electroweak penguins [11] are numerically negligible for the CP-conserving amplitudes and are responsible for the very small shift in the A_2 direction. Therefore, perturbative QCD and factorization lead us from (1) to (4).

Nonfactorizable gluon-condensate corrections, a crucial model-dependent effect, enter at the leading order in the chiral expansion leading to a substantial reduction of the A_2 amplitude (5), as first observed by Pich and de Rafael [12]. Moving the analysis to $O(p^4)$ the chiral loop corrections, computed on the LO chiral Lagrangian via dimensional regularization and minimal subtraction, lead us from (5) to (6), while the corresponding $O(p^4)$ tree-level counterterms calculated in the χQM lead to the point (7). Finally, step (8) represents the inclusion of π-η-η' isospin breaking effects [13].

This model-dependent anatomy shows the relevance of nonfactorizable contributions and higher-order chiral corrections. The suggestion that chiral dynamics may be relevant to the understanding of the $\Delta I = 1/2$ se-

B_i	München 99 $\mu = 1.3$ GeV	Roma 97 $\mu = 2.0$ GeV	Trieste 97 $\mu = 0.8$ GeV
$B_1^{(0)}$	13 (†)	–	9.5
$B_2^{(0)}$	6.1 (†)	–	2.9
$B_1^{(2)} = B_2^{(2)}$	0.48 (†)	–	0.41
B_3	1 (*)	1 (*)	-2.3
B_4	5.2 (* †)	$1 \div 6$ (*)	1.9
$B_5 \simeq B_6$	1.0 ± 0.3 (*)	1.0 ± 0.2	1.6 ± 0.3
$B_7^{(0)} \simeq B_8^{(0)}$	1 (*)	1 (*)	2.5 ± 0.1
$B_9^{(0)}$	7.0 (* †)	1 (*)	3.6
$B_{10}^{(0)}$	7.5 (* †)	1 (*)	4.4
$B_7^{(2)}$	1 (*)	0.6 ± 0.1	0.92 ± 0.02
$B_8^{(2)}$	0.8 ± 0.15 (*)	0.8 ± 0.15	0.92 ± 0.02
$B_9^{(2)}$	0.48 (†)	0.62 ± 0.10	0.41
$B_{10}^{(2)}$	0.48 (†)	1 (*)	0.41
$\hat{B}_K$	0.80 ± 0.15	0.75 ± 0.15	1.1 ± 0.2

Table 12.1: Summary of B factors. Legend: (*) educated guess, (†) derived from the $\Delta I = 1/2$ rule. In the Trieste calculation the $\Delta I = 1/2$ rule is used to constrain the three basic model parameters in terms of which all matrix elements are computed.

lection rule goes back to the work of Bardeen, Buras, and Gérard [14, 15] in the $1/N$ framework using a cutoff regularization. This approach has been recently revived and improved by the Dortmund group, with particular attention to the matching procedure [4, 15]. A pattern similar to that shown in fig. 12.2 for the chiral loop corrections to A_0 and A_2 was previously obtained in a NLO chiral Lagrangian analysis, using dimensional regularization, by Missimer, Kambor, and Wyler [16].

The χQM approach allows us to further investigate the relevance of chiral corrections for each of the effective quark operators of the $\Delta S = 1$ Lagrangian. Fig. 12.3 shows the contributions to the CP-conserving amplitude A_0 of the relevant operators, providing us with a finer (model-dependent) anatomy of the NLO chiral corrections. From fig. 12.3 we notice that, because of the chiral loop enhancement, the Q_6 contribution to A_0 is about 20% of the total amplitude. As we shall see, the $\mathcal{O}(p^4)$ enhancement of the Q_6 matrix element is what drives ε'/ε in the χQM to the 10^{-3} ballpark.

A commonly used way of comparing the estimates of hadronic matrix elements in different approaches is via the so-called B factors, which represent the ratio of the model matrix elements to the corresponding VSA values. However, care must be taken in the comparison of different models due to the scale dependence of the Bs and the values used by different

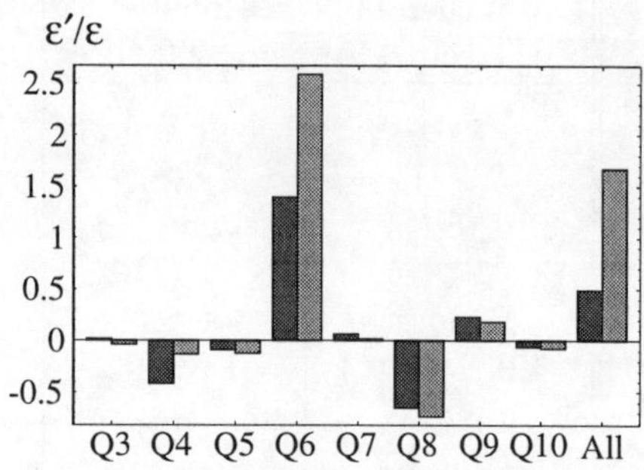

Figure 12.4: Predicting ε'/ε: a (penguin) comparative anatomy of the Munich (dark gray) and Trieste (light gray) results (in units of 10^{-3}).

groups for the parameters that enter the VSA expressions. Table 12.1 reports the B factors used for the predictions shown in fig. 12.1.

An alternative pictorial and synthetic way of analyzing different outcomes for ε'/ε is shown in fig. 12.4, where a "comparative anatomy" of the Trieste and Munich estimates is presented.

From the inspection of the various contributions it is apparent that the final difference on the central value of ε'/ε is almost entirely due to the difference in the Q_6 component. In second order, a larger (negative) contribution of the Q_4 penguin in the Munich calculation goes in the direction of making ε'/ε smaller.

The difference in the Q_4 contribution is easily understood. In the Munich estimate the Q_4 matrix element is obtained using the operatorial relation $\langle Q_4 \rangle = \langle Q_2 \rangle_0 - \langle Q_1 \rangle_0 + \langle Q_3 \rangle$, together with the knowledge acquired on $\langle Q_{1,2} \rangle_0$ from fitting the $\Delta I = 1/2$ selection rule at the charm scale.

As a matter of fact, the phenomenological fit of $\Delta I = 1/2$ rule requires a large value of $\langle Q_2 \rangle_0 - \langle Q_1 \rangle_0$ (which deviates by up to an order of magnitude from the naive VSA estimate). The assumption that $\langle Q_3 \rangle$ is given by its VSA value leads, in the Munich analysis, to a large value of $\langle Q_6 \rangle$: about 5 times larger than its VSA value. On the other hand, in the χQM calculation $\langle Q_3 \rangle$ turns out to have a sign opposite to its VSA expression, in such a way that a smaller value for Q_4 is obtained. A lattice calculation of all gluonic penguins is definitely needed to disentangle such patterns.

At any rate, the main difference between the ε'/ε central values obtained in the Trieste and Munich calculations rests in the Q_6 matrix element. The nature of the difference is apparent in fig. 12.5, where the various penguin contributions to ε'/ε in the Trieste analysis are further separated in LO (dark histograms) and NLO components—chiral loops (gray histograms) and $O(p^4)$ tree-level counterterms (dark histograms).

It is clear that chiral loop dynamics plays a subleading role in the electroweak penguin sector (Q_{8-10}) while enhancing by 60% the Q_6 matrix

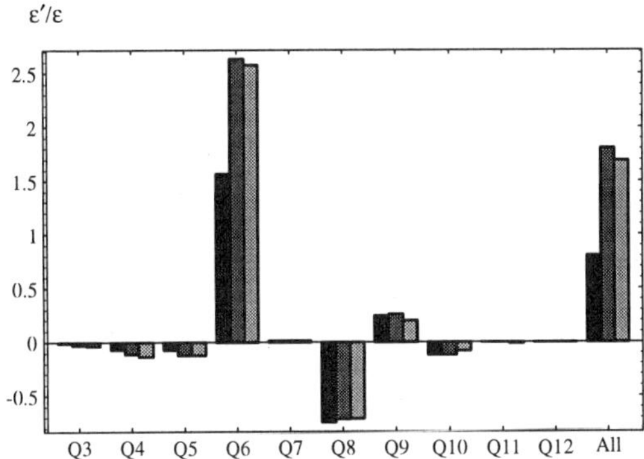

Figure 12.5: Anatomy of ε'/ε (in units of 10^{-3}) within the χQM approach [8]. In black the LO results (which includes the nonfactorizable gluonic corrections), in half-tone the effect of the inclusion of chiral-loop corrections, and in light gray the complete $O(p^4)$ estimate.

element. At $O(p^2)$ the χQM prediction for ε'/ε would just overlap with the Munich estimate once the small effect of the Q_4 operator is taken into account.

The χQM analysis shows that the same dynamics that is relevant to the reproduction of the CP-conserving A_0 amplitude (fig. 12.3) is at work also in the CP-violating sector (gluonic penguins).

In order to ascertain whether the model features represent real QCD effects one should wait for future improvements in lattice calculations. On the other hand, indications for such a dynamics arise from current $1/N$ calculations [4, 17].

The idea of a connection between the $\Delta I = 1/2$ selection rule and ε'/ε is certainly not new [18, 19], although at the GeV scale, where one can trust perturbative QCD, penguins are far from providing the dominant contribution to the CP-conserving amplitudes.

Before concluding, I would like to make a comment on the role of the strange quark mass in the χQM calculation of ε'/ε: in such an approach the basic parameter that enters the relevant penguin matrix elements is the quark condensate and the explicit dependence on m_s appears at the NLO in the chiral expansion. Varying the central value of $\bar{m}_s(m_c)$ from 150 MeV to 130 MeV affects $\langle Q_6 \rangle$ and $\langle Q_8 \rangle$ at the few percent level.

A more sensitive quantity is $\hat{B}_K$, which parameterizes the $\bar{K}-K$ matrix element. This parameter, which equals unity in the VSA, turns out to be quite sensitive to $SU(3)$ breaking effects. Taking $\bar{m}_s(m_c) = 130 \pm 20$ MeV, $\Lambda^{(4)}_{\rm QCD} = 340 \pm 40$ MeV and varying all relevant parameters, the updated χQM result is

$$\hat{B}_K = 1.0 \pm 0.2,$$

to be compared with the value used in the 1997 analysis (table 12.1).

This increases the previous determination of $\mathrm{Im}\,\lambda_t$ by roughly 10% and correspondingly ε'/ε (an updated analysis of ε'/ε in the χQM with Gaussian treatment of experimental inputs is in progress).

I conclude by summarizing with the relevant remarks:

- Phenomenological approaches that embed the $\Delta I = 1/2$ selection rule in $K \to \pi\pi$ decays generally agree with present lattice calculations in the pattern and size of the $I = 2$ components of the $\Delta S = 1$ hadronic matrix elements.

- Concerning the $I = 0$ matrix elements, where lattice calculations suffer from large systematic uncertainties, the $\Delta I = 1/2$ rule forces upon us large deviations from the naive VSA (see table 12.1).

- In the Chiral Quark Model calculation, the fit of the CP-conserving $K \to \pi\pi$ amplitudes, which determines the three basic parameters of the model, feeds down to the penguin sectors showing a substantial enhancement of the Q_6 matrix element, such that $B_6/B_8^{(2)} \approx 2$. This is what drives the ε'/ε prediction in the 10^{-3} ballpark.

- Up to 40% of the present uncertainty in the ε'/ε prediction arises from the uncertainty in the CKM elements $\mathrm{Im}(V_{ts}^* V_{td})$, which is presently controlled by the $\Delta S = 2$ parameter B_K. A better determination of the unitarity triangle from B physics is expected from the B factories and hadronic colliders. From K physics, the rare decay $K_L \to \pi^0 \nu\bar{\nu}$ gives the cleanest "theoretical" determination of $\mathrm{Im}\,\lambda_t$ [2].

- In spite of clever and interesting new-physics explanations for a large ε'/ε [20–23], it is premature to interpret the present "disagreement" between theory and experiment as a signal of physics beyond the standard model. Ungauged systematic uncertainties presently affect all theoretical estimates. Not to forget the long-standing puzzle of the $\Delta I = 1/2$ rule: perhaps an "anomalously" large ε'/ε ($B_6/B_8^{(2)} \approx 2$) is just the CP-violating projection of $A_0/A_2 \approx 20$.

Acknowledgments

My appreciation goes to the organizers of the Kaon 99 Conference for assembling such a stimulating scientific program and for the efficient logistic organization. Finally, I thank J.O. Eeg and M. Fabbrichesi for a most enjoyable and fruitful collaboration.

References

[1] M.S. Sozzi, this volume, chapter 8.

[2] A.J. Buras, this volume, chapter 5.

[3] S. Bertolini, J.O. Eeg, and M. Fabbrichesi, hep-ph/9802405, Rev. Mod. Phys. **72**, 65 (2000).

[4] T. Hambye and P.H. Soldan, this volume, chapter 11.

[5] S. Weinberg, Physica A **96**, 327 (1979); A. Manhoar and H. Georgi, Nucl. Phys. B **234**, 189 (1984).

[6] M. Gell-Mann and A. Pais, Proc. Glasgow Conf. p. 342, 1955.

[7] M. Ciuchini *et al.*, this volume, chapter 28.

[8] S. Bertolini, J.O. Eeg, M. Fabbrichesi, and E.I. Lashin, Nucl. Phys. B **514**, 63 (1998); Nucl. Phys. B **514**, 93 (1998).

[9] M.K. Gaillard and B. Lee, Phys. Rev. Lett. **33**, 108 (1974); G. Altarelli and L. Maiani, Phys. Lett. B **52**, 413 (1974).

[10] M.A. Shifman, A.I. Vainshtein, and V.I. Zacharov, JETP Lett. **22**, 55 (1975); Nucl. Phys. B **120**, 316 (1977); Sov. Phys. JETP **45**, 670 (1977); J. Ellis, M.K. Gaillard, D.V. Nanopoulos, and S. Rudaz, Nucl. Phys. B **131**, 285 (1977), (E) B **132**, 541 (1978); F.J. Gilman and M.B. Wise, Phys. Lett. B **83**, 83 (1979); Phys. Rev. D **20**, 2392 (1979); R.S. Chivukula, J.M. Flynn, and H. Georgi, Phys. Lett. B **171**, 453 (1986).

[11] J. Bijnens and M.B. Wise, Phys. Lett. B **137**, 245 (1984); A.J. Buras and J.M. Gérard, Phys. Lett. B **192**, 156 (1987); M. Lusignoli, Nucl. Phys. B **325**, 33 (1989); J.M. Flynn and L. Randall, Phys. Lett. B **224**, 221 (1989).

[12] A. Pich and E. de Rafael, Nucl. Phys. B **358**, 311 (1991).

[13] B. Holstein, Phys. Rev. D **20**, 1187 (1979); J.F. Donoghue, E. Golowich, B.R. Holstein, and J. Trampetic, Phys. Lett. B **179**, 361 (1986); A.J. Buras and J.M. Gerard, in ref. [11].

[14] W.A. Bardeen, A.J. Buras, and J.M. Gérard, Nucl. Phys. B **293**, 787 (1987).

[15] W.A. Bardeen, this volume, chapter 14.

[16] J. Kambor, J. Missimer, and D. Wyler, Nucl. Phys. B **346**, 17 (1990); Phys. Lett. B **261**, 496 (1991).

[17] J. Bijnens, this volume, chapter 37.

[18] M.A. Shifman, A.I. Vainshtein, and V.I. Zacharov, in ref. [10].

[19] F.J. Gilman and M.B. Wise, in ref. [10].

[20] L. Hall, this volume, chapter 4.

[21] H. Murayama, this volume, chapter 10.

[22] G. Isidori, this volume, chapter 33.

[23] F. Benatti and R. Floreanini, hep-ph/9906272.

13

Prospects for Calculating ϵ_K and ϵ' from Lattice QCD

Rajan Gupta

Abstract

An overview of lattice results for the strange quark mass, B_K, B_6, B_7, and B_8 is presented. I give my assessment of the reliability of various estimates and prospects for future improvements.

13.1 Introduction

The status of CP violation in kaon decays is

$$\begin{aligned} \epsilon &= (2.280 \pm 0.013)10^{-3} e^{i\pi/4} \\ \mathrm{Re}(\epsilon'/\epsilon) &= (21.2 \pm 4.6)10^{-4}, \end{aligned} \tag{13.1}$$

where I have taken the recent world average of ϵ'/ϵ from M. Sozzi's contribution to this volume. The experimental errors on ϵ'/ϵ will decrease once KTeV and NA48 collaborations analyze their full data set. This will provide a unique opportunity to test the standard model. The spotlight is, therefore, now on theory: can one reconcile the two measurements with the predictions of the standard model in which both parameters are governed by the single phase in the CKM matrix, or are these results a window to new physics?

The standard model predictions, evaluated using the effective weak Hamiltonian defined at scale μ, are summarized by Buras in his contribution to this volume:

$$\begin{aligned} \epsilon &= \mathrm{Im}\lambda_t C_\epsilon \hat{B}_K e^{i\pi/4} \{ \mathrm{Re}\lambda_c \left[\eta_1 S_0(x_c) - \eta_3 S_0(x_c, x_t)\right] + \mathrm{Re}\lambda_t \eta_2 S_0(x_t) \} \\ \epsilon' &= \frac{i e^{i(\pi/4+\delta_2-\delta_0)}}{\sqrt{2}} \frac{\mathrm{Im}A_2}{\mathrm{Re}A_0} \left[1 - \epsilon^2\right] \\ &= \mathrm{Im}\lambda_t \frac{G_F e^{i(\pi/4+\delta_2-\delta_0)}}{2\mathrm{Re}A_0} \left[\omega \sum_i y_i \langle Q_i \rangle_0 (1 - \Omega_{\eta+\eta'}) - \sum_i y_i \langle Q_i \rangle_2\right], \end{aligned} \tag{13.2}$$

where $C_\epsilon = 3.78 \times 10^4$, $\omega = \mathrm{Re}A_2/\mathrm{Re}A_0 \approx 1/22$, $\Omega_{\eta+\eta'}$ is the isospin breaking contribution, y_i are the Wilson coefficients, $\langle Q_i\rangle_I = \langle(\pi\pi)_I|Q_i|K\rangle$, and the sum is over all the 4-fermion operators that contribute. Also, I use the convention $\mathrm{Im}A_0 = 0$ and in the last expression neglect the term proportional to ϵ^2. As expected, both quantities are proportional to $\mathrm{Im}\lambda_t \equiv \mathrm{Im}V_{td}V_{ts}^* = A^2\lambda^5\eta$ in the Wolfenstein parameterization.

The equation for ϵ provides a hyberbolic constraint in the $\rho - \eta$ plane provided $\hat{B}_K$ and $|V_{cb}|$ are known. Alternately, a precise determination of $\hat{B}_K$ and $|V_{cb}|$ would fix $\mathrm{Im}\lambda_t$, and the measurements of ϵ'/ϵ could be used to look for new physics. Note that a larger value of $\hat{B}_K$ implies smaller $\mathrm{Im}\lambda_t$, and consequently smaller ϵ'.

For ϵ' we follow the work of Buras *et al.* [1], who use the relations between the various $\langle Q_i\rangle$ and make maximal use of experimental input. Then

$$\left|\frac{\epsilon'}{\epsilon}\right| = \mathrm{Im}\lambda_t\left\{c_0 + \left(c_6 B_6^{1/2} + c_8 B_8^{3/2}\right)\left(\frac{158\ \mathrm{MeV}}{m_d + m_s}\right)^2\right\}. \qquad (13.3)$$

For $\mu = m_c$, $\Lambda^{(4)}_{QCD} = 325 MeV$, and central values of the other parameters, Buras *et al.* [1] get in the NDR scheme

$$\begin{aligned} \mathrm{Im}\lambda_t &\approx 1.29 \times 10^{-4}\,, \\ c_0 &\approx -1.4\,, \\ c_6 &\approx +7.9\,, \\ c_8 &\approx -4.1\,. \end{aligned} \qquad (13.4)$$

From eqs. 13.3 and 13.4 it is clear that there is a strong cancelation between the $\Delta I = 1/2$ (dominated by QCD penguin Q_6) and $\Delta I = 3/2$ (dominated by electroweak penguin Q_8) operators, and the value of the strange quark mass plays a crucial role. For $B_6 = B_8 = 1$ (vacuum saturation approximation (VSA) values currently used as inputs), one needs $m_s + m_d = 70$ MeV at $\mu = m_c$ instead of 158 MeV to get ϵ'/ϵ to $\approx 23 \times 10^{-4}$. A more likely scenario is an enhancement of B_6, a suppression of B_8 and the quark masses, and/or a conspiracy of all other input parameters. It is therefore important to determine all three quantities, m_s, B_6, and B_8, accurately in order to resolve whether or not the standard model predicts the observed value of ϵ'.

There are two omissions from my subsequent discussion of lattice calculations. First, I do not discuss lattice results using domain wall fermions (DWF). Second, G. Martinelli has proposed analyzing B parameters without introducing a dependence on $m_s + m_d$. Recall that the dependence on $m_s + m_d$ is introduced because in VSA $\langle O_6\rangle, \langle O_8\rangle \propto |\langle 0|\mathcal{P}|K\rangle|^2 = M_K^4 f_K^2/(m_s + m_d)^2$. Using B parameters as commonly defined has certain numerical advantages, but it does shift the scale dependence of the operator into a new quantity $(m_s + m_d)$. Which approach is better depends on numerical details, and since at this point I do not have data to

Fermion Action	Z	β	$B_K(\overline{MS}, 2\text{ GeV})$
Staggered [2]	1-loop	$6.0 - 6.4^*$	0.62(2)(2)
Staggered [3]	1-loop	$5.7 - 6.65^*$	0.628(42)
Staggered [4]	1-loop	$5.7 - 6.2^*$	0.573(15)
Wilson [5]	1-loop TI	6.0	0.74(4)(5)
Wilson [6]	1-loop & χWI	$5.9 - 6.5^*$	0.69(7)
Clover [7]	1-loop BPT	6.0	0.65(11)
Clover [7]	Non-pert.	6.0	0.66(11)
TI Clover [8]	1-loop TI	6.0, 6.2	0.72(8)

Table 13.1: Lattice estimates of $B_K(NDR, \mu = 2\text{ GeV})$ for different lattice actions. An asterisk imples that the data were extrapolated to $a = 0$.

make comparative statements, the reader should see Martinelli's writeup (chapter 28).

13.2 B_K

Even though most calculations of B_K have been done in quenched QCD, there are good reasons to believe, as discussed below, that we have a reasonable estimate for the full theory. A summary of results is presented in table 13.1. The most precise calculation in terms of both statistical and systematic errors, is by the JLQCD collaboration. Their result $B_K = 0.628(42)$ when converted to the renormalization group invariant $\hat{B}_K$ is [13]

$$\hat{B}_K = 0.86(6)\,. \tag{13.5}$$

Since this result is for the quenched theory we have to address two associated issues: (i) quenched chiral logs (QCL) and (ii) other effects of quenching. The other remaining systematic error is the use of degenerate quarks for the kaon. The quark mass is typically varied in the range $3m_s - m_s/3$, and the physical kaon is defined in the following two ways: (a) it is assumed to be made up of two degenerate quarks of mass $m_s/2$, or (b) the "light" quark is extrapolated to m_d and the other interpolated to m_s. The issue of quenched chiral logs is relevant to (b). The reason is that for degenerate quarks, quenched QCD and QCD have the same chiral expansion and enhanced logs due to η', an additional Goldstone boson in the quenched theory, vanish [10]. The status on each of these issues and some clarifications on the different published numbers is as follows.

- The parity even part of the 4-fermion operator has two terms, VV and AA. Sharpe [9, 10] has calculated the QCL in these and their lattice volume dependence. The lattice data show the expected behavior, providing a basis for confidence in the CPT (chiral perturbation theory) analyses. The leading chiral log cancels in the sum, $VV + AA$, thus alleviating the uncertainties associated with chiral extrapolations in the quenched theory.

- Estimates of quenching uncertainties provided by CPT are strengthened by the preliminary unquenched calculations [11], and suggest that B_K(full QCD) $\approx 1.05 B_K$(quenched).

- CPT has also been used by Sharpe [10] to estimate the uncertainty associated with using $(m_d \approx m_s)$ rather than the physical ratio $(m_d \approx 0.055 m_s)$. He estimates $B_K(\text{QCD}) \approx 1.05 \pm 0.05 B_K(\text{degenerate})$.

- Finally, it is a fortunate coincidence that the conversion of quenched $B_K(\overline{MS}, \mu = 2 \text{ GeV})$ to $\hat{B}_K$ is very insensitive to whether one uses quenched α_s and anomalous dimensions or those for the full theory.

The success of CPT in estimating errors raises the question whether the systematic shifts due to quenching and degenerate versus physical mass quarks discussed above should be incorporated in the final value of $\hat{B}_K$ or stated as a separate error. Sharpe in [12] includes them and quotes $\hat{B}_K = 0.94 \pm 0.06 \pm 0.14$, where the second error is a very conservative estimate of the systematic uncertainties. In [13], I chose not to include them in the central value and had used a more aggressive estimate of systematic errors in quoting $\hat{B}_K = 0.86 \pm 0.06 \pm 0.06$. Both estimates are based on exactly the same data (JLQCD), and until unquenched data of comparable quality becomes available the choice between them reflects one's taste in the handling of systematic errors.

Finally, in my view, one should not average the numbers given in table 13.1 to get a "best" lattice result. At present, JLQCD's is, by far, the best calculation with respect to lattice size, statistics, and systematics. (The quoted errors in table 13.1 do not always include/address all the systematics uncertainties equally well). What the table does highlight is that all the results agree: a confirmation of the stability of lattice calculations of B_K.

13.3 Light Quark Masses

Since mid-1996 there has been a spurt of activity in the calculation of light quark masses from both lattice QCD and QCD sum-rules. The intriguing possibility first suggested by lattice calculations is that m_u, m_d, and m_s are much lighter than previous estimates based on QCD sum-rules. For a summary of the lattice methodology and results until Oct. 1997 see [14] and also the talk by S. Ryan [15].

Recent quenched lattice results are summarized in table 13.2, and unlike B_K there is no single calculation that is "superior" to the rest. (Unfortunately, once again this is not obvious from quoted errors.) At first glance one sees a significant spread. Focusing attention on m_s extracted by fixing M_K to its physical value, $m_s(M_K)$, the estimates lie between 90 – 115 MeV. A large part of this variation is due to the quantity used to set the lattice scale $1/a$. There is also a large difference between $m_s(M_K)$ and $m_s(M_\phi)$, *i.e.*, different strange mesons give different estimates; and even

	Action	$\bar{m}$	$m_s(M_K)$	$m_s(M_\phi)$	scale $1/a$
Summary 1997 [14]		3.8(1)(3)	99(3)(8)	111(7)(20)	M_ρ
APE 1998 [17]	O(a) SW	4.5(4)	111(12)		M_{K^*}
APE 1999 [18]	O(a) SW	4.8(5)	111(9)		M_{K^*}
CPPACS 1999 [19]	Wilson	4.55(18)	115(2)	143(6)	M_ρ
JLQCD 1999 [20]	Staggered	4.23(29)	106(7)	129(12)	M_ρ
ALPHA-UKQCD 1999 [21]	O(a) SW		97(4)		f_K
RIKEN-BNL 1999 [22]	DWF		95(26)		f_π
QCDSF 1999 [23]	O(a) SW	4.4(2)	105(4)		r_0
QCDSF 1999 [23]	Wilson	3.8(6)	87(15)		r_0

Table 13.2: Recent lattice estimates in MeV of $\bar{m}$ and m_s, in $\overline{MS}$ scheme at 2 GeV. O(a) SW stands for the O(a) Sheikhholeslami-Wohlert action. $r_0 = 0.5$ fermi is the scale from the static $q\bar{q}$ potential.

though neither one reproduces the octet and decuplet baryon mass splittings, $m_s(M_\phi)$, comes much closer [16, 19]. A short explanation of the results is in order.

First, the difference between $m_s(M_K)$ and $m_s(M_\phi)$ and the variation with the observable used to fix $1/a$ are both symptoms of the quenched approximation. The data suggests that it is a $\sim 10\%$ effect, and at present constitutes the biggest uncertainty. Second, the consistency of the results using different actions (from Wilson to domain wall fermions), analyzed using the same states and after an $a = 0$ extrapolation to remove discretization errors, shows that the lattice technology is robust and that we have control over discretization errors. Third, there has been much debate over which renormalization constants (tadpole improved perturbative or from the various nonperturbation methods) to use. The data show that after extrapolation to $a = 0$, the difference is at most a few percent. So the bottom line is that we now have many different methods and consistency checks within the lattice approach for calculating light quark masses and just need the computer power to shed the last approximation—the quenched approximation—to get reliable estimates.

The only unquenched data (albeit for 2 degenerate dynamical flavors) of the modern era (using improved action, nonperturbative renormalization constants, and extrapolation to $a = 0$) are the recent results by the CPPACS collaboration (hep-lat/0004010):

$$\begin{aligned} (m_u + m_d)/2 &= 3.5(3)\,\text{MeV} \\ m_s(M_K) &= 89(7)\,\text{MeV} \\ m_s(M_\phi) &= 99(5)\,\text{MeV}. \end{aligned}$$

It is quite remarkable that $m_s(M_K)$ and $m_s(M_\phi)$ already show consistency. Also, the associated mass splittings in the baryon octet and decuplet are much improved. On the strength of these consistency checks I propose the

Fermion Action	Z	β	$B_7^{3/2}$	$B_8^{3/2}$
Staggered [2]	1-loop TI	6.0, 6.2*	0.62(3)(6)	0.77(4)(4)
Wilson [5]	1-loop TI	6.0	0.58(2)(7)	0.81(3)(3)
SW [7]	1-loop BPT	6.0	0.58(2)	0.83(2)
SW [7]	Non-pert.	6.0	0.72(5)	1.03(3) .
TI SW [8]	1-loop TI	6.0, 6.2	0.72^{+5+2}_{-4-8}	0.80^{+8+1}_{-8-4}

Table 13.3: Lattice estimates of B_7, and B_8 ($NDR, \mu = 2\,\mathrm{GeV}$) for the amplitude $K \to \pi$. An asterisk implies extrapolation to $a = 0$.

following as the current estimate from lattice QCD:

$$m_s(\overline{MS}, \mu = 2GeV) = 85(10)\,\mathrm{MeV}. \tag{13.6}$$

13.4 $\Delta I = 3/2$ Electroweak Penguins: B_7 and B_8

Current lattice calculations of the $\Delta I = 3/2$ amplitude rely on CPT to relate $K \to \pi\pi$ to $K \to \pi$ with degenerate K and π. Under these approximations, the calculation of B_7 and B_8 is equivalent to that of B_K in complexity. Initially, there was a problem of much larger 1-loop renormalization constants. This is now under much better control through the development of better operators and nonperturbative methods. A summary of results is given in table 13.3. All results using perturbative Zs are consistent, confirming that the calculation of the raw correlation functions is under control. The APE calculation [7] using nonpertubative Zs gives a value higher by $\sim 20\%$. However, since almost all the calculations have been done at only one coupling, and anticipating that the extrapolation to the $a = 0$ limit will also be different for the two ways of estimating the Zs, it is too early to choose between the two values. Calculations at other values of the coupling are in progress and I anticipate we will reduce the uncertainty to $< 10\%$ within the year.

The more important issue is whether tree-level CPT is sufficient to relate $K \to \pi\pi$ to $K \to \pi$ with $M_K = M_\pi$. Since a similar situation in B_4 suggests not [24], Golterman and Pallante are doing the needed 1-loop calculations. Thereafter, one has to confront issues of removing the quenched approximation and developing the technology for dealing with the physical case of $K \to \pi\pi$ away from threshold in case 1-loop CPT corrections are very large. Sheer optimism propels me to believe that we will see progress towards realistic estimates of these parameters in the next couple of years.

13.5 Strong Penguin: B_6

Lattice QCD does not yet provide an estimate for B_6. In addition to the issue of using CPT to relate $K \to \pi\pi$ to $K \to \pi$, there is also the problem of mixing with lower-dimension operators and large renormalization factors. There are two calculations under way. The first by Blum using domain wall fermions [25]; and the second by Kilcup and Pekurovsky using staggered fermions [26]. Kilcup *et al.* show that all the needed correlation functions can be calculated with small statistical errors; however, since the 1-loop Zs for staggered fermions are very large ($\sim$ 100%) there are no reliable predictions. One definitely needs nonperturbative calculation of these Zs. It is too early to tell if domain wall fermions will prove to be the method of choice. In short, B_6, which is crucial to understanding both the $\Delta I = 1/2$ rule and ϵ', is still an open problem in lattice QCD.

Acknowledgments

It is a pleasure to thank Bruce and Jon and for organizing such a wonderful conference, and T. Bhattacharya, C. Bernard, and S. Sharpe for their comments.

References

[1] G. Buchalla, A. Buras, and M. E. Lautenbacher, Rev. Mod. Phys. **68**, 1125 (1996); A. Buras, in *Probing the Standard Model of Particle Interactions*, Les Houches session 97, ed. R. Gupta *et al.* (Elsevier, Amsterdam, 1999).

[2] G. Kilcup, R. Gupta, and S. Sharpe, Phys. Rev. D **57**, 1654 (1998).

[3] S. Aoki *et al.*, JLQCD collaboration, Phys. Rev. Lett. **80**, 5271 (1998).

[4] G. Kilcup *et al.*, Nucl. Phys. (Proc. Suppl.) B **53**, 345 (1997).

[5] R. Gupta, T. Bhattacharya, and S. Sharpe, Phys. Rev. D **55**, 4036 (1997).

[6] S. Aoki *et al.*, JLQCD collaboration, hep-lat/9901018.

[7] L. Conti, Phys. Lett. B **421**, 273 (1998).

[8] L. Lellouch and C.D. Lin, Nucl. Phys. (Proc. Suppl.) B **73**, 312 (1999).

[9] S. Sharpe, Phys. Rev. D **46**, 3146 (1992).

[10] S. Sharpe, in TASI 94, ed. J. Donoghue (World Scientific, Singapore 1995).

[11] G. Kilcup *et al.*, Nucl. Phys. (Proc. Suppl.) B **53**, 345 (1997); N. Ishizuka *et al.*, Phys. Rev. Lett. **71**, 24 (1993).

[12] S. Sharpe, Plenary talk at ICHEP98, hep-lat/9811006.

[13] R. Gupta, "Problem of Mass" in *Physics of Mass*, ed. B. Kursunoglu *et al.* (New York, Kluwer Academic/Plenum Publishers, 1998).

[14] T. Bhattacharya and R. Gupta, Nucl. Phys. (Proc. Suppl.) B **63**, 95 (1998).

[15] S. Ryan, this volume, chapter 29.

[16] T. Bhattacharya *et al.*, Phys. Rev. D **53**, 6486 (1996).

[17] D. Becirevic *et al.*, Phys. Lett. B **444**, 401 (1998).

[18] D. Becirevic *et al.*, Phys. Rev. D **61**, 114507 (2000).

[19] S. Aoki *et al.*, CPPACS Collaboration, Phys. Rev. Lett. **84**, 238 (2000).

[20] S. Aoki *et al.*, JLQCD Collaboration, Phys. Rev. Lett. **82**, 4392 (1999).

[21] J. Garden *et al.*, ALPHA-UKQCD Collaboration, Nucl. Phys. B **571**, 237 (2000).

[22] T. Blum *et al.*, RIKEN-BNL Collaboration, Phys. Rev. D **60**, 114507 (1999).

[23] M. Göckeler *et al.*, QCDSF Collaboration, hep-lat/9908005.

[24] M. Golterman and K. Leung, Phys. Rev. D **56**, 2950 (1997).

[25] T. Blum, private communication.

[26] D. Pekurovsky and G. Kilcup, hep-lat/9812019.

14

Weak Matrix Elements in the Large-N_c Limit

William A. Bardeen

Abstract

The matrix elements of the four-quark operators needed to predict many weak interaction processes can be evaluated using the large-N_c limit of quantum chromodynamics. At leading order in the large-N_c expansion, the weak matrix elements of four-quark operators factorize into independent matrix elements of two-quark operators, a common approximation being used today. At next leading order, the weak matrix elements acquire the leading scale and scheme dependence expected for these matrix elements in full QCD. We will discuss methods to evaluate these matrix elements that involve matching perturbative QCD calculations at short distance to nonperturbative hadronic matrix elements at long distance.

The large-N_c expansion for quantum chromodynamics was formulated by 't Hooft [1] and has been used by many authors to study nonperturbative effects in QCD. The large-N_c expansion is based on 't Hooft's observation that the perturbation series could be reorganized by considering the limit, $\alpha_s \to 0, N_c \to \infty, \alpha_s * N_c$ – *fixed*. The leading order of this expansion involves only planar diagrams of quarks and gluons, and all diagrams with internal quark loops are suppressed. This approximation is very similar to the quenched version of QCD used in many lattice computations where quark loops are also suppressed while allowing nonplanar gluon interactions. The large-N_c limit of QCD is nonperturbative as all orders of $\alpha_s * N_c$ must be included at the leading order of the large-N_c expansion. The theory is expected to be a theory of hadronic bound states with color confinement and dynamical chiral symmetry breaking. From the topology of diagrams contributing to the large-N_c limit of QCD, the theory is expected to consist of infinite towers of weakly interacting, color-singlet mesons with all spins and flavors dictated by the quark substructure. The scattering amplitudes are order $1/N_c$ and can be viewed as tree diagrams of the effective meson theory. Higher-order diagrams in the large-N_c expansion involve the insertion of internal quark loops or nonplanar gluon interactions. At the meson level, the higher-order diagrams correspond to a systematic loop expansion of the effective meson theory. The topological structure of these diagrams is very suggestive of a hadronic string picture for large-N_c QCD.

The theoretical description of many weak processes requires knowledge of the hadronic matrix elements of weak currents and chiral densities. In addition, nonleptonic weak decays usually require the knowledge of four-quark operators constructed from products of these currents and densities. The large-N_c expansion can be used to analyze the leading behavior of the matrix elements of these operators. The weak operators are usually written in terms of the products of color-singlet bilinear operators,

$$O_i = (\overline{\psi}\Gamma_i\psi)(\overline{\psi}\Gamma_i\psi) \; . \tag{14.1}$$

At leading order (LO) of the large-N_c expansion, the hadronic matrix elements factorize. Each color-singlet bilinear operator couples independently to the hadrons in the external states:

$$\langle O_i \rangle_{factorized} = \langle (\overline{\psi}\Gamma_i\psi) \rangle \langle (\overline{\psi}\Gamma_i\psi) \rangle . \tag{14.2}$$

Higher-order terms in the large-N_c expansion involve both factorized and nonfactorized contributions to the weak matrix elements. Next-leading-order (NLO) terms include the addition of internal quark loop contributions to the factorized matrix elements and the generation of leading nonfactorizing contributions to the weak matrix elements. These are both one-loop diagrams in the effective meson theory.

At NLO, the nonfactorized amplitudes have a particularly simple structure in the effective meson theory. The matrix element may be written as a momentum integral over a two-current correlation function,

$$\langle O_i \rangle_{nonfactorized} = \int dk \; A_{\Gamma_i\Gamma_i}(k, -k; p_1, \cdots, p_N), \tag{14.3}$$

where k is the momentum flowing through the color-singlet bilinear operators and $p_1, ..., p_N$ are the momenta of the external hadrons. We can use our knowledge of these two-current correlation functions to evaluate the NLO contributions to weak matrix elements [2]. If all the external states are at low energy, the low-momentum part of the integral requires only the low-energy behavior of the two-current correlation function. However, many experiments directly measure these current correlation functions. For meson external states, the lowest energy contributions are summarized in terms an effective chiral Lagrangian. The parameters of the effective chiral Lagrangian are nonperturbative quantities in QCD and have been determined by systematic phenomenological analysis [3]. At higher loop momenta, additional hadronic states must be included such as the vector and axial-vector mesons. In principle, the full intergral can be obtained from the complete tree-level correlation function of the LO meson theory. However, the high-momentum behavior can also be obtained through the use of the operator product expansion,

$$A_{\Gamma_i\Gamma_i}(k,-k;p_1,\cdots,p_N M) \to C_{\Gamma_i\Gamma_i;O_j}(k) * \langle O_j\rangle(p_1,\cdots,p_N), \qquad (14.4)$$

where the coefficient function, $C_{\Gamma_i\Gamma_i;O_j}$, can be computed using the large-N_c version of perturbative QCD, PQCD. It is a remarkable fact that all weak mixing processes are nonleading in the large-N_c expansion. This implies that the coefficient function begins to receive contributions only at NLO. Therefore, the NLO calculation of the current correlation function requires only knowledge of the LO operator matrix element of the operators appearing in eq. (14.4), and these are determined from parameters of the LO effective meson theory.

Since we are able to establish the precise behavior of the integrand appearing in eq. (14.3) both at low energies and at high energies, we can hope to estimate the full integral by interpolating the results at moderate momentum scales. Of course, the accuracy of this interpolation can be improved by including more states or higher derivative terms in the low-energy effective meson theory or by computing higher-order terms in the PQCD expansion of the short-distance theory. In principle, the matching between the effective meson theory and the PQCD expansion of the short-distance theory can be improved to arbitrary accuracy. Comparisons with the conventional definitions of weak matrix elements require knowledge of the particular regularization schemes, NDR or HV, used to define the weak operators in PQCD. Hence, a second short-distance matching must be made between the integral expression of eq. (14.3) and the particular scheme used to reqularize the short-distance behavior. This second matching can be done purely at the quark level as it depends solely on the short-distance physics. Using this short-distance matching, the high-momentum part of the integral of eq. (14.3) can be properly subtracted to obtain the full NLO contribution to the weak matrix elements in any renormalization scheme.

This method can be applied to any of the conventional four-quark operators used to analyze nonleptonic weak interactions: Q_1, Q_2, ..., Q_6, ..., Q_8, ... At NLO in the large-N_c expansion, the weak matrix elements computed by the above method will have the appropriate scale and scheme dependence. These matrix elements can be used with any other analysis of the physical short-distance physics that determines the physical coefficient functions. Scale and scheme dependence will properly cancel between the coefficient functions and the operator matrix elements, at least to NLO in the large-N_c expansion. The precision of the NLO weak matrix elements determined through these methods depends on a number of factors:

- the phenomenological determination of the effective meson theory
  - chiral Lagrangians: $O(p^2)$, $O(p^4)$, $O(p^6)$, ...
  - inclusion of heavy states: vector mesons, scalars, ...
  - models: effective NJL models, chiral quark model, ...

- the long-distance/short-distance matching conditions
- the short-distance expansion of planar QCD
- the scheme-dependent matching conditions of PQCD.

The methods described above have been applied to a number of problems requiring knowledge of hadronic matrix elements. Applications include the $\pi^+ - \pi^0$ electromagnetic mass difference, the $\Delta I = 1/2$ rule in nonleptonic weak decays, the B_K parameter in $K - \bar{K}$ mixing, and the weak matrix elements needed for determining ϵ'/ϵ in the CP-violating kaon decays.

•$\pi^+ - \pi^0$ mass difference. This calculation involves the insertion of explicit one-photon exchange processes that have the same structure as the insertion four-quark operators in weak processes. The inserted vertex is nonlocal due to the photon propagator and does not require the scheme-dependent short-distance matching of the weak matrix elements. The matching between the long-distance meson physics and the short-distance quark physics involving chiral condensates is still required. Using the lowest-order chiral Lagrangian, $O(p^2)$, and the conventional short-distance expansion gives an estimate of the mass difference good to about 15%–20%. Here the matching between the long-distance physics of point-like pions and the short-distance quark physics is determined by the scale where the integrands coincide. This matching occurs in the range of 600–800 MeV although the two different approximations to the integrand have much different energy dependence. We are able to improve our knowledge of the long-distance physics by including the contributions of vector and axial-vector mesons. The matching now becomes excellent at any scale above 600–800 MeV and is related to the known meson saturation of the Weinberg sum rules. The mass difference calculation is now good to about 5% [4].

•$\Delta I = 1/2$ rule in kaon decays. The CP-conserving weak decays of kaons are known to obey the $\Delta I = 1/2$ rule. The $\Delta I = 1/2$ amplitude is enhanced from the factorized matrix element and the $\Delta I = 3/2$ amplitude is suppressed. Part of this enhancement/suppression can be explained in terms of the conventional weak mixing [5] involving weak operators, Q_1 and Q_2 and a charm penguin contribution related to the Q_6 operator. Using the large-N_c approach, the weak matrix elements can be evaluated and additional enhancements/suppressions are observed. Using an $O(p^2)$ chiral Lagrangian, Buras, Bardeen, and Gérard [6] were able to explain about 75% of the observed enhancement of the $\Delta I = 1/2$ amplitude and an additional suppression of the $\Delta I = 3/2$ amplitude. Using an $O(p^4)$ chiral Lagrangian, Hambye *et al.* [7] claim to observe the full enhancement of the $\Delta I = 1/2$ amplitude. Using an extended NJL model to improve the long-distance approximation and the short-distance/long-distance matching conditions, Bijnens *et al.* [8] also claim to see the full enhancement of the $\Delta I = 1/2$ amplitude. These calculations are sensitive to the precise method for calculating the long-distance contributions and the matching procedure used to connect the long- and short-distance calculations. The present calculations do not include the full scheme dependence arising from the short-distance matching conditions.

•$K^0 - \bar{K}^0$ mixing, B_K. The $K^0 - \bar{K}^0$ mixing arises from loops involving the top quark. Integrating out the top quark generates a unique $\Delta S = 2$ four-quark operator. Perturbative QCD can be used to evolve the effective weak operator to a low-energy scale. The large-N_c method can then be used to evaluate the low-energy matrix elements. A number of predictions for the B_K parameter have been made using various approximations for the long distance meson physics:

- $O(p^2)$ chiral Lagrangian [9] $\hat{B}_K \sim 0.7 \pm 0.$
- $O(p^2)$ chiral Lagrangian + vector mesons [10] $\hat{B}_K \sim 0.75 \pm 0.1.$
- ENJL model [8] $\hat{B}_K \sim 0.69 \pm 0.1.$
- $O(p^4)$ chiral Lagrangian [7] $\hat{B}_K \sim 0.6 \pm 0.1.$

• CP violation and ϵ'/ϵ. CP Violation observed at low energy is expected to be generated by loop effects at a high mass scale. Top quark loops contribute to CP violation through complex phases associated with the effective four-quark operators at low energy, particularly the operators Q_6 and Q_8. In the chiral Lagrangian approach, $O(p^4)$ terms are required for the Q_6 matrix element to be nonzero. Leading terms in the nonfactorized amplitudes cancel infrared singularities of the factorized amplitudes. The large-N_c expansion method has been applied to these matrix elements, and the matrix elements of the electropenguin operator, Q_8, were found to be suppressed by the nonfactorizing contributions while the gluopenguin operator, Q_6, may receive a modest enhancement [11]. Both of these effects tend to increase theoretical estimates of ϵ'/ϵ.

The large-N_c expansion permits a consistent evaluation of the weak matrix elements for a number of important physical processes. The method combines our knowledge of perturbative short-distance processes with the nonperturbative contributions contained in the effective meson theories or ENJL models used to describe the long-distance physics. At NLO the method is subject to considerable improvement. The effective meson physics could be extended by considering additional meson states or improved models that evolve the long-distance physics to higher energy scales. This could improve matching of the long-distance and short-distance physics, which is now at a rather crude level for most processes. Also, the scheme dependence of the weak matrix elements requires a specific calculation of the short-distance matching of the momentum integral of the current correlation function to particular regularization scheme in PQCD. At this point, the method is restricted to NLO in the large-N_c expansion as terms $O(1/N_c^2)$ cannot be controlled. At some point, hadronic string theory might be used to obtain a complete picture of the meson amplitudes at higher order. Many of the issues discussed in this short talk will be covered in more detail in other contributions to this conference.

References

[1] G. 't Hooft, Nucl. Phys. B **72**, 461 (1974).

[2] W. Bardeen, A. Buras, and J.-M. Gérard, Nucl. Phys. B **293**, 787 (1987), Phys. Lett. B **192**, 138 (1987), Phys. Lett. B **211**, 343 (1988); W. Bardeen, Nucl. Phys. Proc. Suppl. **7** A, 14 (1989).

[3] J. Gasser and H. Leutwyler, Nucl. Phys. B **250**, 465 (1985); J. Bijnens, G. Colangelo, and G. Ecker, hep-ph/9902437.

[4] W. Bardeen, J. Bijnens, and J.-M. Gérard, Phys. Rev. Lett. **62**, 1343 (1989); J. Bijnens and E. de Rafael, Phys. Lett. B **273**, 483 (1991).

[5] M. Gaillard and B. Lee, Phys. Rev. Lett. **33**, 108 (1974); G. Altarelli and L. Maiani, Phys. Lett. B **52**, 351 (1974); F. Gilman and M. Wise, Phys. Rev. D **20**, 2392 (1979).

[6] W. Bardeen, A. Buras, and J.-M. Gérard, Phys. Lett. B **192**, 138 (1987); A. Buras, in *CP Violation*, ed. C. Jarlskog (World Scientific, Singapore, 1989), p. 575.

[7] T. Hambye, G. Köhler, and P. H. Soldan, hep-ph/9902334.

[8] J. Bijnens and P. Prades, Nucl. Phys. B **444**, 523 (1995); hep-ph/9811472.

[9] W. Bardeen, A. Buras, and J.-M. Gérard, Phys. Lett. B **211**, 343 (1988).

[10] J. Fatelo and J.-M. Gérard, Phys. Lett. B **347**, 136 (1995).

[11] T. Hambye, G. Köhler, E. Paschos, P. Soldan, and W. Bardeen, Phys. Rev. D **58**, 014017 (1998).

Part III

Time Reversal Violation and CPT Studies

With the discovery that not only P and C, but also CP was violated, physicists began to ask if *any* symmetry was respected by the fundamental interactions. In particular, it is a very general feature of quantum field theories [1] that the product CPT, where T stands for time reversal, is valid. The proof relies only on the assumptions of locality (microscopic causality) and Lorentz invariance. Thus, CP violation should imply T violation.

Specific experiments can be performed in order to search directly for T violation. In a four-body decay, with three independent momenta p_1, p_2, and p_3 (where $p_4 = -(p_1 + p_2 + p_3)$ in the center of mass), there are enough independent momenta to construct a T-violating observable just out of momenta. For example, $\epsilon_{ijk} p_{1i} p_{2j} p_{3k}$ is such a quantity. The KTeV group has seen for the first time a very large T-odd asymmetry in the $K_L \to \pi\pi ee$ decay (Ledovskoy) and this has now been confirmed by NA48 (Wronka). KEK E162 (Nomura) has also studied this decay mode. Large CP asymmetries in this decay were originally predicted (Sehgal; Savage), but there is debate as to whether the T-odd effects are T violating or might in principle be CPT violating. (See the contributions of Dalitz (part I) and Kostelecký.)

Other T-violating observables discussed in the present part are the asymmetry between $K^0 \to \bar{K}^0$ and $\bar{K}^0 \to K^0$ transitions (Bloch), transverse muon polarization in the decay $K^+ \to \pi^0 \mu^+ \nu$ (Lim), and permanent electric dipole moments of particles and nuclei (Romalis). A nonzero effect, closely related to the magnitude of CP violation in kaon decays, has been seen in the first of these processes.

There exist direct tests for CPT violation as well. One can examine theories parametrized by a preferred Lorentz frame (Kostelecký), where a host of tests exist in neutral kaon systems and many other arenas. Parametrizations of the neutral kaon system can easily accommodate CPT violation and are subject to numerous tests, as described by Di Falco and Incagli, Bloch, Ryskulov, and Solodov.

References

[1] J. Schwinger, Phys. Rev. **91**, 713 (1953); **94**, 1362 (1954); G. Lüders, Kong. Danske Vid. Selsk., Matt-fys. Medd. **28**, No. 5 (1957); Ann. Phys. (NY) **2**, 1 (1957); J.S. Bell, Proc. Roy. Soc. **A231**, 79 (1955); W. Pauli, in *Niels Bohr and the Development of Physics* (Pergamon, New York, 1955), p. 30.

15

CP and T Violation in the Decay $K_L \to \pi^+\pi^- e^+e^-$ and Related Processes

L. M. Sehgal

Abstract

I review the theoretical basis of the prediction that the decay $K_L \to \pi^+\pi^- e^+e^-$ should show a large CP and T violation, a prediction now confirmed by the KTeV experiment. The genesis of the effect lies in a large violation of CP and T invariance in the decay $K_L \to \pi^+\pi^-\gamma$, which is encrypted in the polarization state of the photon. The decay $K_L \to \pi^+\pi^- e^+e^-$ serves as an analyser of the photon polarization. The asymmetry in the distribution of the angle ϕ between the $\pi^+\pi^-$ and e^+e^- planes is a direct measure of the CP-odd, T-odd component of the photon's Stokes vector. A complete study of the angular distribution can reveal further CP-violating features, which probe the nonradiative (charge-radius and short-distance) components of the $K_L \to \pi^+\pi^- e^+e^-$ amplitude.

Eight years ago, there appeared a report [1] by the E-731 experiment concerning the branching ratio and photon energy spectrum of the decays $K_{L,S} \to \pi^+\pi^-\gamma$. It was found that while the K_S decay could be well reproduced by inner bremsstrahlung from an underlying process $K_S \to \pi^+\pi^-$, the K_L decay contained a mixture of a bremsstrahlung component (IB) and a direct emission component (DE), the relative strength being DE/(DE + IB) = 0.68 for photons above 20 MeV. The simplest matrix element consistent with these features is

$$\mathcal{M}(K_S \to \pi^+\pi^-\gamma) = ef_S\left[\frac{\epsilon\cdot p_+}{k\cdot p_+} - \frac{\epsilon\cdot p_-}{k\cdot p_-}\right] \tag{15.1}$$

$$\mathcal{M}(K_L \to \pi^+\pi^-\gamma) = ef_L\left[\frac{\epsilon\cdot p_+}{k\cdot p_+} - \frac{\epsilon\cdot p_-}{k\cdot p_-}\right] + e\frac{f_{DE}}{{M_K}^4}\epsilon_{\mu\nu\rho\sigma}\epsilon^\mu k^\nu {p_+}^\rho {p_-}^\sigma,$$

where

$$f_L \equiv |f_S| g_{Br},\ g_{Br} = \eta_{+-}e^{i\delta_0(s={M_K}^2)},$$
$$f_{DE} \equiv |f_S| g_{M1},\ g_{M1} = i(0.76)e^{i\delta_1(s)}. \tag{15.2}$$

Here the direct emission has been represented by a CP-conserving magnetic dipole coupling g_{M1}, whose magnitude $|g_{M1}| = 0.76$ is fixed by the empirical ratio DE/IB. The phase factors appearing in g_{Br} and g_{M1} are dictated by the Low theorem for bremsstrahlung, and the Watson theorem for final state interactions. The factor i in g_{M1} is a consequence of CPT invariance [2]. The matrix element for $K_L \to \pi^+\pi^-\gamma$ contains simultaneously electric multipoles associated with bremsstrahlung ($E1$, $E3$, $E5$, ...), which have CP $= +1$, and a magnetic $M1$ multipole with CP $= -1$. It follows that interference of the electric and magnetic emissions should give rise to CP violation.

To understand the nature of this interference, we write the $K_L \to \pi^+\pi^-\gamma$ amplitude more generally as

$$\begin{aligned}\mathcal{M}(K_L \to \pi^+\pi^-\gamma) &= \frac{1}{{M_K}^3}\{E(\omega, \cos\theta)\,[\epsilon\cdot p_+\, k\cdot p_- - \epsilon\cdot p_-\, k\cdot p_+] \\ &\quad + M(\omega, \cos\theta)\epsilon_{\mu\nu\rho\sigma}\epsilon^\mu k^\nu {p_+}^\rho {p_-}^\sigma\}, \qquad (15.3)\end{aligned}$$

where ω is the photon energy in the K_L rest frame, and θ is the angle between π^+ and γ in the $\pi^+\pi^-$ rest frame. In the model represented by eqs. (15.1) and (15.2), the electric and magnetic amplitudes are (up to a common factor of $e\,|f_S|\,M_K$)

$$\begin{aligned}E &= \left(\frac{2M_K}{\omega}\right)^2 \frac{g_{Br}}{1-\beta^2\cos^2\theta} \\ M &= g_{M1}, \qquad (15.4)\end{aligned}$$

where $\beta = (1-4{m_\pi}^2/s)^{1/2}$, $\sqrt{s}$ being the $\pi^+\pi^-$ invariant mass. The Dalitz plot density, summed over photon polarizations, is

$$\frac{d\Gamma}{d\omega\, d\cos\theta} = \frac{1}{512\pi^3}\left(\frac{\omega}{M_K}\right)^3 \beta^3\left(1-\frac{2\omega}{M_K}\right)\sin^2\theta\left[|E|^2+|M|^2\right]. \quad (15.5)$$

Clearly, there is no interference between the electric and magnetic multipoles if the photon polarization is unobserved. Therefore, any CP violation involving the interference of g_{Br} and g_{M1} is hidden in the polarization state of the photon.

The photon polarization can be defined in terms of the density matrix

$$\rho = \begin{pmatrix} |E|^2 & E^*M \\ EM^* & |M|^2 \end{pmatrix} = \frac{1}{2}\left(|E|^2+|M|^2\right)\left[\mathbb{1} + \vec{S}\cdot\vec{\tau}\right], \qquad (15.6)$$

where $\vec{\tau} = (\tau_1, \tau_2, \tau_3)$ denotes the Pauli matrices, and $\vec{S}$ is the Stokes vector of the photon with components

$$\begin{aligned}S_1 &= 2\,\mathrm{Re}\,(E^*M)\,/\,\left(|E|^2+|M|^2\right) \\ S_2 &= 2\,\mathrm{Im}\,(E^*M)\,/\,\left(|E|^2+|M|^2\right) \qquad (15.7) \\ S_3 &= \left(|E|^2-|M|^2\right)/\left(|E|^2+|M|^2\right).\end{aligned}$$

The component S_3 measures the relative strength of the electric and magnetic radiation at a given point in the Dalitz plot. The effects of CP violation reside in the components S_1 and S_2, which are proportional to $\mathrm{Re}(g_{Br}{}^* g_{M1})$ and $\mathrm{Im}(g_{Br}{}^* g_{M1})$, respectively. Of these S_1 is CP-odd, T-odd, while S_2 is CP-odd, T-even. Physically, S_2 is the net circular polarization of the photon: such a polarization is known to be possible in decays like $K_L \to \pi^+\pi^-\gamma$ or $K_{L,S} \to \gamma\gamma$ whenever there is CP violation accompanied by unitarity phases [3]. To understand the significance of S_1, we examine the dependence of the $K_L \to \pi^+\pi^-\gamma$ decay on the angle ϕ between the polarization vector $\vec{\epsilon}$ and the unit vector $\vec{n}_\pi$ normal to the $\pi^+\pi^-$ plane (we choose coordinates such that $\vec{k} = (0, 0, k)$, $\vec{n}_\pi = (1, 0, 0)$, $\vec{p}_+ = (0, p\sin\theta, p\cos\theta)$, and $\vec{\epsilon} = (\cos\phi, \sin\phi, 0)$):

$$\frac{d\Gamma}{d\omega\, d\cos\theta\, d\phi} \sim |E\sin\phi - M\cos\phi|^2 \sim 1 - [S_3\cos 2\phi + S_1\sin 2\phi]\,. \quad (15.8)$$

Thus the CP-odd, T-odd Stokes parameter S_1 appears as a coefficient of the term $\sin 2\phi$. The essential idea of refs. [4, 5] is to use in place of $\vec{\epsilon}$ the vector $\vec{n}_l$ normal to the plane of the Dalitz pair in the reaction $K_L \to \pi^+\pi^-\gamma^* \to \pi^+\pi^- e^+e^-$. This motivates the study of the distribution $d\Gamma/d\phi$ in the decay $K_L \to \pi^+\pi^- e^+e^-$, where ϕ is the angle between the $\pi^+\pi^-$ and e^+e^- planes.

To obtain a quantitative idea of the magnitude of CP violation in $K_L \to \pi^+\pi^-\gamma$, we show in fig. 15.1a the three components of the Stokes vector as a function of the photon energy [6]. These are calculated from the amplitudes (15.4) using weighted averages of $|E|^2$, $|M|^2$, E^*M, and EM^* over $\cos\theta$. The values of S_1 and S_2 are remarkably large, considering that the only source of CP violation is the ϵ-impurity in the K_L wave function ($\epsilon = \eta_{+-}$). Clearly the $1/\omega^2$ factor in E enhances it to a level that makes it comparable to the CP-conserving amplitude M. This is evident from the behavior of the parameter S_3, which swings from a dominant electric behavior at low E_γ ($S_3 \approx 1$) to a dominant magnetic behavior at large E_γ ($S_3 \approx -1$), with a zero in the region $E_\gamma \approx 60\, MeV$. To highlight the difference between the T-odd parameter S_1 and the T-even parameter S_2, we show in fig. 15.1b the behavior of the Stokes parameters in the "hermitian limit": this is the limit in which the T matrix or effective Hamiltonian governing the decay $K_L \to \pi^+\pi^-\gamma$ is taken to be hermitian, all unitarity phases related to real intermediate states being dropped. This limit is realized by taking $\delta_0, \delta_1 \to 0$, and $\arg\epsilon \to \pi/2$. The last of these follows from the fact that ϵ may be written as

$$\epsilon = \frac{\Gamma_{12} - \Gamma_{21} + i\,(M_{12} - M_{21})}{\gamma_S - \gamma_L - 2i\,(m_L - m_S)}, \quad (15.9)$$

where $H_{\text{eff}} = M - i\Gamma$ is the mass matrix of the K^0-$\overline{K}^0$ system. The hermitian limit obtains when $\Gamma_{12} = \Gamma_{21} = \gamma_S = \gamma_L = 0$. As seen from fig. 15.1b, S_2 vanishes in this limit, but S_1 survives, as befits a CP-odd,

T-odd parameter. Fig. 15.1c shows what happens in the CP-invariant limit $\epsilon \to 0$. It is clear that we are dealing here with a dramatic situation in which a CP impurity of a few parts in a thousand in the K_L wave function gives rise to a huge CP-odd, T-odd effect in the photon polarization.

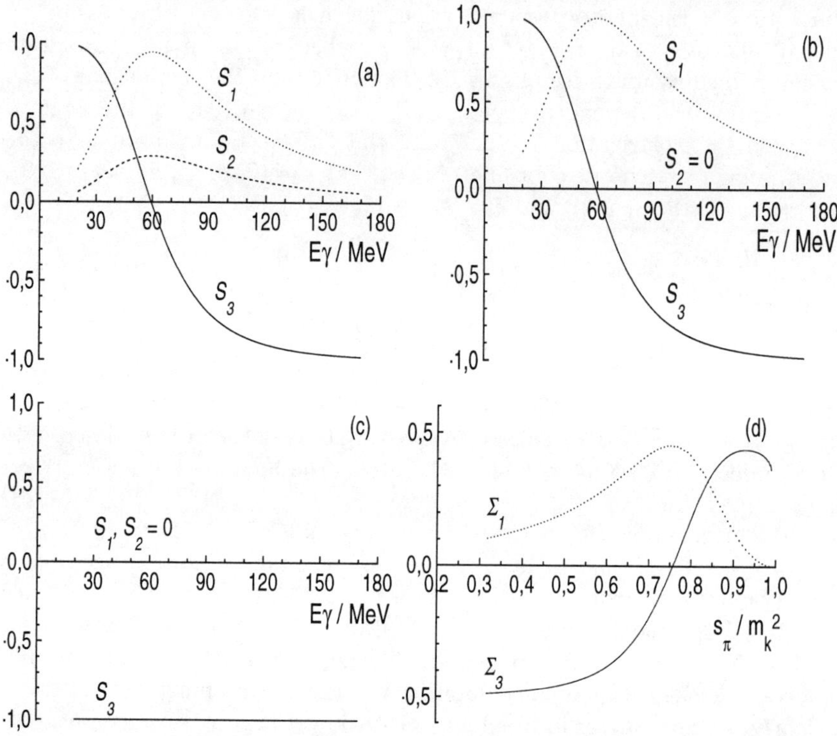

Figure 15.1: (a) Stokes parameters of photon in $K_L \to \pi^+\pi^-\gamma$; (b) hermitian limit $\delta_0 = \delta_1 = 0$, $\arg\epsilon = \pi/2$; (c) CP-invariant limit $\epsilon \to 0$; (d) "Stokes parameters" for $K_L \to \pi^+\pi^-e^+e^-$.

We can now examine how these large CP-violating effects are transported to the decay $K_L \to \pi^+\pi^-e^+e^-$. The matrix element for $K_L \to \pi^+\pi^-e^+e^-$ can be written as [4, 5]

$$\mathcal{M}(K_L \to \pi^+\pi^-e^+e^-) = \mathcal{M}_{br} + \mathcal{M}_{mag} + \mathcal{M}_{CR} + \mathcal{M}_{SD}. \tag{15.10}$$

Here $\mathcal{M}_{br}$ and $\mathcal{M}_{mag}$ are the conversion amplitudes associated with the bremsstrahlung and $M1$ parts of the $K_L \to \pi^+\pi^-\gamma$ amplitude. In addition, we have introduced an amplitude $\mathcal{M}_{CR}$ denoting $\pi^+\pi^-$ production in a $J = 0$ state (not possible in a real radiative decay), as well as an amplitude $\mathcal{M}_{SD}$ associated with the short-distance interaction $s \to d e^+e^-$. The last of these turns out to be numerically negligible because of the smallness of the *CKM* factor $V_{ts}V_{td}{}^*$. The s-wave amplitude $\mathcal{M}_{CR}$, if approximated by the K^0 charge radius diagram, makes a small ($\sim 1\%$) contribution to the decay rate. Thus the dominant features of the decay are due to the conversion amplitude $\mathcal{M}_{br} + \mathcal{M}_{mag}$.

Within such a model, one can calculate the differential decay rate in the form [5]

$$d\Gamma = I(s_\pi,\, s_l,\, \cos\theta_l,\, \cos\theta_\pi,\, \phi)\, ds_\pi\, ds_l\, d\cos\theta_l\, d\cos\theta_\pi\, d\phi. \tag{15.11}$$

Here s_π (s_l) is the invariant mass of the pion (lepton) pair, and θ_π (θ_l) is the angle of the π^+ (l^+) in the $\pi^+\pi^-$ (l^+l^-) rest frame, relative to the dilepton (dipion) momentum vector in that frame. The all-important variable ϕ is defined in terms of unit vectors constructed from the pion momenta $\vec{p_\pm}$ and lepton momenta $\vec{k_\pm}$ in the K_L rest frame:

$$\vec{n}_\pi = (\vec{p}_+ \times \vec{p}_-) \,/\, |\vec{p}_+ \times \vec{p}_-|, \quad \vec{n}_l = \left(\vec{k}_+ \times \vec{k}_-\right) / \left|\vec{k}_+ \times \vec{k}_-\right|,$$
$$\vec{z} = (\vec{p}_+ + \vec{p}_-) \,/\, |\vec{p}_+ + \vec{p}_-|,$$

$$\begin{aligned} \sin\phi &= \vec{n}_\pi \times \vec{n}_l \cdot \vec{z} \quad (CP = -, T = -), \\ \cos\phi &= \vec{n}_l \cdot \vec{n}_\pi \qquad (CP = +, T = +). \end{aligned} \tag{15.12}$$

In ref. [4], an analytic expression was derived for the 3-dimensional distribution $d\Gamma/ds_l\, ds_\pi\, d\phi$, which has been used in the Monte Carlo simulation of this decay. In ref. [5], a formalism was presented for obtaining the fully differential decay function $I(s_\pi,\, s_l,\, \cos\theta_l,\, \cos\theta_\pi, \phi)$.

The principal results of the theoretical model discussed in [4, 5] are as follows:

1. Branching ratio: This was calculated to be [4]

$$\begin{aligned} BR(K_L \to \pi^+\pi^-e^+e^-) &= (1.3\times 10^{-7})_{Br} + (1.8\times 10^{-7})_{M1} \\ &\quad + (0.04\times 10^{-7})_{CR} \\ &\approx 3.1\times 10^{-7}, \end{aligned} \tag{15.13}$$

which agrees well with the result $(3.32 \pm 0.14 \pm 0.28) \times 10^{-7}$ measured in the KTeV experiment [7]. (A preliminary branching ratio 2.9×10^{-7} has been reported by NA48 [8]).

2. Asymmetry in ϕ distribution: The model predicts a distribution of the form

$$\frac{d\Gamma}{d\phi} \sim 1 - (\Sigma_3 \cos 2\phi + \Sigma_1 \sin 2\phi)\,, \tag{15.14}$$

where the last term is CP and T violating, and produces an asymmetry

$$\mathcal{A} = \frac{\left(\int_0^{\pi/2} - \int_{\pi/2}^{\pi} + \int_{\pi}^{3\pi/2} - \int_{3\pi/2}^{2\pi}\right) \frac{d\Gamma}{d\phi} d\phi}{\left(\int_0^{\pi/2} + \int_{\pi/2}^{\pi} + \int_{\pi}^{3\pi/2} + \int_{3\pi/2}^{2\pi}\right) \frac{d\Gamma}{d\phi} d\phi} = -\frac{2}{\pi}\Sigma_1. \tag{15.15}$$

The predicted value [4,5] is

$$|\mathcal{A}| = 15\% \sin(\phi_{+-} + \delta_0(M_K{}^2) - \overline{\delta}_1) \approx 14\% \qquad (15.16)$$

to be compared with the KTeV result [7]

$$|\mathcal{A}|_{KTeV} = (13.6 \pm 2.5 \pm 1.2)\%. \qquad (15.17)$$

The "Stokes parameters" Σ_3 and Σ_1 are calculated to be $\Sigma_3 = -0.133$, $\Sigma_1 = 0.23$. The ϕ distribution measured by KTeV agrees with this expectation (after acceptance corrections made in accordance with the model). It should be noted that the sign of Σ_1 (and of the asymmetry $\mathcal{A}$) depends on whether the numerical coefficient in g_{M1} is taken to be $+0.76$ or -0.76. The data happen to support the positive sign chosen in eq. (15.2).

3. Variation of Stokes parameters with s_π: As shown in fig. 15.1d, the parameters Σ_1 and Σ_3 have a variation with s_π that is in close correspondence with the variation of S_1 and S_3 shown in fig. 15.1b. (Recall that the photon energy E_γ in $K_L \to \pi^+\pi^-\gamma$ can be expressed in terms of s_π: $s_\pi = M_K{}^2 - 2M_K E_\gamma$.) In particular the zero of Σ_3 and the zero of S_3 occur at almost the same value of s_π. This variation with s_π, combined with the low detector acceptance at large s_π, has the consequence of enhancing the measured asymmetry ($23.3 \pm 2.3\%$ in KTeV [7], $20 \pm 5\%$ in NA48 [8]).

4. Generalized Angular Distribution: As shown in ref. [5], a more complete study of the angular distribution of the decay $K_L \to \pi^+\pi^-e^+e^-$ can yield further CP-violating observables, some of which are sensitive to the nonradiative (charge-radius and short-distance) parts of the matrix element. In particular the two-dimensional distribution $d\Gamma/d\cos\theta_l d\phi$ has the form

$$\begin{aligned}\frac{d\Gamma}{d\cos\theta_l d\phi} \quad = \quad & K_1 + K_2\cos 2\theta_l + K_3\sin^2\theta_l\cos 2\phi + K_4\sin 2\theta_l\cos\phi \\ & + K_5\sin\theta_l\cos\phi + K_6\cos\theta_l + K_7\sin\theta_l\sin\phi \\ & + K_8\sin 2\theta_l\sin\phi + K_9\sin^2\theta_l\sin 2\phi. \qquad (15.18)\end{aligned}$$

Considering the behavior of $\cos\theta_l$, $\sin\theta_l$, $\cos\phi$, and $\sin\phi$ under CP and T, the various terms appearing in eq. (15.18) have the following transformation:

	CP	T
K_1, K_2, K_3, K_5	+	+
K_4, K_6	−	+
K_8	+	−
K_7, K_9	−	−

Note that $K_{4,6,8}$ have (CP)(T) = −, a signal that they vanish in the hermitian limit. If only the bremsstrahlung and magnetic dipole terms are retained in the $K_L \to \pi^+\pi^-e^+e^-$ amplitude, one finds $K_4 = K_5 = K_6 = K_7 = K_8 = 0$, the only nonzero coefficients being $K_1 = 1$ (norm),

$K_2 = 0.297$, $K_3 = 0.180$, $K_9 = -0.309$. In this notation, the asymmetry in $d\Gamma/d\phi$ is

$$\mathcal{A} = \frac{2}{\pi}\frac{\frac{2}{3}K_9}{1-\frac{1}{3}K_2} = -14\%.$$

The introduction of a charge-radius term induces a new CP-odd, T-even term $K_4 \approx -1.3\%$, while a short-distance interaction containing an axial vector electron current can induce the CP-odd, T-odd term K_7. The standard model prediction for K_7, however, is extremely small.

We conclude with a list of questions that could be addressed by future research. In connection with $K_{L,S} \to \pi^+\pi^-\gamma$: (i) Is there a departure from bremsstrahlung in $K_S \to \pi^+\pi^-\gamma$ (evidence for direct $E1$)? (ii) Is there a π^+/π^- asymmetry in $K_L \to \pi^+\pi^-\gamma$ (evidence for $E2$)? (iii) Is there a measurable difference between $\eta_{+-\gamma}$ and η_{+-} (existence of direct CP-violating $E1$ in $K_L \to \pi^+\pi^-\gamma$)? With respect to the decay $K_L \to \pi^+\pi^-e^+e^-$: (i) Is there evidence of an s-wave amplitude? (ii) Is there evidence for K_4 or K_7 types of CP violation? On the theoretical front: (i) Can one calculate the s-wave amplitude, and the form factors in $K_L \to \pi^+\pi^-\gamma^*$ [9]? (ii) Can one understand the sign of g_{M1}? (iii) Can one explain why direct $E1$ in $K_S \to \pi^+\pi^-\gamma$ is so small compared to direct $M1$ in $K_L \to \pi^+\pi^-\gamma$ ($|g_{E1}/g_{M1}| < 5\%$)?

References

[1] E.J. Ramberg *et al.*, Fermilab Report No. Fermilab-Conf-91/258, 1991; Phys. Rev. Lett. **70**, 2525 (1993).

[2] T.D. Lee and C.S. Wu, Annu. Rev. Nucl. Sci. **16**, 511 (1966); G. Costa and P.K. Kabir, Nuovo Cimento A **61**, 564 (1967); L.M. Sehgal and L. Wolfenstein, Phys. Rev. **162**, 1362 (1967).

[3] L.M. Sehgal, Phys. Rev. D **4**, 267 (1971).

[4] L.M. Sehgal and M. Wanninger, Phys. Rev. D **46**, 1035 (1992); *ibid.*, D **46**, 5209(E) (1992).

[5] P. Heiliger and L.M. Sehgal, Phys. Rev. D **48**, 4146 (1993).

[6] L.M. Sehgal and J. van Leusen, Phys. Rev. Lett. **83**, 4933 (1999).

[7] KTeV Collaboration, A. Alavi-Harati *et al.*, hep-ex/9908020, submitted to Phys. Rev. Lett.; A. Ledovskoy, this volume, chapter 17.

[8] NA48 Collaboration, S. Wronka, this volume, chapter 18..

[9] M.J. Savage, this volume, chapter 16.

16

$K_L^0 \to \pi^+\pi^-e^+e^-$ in Chiral Perturbation Theory

Martin J. Savage

Kaon decays continue to provide invaluable information about the approximate discrete symmetries of nature. CP violation in $K_L^0 \to \pi\pi$ [1], originating in the $K^0 - \overline{K}^0$ mass matrix, was discovered nearly forty years ago, and direct CP violation in K decays has been unambiguously established [2–4] through a recent remeasurement of $\mathrm{Re}\,(\epsilon'/\epsilon) = (28.0 \pm 3.0 \pm 2.8) \times 10^{-4}$ by the KTeV Collaboration [4]. KTeV has also observed and studied [5] the rare decay $K_L^0 \to \pi^+\pi^-e^+e^-$. A large CP-violating asymmetry, $B_{\mathrm{CP}} = 13.6\pm2.5\pm1.2\%$, constructed from the final state particles was measured [5], consistent with theoretical predictions [6–9]. This decay is dominated by a one-photon intermediate state, $K_L^0 \to \pi^+\pi^-\gamma^* \to \pi^+\pi^-e^+e^-$ and B_{CP} receives a sizable strong interaction enhancement.

A long-standing problem in better understanding K decays and a roadblock to more precisely constraining the standard model of electroweak interactions or uncovering new physics is our present inability to compute the hadronic matrix elements of most electroweak operators to high precision. The lattice provides the only direct method with which to determine these matrix elements; however, it is presently far from being able to compute matrix elements between multihadronic initial and final states. Chiral perturbation theory, χPT, is a framework in which the low-energy strong interactions of the lowest-lying pseudo-Goldstone bosons can be treated in perturbation theory. The external momentum and the light quark mass matrix are treated as small expansion parameters when normalized to the chiral symmetry breaking scale, $\Lambda_\chi \sim 1$ GeV. This article presents the χPT analysis of $K_L^0 \to \pi^+\pi^-e^+e^-$, focusing entirely on the one-photon intermediate state, as shown in fig. 16.1.

The matrix element for $K_L^0 \to \pi^+\pi^-e^+e^-$, assuming CPT invariance, is written in terms of three form factors G, F_+, and F_-,

$$\mathcal{M} = \frac{s_1\, G_F\, \alpha}{4\pi\, f\, q^2}\left[\, i\, G\, \varepsilon_{\mu\lambda\rho\sigma}\, p_+^\lambda\, p_-^\rho\, q^\sigma \; + \; F_+\, p_{+,\mu} \; + \; F_-\, p_{-,\mu}\,\right] \overline{u}(k_-)\gamma^\mu\, v(k_+), \tag{16.1}$$

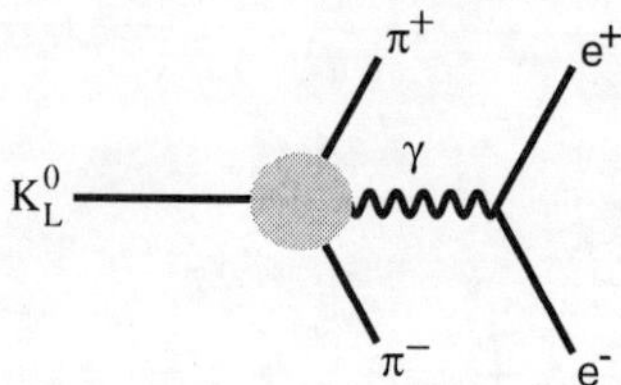

Figure 16.1: The one-photon intermediate state dominates $K_L^0 \to \pi^+\pi^-e^+e^-$. The solid circle denotes the $K_L^0 \to \pi^+\pi^-\gamma^*$ vertex.

where $k_{+,-}$ are the positron and electron momenta respectively, $q = k_+ + k_-$ is the photon momentum, and $p_{+,-}$ are the $\pi^{+,-}$ momenta respectively. G_F is Fermi's coupling constant, s_1 is the sine of the Cabibbo angle, f is the pion decay constant, and α is the electromagnetic fine structure constant. The form factors are functions of hadronic kinematic invariants, *e.g.*, $F_+ = F_+(q^2, q\cdot p_+, q\cdot p_-)$. The smallness of $\mathrm{Re}\,(\epsilon'/\epsilon)$ suggests that to a very good approximation direct CP violation that may contribute to this decay can be neglected. For our purposes the only CP violation that will enter into this decay is due to ϵ, indirect CP violation introduced by the K_L^0 wave function. In terms of the eigenstates of CP, $K_{1,2}$, the K_L^0 wave function is

$$\begin{aligned} |K_L^0\rangle &= |K_2\rangle + \epsilon|K_1\rangle \\ |K_1\rangle &= \frac{1}{\sqrt{2}}\left[\,|K^0\rangle - |\overline{K}^0\rangle\,\right] \\ |K_2\rangle &= \frac{1}{\sqrt{2}}\left[\,|K^0\rangle + |\overline{K}^0\rangle\,\right], \end{aligned} \tag{16.2}$$

where $CP|K_1\rangle = +|K_1\rangle$, $CP|K_2\rangle = -|K_2\rangle$, and $\epsilon = 0.0023\; e^{i44^\circ}$. As direct CP violation is being neglected it is convenient to determine the contributions to G, F_+, and F_- from K_1 and K_2 independently as the two contributions do not interfere in the total decay rate Γ or differential decay rate $d\Gamma/dq^2$. The CP-odd component of the K_L^0 wave function, K_2, gives contributions to the form factors with symmetry properties $G \to +G$, and $F_\pm \to +F_\mp$ under interchange $p_\pm \to p_\mp$, while the contributions from the CP-even component of the K_L^0 wave function, K_1, have symmetry properties $F_\pm \to -F_\mp$ under interchange $p_\pm \to p_\mp$ (where it is understood that the interchange $p_\pm \to p_\mp$ also occurs for the arguments of the form factors).

The Lagrange density that describes the leading order strong and $\Delta S = 1$ weak interactions of the lowest-lying octet of pseudo-Goldstone bosons is

$$\begin{aligned} \mathcal{L} &= \frac{f^2}{8}\mathrm{Tr}\left[D^\mu\Sigma D_\mu\Sigma^\dagger\right] + \lambda\mathrm{Tr}\left[m_q\,\Sigma + \mathrm{h.c.}\right] \\ &+ g_8\,\frac{G_F s_1 f^4}{4\sqrt{2}}\left(\left[D^\mu\Sigma D_\mu\Sigma^\dagger H_w\right] + \mathrm{h.c.}\right), \end{aligned} \tag{16.3}$$

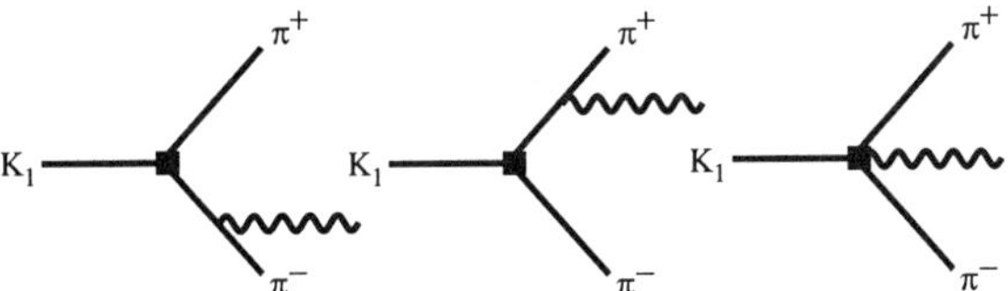

Figure 16.2: The leading order contribution to $K_L^0 \to \pi^+\pi^-\gamma^*$ in χPT. A solid square denotes a weak interaction. Only the K_1 component of the K_L^0 wave function can contribute at tree-level and this contribution is suppressed by a factor of ϵ.

where

$$\Sigma = \exp\left[\frac{2i}{f_\pi}\begin{pmatrix} \frac{1}{\sqrt{2}}\pi^0 + \frac{1}{\sqrt{6}}\eta & \pi^+ & K^+ \\ \pi^- & -\frac{1}{\sqrt{2}}\pi^0 + \frac{1}{\sqrt{6}}\eta & \frac{1}{\sqrt{2}}K_2^0 + \frac{1}{\sqrt{2}}K_1^0 \\ K^- & \frac{1}{\sqrt{2}}K_2^0 - \frac{1}{\sqrt{2}}K_1^0 & -\frac{2}{\sqrt{6}}\eta \end{pmatrix}\right], \tag{16.4}$$

and H_w is a 3×3 matrix with a "1" in the $(1,3)$ entry, inducing a $s \to u$ transition. Octet dominance ($\Delta I = \frac{1}{2}$) has been assumed and thus contributions from the **27** component of the $\Delta S = 1$ Hamiltonian have been neglected. The constant $|g_8| = 5.1$ is fit to the amplitude for $K \to \pi\pi(I = 0)$.

In computing observables in χPT, the external momentum and quark masses are expansion parameters in which the form factors are expanded, *e.g.*, $G = G^{(1)} + G^{(2)} + G^{(3)} + \cdots$. The form factor $G^{(r)}$ is associated with a contribution of order Q^{2r-1}, where $Q = p, m$, the external momenta or meson mass. The same expansion and notation is used for the $F_\pm$ form factors. Unlike the contributions from the K_2 component, contributions from the K_1 component are suppressed by a factor of ϵ. However, the leading order contribution to $K_L^0 \to \pi^+\pi^-\gamma^*$, $r = 1$, is from tree graphs involving the K_1 component, as shown in fig. 16.2. A simple calculation yields

$$\begin{aligned} F_{+,1}^{(1)} &= -\epsilon\, g_8 \frac{32 f_\pi^2 (m_K^2 - m_\pi^2)\pi^2}{q^2 + 2q \cdot p_+} \\ F_{-,1}^{(1)} &= +\epsilon\, g_8 \frac{32 f_\pi^2 (m_K^2 - m_\pi^2)\pi^2}{q^2 + 2q \cdot p_-}, \end{aligned} \tag{16.5}$$

and $G^{(1)} = 0$, which has the correct symmetry under $p_\pm \to p_\mp$ as discussed previously. The subscript on the $F_\pm$ form factors indicates that the contribution comes from the K_1 component. As all constants appearing in eq. (16.5) are determined by other processes; this is a parameter-free leading order prediction. Final state strong interactions that contribute to $F_{\pm,1}$ are important for CP-violating asymmetries such as $B_{\rm CP}$. The leading final state interactions associated with $F_{\pm,1}$ are generated by graphs shown

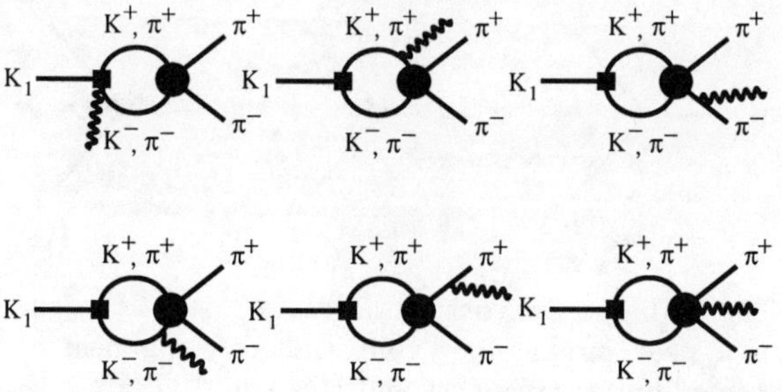

Figure 16.3: The leading final-state interactions in $K_1 \to \pi^+\pi^-\gamma^*$. A solid square denotes a weak interaction and a solid circle denotes a strong interaction. These contribution are proportional to ϵ.

in fig. 16.3. Retaining only the imaginary parts of the graphs, naively enhanced by factors of π over the real parts, we have

$$\mathrm{Im}\left[F_{+,1}^{(2)}\right] = -g_8\ \pi\epsilon\ \frac{\left(m_K^2 - m_\pi^2\right)\left(4m_K^2 - 2m_\pi^2\right)}{q^2 + 2q\cdot p_+}\sqrt{1 - \frac{4m_\pi^2}{m_K^2}}, \qquad (16.6)$$

which is the leading term in building up $e^{i\delta_0}$, where δ_0 is the $I = J = 0$ $\pi\pi$ phase shift evaluated at $s = m_K^2$.

Decay of the K_2 component is described by both the G and $F_{\pm,2}$ form factors starting at $r = 2$, as can be seen from eq. (16.1). At this order in χPT, $G^{(2)}$ is a constant that must be determined from data. The M1 fraction of the decay rate for $K_L^0 \to \pi^+\pi^-\gamma$ is reproduced if $G^{(2)} = 39.3$, where higher-order (momentum-dependent) contributions have been neglected, and $G^{(2)}$ is real. The Dalitz plot for $K_L^0 \to \pi^+\pi^-\gamma$ indicates that there is nonnegligible momentum dependence in G, and therefore higher-order terms will be important [10,11]. This introduces an uncertainty into the prediction of differential rates and CP-violating asymmetries at the order to which we are working. At $r = 3$ there are contributions, not only from loop diagrams, but also from higher-order weak interactions and the Wess-Zumino term. However, as before we are able to compute the leading contribution to the imaginary part of G, which go to build up the final-state interactions, $e^{i\delta_1}$, where δ_1 is the phase shift for $\pi\pi$ scattering in the $I = J = 1$ channel. It is found that

$$\mathrm{Im}\left[G^{(3)}\right] = G^{(2)}\ \frac{s}{48\pi f^2}\left[1 - \frac{4m_\pi^2}{s}\right]^{3/2}, \qquad (16.7)$$

where s is the invariant mass of the $\pi^+\pi^-$ system.

The $F_{\pm,2}$ form factors do not arise only from the charge radius of the K^0 as was assumed in the analyses of [6,7]. In fact, the K^0 charge radius is one of several different types of one-loop graphs arising at $r = 2$ that

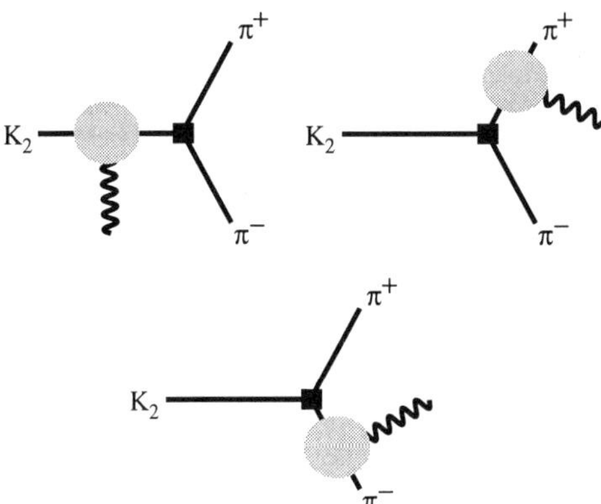

Figure 16.4: Contributions to $K_L^0 \to \pi^+\pi^-\gamma^*$ from the K and π charge radii. The solid square denotes a weak interaction and the lightly shaded circle denotes the sum of one-loop graphs and counterterms that form the charge radius of either the K or the π.

give rise to q^2 dependence in the $F_{\pm,2}$. The diagrams giving contributions from the charge radii of the K^0 and the $\pi^\pm$ are shown in fig. (16.4). The sum of the one-loop diagrams contributing to the K^0 charge radius is finite, while those contributing to the π charge radius are divergent and require the counterterm

$$\mathcal{L} \;=\; -i\,\lambda_{\rm cr}\frac{e}{16\pi^2}\;F^{\mu\nu}\;Tr\left[Q\left(D_\mu\Sigma D_\nu\Sigma^\dagger + D_\mu\Sigma^\dagger D_\nu\Sigma\right)\right], \quad (16.8)$$

where Q is the light-quark electromagnetic charge matrix, and $F^{\mu\nu}$ is the electromagnetic field strength tensor. The coefficient $\lambda_{\rm cr} = -0.91 \pm 0.06$ has been determined from measurements of the π charge radius.

Diagrams that are not charge radius type contributions are shown in fig. 16.5. Analytic expressions for the diagrams shown in fig. 16.5, given in [8,9], are somewhat lengthy and we do not present them here. The sum of the graphs in fig. 16.5 is not finite and the counterterms that enter at this order are described by the Lagrange density [12]

$$\begin{aligned}\mathcal{L} \;=\;\; & i\,g_8\,\frac{G_F\;s_1\;e\;f_\pi^2}{16\sqrt{2}\pi^2}\left[\;a_1\;F^{\mu\nu}\;{\rm Tr}\left[QH_w(\Sigma D_\mu\Sigma^+)(\Sigma D_\nu\Sigma^\dagger)\right]\right.\\ & +a_2\;F^{\mu\nu}\;{\rm Tr}\left[Q(\Sigma D_\mu\Sigma^\dagger)H_w(\Sigma D_\nu\Sigma^\dagger)\right]\\ & +a_3\;F^{\mu\nu}\;{\rm Tr}\left[H_w[Q,\Sigma]D_\mu\Sigma^\dagger\Sigma D_\nu\Sigma^\dagger - H_w D_\mu\Sigma D_\nu\Sigma^\dagger\Sigma[\Sigma^\dagger,Q]\right]\\ & \left.+a_4\;F^{\mu\nu}\;{\rm Tr}\left[H_w\Sigma D_\mu\Sigma^\dagger[Q,\Sigma]D_\nu\Sigma^\dagger\right]\;\right] + h.c., \end{aligned} \quad (16.9)$$

where the constants $a_{1,2,3,4}$ must be determined from data. The combination of counterterms that contributes to $K_L^0 \to \pi^+\pi^-\gamma^*$ is

$$w \;=\; a_3 - a_4 + \frac{1}{6}(a_1 + 2a_2) + \lambda_{\rm cr}, \quad (16.10)$$

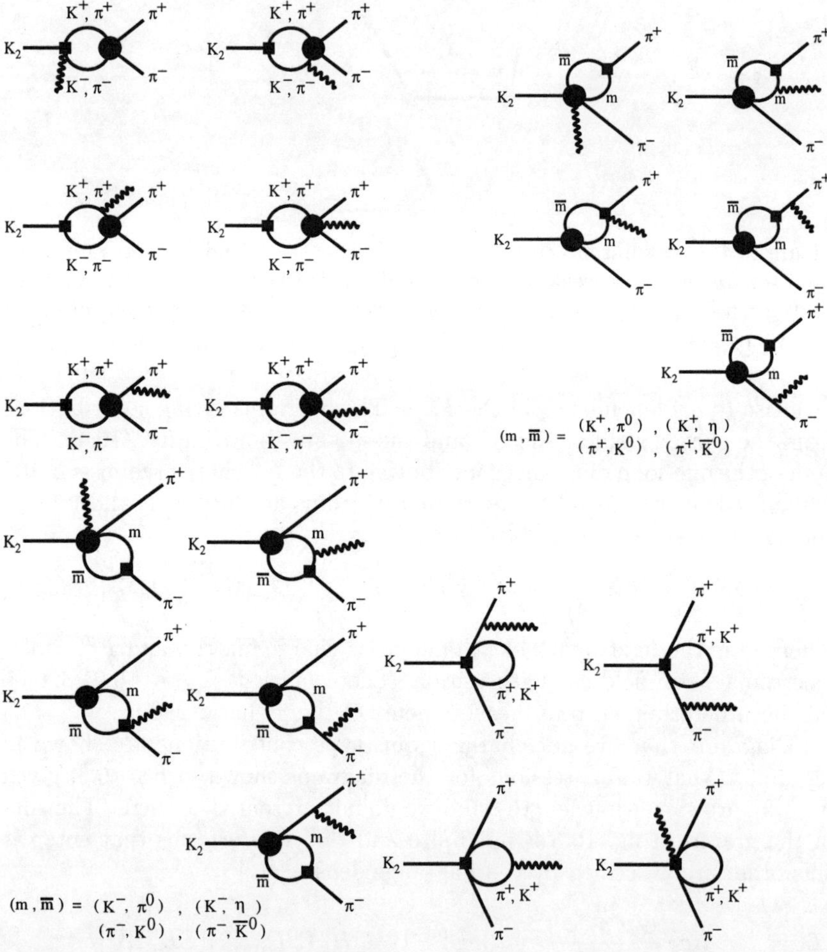

Figure 16.5: One-loop contributions to $K_2 \to \pi^+\pi^-\gamma^*$ from diagrams that do not contribute to the charge radius of the K or π. A solid square denotes a weak interaction and a solid circle denotes a strong interaction.

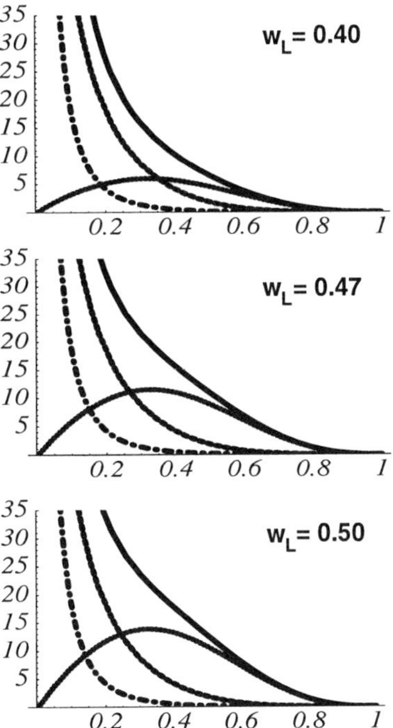

Figure 16.6: The branching fraction $\frac{1}{\Gamma_{\rm tot}}\frac{d\Gamma}{dy}$ verses y, where $y = \sqrt{q^2}/(m_K - 2m_\pi)$. The dot-dashed, dashed, and dotted curves are the contributions from $F_{\pm,1}$, G, and $F_{\pm,2}$ respectively, while the solid curve is the sum of the contributions. The three different plots correspond to the counterterm w_L taking the values 0.40, 0.47, and 0.50 respectively.

while the combination that contributes to $K^+ \to \pi^+e^+e^-$ is

$$w_+ = \frac{2}{3}(a_1 + 2a_2) - 4\lambda_{\rm cr} - \frac{1}{6}\log\left(\frac{m_K^2\, m_\pi^2}{\mu^4}\right) + \frac{1}{3}. \tag{16.11}$$

One has the choice to write the $F_{\pm,2}$ in terms of w or to use the known values of $\lambda_{\rm cr}$ and $a_1 + 2a_2$ and define the finite, μ-independent combination $w_L = a_3 - a_4$ [8]. The value of w_L can be determined from the rate for $K_L^0 \to \pi^+\pi^-e^+e^-$.

The differential decay rate is the incoherent sum of the rates from the three form factors,

$$\frac{d\Gamma}{dq^2} = \frac{d\Gamma_G}{dq^2} + \frac{d\Gamma_{F_1}}{dq^2} + \frac{d\Gamma_{F_2}}{dq^2}, \tag{16.12}$$

due to the symmetry properties of the amplitudes. In fig. (16.6) we have shown the differential branching fraction $\frac{1}{\Gamma_{\rm tot}}\frac{d\Gamma}{dy}$, where $y = \sqrt{q^2}/(m_K - 2m_\pi)$, for different values of w_L, given the central value of $w_+ = 0.89$ [13]

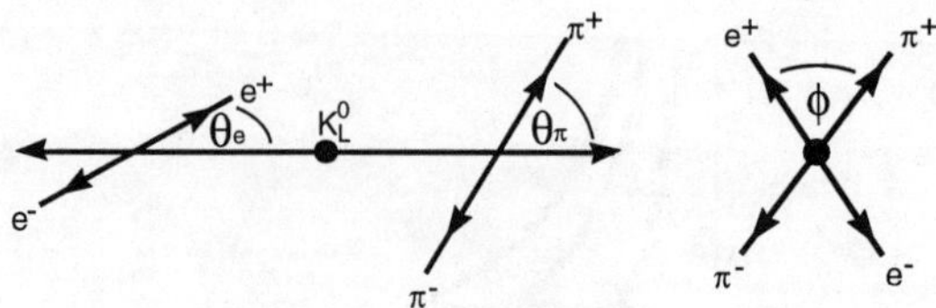

Figure 16.7: Definition of the Pais-Trieman variables, ϕ, θ_e, and θ_π.

and the central value of $\lambda_{\rm cr} = -0.91$. The contribution to the differential rate from $F_{\pm,2}$ vanishes as $q^2 \to 0$, but clearly dominates the high q^2 region (for most values of w_L). Except for the $q^2 \to 0$ region, the contribution from G dominates over the contribution from $F_{\pm,1}$. It is clear that in order to determine w_L a relatively high cut on the e^+e^- invariant mass must be made. To emphasize this point, the branching fraction for $K_L^0 \to \pi^+\pi^-e^+e^-$ with a cut of $q^2_{\rm cut} > (2\ {\rm MeV})^2$ is (using the parameter values already discussed)

$$\mathrm{Br} = \left(16.1 + 10.7 + \left[3.7 - 3.5w_L + 0.8w_L^2\right]\right) 10^{-8}, \tag{16.13}$$

where the first contribution is from G, the second is from F_1, and the third is from F_2. In contrast, the branching fraction with a cut of $q^2_{\rm cut} > (80\ {\rm MeV})^2$ is

$$\mathrm{Br} = \left(0.60 + 0.07 + \left[1.9 - 1.8w_L + 0.4w_L^2\right]\right) 10^{-8}. \tag{16.14}$$

With the presently available branching fraction of $\mathrm{Br} = (3.32\pm0.14\pm0.28)\times 10^{-7}$ from KTeV [5], which has a $q^2_{\rm cut} > (2\ {\rm MeV})^2$ cut, $w_L = 4.7 \pm 0.7$ or -0.6 ± 0.7, but these values depend sensitively upon $G^{(2)}$ and $F_{\pm,1}$ for obvious reasons. Only an analysis of the entire differential spectrum or the shape of the $\pi^+\pi^-$ invariant mass distribution will place more stringent bounds on w_L.

One of the most exciting aspects of $K_L^0 \to \pi^+\pi^-e^+e^-$ is the large value of $B_{\rm CP}$ that is predicted [6–9] and also recently observed by KTeV [5]. $B_{\rm CP}$ is defined to be

$$\begin{aligned} B_{\rm CP} &= \langle \mathrm{Sign}\left[\sin\phi\cos\phi\right]\rangle \\ &= \langle \mathrm{Sign}\left[(n_e\cdot n_\pi)\, n_e \times n_\pi \cdot \left(\frac{p_+ + p_-}{|p_+ + p_-|}\right)\right]\rangle, \end{aligned} \tag{16.15}$$

where ϕ is the Pais-Treiman variable depicted in fig. 16.7, n_e is the normal to the plane formed by the momenta of the e^+e^- pair and n_π is the normal to the plane formed by the momenta of the $\pi^+\pi^-$ pairs. It is integrated over the momenta of the final-state particles with any specified cuts. The integrand that contributes to $B_{\rm CP}$ is proportional to the combination

$$Im\left[G\left(F_{+,1} - F_{-,1}\right)^*\right] \;=\; Im\left[G^{(2)}\left(F_{+,1}^{(1)} - F_{-,1}^{(1)}\right)^*\right.$$

$$+ G^{(2)} \left(F_{+,1}^{(2)} - F_{-,1}^{(2)}\right)^* \qquad (16.16)$$
$$+ G^{(3)} \left(F_{+,1}^{(1)} - F_{-,1}^{(1)}\right)^* + \cdots \Big] .$$

The contribution from Re $\left[G^{(3)}\right]$ has not been computed, and therefore this does not constitute a complete computation of $B_{\rm CP}$ to next-to-leading order. However, the omitted contribution is expected to be small [8, 9]. With a cut of $q_{\rm cut}^2 > (2\ {\rm MeV})^2$ this asymmetry is found to be [8,9]

$$B_{\rm CP} \quad = \quad 9.2\% + 4.2\% + \cdots = 13.4\%, \qquad (16.17)$$

with an uncertainty estimated to be of order $\sim 2\%$ based on the difference between the leading and next-to-leading order contributions. This is in complete agreement with the recent KTeV [5] observation of $B_{\rm CP} = 13.6 \pm 2.5 \pm 1.2\%$ for this invariant mass cut, and consistent with the calculations of [6, 7]. The next-to-leading order contribution of 4.3% is from the final-state interactions associated with $F_{\pm,1}$. It is important to note that $F_{\pm,2}$ does not contribute to $B_{\rm CP}$, and hence the uncertainty in determining w_L does not impact this discussion. As emphasized by Sehgal [14], good agreement between theory and the current experimental value of $B_{\rm CP}$ is obtained within the context of the standard model, with CP violation from ϵ and CPT conservation. Recent discussions of the implication of this observation for T-violating interactions can be found in [15, 16]. While reversing the momenta of the final-state particles does change the sign of $B_{\rm CP}$ (it is T-odd), the initial and final states in the decay have not been interchanged. Therefore, a direct connection to T-violating interactions is absent.

In conclusion, I have presented a systematic analysis of the decay $K_L^0 \to \pi^+\pi^-e^+e^-$ in chiral perturbation up to next-to-leading order. This analysis differs from that of [6,7] in the form of the $K_L^0 \to \pi^+\pi^-\gamma^*$ dependence upon q^2. The size of this contribution is determined by a counterterm, w_L, that presently is only loosely constrained, but could be determined from the existing KTeV data with appropriate kinematic cuts. The large value of the CP asymmetry, $B_{\rm CP}$, that was predicted to arise naturally from ϵ has been confirmed by the KTeV Collaboration [5].

Acknowledgments

I would like to thank Jon Rosner and Bruce Winstein for putting together a very stimulating workshop. This work is supported in part by Department of Energy Grant DE-FG03-97ER41014.

References

[1] J.H. Christenson, J.W. Cronin, V.L. Fitch, and R. Turlay, *Phys. Rev. Lett.* **13**, 138 (1964).

[2] G. Barr *et al.* (NA31 Collaboration), *Phys. Lett.* B **317**, 233 (1993).

[3] L.K. Gibbons *et al.* (E731 Collaboration), *Phys. Rev. Lett.* **70**, 1203 (1993).

[4] A. Alavi-Harati *et al.* (KTeV Collaboration), hep-ex/9905060.

[5] J. Adams *et al.* (KTeV Collaboration), *Phys. Rev. Lett.* **80**, 4123 (1998); A. Ledovskoy (KTeV Collaboration), this volume, chapter 17.

[6] L.M. Sehgal and M. Wanninger, *Phys. Rev.* D **46**, 1035 (1992); *Phys. Rev.* D **46**, 5209 (1992)(E).

[7] P. Heiliger and L.M. Sehgal, *Phys. Rev.* D **48**, 4146 (1993).

[8] J.K. Elwood, M.J. Savage, and M.B. Wise, *Phys. Rev.* D **52**, 5095 (1995); *Phys. Rev.* D **53**, 2855 (1996)(E).

[9] J.K. Elwood, M.J. Savage, J.W. Walden, and M.B. Wise, *Phys. Rev.* D **53**, 4078 (1996).

[10] G. Ecker, H. Neufeld, and A. Pich, *Nucl. Phys.* B **413**, 321 (1994).

[11] G. D'Ambrosio and J. Portoles, *Nucl. Phys.* B **533**, 523 (1998).

[12] G. Ecker, J. Kambor, and D. Wyler, *Nucl. Phys.* B **394**, 101 (1993).

[13] G. Donaldson *et al.*, *Phys. Rev. Lett.* **33**, 554 (1974); *Phys. Rev.* D **14**, 2839 (1976); E.J. Ramberg *et al.*, *Phys. Rev. Lett.* **70**, 2525 (1993).

[14] L.M. Sehgal, this volume, chapter 15.

[15] J. Ellis and N.E. Mavromatos, hep-ph/9903386.

[16] L. Alvarez-Gaume, C. Kounnas, S. Lola, and P. Pavlopoulos, hep-ph/9903458.

17

KTeV Results on $K_L \to \pi^+\pi^-e^+e^-$

A. Ledovskoy[1]

Abstract

We present the first observation of a manifestly CP-violating effect in the $K_L \to \pi^+\pi^-e^+e^-$ decay mode. A large asymmetry was observed in the distribution of these decays in the CP-odd and T-odd angle ϕ between the normals to the decay planes of the e^+e^- and $\pi^+\pi^-$ pairs in the K_L center of mass system. After acceptance corrections according to the model discussed in this report, the asymmetry is found to be 13.6 ± 2.5 (stat) ± 1.2 (syst)%. This is the largest CP-violating effect yet observed and the first such effect observed in an angular variable. A new preliminary branching ratio for this decay is also presented.

Prior to the KTeV experiment, there were several theoretical predictions for the branching ratio and for observation of CP violation in the yet to be observed $K_L \to \pi^+\pi^-e^+e^-$ decay [1–4]. A large asymmetry is predicted in the distribution of CP-odd and T-odd angle ϕ between the normals to the decay planes of the e^+e^- and $\pi^+\pi^-$ pairs in the K_L center of mass system [1]. Talks presented by L. M. Sehgal [5] and M. Savage [6] at this conference gave a detailed theoretical introduction to CP violation aspects of the $K_L \to \pi^+\pi^-e^+e^-$ decay.

The first measurement of the $K_L \to \pi^+\pi^-e^+e^-$ branching ratio [7], reported by the KTeV Collaboration, was based on the observation of 36.6 ± 6.8 events after background subtraction. This corresponds to one day of data taking in 1997. A new preliminary KTeV result for branching ratio is $3.32 \pm 0.14(\text{stat}) \pm 0.28(\text{sys}) \times 10^{-7}$. It is based on 631 events from approximately half of the data accumulated by KTeV during 1997. This measurement is in good agreement with both theoretical predictions [1] and experimental results (published [7, 8] and presented at this conference [9, 10]).

KTeV collected approximately 2000 events of this rare kaon decay during the 1997 run, which made it possible to study various angular distributions and to measure ϕ asymmetry A for the first time. The asymmetry is defined as

[1]For KTeV Collaboration.

$$A \equiv \frac{N_{\sin\phi\cos\phi>0.0} - N_{\sin\phi\cos\phi<0.0}}{N_{\sin\phi\cos\phi>0.0} + N_{\sin\phi\cos\phi<0.0}}. \tag{17.1}$$

The KTeV detector consists of a vacuum decay region, a magnetic spectrometer with a pair of drift chambers on each side of a magnet, a CsI electromagnetic calorimeter, and a muon detector. More detailed description of the KTeV detector can be found in [7,11].

Four-track events with a common vertex and $\pi^+\pi^-e^+e^-$ topology were selected. A track with $0.95 \leq E/p \leq 1.05$ was identified as an electron and with $E/p \leq 0.90$ as a pion, where E is the energy deposited by a particle in the calorimeter and p is the momentum measured by the spectrometer. Various track topology cuts were also applied.

Fig. 17.1 shows the $M_{\pi\pi ee}$ distribution for selected events. The majority of events are $K_L \to \pi^+\pi^-\pi^0_D$ decays with a photon missing and $M_{\pi\pi ee} < M_{kaon}$. There is a noticeable $K_L \to \pi^+\pi^-e^+e^-$ peak above a comparable level of background. The observed ϕ asymmetry for these events as a function of $M_{\pi\pi ee}$ is also shown in fig. 17.1. The asymmetry for background $K_L \to \pi^+\pi^-\pi^0_D$ decays is consistent with zero. The large observed asymmetry in the kaon mass region can be associated only with presence of signal $K_L \to \pi^+\pi^-e^+e^-$ decays. To check that the detector does not introduce significant asymmetry, we reconstructed approximately five million $K_L \to \pi^+\pi^-\pi^0_D$ decays and observed an asymmetry of $-0.02 \pm 0.05\%$.

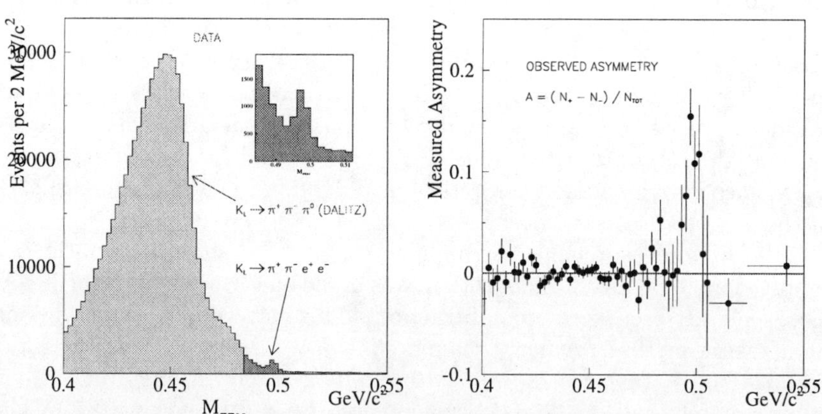

Figure 17.1: $M_{\pi\pi ee}$ distribution of four-track events (left) and observed ϕ asymmetry as a function of $M_{\pi\pi ee}$ (right).

To reduce background arising from other K_L decay modes, several cuts were applied. Four tracks were required to have transverse momentum P_t^2 relative to the direction of the K_L less than 0.6×10^{-4} GeV^2/c^2. Then, all $\pi^+\pi^-e^+e^-$ events were interpreted as $K_L \to \pi^+\pi^-\pi^0_D$ decays. Under this assumption, the momentum squared $P^2_{\pi^0}$ of the assumed π^0 can be calculated in the frame in which the momentum of $\pi^+\pi^-$ is transverse to the K_L direction. $P^2_{\pi^0} \geq 0$ for all of the $K_L \to \pi^+\pi^-\pi^0_D$ decays except for cases where finite detector resolution produces a negative value of this parameter. In contrast, $P^2_{\pi^0} \leq 0$ for most of the $K_L \to \pi^+\pi^-e^+e^-$ decays. We required $P^2_{\pi^0} \leq -0.00625$ GeV^2/c^2 in our analysis. Also, we required e^+e^- invariant mass $M_{e^+e^-} > 2.0$ MeV/c^2 to reduce $K_L \to \pi^+\pi^-\gamma$ background in which the photon converted in the material of the spectrometer. Finally, the reconstructed kaon energy was required to be less than 200 GeV to minimize contamination from $K_S \to \pi^+\pi^-e^+e^-$ decays.

After analysis cuts described above, 1811 $K_L \to \pi^+\pi^-e^+e^-$ signal events above an estimated background of 45 events remained. Fig. 17.2 shows the $M_{\pi\pi ee}$ distribution of this data sample. The $\sin(\phi)\cos(\phi)$ distribution of events with 492 Mev/c^2 $\leq M_{\pi\pi ee} \leq$ 504 MeV/c^2 is also shown in fig. 17.2. The observed asymmetry for this data sample is 23.3 $\pm$ 2.3(stat)%.

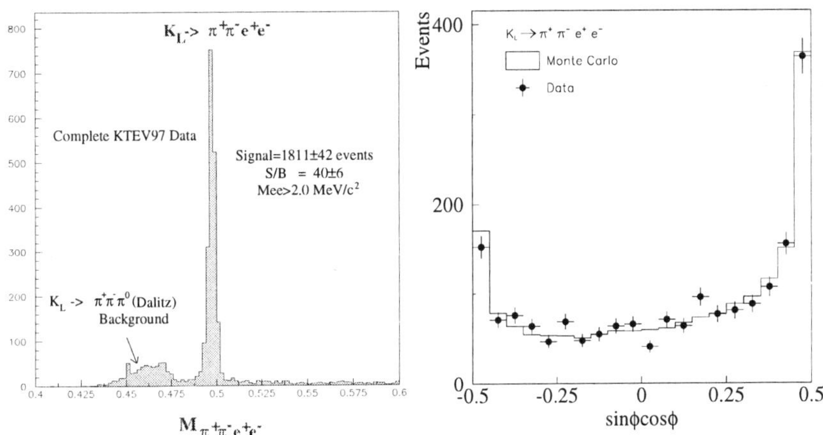

Figure 17.2: $M_{\pi\pi ee}$ distribution of final event sample (left) and $\sin(\phi)\cos(\phi)$ distribution of events with 492 Mev/c^2 $\leq M_{\pi\pi ee} \leq$ 504 MeV/c^2 (right).

The model [1,2] we used for Monte Carlo simulation of $K_L \to \pi^+\pi^-e^+e^-$ includes four amplitudes of photon emission with internal conversion of the photon into an e^+e^- pair: CP-violating decay of the $K_L \to \pi^+\pi^-$ with inner bremsstrahlung, CP-conserving M1 and CP-violating E1 photon emission at the $\pi^+\pi^-$ vertex, and CP-conserving K^0 charge radius process $K_L \to K_S$ via emission of a photon followed by decay of the $K_S \to \pi^+\pi^-$. A detailed presentation of this model was given by L. M. Sehgal [5] at this conference.

Despite very good agreement between this model and our experimental data (see for example fig. 17.2), we incorporated a form factor in the M1 virtual photon emission, similar to ref. [12]:

$$F = \tilde{g}_{M1}[1 + \frac{a_1/a_2}{(M^2_\rho - M^2_K) + 2M_K(E_{e^+} + E_{e^-})}], \tag{17.2}$$

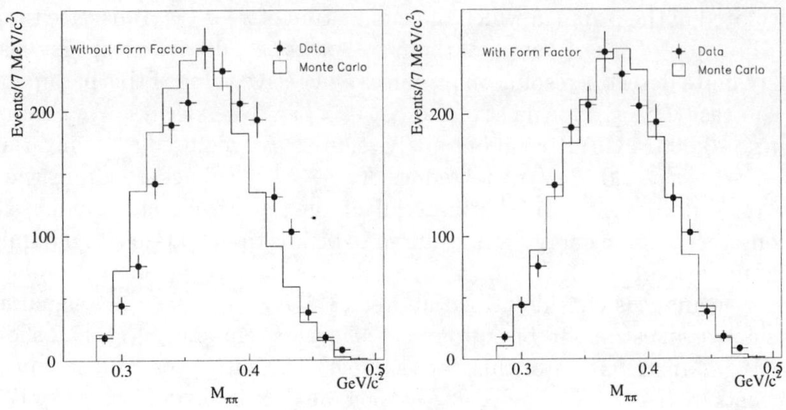

Figure 17.3: Monte Carlo $M_{\pi\pi}$ spectra before (left) and after (right) form factor introduction compared with experimental data.

where M_ρ=770 MeV/c^2. The form factor was necessary to obtain better agreement of Monte Carlo with the $M_{\pi\pi}$ spectrum of the data shown in fig. 17.3.

The ratio a_1/a_2 and $|\tilde{g}_{M1}|$ were determined by fitting the $K_L \to \pi^+\pi^-e^+e^-$ data using the maximum likelihood method. The probability of a given event is based on the $K_L \to \pi^+\pi^-e^+e^-$ matrix element and is a function of five independent variables: ϕ, θ_{e^+} (the angle between the momenta of e^+ and the $\pi^+\pi^-$ system in the e^+e^- cms), θ_{π^+} (the angle between the momenta of π^+ and the e^+e^- system in the $\pi^+\pi^-$ cms), $M_{\pi^+\pi^-}$, and $M_{e^+e^-}$. The likelihood, which is a product of the event probabilities, is a function of two parameters a_1/a_2 and $|\tilde{g}_{M1}|$. The fit gives a result for vector form factor parameters $a_1/a_2 = -0.720\pm0.028\,(\text{stat})\pm0.009\,(\text{syst})\,\text{GeV}^2/\text{c}^2$ and $|\tilde{g}_{M1}| = 1.35^{+0.20}_{-0.17}(\text{stat}) \pm 0.04(\text{syst})$. The average of the form factor F of eq. 17.2 over the range of $E_{e^+} + E_{e^-}$ is found to be 0.84 ± 0.10. This value is in agreement with the constant $|g_{M1}| = 0.76$ used in ref. [1]. The model with the form factor produces an $M_{\pi\pi}$ spectrum in better agreement with experimental data (see fig. 17.3).

Using the model described above, the acceptance-corrected asymmetry in the T-odd angle ϕ is found to be $13.6 \pm 2.5(\text{stat}) \pm 1.2(\text{syst})\%$, consistent with theoretical predictions [1, 5, 6]. This is the largest CP-violating effect yet observed and the first in an angular variable. The KTeV collaboration will have collected 3–6 times more statistics by the end of 1999 which will allow us to further investigate CP and T violation effects in the $K_L \to \pi^+\pi^-e^+e^-$ decay.

Acknowledgments

We thank Fermilab, the U.S. Department of Energy, the U.S. National Science Foundation, and the Ministry of Education and Science of Japan for their support.

References

[1] L.M. Sehgal and M. Wanninger, Phys. Rev. D **46**, 1035 (1992).

[2] P. Heiliger and L.M. Sehgal, Phys. Rev. D **48**, 4146 (1993).

[3] J.K. Elwood, M.B. Wise, and M.J. Savage, Phys. Rev. D **52**, 5095 (1995), hep-ph/9504288.

[4] J.K. Elwood, M.B. Wise, M.J. Savage, and J.W. Walden, Phys. Rev. D **53**, 4078 (1996), hep-ph/9506287.

[5] L.M. Sehgal, this volume, chapter 15.

[6] M. Savage, this volume, chapter 16.

[7] J. Adams *et al.* (KTeV Collaboration]), Phys. Rev. Lett. **80**, 4123 (1998).

[8] Y. Takeuchi *et al.*, Phys. Lett. B **443**, 409 (1998), hep-ex/9810018.

[9] S. Wronka, this volume, chapter 18.

[10] T. Nomura, this volume, chapter 19.

[11] J. Whitmore, this volume, chapter 39.

[12] E.J. Ramberg *et al.* (E731 Collaboration), Phys. Rev. Lett. **70**, 2525 (1993).

18

NA48 Results on $K^0{}_L \rightarrow \pi^+\pi^-e^+e^-$

S. Wronka

Abstract

Preliminary results in $K^0{}_L \rightarrow \pi^+\pi^-e^+e^-$ decay from the full 1998 data set of NA48 experiment at CERN are presented. We find 458±22 events with S/B equal to 37±3. Our result in the asymmetry in ϕ, where ϕ is the angle between $\pi^+\pi^-$ and e^+e^- planes in the kaon rest frame, and branching ratio agree with theoretical predictions.

The amplitude for the $K^0{}_L \rightarrow \pi^+\pi^-e^+e^-$ decay has two distinct components: direct emission, which is CP conserving, and inner bremsstrahlung, which is CP violating. The interference of the CP-even and CP-odd amplitudes gives an asymmetry in ϕ, where ϕ is the angle between $\pi^+\pi^-$ and e^+e^- planes in the kaon rest frame, mostly due to indirect CP violation. The standard definition of that asymmetry is [1, 2]:

$$\mathcal{A} = \frac{\int_0^{\pi/2} \frac{d\Gamma}{d\phi} d\phi + \int_{3/2\pi}^{2\pi} \frac{d\Gamma}{d\phi} d\phi - \int_{\pi/2}^{\pi} \frac{d\Gamma}{d\phi} d\phi - \int_{\pi}^{3/2\pi} \frac{d\Gamma}{d\phi} d\phi}{\int_0^{2\pi} \frac{d\Gamma}{d\phi} d\phi}. \tag{18.1}$$

In this paper we present preliminary results based on the full data set taken in 1998. A detailed description of the NA48 experiment can be found in [3]. To select "good" $\pi^+\pi^-e^+e^-$ events in NA48 first we use a specialized trigger and filter 4 charged track events. We require at least 2 positive and 2 negative tracks in a defined time window and all 6 verticies (combinations of 2 tracks) within a small time and space distance.

If the event succesfuly passes such preliminary selection we identify particles based on their E/p information (where: E is the energy deposition in the krypton calorimeter, and p is the momentum in the spectrometer [3]). (See fig. 18.1 left.) Particles with E/p smaller than 0.85 we classify as pions. Otherwise we treat them as electrons. In each event we require at least 2 pions and 2 electrons with opposite charges. The invariant mass of these events is shown in fig. 18.2 right.

The main background comes from other four-track decays:

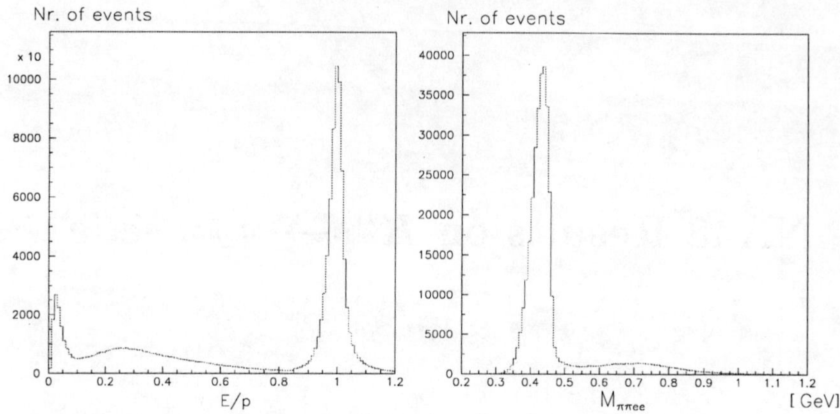

Figure 18.1: Left: E/p ratio for all particles after trigger and filter requirements. Right: Invariant mass of $\pi^+\pi^-e^+e^-$ after particle identification.

$$\begin{array}{ll} K^0{}_L \rightarrow \pi^+\pi^-\gamma & \text{with conversion } \gamma \rightarrow e^+e^-, \\ K^0{}_L \rightarrow \pi^+\pi^-\pi^0_D & \pi^0_D \rightarrow e^+e^-\gamma, \\ \left.\begin{array}{l} K^0{}_L \rightarrow \pi^+\pi^-\gamma \\ K^0{}_L \rightarrow \pi^+\pi^-\pi^0_D \end{array}\right\} & \text{close in time (double } K^0{}_{e3} \text{ event).} \end{array}$$

The huge peak at masses smaller than the $K^0{}_L$ mass (fig. 18.1 right) comes from $K^0{}_L \rightarrow \pi^+\pi^-\pi^0_D$ decays; events with invariant mass bigger than $K^0{}_L$ mass come from double $K^0{}_{e3}$ events. In the $K^0{}_L \rightarrow \pi^+\pi^-\gamma$ background we have four particles in the final state, so it looks exactly like signal on the invariant $\pi^+\pi^-e^+e^-$ mass plot.

To separate signal from background we use the following cuts:

- If the particles come from 4-body decay, their summed P_t calculated relative to the $K^0{}_L$ momentum should be small.

$$P_t = \sqrt{(\sum_{i=1\ldots4} p_{x_i})^2 + (\sum_{i=1\ldots4} p_{y_i})^2} \tag{18.2}$$

  After Monte Carlo studies (fig. 18.2) we use a cut $P_t < 0.01$ GeV.

- To reject $K^0{}_L \rightarrow \pi^+\pi^-\pi^0_D$ background we look at the distribution of $P^2{}_L$ (fig. 18.3), which is equal to the square of the longitudinal momentum of the π^0 in the frame in which the sum of the charged pions' momenta is orthogonal to the kaon momentum:

$$P^2{}_L \equiv \frac{(M_K^2 - M_{\pi^0}^2 - M_{\pi\pi}^2)^2 - 4M_{\pi^0}^2 M_{\pi\pi}^2 - 4M_K^2 (P_T^2)_{\pi\pi}}{4[(P_T^2)_{\pi\pi} + M_{\pi\pi}^2]}. \tag{18.3}$$

  In this analysis we use a cut $P^2{}_L < -0.01\,\text{GeV}^2$.

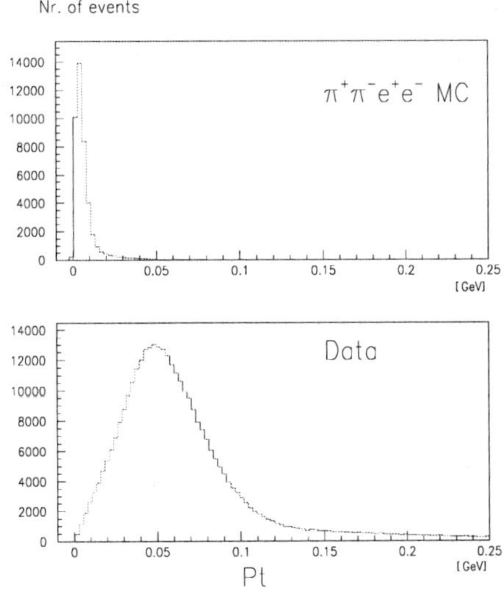

Figure 18.2: P_t calculated relative to the $K^0{}_L$ momentum.

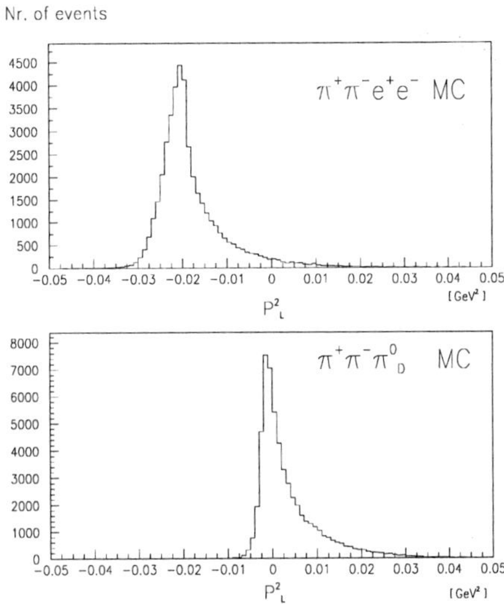

Figure 18.3: Monte Carlo simulation of $P^2{}_L$ distribution.

- Also for $K^0{}_L \rightarrow \pi^+\pi^-\pi^0_D$ rejection we look at the invariant mass of $ee\gamma$. We do not accept events for which this mass is consistent with the π^0 mass (cut between 0.125 GeV < $M_{ee\gamma}$ < 0.145 GeV) (fig. 18.4 left).

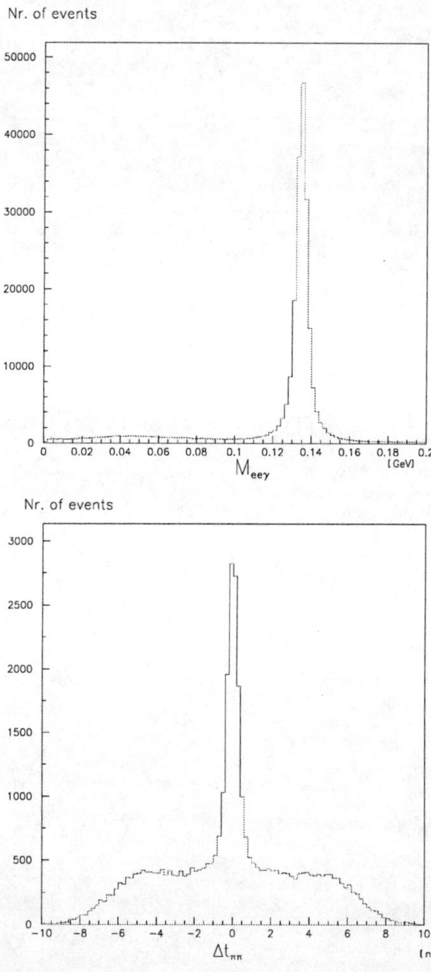

Figure 18.4: Top: Invariant mass of $ee\gamma$ before $K^0{}_\mathrm{L} \to \pi^+\pi^-\pi^0_\mathrm{D}$ rejection cuts. Bottom: The difference between π^+ and π^- times after $K^0{}_\mathrm{L} \to \pi^+\pi^-\pi^0_\mathrm{D}$ rejection cuts.

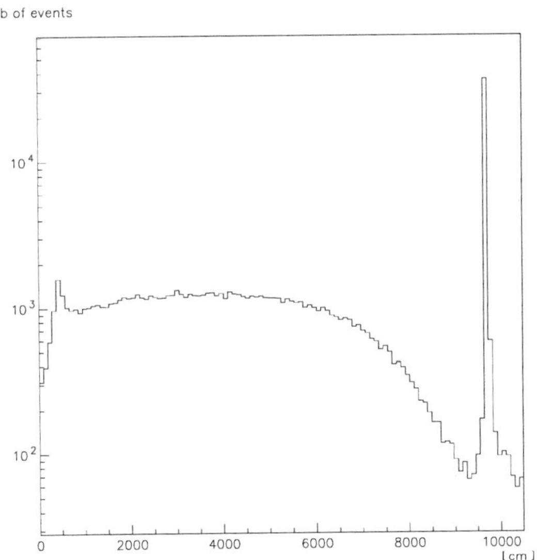

Figure 18.5: Vertex distribution of e^+e^- pairs along the beam.

- Rejection of double ${K^0}_{e3}$ events is done by time cuts. The difference between π^+ and π^- detector times is shown in fig. 18.4 right. We select the events with $-1\,\text{ns} < \Delta t_{\pi\pi} < 1\,\text{ns}$ and require the same for electrons, $-1\,\text{ns} < \Delta t_{ee} < 1\,\text{ns}$.

- We reject ${K^0}_L \to \pi^+\pi^-\gamma$ based on knowledge that the $\gamma \to e^+e^-$ conversion can occur only in matter. The distribution of the vertex position along the beam of e^+e^- pairs is shown in fig. 18.5. A huge peak above 90 m is visible, it is the position of the Kevlar window, which defines the end of the fiducial region [3]. To reject all events which come from γ conversion we require the e^+e^- vertex to be smaller then 87 m. Consequently we have the same requirement for all vertices ($\pi^+\pi^-$, π^+e^-, and π^-e^+).

- Distance from the transverse position of the reconstructed vertex to the beam axis must be smaller than 3 cm.

- The standard "center of gravity" (COG) is calculated for the $\pi^+\pi^-e^+e^-$ system at the krypton calorimeter; we require COG < 3.8 cm from the propagated kaon track position at the calorimeter.

- We reject events with muons in the final state.

During our studies a lot of Monte Carlo simulations (based on the model of L. M. Sehgal and M. Wanninger [1,2]) were done. We generally observe agreement between Monte Carlo and data. However, the observed invariant

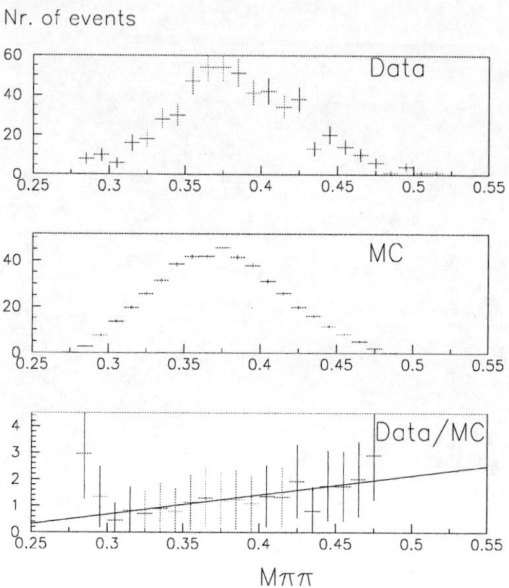

Figure 18.6: Invariant mass of $\pi^+\pi^-$ for the data, Monte Carlo, and Data/Monte Carlo ratio.

$\pi^+\pi^-$ mass from the Monte Carlo does not agree with that of the data, as shown in fig. 18.6. We observe the shift between them, which means that a modification of the theory is necessary [4].

The invariant mass of $\pi^+\pi^-e^+e^-$ after all cuts is presented in fig. 18.7. In the data taken in 1998 we observe 458 ± 22 (statistical error) good events. If we do a fit of the Gauss distribution we get σ=2.1 MeV. The observed background outside the signal peak is flat, so we get the background estimation from a linear fit. Our signal to background ratio is equal to 37 ± 3 (stat.).

For signal events we can plot the $\cos\phi \sin\phi$ distribution, which is presented in fig. 18.8. In our Monte Carlo simulations the generated asymmetry is equal to 14% [1, 2]. After all cuts used in the analysis we get 22%. The result for the 1998 data signal events is equal to 20 ± 5 % (stat.).

To calculate the branching ratio we use ${K^0}_L \to \pi^+\pi^-\pi^0_D$ events for normalization (the same trigger, four charged track events):

$$\mathrm{BR} = \frac{\mathrm{N}_{\pi\pi ee}}{\mathrm{N}_{\pi\pi\pi_D}} \times \frac{\mathrm{Acc}_{\pi\pi\pi_D}}{\mathrm{Acc}_{\pi\pi ee}} \times \mathrm{BR}_{\pi\pi\pi_D}. \tag{18.4}$$

The preliminary result is $(2.90 \pm 0.15) \times 10^{-7}$ (stat. error only).

Our branching ratio and asymmetry agree with theoretical predictions and with the results from KTeV [4]. Analysis will be continued for 1998 data and also for the new data taken in 1999.

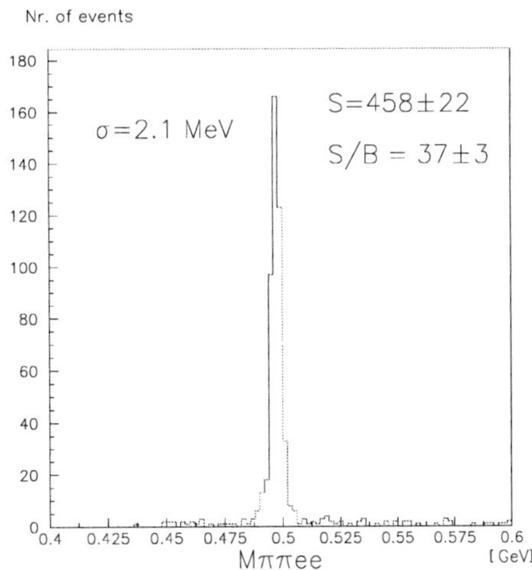

Figure 18.7: Invariant mass of $\pi^+\pi^-e^+e^-$ after all cuts.

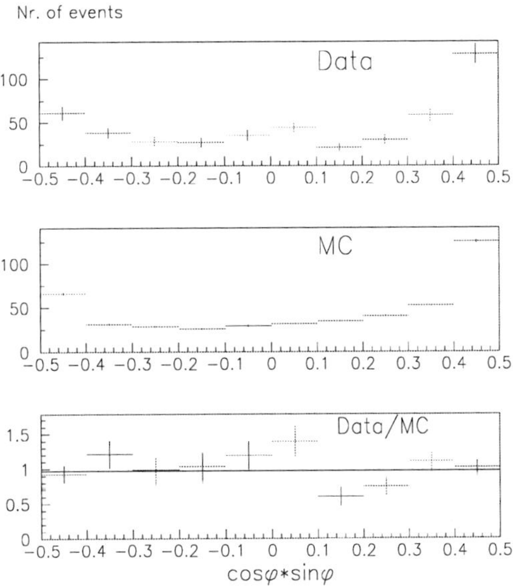

Figure 18.8: Asymmetry for data, Monte Carlo simulation, and Data/Monte Carlo ratio.

References

[1] L.M. Sehgal and M. Wanninger, Phys. Rev. D **46**, 1035 (1992); L.M. Sehgal and M. Wanninger, Phys. Rev. D **46**, 5209(E) (1992); P. Heilger and L.M. Sehgal, Phys. Rev. D **48**, 4146 (1993).

[2] L.M. Sehgal, this volume, chapter 15.

[3] M.S. Sozzi, this volume, chapter 8.

[4] A. Ledovskoy, this volume, chapter 17.

19

KEK E162 Results on $K_L \to \pi^+\pi^-e^+e^-$

Tadashi Nomura

Abstract

Results from KEK E162 are reported in this article. Following the explanation of the experiment itself, the analysis on the decay mode $K_L \to \pi^+\pi^-e^+e^-$ will be explained, based on our published result: $Br(K_L \to \pi^+\pi^-e^+e^-) = (4.4 \pm 1.3 \pm 0.5) \times 10^{-7}$.

19.1 Introduction

KEK E162 is an experiment to study various rare K_L decay modes. It is a collaboration from 3 institutes (Kyoto, KEK, and Hiroshima Inst. Tech.) and 14 members.

At first, let me mention a history of the experiment briefly. The experiment was approved in 1989, aiming to study the direct CP-violating mode $K_L \to \pi^0e^+e^-$. Unique features of our experiment at that time were high-rate capable detectors, such as a drift chamber with 1 GHz pipelined TDC readout, and an electromagnetic calorimeter using pure CsI crystals, which emit fast light output and also enable us to obtain good energy resolution. After the development and the construction of detectors, we did the engineering run to study detector performances, the trigger rate, and so on. Unfortunately, we could not suppress the trigger rate enough because of many neutron backgrounds, and found that it would be impossible to achieve the proposed sensitivity for $K_L \to \pi^0e^+e^-$. In 1996, we decided to change our main purpose to measure the branching ratio of $K_L \to \pi^+\pi^-e^+e^-$, which now becomes a very famous mode as a good testing ground to study CP- and T-violating effects via the ϵ term. After one year of a physics run period, we had finished taking data in June 1997. Results described in this report, which have been already published in 1998 [1], are based on our full data set.

19.2 Description of the Experiment

The experiment was conducted with 12 GeV proton synchrotron at KEK. The primary beam intensity was 10^{12} protons per pulse, 2-second spill every 4-second cycle. The neutral beam was produced at a 6-cm-long copper target with the production angle of 2 degrees. The beam was defined by heavy metal collimators to its size of ± 4 mrad in horizontal and ± 20 mrad in vertical. Following a 4-m-long decay region, the detector area started at 13.5 m from the target. Fig. 19.1 shows our detector system.

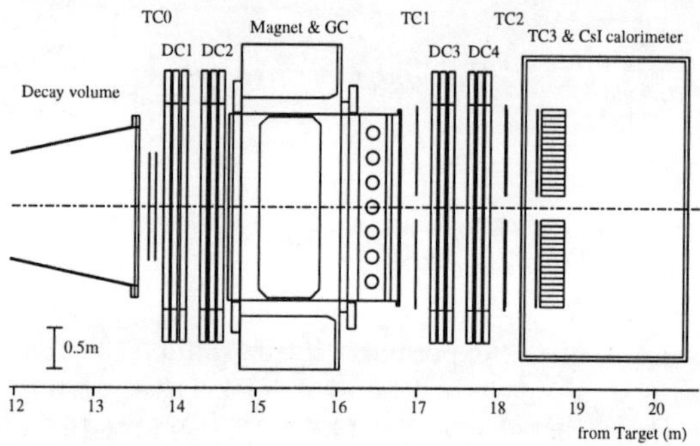

Figure 19.1: Schematic plan view of the KEK-E162 detector.

It was composed with a spectrometer to measure momenta of charged particles, an electromagnetic calorimeter to measure energies of electrons and photons, a gas Cherenkov counter (GC) to identify electrons, and 4 planes of trigger hodoscopes.

The spectrometer consisted of an analyzing magnet and 4 sets of drift chambers. The magnet operated with the field of 0.5 tesla corresponding to an average transverse momentum kick of 136 MeV/c. Each drift chamber had 6 detecting wire planes (X-X′,U-U′,V-V′). Its position resolution was about 140 μm in average, and slightly worse in the beam region, where its counting rate was up to 0.8 MHz per one wire.

Two banks of electromagnetic calorimeters were placed at both sides of the beam hole at the downstream end. Each bank consisted of 15(H)×18(V) crystal blocks, which had the size of 7 cm×7 cm×30 cm (16 X_0) and was read by a 2″ PMT with a transistor-stabilized base. The energy calibration was done by K_{e3} electron samples, and PMT gains were monitored by a Xe flash lamp within 1%. The achieved energy resolution was 3% at 1 GeV and the position resolution was 7 mm. Fig. 19.2 shows the E/p distributions for electrons and pions, where E and p were an energy measured by the calorimeter and a momentum measured by the spectrometer, respectively. It was used to identify particle species in the offline analysis.

Since we wanted to push all the detectors upstream to gain acceptance,

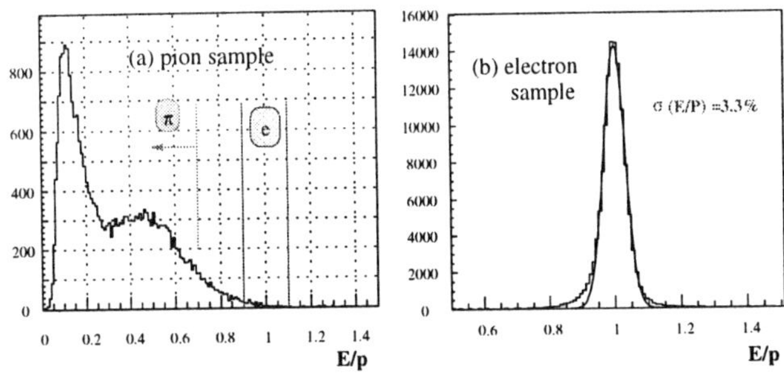

Figure 19.2: The E/p distributions for (a) pions and (b) electrons. The samples were from (a) $K_L \to \pi^+\pi^-\pi^0$ and (b) K_{e3} data set, respectively.

the radiator part of the GC was placed inside the magnet gap. It was filled with pure nitrogen gas at atmospheric pressure and was divided optically into 14 cells, each viewed by 5″ PMT. The efficiency for electrons and the rejection factor for pions are plotted in fig. 19.3.

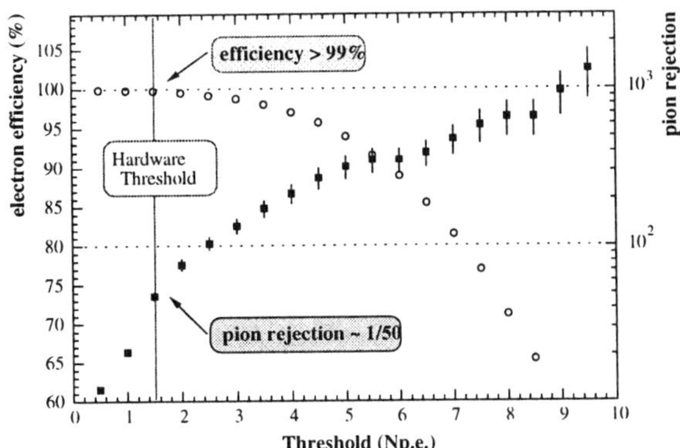

Figure 19.3: The electron efficiency and the π rejection factor of GC.

We obtained an average electron efficiency of 99.7% with a pion rejection factor of 50 in the trigger stage. If we want, we can get a better pion rejection factor by raising an offline threshold, with small loss of electron efficiency.

19.3 Results on $K_L \to \pi^+\pi^-e^+e^-$

19.3.1 Overview

Before the explanation of our analysis procedure, let me remind you of the physics interest of $K_L \to \pi^+\pi^-e^+e^-$. In this process, there exist two dominant intermediate states that have different CP eigenvalues, as shown in fig. 19.4.

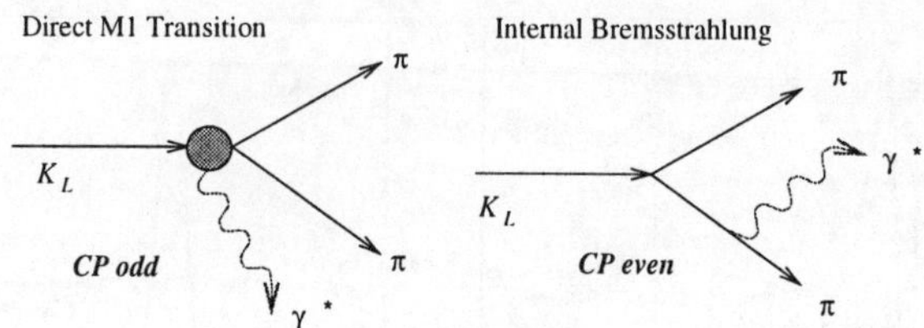

Figure 19.4: The diagrams of two dominant intermediate states, M1 direct transition and internal bremsstrahlung.

If we can observe an interference effect between them, it is evidence of a CP-violating phenomenon. It will be revealed, for example, as an asymmetry in the distribution of the angle between $\pi^+\pi^-$ and e^+e^- planes [2]. Besides, it is found to include a T-violating effect.

Now, let me summarize the strategy of our analysis procedure. We took data for $K_L \to \pi^+\pi^-e^+e^-$ and its normalization mode ($K_L \to \pi^+\pi^-\pi^0$ with π^0 Dalitz decay, $K_L \to \pi^+\pi^-\pi^0_D$) simultaneously under the same trigger conditions. In the offline analysis, we first do some basic selections, common to both modes, including track reconstructions, particle identifications, an event topology check, and background rejections. Then, the way was separated for each mode. In the case of the analysis for the normalization mode, we tried to find a photon from π^0 Dalitz decay. If found, kinematical variables such as the invariant mass were calculated and were examined to determine whether or not the event was consistent with $K_L \to \pi^+\pi^-\pi^0_D$. In the case of the analysis for the target mode, we first reject backgrounds, which mainly stemmed from $K_L \to \pi^+\pi^-\pi^0_D$. After the requirement on kinematical variables and the background subtraction, we finally got the number of observed events. I will explain this strategy.

19.3.2 Basic Selection

In the basic selection that was the first step in the offline analysis, there were four steps. The first step was track reconstruction using the information of drift chambers. We required the number of reconstructed tracks to be at least 4. In addition, they had to have a common vertex in the beam region in the decay volume. Particle species were identified. We tried to find a cluster on the calorimeter at the place where a track injected. If found, we called it as a matched track. The variable E/p, as mentioned above, could be defined in this step. A charged pion was identified as a matched track with its E/p value to be less than 0.7. An electron was required to have GC hits in corresponding cells, and have its energy at least 200 MeV, as well as E/p condition ($0.9 < E/p < 1.1$). The third step was to check an event topology. This means that an event was required to have 4 (and only 4) tracks consistent with $\pi^+\pi^-e^+e^-$. At the final step of the basic selection, we imposed two cuts to reject background events common to both modes. One was the requirement on π momentum balance: $A_{+-} < 0.5$, where $A_{+-} \equiv (p_{\pi^+} - p_{\pi^-})/(p_{\pi^+} + p_{\pi^-})$. It was effective in suppressing backgrounds from nuclear interactions and also those from π-to-μ decays in flight. The other was that we required the invariant mass of e^+e^- to be at least 4 MeV/c^2, to reduce external photon conversion backgrounds.

19.3.3 $K_L \to \pi^+\pi^-\pi^0_D$ as Normalization

Since the $K_L \to \pi^+\pi^-\pi^0_D$ mode has a similar event topology except for a photon, and has a bigger branching ratio, it is suitable for the normalization of $K_L \to \pi^+\pi^-e^+e^-$. Here the expression $K_L \to \pi^+\pi^-\pi^0_D$ means the $K_L \to \pi^+\pi^-\pi^0$ mode with $\pi^0 \to e^+e^-\gamma$.

In the analysis of this mode, we first attempted to find a photon, where a photon was defined as a cluster on the calorimeter that did not match with any reconstructed track. It was also required to have its energy above 200 MeV. If a photon was found successfully, we calculated three kinematical variables: the invariant mass of $e^+e^-\gamma$ ($M_{ee\gamma}$), which had to be M_{π^0}; the invariant mass of $\pi^+\pi^-e^+e^-\gamma$ ($M_{\pi\pi ee\gamma}$), which had to be M_{K_L}; and the θ^2, which should be zero within its resolution, where θ denotes the angle between the K_L momentum vector and the line connecting the target and the decay vertex. As shown in fig. 19.5, their resolutions were found to be 4.6 MeV/c^2 ($\sigma_{M_{\pi^0}}$) and 5.5 MeV/c^2 ($\sigma_{M_{K_L}}$), respectively.

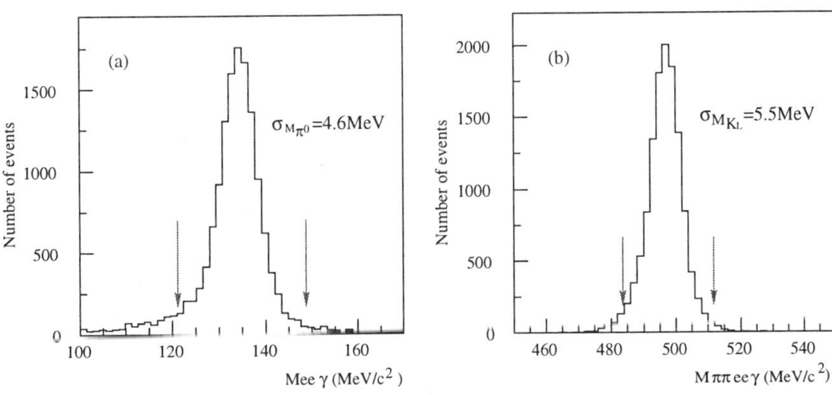

Figure 19.5: The invariant mass distributions of (a) $e^+e^-\gamma$ and (b) $\pi^+\pi^-e^+e^-\gamma$ for $K_L \to \pi^+\pi^-\pi^0_D$ candidates. The mass resolutions were measured to be (a) $\sigma_{M_{\pi^0}} = 4.6\ MeV/c^2$ and (b) $\sigma_{M_{K_L}} - 5.5\ MeV/c^2$, respectively.

To identify the $K_L \to \pi^+\pi^-\pi^0_D$ events, we required events to lie within

3σ of M_{π^0} and M_{K_L}, and to have θ^2 less than 20 mrad^2. After all the cuts, we got 12212 $K_L \to \pi^+\pi^-\pi^0_D$ events.

19.3.4 $K_L \to \pi^+\pi^-e^+e^-$

We now identify the $K_L \to \pi^+\pi^-e^+e^-$ mode. As mentioned, the dominant background at this step came from $K_L \to \pi^+\pi^-\pi^0_D$. At first, we removed events with a photon, which was consistent with a product from $\pi^0 \to e^+e^-\gamma$. Actually, if the invariant mass of $e^+e^-\gamma$ lay within $3\sigma_{M_{\pi^0}}$ of M_{π^0}, we rejected the event without any requirement on $M_{\pi\pi ee\gamma}$. After this cut, there remained backgrounds from $K_L \to \pi^+\pi^-\pi^0_D$ with its photon missed to anywhere. To suppress these backgrounds, we used the following method.

(i) Assume the existence of a photon with $\vec{p}_\gamma$.

(ii) Determine $\vec{p}_\gamma$ by minimizing χ^2_D

$$\chi^2_D(\vec{p}_\gamma) \equiv \left(\frac{M_{ee\gamma} - M_{\pi^0}}{\sigma_{M_{\pi^0}}}\right)^2 + \left(\frac{M_{\pi\pi ee\gamma} - M_{K_L}}{\sigma_{M_{K_L}}}\right)^2 + \left(\frac{\theta}{\sigma_\theta}\right)^2 .$$

(iii) If the minimized χ^2_D value was less than 17, the event was identified as $K_L \to \pi^+\pi^-\pi^0_D$ and thus it was rejected.

Fig. 19.6, which is the distribution of the square root of χ^2_D, illustrates how it worked.

In the figure, the number of K_L in the simulation was normalized to the reconstructed $K_L \to \pi^+\pi^-\pi^0_D$ events, and the branching ratio for $K_L \to \pi^+\pi^-e^+e^-$ was assumed to be given by our result. It can be seen that our data had a structure of the sum of two components: $K_L \to \pi^+\pi^-e^+e^-$ signals and $K_L \to \pi^+\pi^-\pi^0_D$ backgrounds. If we set the cut position to be 17 as above, which was equivalent to a 3σ cut, we can reject 99.7% of $K_L \to \pi^+\pi^-\pi^0_D$ backgrounds, while retaining 92% efficiency for $K_L \to \pi^+\pi^-e^+e^-$.

Fig. 19.7 shows the $M_{\pi\pi ee}$ vs θ^2 scatter plots of the $K_L \to \pi^+\pi^-e^+e^-$ candidate events after the χ^2_D cut, (a) for our data and (b) for the simulation of $K_L \to \pi^+\pi^-\pi^0_D$ (5 times statistics).

In fig. 19.7(a), it is clear that there is a cluster of events in the signal box. We can see a similar structure of backgrounds both in our data and in the simulation (fig. 19.7(b)). The mechanism to produce low-mass backgrounds was found to be the radiation from e^+ and/or e^-, while the high-mass band was the result from π-to-μ decays in flight. The number of events inside the box was 15, which might include backgrounds.

We should estimate the number of backgrounds in the signal box. We projected final events with $|M_{\pi\pi ee} - M_{K_L}| < 3\sigma_{M_{K_L}}$ onto the θ^2-axis. The resultant distribution is shown in fig. 19.8.

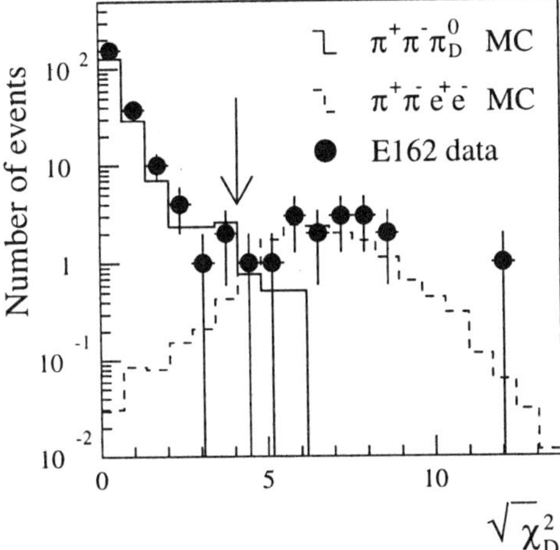

Figure 19.6: The distribution of the square root of χ^2_D. The solid and dashed lines are the results of a Monte Carlo simulation for the $K_L \to \pi^+\pi^-\pi^0_D$ and $K_L \to \pi^+\pi^-e^+e^-$ modes, respectively, and the points indicate our data. The arrow shows the cut position, described in the text.

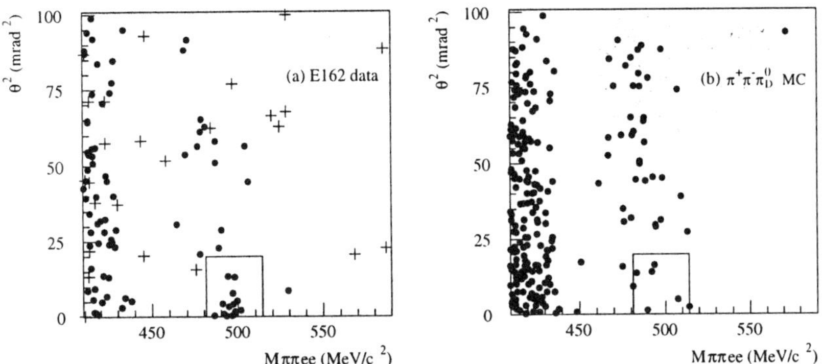

Figure 19.7: (a) The $M_{\pi\pi ee}$ vs θ^2 scatter plot of the $K_L \to \pi^+\pi^-e^+e^-$ candidate events after the χ^2_D cut. The solid dots represent the vacuum data and the plus signs the He data at 1 atm. The box indicates the signal region. (b) The corresponding scatter plot obtained by the $K_L \to \pi^+\pi^-\pi^0_D$ MC simulation. Note that the MC statistics are 5 times as great as the vacuum data.

There is a clear peak at $\theta^2 = 0$, as expected. By comparing our data with the background simulation, we determined the expected number of backgrounds in the signal region. Actually, we rescaled the number predicted by the simulation, with the scale factor by the ratio in the control region. The actual scale factor was 1.15. We found the expected number of backgrounds in the box to be 1.5. Subtracting it from the number of

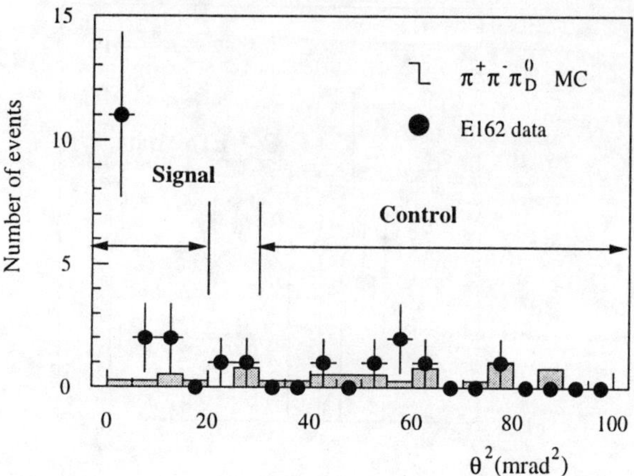

Figure 19.8: The θ^2 projection of the $K_L \to \pi^+\pi^-e^+e^-$ candidate events. The solid line in the figure represents the $K_L \to \pi^+\pi^-\pi^0_D$ MC simulation. Its K_L flux was normalized by the reconstructed $K_L \to \pi^+\pi^-\pi^0_D$ events.

events in the signal region, we determined the number of signal events to be 13.5 ± 4.0, where the error includes uncertainty due to the Monte Carlo statistics as well as the statistical error.

19.3.5 Branching Ratio

The branching ratio was calculated by

$$Br(K_L \to \pi^+\pi^-e^+e^-) = Br(K_L \to \pi^+\pi^-\pi^0) \times Br(\pi^0 \to e^+e^-\gamma) \times \frac{A(\pi^+\pi^-\pi^0_D)}{A(\pi^+\pi^-e^+e^-)} \cdot \frac{\eta(\pi^+\pi^-\pi^0_D)}{\eta(\pi^+\pi^-e^+e^-)} \cdot \frac{N(\pi^+\pi^-e^+e^-)}{N(\pi^+\pi^-\pi^0_D)},$$

where A, η, and N denote acceptance, efficiency, and observed number of events, respectively. The detector acceptances were determined by Monte Carlo simulations. Most of efficiencies were common to both modes, and various uncertainties could be canceled out in the ratio. The largest difference stemmed from the photon finding efficiency in $K_L \to \pi^+\pi^-\pi^0_D$, which was about 80%.

Our systematic errors are summarized in table 19.1, which are divided into three categories. The first one is related to the acceptance-efficiency ratio, the second category is related to the number of observed events, and the third is related to the current experimental errors. Summing up all the uncertainties in quadrature, we found the systematic error to be 11.3%.

We finally arrived at the branching ratio of

$$Br(K_L \to \pi^+\pi^-e^+e^-) = (4.4 \pm 1.3 \pm 0.5) \times 10^{-7},$$

where the first (second) error represents the statistical (systematic) uncertainty.

Source	uncertainty
K_L momentum spectrum	4.8%
Matrix element	3.9%
Others	3.1%
Background subtraction	7.4%
Nuclear interaction	3.6%
Other contamination	1.4%
$Br(K_L \to \pi^+\pi^-\pi^0_D)$	3.1%
Total	11.3%

Table 19.1: Summary of systematic errors in the branching ratio.

19.4 Summary

KEK E162, which aimed to study rare K_L decay, finished taking data in June 1997. All the detectors ran without any failure and kept good performance during the physics run. As for the physics output, the analysis of $K_L \to \pi^+\pi^-e^+e^-$ is reported in this article. Based on 13.5 ± 4.0 events, we determined its branching ratio to be

$$Br(K_L \to \pi^+\pi^-e^+e^-) = (4.4 \pm 1.3 \pm 0.5) \times 10^{-7}.$$

Further studies on CP and T violation in this mode have been reported by KTeV [3] and NA48 [4].

References

[1] Y. Takeuchi *et al.*, Phys. Lett. B **443**, 409 (1998).

[2] L.M. Sehgal and M. Wanninger, Phys. Rev. D **46**, 1035 (1992); **46**, 5209(E) (1992); P. Heilinger and L.M. Sehgal, *ibid.* **48**, 4146 (1993).

[3] S. Ledovskoy, this volume, chapter 17.

[4] S. Wronka, this volume, chapter 18.

20

Direct Observation of Time Reversal Noninvariance in the Neutral-Kaon System

P. Bloch[1]

Abstract

We report on the observation by the CPLEAR experiment of time-reversal symmetry violation through a comparison of the probabilities of $\overline{K}^0$ transforming into K^0 and K^0 into $\overline{K}^0$ as a function of the neutral-kaon eigentime t. The comparison is based on the analysis of the neutral-kaon semileptonic decays. The strangeness of the neutral kaon at time $t = 0$ is tagged by the charge of the associated kaon produced in strong interaction, whereas the strangeness of the kaon at the decay time $t = \tau$ is tagged by the lepton charge in the final state. An average decay-rate asymmetry

$$\left\langle \frac{R(\overline{K}^0_{t=0} \to e^+\pi^-\nu_{t=\tau}) - R(K^0_{t=0} \to e^-\pi^+\overline{\nu}_{t=\tau})}{R(\overline{K}^0_{t=0} \to e^+\pi^-\nu_{t=\tau}) + R(K^0_{t=0} \to e^-\pi^+\overline{\nu}_{t=\tau})} \right\rangle$$
$$= (6.6 \pm 1.3_{\text{stat}} \pm 1.0_{\text{syst}}) \times 10^{-3}$$

was measured over the interval $1\tau_S < \tau < 20\tau_S$, thus leading for the first time to evidence for time reversal noninvariance.

20.1 Introduction

The phenomenology of the neutral kaon system teaches us that the observed CP violation must be accompanied either by time reversal (T) violation or

[1]For the CPLEAR collaboration: University of Athens, Greece; University of Basle, Switzerland; Boston University, USA; CERN, Geneva, Switzerland; LIP and University of Coimbra, Portugal; Delft University of Technology, Netherlands; University of Fribourg, Switzerland; University of Ioannina, Greece; University of Liverpool, UK; J. Stefan Inst. and Phys. Dep., University of Ljubljana, Slovenia; CPPM, IN2P3-CNRS et Université d'Aix-Marseille II, France; CSNSM, IN2P3-CNRS, Orsay, France; Paul Scherrer Institut (PSI), Switzerland; CEA, DSM/DAPNIA, CE-Saclay, France; Royal Institute of Technology, Stockholm, Sweden; University of Thessaloniki, Greece; ETH-IPP Zürich, Switzerland.

by CPT violation, or by both simultaneously. In the past, tests of CPT invariance and phenomenological studies based on the Bell-Steinberger relation have concluded that CP violation is indeed dominantly accompanied by T violation [1]. However, it is only recently that a direct observation of time reversal noninvariance has been performed by the CPLEAR experiment [2].

Following an original proposal of Kabir [3], direct evidence for T violation can be obtained by comparing the rate of transformation of a K^0 into a $\overline{K}^0$ in the course of time with the T (and CP) conjugate transformation rate of a $\overline{K}^0$ into a K^0. Time reversal (T) invariance requires all details of the second process to be deducible from the first; in particular, the probability that a $K^0(t=0)$ is observed as a $\overline{K}^0$ at time τ should be equal to the probability that a $\overline{K}^0(t=0)$ is observed as a K^0 at the same time τ. Any difference of these two probabilities, for example by measuring a nonvanishing value of the A_T asymmetry:

$$A_T = \frac{R(\overline{K}^0 \to K^0) - R(K^0 \to \overline{K}^0)}{R(\overline{K}^0 \to K^0) + R(K^0 \to \overline{K}^0)}. \tag{20.1}$$

is a clear signal of T violation [4].

Experimentally, the measurement of the A_T asymmetry requires the tagging of the neutral kaon strangeness both at production (t=0) and at decay $(t=\tau)$ time. Strangeness tagging at production is the main characteritics of the CPLEAR experiment and will be described in section 20.3. To tag the strangeness of the kaon at the moment of its decay we use semileptonic decays: positive lepton charge is associated to a K^0 and negative lepton charge to a $\overline{K}^0$ ($\Delta S = \Delta Q$ rule). We therefore measure, as a function of time, the decay-rate asymmetry

$$A_L = \frac{R(\overline{K}^0_{t=0} \to e^+\pi^-\nu_{t=\tau}) - R(K^0_{t=0} \to e^-\pi^+\overline{\nu}_{t=\tau})}{R(\overline{K}^0_{t=0} \to e^+\pi^-\nu_{t=\tau}) + R(K^0_{t=0} \to e^-\pi^+\overline{\nu}_{t=\tau})}. \tag{20.2}$$

20.2 Phenomenology of Semileptonic Decay

The A_L asymmetry is only identical with the time reversal asymmetry A_T in the limit of CPT symmetry in the semileptonic decay process and of the validity of the $\Delta S = \Delta Q$ rule. It is therefore important to develop the complete phenomenology of the method to understand how these assumptions enter in the measured asymmetry and, even more, how they can be tested.

The two parameters that describe the neutral-kaon mixing are, respectively, the T and CPT violation parameters:

$$\epsilon = \frac{\Lambda_{\overline{K}^0,K^0} - \Lambda_{K^0,\overline{K}^0}}{2(\lambda_L - \lambda_S)} \qquad \text{and} \qquad \delta = \frac{\Lambda_{\overline{K}^0,\overline{K}^0} - \Lambda_{K^0,K^0}}{2(\lambda_L - \lambda_S)}.$$

Here, Λ_{ij} are the elements and $\lambda_{L,S}$ the eigenvalues of the effective Hamiltonian Λ: $\lambda_{L,S} = m_{L,S} - \frac{i}{2}\Gamma_{L,S}$, where $m_{L,S}$ and $\Gamma_{L,S}$ are the masses and decay widths for the K_L and K_S states, $\Delta m = m_L - m_S$ and $\Delta\Gamma = \Gamma_S - \Gamma_L$. Note that ϵ is T violating by construction. The K_L mixing parameter is defined as $\epsilon_L = \epsilon - \delta$.

The semileptonic decay amplitudes can be written as

$$\langle e^+\pi^-\nu|\Lambda|K^0\rangle = a + b, \qquad \langle e^-\pi^+\overline{\nu}|\Lambda|\overline{K^0}\rangle = a^* - b^*,$$
$$\langle e^-\pi^+\overline{\nu}|\Lambda|K^0\rangle = c + d, \qquad \langle e^+\pi^-\nu|\Lambda|\overline{K^0}\rangle = c^* - d^*.$$

The amplitudes b and d are CPT violating, c and d describe possible violations of the $\Delta S = \Delta Q$ rule, and the imaginary parts are all T violating. The quantities

$$x = \frac{c^* - d^*}{a + b} \qquad \text{and} \qquad \overline{x} = \frac{c^* + d^*}{a - b}$$

describe the violation of the $\Delta S = \Delta Q$ rule in decays into positive and negative leptons, respectively, while

$$y = -\frac{b}{a}$$

describes CPT violation in semileptonic decays in the case where the $\Delta S = \Delta Q$ rule holds. The parameters $x_+ = (x+\overline{x})/2$ and $x_- = (x-\overline{x})/2$ describe the violation of the $\Delta S = \Delta Q$ rule in CPT-conserving and CPT-violating amplitudes, respectively. We assume x, $\overline{x}$ and $y \ll 1$.

The semileptonic decay rates depend on the strangeness of the kaon (K^0 or $\overline{K^0}$) at the production time, $t = 0$, and on the charge of the decay lepton (e^+ or e^-). Only the two rates $R\left[K^0_{t=0} \to e^-\pi^+\overline{\nu}_{t=\tau}\right]$ and $R\left[K^0_{t=0} \to e^+\pi^-\nu_{t=\tau}\right]$ enter in the A_L asymmetry (20.2). These rates can be written as a function of the neutral-kaon decay time, τ, and of the parameters Re(ϵ), Re(y), x, and $\overline{x}$. Neglecting second-order terms, which play a role only for very short lifetime ($\tau \ll 1\tau_S$), we obtain:

$$A_L(\tau) = 4\mathrm{Re}(\epsilon) - 2\mathrm{Re}(y) - 2\mathrm{Re}(x_-) + 2\frac{\mathrm{Re}(x_-)(e^{-\frac{1}{2}\Delta\Gamma\tau} - \cos(\Delta m\tau)) + \mathrm{Im}(x_+)\sin(\Delta m\tau)}{\cosh(\frac{1}{2}\Delta\Gamma\tau) - \cos(\Delta m\tau)}.$$

One observes that:

- For long lifetimes this asymmetry is constant, equal to $4\mathrm{Re}(\epsilon) - 2\mathrm{Re}(y) - 2\mathrm{Re}(x_-)$, and the CPT allowed, $\Delta S = \Delta Q$ violating term ($\mathrm{Im}(x_+)$) does not enter, allowing a measurement free of assumption on the $\Delta S = \Delta Q$ rule.

- In the limit of CPT invariance in the semileptonic decay ($\mathrm{Re}(y) = \mathrm{Re}(x_-) = 0$) A_T and A_L are identical. Without this assumption, the measurement of A_L *alone* can not disentangle between T violation in the mixing and CPT violation in the semileptonic decay. However, we will show that, using a complete set of measurements of kaon decay parameters, CPLEAR gives a bound on the CPT-violating term $\mathrm{Re}(y) + \mathrm{Re}(x_-)$, which makes the observation of T violation unambiguous.

Note that the parameters $\mathrm{Re}(y)$, $\mathrm{Re}(x_-)$ also enter the determination of the parameter $\mathrm{Re}(\epsilon_L)$ from the charge asymmetry δ_l,

$$\delta_l = \frac{R(K_L \to \pi^- l^+ \nu) - R(K_L \to \pi^+ l^- \bar{\nu})}{R(K_L \to \pi^- l^+ \nu) + R(K_L \to \pi^+ l^- \bar{\nu})} = 2\mathrm{Re}(\epsilon_L) - 2\mathrm{Re}(y) - 2\mathrm{Re}(x_-). \tag{20.3}$$

20.3 The CPLEAR Experiment

The CPLEAR experiment produces K^0s and $\overline{K}^0$s through the $p\overline{p}$ annihilation channels

$$p\overline{p} \longrightarrow \begin{cases} K^- \pi^+ K^0 \\ K^+ \pi^- \overline{K}^0. \end{cases}$$

Strangeness conservation in strong interactions enables the initial strangeness of the neutral kaon to be tagged by the charge of the accompanying charged kaon. Full details of the design, operation, and performance of the CPLEAR detector can be found in ref. [5]. Here only a short overview of the detector is presented.

The CPLEAR detector had a cylindrical geometry and was mounted inside a solenoid magnet that produced a field of 0.44 T. It allowed the detection of neutral-kaon decays in the time range $0 < \tau < 20\tau_S$ (τ_S is the K_S mean life). The 200 MeV/c antiprotons from the Low Energy Antiproton Ring (LEAR) of CERN stopped and annihilated inside a gaseous hydrogen target at the center of the detector. Going radially outwards from the target, the detector consisted of a tracking device (two multiwire proportional chambers, six drift chambers, and two layers of streamer tubes), a scintillator-Cherenkov-scintillator sandwich (S1–CE–S2) for particle identification, and a lead/gas-sampling electromagnetic calorimeter. A multilevel trigger system provided fast event selection. The trigger was based on particle identification, event kinematics, and shower counting.

20.4 Event Selection

The desired $p\overline{p}$ annihilations followed by the decay of the neutral kaon into $e\pi\nu$ are first selected by requiring that the events have four charged tracks and zero total charge as well as a good reconstruction quality for each track and vertex, and by identifying one of the decay tracks as an electron or a positron.

The lepton identification is performed with a Neural Network (NN) algorithm [5]. The inputs to the NN algorithm are the momentum of the particle, the energy loss in the two scintillators (S1, S2), the number of

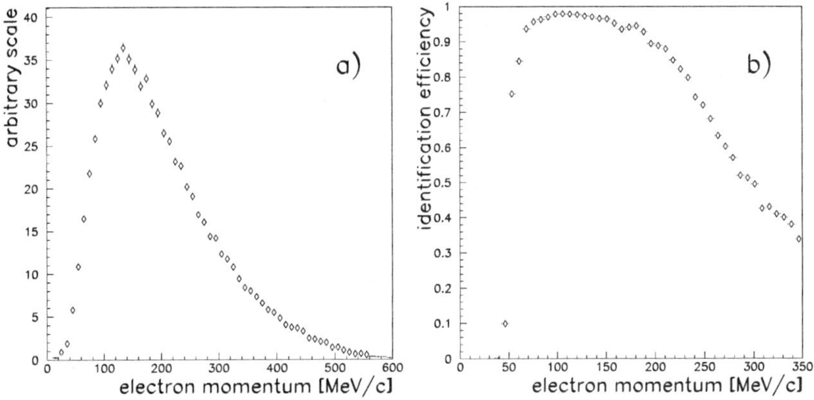

Figure 20.1: a) Expected electron momentum spectrum in semileptonic decays of neutral kaons. b) Electron identification efficiency versus momentum when 2% of pions fake electrons.

photoelectrons in the Cherenkov counter, and the time of flight from the decay vertex to the first scintillator (S1).

The electron momentum spectrum and electron identification efficiency are shown in fig. 20.1. No attempt is made to identify muons with the NN. The probability of identifying a muon as an electron, and thus including a muon event in the electron signal, is $\approx 15\%$.

Finally, the method of kinematic constrained fits (energy-momentum conservation under the semileptonic decay hypothesis and geometric constraints) is used to further reduce the background. The fitted momenta and vertices resulting from the constrained fit determine the neutral-kaon lifetime with a precision that ranges from 0.05 τ_S in the short lifetime region to 0.2–0.3 τ_S in the long one.

A total of 1.3×10^6 $e\pi\nu$ events, with measured decay time $\tau \geq 1\tau_S$, survive the above selection.

From a Monte Carlo simulation we obtain the background-to-signal ratio for the different background channels as a function of the decay time (fig. 20.2).

We have considered contributions from the most abundant neutral-kaon decays. The main background source consists of neutral-kaon two-pion decays and is concentrated at early decay times, while at late decay times there are contributions from $\pi e\nu$ decays where the electron and pion assignments are exchanged, and from $\pi^+\pi^-\pi^0$ decays. Their levels relative to the signal remain below 2%. Fig. 20.2 shows the excellent agreement between real and simulated data. The background simulation is controlled by relaxing some of the selection cuts to increase the background contribution by a large factor. In all cases, data and simulation agree well and we use as a conservative 10% estimate for our systematic error on background level.

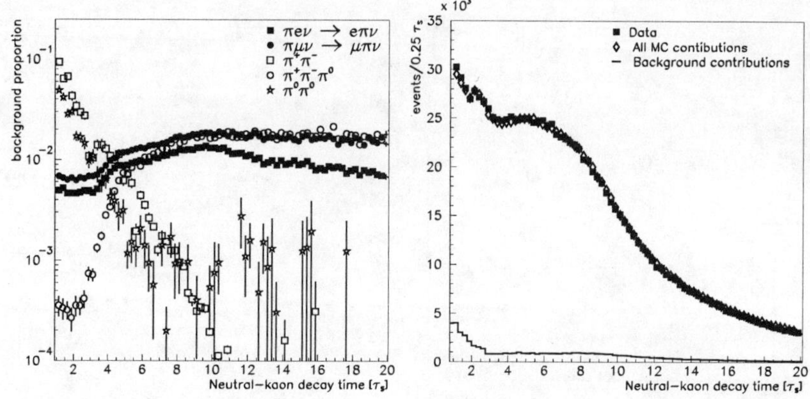

Figure 20.2: Left: proportion of different background channels relative to semileptonic signal. Right: decay time distribution for real and simulated data.

20.5 Analysis

Using an asymmetry has the advantage that the detection efficiencies common to the two processes being compared cancel. Differences in the geometrical acceptances are compensated to first order by frequently reversing the magnetic field. However, different detection probabilities for the charged kaons, pions, and electrons used for tagging the strangeness of the neutral kaon at production and decay times lead to different corrections for each event sample. The corrections are performed on an event-by-event basis via the two normalization factors:

- $\xi = \epsilon(K^+\pi^-)/\epsilon(K^-\pi^+)$, where the efficiencies involved, $\epsilon(K\pi)$, are those of the charged particles at the production vertex (primary vertex normalization);

- $\eta = \epsilon(\pi^+e^-)/\epsilon(\pi^-e^+)$, which takes into account the different detection efficiencies, $\epsilon(\pi e)$, for the particles in the two final states (secondary vertex normalization).

20.5.1 Primary Vertex Normalization

The primary vertex normalization factor ξ differs from unity because of the different strong interaction cross-sections of oppositely charged kaons and pions. These differences are parametrized in terms of transverse and longitudinal momentum of the charged kaon, $p_t(K^\pm)$ and $p_z(K^\pm)$, and of the momentum of the primary pion, $p(\pi^\mp)$, $\xi = \xi(p_t(K^\pm), p_z(K^\pm), p(\pi^\mp))$.

The primary normalization factor ξ is independent of the final state into which the neutral kaon decays. We obtain it with high precision from our data set of $\pi^+\pi^-$ decays [6]. We select events with decay times between 1 and 4 τ_S and calculate the normalization factor ξ by building the ratio of observed K^0 to $\overline{K}^0$ events. In the time interval considered, this ratio can be approximated as

$$\frac{\xi N(K^0_{t=0} \to \pi^+\pi^-_{t=\tau})}{N(\overline{K}^0_{t=0} \to \pi^+\pi^-_{t=\tau})} = (1 - 4\mathrm{Re}(\epsilon_L)) \times \left(1 + 4|\eta_{+-}| \cos(\Delta m\tau - \phi_{+-}) e^{\frac{1}{2}\Gamma_S \tau}\right). \quad (20.4)$$

The oscillating factor on the right-hand side of eq. (20.4) depends on the neutral-kaon parameters η_{+-}, Δm, and $\Gamma_S \equiv 1/\tau_S$, which we take from [7], and, as a result, is known with a precision of $\approx 1 \times 10^{-4}$. The quantity $\mathrm{Re}(\epsilon_L)$ is taken from the semileptonic charge asymmetry δ_l, following eq.(20.3). Experimentally, the charge asymmetry is [7]

$$\delta_l = (3.27 \pm 0.12) \times 10^{-3}. \quad (20.5)$$

The average of ξ over all pion and kaon momenta is $\langle\xi\rangle = 1.12023 \pm 0.00043$. We make the following remarks:

- This normalization procedure does not imply any assumption on direct CPT violation in the $\pi^+\pi^-$ decay (such violation would be implicitly contained in the experimentally measured value of η_{+-}).
- When using δ_l, we have to take into account a possible CPT violation in the semileptonic decay; see eq. (20.3). As a consequence, the asymmetry measured by CPLEAR is not exactly A_L, but $A_L^{exp} = A_L - 2\mathrm{Re}(y) - 2\mathrm{Re}(x_-)$.

20.5.2 Secondary Vertex Normalization

The secondary vertex normalization factor is measured as a function of the momentum of the decay pion (p_π) and electron (p_e).

Pure and unbiased samples of pions and electrons were used to study the difference in the efficiencies. A sample of e^+e^- pairs has been selected from γ conversion data ($K^0 \to 2\pi^0, \pi^0 \to 2\gamma, \gamma \to e^+e^-$), and a sample of π^+ and π^- tracks has been selected from minimum-bias data. The value of η, averaged over the particle momenta, is $\langle\eta\rangle = 1.014 \pm 0.002$.

20.5.3 Regeneration

K^0 and $\overline{K}^0$ are subject to regeneration arising from the forward scattering in the material of the detector. Regeneration corrections are calculated using the regeneration amplitudes measured by CPLEAR [8] and are applied on an event-by-event basis, depending on the momentum of the neutral kaon and on the positions of its production and decay vertices within the detector. This correction results in a positive shift of the asymmetry A_L^{exp} of 0.3×10^{-3}.

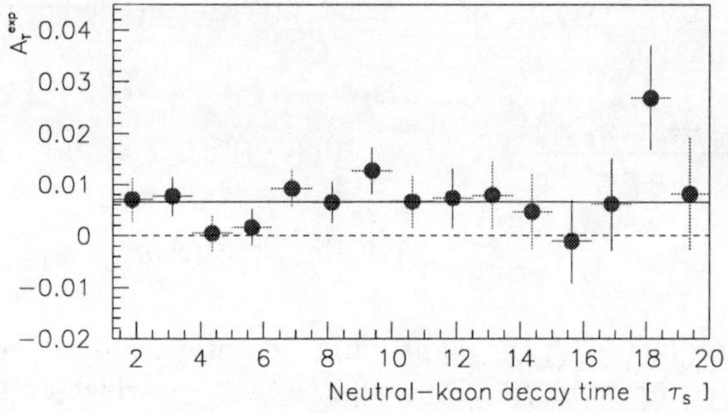

Figure 20.3: The A_L^{exp} asymmetry.

20.6 Results and Discussion

The A_L^{exp} asymmetry is shown in fig. 20.3 for a decay-time interval $1\tau_S \leq \tau \leq 20\tau_S$. The data points scatter around a positive and constant offset from zero, the average being

$$\langle A_L^{\mathrm{exp}} \rangle_{(1-20)\tau_S} = (6.6 \pm 1.3) \times 10^{-3}, \tag{20.6}$$

with $\chi^2/\mathrm{d.o.f.} = 0.84$.

We have investigated the following sources of systematic errors in the measurement of $\langle A_L^{\mathrm{exp}} \rangle$: background level and background asymmetry, normalization factors, decay-time resolution, and regeneration correction. A detailed analysis of the systematic errors can be found in [2]. A summary of the systematic errors for the different parameters is reported in table 20.1.

Source	Known precision	$\langle A_L^{\mathrm{exp}} \rangle$ $[10^{-3}]$	$\mathrm{Im}(x_+)$ $[10^{-3}]$
background level	±10%	±0.03	±0.2
background asymmetry	±1%	±0.02	±0.5
ξ	$\pm 4.3 \times 10^{-4}$	±0.2	±0.1
η	$\pm 2.0 \times 10^{-3}$	±1.0	±0.4
decay-time resolution	10%	negligible	±0.6
regeneration	Ref. [8]	±0.1	±0.1
Total syst.		±1.0	±0.9

Table 20.1: Summary of systematic errors.

20.6.1 *Analysis Assuming CPT Invariance in the Decay*

In the following, we assume CPT invariance in the semileptonic decay amplitudes, ($\mathrm{Re}(y) = 0$ and $x_- = 0$). We can allow for a possible violation of the $\Delta S = \Delta Q$ rule ($x_+ \neq 0$) and fit the data with two free parameters, $\mathrm{Re}(\epsilon)$ and $\mathrm{Im}(x_+)$, both T violating. The result of the fit is

$$\begin{aligned} 4\mathrm{Re}(\epsilon) &= (6.2 \pm 1.4) \times 10^{-3}, \\ \mathrm{Im}(x_+) &= (1.2 \pm 1.9) \times 10^{-3}, \end{aligned}$$

with $\chi^2/\mathrm{d.o.f.} = 0.84$. The errors are statistical only. The correlation coefficient between $4\mathrm{Re}(\epsilon)$ and $\mathrm{Im}(x_+)$ is small, 0.46, since $\mathrm{Im}(x_+)$ is given by the value of the asymmetry for short lifetimes while $4\mathrm{Re}(\epsilon)$ is determined by the long lifetime values.

We observe clear evidence for T violation in the neutral-kaon mixing. $\mathrm{Im}(x_+)$ is compatible with zero. Thus, no T violation is observed in the semileptonic decay amplitude that violates the $\Delta S = \Delta Q$ rule, should this amplitude be different from zero. Note that the systematic errors on $\langle A_L^{\mathrm{exp}} \rangle$ apply as well to the $4\mathrm{Re}(\epsilon)$ term for the case of the two parameters fit.

20.6.2 Relaxing on CPT Invariance in the Decay

In this case, our measurement of $\langle A_L^{\mathrm{exp}} \rangle$ can be considered as a measurement of the parameter sum

$$4(\mathrm{Re}(\epsilon) - \mathrm{Re}(y) - \mathrm{Re}(x_-)) = (6.2 \pm 1.4) \times 10^{-3} \tag{20.7}$$

and can by itself not distinguish between T violation in mixing and CPT violation in the decay. However, the CPLEAR collaboration has recently published a thorough analysis of the neutral kaon decay parameters, using the (unitarity) Bell-Steinberger relation [9]. One of the results of this analysis, which has also been presented at this conference, is a measurement of the sum

$$\mathrm{Re}(y) + \mathrm{Re}(x_-) = (-0.2 \pm 0.3) \times 10^{-3}. \tag{20.8}$$

Reporting this result in eq. (20.7), we see that our asymmetry signals unambiguously a violation of T invariance in the kaon mixing.

20.7 Conclusions

The CPLEAR experiment has, for the first time, directly measured a violation of time reversal (T) invariance by comparing the rates of two time conjugate processes.

Taking this measurement in isolation, this conclusion is valid only in the limit of CPT conservation in the neutral-kaon semileptonic decay. However, combining this measurement with other kaon data and assuming unitarity shows that it is unambiguously due to T violation in the neutral-kaon mixing.

References

[1] See, for instance, T.D. Lee, *Particle Physics and Introduction to Field Theory* (Harwood, Chur, 1981); R.G. Sachs, *The Physics of Time Reversal* (University of Chicago Press, Chicago, 1987); and references therein.

[2] A. Angelopoulos *et al.*, CPLEAR Collaboration, Phys. Lett. B **444**, 43 (1998).

[3] P.K. Kabir, Phys. Rev. D **2**, 540 (1970); A. Aharony, Lett. Nuovo Cimento **3**, 791 (1970).

[4] L. Alvarez-Gaume *et al.*, CERN-TH/99-80

[5] R. Adler *et al.*, CPLEAR Collaboration, Nucl. Instr. and Meth. A **379**, 76 (1996).

[6] R. Adler *et al.*, CPLEAR Collaboration, Phys. Lett. B **363**, 243 (1995).

[7] C. Caso *et al.*, Particle Data Group, Eur. Phys. J. C **3**, 1 (1998).

[8] A. Angelopoulos *et al.*, CPLEAR Collaboration, Phys. Lett. B **413**, 422 (1997).

[9] A. Angelopoulos *et al.*, CPLEAR Collaboration, Phys. Lett. B **456**, 303 (1999).

21

Transverse Muon Polarization in $K^+_{\mu 3}$ Decay

G. Y. Lim[1]

Abstract

KEK-PS E246 aims to search for the violation of time reversal invariance by measuring a nonzero transverse muon polarization, P_T, in $K^+ \to \mu^+\pi^0\nu_\mu$ ($K^+_{\mu 3}$) decay. The experiment started to collect data from early 1996 and successfully completed its first stage in 1998. The result to date, using 3.9 million $K^+_{\mu 3}$ events taken in 1996 and 1997, was given as $P_T = -0.0042 \pm 0.0049(\text{stat}) \pm 0.0009(\text{syst})$ and $\text{Im}\xi = -0.013 \pm 0.016(\text{stat}) \pm 0.003(\text{syst})$.

21.1 Introduction

Time reversal (T) violation is equivalent to CP violation under CPT invariance. In fact, there have recently been important observations of T violation in the neutral kaon system [1], which correspond to ε of CP violation in the standard model. It is argued, on the other hand, that the size of CP violation in the standard model is not enough to explain the observed baryon asymmetry of the universe [2], which implies that there would be other sources of CP violation in addition to that of the standard model. A promising approach to identifying these additional CP violations is to investigate the T-violating transverse muon polarization (P_T) in $K^+ \to \mu^+\pi^0\nu_\mu$ ($K^+_{\mu 3}$) decay. P_T is defined as a triple-vector correlation:

$$P_T = \frac{\vec{\sigma}_\mu \cdot (\vec{p}_{\pi^0} \times \vec{p}_{\mu^+})}{|\vec{p}_{\pi^0} \times \vec{p}_{\mu^+}|}, \tag{21.1}$$

where $\vec{\sigma}_\mu$ is the spin vector of μ^+ and $\vec{p}_{\pi^0}$ and $\vec{p}_{\mu^+}$ are momentum vectors of π^0 and μ^+, respectively. Since the P_T is a T-odd observable and any

[1]Representing the KEK-PS E246 Collaboration formed by KEK, University of Tsukuba, University of Saskatchewan, Virginia Polytechnic Institute and State University, INR/Russia, University of Montreal, University of British Columbia, TRIUMF, National Taiwan University, Yonsei University, Korea University, Princeton University, Osaka University, Tokyo Institute of Technology, and University of Tokyo.

spurious effect from final-state interactions is very small [3], a nonzero value would be evidence for T violation. An important feature here is that there is no standard model contribution to P_T at tree level because only two generations of quarks are involved and the higher-order corrections are also negligible ($\sim 10^{-7}$) [4]. Thus, a nonzero P_T could be a clear signal of new physics beyond the standard model. Recently several models such as multi-Higgs doublet models [5, 6], lepto-quark models [5], and a class of supersymmetric models [7] are suggested and sizable P_T ($\sim 10^{-3}$) is expected.

In the $V - A$ theory, the matrix element of the $K_{\mu 3}$ decay is described as

$$M \propto \frac{G_F}{2} \sin\theta_c [f_+(q^2)(\tilde{p}_K^\lambda + \tilde{p}_\pi^\lambda) + f_-(q^2)(\tilde{p}_K^\lambda - \tilde{p}_\pi^\lambda)] \cdot [\overline{u}_\mu \gamma_\lambda (1-\gamma_5) u_\nu], \quad (21.2)$$

where G_F is the Fermi coupling constant, θ_c is the Cabibbo angle, q^2 is the momentum transfer squared, and $\tilde{p}_K$ and $\tilde{p}_\pi$ are the four momenta of kaon and pion, respectively. T invariance requires that the parameter ξ defined as a ratio between two structure functions $f_+(q^2)$ and $f_-(q^2)$, namely $\xi(q^2) = f_-(q^2)/f_+(q^2)$, should be real. P_T is connected to the imaginary part of ξ with a kinematical factor in the kaon rest frame,

$$P_T = \text{Im}\xi \, \frac{m_\mu}{m_K} \frac{|\vec{p}_\mu|}{[E_\mu + |\vec{p}_\mu| \vec{n}_\mu \cdot \vec{n}_\nu - m_\mu^2/m_K]}. \quad (21.3)$$

Thus, we look for a nonzero T violation parameter Imξ in this P_T experiment. The previous limit comes from a BNL experiment [8] using in-flight decays of an unseparated K^+ beam, which gave $P_T = -0.0031 \pm 0.0053$ and Im$\xi = -0.016 \pm 0.025$.

21.2 Experimental Setup and Method

KEK-PS E246 employs K^+ decay at rest in contrast to previous experiments. The stopped-K^+ method has several advantages especially from the point of view of systematics. First, an apparatus for stopped kaons can be arranged so as to detect π^0s going in any direction, in particular, the detection of π^0s going either in the forward or backward directions along the beam axis. A comparison of these two data groups enables us to reduce any systematic errors significantly. Second, we can deal with perfectly isotropic decays from stopped kaons without any influence from the beam history except for K^+ stopping distribution, the effect of which can be canceled in the first order. Third, the muon polarimeters can be far off the beam and less affected by beam background.

A separated K^+ beam of 660 MeV/c was produced at the 12 GeV proton synchrotron (PS) at the High Energy Accelerator Research Organization (KEK), Japan. The K^+ was distinguished from predominant π^+s ($\pi^+/K^+ \sim 6$) by using a Čerenkov counter. After being slowed down by

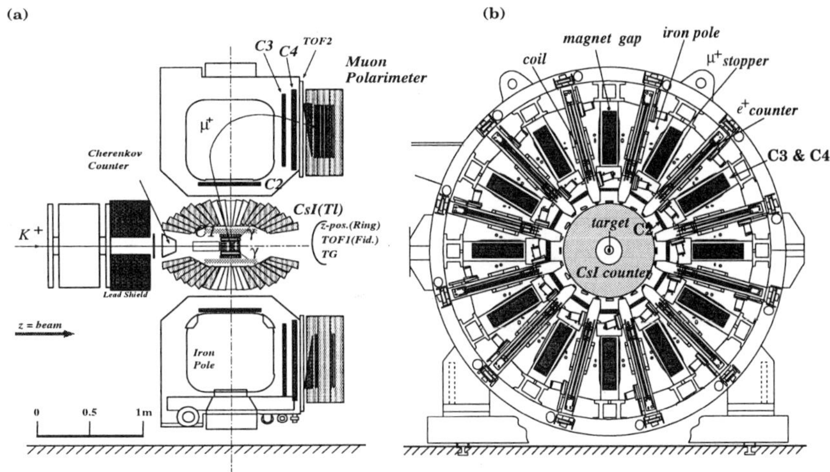

Figure 21.1: E246 experimental setup: (a) side view; (b) end view.

an energy degrader made of BeO, kaons were stopped in a target consisting of 256 scintillating fibers located at the center of a 12-sector iron-core superconducting toroidal spectrometer (fig. 21.1). $K^+_{\mu 3}$ decays having a 3.2% branching ratio were selected and the muon polarization transverse to the decay plane determined by two momentum vectors of muon ($\vec{p}_{\mu^+}$) and pion ($\vec{p}_{\pi^0}$) was measured by means of positron asymmetry in the $\mu^+ \to e^+\nu_e\overline{\nu}_\mu$ decay in the polarimeter.

The $\vec{p}_{\pi^0}$ is determined from two photons detected by a highly segmented CsI(Tl) barrel with silicon PIN photodiode readout [9]. The barrel has 12 holes for the muon to pass into the magnet and the beam entrance and exit holes (fig. 21.2). Since the assembly covers only 75% of 4π, there are a considerable number of events missing one of the photons. Fortunately it is possible to get the direction of the π^0 from only one photon if we require it to have high energy ($E_\gamma > 70$ MeV). The $\vec{p}_{\mu^+}$ is obtained from charged-particle tracking using information recorded at one (C2) and two sets (C3 and C4) of multiwire proportional chambers (MWPCs) at the entrance and exit of the magnetic gap, respectively. In addition, ring counters [10] made of scintillating fibers surround the target to get position information of the track along the beam direction. The tracking showed good momentum resolution ($\sigma_p = 2.6$ MeV$/c$ at 205 MeV$/c$) to reject a predominant background of $K^+ \to \pi^+\pi^0$ ($K^+_{\pi 2}$) decay (fig. 21.3). Positrons from $K^+ \to \pi^0 e^+\nu$ (K^+_{e3}) decay of the same momentum range as that of muons from the $K^+_{\mu 3}$ decay were rejected by time-of-flight (TOF) measurement.

Muons are degraded by a copper block and stopped in a stopper made of pure aluminium plates (fig. 21.4). Pairs of shim plates were used to create a well-defined magnetic field (130 Gauss in average) to hold the muon spin transverse to the decay plane, while the radial (P_L) and axial

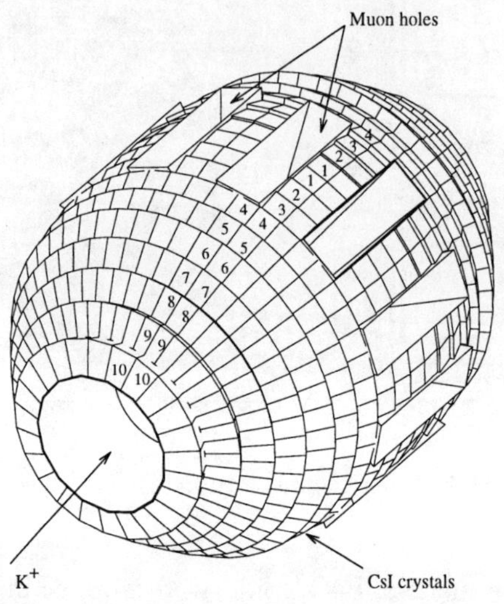

Figure 21.2: Schematic view of the CsI(Tl) barrel.

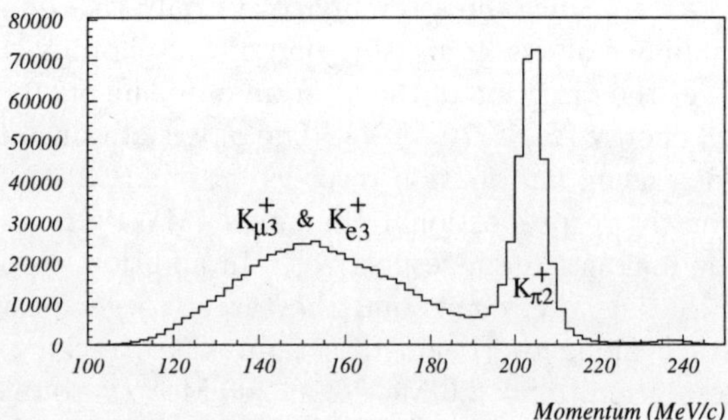

Figure 21.3: Momentum spectrum of charged particles for 2-photon data after correction of energy deposit in the active target.

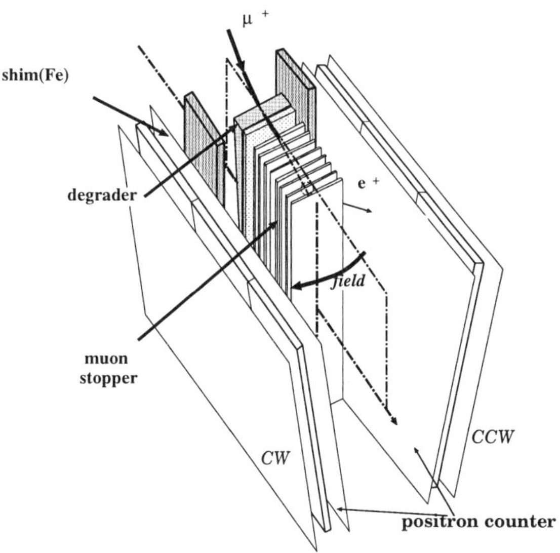

Figure 21.4: Schematic view of one of the polarimeters.

(P_N) in-plane components were precessed. The P_T was deduced from the asymmetry of positron yields detected at the plastic scintillators (positron counters) located at the middle of magnet gaps. In the events that have π^0 moving along the beam direction, the decay plane can be set radially from the detector axis. It means P_T is directed azimuthally in a screw-sense around the detector axis, as shown in fig. 21.5. The asymmetry is defined as a difference in the counting rates of e^+ between the clockwise (cw) and counterclockwise (ccw) side counters sandwiching the muon stopper in the polarimeter.

There are two main techniques to suppress any kinds of spurious asymmetry: summing over all the 12 sectors and forward-backward cancellation. First, the detector has a 12-fold rotational symmetry and one e^+ counter acts as the cw counter in one gap and the ccw counter in the neighboring gap. Thus, summing over the 12 sectors would cancel any non-screw-type biases such as a fake asymmetry from the e^+ counter inefficiency. Furthermore, it cancels out the bias coming from a shift of the K^+ stopping distribution at the target. By summing over the 12 gaps, the P_T is given by

$$\frac{\sum_{i=1}^{12} N_i(cw)}{\sum_{i=1}^{12} N_i(ccw)} \cong 1 \pm 2\alpha\langle\cos\theta_T\rangle P_T, \tag{21.4}$$

where $N(cw)$ and $N(ccw)$ are e^+ counts at the cw and ccw counters for a given gap. The α is an analyzing power and $\langle\cos\theta_T\rangle$ is a geometrical attenuation factor, which are conversion factors to transform the obtained asymmetry into P_T.

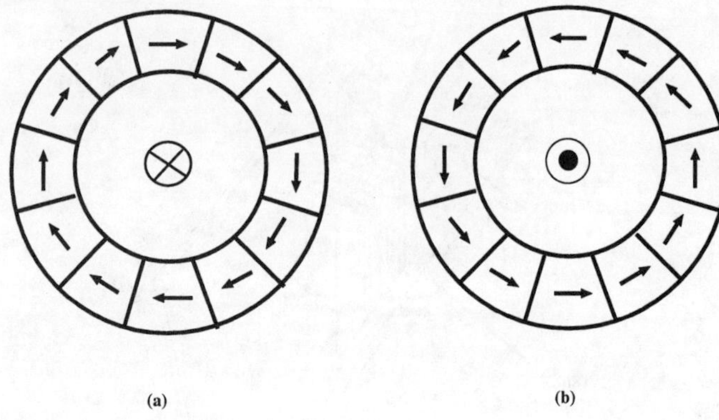

Figure 21.5: Schematic P_T directions in the 12 magnet gaps for (a) forward π^0 events and (b) backward π^0 events.

The second is a comparison between the two data samples according to the direction of π^0, along (forward-going) and the reverse direction (backward-going) to the beam axis. Since the sign of P_T is opposite when the π^0 direction is reversed (fig. 21.5) and any biased asymmetries are likely to be independent of the π^0 directions, their comparison would enable us to reduce the systematic errors significantly. A double ratio can be also formed by these two samples as follows:

$$\frac{[\sum_{i=1}^{12} N_i(cw)/\sum_{i=1}^{12} N_i(ccw)]_{fwd}}{[\sum_{i=1}^{12} N_i(cw)/\sum_{i=1}^{12} N_i(ccw)]_{bwd}} \cong 1 + 4\alpha\langle\cos\theta_T\rangle P_T. \tag{21.5}$$

21.3 Analysis

Data taking started in 1996 and finished its first stage in 1998. Part of the data taken in 1996 and 1997 that corresponds to roughly half of the total statistics have been analyzed. The main object of the data analysis is reconstruction of the $K^+_{\mu 3}$ decay and extraction of the positron time spectrum from the muon decay at the polarimeter. The characteristic muon lifetime of $\tau_\mu = 2.197\mu s$ was correctly observed in the positron time spectrum, as shown in fig. 21.6. The e^+ counts were extracted by integrating the time spectrum from 20 ns to 6.0 μs, subtracting a constant background (BG). The BG was obtained by fitting the time spectrum between 6.0 μs to 19.5 μs to a function $N_0 \exp(-t/\tau_\mu) + BG$.

Two independent analyses (A1 and A2) were performed, enabling us to cross-check the analysis algorithms and to estimate systematic error associated with the analysis. Even though they were done under the same principle of analysis, there were differences in several aspects between the two analyses in detail, especially with regard to the charged-particle tracking and the photon clustering in the CsI calorimeter. The data are divided

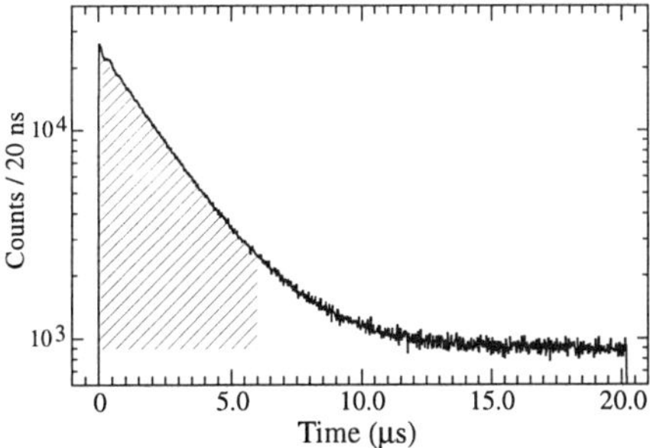

Figure 21.6: Positron time spectrum. The hatched area is the analyzed signal region after constant background subtraction.

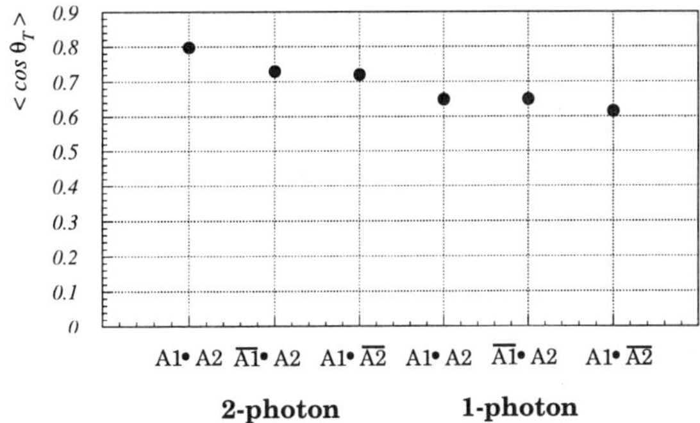

Figure 21.7: Geometrical attenuation factor $\langle \cos \theta_T \rangle$ corrected by the in-plane polarization for each subset.

into 1-photon and 2-photon samples. Each data set is further classified as 'A1·A2' common to both analyses, 'A1·$\overline{\text{A2}}$' selected only in A1, or '$\overline{\text{A1}}$· A2' selected only in A2.

The data quality of all the subsets was tested with a null-asymmetry check and possible decay plane rotation. Since the sign of P_T is reversed for the pions going in opposite directions, there should be no asymmetry for all π^0 directions integrated, if there is no bias in the polarimeter. It is called the null-asymmetry check. Another test is to use the decay plane distribution. Since the decay plane is determined by two vectors, $\vec{p}_{\mu^+}$ and $\vec{p}_{\pi^0}$, a distorted distribution of the decay plane in the data implies nonuniformity of kinematical acceptance of any detector component before the polarimeter, and it would induce an admixture of the in-plane muon polarization. The tests confirmed that there was no significant bias for the 6 data sets. A more important test would be to check the in-plane polarization for all subsets. The in-plane polarization (P_N) is determined by an asymmetry when we take pions going transverse to the beam direction. From the P_N

obtained in the data, we are able to estimate the sensitivity of each subset to P_T. In practice, it was done in terms of the geometrical attenuation factor $\langle\cos\theta_T\rangle$, defined for the angle of the decay plane normal vector relative to the axial direction in the polarimeter, which was evaluated by a Monte Carlo simulation. The corrected $\langle\cos\theta_T\rangle$ by using the observed P_N is shown in fig. 21.7. The 1-photon data sets have smaller $\langle\cos\theta_T\rangle$ because of the angular attenuation of the initial π^0 direction. On the other hand, there was not much difference between the common and uncommon data, which enable us to use all of the subsets to extract P_T.

The transverse polarization was calculated from the obtained asymmetry using eq. (21.5) with the analyzing power and geometrical factor, $\langle\cos\theta_T\rangle$. The analyzing power, $\alpha = 0.198 \pm 0.005$, was determined in an asymmetry measurement for the in-plane component P_N and comparison with a Monte Carlo calculation. Finally, we obtained $P_T = (-4.32 \pm 4.9) \times 10^{-3}$ by summing the 6 data sets with proper statistical weights. The T-violating parameter, Imξ, was extracted from eq. (21.3) as Im$\xi = -0.013 \pm 0.016$ evaluating the kinematical factor with a Monte Carlo calculation.

21.4 Systematic Errors

Estimation of the systematic errors would be the most important task in the measurement of a small effect such as P_T. The major systematic errors in P_T come from admixtures of the in-plane polarization components, P_L and P_N. As described above, the experiment has two powerful techniques, the 12-gap rotational symmetry and the forward-backward cancellation, which suppressed most of the admixtures. Possible admixtures would occur mainly from the misalignment of the detector elements, asymmetry of the magnetic field in the polarimeter, an asymmetric K^+ stopping distribution and background contamination. In order to estimate the systematic errors, the positions of all detector elements were measured precisely, and its effect on P_T was evaluated by using Monte Carlo simulation. Another basic tool is to estimate the forward-backward cancellation power by using the observed data for each item of the systematics. Table 21.1 shows a complete list of systematic errors in E246. The total systematic error, by adding all these contributions in quadrature, was evaluated to be 0.9×10^{-3}.

21.5 Summary

Nonzero transverse muon polarization (P_T) in the $K^+_{\mu 3}$ decay, if observed, will be a clean signal for new physics beyond the standard model. The KEK-PS E246 experiment is measuring P_T using stopped kaons with a potential sensitivity of 10^{-3}. 3.9 million $K^+_{\mu 3}$ events were analyzed from the data of 1996 and 1997 runs, and the first result of E246 is

$$P_T = -0.0042 \pm 0.0049(\text{stat.}) \pm 0.0009(\text{syst.}) \tag{21.6}$$

Source	Cancellation by Σ_{12}	Cancellation fwd/bwd	$\delta P_T \times 10^5$
e^+ counter r-rotation	no	yes	5
e^+ counter z-rotation	no	yes	2
e^+ counter ϕ-offset	no	yes	22
e^+ counter r-offset	yes	yes	< 1
e^+ counter z-offset	yes	yes	< 1
μ^+ counter ϕ-offset	no	yes	< 1
MWPC ϕ-offset (C4)	no	yes	25
CsI(Tl) misalignment	yes	yes	16
$\vec{B}$ offset (ϵ)	no	yes	30
$\vec{B}$ rotation (δ_r)	no	yes	3.7
$\vec{B}$ rotation (δ_z)	no	no	53
K^+ stopping distribution	yes	yes	< 30
Decay plane angle (θ_r)	no	yes	20
Decay plane angle (θ_z)	no	no	9
$K_{\pi 2}$ DIF background	no	yes	6
K^+ DIF background	yes	no	< 19
e^+ spectrum background	yes	yes	8
Analysis	-	-	38
Total			92

Table 21.1: Systematic errors. Σ_{12} and fwd/bwd denote the cancellation capabilities by the azimuthal symmetry of 12 sectors, and by the comparison of π^0 directions, respectively.

and

$$\mathrm{Im}\xi = -0.013 \pm 0.016(\mathrm{stat.}) \pm 0.003(\mathrm{syst.}). \tag{21.7}$$

At this moment, the statistical error is dominant. The data taken in 1998 are now being analyzed, which will improve the statistical sensitivity of Imξ to 0.011 from a naive extrapolation. Additional runs are scheduled in the coming two years, which will further improve this limit.

References

[1] P. Bloch, this volume, chapter 20; A. Ledovskoy, this volume, chapter 17; S. Wronka, this volume, chapter 18.

[2] M. Worah, this volume, chapter 3.

[3] A.R. Zhitnitskii, Yad. Fiz. **31**, 1024 (1980) [Sov. J. Nucl. Phys. **31**, 529 (1980)].

[4] I.I. Bigi, *CP Violation* (Cambridge University Press, 1999).

[5] R. Garisto and G. Kane, Phys. Rev. D **44**, 2038 (1991); G. Bélanger and C.Q. Geng, Phys. Rev. D **44**, 2789 (1991).

[6] M. Kobayashi, T.-T. Lin, and Y. Okada, Prog. Theor. Phys. **95**, 361 (1995).

[7] M. Fabbrichesi and F. Vissani, Phys. Rev. D **55**, 5334 (1997); G.-H. Wu and J.N. Ng, Phys. Lett. B **392**, 93 (1997).

[8] S.R. Blatt *et al.*, Phys. Rev. D **27**, 1056 (1983).

[9] D.V. Dementyev *et al.*, Nucl. Instr. Method A **379**, 499 (1996); Yu. G. Kudenko, O.V. Mineev, and J. Imazato, Nucl. Instr. Method A **411**, 437 (1998).

[10] A.P. Ivashkin *et al.*, Nucl. Instr. Method A **394**, 321 (1997).

22

Limits on T and CP Violation from Permanent Electric Dipole Moments

Michael Romalis

Abstract

Since the discovery of CP violation in K mesons considerable effort has been devoted to searches for T- and CP-violating permanent electric dipole moments (EDM). The limits on the EDMs of the neutron, electron, and ^{199} Hg set particularly stringent and largely orthogonal constraints on CP-violating phases beyond the standard model. Significant improvement in the limits on ^{199}Hg and electron EDMs can be expected in the near future. New techniques may allow further dramatic improvement in the sensitivity.

22.1 Introduction

In order for an elementary particle, atom, or molecule to possess a permanent electric dipole moment (EDM), time-reversal symmetry must be violated. This can be seen by considering the interaction of the electric dipole moment **d** with an electric field **E**

$$H = -\mathbf{d} \cdot \mathbf{E} = -d\frac{\mathbf{S} \cdot \mathbf{E}}{S}, \tag{22.1}$$

where the second equality follows from the fact that the dipole moment, as a vector, must be parallel to the spin of the particle. Since **E** is P-odd, T-even and **S** is P-even, T-odd, the existence of a nonzero electric dipole moment would violate both parity and time reversal. Assuming CPT symmetry, it would also imply violation of CP symmetry.

The EDM operator

$$\mathcal{L} = -d\sigma \cdot E = -d\frac{i}{2}\bar{\psi}\sigma_{\mu\nu}\gamma^5\psi F^{\mu\nu} \tag{22.2}$$

is induced by loop diagrams, the simplest of which are shown in fig. 22.1. In the standard model the contribution of these diagrams cancels since all three generations of quarks must be included to pick up the invariant

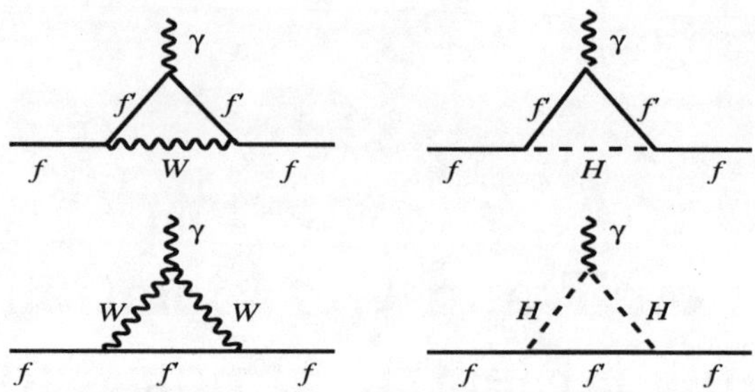

Figure 22.1: One-loop contributions to the EDM.

imaginary part of the CKM matrix. In general, the predictions for EDMs in the standard model are negligible compared with present experimental limits [1].

Beyond the standard model, additional particles can contribute to the loops shown in Figure 22.1, for example, supersymmetric particles, additional Higgs bosons, etc. In general, they can have CP-violating phases that give large one-loop contribution to the EDM. In some cases, two-loop contributions are also significant [2]. Present limits on the EDMs place stringent constraints on such CP-violating effects. For example, new CP-violating phases of SUSY are constrained to be less than 10^{-2} [3]. To express these constraints in a precise form it is necessary to go through atomic, nuclear, and QCD physics that relates the experimental limits to the EDMs of the elementary particles [3, 4].

There are presently three experiments that are most sensitive to new CP-violating effects. They set limits on the EDM of the neutron [5], electron [7], and ^{199}Hg [8]. Significant improvements have been made recently in the electron and ^{199}Hg experiments, and they are expected to give better limits in the near future. Beyond that, a number of new techniques have been proposed to reduce the limits on EDMs by several orders of magnitude.

The rest of this article is organized as follows. In section 22.2 I describe the interpretation of the existing EDM experiments in terms of limits on EDMs of fundamental particles. Section 22.3 is devoted to the review of the existing experiments and a brief description of new experimental proposals.

22.2 Limits on EDMs of elementary particles

The most sensitive EDM experiments set a limit on the EDM of a composite object, such as the neutron, an atom, or a molecule. To be compared with predictions for new CP-violating effects, these limits must be translated into limits on EDMs of quarks and leptons.

Naively, the EDM of the neutron can be easily expressed in terms of

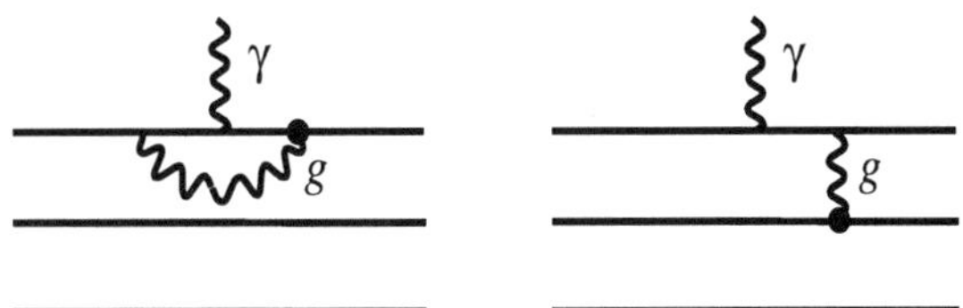

Figure 22.2: Contributions of the chromo-EDM operator (shown by a dark blob) to the EDM of the neutron.

the quark EDMs. In the nonrelativistic SU(6) model

$$d_n = \frac{4}{3}d_d - \frac{1}{3}d_u. \tag{22.3}$$

The SU(6) model is expected to be reliable in this case [9] and agrees relatively well with a lattice QCD calculation [10]:

$$d_n = 0.8d_d - 0.23d_u. \tag{22.4}$$

However, one must also include the contribution of the chromo-EDM operator

$$\mathcal{L} = -\tilde{d}\frac{i}{2}\bar{\psi}\sigma_{\mu\nu}\gamma^5\lambda^a\psi G^{a\,\mu\nu}, \tag{22.5}$$

where $\tilde{d}$ denotes the chromo-EDM of the quarks. This operator contributes to the neutron EDM through the diagrams shown in fig. 22.2. Its contribution is comparable to the usual EDM operator (eq. (22.2)), but it is much more difficult to estimate. Naive dimensional analysis is usually used [3]

$$d_n^C = \frac{eg_s}{4\pi}(O(1)\tilde{d}_u + O(1)\tilde{d}_d), \tag{22.6}$$

while the chiral perturbation analysis gives [11]

$$d_n^C = e(0.7\tilde{d}_u + 0.7\tilde{d}_d + 0.1\tilde{d}_s). \tag{22.7}$$

The uncertainty in these estimates is on the order of 100%. Thus, while one can set a rough limit on the size of the quark EDMs from the neutron EDM, the exact linear combination constrained by the neutron limit cannot be calculated reliably [4, 12].

The interpretation of the electron EDM limit is, by far, the cleanest. The experiment measures the EDM of Tl atoms, which is proportional to the electron EDM

$$d_{Tl} = (-585 \pm 50)d_e, \tag{22.8}$$

where the uncertainty comes entirely from atomic calculations [13].

To relate the limit on the atomic EDM of ^{199}Hg to elementary particles one must go through several steps. The atomic EDM is sensitive to the so-called Schiff moment of the nucleus S,

$$d_{^{199}Hg} = R_A\, S. \tag{22.9}$$

Experiment	Present limit
n (Grenoble) [5]	$\|d_n\| < 6.3 \times 10^{-26}$ e cm (90% C.L.)
	$\left\|0.8d_d - 0.2d_u + e(0.7\tilde{d}_d + 0.7\tilde{d}_u)\right\|$ $< 6.3 \times 10^{-26}$ e cm
Tl (Berkeley) [7]	$\|d_{Tl}\| < 4 \times 10^{-27}$ e cm (95% C.L.)
	$\left\|d_e \frac{m_d}{m_e}\right\| < 7.6 \times 10^{-26}$ e cm
^{199}Hg (Seattle) [8]	$\|d_{Hg}\| < 8.7 \times 10^{-28}$ e cm (95% C.L.)
	$e\left\|\tilde{d}_d - \tilde{d}_u\right\| < 2.7 \times 10^{-26}$e cm

Table 22.1: Present experimental limits on EDMs and their implications for EDMs of elementary particles. The electron EDM is scaled by the quark–to–electron mass ratio for comparison purposes, since most theories predict an EDM proportional to the mass.

The factor R_A depends on the atomic wave function at the nucleus and has been calculated with an accuracy of 30% [14]. The Schiff moment is proportional to the difference between the charge distribution and the electric dipole moment distribution of the nucleus. It can be induced by the electric dipole moments of the nucleons and CP-violating nucleon-nucleon interactions. It turns out that the second mechanism dominates and one can write

$$S = R_N \xi_{np}, \tag{22.10}$$

where ξ_{np} is the strength of CP-violating nuclear operator $G_F(\bar{p}p)(\bar{n}i\gamma^5 n)$ and R_N depends on the nuclear structure. It has been calculated for square-well and Woods-Saxon potentials with an estimated accuracy of 30% [15]. The CP-violating n-p interaction is dominated by π_0 exchange, which is proportional to the chromo-EDMs of the quarks [4]

$$\xi_{np} = G_F^{-1} \frac{3g_{\pi pp} m_0^2}{f_\pi m_\pi^2} (\tilde{d}_d - \tilde{d}_u). \tag{22.11}$$

Even though the calculations for ^{199}Hg involve several steps, the relative contribution of different quarks is fairly certain and the accuracy of the absolute scaling parameter can be improved by more detailed atomic and nuclear structure calculations.

I summarize the experimental limits and limits on elementary particle EDMs in table 22.1. As can be seen all three limits are comparable to each other and constrain a different linear combination of EDMs. This allows one to exclude the possibility of an accidental cancellation of different contributions [16].

22.3 EDM Experiments

All EDM experiments performed to date are based on detecting an energy difference produced by the electric dipole moment interacting with an electric field, eq. (22.1). The particles are spin polarized, placed in parallel electric and magnetic fields, and their Larmor precession frequency

$$\omega_L = \frac{2dE + 2\mu B}{\hbar} \tag{22.12}$$

is measured. Then the electric field is reversed and the frequency is measured again. A change of the precession frequency correlated with the reversal of the electric field would indicate the presence of an EDM. The ultimate sensitivity of such method is limited by the energy uncertainty principle. If the spin coherence time or measurement time is τ, then the Larmor precession frequency of a single particle can be measured with an uncertainty

$$\delta\omega = \frac{1}{\tau}. \tag{22.13}$$

If the measurement is performed on N particles simultaneously and repeated many times for a total time $T \gg \tau$, the statistical uncertainty is given by

$$\delta d = \frac{\hbar}{2E\sqrt{N\,\tau\,T}}. \tag{22.14}$$

Thus, in all experiments one tries to increase E, N, and τ, while eliminating any changes of the magnetic field that might be correlated with the electric field.

22.3.1 Neutron EDM Experiment

The latest neutron EDM experiments have been done at nuclear reactors in Grenoble, France, and Gatchina, Russia. The experiments utilize ultracold neutrons, with temperature on the order of 1 mK. Such neutrons are reflected from many materials that present a repulsive potential barrier on the order of 200neV. The neutrons are extracted from the reactor at room temperature and moderated to 30 K in liquid D_2. They are polarized by passing through a magnetized foil that transmits neutrons in one spin orientation and reflects the other. About 10^4 neutrons with an energy less than 200 neV are stored in a 20 liter cell for a period of 100 sec. By applying a $\pi/2$ RF pulse the neutrons are made to precess in a magnetic field of 10 mG and an electric field of 10 kV/cm. After 100 sec of free precession a second $\pi/2$ RF pulse, coherent with the first, is applied, and the number of neutrons in each of the spin states is counted using the magnetized foil as an analyzer. By varying the RF frequency relative to the Larmor precession frequency of the neutrons one can observe the Ramsey fringes. The sign of the electric field is periodically reversed and any shift of the Ramsey fringes correlated with the electric field would indicate the presence of an

EDM. The apparatus is placed inside multilayer magnetic shields and the magnetic fields inside are monitored using alkali metal magnetometers.

In 1990 the Grenoble group obtained a result [17] $d_n = -(3.4 \pm 2.6 \pm 4.1) \times 10^{-26}$e cm, which was limited by systematic errors due to changes in the magnetic field correlated with the reversal of the electric field. To address this problem they have added a ^{199}Hg comagnetometer in the new version of the experiment. Polarized ^{199}Hg atoms are introduced into the storage cell together with the neutrons, and their Larmor frequency is measured. Since the limit on the ^{199}Hg EDM is much smaller that the neutron limit, any changes of the ^{199}Hg Larmor frequency must be due to changes of the magnetic field. Using the ^{199}Hg comagnetometer they have recently obtained a result [5] $d_n = (1.9 \pm 5.4) \times 10^{-26}$e cm, limited by statistical error. They have combined this number with their 1990 result to set a new limit $|d_n| < 6.3 \times 10^{-26}$ e cm (90% CL). The validity of this procedure was recently criticized in [6].

The Gatchina experiment does not use a comagnetometer; however, it has two separate neutron storage cells with oppositely directed electric fields located next to each other. The sign of the EDM shift is opposite in the two cells, while many of the magnetic field effects are the same. The last experimental result published by the Gatchina group [18] is $d_n = +(2.6 \pm 4.2 \pm 1.6) \times 10^{-26}$e cm, comparable to the Grenoble result. To address some of the objections raised in [6] the limit given in [5] can be interpreted as a world average.

22.3.2 Electron EDM Experiment

The search for the EDM of the electron is done using heavy atoms with an unpaired electron, such as Tl or Cs. If the interactions of the electron were purely electrostatic and nonrelativistic, then no EDM could be observed, since the average electric field seen by the electron must be zero. However, as was pointed out by Sandars [19], spin-orbit interaction and relativistic corrections can actually enhance the atomic EDM relative to the electron EDM by a factor on the order of $\alpha^2 Z^3$.

The Berkeley EDM experiment uses an atomic beam of Tl atoms and the classic Ramsey separated field technique. To achieve high statistical sensitivity the atomic beams are very intense, with a counting rate of 10^{10} atoms/sec. The interaction region is 1 m long with an electric field of 10^5 V/cm. Atomic state preparation and detection are done using lasers. The dominant systematic effect is due to the magnetic field that the atoms see as they move through the electric field:

$$\mathbf{B}_m = \frac{\mathbf{v} \times \mathbf{E}}{c}. \tag{22.15}$$

This magnetic field is odd in E and can produce a false signal. To reduce the size of the effect, two counterpropagating atomic beams are used in the experiment. The velocity of the beams is aligned perpendicular to the direction of the **E** and **B** fields to eliminate the projection of the $\mathbf{B}_m$ field on the **B** field. The last result [7] $d_{Tl} = (-1.05 \pm 0.7 \pm 0.59) \times 10^{-24}$e cm was limited by magnetic field noise and several systematic effects.

The Berkeley group is now working on the new version of the experiment, which incorporates several major improvements. To reduce the magnetic field noise, the experiment is now using two pairs of counterpropagating Tl beams located on the opposite sides of the electric field plate. In addition, two pairs of counterpropagating Na beams have been added. Since Na atoms are much less sensitive to the electron EDM, they serve as a comagnetometer and a diagnostic tool for the systematic effects. The Tl and Na beams emerge from the same oven slits and thus follow very similar paths through the apparatus. The beam fluxes and the detection efficiency have also been increased, giving a total reduction of the statistical noise by a factor of 12. Better control of the magnetic field and its gradients combined with the diagnostic power of the Na beams allows reduction of the systematic errors by a comparable amount. A significant improvement in the limit on the electron EDM can be expected in the near future.

22.3.3 ^{199}Hg EDM Experiment

The Seattle ^{199}Hg experiment uses glass cells filled with 10^{14} atoms of ^{199}Hg and a buffer gas mixture. An electric field of 10kV/cm is applied using a conductive film deposited on the inside surfaces of the cell. The walls of the cells are coated with a paraffin film to increase the spin relaxation time of ^{199}Hg to 100 sec. The atoms are polarized by optical pumping using 254 nm line and the Larmor precession of the atoms is measured by monitoring the transmission of the light through the cell. The experiment uses two cells with oppositely directed electric fields to reduce magnetic field noise. The last experimental result [8] $d_{Hg} = (-1.1 \pm 2.4 \pm 3.6) \times 10^{-28}$e cm was limited by statistical noise and statistically limited tests for possible systematic effects.

We are presently working on a new version of the experiment that incorporates a number of improvements. For optical pumping and detection we now use a tunable UV laser instead of a microwave lamp. This allows us to set the frequency of the laser close to the optical absorption line for optical pumping and away from the absorption line for monitoring the precession of the spins using optical rotation of plane-polarized light. With this technique we can increase the polarization and density of ^{199}Hg atoms and reduce spin relaxation due to absorption of the light. Moreover, the systematic effects due to the Stark shift of the absorption line in the electric field are substantially reduced. These improvements, combined with higher detector efficiencies and longer spin relaxation times allowed us to reduce the statistical noise by a factor of 50. One of the main systematic effects is due to the magnetic fields produced by the leakage currents flowing on the surfaces of the cell, which are correlated with the reversal of the electric field. These fields are reduced to an acceptable level by keeping the leakage currents below 1 pA. Our goal is to improve the limit on ^{199}Hg EDM by a factor of 10.

22.3.4 New Experimental Techniques

Several new experimental techniques have been proposed to search for an EDM in various systems. To improve the limit on the neutron EDM R. Golub and S. Lamoreaux [20] have proposed to use ultra-cold neutrons produced by down-scattering from phonons in superfluid He. This mechanism allows one to increase the density of the neutrons by a factor of 100. A small amount of spin-polarized ^{3}He will be added to superfluid ^{4}He to serve several purposes. ^{3}He can act as a spin polarizer and analyzer for the neutrons since the cross-section of the process ^{3}He+n→p+T is highly spin dependent. The Larmor precession of the neutrons can be detected by observing oscillations in the scintillation rate of the reaction products in ^{4}He. In addition, ^{3}He can also act as a comagnetometer. It is expected that the electric field can be increased by a factor of 5 due to good dielectric properties of liquid He, and the neutron storage time can be increased by a factor of 8, resulting in overall improvement of sensitivity by a factor of 100. The experiment is in the initial design and testing stages at Los Alamos.

To improve the limit on the electron EDM several groups are exploring the possibility of using laser-cooled atoms. With cold atoms held in an optical trap one can achieve coherence times up to 100 sec, aXe factor of 3×10^4 longer than the transit time in the Tl EDM experiment, and increase the number of atoms by a factor of 10. In addition, most systematic effects limiting the Berkeley experiment will be eliminated. Using Cs atoms, which are a factor of 5 less sensitive to the electron EDM, the statistical sensitivity can be improved by factor of 100. The frequency shifts due to cold collisions [21] and interaction with the trapping light [22] appear to be manageable. D. Heinzen, S. Chu, and D. Weiss are independently working on preliminary experiments.

A method to improve the limit on the electron EDM using a PbO molecule has been proposed by D. DeMille. Very high EDM sensitivity can be achieved in polar molecules by looking for the interaction of the electric dipole moment with the interatomic electric field, which is on the order of 10^9 V/cm. The interatomic field can be aligned by applying a relatively small external electric field. Several EDM experiments with TlF have been performed [23] and an experiment with YbF is underway at Sussex. These experiments are mostly limited by a low counting rate. The PbO experiment will be done in a metastable excited state of the molecule, which can be polarized in an electric field of only 5V/cm. This field can be easily applied in a vapor cell with a high density of PbO, resulting in a much higher counting rate. Depending on the excitation and detection efficiency, an improvement by a factor of 100 to 10^4 is possible. Many systematic effects can be eliminated by switching between two different excited states of the molecule that have the opposite sign of the EDM shift. Preliminary experiments are under way at Yale.

To improve the limit on EDM of a diamagnetic atom we propose to use liquid ^{129}Xe. With liquid ^{129}Xe one can increase the number of atoms by 10^8, the electric field by a factor of 10, and the spin lifetime by a factor of 10 compared with the ^{199}Hg experiment. After taking into account a factor of 10 lower sensitivity of ^{129}Xe to CP-violating interactions, one can still improve the limit by a factor of 10^4. Bulk quantities of polarized liquid ^{129}Xe can be readily produced using existing techniques of optical pumping. The Larmor precession of liquid ^{129}Xe can be detected with sufficient sensitivity using a SQUID magnetometer. To reduce the magnetic field self-interaction effects [24] liquid ^{129}Xe will be placed in a spherical cell. Preliminary experiments are being started in Seattle.

22.4 Conclusions

Present limits on the EDM of the neutron, electron and ^{199}Hg set stringent constraints on new sources of CP violation. The limits from the three experiments are comparable and largely orthogonal, making accidental cancellations highly unlikely. The limits on the electron and ^{199}Hg EDM are likely to be improved in the near future. Several new techniques have been proposed, promising improvements by factors of 100 to 10^4.

References

[1] I.B. Khriplovich and S.K. Lamoreaux, *CP Violation without Strangeness*, (Springer, Berlin, 1997).

[2] S.M. Barr and A. Zee, Phys. Rev. Lett. **65**, 21 (1990).

[3] S.M. Barr, Int. J. Mod. Phys. **8**, 209 (1993).

[4] T. Falk, K.A. Olive, M. Pospelov, and R. Roiban, Nucl. Phys. B **560**, 3 (1999).

[5] P.G. Harris *et al.*, Phys. Rev. Lett **82**, 904 (1999). For criticism of the new limit see [6].

[6] S.K. Lamoreaux and R. Golub, Phys. Rev. D **61**, 051301 (2000).

[7] E.D. Commins, S. B. Ross, D. DeMille, and B.C. Regan, Phys. Rev. A **50**, 2960 (1994).

[8] J.P. Jacobs, W.M. Klipstein, S.K. Lamoreaux, B.R. Heckel, and E.N. Fortson, Phys. Rev. A **52**, 3521 (1995).

[9] C. Hamzaoui, M. Pospelov, and R. Roiban, Phys. Rev. D **56**, 4295 (1997).

[10] S. Aoki *et al.*, Phys. Rev. D **56**, 433 (1997).

[11] V.M. Khatsimovsky and I.B. Khriplovich, Phys. Lett. B **296**, 219 (1994).

[12] A. Bartl, T. Gajdosik, W. Porod, P. Stockinger, and H. Stremnitzer, Phys. Rev. D **60**, 073003 (1999).

[13] Z.W. Liu and H.P. Kelley, Phys. Rev. A **45**, 4210 (1992).

[14] V.V. Flambaum, I.B. Khriplovich, and O.P. Sushkov, Phys. Lett. B **162**, 213 (1985).

[15] O.P. Sushkov, V.V. Flambaum, and I.B. Khriplovich, Sov. Phys. JETP **60**, 873 (1984).

[16] T. Ibrahim and P. Nath, Phys. Lett. B **418**, 98 (1998).

[17] K.F. Smith *et al.*, Phys. Lett. B **234**, 191 (1990).

[18] I.S. Altarev *et al.*, Phys. Lett. B **276**, 242 (1992).

[19] P.G.H. Sandars, Phys. Lett. **14**, 194 (1965).

[20] R. Golub and S.K. Lamoreaux, Phys. Rep. **237**, 1 (1994).

[21] M. Bijlsma, B.J. Verhaar, and D.J. Heinzen, Phys. Rev. A **49**, R4285 (1994).

[22] M.V. Romalis and E.N. Fortson, Phys. Rev. A **59**, 4547 (1999).

[23] D. Cho, K. Sangster, and E.A. Hinds, Phys. Rev. Lett. **63**, 2559 (1989).

[24] M.V. Romalis and W. Happer, Phys. Rev. A **60**, 1385 (1999).

23

Bounds on CPT and Lorentz Violation from Experiments with Kaons

V. Alan Kostelecký

Abstract

Possible signals for indirect CPT violation arising in experiments with neutral kaons are considered in the context of a general CPT- and Lorentz-violating standard-model extension. Certain CPT observables can depend on the meson momentum and exhibit sidereal variations in time. Any leading-order CPT violation would be controlled by four parameters that can be separately constrained in appropriate experiments. Recent experiments bound certain combinations of these parameters at the level of about 10^{-20} GeV.

Experiments using neutral-meson oscillations can place constraints of remarkable precision on possible violations of CPT invariance. For kaons, recent results [1–3] bound the CPT figure of merit $r_K \equiv |m_K - m_{\overline{K}}|/m_K$ to less than a part in 10^{18}. Other experiments [4–6] are expected to improve this bound in the near future. Experiments with neutral-B mesons [7, 8] have also placed high-precision constraints on possible CPT violation, and the B and charm factories should produce additional bounds on the heavy neutral-meson systems.

A purely phenomenological treatment of possible CPT violation in the kaon system has been known for some time [9]. In this approach, a complex phenomenological parameter δ_K allowing for indirect CPT violation is introduced in the standard relationships between the physical meson states and the strong-interaction eigenstates. No information about δ_K itself can be obtained within this framework. However, over the past ten years a plausible theoretical framework allowing the possibility of CPT violation has been developed. It involves the notion of spontaneous breaking of CPT and Lorentz symmetry in a fundamental theory [10], perhaps arising at the Planck scale from effects in a quantum theory of gravity or string theory, and it is compatible both with established quantum field theory and with present experimental constraints. At low energies, a general CPT- and

Lorentz-violating standard-model extension emerges that preserves gauge invariance and renormalizability [11, 12] and that provides an underlying basis for the phenomenology of CPT violation in the kaon system. The resulting situation is comparable to that for conventional CP violation, where the nonzero value of the phenomenological parameter ϵ_K for T violation in the kaon system can in principle be calculated from the usual standard model of particle physics [13, 14].

In this talk, the primary interest is in the application of the standard-model extension to CPT tests with kaons. However, the standard-model extension also provides a quantitative microscopic framework for CPT and Lorentz violation that can be used to evaluate and compare a wide variety of other experiments [15]. These include tests with heavy neutral-meson systems [7,8,11,16,17], studies of fermions in Penning traps [18–22], constraints on photon birefringence and radiative QED effects [12,23–25], hydrogen and antihydrogen spectroscopy [26, 27], clock-comparison experiments [28, 29], measurements of muon properties [30], cosmic-ray and neutrino tests [31], and baryogenesis [32].

Developing a plausible theoretical framework for CPT violation without radical revisions of established quantum field theory is a difficult proposition [15, 33]. It is therefore perhaps to be expected that in the context of the standard-model extension the parameter δ_K displays features previously unexpected, including dependence on momentum magnitude and orientation. The implications include, for instance, time variations of the measured value of δ_K with periodicity of one sidereal (not solar) day [17].

First, consider some general theoretical features relevant for oscillations in any neutral-meson system. Denote generically the strong-interaction eigenstate by P^0, where P^0 is one of K^0, D^0, B_d^0, B_s^0, and denote the opposite-flavor antiparticle by $\overline{P^0}$. Then a neutral-meson state is a linear combination of the Schrödinger wave function for P^0 and $\overline{P^0}$. The time evolution of the associated two-component object Ψ state is given [9] in terms of a 2×2 effective Hamiltonian Λ by $i\partial_t\Psi = \Lambda\Psi$. The physical propagating states are the eigenstates P_S and P_L of Λ. They have eigenvalues $\lambda_S \equiv m_S - \frac{1}{2}i\gamma_S$ and $\lambda_L \equiv m_L - \frac{1}{2}i\gamma_L$, respectively, where m_S, m_L are the propagating masses and γ_S, γ_L are the associated decay rates. Flavor oscillations between P^0 and $\overline{P^0}$ are controlled by the off-diagonal components of Λ, while indirect CPT violation [34] occurs if and only if the diagonal elements of Λ have a nonzero difference $\Lambda_{11} - \Lambda_{22} \neq 0$. Writing Λ as $\Lambda \equiv M - \frac{1}{2}i\Gamma$, where M and Γ are hermitian, the condition for CPT violation becomes $\Delta M - \frac{1}{2}i\Delta\Gamma \neq 0$, where $\Delta M \equiv M_{11} - M_{22}$ and $\Delta\Gamma \equiv \Gamma_{11} - \Gamma_{22}$.

A perturbative calculation in the general standard-model extension provides the dominant CPT-violating contributions to Λ [10]. It turns out that the hermiticity of the perturbing Hamiltonian enforces $\Delta\Gamma = 0$ at leading order. The leading-order signal therefore arises in the difference ΔM, and so the standard figure of merit

$$r_P \equiv \frac{|m_P - m_{\overline{P}}|}{m_P} = \frac{|\Delta M|}{m_P} \tag{23.1}$$

provides a complete description of the magnitude of the dominant CPT-violating effects. An explicit expression for ΔM in terms of quantities in the standard-model extension is known [11, 17]. For several reasons, its form turns out to be relatively simple,

$$\Delta M \approx \beta^\mu \Delta a_\mu. \tag{23.2}$$

Here, $\beta^\mu = \gamma(1, \vec{\beta})$ is the four-velocity of the meson state in the observer frame and Δa_μ is a combination of CPT- and Lorentz-violating coupling constants for the two valence quarks in the P^0 meson. Note that the oscillation experiments considered here provide the only known sensitivity to Δa_μ. Note also that the velocity dependence and the corresponding momentum dependence of ΔM are compatible with the anticipated substantial modifications to standard physics if the CPT theorem is violated.

The experimental implications of momentum dependence in observables for CPT violation are substantial. Effects can be classified according to whether they arise primarily from a dependence on the magnitude of the boost or from the variation with its direction [17]. The dependence on momentum magnitude implies the possibility of increasing the CPT reach by changing the meson boost and even the possibility of increasing sensitivity by restricting attention to a momentum subrange in a given data set. The dependence on momentum direction implies variation of observables with the beam direction for collimated mesons, variation with the meson angular distribution for other situations, and sidereal effects arising from the rotation of the earth relative to the constant 3-vector $\Delta\vec{a}$. In actual experiments the momentum and angular dependences are frequently used to determine detector properties and experimental systematics, so there is a definite risk of cancelling or averaging away CPT-violating effects. However, the detection of a momentum dependence in observables would be a unique feature of CPT violation. There are also new possibilities for data analysis. For instance, measurements of an observable can be binned according to sidereal time to search for possible time variations as the earth rotates.

The above discussion holds for any neutral-meson system. For definiteness, the remainder of this talk considers the special case of kaons. The parameter δ_K, which is effectively a phase-independent quantity, can be defined through the relationship between the eigenstates of the strong interaction and those of the effective Hamiltonian:

$$\begin{aligned} |K_S\rangle &= \frac{(1+\epsilon_K+\delta_K)|K^0\rangle + (1-\epsilon_K-\delta_K)|\overline{K^0}\rangle}{\sqrt{2(1+|\epsilon_K+\delta_K|^2)}}, \\ |K_L\rangle &= \frac{(1+\epsilon_K-\delta_K)|K^0\rangle - (1-\epsilon_K+\delta_K)|\overline{K^0}\rangle}{\sqrt{2(1+|\epsilon_K-\delta_K|^2)}}. \end{aligned} \tag{23.3}$$

Assuming that all CP violation is small, δ_K is in general given as

$$\delta_K \approx \Delta\Lambda/2\Delta\lambda, \tag{23.4}$$

where $\Delta\lambda \equiv \lambda_S - \lambda_L$ is the eigenvalue difference of Λ. In terms of the mass and decay-rate differences $\Delta m \equiv m_L - m_S$ and $\Delta\gamma \equiv \gamma_S - \gamma_L$, it follows that $\Delta\lambda = -\Delta m - \frac{1}{2}i\Delta\gamma = -i\Delta m e^{-i\hat{\phi}}/\sin\hat{\phi}$, where $\hat{\phi} \equiv \tan^{-1}(2\Delta m/\Delta\gamma)$.

In the context of the standard-model extension, the above expressions show that a meson with velocity $\vec{\beta}$ and corresponding boost factor γ displays CPT-violating effects given by

$$\delta_K \approx i \sin\hat{\phi}\, e^{i\hat{\phi}} \gamma(\Delta a_0 - \vec{\beta}\cdot\Delta\vec{a})/\Delta m. \tag{23.5}$$

The conventional figure of merit r_K becomes

$$\begin{aligned} r_K &\equiv \frac{|m_K - m_{\overline{K}}|}{m_K} \approx \frac{2\Delta m}{m_K \sin\hat{\phi}}|\delta_K| \\ &\approx \frac{|\beta^\mu \Delta a_\mu|}{m_K}. \end{aligned} \tag{23.6}$$

After substitution for the known experimental values [35] for Δm, m_K, and $\sin\hat{\phi}$, this gives

$$r_K \simeq 2 \times 10^{-14}|\delta_K| \simeq 2\left|\beta^\mu \frac{\Delta a_\mu}{1\ \text{GeV}}\right|. \tag{23.7}$$

A constraint on $|\delta_K|$ of about 10^{-4} corresponds to a limit on $|\beta^\mu\Delta a_\mu|$ of about 10^{-18} GeV.

The dependence of the eigenfunctions and eigenvalues of Λ on M_{11} and M_{22} raises the possibility of leading-order momentum dependence in the parameter ϵ_K, in the masses and decay rates m_S, m_L, γ_S, γ_L, and in various associated quantities such as Δm, $\Delta\gamma$, $\hat{\phi}$. However, this possible dependence is in fact absent because the CPT-violating contribution from M_{22} is the negative of that from M_{11}, and only δ_K is sensitive to ΔM at leading order. Thus, for example, the usual parameter ϵ_K for indirect T violation is independent of momentum in the present framework [36].

The expressions obtained above can be viewed as defined in the laboratory frame. To exhibit the time dependence of δ_K arising from the rotation of the earth, a different and nonrotating frame is useful [29]. A basis $(\hat{X}, \hat{Y}, \hat{Z})$ for this frame can be introduced in terms of celestial equatorial coordinates. The $\hat{Z}$ axis is defined as the rotation axis of the earth, while $\hat{X}$ has declination and right ascension 0° and $\hat{Y}$ has declination 0° and right ascension 90°. This provides a right-handed orthonormal basis that is independent of any particular experiment. Denote the spatial basis in the laboratory frame as $(\hat{x}, \hat{y}, \hat{z})$, where $\hat{z}$ and $\hat{Z}$ differ by a nonzero angle given by $\cos\chi = \hat{z}\cdot\hat{Z}$. Then, $\hat{z}$ precesses about $\hat{Z}$ with the earth's sidereal frequency Ω. A convenient choice of $\hat{z}$ axis is often along the beam direction. If the origin of time $t = 0$ is taken such that $\hat{z}(t = 0)$ is in the first quadrant of the $\hat{X}$-$\hat{Z}$ plane and if $\hat{x}$ is defined perpendicular to $\hat{z}$ and lies in the $\hat{z}$-$\hat{Z}$ plane for all t, then a right-handed orthonormal basis can be completed with $\hat{y} := \hat{z} \times \hat{x}$. It follows that $\hat{y}$ lies in the plane of the earth's equator and is perpendicular to $\hat{Z}$ at all times. Disregarding relativistic effects due to the rotation of the earth, a nonrelativistic transformation (given by eq. (16) of ref. [29]) provides the conversion between the two bases.

Using the above results, one can obtain in the nonrotating frame an expression for the parameter δ_K in the general case of a kaon with three-velocity $\vec{\beta} = \beta(\sin\theta\cos\phi, \sin\theta\sin\phi, \cos\theta)$. Here, θ and ϕ are standard spherical polar coordinates specified in the laboratory frame about the $\hat{z}$ axis. If $\hat{z}$ coincides with the beam axis, the spherical polar coordinates can be taken as the usual polar coordinates for a detector. One finds

$$\begin{aligned}
\delta_K(\vec{p}, t) \quad = \quad & \frac{i \sin\hat{\phi}\, e^{i\hat{\phi}}}{\Delta m} \gamma(\vec{p}) \times \\
& \Big[\Delta a_0 + \beta(\vec{p})\Delta a_Z(\cos\theta\cos\chi - \sin\theta\cos\phi\sin\chi) \\
& + \beta(\vec{p})\,(-\Delta a_X \sin\theta\sin\phi \\
& \qquad + \Delta a_Y(\cos\theta\sin\chi + \sin\theta\cos\phi\cos\chi))\sin\Omega t \\
& + \beta(\vec{p})\,(\Delta a_X(\cos\theta\sin\chi + \sin\theta\cos\phi\cos\chi) \\
& \qquad + \Delta a_Y \sin\theta\sin\phi)\cos\Omega t\Big], \qquad (23.8)
\end{aligned}$$

where $\gamma(\vec{p}) = \sqrt{1 + |\vec{p}|^2/m_K^2}$ and $\beta(\vec{p}) = |\vec{p}|/m\gamma(\vec{p})$, as usual. This expression has direct implications for experiment. For example, the complex phase of δ_K is $i\exp(i\hat{\phi})$, independent of momentum and time. The real and imaginary parts of δ_K therefore exhibit the same momentum and time dependence, and so $\mathrm{Re}\,\delta_K$ and $\mathrm{Im}\,\delta_K$ scale proportionally when a meson is boosted. Another property of eq. (23.8) is the variation of the CPT-violating effects with the meson boost. For example, if $\Delta a_0 = 0$ in the laboratory frame then there is no CPT violation for a meson at rest but effects appear when the meson is boosted. In contrast, for the case where $\Delta\vec{a} = 0$ in the laboratory frame, CPT violation is enhanced by the boost factor γ relative to a meson at rest. Other implications follow from the angular dependence in eq. (23.8) and from the variation of δ_K with sidereal time t. For example, under some circumstances all CPT violation can average to zero if, as usual, neither angular separation nor time binning are performed.

The momentum and time dependence given by eq. (23.8) implies that the experimental setup and data-taking procedure affect the CPT reach. Space restrictions here preclude consideration of all the different classes of scenario realized in practice. Instead, attention is restricted here to a single one, typified by the E773 and KTeV experiments [1,37]. This class of experiment, which involves highly collimated uncorrelated kaons having nontrivial momentum spectrum and large mean boost, is particularly relevant here because the KTeV collaboration announced at this conference the first constraints on the sidereal-time dependence of CPT observables in the kaon system [2]. A discussion of some issues relevant to other types of experiment can be found in ref. [17].

The KTeV experiment involves kaons with $\beta \simeq 1$ and average boost factor $\overline{\gamma}$ of order 100. For this case, $\hat{z} \cdot \hat{Z} = \cos\chi \simeq 0.6$. In all experiments with boosted collimated kaons, eq. (23.8) simplifies because the kaon three-velocity in the laboratory frame can be taken as $\vec{\beta} = (0, 0, \beta)$. The expression for δ_K becomes

$$\begin{aligned} \delta_K(\vec{p}, t) \quad = \quad & \frac{i \sin\hat{\phi}\; e^{i\hat{\phi}}}{\Delta m} \gamma \times \\ & [\Delta a_0 + \beta \Delta a_Z \cos\chi + \beta \sin\chi (\Delta a_Y \sin \Omega t + \Delta a_X \cos \Omega t)]. \end{aligned} \tag{23.9}$$

In this equation, each of the four components of Δa_μ has momentum dependence through the boost factor γ. However, only the coefficients of Δa_X and Δa_Y vary with sidereal time.

To gain insight into the implications of eq. (23.9), consider first a conventional analysis that seeks to constrain the magnitude $|\delta_K|$ but disregards the momentum and time dependence. Assuming the experiment is performed over an extended time period, as is typically the case, the relevant quantity is the time and momentum average of eq. (23.9):

$$|\overline{\delta_K}| = \frac{\sin\hat{\phi}}{\Delta m} \overline{\gamma} (\Delta a_0 + \overline{\beta} \Delta a_Z \cos\chi), \tag{23.10}$$

where $\overline{\beta}$ and $\overline{\gamma}$ are appropriate averages of β and γ, respectively, taken over the momentum spectrum of the data. Substitution of the experimental quantities and the current constraint on $|\delta_K|$ from this class of experiment permits the extraction of a bound on a combination of Δa_0 and Δa_Z [17]:

$$|\Delta a_0 + 0.6 \Delta a_Z| \lesssim 10^{-20} \text{ GeV}. \tag{23.11}$$

The ratio of this to the kaon mass compares favorably with the ratio of the kaon mass to the Planck scale. Note that the CPT reach of this class of experiments is some two orders of magnitude greater than might be inferred from the bound on r_K, due to the presence of the boost factor $\overline{\gamma} \simeq 100$.

In experiments with kaon oscillations, the bounds obtained on δ_K are extracted from measurements on other observables including, for instance, the mass difference Δm, the K_S lifetime $\tau_S = 1/\gamma_S$, and the ratios η_{+-}, η_{00} of amplitudes for 2π decays. The latter are defined by

$$\begin{aligned} \eta_{+-} \quad & \equiv \quad \frac{A(K_L \to \pi^+\pi^-)}{A(K_S \to \pi^+\pi^-)} \equiv |\eta_{+-}| e^{i\phi_{+-}} \approx \epsilon + \epsilon', \\ \eta_{00} \quad & \equiv \quad \frac{A(K_L \to \pi^0\pi^0)}{A(K_S \to \pi^0\pi^0)} \equiv |\eta_{00}| e^{i\phi_{00}} \approx \epsilon - 2\epsilon'. \end{aligned} \tag{23.12}$$

Adopting the Wu-Yang phase convention [38], it follows that $\epsilon \approx \epsilon_K - \delta_K$ [39, 40]. Experimentally, it is known that $|\epsilon| \simeq 2 \times 10^{-3}$ [35] and that $|\epsilon'| \simeq 6 \times 10^{-6}$ [41]. Since δ_K is bounded only to about 10^{-4} it is acceptable at present to neglect ϵ', equivalent to assuming the hierarchy $|\epsilon_K| > |\delta_K| > |\epsilon'|$. Noting that the phases of ϵ_K and δ_K differ by 90° [42] then gives

$$\begin{aligned} |\eta_{+-}|e^{i\phi_{+-}} &\approx |\eta_{00}|e^{i\phi_{00}} \approx \epsilon \approx \epsilon_K - \delta_K \\ &\approx (|\epsilon_K| + i|\delta_K|)e^{i\hat{\phi}}. \end{aligned} \tag{23.13}$$

This implies

$$\begin{aligned} |\eta_{+-}| &\approx |\eta_{00}| \approx |\epsilon_K|(1 + O(|\delta_K/\epsilon_K|^2), \\ \phi_{+-} &\approx \phi_{00} \approx \hat{\phi} + |\delta_K/\epsilon_K|, \end{aligned} \tag{23.14}$$

which shows that leading-order momentum and time dependences in measured quantities appear only in the phases ϕ_{+-} and ϕ_{00}. The momentum and time dependences are absent or suppressed in other observables, including $|\eta_{+-}|$, $|\eta_{00}|$, ϵ', Δm, $\hat{\phi}$, and $\tau_S = 1/\gamma_S$.

Substituting for δ_K in ϕ_{+-} and ϕ_{00} yields expressions displaying explicitly the time and momentum dependences:

$$\begin{aligned} \phi_{+-} \approx \phi_{00} &\approx \hat{\phi} + \frac{\sin\hat{\phi}}{|\eta_{+-}|\Delta m}\gamma[\Delta a_0 + \beta\Delta a_Z \cos\chi \\ &\quad +\beta \sin\chi(\Delta a_Y \sin\Omega t + \Delta a_X \cos\Omega t)]. \end{aligned} \tag{23.15}$$

Since the coefficients of each of the four components Δa_0, Δa_X, Δa_Y, Δa_Z are all distinct, this equation shows that in principle each component can be independently bounded in the class of experiments involving collimated kaons with a nontrivial momentum spectrum. Thus, binning in time and fitting to sine and cosine terms would allow independent constraints on Δa_X and Δa_Y, while a time-averaged analysis would permit the extraction of Δa_0 and Δa_Z provided the momentum spectrum includes a significant range of $\vec{\beta}$. Note, however, that the latter separation is unlikely to be possible at experiments with high mean boost because then $\beta \simeq 1$ over much of the momentum range.

A constraint $A_{+-} \lesssim 0.5°$ on the amplitude A_{+-} of time variations of the phase ϕ_{+-} with sidereal periodicity was announced at this conference [2]. Eq. (23.15) shows that A_{+-} is given by

$$A_{+-} = \beta\gamma\frac{\sin\hat{\phi}\sin\chi}{|\eta_{+-}|\Delta m}\sqrt{(\Delta a_X)^2 + (\Delta a_X)^2}. \tag{23.16}$$

Substitution for known quantities and for the experimental constraint on A_{+-} places the bound

$$\sqrt{(\Delta a_X)^2 + (\Delta a_Y)^2} \lesssim 10^{-20} \text{ GeV} \tag{23.17}$$

on the relevant parameters for CPT violation. Like the bound (23.11), the ratio of this bound to the kaon mass compares favorably with the ratio of the kaon mass to the Planck scale. Note that the bounds (23.11) and (23.17) represent independent constraints on possible CPT violation. Note also that in principle a constraint on the phase of the sidereal variations of ϕ_{+-}, determined by the ratio $\Delta a_Y/\Delta a_X$, would permit the separation of Δa_X and Δa_Y.

The examples discussed in this talk show that the study of momentum and time dependence in CPT observables is necessary to obtain the full CPT reach in a given experiment. Additional interesting results would emerge from careful analyses for experiments other than the ones considered here. Moreover, although emphasis has been given to the kaon system, related analyses in other neutral-meson systems would be well worth pursuing.

References

[1] E773 Collaboration, B. Schwingenheuer *et al.*, Phys. Rev. Lett. **74**, 4376 (1995). See also E731 Collaboration, L.K. Gibbons *et al.*, Phys. Rev. D **55**, 6625 (1997); R. Carosi *et al.*, Phys. Lett. B **237**, 303 (1990).

[2] Y.B. Hsiung, this volume, chapter 6.

[3] P. Bloch, this volume, chapter 25.

[4] P. Franzini, in G. Diambrini-Palazzi, C. Cosmelli, L. Zanello, eds., *Phenomenology of Unification from Present to Future* (World Scientific, Singapore, 1998); P. Franzini and J. Lee-Franzini, Nucl. Phys. Proc. Suppl. **71**, 478 (1999).

[5] A. Antonelli, this volume, chapter 7; S. Di Falco and M. Incagli, this volume, chapter 24.

[6] See, for example, C. Bhat *et al.*, preprint FERMILAB-P-0894 (1998).

[7] OPAL Collaboration, R. Ackerstaff *et al.*, Z. Phys. C **76**, 401 (1997).

[8] DELPHI Collaboration, M. Feindt *et al.*, preprint DELPHI 97-98 CONF 80 (1997).

[9] See, for example, T.D. Lee and C.S. Wu, Annu. Rev. Nucl. Sci. **16**, 511 (1966).

[10] V.A. Kostelecký and S. Samuel, Phys. Rev. Lett. **63**, 224 (1989); *ibid.*, **66**, 1811 (1991); Phys. Rev. D **39**, 683 (1989); *ibid.*, **40**, 1886 (1989); V.A. Kostelecký and R. Potting, Nucl. Phys. B **359**, 545 (1991); Phys. Lett. B **381**, 89 (1996).

[11] V.A. Kostelecký and R. Potting, Phys. Rev. D **51**, 3923 (1995).

[12] D. Colladay and V.A. Kostelecký, Phys. Rev. D **55**, 6760 (1997); *ibid.*, **58**, 116002 (1998).

[13] The discrete symmetries C, P, T and their combinations are discussed in, for example, R.G. Sachs, *The Physics of Time Reversal* (University of Chicago Press, Chicago, 1987).

[14] For a review, see B. Winstein and L. Wolfenstein, Rev. Mod. Phys. **65**, 1113 (1993).

[15] See, for example, V.A. Kostelecký, ed., *CPT and Lorentz Symmetry* (World Scientific, Singapore, 1999).

[16] V.A. Kostelecký and R. Potting, in D.B. Cline, ed., *Gamma Ray–Neutrino Cosmology and Planck Scale Physics* (World Scientific, Singapore, 1993) (hep-th/9211116); D. Colladay and V.A. Kostelecký, Phys. Lett. B **344**, 259 (1995); Phys. Rev. D **52**, 6224 (1995); V.A. Kostelecký and R. Van Kooten, Phys. Rev. D **54**, 5585 (1996).

[17] V.A. Kostelecký, Phys. Rev. Lett. **80**, 1818 (1998); Phys. Rev. D **61**, 16002 (2000).

[18] P.B. Schwinberg, R.S. Van Dyck, Jr., and H.G. Dehmelt, Phys. Lett. A **81**, 119 (1981); Phys. Rev. D **34**, 722 (1986); L.S. Brown and G. Gabrielse, Rev. Mod. Phys. **58**, 233 (1986); R.S. Van Dyck, Jr., P.B. Schwinberg, and H.G. Dehmelt, Phys. Rev. Lett. **59**, 26 (1987); G. Gabrielse *et al.*, *ibid.*, **74**, 3544 (1995).

[19] R. Bluhm, V.A. Kostelecký, and N. Russell, Phys. Rev. Lett. **79**, 1432 (1997); Phys. Rev. D **57**, 3932 (1998).

[20] G. Gabrielse *et al.*, in ref. [15]; Phys. Rev. Lett. **82**, 3198 (1999).

[21] H. Dehmelt *et al.*, Phys. Rev. Lett. **83**, 4694 (1999).

[22] R. Mittleman, I. Ioannou, and H. Dehmelt, in ref. [15]; R. Mittleman *et al.*, Phys. Rev. Lett. **83**, 2116 (1999).

[23] S.M. Carroll, G.B. Field, and R. Jackiw, Phys. Rev. D **41**, 1231 (1990).

[24] R. Jackiw and V.A. Kostelecký, Phys. Rev. Lett. **82**, 3572 (1999).

[25] M. Pérez-Victoria, Phys. Rev. Lett. **83**, 2518 (1999); J.M. Chung, Phys. Lett. B **461**, 138 (1999).

[26] M. Charlton *et al.*, Phys. Rep. **241**, 65 (1994); J. Eades, ed., *Antihydrogen* (J.C. Baltzer, Geneva, 1993).

[27] R. Bluhm, V.A. Kostelecký, and N. Russell, Phys. Rev. Lett. **82**, 2254 (1999).

[28] V.W. Hughes, H.G. Robinson, and V. Beltran-Lopez, Phys. Rev. Lett. **4**, 342 (1960); R.W.P. Drever, Philos. Mag. **6**, 683 (1961); J.D. Prestage *et al.*, Phys. Rev. Lett. **54**, 2387 (1985); S.K. Lamoreaux *et al.*, Phys. Rev. Lett. **57**, 3125 (1986); Phys. Rev. A **39**, 1082 (1989); T.E. Chupp *et al.*, Phys. Rev. Lett. **63**, 1541 (1989); C.J. Berglund *et al.*, Phys. Rev. Lett. **75**, 1879 (1995).

[29] V.A. Kostelecký and C.D. Lane, Phys. Rev. D **60**, 116010 (1999); J. Math. Phys. (N.Y.) **40**, 6245 (1999).

[30] R. Bluhm, V.A. Kostelecký, and C.D. Lane, Phys. Rev. Lett. **84**, 1098 (2000).

[31] S. Coleman and S. Glashow, Phys. Rev. D **59**, 116008 (1999).

[32] O. Bertolami *et al.*, Phys. Lett. B **395**, 178 (1997).

[33] The possibility that unconventional quantum mechanics in the kaon system might generate CPT violation is discussed in J. Ellis *et al.*, Phys. Rev. D **53**, 3846 (1996), where it is shown that the resulting effects can be separated from δ_K.

[34] Direct CPT violation in the decay amplitudes is neglected in this talk because it is expected to be unobservable in the standard-model extension [11].

[35] *Review of Particle Properties,* Eur. Phys. J. C **3**, 1 (1998).

[36] The possibility of a relatively large momentum dependence for T violation is considered in, for example, J.S. Bell and J.K. Perring, Phys. Rev. Lett. **13**, 348 (1964); S.H. Aronson *et al.*, Phys. Rev. D **28**, 495 (1983).

[37] J. Adams *et al.*, Phys. Rev. Lett. **79**, 4093 (1997).

[38] T.T. Wu and C.N. Yang, Phys. Rev. Lett. **13**, 380 (1964).

[39] V.V. Barmin et al., Nucl. Phys. B **247** 293 (1984).

[40] N.W. Tanner and R.H. Dalitz, Ann. Phys. **171**, 463 (1986).

[41] KTeV Collaboration, A. Alavi-Harati *et al.*, preprint EFI 99-25 (1999). See also NA31 Collaboration, G.D. Barr *et al.*, Phys. Lett. B **317**, 233 (1993); E731 Collaboration, L.K. Gibbons *et al.*, Phys. Rev. Lett. **70**, 1203 (1993).

[42] See, for example, C.D. Buchanan et al., Phys. Rev. D **45**, 4088 (1992).

24

CPT Studies with KLOE

S. Di Falco and M. Incagli[1]

Abstract

Several elegant CPT tests through interferometric studies can be performed with the KLOE experiment currently running at the Frascati ϕ factory DAΦNE.

In particular, since K_S and K_L are always produced in pairs and with opposite momentum, the identification of a K_L (K_S) decay provides a very clean tagging for the K_S (K_L).

The existence of a tagged K_S beam gives to KLOE the unique opportunity to directly measure the semileptonic asymmetry of the K_S and the CPT-violating parameter in the kaon mass matrix $\mathcal{R}(\delta_K)$.

A very preliminary Monte Carlo study shows that a statistical error of $\mathcal{O}(2 \times 10^{-4})$ on the $\mathcal{R}(\delta_K)$ can be reached.

24.1 Introduction

A difference between the integrated semileptonic asymmetries

$$A_{\ell,S} = \frac{\Gamma(K_S \to \ell^+ \nu_\ell \pi^-) - \Gamma(K_S \to \ell^- \bar{\nu}_\ell \pi^+)}{\Gamma(K_S \to \ell^+ \nu_\ell \pi^-) + \Gamma(K_S \to \ell^- \bar{\nu}_\ell \pi^+)}$$

[1]For the KLOE Collaboration: M. Adinolfi, A. Aloisio, F. Ambrosino, A. Andryakov, A. Antonelli, C. Bacci, A. Bankamp, G. Barbiellini, G. Bencivenni, S. Bertolucci, C. Bini, C. Bloise, V. Bocci, F. Bossi, P. Branchini, G. Cabibbo, R. Caloi, P. Campana, G. Capon, G. Carboni, A. Cardini, G. Cataldi, F. Ceradini, F. Cervelli, F. Cevenini, G. Chiefari, P. Ciambrone, S. Conticelli, E. De Lucia, G. De Robertis, P. De Simone, G. De Zorzi, S. Dell'Agnello, A. Denig, A. Di Domenico, S. Di Falco, A. Doria, E. Drago, O. Erriquez, A. Farilla, G. Felici, A. Ferrari, M. L. Ferrer, G. Finocchiaro, C. Forti, G. Foti, A. Franceschi, P. Franzini, M. L. Gao, C. Gatti, P. Gauzzi, S. Giovannella, V. Golovatyuk, E. Gorini, F. Grancagnolo, E. Graziani, P. Guarnaccia, X. Huang, M. Incagli, L. Ingrosso, Y. Y. Jiang, W. Kim, W. Kluge, V. Kulikov, F. Lacava, G. Lanfranchi, J. Lee-Franzini, T. Lomtadze, C. Luisi, C. S. Mao, A. Martini, W. Mei, L. Merola, R. Messi, S. Miscetti, S. Moccia, M. Moulson, S. Mueller, F. Murtas, M. Napolitano, A. Nedosekin, L. Pacciani, P. Pagès, M. Palutan, L. Paoluzi, E. Pasqualucci, L. Passalacqua, A. Passeri, V. Patera, E. Petrolo, D. Picca, G. Pirozzi, L. Pontecorvo, M. Primavera, F. Ruggieri, P. Santangelo, E. Santovetti, G. Saracino, R. D. Schamberger, B. Sciascia, A. Sciubba, F. Scuri, I. Sfiligoi, T. Spadaro, E. Spiriti, C. Stanescu, L. Tortora, P. Valente, G. Venanzoni, S. Veneziano, Y. Wu.

$$A_{\ell,L} = \frac{\Gamma(K_L \to \ell^+\nu_\ell\pi^-) - \Gamma(K_L \to \ell^-\bar{\nu}_\ell\pi^+)}{\Gamma(K_L \to \ell^+\nu_\ell\pi^-) + \Gamma(K_L \to \ell^-\bar{\nu}_\ell\pi^+)} \quad (24.1)$$

where $\ell = e, \mu$, is a clean evidence of CPT violation in the mass matrix of the kaons, *i.e.*, of the δ_K parameter defined by:

$$|K_S\rangle \simeq \frac{(1+\varepsilon_K+\delta_K)|K^0\rangle + (1-\varepsilon_K-\delta_K)|\bar{K}^0\rangle}{\sqrt{2}}$$

$$|K_L\rangle \simeq \frac{(1+\varepsilon_K-\delta_K)|K^0\rangle + (1-\varepsilon_K+\delta_K)|\bar{K}^0\rangle}{\sqrt{2}}. \quad (24.2)$$

By the way, this kind of measurement has never been performed up to now because of the rarity of the semileptonic decays of the K_S ($\mathcal{BR} \sim 10^{-4}$)[2] and the consequent critical nature of the background rejection.

The unique way to overcome the problem is the realization of an intense and pure K_S beam. This cannot be done at a fixed-target experiment where the contamination due to K_L is always present; a high-luminosity ϕ factory like DAΦNE at Frascati is the ideal solution.

A ϕ factory is an e^+e^- circular collider with a c.o.m. energy $\sqrt{s} \simeq M_\phi \simeq 1020\ MeV$. The ϕ meson is produced at the resonance with a cross section $\sigma \sim 4.4\,\mu$b.

When the design luminosity will be reached, DAΦNE will be able to collect an integrated luminosity $L = 5\,\text{fb}^{-1}/\text{year}$ corresponding to $\sim 2.2\ 10^{10}\phi$ mesons/year. Almost one-third of these mesons decay in K_S-K_L pairs (see table 24.1).

ϕ decays	
K^+K^-	49.1 ± 0.8 %
$K_S\ K_L$	34.1 ± 0.6 %
$\rho\pi^0$, $\pi^+\pi^-\pi^0$	15.5 ± 0.7 %

Table 24.1: Main decays channels for the ϕ meson.

The momentum of the neutral kaons is $\sim 110\,\text{MeV/c}$, corresponding to a mean decay path of

$$\lambda_S = \beta\gamma c\tau_S = 0.56\,\text{cm}, \quad \lambda_L = \beta\gamma c\tau_L = 354\,\text{cm}$$

Since K_S and K_L are always produced in pairs, the identification of a K_L (K_S) decay provides also a very clean tagging for the K_S (K_L).

For example, K_L *tagging* is provided either by the identification of a $K_S \to \pi^+\pi^-$ or a $K_S \to \pi^0\pi^0$ decay.[3]

[2]The decay $K_S \to e\nu\pi$ has been observed only recently by CMD-2 at VEPP-2M with a few tens of events [1].

[3]The signatures of this decays are respectively two charged tracks or four photons coming out from the interaction region and having the invariant mass and the momentum expected for the K_S.

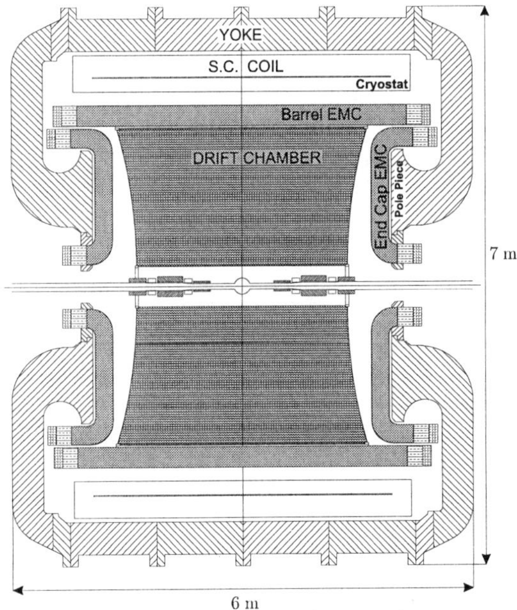

Figure 24.1: Section view of KLOE detector. The shaded areas close to the beam pipe are the six (three on each side) permant focusing quadrupoles installed inside the detector.

On the other side the K_S *tagging* is provided either by the identification of a $K_L \to$ (charged particles) decay, *i.e.*, a "V" inside the drift chamber not connected with the interaction region, or by the identification of a K_L interaction in the calorimeter, *i.e.*, a nonelectromagnetic cluster with the time of flight expected for the kaon.

Using all these tagging methods together $\sim$ 88% of all $K_S K_L$ events can be identified.[4]

The main inefficiencies come from the loss of the particles that hit the low β quadrupoles close to the interaction region (see fig. 24.1) or from the *punch-through* of the K_Ls across the calorimeter.

The tagging algorithms described above have been succesfully tested on the first data acquired by KLOE confirming the purity and the elegance of the K-K pairs produced by DAΦNE.

Besides having the possibility of a pure tagging, in particular of the K_S beam, in a ϕ factory the two neutral kaons are produced in a pure antisymmetric quantum state having the quantum numbers of the ϕ meson ($J^{PC} = 1^{--}$):

$$|i\rangle \;=\; \frac{1}{\sqrt{2}}\{|\bar{K}^0(\vec{p})\rangle|K^0(-\vec{p})\rangle - |\bar{K}^0(-\vec{p})\rangle|K^0(\vec{p})\rangle\} =$$

[4]The trigger efficiency is also included.

$$\simeq \frac{1}{\sqrt{2}}\{|K_S(\vec{p})\rangle|K_L(-\vec{p})\rangle - |K_S(-\vec{p})\rangle|K_L(\vec{p})\rangle\}, \quad (24.3)$$

where the second equivalence holds if terms $\mathcal{O}(\varepsilon^2)$ are neglected.

This allows for the study of interferometric patterns from which CP and CPT measurements are obtained. They will be described in section 24.3.

24.2 The KLOE Detector

The KLOE detector is described in detail elsewhere (see [2]). In this section only the main characteristics will be underlined.

The detector (see fig. 24.1) has two main subsystems: the large cylindrical He-based drift chamber, with a r/ϕ resolution better than 200 μm, which ensures a very good momentum resolution of $\sigma_{p_\perp}/p_\perp \simeq 0.5\%$ in a magnetic field of ≈ 0.6 T, surrounded by the fine sampling lead/scintillating fibers electromagnetic calorimeter with an excellent time resolution of 55 ps/$\sqrt{E(\mathrm{GeV})}$ and good energy resolution of $4.7\%/\sqrt{E(\mathrm{GeV})}$, high hermeticity and full efficiency for photons down to 20 MeV.

The data taking has started in spring 1999 at the initial luminosity of $\sim 10^{30}$ cm^{-2} s^{-1} that will be raised by a factor of 30 allowing us to collect, in the near future, an integrated luminosity of 100 pb^{-1} corresponding to 1/100 of the final KLOE statistics. In terms of ϕ decays, this integrated luminosity corresponds to 2×10^8 K_SK_L pairs produced, to $\sim 8\times 10^6$ $K^L_{\mu 3}$ and $\sim 12\times 10^6$ K^L_{e3} decays reconstructed[5] and to $\sim 7\times 10^4$ $K^S_{\mu 3}$ and $\sim 9\times 10^4$ K^S_{e3} decays reconstructed.[6]

24.3 Interferometry at KLOE

Since K_S and K_L are produced in a pure quantum state (eq. (24.3)), several tests of CPT can be performed using *interference patterns.*

The probability of having one kaon decaying into the final state f_1 at time t_1 and the other decaying into the final state f_2 at time t_2 is given by (see [3]):

$$\begin{aligned} I(f_1,t_1;f_2,t_2) &= \frac{\langle f_1|H_w|K_S\rangle\langle f_2|H_w|K_S\rangle}{\sqrt{2}} \\ &\quad \times\{\eta_2 e^{-i(m_St_1+m_Lt_2)}e^{-(\Gamma_St_1+\Gamma_Lt_2)/2} \\ &\quad - \eta_1 e^{-i(m_Lt_1+m_St_2)}e^{-(\Gamma_Lt_1+\Gamma_St_2)/2}\} \end{aligned} \quad (24.4)$$

with $\eta_i = \frac{\langle \mathrm{f_i}\, |K_L\rangle}{\langle \mathrm{f_i}\, |K_S\rangle} = |\eta_i|e^{i\phi_i}$. Integrating over $T = t_1 + t_2$, the integrated probability as function of the time difference $\Delta t = t_1 - t_2$ is

$$I(f_1,f_2;\Delta t>0) = \frac{1}{2\Gamma}|\langle f_1|H_w|K_S\rangle\langle f_2|H_w|K_S\rangle|$$

[5] A reconstrucion efficiency of 21% is included. This term is dominated by the geometrical efficiency ($\lambda_L = 3.5$ m).

[6] The reconstruction efficiency for these decays is at the level of 90%.

	f_1	f_2	Quantity
identical final states	f	f	$\Gamma_S, \Gamma_L, \Delta m$
quasi-identical final states	$\pi^+\pi^-$ $\pi^+ l^- \nu$	$\pi^0\pi^0$ $\pi^- l^+ \nu$	$\mathcal{R}\frac{\epsilon'}{\epsilon}, \mathcal{I}\frac{\epsilon'}{\epsilon}$ δ_K
different final states	$\pi\pi$	K_{l3}	$A_L, \Delta m, \eta_{\pi\pi}, \phi_{\pi\pi}$

Table 24.2: Main parameters measurable from the interference patterns given by the decays of the two kaons.

$$\times\{|\eta_1|^2 e^{-\Gamma_L|\Delta t|} + |\eta_2|^2 e^{-\Gamma_S|\Delta t|} - 2|\eta_1||\eta_2| e^{-\Gamma|\Delta t|/2}\cos(\Delta m|\Delta t| + \phi_2 - \phi_1)\}, \tag{24.5}$$

where $\Gamma = \Gamma_L + \Gamma_S$. The corresponding intensity for $\Delta t < 0$ is given changing the pedices 1 and 2.

Scanning all the possible combinations for the two final states a lot of CP- or CPT-violating parameters can be directly measured. A summary is reported in table 24.2.

In the table the final states are divided in three categories, following the suggestion of ref. [3]. In the first category identical final states are compared, providing only kinematical informations. The second category is the most interesting for CP and CPT studies, in which direct information on ϵ' and δ is obtained. The third category includes disparate final states, and it essentially provides information on the K_L leptonic asymmetry.

The interference pattern resulting from the semileptonic decay of the two neutral kaons is particularly relevant for CPT violation analysis, and it is illustrated in fig. 24.2. The central bump is due to the fact that the two amplitudes are almost opposite, therefore the antisymmetry of the initial state turns out into a constructive interference.

The picture is not symmetric around 0, and the asymmetry of this interference pattern provides information about the real and imaginary parts of the δ parameter. For example, two simple integrated asymmetries can be defined:

$$A^{\mathcal{R}}_{CPT} \equiv \frac{\int_0^{125cm} I(\Delta)d\Delta - \int_{-125cm}^{0} I(\Delta)d\Delta}{\int_0^{125cm} I(\Delta)d\Delta + \int_{-125cm}^{0} I(\Delta)d\Delta} \simeq -4\,\mathcal{R}\delta, \tag{24.6}$$

$$A^{\mathcal{I}}_{CPT} \equiv \frac{\int_0^{3cm} I(\Delta)d\Delta - \int_{-3cm}^{0} I(\Delta)d\Delta}{\int_0^{3cm} I(\Delta)d\Delta + \int_{-3cm}^{0} I(\Delta)d\Delta} \simeq -(\mathcal{I}\delta_K + 2\mathcal{R}\delta_K), \tag{24.7}$$

where $\Delta = \beta\gamma c\Delta t$ is the distance between the two decay vertices expressed in cm.

Therefore the asymptotic behavior of the distribution provides information on $\mathcal{R}\delta$, while the imaginary part of δ contributes to the asymmetry in the central region.

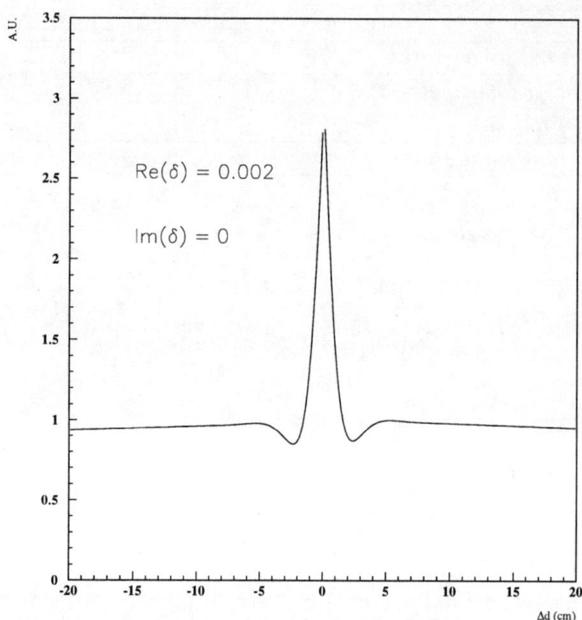

Figure 24.2: Interference pattern for the semileptonic decay of the two neutral kaons. The asymptotic behavior of the distribution provides information on $\mathcal{R}\delta$, while the imaginary part of δ contributes to the asymmetry in the central region.

24.4 The Semileptonic Asymmetries

A particularly relevant decay that will be studied in KLOE is the semileptonic K_S decay. In fact, without assuming the CPT theorem and the $\Delta S = \Delta Q$ rule, the semileptonic amplitudes of the strangeness eigenstate kaons can be written as

$$\begin{aligned} A(K^0 \to \ell^+) &= a + b \quad & A(\bar{K}^0 \to \ell^-) &= a^* - b^*, \\ A(K^0 \to \ell^-) &= c + d \quad & A(\bar{K}^0 \to \ell^+) &= c^* - d^*, \end{aligned} \tag{24.8}$$

where the amplitudes c and d describe the violation of the $\Delta S = \Delta Q$ rule, while $b \neq 0$ implies a "direct" CPT violation.

The behavior of these amplitudes under the CP, T, and CPT symmetries is summarized in table 24.3. Using (24.1), (24.2), and (24.8) the semileptonic asymmetries can be written as

$$\begin{aligned} A_{\ell,S} &= \frac{\Gamma_S^{\ell^+} - \Gamma_S^{\ell^-}}{\Gamma_S^{\ell^+} + \Gamma_S^{\ell^-}} = 2\mathcal{R}\varepsilon_K + 2\mathcal{R}\delta_K + 2\mathcal{R}(\frac{b}{a}) - 2\mathcal{R}(\frac{d^*}{a}), \\ A_{\ell,L} &= \frac{\Gamma_L^{\ell^+} - \Gamma_L^{\ell^-}}{\Gamma_L^{\ell^+} + \Gamma_L^{\ell^-}} = 2\mathcal{R}\varepsilon_K - 2\mathcal{R}\delta_K + 2\mathcal{R}(\frac{b}{a}) + 2\mathcal{R}(\frac{d^*}{a}). \end{aligned} \tag{24.9}$$

The semileptonic asymmetry for the K_L has already been measured with good accuracy ($A_{\ell,L} = (3.27 \pm 0.12) \cdot 10^{-3}$) [4], and further improve-

	$\mathcal{R}(a,c)$	$\mathcal{I}(a,c)$	$\mathcal{R}(b,d)$	$\mathcal{I}(b,d)$
CP	+	−	−	+
T	+	−	+	−
CPT	+	+	−	−

Table 24.3: Behavior of the semileptonic amplitudes described in 24.8 under CP, T, and CPT symmetries: a and c are CPT conserving, b and d describe direct CPT violation.

ments are expected both from KLOE and from the fixed target experiments. The semileptonic asymmetry for the K_S has never been measured up to now and will be measured for the first time by KLOE.

Combining the two asymmetries, we get:

$$\begin{aligned} A_{\ell,S} + A_{\ell,L} &= 4\,\mathcal{R}\varepsilon_K + 4\,\mathcal{R}(\frac{b}{a}), \\ A_{\ell,S} - A_{\ell,L} &= 4\,\mathcal{R}\delta_K - 4\,\mathcal{R}(\frac{d^*}{a}). \end{aligned} \tag{24.10}$$

In particular from the second of the (24.10) it's possible to get a clean "direct" measure of the indirect CPT violation disentangled from both the CP violation and the direct CPT violation. Nonetheless it's not possible to disentangle the indirect CPT violation from the amplitude violating the $\Delta S = \Delta Q$ rule using only the neutral kaons.[7]

24.5 Semileptonic Asymmetries with KLOE

The huge amount of kaon decays that can be observed at KLOE will allow for an accurate determination of A_L, while the efficient and high-purity tagged K_S beam will give the opportunity to measure for the first time the branching ratio of $K_S \to \ell\nu\pi$ and the A_S asymmetry. The measurement of the two asymmetries will thus allow a test of CPT by measuring the δ parameter.

The two problems that must be faced in order to determine A_S are to isolate the $K_S \to \ell\nu\pi$ and to correctly identify the lepton between the two charged tracks connected to the observed vertex.

The semileptonic decay is in part discriminated from the much more copious two-body decay by kinematical cuts on the mass and on the total momentum of the two charged tracks, assumed to be pions. This cuts out 83% of the two-body decays while retaining all the signal.

The correct assignment of the lepton charge can be done using calorimeter information or using kinematical information. In this second technique the *neutrino 4-momentum* is evaluated assigning the pion mass alterna-

[7]The $\Delta S = \Delta Q$ rule itself can be tested at KLOE using the "flavor tag" provided by the charge exchange reactions of the copious $K^{\pm}$.

tively to each one of the two tracks:

$$p_\nu = p_\pi + p_e - p_S \equiv (E_\pi + E_e - E_S \; ; \; \vec{p}_\pi + \vec{p}_e - \vec{p}_S) \tag{24.11}$$

with:

$$\begin{array}{lcccc} \text{“}E_\pi\text{”} & = & \sqrt{m_\pi^2 + |\vec{p}_1|^2} & ; & \sqrt{m_\pi^2 + |\vec{p}_2|^2} \\ \text{“}E_e\text{”} & = & |\vec{p}_2| & ; & |\vec{p}_1| \end{array} \tag{24.12}$$

The combination providing the smallest value of the *neutrino mass* is selected.

The resulting distribution is plotted in fig. 24.3. The distribution has a long tail. The shaded histogram isolates events in which the K_L has been correctely identified showing that the long tail is actually due to events in which the K_L *vertex* has been misidentified, due to kinks, decays in flights, spiralizing tracks, and so on. A cut at 42 MeV on the *neutrino mass* cleans the sample with an efficiency of $\epsilon_{CUT} = 51\%$ on signal events and of $\epsilon_{CUT}^{bck} = 24\%$ on background events.

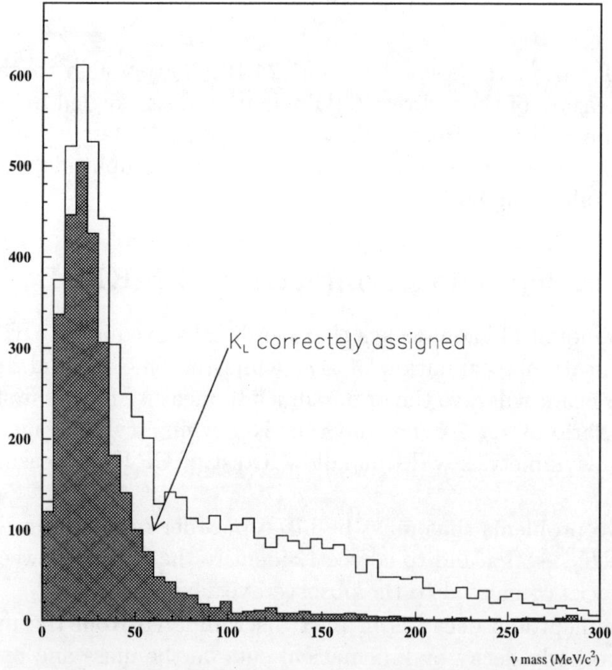

Figure 24.3: Neutrino mass reconstructed by imposing 4-momentum conservation. The events in the shaded histogram are the ones in which the K_L has been correctely identified.

With these cuts the *correct charge assignment probability* is $d = 0.83$ corresponding to a *dilution factor* $D = 2d - 1 = 0.66$. This dilution enters in the evaluation of the statistical accuracy on the measurement of the semileptonic asymmetry as follows:

$$\delta A_{STAT} = \frac{1}{\sqrt{\epsilon D^2 N}}\sqrt{1+\frac{B}{S}}, \tag{24.13}$$

where $S = \epsilon N$ are the signal events after selection.

Putting all efficiencies together one has:

$$\epsilon = \epsilon_{TRIG} \times \epsilon_{TAG} \times \epsilon_{VTX} \times \epsilon_{KINE} \times \epsilon_{CUT} \sim 0.33 \tag{24.14}$$

where:

- $\epsilon_{TRIG} \sim 95\%$: trigger efficiency
- $\epsilon_{TAG} \sim 90\%$: probability of *tagging* the event
- $\epsilon_{VTX} \sim 80\%$: probability of finding the K_S vertex; this value is dominated by geometrical inefficiencies (pions that hit the quadrupoles)
- $\epsilon_{KINE} \sim 99\%$: kinematical cuts
- $\epsilon_{CUT} \sim 51\%$: analysis cuts on neutrino quantities

The number of semileptonic decays expected for an integrated luminosity of $L\,\text{pb}^{-1}$ is:

$$N = L \times \sigma_\phi(\text{pb}) \times \text{BR}(\phi \to K_L K_S) \times \text{BR}(K_S \to \pi e \nu) \sim 1000 \cdot L. \tag{24.15}$$

Putting the numbers all together, the statistical error is:

$$\delta A_S^{stat} = \frac{1}{\sqrt{\epsilon D^2 N}} \sim \frac{0.8 \cdot 10^{-1}}{L(pb^{-1})} = \begin{cases} 0.8 \cdot 10^{-2} & L = 100\,\text{pb}^{-1} \\ 0.8 \cdot 10^{-3} & L = 10\,\text{fb}^{-1}. \end{cases} \tag{24.16}$$

Since the main statistical limit on the measurement of $\mathcal{R}\delta_K$ using the semileptonic asymmetries will be dominated by A_S, this quantity will have a statistical accuracy of $\sim 2 \times 10^{-4}$ once the KLOE full statistics are collected.

However, this result is diluted by the huge background due to $K_S \to \pi^+\pi^-$ decays. In fact the ratio B/S is ~ 1000 before applying cuts. Applying the kinematical and analysis cuts it becomes

$$\frac{B}{S} = \frac{\epsilon^{bck}_{KINE}\epsilon^{bck}_{CUTS}N^{bck}}{\epsilon_{KINE}\epsilon_{CUTS}N} = \frac{0.17 * 0.24}{0.99 * 0.51} \cdot 1000 = 81, \tag{24.17}$$

from which $\sqrt{1+B/S} \sim 9$!

Improvements in the B/S ratio are also expected from applying calorimeter information.

24.6 Conclusions

The possibility of having a clean, tagged, intense beam of K_Ss and the excellent performances of the KLOE detector will allow many measurements in the leptonic sector that were not available so far, such as the $K_S \to \ell\nu_\ell\pi$ branching ratio measurement, to an accuracy of 10^{-3}, and the $A_{S,\ell}$ charge asymmetry to a level of 8×10^{-4}. Together with the $A_{L,\ell}$ determination this will allow one to test CPT violation by measuring the δ parameter to an accuracy of 2×10^{-4}, albeit assuming the $\Delta S = \Delta Q$ rule.

The first months of running of KLOE at the initial DAΦNE luminosity will already give preliminary results with 1/100 of the full statistics, *i.e.*, with a factor 10 in the statistical error.

References

[1] R. R. Akhmetshin *et al.*, Preprint Budker INP 99-11 (1999).

[2] A. Antonelli *et al.*, the KLOE Collaboration, *The KLOE Detector: Technical Proposal*, LNF-93/002(IR) (1993).

[3] C.D. Buchanan *et al.*, Phys. Rev. D **45**, 4088 (1992).

[4] C. Caso *et al.*, Particle Data Group, Eur. Phys. Soc. J. C **3**, (1998).

25

Tests of CPT Invariance in the CPLEAR Experiment

P. Bloch[1]

Abstract

We present recent results on CPT invariance obtained by the CPLEAR Collaboration. Using the full sample of neutral-kaon semileptonic decays we have improved by two orders of magnitude the limit currently available for the CPT violation parameter Re(δ), Re(δ) $= (3.0 \pm 3.3_{\rm stat} \pm 0.6_{\rm syst}) \times 10^{-4}$. Data from the CPLEAR experiment, together with the most recent world averages for some of the neutral-kaon parameters, were constrained with the Bell–Steinberger relation, allowing the CPT-violation parameter Im(δ) to be determined with an increased accuracy. Combining these two results, we have evaluated the K^0 - $\overline{K}^0$ mass and width difference. We have obtained $(\mathrm{M}_{K^0K^0} - \mathrm{M}_{\overline{K}^0\overline{K}^0}) = (-1.5 \pm 2.0) \times 10^{-18}$GeV and $(\Gamma_{K^0K^0} - \Gamma_{\overline{K}^0\overline{K}^0}) = (3.9 \pm 4.2) \times 10^{-18}$GeV. Moreover, our analysis allows most of the CPT-violation parameters for the neutral-kaon decays to $e\pi\nu$ and $\pi\pi$ to be determined. All our results are consistent with CPT invariance.

25.1 Introduction

The CPT theorem [1], which is based on general principles of relativistic quantum field theories, states that any order of the triple product of the discrete symmetries C, P, and T represents an exact symmetry. The theorem predicts that particles and antiparticles have equal masses, lifetimes, charge-to-mass ratios, and gyromagnetic ratios. The CPT symmetry has

[1]For the CPLEAR Collaboration: University of Athens, Greece; University of Basle, Switzerland; Boston University, USA; CERN, Geneva, Switzerland; LIP and University of Coimbra, Portugal; Delft University of Technology, Netherlands; University of Fribourg, Switzerland; University of Ioannina, Greece; University of Liverpool, UK; J. Stefan Inst. and Phys. Dep., University of Ljubljana, Slovenia; CPPM, IN2P3-CNRS et Université d'Aix-Marseille II, France; CSNSM, IN2P3-CNRS, Orsay, France; Paul Scherrer Institut (PSI), Switzerland; CEA, DSM/DAPNIA, CE-Saclay, France; Royal Institute of Technology, Stockholm, Sweden; University of Thessaloniki, Greece; ETH-IPP Zürich, Switzerland.

been tested in a variety of experiments (see for example ref. [2]) and remains to date the only combination of C, P, T that is observed as an exact symmetry in nature.

The neutral-kaon system, where CP and T violation have been observed, is a very sensitive place to look for departure from CPT invariance.

The best test of the CPT symmetry comes from the limit on the mass difference between K^0 and $\overline{K}^0$. However, all previous analyses contain the assumption of CPT invariance in the decays of the neutral kaons and/or neglect some of the contributions of decay channels other than the two-pion decay. In the present analysis we overcome these limitations. Using the strangeness-tagged CPLEAR data, in particular the semileptonic decays, we are able to give, for the first time, bounds on both the mass and width difference between K^0 and $\overline{K}^0$.

25.2 Neutral Kaon Phenomenology

A neutral-kaon state can be written as a superposition of $|K^0\rangle$ and $|\overline{K}^0\rangle$, the eigenstates of the strong and electromagnetic interactions, with strangeness $+1$ and -1, respectively

$$|\Psi(t)\rangle = \alpha(t)|K^0\rangle + \beta(t)|\overline{K}^0\rangle \ . \tag{25.1}$$

As weak interactions do not conserve strangeness $|K^0\rangle$ and $|\overline{K}^0\rangle$ undergo strangeness oscillations. The time evolution of the state in eq. (25.1) is described by [3]

$$\begin{aligned} \frac{d}{dt}\Psi &= -i\Lambda\Psi, \qquad (25.2) \\ \Lambda &\equiv \mathrm{M} - \frac{i}{2}\Gamma, \\ \begin{pmatrix} \Lambda_{K^0K^0} & \Lambda_{K^0\overline{K}^0} \\ \Lambda_{\overline{K}^0K^0} & \Lambda_{\overline{K}^0\overline{K}^0} \end{pmatrix} &\equiv \begin{pmatrix} \mathrm{M}_{K^0K^0} & \mathrm{M}_{K^0\overline{K}^0} \\ \mathrm{M}_{\overline{K}^0K^0} & \mathrm{M}_{\overline{K}^0\overline{K}^0} \end{pmatrix} - \frac{i}{2}\begin{pmatrix} \Gamma_{K^0K^0} & \Gamma_{K^0\overline{K}^0} \\ \Gamma_{\overline{K}^0K^0} & \Gamma_{\overline{K}^0\overline{K}^0} \end{pmatrix}, \end{aligned}$$

where M and Γ are Hermitian matrices known as the mass and decay matrices. The corresponding eigenvalues are $\lambda_{L,S} = m_{L,S} - \frac{i}{2}\Gamma_{L,S}$. The symmetry properties of the matrix elements are shown in table 25.1. For this

Symmetry	Matrix Elements Property
CPT invariance	$\Lambda_{K^0K^0} = \Lambda_{\overline{K}^0\overline{K}^0}$
T invariance	$\lvert\Lambda_{K^0\overline{K}^0}\rvert = \lvert\Lambda_{\overline{K}^0K^0}\rvert$
CP invariance	$\Lambda_{K^0K^0} = \Lambda_{\overline{K}^0\overline{K}^0}$ and $\lvert\Lambda_{K^0\overline{K}^0}\rvert = \lvert\Lambda_{\overline{K}^0K^0}\rvert$

Table 25.1: The symmetry of the matrix elements under the assumption of CPT, T, and CP invariance, respectively.

analysis, we use a T-violation parameter that has exactly the superweak phase $\phi_{\rm sw} = \tan^{-1}(2\Delta m/\Delta\Gamma)$ and is defined as

$$\epsilon_{\rm T} = \sin(\phi_{\rm sw}) \frac{|\Lambda_{K^0\overline{K^0}}|^2 - |\Lambda_{\overline{K^0}K^0}|^2}{\Delta\Gamma\Delta m} \exp\Big(i\phi_{\rm sw}\Big), \tag{25.3}$$

where we introduced $\Delta m = m_{\rm L} - m_{\rm S}$, $\Delta\Gamma = \Gamma_{\rm S} - \Gamma_{\rm L}$.

The CPT-violation parameter δ is defined as

$$\delta = \frac{\Lambda_{\overline{K^0}\overline{K^0}} - \Lambda_{K^0K^0}}{2(\lambda_L - \lambda_S)} = \delta_{||} \exp\Big(i\phi_{\rm sw}\Big) + \delta_\perp \exp\Big(i(\phi_{\rm sw} + \frac{\pi}{2})\Big), \tag{25.4}$$

where the projections $\delta_{||}$ and $\delta_\perp$ are respectively

$$\delta_{||} = \frac{1}{4}\frac{\Gamma_{K^0K^0} - \Gamma_{\overline{K^0}\overline{K^0}}}{\sqrt{\Delta m^2 + (\frac{\Delta\Gamma}{2})^2}} \quad \text{and} \quad \delta_\perp = \frac{1}{2}\frac{\mathrm{M}_{K^0K^0} - \mathrm{M}_{\overline{K^0}\overline{K^0}}}{\sqrt{\Delta m^2 + (\frac{\Delta\Gamma}{2})^2}}. \tag{25.5}$$

In turn, the parameters $\delta_{||}$ and $\delta_\perp$ are expressed as functions of measureable quantities, Re(δ), Im(δ) and $\phi_{\rm sw}$, as

$$\begin{aligned} \delta_{||} &= \mathrm{Re}(\delta)\cos(\phi_{\rm sw}) + \mathrm{Im}(\delta)\sin(\phi_{\rm sw}), \\ \delta_\perp &= -\mathrm{Re}(\delta)\sin(\phi_{\rm sw}) + \mathrm{Im}(\delta)\cos(\phi_{\rm sw}). \end{aligned} \tag{25.6}$$

In the decay to $\pi\pi$ ($\pi^+\pi^-$ or $\pi^0\pi^0$), the observables of CP violation are the amplitude ratios

$$\eta_{\pi\pi} = \frac{\langle\pi\pi|\mathrm{T}|K_L\rangle}{\langle\pi\pi|\mathrm{T}|K_S\rangle} = |\eta_{\pi\pi}| \times e^{i\phi_{\pi\pi}}. \tag{25.7}$$

One can show that

$$\eta_{+-} = \varepsilon + \varepsilon' \qquad \text{and} \qquad \eta_{00} = \varepsilon - 2\varepsilon' \tag{25.8}$$

with

$$\varepsilon = \epsilon_{\rm T} - \delta + i\Delta\phi + \frac{\mathrm{Re}(B_0)}{\mathrm{Re}(A_0)}.$$

Let us discuss the various contributions to $\eta_{\pi\pi}$:

1. ϵ_T and δ have already been defined previously.

2. ε' parametrizes the direct CP violation in the 2π decay. It is given by

$$\varepsilon' = \frac{1}{\sqrt{2}} \exp^{i(\delta_2 - \delta_0)} \left[i\mathrm{Im}(\frac{A_2}{A_0}) + \frac{\mathrm{Re}(A_2)}{\mathrm{Re}(A_0)} \left\{ \frac{\mathrm{Re}(B_2)}{\mathrm{Re}(A_2)} - \frac{\mathrm{Re}(B_0)}{\mathrm{Re}(A_0)} \right\} \right] \tag{25.9}$$

where A_0 (resp B_0) and A_2 (resp B_2) are the CPT-conserving (resp CPT-violating) decay amplitudes with isospin 0 and 2 . δ_0 and δ_2 are the corresponding strong interaction phase shifts.

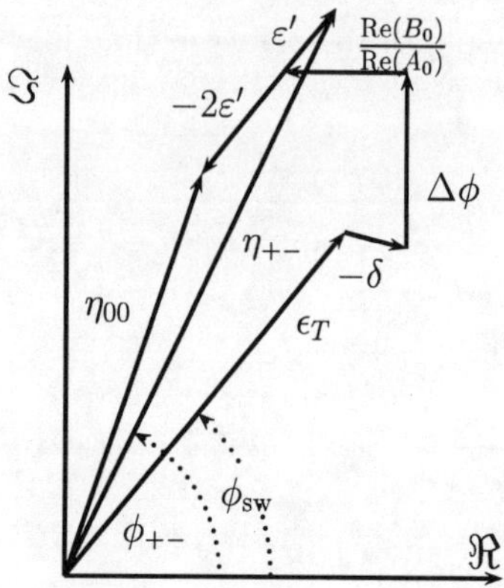

Figure 25.1: T, CP, and CPT violation parameters in 2π decay.

Since experimentally $\delta_2 - \delta_0 = -42° \pm 4°$ [4], the CPT-conserving part of ε' is almost parallel to ϵ_T, and $\varepsilon'_\perp$, the component of ε' that is orthogonal to ϵ_T, is CPT violating. Note that $\varepsilon'_\perp$ can be experimentally constrained by a measurement of the phase difference $\phi_{+-} - \phi_{00}$, using the relation

$$\varepsilon'_\perp = \frac{1}{3}(\phi_{+-} - \phi_{00})\,|\eta_{+-}|. \tag{25.10}$$

3. $\Delta\phi = \frac{1}{2}\left[\varphi_\Gamma - \arg(A_0^*\overline{A}_0)\right]$ is the phase difference between the I=0 component of the decay amplitude and the matrix element $\Gamma_{K^0\overline{K^0}}$. This phase difference, expected to be extremely small in the standard model, can be approximated by the sum of 3 terms:

$$\begin{aligned}\Delta\phi &= |\frac{A_2}{A_0}|^2 \arg\left(\frac{A_2}{A_0}\right) + \frac{\Gamma_L}{\Gamma_S}\Big[4\mathrm{BR}(K_L \to \mathrm{l}^+\pi^-\nu) \times \mathrm{Im}(x_+) \\ &\quad - \mathrm{BR}(K_L \to 3\pi) \times \mathrm{Im}(\epsilon + \delta - \eta_{+-0})\Big]\end{aligned} \tag{25.11}$$

The first term (contribution of the $I = 2$ amplitude) can be bounded, using the measured value (from charged kaon decays) of $|A_2/A_0|$ and the value of $\mathrm{Re}(\varepsilon'/\varepsilon)$, to be less than 10^{-6}. The second and third terms are related to a possible contribution of the $\Delta S = -\Delta Q$ amplitude in the semileptonic decay and of direct CP violation in the 3π sector, respectively.

4. $\mathrm{Re}(B_0)/\mathrm{Re}(A_0)$ is the ratio of the CPT-violating to the CPT-allowed amplitude in the isospin I=0 component of the decay amplitude.

The relation between all these parameters is illustrated in fig. 25.1. It appears clearly that

- If we can neglect $\Delta\phi$,[2] the equality of ϕ_{+-}, ϕ_{00}, and $\phi_{\rm sw}$ is *required* by CPT invariance. Indeed, the compilation of the available data [2] gives $\phi_{+-} = 43.5^\circ \pm 0.6^\circ$, $\phi_{\rm sw} = 43.46^\circ \pm 0.08^\circ$, and $\phi_{00} - \phi_{+-} = -0.3^\circ \pm 0.8^\circ$, in excellent agreement with CPT conservation.

- However, the equality of these phases is not *sufficient* to prove CPT invariance. Spurious cancellations can happen between δ and B_0 [eq. (25.8)] or between B_0 and B_2 [eq. (25.9)].

It is therefore essential to make additional measurements in other decay channels to fully constrain the system. This is the main purpose of the Re(δ) measurement, which will be described in the next section.

25.3 Direct Measurement of the CPT Parameter Re(δ)

The measurement of Re(δ) has been performed in CPLEAR with semileptonic decays using an experimental method very similar to the one used for the T-violation analysis [5], which has been presented in chapter 20. The phenomenology of neutral-kaon semileptonic decays can therefore be found earlier in this volume. Similarly, the experimental conditions of the measurement have already been exposed and will not be repeated. More details can also be found in [6].

To extract the CPT-violation parameter Re(δ) in an optimal way, CPLEAR builds the time-dependent decay-rate asymmetry A_δ, defined as

$$A_\delta(\tau) \equiv \frac{\overline{R}_+ - R_-(1+4\mathrm{Re}(\epsilon_L))}{\overline{R}_+ + R_-(1+4\mathrm{Re}(\epsilon_L))} + \frac{\overline{R}_- - R_+(1+4\mathrm{Re}(\epsilon_L))}{\overline{R}_- + R_+(1+4\mathrm{Re}(\epsilon_L))}, \tag{25.12}$$

where we defined

$$\begin{aligned} R_+(\tau) &\equiv R\left[K^0_{t=0} \to e^+\pi^-\nu_{t=\tau}\right], & \overline{R}_-(\tau) &\equiv R\left[\overline{K}^0_{t=0} \to e^-\pi^+\overline{\nu}_{t=\tau}\right], \\ R_-(\tau) &\equiv R\left[K^0_{t=0} \to e^-\pi^+\overline{\nu}_{t=\tau}\right], & \overline{R}_+(\tau) &\equiv R\left[\overline{K}^0_{t=0} \to e^+\pi^-\nu_{t=\tau}\right], \end{aligned}$$

and $\mathrm{Re}(\epsilon_L) = \mathrm{Re}(\epsilon_T) - \mathrm{Re}(\delta)$. Eq. (25.12) can be written, to first order in the small parameters, as

$$\begin{aligned} A_\delta(\tau) &= 2\frac{\mathrm{Im}(x_+)e^{-\frac{1}{2}(\Gamma_S+\Gamma_L)\tau}\sin(\Delta m\tau) + \mathrm{Re}(x_-)\mathrm{E}_-(\tau)}{\mathrm{E}_+(\tau) - e^{-\frac{1}{2}(\Gamma_S+\Gamma_L)\tau}\cos(\Delta m\tau)} \\ &\quad + \Big(-4\mathrm{Re}(\delta)\mathrm{E}_-(\tau) - 2\mathrm{Re}(x_-)\mathrm{E}_-(\tau) \\ &\qquad + [2\mathrm{Im}(x_+) + 4\mathrm{Im}(\delta)]e^{-\frac{1}{2}(\Gamma_S+\Gamma_L)\tau}\sin(\Delta m\tau)\Big)\Big/ \end{aligned}$$

[2]The measurements of CPLEAR in semileptonic and 3π sectors have allowed one to put good limits on $\Delta\phi$. If one assumes there is no $I = 3$ decay amplitude in the three-pion decay so that $\eta_{000} = \eta_{+-0}$ we obtain $\Delta\phi = (-1.2 \pm 8.5) \times 10^{-6}$, making the contribution of $\Delta\phi$ negligible in the analysis. If, however, one uses the measured value for η_{000}, the error increases by an order of magnitude and becomes dominant in (25.8).

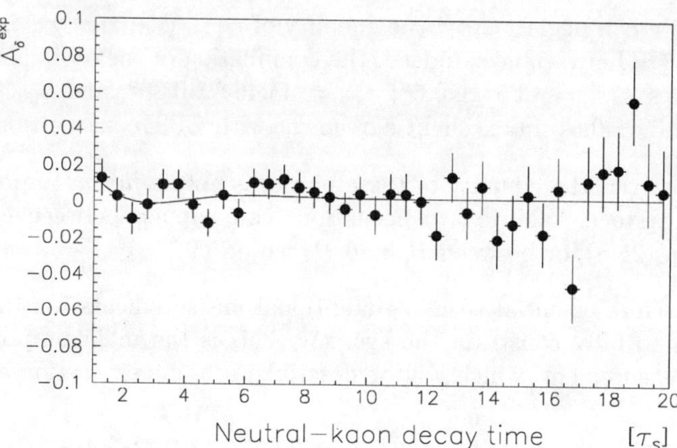

Figure 25.2: The asymmetry A_δ^{exp} versus the neutral-kaon decay time (in units of τ_S). The solid line represents the result of the fit.

$$\left(\mathrm{E}_+(\tau) + e^{-\frac{1}{2}(\Gamma_S+\Gamma_L)\tau}\cos(\Delta m\tau)\right)$$
$$+ 4\mathrm{Re}(\delta). \tag{25.13}$$

Note that

- the parameter y describing direct CPT violation in semileptonic decay cancels out,
- the factor $(1 + 4\mathrm{Re}(\epsilon_L))$ is absorbed in the CPLEAR primary vertex normalization factor, so that the A_δ measurement does not require any external input (this is different from the T-violation analysis where we used δ_L, the K_L charge asymmetry),
- all the terms including parameters other than $\mathrm{Re}(\delta)$ are damped after a few lifetimes.

Fig. 25.2 shows the data A_δ^{exp} for A_δ asymmetry. A fit using eq. (25.13) yields

$$\mathrm{Re}(\delta) = (\ \ 3.0 \pm 3.3_{\mathrm{stat}} \pm 0.6_{\mathrm{syst}}) \times 10^{-4}.$$

The value of $\mathrm{Re}(\delta)$ is compatible with zero and ≈50 times more accurate than in previous measurements. The fit yields also

$$\begin{aligned}
\mathrm{Im}(\delta) &= (-1.5 \pm 2.3_{\mathrm{stat}} \pm 0.3_{\mathrm{syst}}) \times 10^{-2}, \\
\mathrm{Re}(x_-) &= (0.2 \pm 1.3_{\mathrm{stat}} \pm 0.3_{\mathrm{syst}}) \times 10^{-2}, \\
\mathrm{Im}(x_+) &= (1.2 \pm 2.2_{\mathrm{stat}} \pm 0.3_{\mathrm{syst}}) \times 10^{-2}.
\end{aligned}$$

25.4 Fit of Data Including the Bell-Steinberger Relation; Measurement of Im(δ)

Data from the CPLEAR experiment, together with the most recent world averages for some of the neutral-kaon parameters, can be constrained with the Bell–Steinberger (or unitarity) relation, allowing the T-violation parameter Re(ϵ_T) and the CPT-violation parameter Im(δ) of the neutral-kaon mixing matrix to be determined with an increased accuracy. In the $K_S - K_L$ basis the Bell–Steinberger relation [7] can be written as

$$-i(\lambda_{\rm L}^* - \lambda_{\rm S})\langle K_L|K_S\rangle = \sum \langle f|\Lambda|K_L\rangle^* \langle f|\Lambda|K_S\rangle, \tag{25.14}$$

where we sum over all the final states f. The above equation becomes

$$\mathrm{Re}(\epsilon_T) - i\mathrm{Im}(\delta) = \frac{1}{2(i\Delta m + \frac{1}{2}(\Gamma_L + \Gamma_S))} \times \sum \mathrm{A}_{f_L}\mathrm{A}^*_{f_S} \tag{25.15}$$

with

$$\mathrm{A}_{f_L} = \langle f|\Lambda|K_L\rangle, \qquad \mathrm{A}_{f_S} = \langle f|\Lambda|K_S\rangle,$$

$$\begin{aligned}\sum \mathrm{A}_{f_L}\mathrm{A}^*_{f_S} &= \sum(|\mathrm{A}_S|^2\eta_{\pi\pi}) + \sum(|\mathrm{A}_L|^2\eta^*_{\pi\pi\pi}) \\ &\quad + 2\big[\mathrm{Re}(\epsilon_T) - \mathrm{Re}(y) - i(\mathrm{Im}(x_+) + \mathrm{Im}(\delta))\big]|f_{\pi\ell\nu}|^2,\end{aligned}$$

and

$$|\mathrm{A}_S|^2 = \mathrm{BR}^S{}_{\pi\pi}\Gamma_S, \qquad |\mathrm{A}_L|^2 = \mathrm{BR}^L{}_{\pi\pi\pi}\Gamma_L, \qquad |f_{\pi\ell\nu}|^2 = \mathrm{BR}^L{}_{\pi\ell\nu}\Gamma_L.$$

where BR stands for branching ratio, with the upper index referring to the decay particle and the lower index to the final state, $\eta_{\pi\pi}$ and $\eta_{\pi\pi\pi}$ are the CP-violation parameters when neutral kaons decay to two and three pions, respectively, and ℓ denotes electrons and muons. The radiative modes, like $\pi^+\pi^-\gamma$, are included in the corresponding parent modes [2]. Channels with $\mathrm{BR}^{\rm S}_f$ (or $\mathrm{BR}^{\rm L}_f \times \Gamma_{\rm L}/\Gamma_{\rm S}$) $< 10^{-5}$ do not contribute to eq. (25.15) within the accuracy of the present analysis.

Table 25.2 summarizes the experimental values of the parameters entered to evaluate eq. 25.15. The values of ϕ_{+-} and Δm in table 25.2 result from experiments that do not assume CPT invariance in the decay (regeneration experiments). We note that experimental data exist for all the parameters related to two- and three-pion decays, $\eta_{\pi\pi}$ and $\eta_{\pi\pi\pi}$. For the semileptonic decays we lack the measurement of the parameter Re(y), while for the parameter Im(x_+) the measurement comes from CPLEAR [5].

As a result, we obtain

$$\begin{aligned}\mathrm{Im}(x_+)&=(-2.0 \pm 2.7) \times 10^{-3}, &\quad \mathrm{Re}(y)&=(0.3 \pm 3.1) \times 10^{-3},\\ \mathrm{Re}(\delta)&=(2.4 \pm 2.8) \times 10^{-4}, &\quad \mathrm{Re}(x_-)&=(-0.5 \pm 3.0) \times 10^{-3},\end{aligned}$$

and

$$\mathrm{Re}(\epsilon_T) = (164.9 \pm 2.5) \times 10^{-5}, \qquad \mathrm{Im}(\delta) = (\;\;2.4 \pm 5.0) \times 10^{-5}.$$

Parameter	Value	
$\|\eta_{+-}\|$	$(2.283 \pm 0.025) \times 10^{-3}$	refs. [2,8]
ϕ_{+-}	$43.6° \pm 0.6°$	see text
Δm	$(530.2 \pm 1.5) \times 10^{7} \hbar s^{-1}$	see text
$\Delta\phi$	$-0.3° \pm 0.8°$	ref. [9]
$r = \frac{\|\eta_{00}\|}{\|\eta_{+-}\|}$	0.9930 ± 0.0020	ref. [2]
$\mathrm{Re}\eta_{+-0}$	$(-2 \pm 8) \times 10^{-3}$	ref. [10]
$\mathrm{Im}\eta_{+-0}$	$(-2 \pm 9) \times 10^{-3}$	ref. [10]
$\mathrm{Re}\eta_{000}$	0.08 ± 0.11	refs. [11,12]
$\mathrm{Im}\eta_{000}$	0.07 ± 0.16	refs. [11,12]
$\mathrm{BR}^{\mathrm{S}}{}_{\pi^0\pi^0}$	$(31.39 \pm 0.28)\%$	ref. [2]
$\mathrm{BR}^{\mathrm{L}}{}_{\pi^+\pi^-\pi^0}$	$(12.56 \pm 0.20)\%$	ref. [2]
$\mathrm{BR}^{\mathrm{L}}{}_{\pi^0\pi^0\pi^0}$	$(21.12 \pm 0.27)\%$	ref. [2]
$\mathrm{BR}^{\mathrm{L}}{}_{\pi\ell\nu}$	$(65.95 \pm 0.37)\%$	ref. [2]
τ_{S}	$(0.8934 \pm 0.0008) \times 10^{-10} s$	ref. [2]
τ_{L}	$(5.17 \pm 0.04) \times 10^{-8} s$	ref. [2]

Table 25.2: Experimental status of the neutral-kaon system parameters.

The correlation coefficients between the parameters are shown in Table 25.3.

The error on $\mathrm{Re}(\epsilon_T)$ and $\mathrm{Im}(\delta)$ is dominated by the error on η_{000}. If we assume that $\eta_{+-0} = \eta_{000}$, our analysis yields

$$\mathrm{Re}(\epsilon_T) = (165.0 \pm 1.9) \times 10^{-5}, \qquad \mathrm{Im}(\delta) = (\,-0.5 \pm 2.0) \times 10^{-5},$$

thus reducing the error on $\mathrm{Im}(\delta)$ by a factor of two. Table 25.3 shows a strong anticorrelation between the values of $\mathrm{Re}(x_-)$ and $\mathrm{Re}(y)$ given by the fit. If we consider their sum we find

$$\mathrm{Re}(y + x_-) = (-0.2 \pm 0.3) \times 10^{-3}.$$

We note that this quantity appears in the asymptotic value of the time-reversal asymmetry measured by CPLEAR. The present result confirms that the possible contribution to this asymmetry arising from CPT-violating decay amplitudes is negligible.

	$\mathrm{Im}(x_+)$	$\mathrm{Re}(y)$	$\mathrm{Re}(\delta)$	$\mathrm{Re}(x_-)$	$\mathrm{Re}(\epsilon_T)$	$\mathrm{Im}(\delta)$
$\mathrm{Im}(x_+)$	-	−0.624	−0.555	0.651	−0.142	0.075
$\mathrm{Re}(y)$		-	0.279	−0.997	−0.159	−0.075
$\mathrm{Re}(\delta)$			-	−0.349	0.039	−0.051
$\mathrm{Re}(x_-)$				-	0.060	0.109
$\mathrm{Re}(\epsilon_T)$					-	−0.256
$\mathrm{Im}(\delta)$						-

Table 25.3: The correlation coefficients for the parameters $\mathrm{Im}(x_+)$, $\mathrm{Re}(y)$, $\mathrm{Re}(\delta)$, $\mathrm{Re}(x_-)$, $\mathrm{Re}(\epsilon_T)$ and $\mathrm{Im}(\delta)$.

25.5 K^0 and $\overline{K}^0$ Mass and Width Difference

25.5.1 CPT Test Based on the δ Parameter

Introducing the measured values of Re(δ) and Im(δ) in eqs. 25.5 and 25.6, we can now compute both the K^0 and $\overline{K}^0$ mass and width differences. We obtain

$$\begin{aligned} \Gamma_{K^0K^0} - \Gamma_{\overline{K}^0\overline{K}^0} &= (3.9 \pm 4.2) \times 10^{-18}\text{GeV}, \\ \text{M}_{K^0K^0} - \text{M}_{\overline{K}^0\overline{K}^0} &= (-1.5 \pm 2.0) \times 10^{-18}\text{GeV}, \end{aligned}$$

with a correlation coefficient of -0.95. Fig. 25.3 shows the error ellipses corresponding to 1σ, 2σ, and 3σ. Our result on the mass difference is a factor of two better than the one obtained with a similar calculation in ref. [13]. For the sake of comparison with PDG [2], we ignore contributions

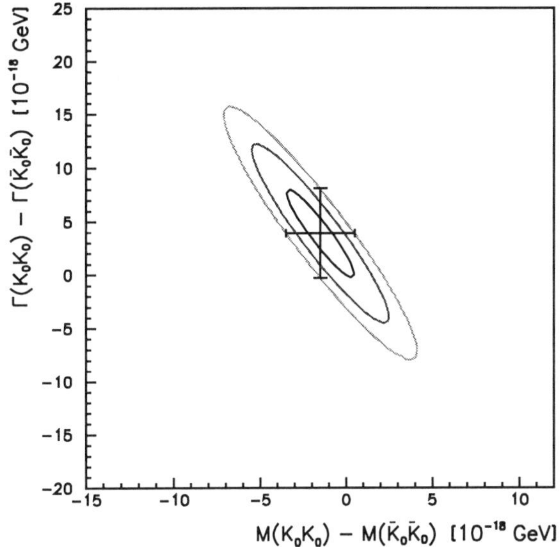

Figure 25.3: The $K^0 - \overline{K}^0$ decay-width versus mass difference. The one-, two-, and three-sigma ellipses are also shown.

other than those of the two-pion decay channel and assume CPT-invariant decay amplitudes so that $\Gamma_{K^0K^0} = \Gamma_{\overline{K}^0\overline{K}^0}$. We then obtain

$$\begin{aligned} \text{M}_{K^0K^0} - \text{M}_{\overline{K}^0\overline{K}^0} &= (-0.4 \pm 2.4) \times 10^{-19}, \text{ and also} \\ |\text{M}_{K^0K^0} - \text{M}_{\overline{K}^0\overline{K}^0}| &\leq 4.0 \times 10^{-19} \text{ (90\% CL)}. \end{aligned}$$

25.5.2 CPT-Violating Amplitudes in 2π Decay

The measurement of Re(δ) and Re(ϵ_T) can also be used, by projecting eq. (25.8) to the real axis, to give an estimate of Re(B_0)/Re(A_0). We obtain

$$\frac{\mathrm{Re}(B_0)}{\mathrm{Re}(A_0)} = (2.6 \pm 2.9) \times 10^{-4}.$$

Introducing this result in eq. (25.9), together with the measured values

$$\begin{aligned} |\eta_{+-}| &= (2.283 \pm 0.025) \times 10^{-3} \text{ (see refs. [2,8])}, \\ \phi_{00} - \phi_{+-} &= (-0.3 \pm 0.8)^\circ \text{ (see ref. [2])}, \\ \frac{1}{\sqrt{2}} \frac{\mathrm{Re}(A_2)}{\mathrm{Re}(A_0)} &\sim \frac{1}{\sqrt{2}} \left|\frac{A_2}{A_0}\right| = 0.0317 \pm 0.0001 \text{ (see ref. [14])}, \end{aligned}$$

we obtain

$$\frac{\mathrm{Re}(B_2)}{\mathrm{Re}(A_2)} - \frac{\mathrm{Re}(B_0)}{\mathrm{Re}(A_0)} = (-1.3 \pm 3.4) \times 10^{-4}.$$

25.6 Conclusions

The precise measurement of the CPT-violation parameter $\mathrm{Re}(\delta)$ by CPLEAR has opened the possibility to study CPT violation in the neutral-kaon system in a general way, without assuming CPT in the decay processes. Combining the CPLEAR results with other neutral-kaon data, we have obtained for the first time a measurement of the K^0 and $\overline{K}^0$ width difference and of most of the CPT-violating amplitudes in semileptonic and 2π decay channels. All our results are compatible with CPT invariance.

References

[1] J.S. Bell, Proc. Royal Soc. A **231**, 479 (1955); G.L. Lüders, Ann. Phys. **2**, 1 (1957); R. Jost, Helv. Phys. Acta **30**, 409 (1957).

[2] C. Caso *et al.*, Particle Data Group, Eur. Phys. J. C **3**, 1 (1998).

[3] See for example: T.T. Wu and C.N. Yang, Phys. Rev. Lett. **13**, 380 (1964); T.D. Lee and C.S. Wu, Ann. Rev. Nucl. Sci. **16**, 511 (1966); C.D. Buchanan *et al.*, Phys. Rev. D **45**, 4088 (1992); L. Maiani, "CP and CPT Violation in Neutral Kaon Decays," in *The Second DAΦNE Physics Handbook*, ed. L. Maiani *et al.* (INFN, Frascati, 1995), p. 3.

[4] E. Chell and M.G. Olsson, Phys. Rev. D **48**, 4076 (1993).

[5] A. Angelopoulos *et al.*, CPLEAR Collaboration, Phys. Lett. B **444**, 43 (1998).

[6] A. Angelopoulos *et al.*, CPLEAR Collaboration, Phys. Lett. B **444**, 52 (1998).

[7] J.S. Bell and J. Steinberger, "Weak Interactions of Kaons," in *Proc. of the Oxford Int. Conf. on Elementary Particles*, ed. R.G. Moorhouse *et al.* (Rutherford Laboratory, Chilton, England, 1965), p. 195.

[8] A. Apostolakis *et al.*, CPLEAR Collaboration, Phys. Lett. B **458**, 545 (1999).

[9] B. Schwingenheuer *et al.*, Phys. Rev. Lett. **74**, 4376 (1995); R. Carosi *et al.*, Phys. Lett. B **237**, 303 (1990).

[10] R. Adler *et al.*, CPLEAR Collaboration, Phys. Lett. B **407**, 193 (1997).

[11] A. Angelopoulos *et al.*, CPLEAR Collaboration, Phys. Lett. B **425**, 391 (1998).

[12] V.V. Barmin *et al.*, Phys. Lett. B **128**, 129 (1983).

[13] E. Shabalin, Phys. Lett. B **369**, 335 (1996).

[14] T.J. Devlin and J.O. Dickey, Rev. Mod. Phys. **51**, 237 (1979).

26

Observation of $K_S^0 \to \pi e \nu$

N. M. Ryskulov[1]

Abstract

The decay $K_S^0 \to \pi e \nu$ has been observed by the CMD-2 detector at the e^+e^- collider VEPP-2M at Novosibirsk. Of 6 million produced $K_L^0 K_S^0$ pairs, 75 ± 13 events of the $K_S^0 \to \pi e \nu$ decay were selected. The corresponding branching ratio is $\mathrm{B}(K_S^0 \to \pi e \nu) = (7.2 \pm 1.4) \times 10^{-4}$. This result is consistent with the evaluation of $\mathrm{B}(K_S^0 \to \pi e \nu)$ from the K_L^0 semileptonic rate and K_S^0 lifetime assuming $\Delta S = \Delta Q$.

While semileptonic decays of the K_L^0 have been well measured, the information on similar decays of K_S^0 is scarce. The only measurement of the $K_S \to \pi^\pm e^\mp \nu$ performed long ago assumed $\Delta S = \Delta Q$ and has low accuracy [1]. The Review of Particle Physics evaluates the corresponding decay rate indirectly, using the K_L^0 measurements and assuming that $\Delta S = \Delta Q$ so that $\Gamma(K_S^0 \to \pi^\pm e^\mp \nu) = \Gamma(K_L^0 \to \pi^\pm e^\mp \nu)$ [2].

We present results of the direct measurement of the branching ratio for the $K_S^0 \to \pi e \nu$ using the unique opportunity to study events containing a pure $K_L^0 K_S^0$ state produced in the reaction $e^+e^- \to \phi \to K_L^0 K_S^0$. The data were collected during the period of 1993–1998 with the CMD-2 detector [3,4].

The CMD-2 is a general-purpose detector consisting of a drift chamber (DC) and proportional Z-chamber (ZC) used for the trigger, both inside a thin ($0.4X_0$) superconducting solenoid with a field of 1 T. Outside the field, there is a barrel (CsI) calorimeter and a muon range system. The CsI calorimeter covers polar angles from 0.8 to 2.3 radians. The vacuum

[1]On behalf of R. R. Akhmetshin, E. V. Anashkin, M. Arpagaus, V. M. Aulchenko, V. Sh. Banzarov, L. M. Barkov, S. E. Baru, N. S. Bashtovoy, A. E. Bondar, D. V. Bondarev, A. V. Bragin, D. V. Chernyak, A. G. Chertovskikh, A. S. Dvoretsky, S. I. Eidelman, G. V. Fedotovich, N. I. Gabyshev, A. A. Grebeniuk, D. N. Grigoriev, P. M. Ivanov, S. V. Karpov, B. I. Khazin, I. A. Koop, L. M. Kurdadze, A. S. Kuzmin, I. B. Logashenko, P. A. Lukin, K. Yu. Mikhailov, I. N. Nesterenko, V. S. Okhapkin, E. A. Perevedentsev, A. S. Popov, T. A. Purlatz, N. I. Root, A. A. Ruban, N. M. Ryskulov, A. G. Shamov, Yu. M. Shatunov, A. I. Shekhtman, B. A. Shwartz, V. A. Sidorov, A. N. Skrinsky, V. P. Smakhtin, I. G. Snopkov, E. P. Solodov, P. Yu. Stepanov, A. I. Sukhanov, J. A. Thompson, V. M. Titov, Yu. Y. Yudin, and S. G. Zverev.

beam pipe with a radius of 1.8 cm is placed inside the DC and K_S^0 mesons with the decay length $\lambda = 0.6$ cm decay within it. The DC has momentum resolution of 3% for 200 MeV/c charged particles. The CsI calorimeter with $6 \times 6 \times 15\,\text{cm}^3$ crystals is placed at a distance of 40 cm from the beam axis, and about a half of K_L^0 mesons with the decay length $\lambda = 3.3$ m have interactions within CsI crystals. The energy resolution for photons in the CsI calorimeter is about 8%. Charged particles from the neutral kaon decays have momenta less than 290 MeV/c and stop within the CsI crystals.

K_S^0 decays can be tagged using the presence of the second vertex with two charged particles at a distance from the e^+e^- interaction region or the CsI cluster from K_L^0 interactions in CsI. The most probable decay channel $K_S^0 \to \pi^+\pi^-$ was used for the normalization of the semileptonic $K_S^0 \to \pi e\nu$ decay. Both channels have a vertex with two charged particles near the beam axis.

To identify electrons in the decay under study, we are using the difference between measured momentum and energy loss in the detector material for stopped particles. The basic parameter used for charged particle identification was

$$DPE = P_{particle} - E_{loss} - E_{cluster},$$

where $P_{particle}$ is the particle momentum measured in the DC, E_{loss} is the ionization energy loss (about 10 MeV) in the material in front of the CsI calorimeter, $E_{cluster}$ is the energy deposition in the CsI cluster matched with a particle track. CsI clusters that do not match any track are further referred to as photons. Electrons must have $DPE = 0$ if the resolution of the detector is ideal and the leakage of showers in the CsI calorimeter is negligible. On the other hand, pions and positive muons have a broad distribution displaced from zero. Negative muons have a sharp peak displaced from zero as the energy of the CsI cluster is equal to the difference between the muon kinetic energy and E_{loss}. Pions from the decay $K_S^0 \to \pi^+\pi^-$ were used to obtain the distribution over this parameter for charged pions in the momentum range 160–200 MeV/c. This distribution together with the fitting function is shown in fig. 26.1. For electrons and muons the distribution over this parameter was obtained from experimental data for reactions $e^+e^- \to e^+e^-, \mu^+\mu^-$ at the beam energy of 195 MeV. At this energy particle momenta are 195 MeV/c for electrons and 164 MeV/c for muons. The DPE distribution for electrons as well as the fitting function are shown in fig. 26.2. The same distribution for muons (fig. 26.3 and fig. 26.4) overlaps with the distribution for pions and this is properly taken into account.

Some kinematic features for the decay mode $K_S^0 \to \pi e\nu$ are:

- the opening angle between two tracks is between 0 and π (fig. 26.5)
- the momenta of charged particles are less than 290 MeV/c

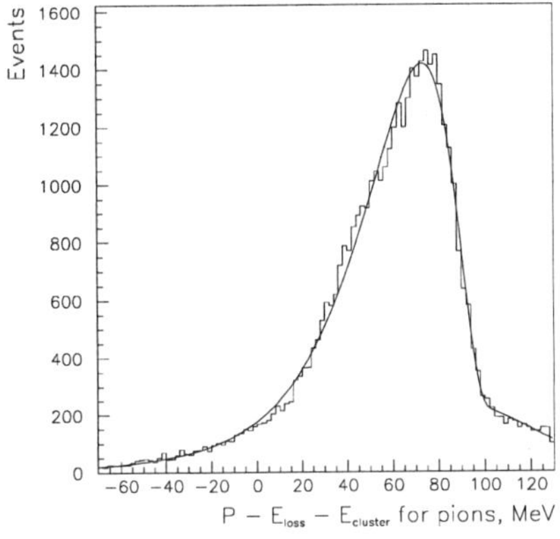

Figure 26.1: DPE distribution for pions with momenta 160–200 MeV/c in $K_S^0 \to \pi^+\pi^-$ decay.

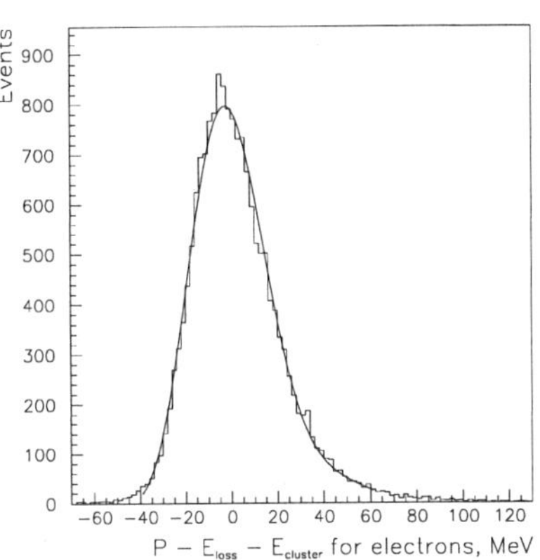

Figure 26.2: DPE distribution for collinear electrons with momentum 195 MeV/c.

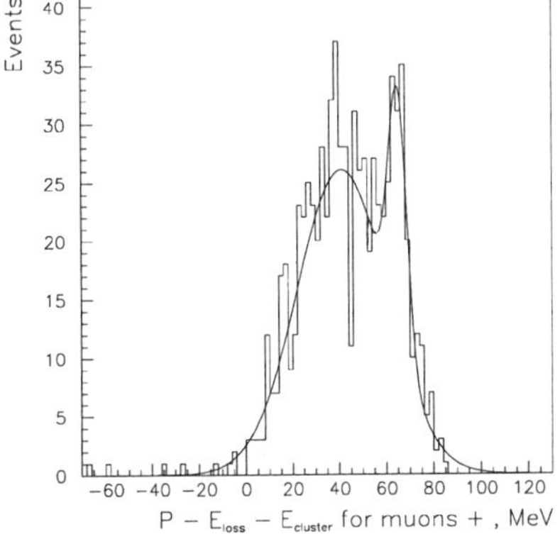

Figure 26.3: DPE distribution for positive muons with momentum 164 MeV/c.

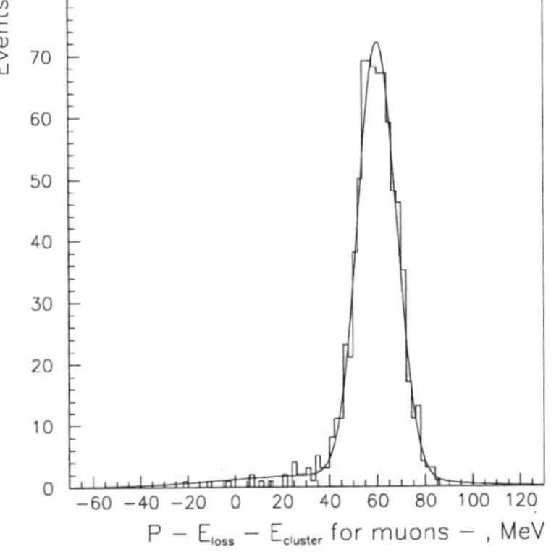

Figure 26.4: DPE distribution for negative muons with momentum 164 MeV/c.

- the total energy of charged particles (assuming that both particles are charged pions) is between 330 and 550 MeV (fig. 26.6).

The same parameters for the decay mode $K_S^0 \to \pi^+\pi^-$ are.

- the opening angle between two tracks is more than 2.6 radians (fig. 26.7)
- the pion momenta are between 160 and 270 MeV/c
- the total energy of charged particles (assuming that both particles are charged pions) is equal to the beam energy (between 508 and 512 MeV) (fig. 26.8).

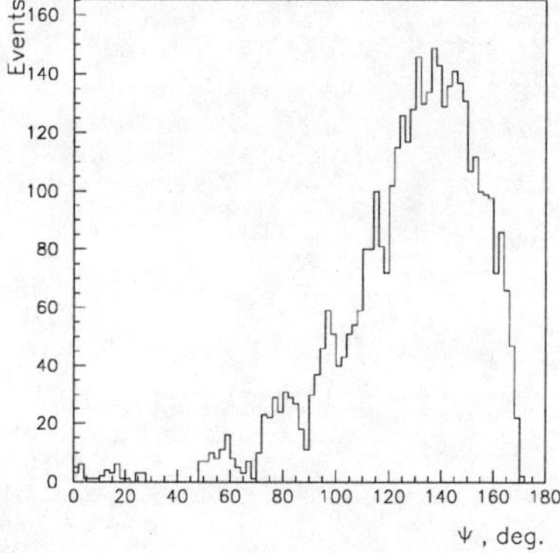

Figure 26.5: Distribution over the opening angle between two tracks in $K_S^0 \to \pi e \nu$ decay (Monte Carlo simulation).

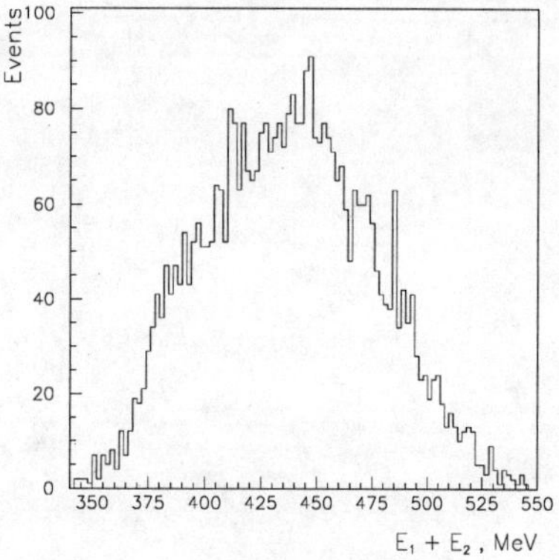

Figure 26.6: Distribution over the sum energy of charged particles in $K_S^0 \to \pi e \nu$ decay (Monte Carlo simulation).

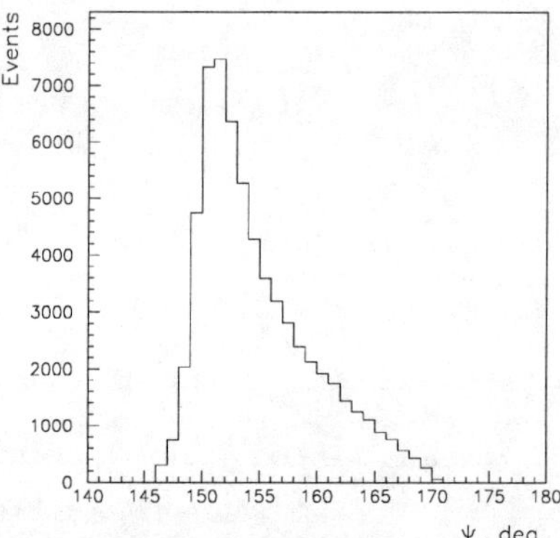

Figure 26.7: Distribution over the opening angle between two tracks in $K_S^0 \to \pi^+\pi^-$ decay.

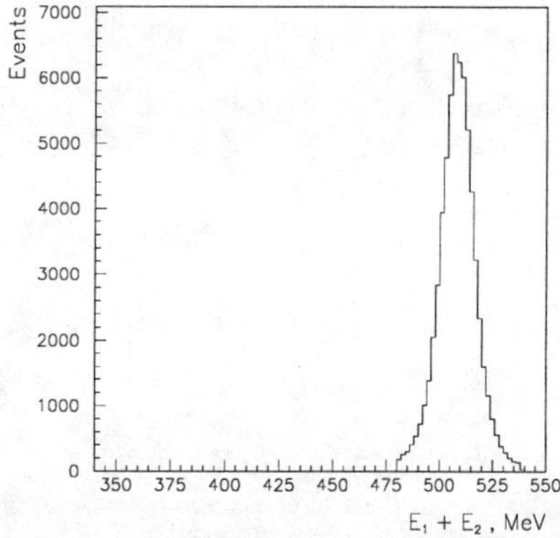

Figure 26.8: Distribution over the sum energy of charged particles in $K_S^0 \to \pi^+\pi^-$ decay.

The selection criteria for both modes of K_S^0 decay were:

- one or two vertices are found in the event
- two minimum ionizing tracks with the opposite charge sign are reconstructed from the first vertex (nearest to the beam) and there is no other track with distance to the beam less than 1.4 cm
- the distance from the first vertex to the beam is between 0.2 cm and 1.4 cm (this cut rejects background from the beam region and material of the beam pipe)

- the distance from the first vertex to the interaction point along the beam direction is less than 7 cm
- each charged particle at the first vertex has a momentum between 90 and 270 MeV/c since particles with a momentum less than 90 MeV/c cannot reach the CsI calorimeter in the magnetic field of the detector
- each track from the first vertex crosses all sensitive layers in the DC in the radial direction and therefore has a polar angle θ between 0.87 and 2.27 radians
- each charged particle at the first vertex fires the ZC and does not fire the muon range system
- the azimuthal angle difference between two tracks at the first vertex ($\Delta\phi$) is between 0.17 and 2.97 radians
- the azimuthal angle difference ($\Delta\phi$) between the plane "the first vertex (K_S^0)—the beam axis" and a photon with the energy greater than 50 MeV (supposedly the K_L^0 cluster) or the second vertex in DC (supposedly the K_L^0 decay in DC) is within ± 0.5 radian
- There are no photons with the energy greater than 15 MeV outside the direction between "the first vertex—the beam axis" ± 1 radian in the ϕ plane (this cut rejects background from processes with the neutral pions).

To select the decay mode $K_S^0 \to \pi e \nu$ the following criteria for the first vertex were additionally used:

- the opening angle between two tracks is between 0.35 and 2.50 radians
- the total energy of charged particles (assuming that both particles are charged pions) is between 300 and 470 MeV
- the DPE parameter corresponding to the charged particle at the first vertex is included in a histogram when this particle track matches the CsI cluster independently of the matching conditions of the other track
- the invariant mass squared of the assumed neutrino is greater than $-10000\ MeV^2/c^4$ and less than $6000\ MeV^2/c^4$.

To select the decay mode $K_S^0 \to \pi^+\pi^-$ the following criteria for the first vertex were additionally used:

- the opening angle between two tracks is more than 2.55 radians
- the total energy of charged particles (assuming that both particles are charged pions) is between 480 and 540 MeV

- the pion momenta are between 140 and 270 MeV/c
- the average momentum of two charged pions is between 190 and 230 MeV/c
- the ratio of the smaller momentum to the larger one is more than 0.58
- the angle between the vector sum of momenta and the direction "the beam axis → the first vertex" is less than $\pi/2$
- each charged particle at the first vertex has a matched CsI cluster
- the invariant mass squared of the assumed photon is greater than $-10000\ MeV^2/c^4$ and less than $6000\ MeV^2/c^4$.

The DPE distribution for events selected as candidates for the decay $K_S^0 \to \pi e\nu$ is shown in fig. 26.9. The data were fit using the DPE distribution of e, μ, and π measured in experiment. The result of the fit for the number of the electrons is $N_e = 83.5 \pm 12.7$. The number of mesons is equal to $N_m = 354 \pm 21$. The distribution over the distance between the vertex and beam axis for these events is consistent with that for $K_S^0 \to \pi^+\pi^-$ decays.

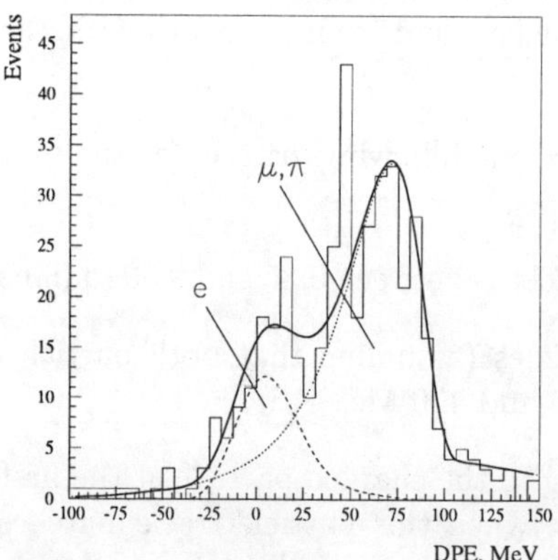

Figure 26.9: DPE distribution for charged particles in K_S^0 decays: dashed line—electrons; dotted line—muons and pions.

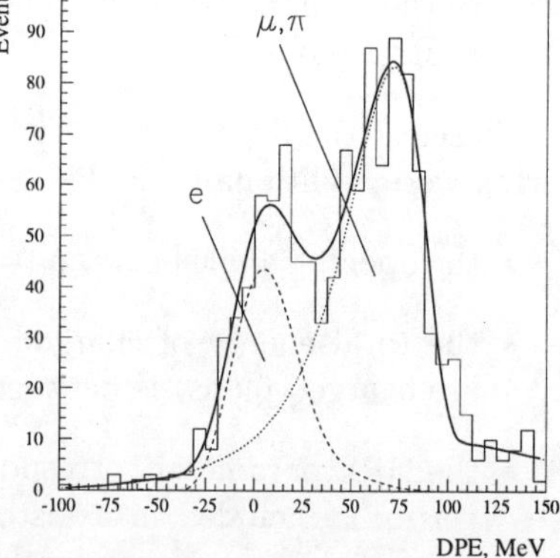

Figure 26.10: DPE distribution for charged particles in K_L^0 decays: dashed line—electrons; dotted line—muons and pions.

The main background for the $K_S^0 \to \pi e\nu$ decay mode after applying the above cuts comes from the decays $K_S^0 \to \pi^+\pi^-\gamma$, $K_S^0 \to \pi\mu\nu$, and $K_L^0 \to \pi e\nu$. The former two processes are taken into account while fitting the histogram over DPE (the fit has two free parameters—the number of electrons N_e and that of mesons $N_\pi + N_\mu$). To take into account the background from the latter process, the same procedure was applied to events with a distance from the first vertex to the beam axis between 3 and 7 cm. The resulting number of electrons for these events is 24.2 ± 6.1. Taking into account the dependence of the efficiency of vertex reconstruction on the distance from the beam axis as well as the ratio of the distance intervals for K_L^0 and K_S^0 decays it was found that the contribution of the K_L^0 decays is equal to 8.6 ± 2.2. Thus, the number of electrons, and correspondingly the number of events of the $K_S^0 \to \pi e \nu$ decay, is equal to

$$N_e = 75 \pm 13.$$

To illustrate the correctness of the identification based on the DPE parameter, we applied the same procedure to look for events of the decay $K_L^0 \to \pi e \nu$. Events were selected in which there were a $K_S^0 \to \pi^+\pi^-$ decay near the beam axis and a second vertex in DC at a long distance from the beam axis. The ratio of the number of electrons N_e to the number of pions and muons $N_\mu + N_\pi$ can be calculated from the branching ratios of the main decay modes of the K_L^0 and should be equal to 0.33 [2]. The DPE distribution for events selected as candidates for the decays of K_L^0 meson is shown in fig. 26.10. The results of the fit are: $N_e = 300 \pm 22$, $N_m = 887 \pm 34$ and their ratio is 0.34 ± 0.03 in agreement with the estimate above.

Under the applied cuts the number of $K_S^0 \to \pi^+\pi^-$ detected decays equals $N_{\pi\pi} = 178110$ (of about 6 million produced $K_S^0 K_L^0$ pairs). Applying the selection criteria above to the events from simulation, one finds that the ratio of the detection efficiency for the normalization process to one for the process under study should be $\varepsilon_{rel} = 2.49 \pm 0.12$. The simulation of the $K_S^0 \to \pi e \nu$ decay was performed using the same Dalitz plot as for the K_L^0 decay. Then one would expect for the branching ratio

$$B(K_S^0 \to \pi e \nu) = (N_e \cdot \varepsilon_{rel}/N_{\pi\pi}) \cdot B(K_S^0 \to \pi^+\pi^-).$$

From 75 ± 13 events observed by us the following result was obtained for the branching ratio:

$$B(K_S^0 \to \pi e \nu) = (7.2 \pm 1.4) \times 10^{-4}.$$

The quoted error contains the statistical error and the systematic uncertainty (5% from the simulation detection efficiency and 5% from the selection criteria) added in quadrature. This result is consistent with the previous determination of B($K_S^0 \to \pi e \nu$) [1] as well as with the Review of Particle Physics evaluation [2].

Acknowledgments

This work is supported in part by the Russian Foundation for Basic Research under grant RFBR-98-02-17851 and the US DOE grant DE-FG02-91ER40646.

References

[1] B. Aubert *et al.*, Phys. Lett. **17**, 59 (1965).

[2] C. Caso *et al.*, Eur. Phys. J. C **3**, 1 (1998).

[3] G.A. Aksenov *et al.*, Preprint BudkerINP 85-118. Novosibirsk, 1985.

[4] E.V. Anashkin *et al.*, ICFA Instr. Bulletin **5**, 18 (1988).

27

Studies of Kaon Decays at the ϕ Resonance

E. P. Solodov

Abstract

An integrated luminosity $\approx 14\,\text{pb}^{-1}$ has been collected with the CMD-2 detector at the VEPP-2M collider for ϕ meson study ($\approx 20 \times 10^6 \phi$'s). The ϕ meson is a good source of the neutral and charged kaons and their decays in flight are detected with complete reconstruction in the CMD-2 detector. The latest analysis of the $K_S^0 K_L^0$ coupled decays and charged kaon decays are presented in this paper.

27.1 Introduction

The neutral- and charged-kaon pairs are the main decay channels of the ϕ resonance, and a few million of these decays have been detected with the CMD-2 detector [1,2] at the VEPP-2M [3] electron-positron collider at the Budker Institute of Nuclear Physics in Novosibirsk. The common decay of the ϕ into two slow kaons and the good 4π detector makes it possible to study kaon decays and interactions using the recoil particle as a tag.

Some preliminary results in the $K_S^0 K_L^0$ coupled decays study with the CMD-2 detector, based on a relatively small data sample have been published [4,5].

All major decay modes of charged kaons are well detected in the CMD-2 detector, and their relative rates can be measured with relatively low systematic errors. These branching ratios were measured by optical spark chamber experiments [6] in the 1970s and were not in good agreement with each other. Moreover, the K_{e3} decay is very important for the precise determination of V_{us} in the CKM matrix.

The CMD-2 detector has been described in more detail elsewhere [1,2]. A 3.4 cm diameter vacuum beam pipe is made of Be with a 0.077 cm wall thickness and may be considered as a target for studies of the kaon nuclear interaction.

In this paper we present the latest results in the study of coupled $K_S^0 K_L^0$ decays and interactions, including better measurement of the regeneration cross section and preliminary results on measurements of charged kaon decay rates. The latter is based on about 20% of available data.

27.2 Study of $K_S^0 K_L^0$ Coupled Decays

Candidates were selected from a sample in which two vertices, each with two oppositely charged tracks, were found within 15 cm from the beam axis and all tracks were reconstructed.

The cuts (470 MeV/c^2 < M_{inv} < 525 MeV/c^2 and 80 MeV/c < P_{mis} < 140 MeV/c with an additional requirement to have another reconstructed vertex in the P_{mis} direction) select $K_S^0 \to \pi^+\pi^-$ events at one vertex. In this case the other vertex is expected to be a K_L^0. With these conditions about 14000 events with double vertices have been selected.

Fig. 27.1a shows the decay length distribution for the selected K_S^0s compared to calculation expected from the K_S^0 decay length of 0.54 ± 0.01 cm averaged over beam energy. The decay radius distribution for the K_L^0s with two charged tracks in the final state is shown in fig. 27.1b.

At the Be beam pipe position (1.7 cm) a sharp peak with 562 ± 42 events is seen followed by a bump with about 1100 events corresponding respectively to K_L^0 interactions with nuclei in the Be tube and in the DC material (Cu-Ti wires and Ar gas). The remaining events represent K_L^0 decaying in flight. A dashed line shows the expected decay distribution for uniform reconstruction efficiency.

To select candidates for $K_L^0 \to \pi^+\pi^-$, an additional cut requiring the invariant mass of two tracks from a K_L^0 vertex to be in the range of 470–525 MeV/c^2 was applied. The distribution obtained is presented in fig. 27.1d together with the fit function describing the remaining K_L^0 semileptonic decays and a peak with K_S^0 decay length. The latter is interpreted as regeneration of K_L^0 into K_S^0 on the Be pipe. The number of events under the peak was found to be 238 ± 20. One can apply stronger requirements for these events to satisfy $K_L^0 \to \pi^+\pi^-$ kinematics within detector resolution, *i.e.*, 80 MeV/c < P_{mis} < 140 MeV/c and K_S^0 vertex being in the P_{mis} direction. This selection is illustrated in fig. 27.1d by the histogram. The peak at the Be tube survives with 91 ± 12 events and 180 K_L^0 decays in flight and interactions in the DC material remain. About 40 CP violation decays of $K_L^0 \to \pi^+\pi^-$ are expected but cannot be identified because of K_L^0 semileptonic decay backgrounds and nuclear interactions. Better DC resolution, as in the KLOE detector for the dedicated DAΦNE ϕ factory, is needed (see A. Antonelli, this volume, chapter 7). Using simulated efficiencies for estimation of the full number of K_L^0 passing through the beryllium pipe, the following cross sections for regeneration and visible inelastic scattering have been

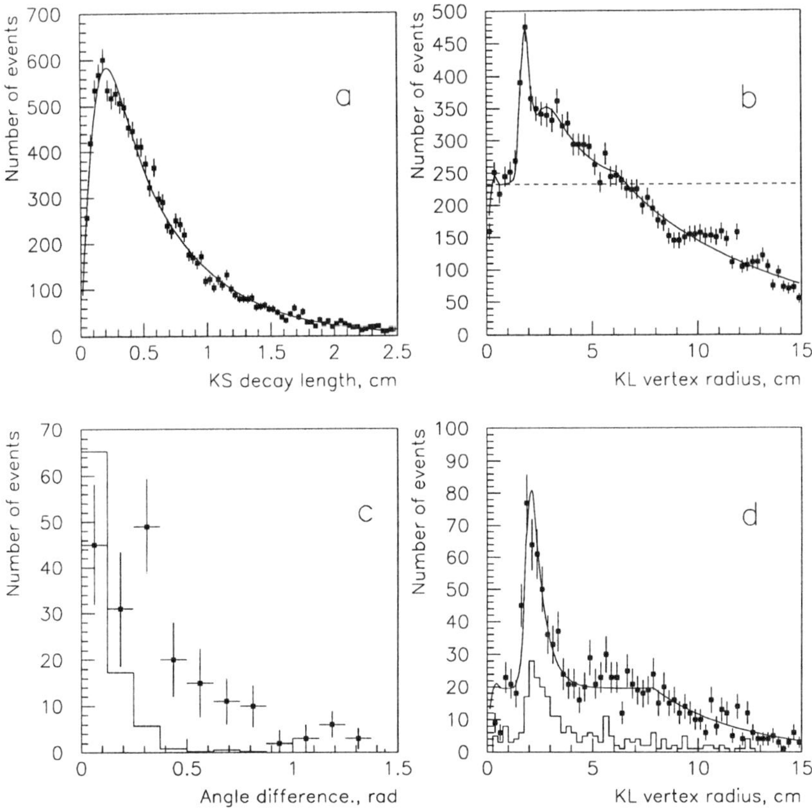

Figure 27.1: a. Decay length for K_S^0. b. Decay radius for K_L^0. c. Projected angular distribution for "tube" events and for K_S^0 two pion decays (histogram). d. Decay radius for K_L^0s after M_{inv} cut and after $K_L^0 \to \pi^+\pi^-$ selecting cut (histogram).

obtained:

$$\begin{aligned}\sigma_{reg}^{Be} &= 55.1 \pm 5.9 \pm 5.0\,\text{mb}.\\ \sigma_{inel}^{vis} &= 72 \pm 9\,\text{mb}.\end{aligned}$$

The sources of the inelastic scattering events are the reactions with Σ and Λ production. To estimate the total cross section, the relative weight of these reactions was taken as 0.21 from the CERN GEANT code (NUCRIN). With the ratio $\sigma_{inel}/\sigma_{tot} = 0.52$ taken from [7], one can estimate $\sigma_{tot}^{Be} = (580 \pm 72 \pm 174)$ mb. A comparison of this cross section with other experiments and calculations is shown in fig. 27.2.

The histogram in fig. 27.1c shows the angular distribution (in the $r - \phi$ plane) for the K_S^0 regenerated at the beam pipe (events at 1.5–4.0 cm) after subtraction of the background from the semileptonic decays of K_L^0 (events

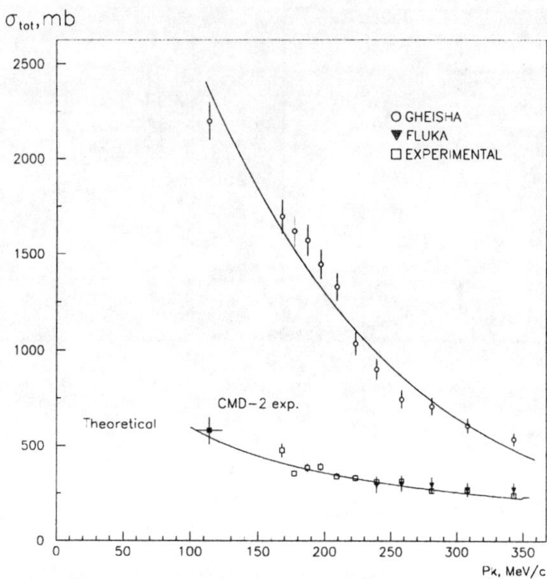

Figure 27.2: Comparison of total nuclear cross section for K_L with different simulation packages. Experimental points at higher momenta are shown. The theoretical line represents calculations from [7].

at 4.0–6.5 cm). The obtained angular distribution is wider than in the case of coherent regeneration, which can be illustrated by the distribution shown by the histogram for original K_S^0 decays at the same distance. There is no evidence for a coherent contribution to the regenerated events.

The data were obtained at different energies around the ϕ. The regeneration cross section can be calculated for different kaon momenta. For the average momenta (105 ± 2), (110 ± 2), and (115 ± 2) MeV/c the cross sections (51.9 ± 8.5), (63.5 ± 7.4), and (48.0 ± 10.5) mb have been obtained respectively.

In fig. 27.3a the measured regeneration cross section is plotted together with the theoretical calculations [7] for Be and Cu. The comparison of the calculated regeneration cross sections for these two different materials shows, that at momenta below 200 MeV/c one cannot scale them by a simple $A^{2/3}$ dependence. The experimental angular distribution of the regenerated K_S^0 after background subtraction is presented in fig. 27.3b together with a fit function and theoretical prediction [7] and seems to be in good agreement.

The obtained regeneration cross section for Be for low momentum kaons makes it possible to estimate the regeneration background for the KLOE experiments at DAΦNE and helps to subtract it.

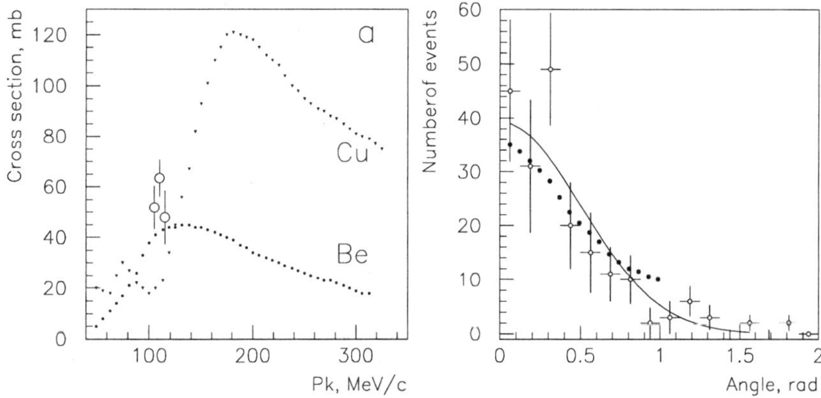

Figure 27.3: a. Experimental regeneration cross section and theoretical calculations for Be and Cu. b. Projected angular distribution of the regenerated K_S^0 with fit function (solid line) and theoretical prediction (dots);

27.3 Charged Kaon Decays Study

With 120 MeV/c momentum the charged kaons from ϕ decays have about 50 cm decay length. Within a 9 cm radius about 6% of charged kaon decays can be completely reconstructed.

Candidates were selected by the presence of two charged tracks in the DC—one track coming from the interaction point with 120 MeV/c momentum and high dE/dx and the second track with low dE/dx signal, acollinearity angle greater than 0.15 radian, and impact parameter greater than 0.2 cm. The decay point was reconstructed by prolongation of the kaon track to the intersection with the second track, which was assumed to be from decay of another kaon. Under these conditions about 50000 tagged kaon decays were selected with decay radius less than 9 cm.

Most of the major charged kaon decays have one charged particle in the final state with some number of neutrals (neutrino and π^0). Only $K^\pm \to \pi^\pm\pi^+\pi^-$ with 5.59±0.05% [6] decay probability has three charged particles in the final state. But this decay is also present in the selected sample when two pions escape detection.

The momentum and the direction of the decayed kaon are completely determined by the tagging kaon, and the missing mass can be calculated assuming the decay particle to be a pion. Fig. 27.4 presents the distribution of calculated squared missing masses in the detected two-track systems. Peaks correspond to the two body decays $K^\pm \to \mu^\pm\nu$ and $K^\pm \to \pi^\pm\pi^0$. Using the number of events under the peaks and with the efficiencies obtained by simulation the following ratio has been obtained:

$$\Gamma(K^\pm \to \pi^\pm\pi^0)/\Gamma(K^\pm \to \mu^\pm\nu) = 0.340 \pm 0.008 \pm 0.017.$$

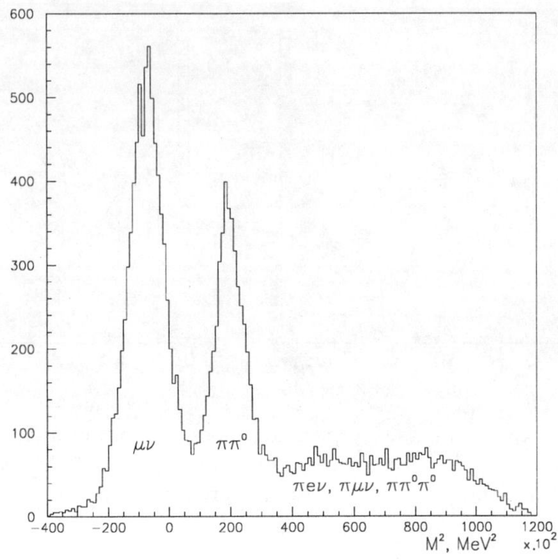

Figure 27.4: Missing mass distribution for selected events.

This result should be compared with one listed in the PDG tables [6]:

$$\Gamma(K^\pm \to \pi^\pm\pi^0)/\Gamma(K^\pm \to \mu^\pm\nu) = 0.3316 \pm 0.0032.$$

The broad distribution in fig. 27.4 represents the three body decays of charged kaons to $\pi^0\mu\nu$, $\pi^0 e\nu$, and $\pi\pi^0\pi^0$. The reconstruction of π^0's from detected photons can help to separate the channels mentioned above. In the present analysis a different calorimeter response for electrons and mesons was used. The charged particles from three body decays have low momenta and stop in the barrel CsI calorimeter. It was found that energy deposition for mesons is close to their kinetic energy and a parameter

$$\text{DPE} = P \cdot c - \Delta E - E_{cluster}$$

has weak dependence on momentum and gives a relatively sharp peak at 90 MeV. In this parameter P is the particle momentum measured in the DC, ΔE is the ionization energy loss in material in front of the CsI calorimeter, and $E_{cluster}$ is the energy deposition in the CsI cluster matched with the particle track. The shape of the DPE was studied with pions and muons from two body decays, and the functions fitting DPE for positive and negative mesons were found. The average value of the DPE parameter for electrons is close to zero and electrons could be separated from mesons in the DPE distribution.

Figs. 27.5a and b present the DPE distribution for negative and positive particles for events with squared missing mass $M_{miss} > 3.5 \times 10^4$, in

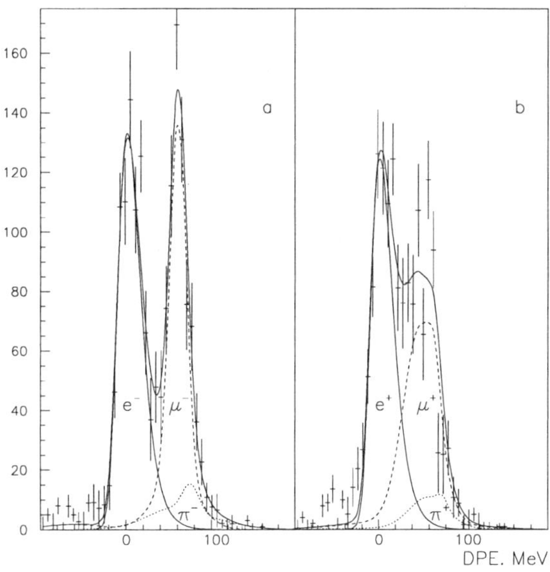

Figure 27.5: The DPE parameter distributions for negative (a) and positive (b) decay particles.

order to select three body decays. Lines present fit functions used for the separation of electrons from mesons. Electrons from semileptonic decays are relatively well separated and 1503±52 events have been found. Muon and pion distributions overlap and more sophisticated analysis including information from detected photons is under preparation now to separate all decay channels.

With the number of electrons found the following relative decay rate has been measured:

$$\frac{\Gamma(K^\pm \to e^\pm \pi^0 \nu)}{\Gamma(K^\pm \to \mu^\pm \nu) + \Gamma(K^\pm \to \pi^\pm \pi^0)} = 0.0578 \pm 0.0020 \pm 0.0020.$$

This result should be compared to the values 0.0601 ± 0.0015 (average) or 0.0570 ± 0.0008 (fit with $S = 1.4$) listed in the tables [6].

Only about 20% of available data are analyzed. The systematic errors shown here are dominated by uncertainty in the acceptance obtained from simulation. Data analysis is in progress and an overall error in the K_{e3} decay rate of order of 2% is expected.

27.4 Acknowledgements

This work is supported in part by the grant INTAS 96-0624.

References

[1] G.A. Aksenov *et al.*, Preprint BudkerINP 85-118, Novosibirsk, 1985.

[2] E.V. Anashkin *et al.*, ICFA Instrumentation Bulletin **5**, 18 (1988).

[3] V.V. Anashin *et al.*, Preprint Budker INP 84-114, Novosibirsk, 1984.

[4] R.R. Akhmetshin *et al.*, Phys. Lett. B **398**, 423 (1997).

[5] E.P. Solodov (CMD-2 Collaboration), Proceedings of the 29th International Conference on High Energy Physics, June 1998, Vancouver, Canada.

[6] Particle Data Group, C. Caso *et al.*, European Physical Journal C **3**, 1 (1998).

[7] R. Baldini and A. Michetti, Preprint LNF-96/008 (1996).

Part IV

Theoretical Topics in Kaon Physics

Basic to any approach to the strong interaction problems inherent in kaon physics is quantum chromodynamics, or QCD, the theory of interacting quarks and gluons. However, the low energies associated with kaon decays make it crucial to extend the range of validity of QCD beyond that of perturbation theory, which has been applied most successfully at high energies where the strong coupling constant, α_s, is sufficiently small to allow a perturbative expansion.

A key approach to nonperturbative (strongly interacting) QCD is to simulate its behavior on a lattice of space-time points. Recent advances in computing power have permitted great strides in lattice QCD. Nonetheless, limitations are still formidable, particularly those associated with the neglect of production of light quark-antiquark pairs. It has been possible in the past couple of years to go beyond this *quenched* approximation to lattice QCD. Martinelli and collaborators and Ryan review some progress on lattice results; see also Gupta's contribution to part II.

Another approach to extending the domain of QCD's validity is to imagine the number of colors, N_c, being taken to infinity in such a way that $N_c\alpha_s$ remains fixed [1]. In this limit only planar Feynman diagrams containing quarks and gluons survive, and resonances become stable. Progress in large-N_c QCD is described by de Rafael, as well as by Hambye and Soldan and by Bardeen in part II.

Chiral perturbation theory relies on a systematic expansion of low-energy amplitudes in powers of momenta (scaled by f_π, the pion decay constant) to relate various kaon decay processes to one another. See, for example, the discussions by Savage in part III and Bijnens in part V.

Parametrizations of weak decay amplitudes according to their flavor structure are useful in identifying regularities when several different sorts of contributions are present. These can include "penguin" amplitudes (whose name is associated with the shape of their Feynman graphs, with some stretch of the imagination). Ciuchini and his collaborators discuss such amplitudes, with reference to charmed quarks on the internal lines of the graphs.

General techniques associated with properties of scattering amplitudes and the unitarity of the S matrix can also provide insight for weak decays into final states consisting exclusively of strongly interacting particles, as pointed out by Petrov.

References

[1] G. 't Hooft, Nucl. Phys. B **72**, 461 (1974).

28

ε'/ε from Lattice QCD

M. Ciuchini, E. Franco, L. Giusti, V. Lubicz, and G. Martinelli

Abstract

Lattice calculations of matrix elements relevant for kaon decays, and in particular for ε'/ε, are reviewed. The rôle of the strange quark mass is also discussed. A comparison with other non-perturbative approaches used to compute kaon decay amplitudes is made.

28.1 Introduction

Theoretical predictions for nonleptonic decays are obtained by introducing an effective low-energy Hamiltonian expressed in terms of local operators and of the corresponding Wilson coefficients. The latter can be computed in perturbation theory, whereas the matrix elements of the operators have to be evaluated within some nonperturbative approach. For kaon decays, the Wilson coefficients are known at the next-to-leading order accuracy [1–8] and the main uncertainties come from the calculation of the matrix elements. In this talk we review the present status of lattice computations of matrix elements that are relevant in kaon decays. We focus, in particular, on those which enter the calculation of ε'/ε and also make a comparison with other methods, namely the Chiral Quark Model (χQM) and the large N expansion [9–14]. For a more general discussion of the theory of CP violation in kaon decays see [15] and references therein.

Some general remarks are necessary before entering a more detailed discussion. Given the large numerical cancellations that may occur in the theoretical expression of ε'/ε, a solid prediction should avoid the "Harlequin procedure." This procedure consists in patching together B_6 from the χQM, B_8 from the $1/N$ expansion, $m_s^{\overline{MS}}$ from the lattice, etc., or any other combination/average of different methods. All these quantities are indeed strongly correlated (for example B_6 and B_8 in the $1/N$ expansion or B parameters and quark masses in the lattice approach [16]) and should be consistently computed within each given theoretical framework. Unfortunately, none of the actual nonperturbative methods is in the position to avoid completely the Harlequin procedure, not even for the most important input parameters only. The second important issue is the consistency of the renormalization procedure adopted in the perturbative calculation of the Wilson coefficients and in the nonperturbative computation of the operator matrix elements. This problem is particularly serious for the χQM and the $1/N$ expansion and will be discussed when we compare the lattice approach to these methods. We will address, in particular, the problem of the quadratic divergences appearing in the $1/N$ expansion. This is an important issue, since the authors of refs. [13, 14] find that these divergences provide the enhancement necessary to explain the large values of $\mathrm{Re}\, A_0$ and of ε'/ε.

Schematically, ε' can be cast in the form

$$\varepsilon' = \frac{\exp(i\pi/4)}{\sqrt{2}} \frac{\omega}{\mathrm{Re}\, A_0} \times \left[\omega^{-1}\, \mathrm{Im}\, A_2 - (1 - \Omega_{IB})\, \mathrm{Im}\, A_0\right], \tag{28.1}$$

where $\omega = \mathrm{Re}\, A_2/\, \mathrm{Re}\, A_0$ and $\mathrm{Re}\, A_0$ are taken from experiments, and Ω_{IB} is a correcting factor, estimated in refs. [17–19], due to isospin-breaking effects. Using the operator product expansion, the $K \to \pi\pi$ amplitudes $\mathrm{Im}\, A_2$ and $\mathrm{Im}\, A_0$ are computed from the matrix elements of the effective Hamiltonian, expressed in terms of Wilson coefficients and renormalized operators

$$\langle \pi\pi | \mathcal{H}^{\Delta S=1} | K^0 \rangle_{I=0,2} = \sum_i \langle \pi\pi | Q_i(\mu) | K^0 \rangle_{I=0,2}\, C_i(\mu), \tag{28.2}$$

where the sum is over a complete set of operators, which depend on the renormalization scale μ. Wilson coefficients and matrix elements of the operators $Q_i(\mu)$, appearing in the effective Hamiltonian, separately depend on the choice of the renormalization scale and scheme. This dependence cancels in physical quantities, such as $\mathrm{Im}\, A_2$ and $\mathrm{Im}\, A_0$, up to higher-order corrections in the perturbative expansion of the Wilson coefficients. For this crucial cancellation to take place, the nonperturbative method used to compute hadronic matrix elements must allow a definition of the renormalized operators consistent with the scheme used in the calculation of the Wilson coefficients.

So far, lattice QCD is the only nonperturbative approach in which both the scale and scheme dependence can be consistently accounted for, using either lattice perturbation theory or nonperturbative renormalization techniques [20, 21]. This is the main reason why the authors of Refs. [8, 22–24] have followed this approach over the years.

There is a general consensus [15] that the largest contributions are those coming from Q_6 and Q_8 (for $\mathrm{Im}\, A_2$), with opposite sign, and sizeable contributions may come from Q_3, Q_4 and Q_9 (for $\mathrm{Im}\, A_2$) in the presence of large cancellations between Q_6 and Q_8, *i.e.*, when the prediction for ε'/ε is of $\mathcal{O}(10^{-4})$.[1] For this reason the following discussion, and the comparison

[1] The operators $Q_{3,4,5,6}$ only contribute to the $I = 0$ amplitudes.

with other calculations, will be focused on the determination, and errors, of the matrix elements of the two most important operators.

28.2 Matrix Elements from Lattice QCD

The evaluation of physical $K \to \pi\pi$ matrix elements on the lattice relies on the use of chiral perturbation theory (χPT): so far only $\langle\pi|Q_i(\mu)|K\rangle$ and $\langle\pi(\vec{p}=0)\pi(\vec{q}=0)|Q_i(\mu)|K\rangle_{I=2}$ (with the two pions at rest) have been computed for a variety of operators. The physical matrix elements are then obtained by using χPT at the lowest order. This is a consequence of the difficulties in extracting physical multiparticle amplitudes in Euclidean space-time [25]. Proposals to overcome this problem have been presented, at the price of introducing some model dependence in the lattice results [26]. The use of χPT implies that large systematic errors may occur in the presence of large corrections from higher-order terms in the chiral expansion and/or from FSI. This problem is common to all approaches: if large higher-order terms in the chiral expansion are indeed present and important, any method aiming to have these systematic errors under control must be able to reproduce the FSI phases of the physical amplitudes. The approaches of refs. [9,10] and [12–14,27], however, give FSI smaller than their physical values.

28.2.1 $I = 2$ Matrix Element of Q_8

There exists a large set of quenched calculations of $\langle Q_8\rangle_2$ performed with different formulations of the lattice fermion actions (staggered, Wilson, tree-level improved, tadpole improved) and renormalization techniques (perturbative, boosted perturbative, nonperturbative), at several values of the inverse lattice spacing $a^{-1} = 2 \div 3$ GeV [16,21,28–30]. All these calculations, usually expressed in terms of $B_8^{(3/2)}$, give consistent results within 20% uncertainty. Among the results, in the numerical estimates presented in sec. 28.5, we have taken the central value from the recent calculation of ref. [16], where the matrix elements $\langle Q_8\rangle_2$ and $\langle Q_7\rangle_2$ have been computed without any reference to the quark masses, and inflated the errors to account from the uncertainty due to the quenched approximation (unquenched results are expected very soon) and the lack of extrapolation to zero lattice spacing. For a discussion on the rôle of quark masses see the end of this section.

28.2.2 Matrix Element of Q_6

For $\langle Q_6\rangle_0$ from the lattice, the situation appears worse today than a few years ago when the calculations of refs. [8,22,23] were performed:

i) Until 1997, the only existing lattice result, obtained with staggered fermions (SF) without NLO lattice perturbative corrections, was $B_6 = 1.0 \pm 0.2$ [31]. This is the value used in previous analyses [8,22,23].

ii) With SF even more accurate results have been quoted recently, namely $B_6 = 0.67 \pm 0.04 \pm 0.05$ (quenched) and $B_6 = 0.76 \pm 0.03 \pm 0.05$ (with $n_f = 2$) [32].

iii) $\mathcal{O}(\alpha_s)$ corrections, necessary to match lattice operators to continuum ones at the NLO, are so huge for Q_6 in the case of SF (in the neighborhood of -100% [33]) as to make all the above results unreliable. Note, however, that the corrections tend to diminish the value of $\langle Q_6 \rangle_0$.

iv) The latest lattice results for this matrix element, computed with domain-wall fermions [34] from $\langle \pi | Q_6 | K \rangle$, are absolutely surprising: $\langle Q_6 \rangle_0$ has a sign opposite to what expected in the VSA, and to what is found with the χQM and the $1/N$ expansion. Moreover, the absolute value is so large as to give $\varepsilon'/\varepsilon \sim -120 \times 10^{-4}$. Were this confirmed, even the conservative statement by Andrzej Buras [15], namely *that certain features present in the standard model are confirmed by the experimental results; indeed, the sign and the order of magnitude of* ε'/ε *predicted by the standard model turn out to agree with the data* would be too optimistic. In order to reproduce the experimental number, $\varepsilon'/\varepsilon \sim 20 \times 10^{-4}$, not only new physics is required, but a large cancellation should also occur between the standard model and the new physics contributions. Since this result has been obtained with domain-wall fermions, a lattice formulation for which numerical studies started very recently, and no details on the renormalization and subtraction procedure have been given, we consider it premature to use the value of the matrix element of ref. [34] in phenomenological analyses. Hopefully, new lattice calculations will clarify this fundamental issue.

28.2.3 B Parameters and Quark Masses

Following the common lore, matrix elements of weak four-fermion operators are given in terms of the so-called B parameters, which measure the deviation of their values from those obtained in the vacuum saturation approximation (VSA). A classical example is provided by the matrix element of the $\Delta S = 2$ left-left operator $Q^{\Delta S=2} = \bar{s}\gamma_\mu(1-\gamma_5)d\, \bar{s}\gamma^\mu(1-\gamma_5)d$ relevant to the prediction of the CP-violation parameter ε

$$\langle \bar{K}^0 | Q^{\Delta S=2} | K^0 \rangle = \frac{8}{3} M_K^2 f_K^2 B_K . \tag{28.3}$$

VSA values and B parameters are also used for matrix elements of operators entering the expression of ε'/ε, in particular $Q_6 = \bar{s}_\alpha \gamma_\mu (1-\gamma_5) d_\beta \times \sum_q \bar{q}_\beta \gamma_\mu (1+\gamma_5) q_\alpha$ and $Q_8 = (3/2) \bar{s}_\alpha \gamma_\mu (1-\gamma_5) d_\beta \sum_q e_q \bar{q}_\beta \gamma_\mu (1+\gamma_5) q_\alpha$

$$\begin{aligned}
\langle \pi\pi | Q_6(\mu) | K^0 \rangle_{I=0} &= -4 \left[\frac{M_{K^0}^2}{m_s(\mu) + m_d(\mu)} \right]^2 (f_K - f_\pi)\, B_6(\mu) \\
\langle \pi\pi | Q_8(\mu) | K^0 \rangle_{I=2} &= \sqrt{2} f_\pi \left[\left(\frac{M_{K^0}^2}{m_s(\mu) + m_d(\mu)} \right)^2 \right.
\end{aligned}$$

$$- \frac{1}{6}\left(M_K^2 - M_\pi^2\right)\Big] \; B_8^{(3/2)}(\mu). \tag{28.4}$$

Since in the VSA and in the $1/N$ expansion the expression of the matrix elements is quadratic in $m_s + m_d$, predictions for the physical amplitudes are heavily affected by the specific value that we assume for this quantity. Unlike f_K and M_K, quark masses are not directly measured by experiments and the present accuracy in their determination is still rather poor [35,36]. Therefore, the "conventional" parametrization induces a large systematic uncertainty in the prediction of the physical amplitudes of $\langle Q_6\rangle_{I=0}$ and $\langle Q_8\rangle_{I=2}$ (and of any other left-right operator). Moreover, whereas for $Q^{\Delta S=2}$ we introduce $\hat{B}_K$ as an alias of the matrix element, by using (28.4) we replace each of the matrix elements with 2 unknown quantities, *i.e.*, the B parameter and $m_s + m_d$. Finally, in many phenomenological analyses, the values of the B parameters of $\langle Q_6\rangle_{I=0}$ and $\langle Q_8\rangle_{I=2}$ and of the quark masses are taken by independent lattice calculations, thus increasing the spread of the theoretical predictions. All this can be avoided in the lattice approach, where matrix elements can be computed from first principles. In ref. [16] a new parametrization of the matrix elements in terms of well-known experimental quantities, without any reference to the strange (down) quark mass, has been introduced. This results in a determination of physical amplitudes with smaller systematic errors. The interested reader can refer to [16] for details.

Before ending this discussion, we wish to illustrate the correlation existing between the B parameters and the quark masses in lattice calculations. On the lattice, quark masses are often extracted from the matrix elements of the (renormalized) axial current (A_μ) and pseudoscalar density ($P(\mu)$) (for simplicity we assume degenerate quark masses)

$$m(\mu) \equiv \frac{1}{2}\frac{\langle\alpha|\partial_\mu A_\mu|\beta\rangle}{\langle\alpha|P(\mu)|\beta\rangle}\,, \tag{28.5}$$

where α and β are physical states (typically α is the vacuum state and β the one-pseudoscalar meson state) and $m(\mu)$ and $P(\mu)$ are renormalized in the same scheme. On the other hand, the B parameters of Q_6 and Q_8 are obtained (schematically) from the ratio of the following matrix elements, evaluated using suitable ratios of correlation functions:[2]

$$B_{6,8}(\mu) \propto \frac{\langle\pi|Q_{6,8}(\mu)|K\rangle}{\langle\pi|P_\pi(\mu)|0\rangle\langle 0|P_K(\mu)|K\rangle}\,, \tag{28.6}$$

where P_π and P_K are the pseudoscalar densities with the flavor content of the pion or kaon, respectively. Eqs. (28.5) and (28.6) demonstrate the strong correlation existing between B parameters and quark masses: large values of the matrix elements of $P(\mu)$ correspond, at the same time, to

[2]See, for example, ref. [21]. For simplicity the superscript (3/2) in B_8 is omitted.

small values of $m(\mu)$ and $B_{6,8}(\mu)$. Physical amplitudes, instead, behave as

$$\langle Q_{6,8}\rangle = \text{const.} \times \frac{B_{6,8}(\mu)}{m(\mu)^2}, \tag{28.7}$$

where "const." is a constant that may be expressed in terms of measurable quantities (specifically M_K and f_K) only. From eqs. (28.5) and (28.6), we recognize that the dependence on $\langle P(\mu)\rangle$ cancels in the ratio $B_{6,8}/m(\mu)^2$, appearing in the physical matrix elements.

Previous lattice studies preferred to work with B parameters because these are dimensionless quantities, not affected by the uncertainty due to the calibration of the lattice spacing. This method can still be used, provided that quark masses and the B parameters from the same simulation are presented together (alternatively one can give directly the ratio $B_{6,8}/m(\mu)^2$). In ref. [16], two possible definitions of dimensionless "B parameters," which can be directly related to physical matrix elements without using the quark masses, have been proposed.

28.2.4 The Strange Quark Mass

Although in lattice calculations of matrix elements any reference to quark masses can be avoided, these are fundamental parameters of the standard model and are used in the large-N expansion. Here we would like to add only a few remarks to ref. [36], where this subject has been reviewed.

First, lattice calculations that use nonperturbative renormalization methods obtain the quark masses without errors coming from the truncation of the perturbative series (typically in the RI-MOM or the Schrödinger renormalization schemes; for a complete set of references see ref. [36]). The conversion of these results to the "standard" $\overline{MS}$ scheme can be done at the N^{3}LO. Differences between NLO, N^{2}LO, and N^{3}LO are important, $\sim 6 \div 10$ MeV for $m_s^{\overline{\text{MS}}}$, as demonstrated by

$$\begin{array}{lcllll} & & \text{NLO} & \text{N}^2\text{LO} & \text{N}^3\text{LO} & \\ m_\ell^{\overline{\text{MS}}}(2\,\text{GeV}) & = & \{5.2(5); & 4.9(5); & 4.8(5)\}\,\text{MeV} & \\ m_s^{\overline{\text{MS}}}(2\,\text{GeV}) & = & \{120(9); & 114(9); & 111(9)\}\,\text{MeV} & \end{array} \tag{28.8}$$

taken from ref. [37]. Therefore, when confronting results from different calculations it is necessary to specify the order at which the results have been obtained. In table 1 of ref. [36], results obtained with perturbation theory at NLO or with nonperturbative methods at the N^{3}LO are directly compared. This, for the reasons discussed above, is misleading. We also note that in most of the phenomenological applications, for example with QCD sum rules, the theoretical expressions are known only at the NLO and, for consistency, quark masses at the same level of accuracy should be used.

By comparing the results obtained with the nonperturbatively improved action at $\beta = 6.2$ (corresponding to $a^{-1} \sim 2.6$ GeV, which is the value used by the APE Collaboration) with those extrapolated to the continuum (table 2 of ref. [38]), one finds the discretization errors at this value of the lattice spacing and with this action to be $3 \div 4\%$. This is much smaller than the 15% quoted in [36].

The large reduction of the value of the masses in the unquenched case, found by the CP-PACS Collaboration, is not confirmed by other lattice calculations by the MILC [39] and APE [40] Collaborations and it is at variance with the bounds of ref. [41]. We think that further investigation is required on this important point.

28.3 Renormalization Group Invariant Operators

Wilson coefficients and renormalized operators are usually defined in a given scheme (HV, NDR, RI), at a fixed renormalization scale μ and depend on the renormalization scheme and scale. This is a source of confusion in the literature. Quite often, for example, one finds comparisons of B parameters computed in different schemes. Incidentally, we note that the NDR scheme used in the lattice calculation of ref. [29] differs from the standard NDR scheme of refs. [3–8]; on the other hand, the HV scheme of ref. [3] is not the same as the HV scheme of ref. [7]. In some cases, the differences between different schemes may be numerically large, *e.g.*, $B_8^{(3/2)HV} \sim 1.3\, B_8^{(3/2)NDR}$ at $\mu \sim 2$ GeV. To avoid all these problems, it is convenient to introduce a renormalization group invariant (RGI) definition of Wilson coefficients and composite operators that generalizes what is usually done for B_K using the RGI B parameter $\hat{B}_K$. The idea is very simple. Physical amplitudes can be written as

$$\langle F|\mathcal{H}|I\rangle = \langle F|\vec{Q}(\mu)|I\rangle \cdot \vec{C}(\mu), \tag{28.9}$$

where $\vec{Q}(\mu) \equiv (Q_1(\mu), Q_2(\mu), \ldots, Q_N(\mu))$ is the operator basis and $\vec{C}(\mu)$ the corresponding Wilson coefficients, represented as a column vector. $\vec{C}(\mu)$ is expressed in terms of its counterpart, computed at a large scale M, through the renormalization-group evolution matrix $\hat{W}[\mu, M]$

$$\vec{C}(\mu) = \hat{W}[\mu, M]\vec{C}(M). \tag{28.10}$$

The initial conditions for the evolution equations, $\vec{C}(M)$, are obtained by perturbative matching of the full theory, which includes propagating heavy-vector bosons (W and Z^0), the top quark, SUSY particles, etc., to the effective theory where the W, Z^0, the top quark, and all the heavy particles have been integrated out. In general, $\vec{C}(M)$ depends on the scheme used to define the renormalized operators. It is possible to show that $\hat{W}[\mu, M]$ can be written in the form

$$\hat{W}[\mu, M] = \hat{M}[\mu]\hat{U}[\mu, M]\hat{M}^{-1}[M], \tag{28.11}$$

with

$$\hat{U}[\mu, M] = \left[\frac{\alpha_s(M)}{\alpha_s(\mu)}\right]^{\hat{\gamma}_Q^{(0)T}/2\beta_0}, \quad \hat{M}[\mu] = \hat{1} + \frac{\alpha_s(\mu)}{4\pi}\hat{J}[\lambda(\mu)], \tag{28.12}$$

where $\hat{\gamma}_Q^{(0)T}$ is the leading order anomalous dimension matrix and $\hat{J}[\lambda(\mu)]$ can be obtained by solving the renormalization group equations (RGE) at the NLO. By defining

$$\hat{w}^{-1}[\mu] \equiv \hat{M}[\mu]\,[\alpha_s(\mu)]^{-\hat{\gamma}_Q^{(0)T}/2\beta_0}, \tag{28.13}$$

we get

$$\hat{W}[\mu, M] = \hat{w}^{-1}[\mu]\hat{w}[M]. \tag{28.14}$$

The effective Hamiltonian (28.9) can then be written as

$$\begin{aligned} \mathcal{H} &= \vec{Q}(\mu)\cdot\vec{C}(\mu) = \vec{Q}(\mu)\hat{W}[\mu, M]\vec{C}(M) \\ &= \vec{Q}(\mu)\hat{w}^{-1}[\mu]\cdot\hat{w}[M]\vec{C}(M) = \vec{Q}^{RGI}\cdot\vec{C}^{RGI}, \end{aligned} \tag{28.15}$$

with

$$\vec{C}^{RGI} = \hat{w}[M]\vec{C}(M), \quad \vec{Q}^{RGI} = \vec{Q}(\mu)\cdot\hat{w}^{-1}[\mu]. \tag{28.16}$$

$\vec{C}^{RGI}$ and $\vec{Q}^{RGI}$ are scheme and scale independent at the order at which the Wilson coefficients have been computed.

28.4 Comparison with Other Methods

In this section, we briefly discuss the relevant aspects that distinguish the lattice approach from others that have been used in the literature to predict ε'/ε.

The original approach of the Munich group was to extract the values of the relevant matrix elements from experimental measurements [4,6]. This method guarantees the consistency of the operator matrix elements with the corresponding Wilson coefficients. Unfortunately, with the Munich method it is impossible to get the two most important contributions, namely those corresponding to $\langle Q_6\rangle_0$ and $\langle Q_8\rangle_2$. For this reason, "guided by the results presented above and biased to some extent by the results from the large-N approach and lattice calculations," the authors of ref. [42] have taken $B_6 = 1.0 \pm 0.3$ and $B_8^{(3/2)} = 0.8 \pm 0.2$, and $m_s^{\overline{MS}} = 110 \pm 20$ MeV at $\mu = 1.3$ GeV. These values, if assumed to hold in the HV regularization, are very close to those used in ref. [24]. They do not come, however, from a calculation consistently made within a given theoretical approach (large-N expansion, χQM, or lattice for example).

The $1/N$ expansion and the χQM are effective theories. To be specific, in the framework of the $1/N$ expansion the starting point is given by the chiral Lagrangian for pseudoscalar mesons expanded in powers of masses and momenta. At the leading order in $1/N$, local four-fermion operators can be written in terms of products of currents and densities, which are expressed in terms of the fields and coupling of the effective theory. In higher orders, a (hard) cutoff, Λ_c, must be introduced to compute the relevant loop diagrams. The cutoff is usually identified with the scale at which the short-distance Wilson coefficients must be evaluated.

Divergences appearing in factorizable contributions can be reabsorbed in the renormalized coupling of the effective Lagrangian and in the quark masses; nonfactorizable corrections constitute the part that should be matched to the short-distance coefficients. By using the intermediate color-singlet boson method, the authors of refs. [11–14, 43] claim to be able to perform a consistent matching, including the finite terms, of the matrix elements of the operators in the effective theory to the corresponding Wilson coefficients. It is precisely this point that, in our opinion, has never been demonstrated in a convincing way. If the matching is "consistent," then it should be possible to show analytically that the cutoff dependence of the matrix elements computed in the $1/N$ expansion cancels that of the Wilson coefficients, at least at the order in $1/N$ at which they are working. Moreover, if finite terms are really under control, it should be possible to tell whether the coefficients should be taken in HV, NDR, or any other renormalization scheme.

The fact that in higher orders even quadratic divergences appear, with the result that the logarithmic divergences depend now on the regularization, makes the matching even more problematic. Theoretically, we cannot imagine any mechanism to cancel the cutoff dependence of the physical amplitude in the presence of quadratic divergences, which should, in our opinion, disappear in any reasonable version of the effective theory. It is also important to show (and to our knowledge it has never been done) that the numerical results for the matrix elements are stable with respect to the choice of the ultraviolet cutoff. This would also clarify the issue of the routing of the momenta in divergent integrals. For example, the matrix elements in the meson theory could be computed in some lattice regularization.

28.5 Numerical Results

As discussed above, all the methods used in the calculation of ε'/ε are not completely satisfactory and in general suffer from large theoretical uncertainties.

The lattice approach can, in principle, compute the relevant matrix elements without any model assumption (at least at the lowest order in the chiral expansion), and with operators consistently defined to match the Wilson coefficients of the effective Hamiltonian. In spite of these advantages the lattice results for $\langle Q_6 \rangle_0$ are inconclusive, as discussed before. Regarding the surprising result of ref. [34], we think that further scrutiny and confirmation from other calculations are needed before using it in a phenomenological analysis.

In the absence of any definite result for $\langle Q_6\rangle_0$ from the lattice, the authors of ref. [24] have assumed for this matrix element the value

$$\langle Q_6\rangle_0 \equiv \langle\pi\pi|Q_6^{HV}|K^0\rangle_{I=0} = -0.4 \pm 0.4\,\text{GeV}^3, \tag{28.17}$$

and

$$\langle Q_5\rangle_0 = 1/3\langle Q_6\rangle_0, \tag{28.18}$$

at a scale $\mu = 2$ GeV. The value of the matrix element in eq. (28.17) corresponds to $B_6 = 1.0 \pm 1.0$ for a "conventional" mass fixed to $m_s^{\overline{MS}} + m_d^{\overline{MS}} = 130$ MeV.

For $\langle Q_{7,8}\rangle_2$, the values of ref. [16] (obtained with an improved action using nonperturbatively renormalized operators at $\mu = 2$ GeV) have been used, namely

$$\langle Q_7\rangle_2 \equiv \langle\pi\pi|Q_7^{HV}|K^0\rangle_{I=2} = 0.18 \pm 0.06\,\text{GeV}^3, \tag{28.19}$$

$$\langle Q_8\rangle_2 \equiv \langle\pi\pi|Q_8^{HV}|K^0\rangle_{I=2} = 0.62 \pm 0.12\,\text{GeV}^3, \tag{28.20}$$

where the superscript HV denotes the t'Hooft-Veltman renormalization scheme.

By varying the input parameters as described in ref. [24] and by weighting the Monte Carlo events with the experimental constraints, the prediction for ε'/ε is

$$\varepsilon'/\varepsilon = (3.6^{+6.7}_{-6.3} \pm 0.5) \times 10^{-4}, \tag{28.21}$$

where the third error on ε'/ε is an estimate of the residual scheme dependence due to unknown higher-order corrections in the perturbative expansion.[3] Given the large theoretical uncertainties, and taking into account some differences in the calculation of this quantity (choice of the renormalization scale, values of several B parameters, etc.), the result in eq. (28.21) is in substantial agreement, though slightly lower, with the recently upgraded evaluation of ref. [42]: $\varepsilon'/\varepsilon = (7.7^{+6.0}_{-3.5}) \times 10^{-4}$ and $\varepsilon'/\varepsilon = (5.2^{+4.6}_{-2.7}) \times 10^{-4}$ in NDR and in HV respectively. It is also very close to previous estimates of the Rome [22, 23] and Munich [4, 6] groups. This agreement is not surprising since the two groups use very similar inputs for the matrix elements and the experimental parameters have only slightly changed in the last few years. The crucial question, namely a quantitative determination of $\langle Q_6\rangle_0$, remains unfortunately still unsolved.

All the above results are, however, much lower than the recent measurements of KTeV, $\text{Re}(\varepsilon'/\varepsilon) = (28.0 \pm 4.1) \times 10^{-4}$, of NA48, $\text{Re}(\varepsilon'/\varepsilon) = (18.5 \pm 7.3) \times 10^{-4}$, or than the present world average $\text{Re}(\varepsilon'/\varepsilon)_{WA} = (21.2 \pm 4.6) \times 10^{-4}$, determined by the results of refs. [44–47].

By scanning various input parameters (in the conventional approach B_6, $B_8^{(3/2)}$, $\alpha_s(M_Z)$, Im λ_t, etc.) and in particular by choosing them close

[3]The value in eq. 28.21 is slightly different from that presented at the the conference and quoted by ref. [15]. The reason is that the final analysis of ref. [16] found for $\langle Q_8\rangle_{I=2}$ a value larger by about 15% than the preliminary one.

to their extreme values it is possible to obtain $\varepsilon'/\varepsilon \sim 20 \times 10^{-4}$. This also gives the impression of a better agreement (lesser disagreement) between the theoretical predictions and the data. For example, in ref. [24] the scanning gives $-11 \times 10^{-4} \leq \varepsilon'/\varepsilon \leq 27 \times 10^{-4}$. In spite of the fact that the experimental world average is compatible with the "scanned" range above, a conspiracy of several inputs in the same direction is necessary in order to get a large value of ε'/ε. For central values of the parameters, the predictions are, in general, much lower than the experimental results. For this reason, barring the possibility of new physics effects [48], we believe that an important message is arriving from the experimental results:

Penguin contractions (or eye-diagrams, not to be confused with penguin operators), which are usually neglected within factorization, give contributions that make the matrix elements definitely larger than their factorized values.

This implies that the "effective" B parameters of the relevant operators, specifically those relative to the matrix elements of Q_1 and Q_2 for $Re(A_0)$ and of Q_6 for ε'/ε, are much larger than 1. This interpretation would provide a unique dynamical mechanism to explain both the $\Delta I = 1/2$ rule and a large value of ε'/ε [49]. Large contributions from penguin contractions are actually found by calculations performed in the framework of the χQM [9, 10] or the $1/N$ expansion [11, 13, 14, 27]. It is very important that these indications find quantitative confirmation in other approaches, for example in lattice QCD calculations. Note that naïve explanations of the large value of ε'/ε, such as a very low value of $m_s^{\overline{MS}}$, would leave the $\Delta I = 1/2$ rule unexplained.

Finally, one may try to quantify the amount of enhancement required for the matrix element of Q_6 in order to explain the experimental value of ε'/ε. A fit of $\langle Q_6 \rangle_0$ to $\text{Re}(\varepsilon'/\varepsilon)_{WA}$ gives $\langle Q_6 \rangle_0 = -1.2^{+0.25}_{-0.21} \pm 0.15$ GeV^3, about 2.5 times larger than the central value used in our analysis.

References

[1] G. Altarelli, G. Curci, G. Martinelli, and S. Petrarca, Nucl. Phys. B **187**, 461 (1981).

[2] A.J. Buras and P.H. Weisz, Nucl. Phys. B **333**, 66 (1990).

[3] A.J. Buras, M. Jamin, M.E Lautenbacher, and P.H. Weisz, Nucl. Phys. B **370**, 69 (1992), Addendum, Nucl. Phys. B **375**, 501 (1992).

[4] A. Buras, M. Jamin, and M.E. Lautenbacher, Phys. Lett. B **389**, 749 (1996).

[5] A.J. Buras, M. Jamin, and M.E. Lautenbacher, Nucl. Phys. B **400**, 37 (1993) and B **400**, 75 (1993).

[6] A. Buras, M. Jamin, and M.E. Lautenbacher, Nucl. Phys. B **408**, 209 (1993).

[7] M. Ciuchini, E. Franco, G. Martinelli, and L. Reina, Nucl. Phys. B **415**, 403 (1994).

[8] M. Ciuchini *et al.*, Z. Phys. C **68**, 239 (1995).

[9] S. Bertolini, J.O. Eeg, and M. Fabbrichesi, Nucl. Phys. B **476**, 225 (1996).

[10] S. Bertolini, J.O. Eeg, M. Fabbrichesi, and E.I. Lashin, Nucl. Phys. B **514**, 93 (1998).

[11] J. Heinrich, E.A. Paschos, J.M. Schwarz, and Y.L. Wu, Phys. Lett. B **279**, 140 (1992); E.A. Paschos, review presented at the 27th Lepton-Photon Symposium, Beijing, China (1995).

[12] T. Hambye *et al.*, Phys. Rev. D **58**, 014017 (1998).

[13] T. Hambye, G.O. Köhler, and P.H. Soldan, hep-ph/9902334.

[14] T. Hambye, G.O. Köhler, E.A. Paschos, and P.H. Soldan, hep-ph/9906434.

[15] A. Buras, this volume, chapter 5, hep-ph/9908395.

[16] A. Donini, V. Giménez, L. Giusti, and G. Martinelli, BUHEP-99-23; L. Giusti, presented at Latt99, 29 June–3 July 1999, Pisa, Italy, to appear in the Proceedings, hep-lat/9909041.

[17] J.F. Donoghue, E. Golowich, B.R. Holstein, and J. Trampetic, Phys. Lett. B **179**, 361 (1986).

[18] A.J. Buras and J.M. Gérard, Phys. Lett. B **192**, 156 (1987).

[19] M. Lusignoli, Nucl. Phys. B **325**, 33 (1989).

[20] G. Martinelli, C. Pittori, C.T. Sachrajda, M. Testa, and A. Vladikas, Nucl. Phys. B **445**, 81 (1995).

[21] G. Martinelli *et al.*, Nucl. Phys. B **445**, 81 (1995); A. Donini *et al.*, Phys. Lett. B **360**, 83 (1996); M. Crisafulli *et al.*, Phys. Lett. B **369**, 325 (1996); L. Conti *et al.*, Phys. Lett. B **421**, 273 (1998); C. R. Allton et al., Phys. Lett. B **453**, 30 (1999).

[22] M. Ciuchini, E. Franco, G. Martinelli, and L. Reina, Phys. Lett. B **301**, 263 (1993).

[23] M. Ciuchini, Nucl. Phys. (Proc. Suppl.) **59**, 149 (1997).

[24] M. Ciuchini, E. Franco, L. Giusti, V. Lubicz, and G. Martinelli, in preparation.

[25] L. Maiani and M. Testa, Phys. Lett. B **245**, 585 (1990).

[26] M. Ciuchini, E. Franco, G. Martinelli, and L. Silvestrini, Phys. Lett. B **380**, 353 (1996).

[27] W.A. Bardeen, A.J. Buras, and J.M. Gérard, Phys. Lett. B **180**, 133 (1986); Nucl. Phys. B **293**, 787 (1987); Phys. Lett. B **192**, 138 (1987).

[28] G. Kilcup, R. Gupta, and S. Sharpe, Phys. Rev. D **57**, 1654 (1998).

[29] T. Bhattacharaya, R. Gupta, and S. Sharpe, Phys. Rev. D **55**, 4036 (1997).

[30] L. Lellouch and D. Lin, Nucl. Phys. (Proc. Suppl.) **73**, 314 (1999).

[31] G. Kilcup, Nucl. Phys. B (Proc. Suppl) **20**, 417 (1991); S. Sharpe, Nucl. Phys. B (Proc. Suppl) **20**, 429 (1991); S. Sharpe *et al.*, Phys. Lett. B **192**, 149 (1987).

[32] D. Pekurovsky and G. Kilcup, hep-lat/9709146 and hep-lat/9812019.

[33] S. Sharpe and A. Patel, hep-lat/9310004; N. Ishizuka and Y. Shizawa, Phys. Rev. D **49**, 3519 (1994).

[34] T. Blum *et al.*, BNL-66731, hep-lat/9908025.

[35] R. D. Kenway, Nucl. Phys. (Proc. Suppl.) **73**, 16 (1999); S. R. Sharpe, talk given at 29th International Conference on High-Energy Physics (ICHEP 98), Vancouver, Canada 1998, hep-lat/9811006; V. Lubicz, Nucl. Phys. (Proc. Suppl.) **74**, 291 (1999).

[36] S. Ryan, this volume, chapter 29, hep-ph/9908386.

[37] D. Becirevic, V. Giménez, V. Lubicz, and G. Martinelli, hep-lat/9909082.

[38] J. Garden *et al.*, ALPHA and UKQCD Collaboration, DESY-99-075, hep-lat/9906013.

[39] C. Bernard, private comunication.

[40] V. Giménez, presented at Latt99, 29 June–3 July 1999, Pisa, Italy, to appear in the Proceedings.

[41] L. Lellouch, E. De Rafael, and J. Taron, Phys. Lett. B **414**, 195 (1997).

[42] S. Bosch *et al.*, hep-ph/9908408.

[43] J. Bijnens and J. Prades, JHEP **01**, 023 (1999).

[44] KTeV Collaboration, A. Alavi-Harati *et al.*, Phys. Rev. Lett. **83**, 22 (1999).

[45] M.S. Sozzi, this volume, chapter 8.

[46] NA31 Collaboration, H. Burkhardt *et al.*, Phys. Lett. B **206**, 169 (1988); G.D. Barr *et al.*, Phys. Lett. B **317**, 233 (1993).

[47] E731 Collaboration, L.K. Gibbons *et al.*, Phys. Rev. Lett. **70**, 1203 (1993).

[48] H. Murayama, this volume, chapter 10, hep-ph/9908442.

[49] M. Ciuchini *et al.*, this volume, chapter 31.

29

Lattice Determinations of the Strange Quark Mass

Sinéad Ryan

29.1 Introduction

The importance of the strange quark mass, as a fundamental parameter of the standard model and as an input to many interesting quantities, has been highlighted in many reviews, *e.g.*, in ref. [1]. A first-principles calculation of m_s is possible in lattice QCD but to date there has been a rather large spread in values from lattice calculations. This review aims to clarify the situation by explaining the particular systematic errors and their effects and illustrating the emerging consenus.

In addition, a discussion of the strange quark mass is timely given the recent results from KTeV [2] and NA48 [3] for ϵ'/ϵ, which firmly establish direct CP violation in the standard model and when combined with previous measurements give a world average $\epsilon'/\epsilon = (21.2 \pm 2.8) \times 10^{-4}$. This is in stark disagreement with the theoretical predictions that favor a low ϵ'/ϵ [4].

Although in principle ϵ'/ϵ does not depend directly on m_s in practice it has been an input in current phenomenological analyses. This dependence arises because the matrix elements of the gluonic, $\langle Q_6 \rangle_0$, and electroweak, $\langle Q_8 \rangle_2$, penguin operators[1] are of the form $\langle \pi\pi | Q_i | K \rangle$ and final-state interactions make them notoriously difficult to calculate directly. They have been, therefore, parameterized in terms of bag parameters, $\mathcal{B}_i$, the strange quark mass, m_s and the top quark mass, m_t, as discussed in detail in ref. [4]. A recent review of lattice calculations of the matrix elements is in ref. [5]. In this talk I will focus on some recent and careful lattice determinations of m_s, illustrating the reasons for the large spread in earlier results.

29.2 The Strange Quark Mass from Lattice QCD

In lattice QCD, m_s is determined in two ways, each of which relies on calculations of experimentally measured quantities to fix the lattice bare

[1] Keeping only the numerically dominant contributions for simplicity.

coupling and quark masses. The 1P-1S charmonium splitting, M_ρ, and r_0 are some of the parameters typically chosen to fix the inverse lattice spacing, a^{-1}. To determine m_s either M_K or M_ϕ is used. It is an artifact of the quenched approximation that m_s depends on the choice of input parameters, so that some of the spread in answers from lattice QCD can be attributed to different choices here. Naturally, some quantities are better choices than others, being less sensitive to quenching or having smaller systematic errors.

The quark mass can be determined from hadron spectroscopy, using chiral perturbation theory to match a lattice calculation of M_K (or M_ϕ) to its experimental value with

$$M_{PS}^2 = B_{PS}\frac{(m_i + m_j)}{2} + \ldots \quad \textbf{or} \quad M_V = A_V + B_V\frac{(m_i + m_j)}{2} + \ldots \tag{29.1}$$

This is the hadron spectrum or vector Ward identity (VWI) method.

Alternatively, one can use the axial Ward identity (AWI): $\partial_\mu A_\mu(x) = (m_i + m_j)P(x)$ imposed at quark masses to correspond to either the experimentally measured M_K or M_ϕ determines m_s.

The lattice bare masses and matrix elements are related to their continuum counterparts, in (say) the $\overline{MS}$ scheme, by the renormalization coefficients, Z_s or $Z_{(A,P)}$, calculated perturbatively or nonperturbatively,

$$m_s^{\overline{MS}}(\mu) = Z_s^{-1}(\mu, ma)m_q^0, \qquad (m_s + \overline{m})^{\overline{MS}}(\mu) = \frac{Z_A(ma)}{Z_P(\mu, ma)}\frac{\langle \partial_\mu A_\mu J(0)\rangle}{\langle P(x)J(0)\rangle}.$$

m_s has been calculated in all three lattice fermion formalisms: Wilson, staggered and domain wall. Although the domain wall fermion results are extremely interesting, since this approach has the good flavor structure of Wilson fermions while preserving chiral symmetry, the results for m_s are still preliminary so I will focus on results with Wilson and staggered fermions. A description of the domain wall formalism and results can be found in ref. [8].

Comparing results from these different methods provides a nice check of lattice calculations.

29.3 Main Uncertainties in the Calculation

The difference in early lattice results can be understood in terms of the treatment of systematic uncertainties in these particular calculations. The largest of these are discretization errors, calculation of renormalization coefficients and the quenched approximation.

29.3.1 Discretization Errors

The Wilson action has discretization errors of $\mathcal{O}(a)$, so for a reliable result one needs fine lattices and a continuum extrapolation, $a \to 0$. See fig. 29.1 for the CP-PACS Collaboration's quenched Wilson results [9]. The Sheikholeslami-Wohlert (SW) clover action includes a term $\sim c_{SW}\overline{\Psi}\sigma_{\mu\nu} \times F_{\mu\nu}\Psi$ and discretization errors start at $\mathcal{O}(\alpha_s a)$, when c_{SW} is determined perturbatively. The remaining a dependence must be removed by continuum extrapolation, but the slope of the extrapolation is milder [6]. A nonperturbative determination of c_{SW} [10] gives an $\mathcal{O}(a)$-improved action, which should futher reduce the lattice spacing dependence. Recent results from the APE, ALPHA/UKQCD, and QCDSF Collaborations use this approach [12,14,15]. The latter two groups include continuum extrapolations and find significant a dependence ($\approx$ 15% between the finest lattice and $a = 0$ as found by ALPHA/UKQCD). In the case of the more commonly used VWI approach the slope of the extrapolation in a is positive and therefore m_s at finite lattice spacing is too high, even with improvement.

The staggered fermion action is $\mathcal{O}(a)$ improved so the lattice spacing dependence should be mild.

29.3.2 Renormalization Coefficients

Z_S and $Z_{(A,P)}$ can be determined perturbatively or nonperturbatively. A nonperturbative calculation is preferable as it removes any perturbative ambiguity. This was pioneered by the ALPHA and APE groups [10,11].

For Wilson fermions perturbative corrections are smaller and therefore more reliable in the VWI approach (*i.e.*, for Z_S) than in the AWI approach. In ref. [12] the difference between nonperturbative results and boosted perturbation theory is $\sim$ 10% for Z_S and $\sim$ 30% for Z_P at $a^{-1} \sim 2.6$ GeV. For staggered fermions the perturbative coefficients are large and positive so the results are unreliable and nonperturbative renormalization is essential. The perturbative staggered results are therefore too low and this effect combined with the too-high values of m_s from Wilson results at finite lattice spacing explains much of the spread in lattice results.

29.3.3 Quenching

Most calculations are done in the quenched approximation—neglecting internal quark loops—as a computational expedient. An estimate of this approximation, based on phenomenological arguments, was made in [6]. The authors estimated that unquenching lowers m_s by $\approx 20 - 40\%$. They also argued that M_K rather than M_ϕ is a better choice of input parameter since it is less sensitive to quenching. Unquenched calculations by CP-PACS have shown that these estimates were of the correct size and sign [17].

A number of clear trends are therefore identified:

- There is significant a dependence in the Wilson action results, which raises m_s at finite lattice spacing. Although this is milder for the improved actions it is still present, as pointed out in refs. [6,7,15].

- Using perturbative improvement, the VWI and AWI methods differ at finite lattice spacing but agree after continuum extrapolation. This indicates the methods have discretization errors larger than the perturbative uncertainty. Nonperturbative renormalization has a larger effect on AWI results, bringing them into agreement with VWI results at finite lattice spacing. However, discretization errors remain a significant uncertainty and without a continuum extrapolation lead to an overestimate of m_s.
- Perturbative renormalization of staggered fermions results in an underestimate of m_s. Nonperturbative renormalization is essential.
- A lower value of m_s is expected from an unquenched calculation.

29.4 Recent Results for m_s

The systematic uncertainties in the lattice determination of m_s are now well understood. Some recent results that I believe provide a definitive value of m_s in quenched QCD and an unquenched result are now discussed.

29.4.1 Quenched Results

Table 29.1 compares a number of recent calculations of m_s. The JLQCD, ALPHA/UKQCD, and QCDSF Groups have removed all uncertainties within the quenched approximation. JLQCD use staggered fermions and nonperturbative renormalization [13]. They observe mild a dependence, as expected, and take the continuum limit. The effect of nonperturbative renormalization is considerable, again as expected: $\sim +18\%$ when compared to the perturbative result.

The ALPHA/UKQCD [15] and QCDSF [14] Collaborations use a nonperturbatively improved SW action and renormalization and include a continuum extrapolation. This explains the difference between their results and that of the APE Group (which has not been extrapolated to $a = 0$). Interestingly, ALPHA/UKQCD, QCDSF, and the Fermilab [6] and LANL [7] results for m_s are in agreement. The difference in analyses is nonperturbative versus perturbative renormalization, indicating that the perturbative result for the VWI method is reliable (for Wilson fermions).

29.4.2 An Unquenched Result

There are a number of new preliminary unquenched calculations of m_s [16]. However, CP-PACS have recently completed their analysis [17], shown in fig. 29.1, so I will concentrate on this. Since unquenching requires a huge increase in computing time it is prudent to use coarser (less time consuming) lattice spacings. This in turn requires improved actions to control the discretization effects. CP-PACS use a perturbatively improved quark and gluon action and extrapolate to the continuum limit. The perturbative

GROUP	lattice spacings	$a \to 0$	$m_s^{\overline{MS}}(2\text{GeV})$
APE '98 [12]	2	no	121(13)
FNAL '96 [6]	3	yes	95(16)
LANL '96 [7]	3	yes	100(21)(10)
JLQCD '99 [13]	4	yes	106(7.1)
QCDSF '99 [14]	3	yes	105(4)
ALPHA/UKQCD '99 [15]	4	yes	97(4)

Table 29.1: Quenched lattice results. The APE result is obtained at $a^{-1} =$ 2.6 GeV.

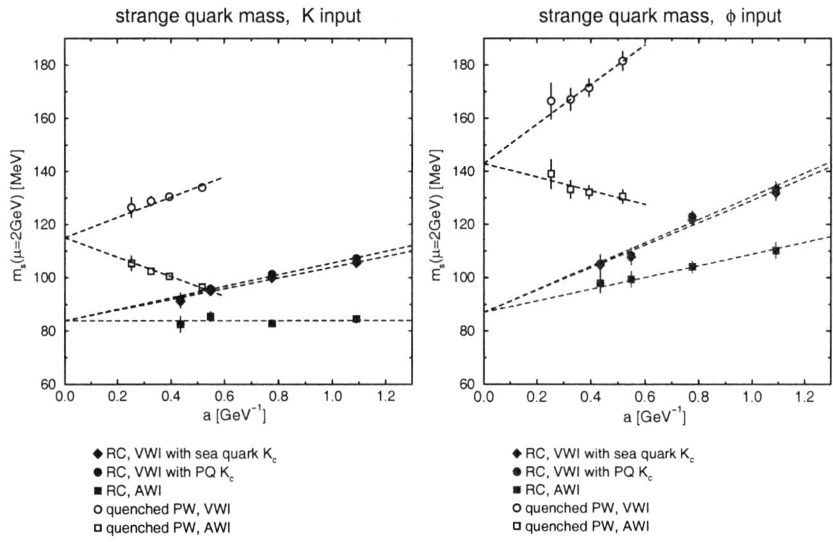

Figure 29.1: RC is the RG-improved gluon and Clover action (with two definitions of quark mass in the chiral limit), PW are the quenched Wilson results already discussed.

renormalization is reliable with a remaining perturbative error of $\mathcal{O}(\alpha_s^2)$, for Wilson fermions. The final result is $m_s^{\overline{MS}}(2\,\text{GeV}) = 84(7)$ MeV. Although this result disagrees with bounds derived from the positivity of the spectral function [18] it remains unclear at what scale, μ, perturbative QCD and thus the bound itself become reliable. CP-PACS conclude that unquenching lowers m_s—compare the filled and open symbols in fig. 29.1. As in the quenched case the VWI and AWI methods differ at finite lattice spacing but extrapolate to the same result—compare the ○ and □ symbols. Finally, the strange quark mass obtained from the K and ϕ mesons yields consistent continuum values in full QCD: 84(7) MeV and 87(11) MeV respectively.

29.5 Conclusions

There has been much progress this year in lattice calculations of m_s. Current computing power and theoretical understanding are sufficient to determine m_s to great precision. A calculation removing *all* uncertainties would include unquenched simulations, a continuum extrapolation, and nonperturbative renormalization and can be done in the short term. Simulations at $n_f = 2$ and 4 with an interpolation to $n_f = 3$ are also possible.

Finally, I look at the implications for ϵ'/ϵ from current theoretical calculations given the recent lattice calculations of m_s. The dependence is shown in fig. 29.2 from the analytic expression

$$\epsilon'/\epsilon = \mathrm{Im}\lambda_t \cdot \left[c_0 + \left(c_6 \mathcal{B}_6^{(1/2)} + c_8 \mathcal{B}_8^{(3/2)} \right) \left(\frac{M_K}{m_s(m_c) + m_d(m_c)} \right)^2 \right] \quad (29.2)$$

and input from lattice calculations for $\mathcal{B}_i$ [19]. The values of other standard model parameters are from ref. [20]. The lines represent the effect of

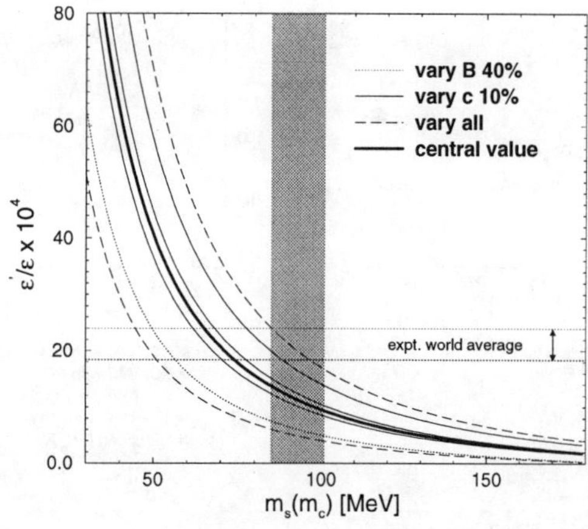

Figure 29.2: ϵ'/ϵ as a function of m_s from Equation. 29.2.

varying the bag parameters and/or the Wilson coefficients and the band is the unquenched m_s from CP-PACS, run to m_c. Further reducing the uncertainty on m_s is more straightforward than for the $\mathcal{B}_i$ and can constrain theoretical calculations of ϵ'/ϵ. Clearly the lower values of m_s give higher ϵ'/ϵ values, in better agreement with experiment!

References

[1] G. Martinelli, Nucl. Phys. (Proc. Suppl.) **73**, 58 (1999).

[2] KTeV Collaboration, Phys. Rev. Lett. **83**, 22 (1999).

[3] M.S. Sozzi, this volume, chapter 8.

[4] G. Buchalla, A. Buras, and M. Lautenbacher, Rev. Mod. Phys. **68**, 1125 (1996); M. Ciuchini *et al.*, Z. Phys. C **68**, 239 (1995); S. Bertolini, J.O. Eeg, and M. Fabbrichesi, hep-ph/9802405.

[5] M. Ciuchini *et al.*, this volume, chapter 28.

[6] B. Gough *et al.*, Phys. Rev. Lett. **79**, 1622 (1997).

[7] T. Bhattacharya and R. Gupta, Phys. Rev. D **55**, 7203 (1997).

[8] T. Blum *et al.*, Phys. Rev. D **60**, 114507 (1999); and M. Wingate, Nucl. Phys. B (Proc. Suppl.) **83–84**, 221 (2000).

[9] CP-PACS Collaboration, S. Aoki *et al.*, Phys. Rev. Lett. **84**, 238 (2000).

[10] M. Lüscher, S. Sint, R. Sommer, and P. Weisz, Nucl. Phys. B **478**, 365 (1996).

[11] G. Martinelli *et al.*, Nucl. Phys. B **445**, 81 (1995).

[12] D. Becirevic *et al.*, Phys. Lett. B **444**, 401 (1998).

[13] JLQCD Collaboration, S. Aoki *et al.*, Phys. Rev. Lett. **82**, 4392 (1999).

[14] M. Göckeler *et al.*, hep-lat/9908005.

[15] J. Garden *et al.*, Nucl. Phys. B **571**, 237 (2000).

[16] MILC Collaboration and APE Collaboration, private communication from G. Martinelli.

[17] CP-PACS Collaboration, R. Burkhalter *et al.*, private communication, and Nucl. Phys. B (Proc. Suppl.) **83–84**, 176 (2000).

[18] L. Lellouch, E. de Rafael, and J. Taron, Phys. Lett. B **414**, 195 (1997).

[19] R. Gupta, hep-ph/98041412.

[20] A. Buras, M. Jamin, and M. Lautenbacher, Phys. Lett. B **389**, 749 (1996).

30

Large-N_c QCD and Weak Matrix Elements

Eduardo de Rafael

Abstract

I report on recent progress [1, 2] in calculating electroweak processes within the framework of QCD in the $1/N_c$ expansion.

30.1 Introduction

In the standard model, the physics of nonleptonic K decays is described by an effective Lagrangian which is the sum of four-quark operators modulated by c-number coefficients (Wilson coefficients). This effective Lagrangian results from integrating out the fields in the standard model with heavy masses ($Z^0, W^{\pm}, t, b$, and c), in the presence of the strong interactions evaluated in perturbative QCD (pQCD) down to a scale μ below the mass of the charm quark M_c. The scale μ has to be large enough for the pQCD evaluation of the c-number coefficients to be valid and, therefore, it is much larger than the scale at which an effective Lagrangian description in terms of the Nambu-Goldstone degrees of freedom (K, π, and η) of the spontaneous $SU(3)_L \times SU(3)_R$ symmetry breaking (SχSB) is appropriate. Furthermore, the evaluation of the coupling constants of the low-energy effective chiral Lagrangian cannot be made within pQCD because at scales $\mu \lesssim 1\,\mathrm{GeV}$ we enter a regime where SχSB and confinement take place and the dynamics of QCD is then fully governed by nonperturbative phenomena.

The structure of the low-energy effective Lagrangian, in the absence of virtual electroweak interactions, is well-known [3]:

$$\mathcal{L}_{\text{eff}} = \frac{1}{4} f_\pi^2 \,\mathrm{tr} D_\mu U D^\mu U + \cdots + L_{10}\,\mathrm{tr} U^\dagger F_R^{\mu\nu} U F_{L\mu\nu} + \cdots . \tag{30.1}$$

Here the unitary matrix U collects the meson fields (K, π and η), and F_L, (F_R) denote field-strength tensors associated with external gauge field sources. The dots indicate other terms with the same chiral power counting

$\mathcal{O}(p^4)$ as the L_{10} term and higher-order terms. The important point that I wish to emphasize here is that *the coupling constants of this effective Lagrangian correspond to coefficients of the Taylor expansion in powers of momenta (and quark masses), of specific QCD Green's functions of color singlet quark-currents.* Let us consider as an example, and in the chiral limit where the light quark masses are set to zero, the two-point function ($Q^2 = -q^2$; $L^\mu = \bar{q}\gamma^\mu \frac{1}{2}(1-\gamma_5)q$; $R^\mu = \bar{q}\gamma^\mu \frac{1}{2}(1+\gamma_5)q$)

$$\Pi_{LR}^{\mu\nu}(q) = 2i \int d^4x\, e^{iq\cdot x} \langle 0|\mathrm{T}\left(L^\mu(x) R^\nu(0)^\dagger\right)|0\rangle = (q^\mu q^\nu - g^{\mu\nu}q^2)\Pi_{LR}(Q^2). \tag{30.2}$$

For Q^2 small, $-Q^2\Pi_{LR}(Q^2) = f_\pi^2 + 4L_{10}\, Q^2 + \mathcal{O}(Q^4)$, clearly showing the correspondence stated above.

In the presence of virtual electroweak interactions there appear new couplings in the low-energy effective Lagrangian, such as the term

$$e^2 C\, \mathrm{tr}\left(Q_R U Q_L U^\dagger\right) = -2e^2 C \frac{1}{f_\pi^2}(\pi^+\pi^- + K^+K^-) + \cdots, \tag{30.3}$$

where $Q_R = Q_L = \mathrm{diag}[2/3, -1/3, -1/3]$, showing that, in the presence of the electroweak interactions, the charged pion and kaon fields become massive. The basic complication in evaluating coupling constants like C in eq. (30.3), which originate in loops with electroweak gauge fields, is that *they correspond to integrals over all values of the Euclidean momenta of specific combinations of QCD Green's functions of color singlet quark-currents.* In our particular example, it can be shown [1,4] that

$$C = \frac{-1}{8\pi^2}\frac{3}{4}\int_0^\infty dQ^2\, Q^2 \left(1 - \frac{Q^2}{Q^2 + M_Z^2}\right)\Pi_{LR}(Q^2), \tag{30.4}$$

with Q the Euclidean momentum of the virtual gauge field; the first term in the parenthesis is the well-known [4] contribution from electromagnetism; the second term is the one induced by the weak neutral current [1]. It is clear that the evaluation of coupling constants of this type represents a rather formidable task. As we shall see below, it is possible, however, to proceed further within the framework of the $1/N_c$ expansion in QCD [5].

30.2 Large-N_c QCD and the OPE

In the limit where the number of colors N_c becomes infinite, with $\alpha_s \times N_c$ fixed, the QCD spectrum reduces to an infinite number of zero-width mesonic resonances, and the leading large-N_c contribution to an n-point correlator is given by all the possible tree-level exchanges of these resonances in the various channels. In this limit, the analytical structure of an n-point function is very simple: the singularities in each channel consist only of a succession of *simple poles.* For example, in the case of Π_{LR} in eq. (30.2),

$$-Q^2\Pi_{LR}(Q^2) = f_\pi^2 + \sum_A f_A^2 M_A^2 \frac{Q^2}{M_A^2 + Q^2} - \sum_V f_V^2 M_V^2 \frac{Q^2}{M_V^2 + Q^2}, \tag{30.5}$$

where the sums extend over all vector (V) and axial-vector (A) states. Furthermore, in the chiral limit, the operator product expansion (OPE) applied to the correlation function $\Pi_{LR}(Q^2)$ implies

$$\lim_{Q^2\to\infty} Q^2\Pi_{LR}(Q^2) \to 0, \qquad \lim_{Q^2\to\infty} Q^4\Pi_{LR}(Q^2) \to 0, \tag{30.6}$$

and [6]

$$\lim_{Q^2\to\infty} Q^6\Pi_{LR}(Q^2) = -4\pi^2 \left(\frac{\alpha_s}{\pi} + \mathcal{O}(\alpha_s^2)\right) \langle\bar{\psi}\psi\rangle^2 . \tag{30.7}$$

The first two relations result in the two Weinberg sum rules

$$\sum_V f_V^2 M_V^2 - \sum_A f_A^2 M_A^2 = f_\pi^2 \quad \text{and} \quad \sum_V f_V^2 M_V^4 - \sum_A f_A^2 M_A^4 = 0 . \tag{30.8}$$

There is in fact a new set of constraints that emerge in the large-N_c limit that relate order parameters of the OPE to couplings and masses of the narrow states. In our example, we have from eqs. (30.5) and (30.7) that

$$\sum_V f_V^2 M_V^6 - \sum_A f_A^2 M_A^6 = -4\pi^2 \left(\frac{\alpha_s}{\pi} + \mathcal{O}(\alpha_s^2)\right) \langle\bar{\psi}\psi\rangle^2. \tag{30.9}$$

On the other hand, the coupling constants of the low-energy Lagrangian in the strong interaction sector are also related to couplings and masses of the narrow states of the large-N_c QCD spectrum; *e.g.*,

$$-4L_{10} = \sum_V f_V^2 - \sum_A f_A^2 . \tag{30.10}$$

It is to be remarked that the convergence of the integral in eq. (30.4) in the large-N_c limit is guaranteed by the two Weinberg sum rules in eqs. (30.8). However, in order to obtain a numerical estimate, and in the absence of an explicit solution of QCD in the large-N_c limit, one still needs to make further approximations. Partly inspired by the phenomenological successes of "vector meson dominance" in predicting, *e.g.*, the low-energy constants of the effective chiral Lagrangian [7], we have recently proposed [8] to consider the approximation to large-N_c QCD that restricts the hadronic spectrum to a minimal pattern, compatible with the short-distance properties of the QCD Green's functions that govern the observable(s) one is interested in. In the channels with J^P quantum numbers 1^- and 1^+ this minimal pattern, in the cases we have discussed so far, is the one with a spectrum that consists of a hadronic lowest energy narrow state and treats the rest of the narrow states as a large-N_c pQCD continuum, the onset of the continuum being fixed by consistency constraints from the OPE, like the absence of dimension $d = 2$ operators. We call this the *lowest meson dominance* (LMD) approximation to large-N_c QCD. The basic observation here is that *order parameters of SχSB in QCD have a smooth behavior at short distances.* For example, in the case of the function Π_{LR}

and, therefore, the coupling C, this is reflected by the fact that (in the chiral limit) the pQCD continuum contributions in the V sum and the A sum in eq. (30.5) cancel each other. The evaluation of the constant C in eq. (30.4) in this approximation corresponds to a mass difference $\Delta m_\pi = 4.9\,\text{MeV}$, remarkably close to the experimental result: $\Delta m_\pi|_{\text{exp.}} = 4.59\,\text{MeV}$.

30.3 Electroweak Penguin Operators

Within the framework discussed above, we have also shown [1] that the $K \to \pi\pi$ matrix elements of the four-quark operator

$$Q_7 = 6(\bar{s}_L \gamma^\mu d_L) \sum_{q=u,d,s} e_q(\bar{q}_R \gamma_\mu q_R), \tag{30.11}$$

generated by the electroweak penguinlike diagrams of the standard model, can be calculated to first nontrivial order in the chiral expansion and in the $1/N_c$ expansion. What is needed here is the bosonization of the operator Q_7 to next-to-leading order in the $1/N_c$ expansion. The problem turns out to be entirely analogous to the bosonization of the operator $Q_{LR} \equiv (\bar{q}_L \gamma^\mu Q_L q_L)(\bar{q}_R \gamma^\mu Q_R q_R)$, which governs the electroweak $\pi^+ - \pi^0$ mass difference discussed above. Because of the LR structure, the factorized component of Q_7, which is leading in $1/N_c$, cannot contribute to order $\mathcal{O}(p^0)$ in the low-energy effective Lagrangian. The first $\mathcal{O}(p^0)$ contribution from this operator is next-to-leading in the $1/N_c$ expansion and is given by an integral, $[\left(\lambda_L^{(23)}\right)_{ij} = \delta_{i2}\delta_{3j}\ (i,j = 1,2,3)]$,

$$Q_7 \to -3ig_{\mu\nu} \int \frac{d^4q}{(2\pi)^4} \Pi_{LR}^{\mu\nu}(q)\ \text{tr}\left(U\lambda_L^{(23)}U^\dagger Q_R\right), \tag{30.12}$$

involving the *same* two-point function as in eq. (30.2). Although the resulting B factors of $\Delta I = 1/2$ and $\Delta I = 3/2$ transitions are found to depend only logarithmically on the matching scale μ, their actual numerical values turn out to be rather sensitive to the precise choice of μ in the GeV region. Furthermore, because of the normalization to the vacuum saturation approximation (VSA) inherent in the (rather disgraceful) conventional definition of B factors, there appears a spurious dependence on the light quark masses as well. In fig. 30.1 we show our prediction for the ratio

$$\widetilde{B}_7^{(3/2)} \equiv \frac{\langle \pi^+|Q_7^{(3/2)}|K^+\rangle}{\langle \pi^+|Q_7^{(3/2)}|K^+\rangle_0^{\text{VSA}}}, \tag{30.13}$$

versus the matching scale μ defined in the $\overline{MS}$ scheme. This is the ratio considered in recent lattice QCD calculations [9]. [In fact, the lattice definition of $\widetilde{B}_7^{(3/2)}$ uses a current algebra relation between the $K \to \pi\pi$ and the $K \to \pi$ matrix elements that is only valid at order $\mathcal{O}(p^0)$ in the chiral expansion.] In eq. (30.13), the matrix element in the denominator is evaluated in the chiral limit, as indicated by the subscript "0."

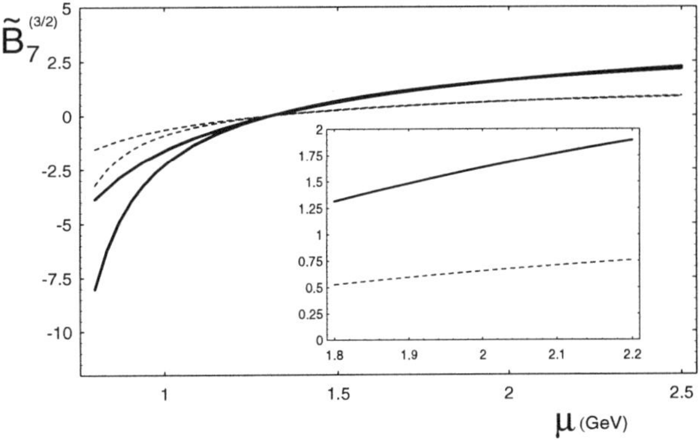

Figure 30.1: The $\tilde{B}_7^{(3/2)}$ factor in eq. (30.13) versus μ in GeV. Solid lines correspond to $(m_s+m_d)(2\,\mathrm{GeV}) = 158\,\mathrm{MeV}$; dashed lines to $(m_s+m_d)(2\,\mathrm{GeV}) = 100\,\mathrm{MeV}$.

30.4 Decay of Pseudoscalars into Lepton Pairs

The processes $\pi \to e^+e^-$ and $\eta \to l^+l^-$ $(l = e, \mu)$ are dominated by the exchange of two virtual photons. It is then useful to consider the ratios $(P = \pi^0, \eta)$

$$R(P \to \ell^+\ell^-) = \frac{Br(P \to \ell^+\ell^-)}{Br(P \to \gamma\gamma)} = 2\left(\frac{\alpha m_\ell}{\pi M_P}\right)^2 \beta_\ell(M_P^2)\,|\mathcal{A}(M_P^2)|^2, \tag{30.14}$$

with $\beta_\ell(s) = \sqrt{1 - 4m_\ell^2/s}$. To lowest order in the chiral expansion, the unknown dynamics in the amplitude $\mathcal{A}(M_P^2)$ depends entirely on a low-energy coupling constant χ. We have recently shown [2] that this constant can be expressed as an integral over the three-point function

$$\int d^4x \int d^4y\, e^{iq_1\cdot x} e^{iq_2\cdot y} < 0\,|\,T\{j_\mu^{\mathrm{em}}(x) j_\nu^{\mathrm{em}}(y) P^3(0)\}\,|\,0 > \\ = \frac{2}{3}\,\epsilon_{\mu\nu\alpha\beta}\, q_1^\alpha q_2^\beta\, \mathcal{H}(q_1^2, q_2^2, (q_1+q_2)^2), \tag{30.15}$$

involving the electromagnetic current j_μ^{em} and the density current $P^3 = \frac{1}{2}(\bar{u}i\gamma_5 u - \bar{d}i\gamma_5 d)$. More precisely, ($d$= space-time dimension),

$$\frac{\chi(\mu)}{32\pi^4}\frac{<\bar{\psi}\psi>}{F_\pi^2} = -\left(1 - \frac{1}{d}\right)\int \frac{d^dq}{(2\pi)^d}\left(\frac{1}{q^2}\right)^2 \\ \times \lim_{(p'-p)^2 \to 0} (p'-p)^2 \left[\mathcal{H}(q^2, q^2, (p'-p)^2) - \mathcal{H}(0, 0, (p'-p)^2)\right]. \tag{30.16}$$

The evaluation of this coupling in the LMD approximation to large-N_c QCD, which we have discussed above, leads to the result $\chi^{\mathrm{LMD}}(\mu = M_V) = 2.2 \pm 0.9$. The corresponding branching ratios are shown in table 30.1.

R	LMD	Experiment
$R(\pi^0 \to e^+e^-) \times 10^8$	6.2 ± 0.3	7.13 ± 0.55 [10]
$R(\eta \to \mu^+\mu^-) \times 10^5$	1.4 ± 0.2	1.48 ± 0.22 [11]
$R(\eta \to e^+e^-) \times 10^8$	1.15 ± 0.05	?

Table 30.1: Ratios $R(P \to \ell^+\ell^-)$ in eq. (30.14) obtained within the LMD approximation to large-N_C QCD and the comparison with available experimental results.

It was shown in ref. [12] that, when evaluated within the chiral $U(3)$ framework and in the $1/N_c$ expansion, the $|\Delta S| = 1$ $K_L^0 \to \ell^+\ell^-$ transitions can also be described by an expression similar to eq. (30.14) with an effective constant $\chi_{K_L^0}$ containing an additional piece from the short-distance contributions [13]. The most accurate experimental determination [14] gives $Br(K_L^0 \to \mu^+\mu^-) = (7.18 \pm 0.17) \times 10^{-9}$. In the framework of the $1/N_c$ expansion and using the experimental branching ratio [11] $Br(K_L^0 \to \gamma\gamma) = (5.92 \pm 0.15) \times 10^{-4}$, this leads to a unique solution for an *effective* $\chi_{K_L^0} = 5.17 \pm 1.13$. Furthermore, following ref. [12], $\chi_{K_L^0} = \chi - \mathcal{N}\ \delta\chi_{SD}$ where $\mathcal{N} = (3.6/g_8 c_{\text{red}})$ normalizes the $K_L^0 \to \gamma\gamma$ amplitude. The coupling g_8 governs the $\Delta I = 1/2$ rule, the constant c_{red} is defined in ref. [12] and $\delta\chi_{SD}^{\text{Standard}} = (+1.8 \pm 0.6)$ is the short-distance contribution in the standard model [13]. Therefore, a test of the *short-distance* contribution to this process completely hinges on our understanding of the *long-distance* constant $\mathcal{N}$ and therefore of the $\Delta I = 1/2$ rule in the $1/N_c$ expansion. Moreover, c_{red} is regrettably very unstable in the chiral and large-N_c limits, a behavior that surely points towards the need to have higher-order corrections under control. The analysis of ref. [12] uses $c_{\text{red}} \simeq +1$ and $g_8 \simeq 3.6$, where these numbers are obtained phenomenologically by requiring agreement with the two-photon decay of K_L^0, π^0, η, and η' as well as $K \to 2\pi, 3\pi$. Should we use these values of c_{red} and g_8 with our result $\chi^{\text{LMD}}(\mu = M_V) = 2.2 \pm 0.9$, we would obtain $\chi_{K_L^0} = 0.4 \pm 1.1$, corresponding to a ratio $R(K_L^0 \to \mu^+\mu^-) = (2.24 \pm 0.41) \times 10^{-5}$, which is 2.5σ above the experimental value $R(K_L^0 \to \mu^+\mu^-) = (1.21 \pm 0.04) \times 10^{-5}$.

Acknowledgments

It is a pleasure to thank my colleagues Marc Knecht, Santi Peris, and Michel Perrottet for a very pleasant collaboration. This research was supported, in part, by TMR, EC-Contract No. ERBFMRX-CT980169.

References

[1] M. Knecht, S. Peris, and E. de Rafael, Phys. Lett. B **443**, 255 (1998); *ibid.* Phys. Lett. B **457**, 227 (1999).

[2] M. Knecht, S. Peris, M. Perrottet, and E. de Rafael, Phys. Rev. Lett. **83**, 5230 (1999).

[3] S. Weinberg, Phys. Rev. **18**, 507 (1967); J. Gasser and H. Leutwyler, Nucl. Phys. B **250**, 465 (1984).

[4] T. Das *et al.*, Phys. Rev. Lett. **18**, 759 (1967) .

[5] G. t'Hooft, Nucl. Phys. B **72**, 461 (1974); *ibid.* B **75**, 461 (1974); E. Witten, Nucl. Phys. B **160**, 57 (1979); M. Knecht and E. de Rafael, Phys. Lett. B **424**, 335 (1998).

[6] M.A. Shifman, A.I. Vainshtein, and V.I. Zakharov, Nucl. Phys. B **147**, 385, 447 (1979).

[7] G. Ecker, J. Gasser, A. Pich, and E. de Rafael, Nucl. Phys. B **321**, 311 (1989); G. Ecker, J. Gasser, H. Leutwyler, A. Pich, and E. de Rafael, Phys. Lett. B **223**, 425 (1989).

[8] S. Peris, M. Perrottet, and E. de Rafael, JHEP **05**, 011 (1998). See also M. Golterman and S. Peris, hep-ph/9908252.

[9] L. Conti *et al.*, Phys. Lett. B **421**, 273 (1998); C.R. Allton *et al.*, hep-lat/9806016; L. Lellouch and C.-J. David Lin, hep-lat/9809142.

[10] A. Alavi-Harati *et al.*, hep-exp/9903007, and references therein.

[11] C. Caso *et al.*, Eur. Phys. J. C **3**, 1 (1998).

[12] D. Gómez Dumm and A. Pich, Phys. Rev. Lett. **80**, 4633 (1998).

[13] G. Buchalla and A.J. Buras, Nucl. Phys. B **412**, 106 (1994); A. J. Buras and R. Fleischer, in *Heavy Flavors II*, ed. A. J. Buras and M. Lindner, hep-ph/9704376.

[14] D. Ambrose, this volume, chapter 41.

31

Penguin Amplitudes: Charming Contributions

M. Ciuchini, E. Franco, G. Martinelli, and L. Silvestrini

Abstract

We briefly introduce the Wick-contraction parametrization of hadronic matrix elements and discuss some applications to B and K physics.

In spite of the progress in nonperturbative techniques, the computation of hadronic matrix elements is still an open problem, particularly when the final state contains more than one meson. In this case, methods based on Euclidean field theory, such as QCD sum rules or lattice QCD, have serious difficulties in computing physical amplitudes [1, 2]. Besides the standard parametrization of hadronic matrix elements in terms of B parameters, it is useful for phenomenological studies to introduce a different parametrization based on the contractions of quark fields inside the matrix element. In the following, we briefly discuss the Wick-contraction parametrization introduced within the framework of nonleptonic B decays in ref. [3].

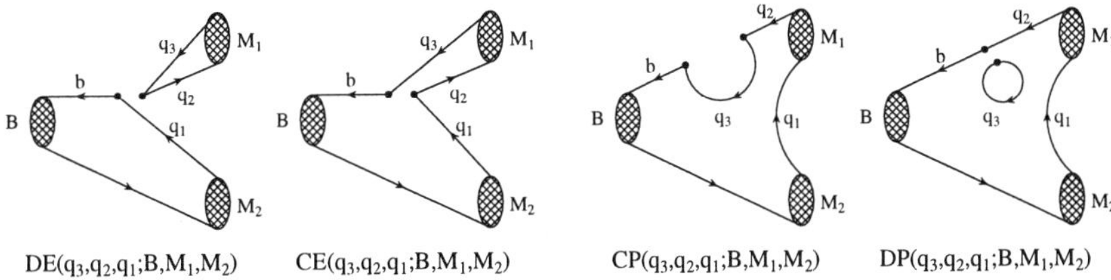

Figure 31.1: Emission and penguin Wick-contraction topologies. The double dots denote the operator insertion.

To be concrete, let us illustrate how this parametrization works in a few examples taken from B physics. Consider the Cabibbo-allowed decay $B^+ \to \bar{D}^0\pi^+$. Only two operators of the $\Delta B = 1$ effective weak Hamiltonian contribute to this amplitude, namely

$$\langle \bar{D}^0\pi^+|Q_1^{\Delta C=1}|B^+\rangle \;=\; \langle \bar{D}^0\pi^+|\bar{b}\gamma_\mu(1-\gamma_5)d\;\bar{u}\gamma^\mu(1-\gamma_5)c|B^+\rangle,$$

$$\langle \bar{D}^0\pi^+|Q_2^{\Delta C=1}|B^+\rangle = \langle \bar{D}^0\pi^+|\bar{b}\gamma_\mu(1-\gamma_5)c\,\bar{u}\gamma^\mu(1-\gamma_5)d|B^+\rangle . \quad (31.1)$$

In this particularly simple example, the quark fields in the operators can be contracted only according to the emission topologies DE and CE, shown in fig. 31.1. In the absence of a method for computing them, these contractions can be taken as complex parameters in phenomenological studies. The matrix elements can be rewritten as

$$\begin{aligned} \langle \bar{D}^0\pi^+|Q_1^{\Delta C=1}|B^+\rangle &= CE_{LL}(d,u,c;B^+,\bar{D}^0,\pi^+) \\ &\quad +DE_{LL}(c,u,d;B^+,\pi^+,\bar{D}^0), \\ \langle \bar{D}^0\pi^+|Q_2^{\Delta C=1}|B^+\rangle &= CE_{LL}(c,u,d;B^+,\bar{D}^0,\pi^+) \\ &\quad +DE_{LL}(d,u,c;B^+,\pi^+,\bar{D}^0). \end{aligned} \quad (31.2)$$

The subscript LL refers to the Dirac structure of the inserted operators. In general there are 14 different topologies [3–5]. Of course, in order to be predictive, one needs to introduce relations among different parameters given by dynamical assumptions based on flavor symmetries, chiral properties, heavy quark expansion, $1/N$ expansion, etc. This approach proves particularly useful for studying the $\Delta S = 1$ B decays. For instance, let us consider the decay $B^+ \to K^+\pi^0$. Its amplitude receives contributions from all the operators of the $\Delta B = 1$, $\Delta S = 1$ effective Hamiltonian. We consider only the matrix elements of operators that are both proportional to the largest Wilson coefficients C_1 and C_2 and leading order in the Cabibbo angle. They read

$$\begin{aligned} \langle K^+\pi^0|Q_1^c|B^+\rangle &= \langle K^+\pi^0|\bar{b}\gamma_\mu(1-\gamma_5)s\,\bar{c}\gamma^\mu(1-\gamma_5)c|B^+\rangle \\ &= DP_{LL}(c,s,u;B^+,K^+,\pi^0), \\ \langle K^+\pi^0|Q_2^c|B^+\rangle &= \langle K^+\pi^0|\bar{b}\gamma_\mu(1-\gamma_5)c\,\bar{c}\gamma^\mu(1-\gamma_5)s|B^+\rangle \\ &= CP_{LL}(c,s,u;B^+,K^+,\pi^0). \end{aligned} \quad (31.3)$$

The penguin contractions CP and DP are shown in fig. 31.1. We stress the difference between penguin operators, which we have neglected here, and penguin contractions, which can contribute to the matrix elements of any operator. This kind of nonperturbative contribution, called "charming penguins" in ref. [3], could dominate $\Delta S = 1$, $\Delta C = 0$ B decays because other contributions are either proportional to the small Wilson coefficients C_3–C_{10} or are doubly Cabibbo suppressed, as in the case of the factorizable emission topologies of $Q^u_{1,2}$. A detailed analysis of nonleptonic B decays in this framework can be found in refs. [4, 6, 7]. The presence of "charming penguin" contributions is likely to make the naïve factorization approach fail in describing this class of decays.

A different but related parametrization of hadronic matrix elements has been recently proposed in ref. [5]. In this approach, the parameters are the suitable combinations of Wick contractions and Wilson coefficients that are renormalization scale and scheme independent. In this way, the relations among contractions enforced by the renormalization group equations are explicit. Besides, the phenomenological determination of the parameters do not depend on the choice of the Wilson coefficients. On the other hand, imposing relations among parameters based on dynamical assumptions may be more involved.

Matrix-element parametrizations are less useful when applied to K decays, because there are few decay channels to fix the parameters and test the assumptions.[1] In addition, chiral relations allow the connection between matrix elements with one pion and those with two or more pions in the final states, the former being calculable with lattice QCD. However, a reliable lattice determination of $\langle\pi\pi|Q_6|K\rangle$, the dominant contribution to ε'/ε [8], is currently missing [9].

Let us apply the Wick-contraction parametrization to $K \to \pi\pi$ and verify whether there could be a connection between the long-standing problem of the $\Delta I = 1/2$ rule and a large value of the matrix element of Q_6, as suggested by the recent measurement of ε'/ε. In terms of the Buras-Silvestrini parameters [5], the amplitudes $K \to \pi\pi$ with definite isospin are

$$
\begin{aligned}
\mathrm{Re}A_2 &\sim \frac{1}{3}\left(E_1 + E_2\right), \\
\mathrm{Re}A_0 &\sim \left(-\frac{2}{3}E_1 + \frac{1}{3}E_2 - A_2 + P_1' + P_3'\right), \\
\mathrm{Im}A_0 &\sim -\left(P_1 + P_3\right),
\end{aligned}
\tag{31.4}
$$

where $E_{1,2}$ are the emission parameters, A_2 is built with annihilations and

$$
\begin{aligned}
P_1 \equiv\ & \sum_{i=2}^{5}\Big(y_{2i-1}\langle Q_{2i-1}\rangle_{CE} + y_{2i}\langle Q_{2i}\rangle_{DE}\Big) \\
& + \sum_{i=3}^{10}\Big(y_i\langle Q_i\rangle_{CP} + y_i\langle Q_i\rangle_{DP}\Big) \\
& + \sum_{i=2}^{5}\Big(y_{2i-1}\langle Q_{2i-1}\rangle_{CA} + y_{2i}\langle Q_{2i}\rangle_{DA}\Big), \\
P_3 \equiv\ & \sum_{i=2}^{5}\Big(y_{2i-1}\langle Q_{2i-1}\rangle_{DA} + y_{2i}\langle Q_{2i}\rangle_{CA}\Big) \\
& + \sum_{i=3}^{10}\Big(y_i\langle Q_i\rangle_{CPA} + y_i\langle Q_i\rangle_{DPA}\Big),
\end{aligned}
\tag{31.5}
$$

are the penguinlike parameters. The notation $\langle Q_i\rangle_{CE}$ refers to the connected emission with the insertion of the operator Q_i, etc.

Neglecting annihilations, as suggested by large-N counting or by CPS + chiral symmetries [10], we are left with four parameters and three measured

[1] Indeed, in the case of $K \to \pi\pi$, there are only two complex amplitudes corresponding to the $\pi\pi(I = 0, 2)$ final states.

quantities. It is unlikely that ReA_0 is dominated by emissions, since the large ratio ReA_0/ReA_2 would require large cancellations between E_1 and E_2; see eq. (31.4). Therefore, in the most natural scenario, both ReA_0 and ImA_0 are dominated by penguin parameters. Notice that P_1 and P_1' are different, so that no parametric relation between ReA_0 and ImA_0 can be established. However, the following relation holds:

$$P_1' = z_1 \langle Q_1 \rangle_{DP} + z_2 \langle Q_2 \rangle_{CP} + P_1(y \to z), \tag{31.6}$$

where y_i and z_i are the Wilson coefficients of the 3-flavor effective weak Hamiltionian. Given this relation, we can envisage a dynamical mechanism to connect the two parameters. Let us assume that $P_1' \gg P_1(y \to z)$ and that $\langle Q_{1,2} \rangle_{DP}$ and $\langle Q_{5,6} \rangle_{DP}$ share the same enhancement.[2] Arguments may be provided to assume that [11]

$$f = \langle Q_1 \rangle_{DP} \sim 1/N_c \langle Q_2 \rangle_{CP} \sim -\langle Q_5 \rangle_{DP} \sim -1/N_c \langle Q_6 \rangle_{CP}. \tag{31.7}$$

By using the experimental value of ReA_0 and factorizing the emission contractions, we extract f, from which we derive

$$B_1 = -9, \; B_2 = 7.5, \; B_5 = B_6 = 1.5. \tag{31.8}$$

It is interesting that the same mechanism enhances the B parameters entering ReA_0 by a factor of ~ 10 and those of ImA_0 by a factor of ~ 2, as required by the theoretical calculations to explain the experimental data.

Alternatively, we could assume $P_1' \sim P_1(y \to z)$ in eq. (31.6), namely that everything comes from the penguin operators Q_5 and Q_6. This is the old suggestion of SVZ [12]. Using perturbative coefficients, it is possible to show that this requires $B_6 \sim 20$ in order to fit (ReA_0)$_{\text{exp}}$. Such a large value is excluded by the measurement of ε'/ε.

To summarize, a connection between the enhancement of ReA_0 and a large value of ε'/ε cannot be established without some assumption on the long-distance dynamics. We have presented a simple example, which assumes penguin-contraction dominance, that shows the correct pattern of enhancements. In this respect, models could give some insight [13], but quantitative predictions may prove hard to produce. Hopefully, nonperturbative renormalization and new computing techniques will help overcoming the problems which prevent the lattice computation of ReA_0 and B_6 [14].

Acknowledgments

M. C. and L. S. thank A. Buras for useful discussions and excellent steaks, beer, and particularly desserts. G. M. looks forward to acknowledging the same in the future.

[2]We found that these two assumptions are compatible.

References

[1] B.Y. Blok and M.A. Shifman, Sov. J. Nucl. Phys. **45**, 522 (1987).

[2] L. Maiani and M. Testa, Phys. Lett. B **245**, 585 (1990).

[3] M. Ciuchini, E. Franco, G. Martinelli, and L. Silvestrini, Nucl. Phys. B **501**, 271 (1997).

[4] M. Ciuchini *et al.*, Nucl. Phys. B **512**, 3 (1998).

[5] A.J. Buras and L. Silvestrini, hep-ph/9812392.

[6] M. Ciuchini *et al.*, Nucl. Instrum. Meth. A **408**, 28 (1998).

[7] M. Ciuchini, R. Contino, E. Franco, and G. Martinelli, Eur. Phys. J. C **9**, 43 (1999).

[8] A.J. Buras, this volume, chapter 5.

[9] D. Pekurovsky and G. Kilcup, hep-lat/9812019.

[10] C. Bernard, T. Draper, A. Soni, H.D. Politzer, and M.B. Wise, Phys. Rev. D **32**, 2343 (1985).

[11] M. Ciuchini, E. Franco, G. Martinelli, and L. Silvestrini, in preparation.

[12] A.I. Vainshtein, V.I. Zakharov, and M.A. Shifman, Sov. Phys. JETP **45**, 670 (1977).

[13] S. Bertolini, J.O. Eeg, M. Fabbrichesi, and E.I. Lashin, Nucl. Phys. B **514**, 63 (1998); *ibid.* **514**, 93 (1998); T. Hambye, G.O. Kohler, and P.H. Soldan, hep-ph/9902334; T. Hambye, G.O. Kohler, E.A. Paschos, and P.H. Soldan, hep-ph/9906434; H.-Y. Cheng, hep-ph/9906403; A.A. Belkov, G. Bohm, A.V. Lanyov, and A.A. Moshkin, hep-ph/9907335; J. Bijnens, this volume, chapter 37.

[14] T. Blum, private communication.

32

Final State Interactions: From Strangeness to Beauty

Alexey A. Petrov

Abstract

I give a brief review of final state interactions (FSI) in meson decays. I describe possible effects of FSI in K, D, and B systems, paying particular attention to the description of the heavy meson decays. Available theoretical methods for dealing with the effects of FSI are discussed.

32.1 Motivation

Final state interactions (FSI) play an important role in meson decays. The effect of FSI might be significant, especially if one is interested in rare decays, where the presence of the stronger channel, coupled to the channel of interest via the final state interaction, might significantly affect the prediction in which FSIs are not accounted for. This obvious observation, of course, does not exhaust the list of the motivations for better understanding of FSI. Many important observables that are sensitive to new physics could also receive contributions from the final state rescatterings. An excellent example is provided by the T-violating lepton polarizations in K decays (such as $K \to \pi l \nu$ and $K \to \gamma l \nu$) that are not only sensitive to new physics but could also be induced by the electromagnetic FSI. However, the most profound effect of FSI is in the decays of B and D mesons used for studies of direct CP violation, where one compares the rates of a B or D meson decay with the charged conjugated process [1]. The corresponding asymmetries, in order to be nonzero, require two different final states produced by different weak amplitudes that can go into each other by a strong interaction rescattering and therefore depend on both weak CKM phase and strong rescattering phase provided by the FSI. Thus, FSI directly affect the asymmetries, and their size can be interpreted in terms of fundamental parameters *only* if these FSI phases are calculable. In all of these examples FSI complicates the interpretation of experimental observables in terms of fundamental parameters [2, 3]. In this talk I review the progress in understanding of FSI in meson decays.

The difference of the physical picture at the energy scales relevant to K, D, and B decays calls for a specific descriptions for each class of decays. For instance, the relevant energy scale in K decays is $m_K \ll 1$ GeV. With such a low energy release only a few final state channels are available. This significantly simplifies the theoretical understanding of FSI in kaon decays. In addition, chiral symmetry can also be employed to assist the theoretical description of FSI in K decays. In D decays, the relevant scale is $m_D \sim 1$ GeV. This region is populated by the light quark resonances, so one might expect their significant influence on the decay rates and CP-violating asymmetries. No model-independent description of FSI is available, but it is hinted experimentally that the number of available channels is still limited, allowing for a modeling of the relevant QCD dynamics. Finally, in B decays, where the relevant energy scale $m_B \gg 1$ GeV is well above the resonance region, the heavy quark limit might prove useful.

32.2 Some Formal Aspects of FSI

Final state interactions in $A \to f$ arise as a consequence of the unitarity of the $\mathcal{S}$ matrix, $\mathcal{S}^\dagger \mathcal{S} = 1$, and involve the rescattering of physical particles in the final state. The $\mathcal{T}$ matrix, $\mathcal{T} = i\,(1 - \mathcal{S})$, obeys the optical theorem:

$$\mathcal{D}isc\ \mathcal{T}_{A\to f} \equiv \frac{1}{2i}\left[\langle f|\mathcal{T}|A\rangle - \langle f|\mathcal{T}^\dagger|A\rangle\right] = \frac{1}{2}\sum_i \langle f|\mathcal{T}^\dagger|i\rangle\langle i|\mathcal{T}|A\rangle, \quad (32.1)$$

where $\mathcal{D}isc$ denotes essentially the imaginary part. Using CPT in the form $\langle \bar{f}|\mathcal{T}|\bar{A}\rangle^* = \langle \bar{A}|\mathcal{T}^\dagger|\bar{f}\rangle = \langle f|\mathcal{T}^\dagger|A\rangle$ this can be tranformed into the more intuitive form

$$\langle \bar{f}|\mathcal{T}|\bar{A}\rangle^* = \sum_i \langle f|\mathcal{S}^\dagger|i\rangle\langle i|\mathcal{T}|A\rangle. \quad (32.2)$$

Here, the states $|i\rangle$ represent all possible final states (including $|f\rangle$) that can be reached from the state $|A\rangle$ by the weak transition matrix $\mathcal{T}$. The right-hand side of eq. (32.2) can then be viewed as a weak decay of $|A\rangle$ into $|i\rangle$ followed by a strong rescattering of $|i\rangle$ into $|f\rangle$. Thus, we identify $\langle f|\mathcal{S}^\dagger|i\rangle$ as a FSI rescattering of particles. Notice that if $|i\rangle$ is an eigenstate of $\mathcal{S}$ with a phase $e^{2i\delta}$, we have

$$\langle \bar{i}|\mathcal{T}|\bar{A}\rangle^* = e^{-2i\delta_i}\langle i|\mathcal{T}|A\rangle, \quad (32.3)$$

which implies equal rates for the charge-conjugated decays.[1] Also

$$\langle \bar{i}|\mathcal{T}|\bar{A}\rangle = e^{i\delta}T_i, \quad \langle i|\mathcal{T}|A\rangle = e^{i\delta}T_i^*. \quad (32.4)$$

[1] This fact will be important in the studies of CP-violating asymmetries as no CP asymmetry is generated in this case.

The matrix elements T_i are assumed to be the "bare" decay amplitudes and have no rescattering phases. This implies that these transition matrix elements between charge-conjugated states are just the complex conjugated ones of each other. eq. (32.4) is known as Watson's theorem [4]. Yet, a quite complicated problem of separating "true bare" amplitudes from the "FSI" ones (known as the Omnès problem) still exists.

32.2.1 K Decays

The low scale associated with K decays suggests an effective theory approach of integrating out heavy particles and making use of the chiral symmetry of QCD. This theory has been known for a number of years as chiral perturbation theory (χPT), which makes use of the fact that kaons and pions are the Goldstone bosons of chiral $SU(3)_L \times SU(3)_R$ broken down to $SU(3)_V$, and are the only relevant degrees of freedom at this energy. χPT allows for a consistent description of the strong and electromagnetic FSI in kaon system.

The discussion of strong FSI is naturally included in the χPT calculations of kaon decays processes at one or more loops [5]. In addition, the kaon system is rather unique for its sensitivity to the electromagnetic final state interaction effects. Normally, one expects this class of corrections to be negligibly small. However, in some cases they are still very important. For instance, it is known that in nonleptonic K decays the $\Delta I = 1/2$ isospin amplitude is enhanced compared to the $\Delta I = 3/2$ amplitude by approximately a factor of 22. Since electromagnetism does not respect isospin symmetry, one might expect that electromagnetic FSI might contribute to the $\Delta I = 3/2$ amplitude at the level of $22/137 \sim 20\%$! Of course, some cancelations might actually lower the impact of this class of FSI [6].

There is a separate class of observables that is directly affected by electromagnetic FSI. It includes the T-violating transverse lepton polarizations in the decays $K \to \pi l \nu$ and $K \to l \nu \gamma$

$$P_l^{\perp} = \frac{\vec{s}_l \cdot (\vec{p}_i \times \vec{p}_l)}{|\vec{p}_i \times \vec{p}_l|}, \qquad (32.5)$$

where $i = \gamma, \pi$. These polarizations are usually very small in the standard model, so their observation implies an effect induced by new physics.

A number of parameters of various extensions of the standard model can be constrained via these measurements [7]. It is, however, important to realize that the polarizations as high as $10^{-3}(10^{-6})$ could be generated by the electromagnetic rescattering of the final state lepton and pion [8] or due to other intermediate states. These corrections have been estimated for a number of experimentally interesting final states.

32.2.2 D Decays

The relatively low mass of the charm quark puts the D mesons in the region populated by the higher excitations of the light quark resonances. It is therefore natural to assume that the final state rescattering is dominated by the intermediate resonance states [9]. Unfortunately, no model-independent description exists at this point, but the wealth of experimental results allows for the introduction of testable models of FSI [10]. These models are very important in the studies of direct CP-violating asymmetries

$$A_{CP} = \frac{\Gamma(D \to f) - \Gamma(\bar{D} \to \bar{f})}{\Gamma(D \to f) + \Gamma(\bar{D} \to \bar{f})} \sim \sin\theta_w \sin\delta_s, \tag{32.6}$$

which explicitly depend on the values of both weak (θ_w) and strong (δ_s) phases. In most models of FSI in D decay, the phase δ_s is generated by the width of the nearby resonance and by calculating the imaginary part of loop integral with the final state particles coupled to the nearby resonance.

It is important to realize that the large final state interactions and the presence of the nearby resonances in the D system has an immediate impact on the $D - \bar{D}$ mixing parameters. It is well known that the short-distance contribution to Δm_D and $\Delta\Gamma$ is very small, of the order of 10^{-18} GeV. Nearby resonances can enhance them by one or two orders of magnitude [11]. In addition, they provide a source for quark-hadron duality violations, as they populate the gap between the QCD scale and the scale set by the mass of the heavy quark normally required for the application of heavy quark expansions.

32.2.3 B Decays

In the B system, where the density of the available resonances is large due to the increased energy, a different approach must be employed. One can use the fact that the b quark mass is large compared to the QCD scale and investigate the behavior of final state phases in the $m_b \to \infty$ limit.

Significant energy release available in B decays allows the study of inclusive quantities, for instance inclusive CP-violating asymmetries of the form of eq. (32.6). There, one can use duality arguments to calculate final state phases for charmless B decays using perturbative QCD [12]. Indeed, the $b \to c\bar{c}s$ process, with subsequent final state rescattering of the two charmed quarks into the final state (penguin diagram), does the job, as for the energy release of the order $m_b > 2m_c$ available in b decay, the rescattered c quarks can go on-shell generating a CP-conserving phase and thus $\mathcal{A}_{\rm CP}^{dir}$, which is usually quite small for the experimentally feasible decays, $\mathcal{O}(1\%)$. It is believed that larger asymmetries can be obtained in exclusive decays. However, a simple picture is lost because of the absence of the duality argument.

It is known that scattering of high-energy particles may be divided into "soft" and "hard" parts. Soft scattering occurs primarily in the forward direction with a limited transverse momentum distribution that falls exponentially with a scale of order 0.5 GeV. At higher transverse momentum one encounters the region of hard scattering, which can be described by perturbative QCD. Since final state particles in B decay fly back-to-back, one faces the difficulty of separating the two. It might prove useful to employ unitarity in trying to describe FSI in exclusive B decays.

It is easy to investigate first the *elastic* channel. The inelastic channels have to share a similar behavior in the heavy quark limit, as the unitarity of the $\mathcal{S}$ matrix requires that they are indeed the dominant contributors to soft rescattering. The elastic channel is convenient because of the optical theorem, which connects the forward (imaginary) invariant amplitude $\mathcal{M}$ to the total cross section,

$$\mathcal{I}m\ \mathcal{M}_{f\to f}(s,\ t=0) = 2k\sqrt{s}\sigma_{f\to \text{all}} \sim s\sigma_{f\to \text{all}}, \tag{32.7}$$

where s and t are the usual Mandelstam variables. The asymptotic total cross sections are known experimentally to rise slowly with energy and can be parameterized by the form [13] $\sigma(s) = X\,(s/s_0)^{0.08} + Y\,(s/s_0)^{-0.56}$, where $s_0 = \mathcal{O}(1)$ GeV is a typical hadronic scale. Considering only the imaginary part of the amplitude, and building in the known exponential fall-off of the elastic cross section in t ($t < 0$) [14] by writing

$$i\mathcal{I}m\ \mathcal{M}_{f\to f}(s,t) \simeq i\beta_0 \left(\frac{s}{s_0}\right)^{1.08} e^{bt}, \tag{32.8}$$

one can calculate its contribution to the unitarity relation for a final state $f = ab$ with kinematics $p'_a + p'_b = p_a + p_b$ and $s = (p_a + p_b)^2$:

$$\begin{aligned} \mathcal{D}isc\ \mathcal{M}_{B\to f} &= \frac{-i}{8\pi^2}\int \frac{d^3p'_a}{2E'_a}\frac{d^3p'_b}{2E'_b}\delta^{(4)}(p_B - p'_a - p'_b) \\ &\qquad \times \mathcal{I}m\ \mathcal{M}_{f\to f}(s,t)\mathcal{M}_{B\to f} \\ &= -\frac{1}{16\pi}\frac{i\beta_0}{s_0 b}\left(\frac{m_B^2}{s_0}\right)^{0.08} \mathcal{M}_{B\to f}, \end{aligned} \tag{32.9}$$

where $t = (p_a - p'_a)^2 \simeq -s(1-\cos\theta)/2$, and $s = m_B^2$.

One can refine the argument further, since the phenomenology of high-energy scattering is well accounted for by Regge theory [14]. In the Regge model, scattering amplitudes are described by the exchanges of Regge trajectories (families of particles of differing spin) with the leading contribution given by the Pomeron exchange. Calculating the Pomeron contribution to the elastic final state rescattering in $B \to \pi\pi$ one finds [15]

$$\mathcal{D}isc\ \mathcal{M}_{B\to\pi\pi}|_{\text{Pomeron}} = -i\epsilon\mathcal{M}_{B\to\pi\pi}, \qquad \epsilon \simeq 0.21. \tag{32.10}$$

It is important that the Pomeron-exchange amplitude is seen to be almost purely imaginary. However, of chief significance is the identified weak dependence of ϵ on m_B: the $(m_B^2)^{0.08}$ factor in the numerator is attenuated by the $\ln(m_B^2/s_0)$ dependence in the effective value of b.

The analysis of the elastic channel suggests that, at high energies, FSI phases are *mainly generated by inelastic effects*, which follows from the fact that the high-energy cross section is mostly inelastic. This also follows from the fact that the Pomeron elastic amplitude is almost purely imaginary. Since the study of elastic rescattering has yielded a $\mathcal{T}$ matrix element $\mathcal{T}_{ab\to ab} = 2i\epsilon$, *i.e.*, $\mathcal{S}_{ab\to ab} = 1 - 2\epsilon$, and since the constraint of unitarity of the $\mathcal{S}$ matrix implies that the off-diagonal elements are $\mathcal{O}(\sqrt{\epsilon})$, with ϵ approximately $\mathcal{O}(m_B^0)$ in powers of m_B and numerically $\epsilon < 1$, then the inelastic amplitude must also be $\mathcal{O}(m_B^0)$ with $\sqrt{\epsilon} > \epsilon$. Similar conclusions follow from the consideration of the final state unitarity relations. This complements the old Bjorken picture of heavy meson decay (the dominance of the matrix element by the formation of the small hadronic configuration that grows into the final state pion "far away" from the point where it was produced and does not interact with the soft gluon fields present in the decay; see also [16] for the discussion) by allowing for the rescattering of multiparticle states, production of which is favorable in the $m_b \to \infty$ limit, into the two body final state. Analysis of the final-state unitarity relations in their general form is complicated due to the many contributing intermediate states present at the B mass, but we can illustrate the systematics of inelastic scattering in a two-channel model. It involves a two-body final state f_1 undergoing elastic scattering and a final state f_2 that represents "everything else." As before, the elastic amplitude is purely imaginary, and the scattering can be described in the one-parameter form

$$S = \begin{pmatrix} \cos 2\theta & i\sin 2\theta \\ i\sin 2\theta & \cos 2\theta \end{pmatrix}, \qquad T = \begin{pmatrix} 2i\sin^2\theta & \sin 2\theta \\ \sin 2\theta & 2i\sin^2\theta \end{pmatrix}, \tag{32.11}$$

where we identify $\sin^2\theta \equiv \epsilon$. The unitarity relations become

$$\begin{aligned} \mathcal{D}isc\ \mathcal{M}_{B\to f_1} &= -i\sin^2\theta \mathcal{M}_{B\to f_1} + \frac{1}{2}\sin 2\theta \mathcal{M}_{B\to f_2}, \\ \mathcal{D}isc\ \mathcal{M}_{B\to f_2} &= \frac{1}{2}\sin 2\theta \mathcal{M}_{B\to f_1} - i\sin^2\theta \mathcal{M}_{B\to f_2}. \end{aligned} \tag{32.12}$$

Denoting $\mathcal{M}_1^0$ and $\mathcal{M}_2^0$ to be the decay amplitudes in the limit $\theta \to 0$, an exact solution to eq. (32.12) is given by

$$\mathcal{M}_{B\to f_1} = \cos\theta \mathcal{M}_1^0 + i\sin\theta \mathcal{M}_2^0, \qquad \mathcal{M}_{B\to f_2} = \cos\theta \mathcal{M}_2^0 + i\sin\theta \mathcal{M}_1^0. \tag{32.13}$$

Thus, the phase is given by the inelastic scattering with a result of order

$$\mathcal{I}m\ \mathcal{M}_{B\to f}/\mathcal{R}e\ \mathcal{M}_{B\to f} \sim \sqrt{\epsilon}\ \left(\mathcal{M}_2^0/\mathcal{M}_1^0\right). \tag{32.14}$$

Clearly, for physical B decay, we no longer have a simple one-parameter $\mathcal{S}$ matrix, and, with many channels, cancellations or enhancements are possible for the sum of many contributions. However, the main feature of the above result is expected to remain: Inelastic channels cannot vanish and provide the FSI phase, which is systematically of order $\sqrt{\epsilon}$ and thus does not vanish in the large m_B limit.

A contrasting point of view is taken in a recent calculation [17] that claims that the cancellation among multiparticle states actually occurs. The argument is based on the perturbative factorization of currents (*i.e.*, the absence of infrared singularities, etc.) in the matrix elements of four quark operators in the Bjorken setup and indicates that leading corrections to the factorization result are suppressed by $1/m_b$ (see, however, [18]). Note, however, that almost all of the important long-distance final state rescattering effects involve exchange of global quantum numbers, such as charge or strangeness, and thus are suppressed by $\approx 1/m_B$ [2, 19]. This is easy to see in the Regge description of FSI where this exchange is mediated by the suppressed ρ or higher lying trajectories.

(i) Bounds on the FSI Corrections. In view of the large theoretical uncertainties [20] involved in the calculation of the FSI contributions, it would be extremely useful to find a phenomenological method by which to bound the magnitude of the FSI contribution. The observation of a larger asymmetry would then be a signal for new physics. Here the application of flavor $SU(3)$ flavor symmetry provides powerful methods to obtain a direct upper bound on the FSI contribution [19]. The simplest example involves bounding FSI in $B \to \pi K$ decays using $B^\pm \to K^\pm K$ transitions [2].

(ii) Direct Observation. Another interesting way of studying FSI involves rare weak decays for which the direct amplitude $A(B \to f)$ is suppressed compared to $A(B \to i)$. They offer a tantalizing possibility of the *direct observation* of the effects of FSI.

One of the possibilities involves dynamically suppressed decays that proceed via weak annihilation diagrams. It has been argued that final state interactions, if large enough, can modify the decay amplitudes, violating the expected hierarchy of amplitudes. For instance, it was shown [21] that the rescattering from the dominant tree-level amplitude leads to the suppression of the weak annihilation amplitude by only $\lambda \sim 0.2$ compared to $f_B/m_B \sim \lambda^2$ obtained from the naive quark diagram estimate.

Alternatively, one can study OZI-violating modes, *i.e.*, the modes that cannot be realized in terms of quark diagrams without annihilation of at least one pair of the quarks, like $\overline{B}^0_d \to \phi\phi, D^0\phi$ and $J/\psi\phi$. The unitarity argument implies that this decay can also proceed via the OZI-allowed weak transition followed by final state rescattering into the final state under consideration [22, 23]. In B decays these OZI-allowed steps involve multiparticle intermediate states and might provide a source for violation of the OZI rule. For instance, the FSI contribution can proceed via $\overline{B}^0_d \to \eta^{(\prime)}\eta^{(\prime)} \to \phi\phi$, $\overline{B}^0_d \to D^{*0}\eta^{(\prime)} \to D^0\phi$ and $\overline{B}^0_d \to \psi'\eta^{(\prime)} \to J/\psi\phi$. The intermediate state also includes additional pions. The weak decay into the intermediate state occurs at tree level, through the $(u\bar{u} + d\bar{d})/\sqrt{2}$ component of the $\eta^{(\prime)}$ wave function, whereas the strong scattering into the final state involves the $s\bar{s}$ component. Hence the possibility of using these decay modes as direct probes of the FSI contributions to B decay amplitudes. It is, however, possible to show that there exist strong cancellations [22] among various *two-body* intermediate channels. In the example of $\overline{B}^0_d \to \phi\phi$, the cancellation among η and η' is almost complete, so the effect is of the second order in the $SU(3)$-breaking corrections:

$$Disc\, \mathcal{M}_{B\to\phi\phi} = O(\delta^2, \Delta^2, \delta\Delta) f_\eta F_0 A, \quad \delta = f_{\eta'} - f_\eta, \; \Delta = F_0' - F_0, \quad (32.15)$$

with $A \sim s^{\alpha_0 - 1} e^{i\pi\alpha_0/2}/8b$. This implies that the OZI-suppressed decays provide an excellent probe of the multiparticle FSI. Given the very clear signature, these decay modes could be probed at the upcoming B factories.

32.3 Outlook

One of the main goals of physics of CP violation and meson decay is to correctly extract the underlying parameters of the fundamental Lagrangian that are responsible for these phenomena. The understanding of final state interactions is very important for the success of this program.

References

[1] M. Gronau, this volume, chapter 48.

[2] A.F. Falk, A.L. Kagan, Y. Nir, and A.A. Petrov, Phys. Rev. D **57**, 4290 (1998); M. Neubert, Phys. Lett. B **424**, 152 (1998); D. Atwood and A. Soni, Phys. Rev. D **58**, 036005 (1998).

[3] J. Donoghue, E. Golowich, and A. Petrov, Phys. Rev. D **55**, 2657 (1997).

[4] K.M. Watson, Phys. Rev. **88**, 1163 (1952).

[5] M. Savage, this volume, chapter 16; J. Bijnens, this volume, chapter 37.

[6] V. Cirigliano, J. Donoghue, and E. Golowich, Phys. Lett. B **450**, 241 (1999).

[7] M. Kobayashi *et al.*, Prog. Theor. Phys. **95**, 361 (1996); C.H. Chen, *et al.*, Phys. Rev. D **56** (1997) 6856; G. Wu and J.N. Ng, Phys. Rev. D **55**, 2806 (1997).

[8] A.R. Zhitnitsky, Sov. J. Nucl. Phys. **31**, 529 (1980).

[9] F. Buccella *et al.*, Phys. Lett. B **379**, 249 (1996).

[10] J.L. Rosner, hep-ph/9903543, hep-ph/9905366.

[11] E. Golowich and A.A. Petrov, Phys. Lett. B **427**, 172 (1998).

[12] M. Bander, D. Silverman, and A. Soni, Phys. Rev. Lett. **43**, 242 (1979); J.-M. Gérard and W.-S. Hou, Phys. Rev. D **43**, 2909 (1991); Yu. Dokshitser and N. Uraltsev, JETP Lett. **52** (10), 1109 (1990).

[13] C. Caso *et al.* (PDG), Eur. Phys. J. C **3**, 1 (1998).

[14] P.D.B. Collins, *Introduction to Regge Theory* (Cambridge University Press, Cambridge, England, 1977).

[15] J.F. Donoghue, E. Golowich, A.A. Petrov, and J.M. Soares, Phys. Rev. Lett. **77**, 2178 (1996).

[16] J.F. Donoghue and A.A. Petrov, Phys. Lett. B **393**, 149 (1997).

[17] M. Beneke *et al.*, hep-ph/9905312.

[18] N. Isgur and C.H. Llewellyn Smith, Phys. Lett. B **217**, 535 (1989).

[19] A. Buras and R. Fleischer, hep-ph/9810260; R. Fleischer, Eur. Phys. J. C **6**, 451 (1999); M. Suzuki, Phys. Rev. D **58**, 111504 (1998); M. Gronau and D. Pirjol, hep-ph/9902482; D. Pirjol, hep-ph/9908306.

[20] M. Gronau and J.L. Rosner, Phys. Rev. D **57**, 6843 (1998).

[21] B. Blok, M. Gronau, and J.L. Rosner, Phys. Rev. Lett. **78**, 3999 (1997).

[22] J.F. Donoghue, E. Golowich, A.A. Petrov, and J.M. Soares, unpublished; A.A. Petrov, talk given at DPF99, hep-ph/9903366.

[23] P. Geiger and N. Isgur, Phys. Rev. Lett. **67**, 1066 (1991); H. Lipkin, Nucl. Phys. B **291**, 720 (1987); J. Donoghue, Phys. Rev. D **33**, 1516 (1986).

Part V

Rare Kaon Decays

The Glashow-Weinberg-Salam electroweak theory could have entailed large flavor-changing *neutral* currents (FCNC) unless some new physics, such as a fourth (charmed) quark, was postulated [1]. Taking this quark seriously, Gaillard and Lee [2] calculated the rates for a number of rare processes, such as $K_L \to \gamma\gamma$, $K_L \to \mu^+\mu^-$, $K^+ \to \pi^+\nu\bar{\nu}$, $K_L \to \pi^0 e^+e^-$, $K_L \to \pi^0\nu\bar{\nu}$, and so on. In this volume, Isidori gives a contemporary view of the ability of rare kaon decays to search for new physics.

With the discovery of charm, and subsequently of bottom and top in accord with Kobayashi and Maskawa's prediction, the stage was set for accurate predictions of rates for a number of the above processes. The rates for $K_L \to \gamma\gamma$ and $K_L \to \mu^+\mu^-$ would have been substantially affected if the cancellation of FCNC postulated by Glashow, Iliopoulos, and Maiani (GIM) [1] due to the charmed quark were not present. A number of other processes are sensitive to those elements of the Cabibbo-Kobayashi-Maskawa (CKM) matrix describing the coupling of light quarks to the top quark. For example, the element V_{td} plays a key role in the decay $K^+ \to \pi^+\nu\bar{\nu}$. Based on the single event for this process seen by Experiment E787 at Brookhaven National Laboratory, as described by Redlinger in this volume, the experimental branching ratio is only slightly above the theoretical expectation, and may well be compatible with it.

A number of processes that are not expected to occur in the standard electroweak theory can be probed with remarkable sensitivity in rare kaon decays. These include $K_L \to \mu e$ (Molzon) and $K^+ \to \pi^+\mu^+e^-$ (Zeller). Others are expected to occur with very low branching ratios. Some have already been seen, including the event of $K^+ \to \pi^+\nu\bar{\nu}$ mentioned above and the decays $K^+ \to \pi^+e^+e^-$ and $K^+ \to \pi^+\mu^+\mu^-$ (Zeller). In recent years, the high-energy experiments desgined to study ϵ'/ϵ have also made substantial studies in the area of rare kaon decays, particularly those with γ's and π^0's in the final state and other modes studied by CERN NA48 (Köpke), Fermilab E799 (Whitmore), and Brookhaven (Komatsubara and Ambrose). Some are within reach of experiments foreseen for the next few years. A theoretical perspective on some of these decays from the standpoint of chiral perturbation theory is given by Bijnens. One expects to learn a great deal about the structure of the CKM matrix, and possibly about new physics, from forthcoming experiments in this area.

References

[1] S.L. Glashow, J. Iliopoulos, and L. Maiani, Phys. Rev. D **2**, 1285 (1970).

[2] M.K. Gaillard and B.W. Lee, Phys. Rev. D **10**, 897 (1974).

33

Standard Model vs. New Physics in Rare Kaon Decays

Gino Isidori

Abstract

We present a brief overview of rare K decays, emphasizing the different role of standard model and possible new physics contributions in various channels.

Being sensitive to flavor dynamics from few MeV up to several TeV, rare kaon decays provide a powerful tool to test the standard model (SM) and to search for new physics (NP). In the following we shall outline some of the most interesting aspects of these decays, starting with the most rare ones, strongly sensitive to NP effects, moving toward processes that are more and more dominated by low-energy dynamics.

33.1 Lepton-Flavor Violating Modes

Decays like $K_L \to \mu e$ and $K \to \pi\mu e$ are completely forbidden within the SM, where lepton flavor is conserved, but are also absolutely negligible if we simply extend the model by including only Dirac-type neutrino masses. A positive evidence of any of these processes would therefore unambiguously signal NP, calling for nonminimal extensions of the SM. Moreover, as long as the final state contains at most one pion in addition to the lepton pair, the experimental information on the decay rate can be easily translated into a precise information on the short-distance amplitude $s \to d\mu e$. In this respect we stress that $K_L \to \mu e$ and $K \to \pi\mu e$ provide a complementary information: the first mode is sensitive to pseudoscalar and axial-vector $s \to d$ couplings, whereas the second one is sensitive to scalar, vector, and tensor structures.

In exotic scenarios, like R-parity-violating SUSY or models with leptoquarks, the $s \to d\mu e$ amplitude can be generated already at tree level. In this case naive power counting suggests that limits on $B(K_L \to \mu e)$ or $B(K \to \pi\mu e)$ at the level of 10^{-11} probe NP scales of the order of 100 TeV [1]. On the other hand, in more "conservative" scenarios where the

$s \to d\mu e$ transition can occur only at the one-loop level, it is more appropriate to say that the scale probed is around the (still remarkable !) value of 1 TeV. An interesting example of the second type of scenario is provided by left-right models with heavy Majorana neutrinos [2].

33.2 $K \to \pi\nu\bar{\nu}$

These decays are particularly fascinating since on one side, within the SM, their small but nonnegligible rates are calculable with high accuracy in terms of the less known Cabibbo-Kobayashi-Maskawa (CKM) angles [3]. On the other side, the flavor-changing neutral-current (FCNC) nature implies a strong sensitivity to possible NP contributions, even at very high energy scales.

Within the SM the $s \to d\nu\bar{\nu}$ amplitude is generated only at the quantum level, through Z penguin and W box diagrams. Separating the contributions to the amplitude according to the intermediate up-type quark running inside the loop, one can write

$$\mathcal{A}(s \to d\nu\bar{\nu}) = \sum_{q=u,c,t} V_{qs}^* V_{qd} \mathcal{A}_q \sim \begin{cases} \mathcal{O}(\lambda^5 m_t^2) + i\mathcal{O}(\lambda^5 m_t^2) & (q=t) \\ \mathcal{O}(\lambda m_c^2) \ + i\mathcal{O}(\lambda^5 m_c^2) & (q=c) \\ \mathcal{O}(\lambda \Lambda_{\mathrm{QCD}}^2) & (q=u), \end{cases} \tag{33.1}$$

where V_{ij} denote the elements of the CKM matrix. The hierarchy of these elements [4] would favor up- and charm-quark contributions; however, the hard GIM mechanism of the parton-level calculation implies $\mathcal{A}_q \sim m_q^2/M_W^2$, leading to a completely different scenario. As shown on the r.h.s. of (33.1), where we have employed the standard phase convention ($\Im V_{us} = \Im V_{ud} = 0$) and expanded the CKM matrix in powers of the Cabibbo angle ($\lambda = 0.22$) [4], the top-quark contribution dominates both real and imaginary parts.[1] This structure implies several interesting consequences for $\mathcal{A}(s \to d\nu\bar{\nu})$: it is dominated by short-distance dynamics and therefore calculable with high precision in perturbation theory; it is very sensitive to V_{td}, which is one of the less constrained CKM matrix elements; it is likely to have a large CP-violating phase; it is very suppressed within the SM and thus very sensitive to possible NP effects.

The short-distance contributions to $\mathcal{A}(s \to d\nu\bar{\nu})$, within the SM, can be efficiently described by means of a single effective dimension-6 operator: $O_{LL}^{\nu} = (\bar{s}_L \gamma^\mu d_L)(\bar{\nu}_L \gamma_\mu \nu_L)$. The Wilson coefficient of this operator has been calculated by Buchalla and Buras including next-to-leading-order QCD corrections [5] (see also [6, 7]), leading to a very precise description of the partonic amplitude. Moreover, the simple structure of O_{LL}^{ν} has two major advantages:

- The relation between partonic and hadronic amplitudes is quite accurate, since the hadronic matrix elements of the $\bar{s}\gamma^\mu d$ current between

[1] The Λ_{QCD}^2 factor in the last line of (33.1) follows from a naive estimate of long-distance effects.

a kaon and a pion are related by isospin symmetry to those entering K_{l3} decays, which are experimentally well known.

- The lepton pair is produced in a state of definite CP and angular momentum, implying that the leading SM contribution to $K_L \to \pi^0 \nu\bar{\nu}$ is CP violating.

33.2.1 SM Uncertainties

The dominant theoretical error in estimating $B(K^+ \to \pi^+ \nu\bar{\nu})$ is due to the uncertainty of the QCD corrections to the charm contribution (see [7] for an updated discussion), which can be translated into a 5% error in the determination of $|V_{td}|$ from $B(K^+ \to \pi^+ \nu\bar{\nu})$. This uncertainty can be considered as generated by "intermediate-distance" dynamics; genuine long-distance effects associated with the up quark have been shown to be substantially smaller [8].

The case of $K_L \to \pi^0 \nu\bar{\nu}$ is even more clean from the theoretical point of view [9]. Indeed, because of the CP structure, only the imaginary parts in (33.1)—where the charm contribution is absolutely negligible—contribute to $\mathcal{A}(K_2 \to \pi^0 \nu\bar{\nu})$. Thus the dominant direct-CP-violating component of $\mathcal{A}(K_L \to \pi^0 \nu\bar{\nu})$ is completely saturated by the top contribution, where the QCD uncertainties are very small (around 1%). Intermediate and long-distance effects in this process are confined to the indirect-CP-violating contribution [10] and to the CP-conserving one [11], which are both extremely small. Taking into account also the isospin-breaking corrections to the hadronic matrix element [12], one can therefore write a very accurate expression (with a theoretical error around 1%) for $B(K_L \to \pi^0 \nu\bar{\nu})$ in terms of short-distance parameters [7,10]:

$$B(K_L \to \pi^0 \nu\bar{\nu})_{\rm SM} = 4.25 \times 10^{-10} \left[\frac{\overline{m}_t(m_t)}{170 \text{ GeV}}\right]^{2.3} \left[\frac{\Im \lambda_t}{\lambda^5}\right]^2 . \qquad (33.2)$$

The high accuracy of the theoretical predictions of $B(K^+ \to \pi^+ \nu\bar{\nu})$ and $B(K_L \to \pi^0 \nu\bar{\nu})$ in terms of the modulus and the imaginary part of $\lambda_t = V_{ts}^* V_{td}$ could clearly offer the possibility of very interesting tests of the CKM mechanism. Indeed, a measurement of both channels would provide two independent pieces of information on the unitarity triangle, which can be probed also by B-physics observables. In particular, as emphasized in [10], the ratio of the two branching ratios could be translated into a clean and complementary determination of $\sin(2\beta)$.

Taking into account all the indirect constraints on V_{ts} and V_{td} obtained within the SM, the present range of the SM predictions for the two branching ratios reads [7]:

$$B(K^+ \to \pi^+ \nu\bar{\nu})_{\rm SM} = (0.82 \pm 0.32) \times 10^{-10}, \qquad (33.3)$$

$$B(K_L \to \pi^0 \nu\bar{\nu})_{\rm SM} = (3.1 \pm 1.3) \times 10^{-11}. \qquad (33.4)$$

Moreover, as pointed out recently in [7], a stringent and theoretically clean upper bound on $B(K^+ \to \pi^+ \nu\bar{\nu})_{\rm SM}$ can be obtained using only the experimental information on $\Delta M_{B_d}/\Delta M_{B_s}$ to constrain $|V_{td}/V_{ts}|$. In particular, using $(\Delta M_{B_d}/\Delta M_{B_s})^{1/2} < 0.2$ it is found that

$$B(K^+ \to \pi^+ \nu\bar{\nu})_{\rm SM} < 1.67 \times 10^{-10}, \tag{33.5}$$

which represents a very interesting challenge for the BNL-E787 experiment [13].

33.2.2 Beyond the SM: General Considerations

As far as we are interested only in $K \to \pi\nu\bar{\nu}$ decays, we can roughly distinguish the extensions of the SM into two big groups: those involving new sources of quark-flavor mixing (like generic SUSY extensions of the SM, models with new generations of quarks, etc.) and those where the quark mixing is still ruled by the CKM matrix (like the 2-Higgs-doublet model of type II, constrained SUSY models, etc.). In the second case NP contributions are typically smaller than SM ones at the amplitude level (see, *e.g.*, [14, 15] for some recent discussions). On the other hand, in the first case it is possible to overcome the $\mathcal{O}(\lambda^5)$ suppression of the dominant SM amplitude. If this is the case, it is then easy to generate sizable enhancements of $K \to \pi\nu\bar{\nu}$ rates (see, *e.g.*, [16, 17]).

Concerning $K_L \to \pi^0\nu\bar{\nu}$, it is worthwhile to emphasize that if lepton flavor is not conserved [18, 19] or right-handed neutrinos are involved [20], then new CP-conserving contributions could in principle arise.

Interestingly, despite the variety of NP models, it is possible to derive a model-independent relation among the widths of the three neutrino modes [18]. Indeed, the isospin structure of any $s \to d$ operator bilinear in the quark fields implies

$$\Gamma(K^+ \to \pi^+ \nu\bar{\nu}) = \Gamma(K_L \to \pi^0 \nu\bar{\nu}) + \Gamma(K_S \to \pi^0 \nu\bar{\nu}), \tag{33.6}$$

up to small isospin-breaking corrections, which then leads to

$$B(K_L \to \pi^0 \nu\bar{\nu}) < \frac{\tau_{K_L}}{\tau_{K^+}} B(K^+ \to \pi^+ \nu\bar{\nu}) \simeq 4.2 B(K^+ \to \pi^+ \nu\bar{\nu}). \tag{33.7}$$

Any experimental limit on $B(K_L \to \pi^0\nu\bar{\nu})$ below this bound can be translated into nontrivial dynamical information on the structure of the $s \to d\nu\bar{\nu}$ amplitude.

33.2.3 SUSY Contributions and the $Z\bar{s}d$ Vertex

We will now discuss in more detail the possible modifications of $K \to \pi\nu\bar{\nu}$ decays in the framework of a generic low-energy supersymmetric extension of the SM, which represents a very attractive possibility from the theoretical point of view [21]. Similarly to the SM, also in this case FCNC amplitudes are generated only at the quantum level, provided we assume unbroken R parity and minimal particle content. However, in addition to the standard penguin and box diagrams, also their corresponding superpartners, generated by gaugino-squarks loops, play an important role. In particular, the chargino-up-squarks diagrams provide the potentially dominant non-SM effect to the $s \to d\nu\bar{\nu}$ amplitude [22]. Moreover, in the limit where the average mass of SUSY particles is substantially larger than M_W, the penguin diagrams tend to dominate over the box ones and the dominant SUSY effect can be encoded through an effective $Z\bar{s}d$ coupling [16, 23, 24].

The flavor structure of a generic SUSY model is quite complicated and a convenient model-independent parameterization of the various flavor-mixing terms is provided by the so-called mass-insertion approximation [25]. This consists of choosing a simple basis for the gauge interactions and, in that basis, to perform a perturbative expansion of the squark mass matrices around their diagonal. Employing a squark basis where all quark-squark-gaugino vertices involving down-type quarks are flavor diagonal, it is found that the potentially dominant SUSY contribution to the $Z\bar{s}d$ vertex arises from the double mixing $(\tilde{u}^d_L - \tilde{t}_R) \times (\tilde{t}_R - \tilde{u}^s_L)$ [16]. Indirect bounds on these mixing terms dictated by vacuum stability, neutral-meson mixing, and $b \to s\gamma$ leave open the possibility of large effects [16]. More stringent constraints can be obtained employing stronger theoretical assumptions on the flavor structure of the SUSY model [23]. However, the possibility of sizable modifications of $K \to \pi\nu\bar{\nu}$ widths (including enhancements of more than one order of magnitude in the case of $K_L \to \pi^0\nu\bar{\nu}$) cannot be excluded *a priori*.

Interestingly a nonstandard $Z\bar{s}d$ vertex can be generated also in non-SUSY extensions of the SM (see, *e.g.*, [26]). It is therefore useful trying to constrain this scenario in a model-independent way. At present the best direct limits on the $Z\bar{s}d$ vertex are dictated by $K_L \to \mu^+\mu^-$ [16, 18, 24], bounding the real part of the coupling, and $\Re(\epsilon'/\epsilon)$ [24], constraining the imaginary one. Unfortunately in both cases the bounds are not very accurate, being affected by sizable hadronic uncertainties. Concerning ϵ'/ϵ, it is worthwhile to mention that the nonstandard $Z\bar{s}d$ vertex could provide an explanation for the apparent discrepancy between $(\epsilon'/\epsilon)_{\rm exp}$ and $(\epsilon'/\epsilon)_{\rm SM}$ [23, 27], even if it is certainly too early to make a definite statement in this respect [28]. In the future the situation could become much clearer with precise determinations of both the real and the imaginary parts of the $Z_{\bar{s}d}$ coupling by means of $\Gamma(K^+ \to \pi^+\nu\bar{\nu})$ and $\Gamma(K_L \to \pi^0\nu\bar{\nu})$. Note that if we only use the present constraints from $K_L \to \mu^+\mu^-$ and $\Re(\epsilon'/\epsilon)$, we cannot exclude enhancements up to one order of magnitude for $\Gamma(K_L \to \pi^0\nu\bar{\nu})$ and up to a factor ~ 3 for $\Gamma(K^+ \to \pi^+\nu\bar{\nu})$ [23, 24].

33.3 $K \to \pi\ell^+\ell^-$ and $K \to \ell^+\ell^-$

Similarly to $K \to \pi\nu\bar{\nu}$ decays, the short-distance contributions to $K \to \pi\ell^+\ell^-$ and $K \to \ell^+\ell^-$ are calculable with high accuracy and are potentially sensitive to NP effects. However, in these processes the size of long-distance contributions is usually much larger due to the presence of electromagnetic interactions. Only in few cases (mainly in CP-violating observables), long-distance contributions are suppressed and it is possible to extract the interesting short-distance information.

33.3.1 $K \to \pi \ell^+ \ell^-$

The single-photon exchange amplitude, dominated by long-distance dynamics, provides the largest contribution to the CP-allowed transitions $K^+ \to \pi^+ \ell^+ \ell^-$ and $K_S \to \pi^0 \ell^+ \ell^-$. The former has been observed, both in the electron and in the muon mode, whereas only an upper bound of about 10^{-6} exists on $B(K_S \to \pi^0 e^+ e^-)$ [29]. This amplitude can be described in a model-independent way in terms of two form factors, $W_+(z)$ and $W_S(z)$, defined by [30]

$$i \int d^4x e^{iqx} \langle \pi(p) | T \left\{ J^\mu_{\text{elm}}(x) \mathcal{L}_{\Delta S=1}(0) \right\} | K_i(k) \rangle =$$
$$\frac{W_i(z)}{(4\pi)^2} \left[z(k+p)^\mu - (1 - r_\pi^2) q^\mu \right], \tag{33.8}$$

where $q = k - p$, $z = q^2/M_K^2$, and $r_\pi = M_\pi/M_K$. The two form factors are nonsingular at $z = 0$ and, due to gauge invariance, vanish to lowest order in chiral perturbation theory (CHPT) [31]. Beyond lowest order one can identify two separate contributions to the $W_i(z)$: a nonlocal term, $W_i^{\pi\pi}(z)$, due to the $K \to 3\pi \to \pi\gamma^*$ scattering, and a local term, $W_i^{\text{pol}}(z)$, that encodes the contributions of unknown low-energy constants (to be determined by data) [30]. At $\mathcal{O}(p^4)$ the local term is simply a constant, whereas at $\mathcal{O}(p^6)$ also a term linear in z arises. We note, however, that already at $\mathcal{O}(p^4)$ chiral symmetry alone does not help to relate W_S and W_+, or K_S and K^+ decays [31].

Recent results on $K^+ \to \pi^+ e^+ e^-$ and $K^+ \to \pi^+ \mu^+ \mu^-$ by BNL-E865 [32] indicate very clearly that, owing to a large linear slope, the $\mathcal{O}(p^4)$ expression of $W_+(z)$ is not sufficient to describe experimental data. This should not be considered a failure of CHPT, but rather as an indication that large $\mathcal{O}(p^6)$ contributions are present in this channel.[2] Indeed the $\mathcal{O}(p^6)$ expression of $W_+(z)$ seems to fit data well. Interestingly, this is due not only to a new free parameter appearing at $\mathcal{O}(p^6)$, but also to the presence of the nonlocal term. The evidence of the latter provides a real significant test of the CHPT approach.

In the $K_L \to \pi^0 \ell^+ \ell^-$ decay the long-distance part of the single-photon exchange amplitude is forbidden by CP invariance but it contributes to the processes via K_L-K_S mixing, leading to

$$B(K_L \to \pi^0 e^+ e^-)_{\text{CPV-ind}} = 3 \times 10^{-3}\, B(K_S \to \pi^0 e^+ e^-). \tag{33.9}$$

[2] This should not be surprising since in this mode sizable next-to-leading order contributions could arise due to vector-meson exchange.

On the other hand, the direct-CP-violating part of the decay amplitude is very similar to the one of $K_L \to \pi^0 \nu\bar{\nu}$ but for the fact that it receives an additional short-distance contribution due to the photon penguin. Within the SM, this theoretically clean part of the amplitude leads to [33]

$$B(K_L \to \pi^0 e^+ e^-)^{\rm SM}_{\rm CPV-dir} = 0.67 \times 10^{-10} \left[\frac{\overline{m}_t(m_t)}{170~\rm GeV}\right]^2 \left[\frac{\Im\lambda_t}{\lambda^5}\right]^2, \tag{33.10}$$

and, similarly to the case of $B(K_L \to \pi^0 \nu\bar{\nu})$, it could be substantially enhanced by SUSY contributions [16,23]. The two CP-violating components of the $K_L \to \pi^0 e^+ e^-$ amplitude will in general interfere. Given the present uncertainty on $B(K_S \to \pi^0 e^+ e^-)$, at the moment we can only set the rough upper limit

$$B(K_L \to \pi^0 e^+ e^-)^{\rm SM}_{\rm CPV-tot} \lesssim \text{few} \times 10^{-11} \tag{33.11}$$

on the sum of all the CP-violating contributions to this mode [30]. We stress, however, that the phases of the two CP-violating amplitudes are well known. Thus if $B(K_S \to \pi^0 e^+ e^-)$ will be measured, it will be possible to determine the interference between direct and indirect CP-violating components of $B(K_L \to \pi^0 e^+ e^-)_{\rm CPV}$ up to a sign ambiguity. Finally, it is worth noting that evidence for $B(K_L \to \pi^0 e^+ e^-)_{\rm CPV}$ above the 10^{-10} level, possible within specific supersymmetric scenarios [23], would be a clear signal of physics beyond the SM.

An additional contribution to $K_L \to \pi^0 \ell^+ \ell^-$ decays is generated by the CP-conserving processes $K_L \to \pi^0 \gamma\gamma \to \pi^0 \ell^+ \ell^-$ [34]. This however does not interfere with the CP-violating amplitude and, as we shall discuss in the next section, it is quite small ($\lesssim 4 \times 10^{-12}$) in the case of $K_L \to \pi^0 e^+ e^-$.

33.3.2 $K_L \to l^+ l^-$

The two-photon intermediate state plays an important role in $K_L \to \ell^+ \ell^-$ transitions. This is by far the dominant contribution in $K_L \to e^+ e^-$, where the dispersive integral of the $K_L \to \gamma\gamma \to l^+ l^-$ loop is dominated by the term proportional to $\log(m_K^2/m_e^2)$. The presence of this large logarithm implies also that $\Gamma(K_L \to e^+ e^-)$ can be estimated with relatively good accuracy in terms of $\Gamma(K_L \to \gamma\gamma)$, yielding the prediction $B(K_L \to e^+ e^-) \sim 9 \times 10^{-12}$ [35], which recently seems to have been confirmed by the four events observed at BNL-E871 [36].

More interesting from the short-distance point of view is the case of $K_L \to \mu^+ \mu^-$. Here the two-photon long-distance amplitude is not enhanced by large logs and it is almost comparable in size with the short-distance one, sensitive to $\Re V_{td}$ [5]. Unfortunately the dispersive part of the two-photon contribution is much more difficult to estimate in this case, owing to its stronger sensitivity to the $K_L \to \gamma^* \gamma^*$ form factor. Despite the precise experimental determination of $B(K_L \to \mu^+ \mu^-)$, the present constraints on $\Re V_{td}$ from this observable are not very stringent [37]. Nonetheless, the measurement of $B(K_L \to \mu^+ \mu^-)$ is still useful to put significant bounds on possible NP contributions. Moreover, we stress that the uncertainty of the $K_L \to \gamma^*\gamma^* \to \mu^+\mu^-$ amplitude could be partially decreased in the future by precise experimental information on the form factors of $K_L \to \gamma\ell^+\ell^-$ and $K_L \to e^+e^-\mu^+\mu^-$ decays, especially if these would be consistent with the general parameterization of the $K_L \to \gamma^*\gamma^*$ vertex proposed in [37].

33.4 Two-Photon Processes

$K \to \pi\gamma\gamma$ and $K \to \gamma\gamma$ decays are completely dominated by long-distance dynamics and therefore not particularly useful to search for NP. However, these modes are interesting on one hand to perform precision tests of CHPT, and on the other to estimate long-distance corrections to the $\ell^+\ell^-$ channels (see, *e.g.*, [38] and references therein).

Among the CHPT tests, an important role is played by $K_S \to \gamma\gamma$. The first nonvanishing contribution to this process arises at $\mathcal{O}(p^4)$ and, being generated only by a finite loop amplitude, is completely determined [39]. Since in this channel vector meson exchange contributions are not allowed, and unitarity corrections are automatically included in the $\mathcal{O}(p^2)$ coupling [38], we expect that the $\mathcal{O}(p^4)$ result provides a good approximation to the full amplitude. This is confirmed by present data [29], but a more precise determination of the branching ratio is needed in order to perform a more stringent test.

Similarly to the $K_S \to \gamma\gamma$ case, the leading nonvanishing contribution to $K_L \to \pi^0\gamma\gamma$ also arises only at $\mathcal{O}(p^4)$ and is completely determined [40]. However, in this case large $\mathcal{O}(p^6)$ corrections can be expected as a result of both unitarity corrections and vector meson exchange contributions. Indeed the $\mathcal{O}(p^4)$ prediction for $B(K_L \to \pi^0\gamma\gamma)$ turns out to be substantially smaller (more than a factor 2) than the experimental findings [38]. After the inclusion of unitarity corrections and vector meson exchange contributions, both spectrum and branching ratio of this decay can be expressed in terms of a single unknown coupling: a_V [41]. The recent KTeV measurement [42] has shown that the determination of a_V from both spectrum and branching ratio of $K_L \to \pi^0\gamma\gamma$ leads to the same value, $a_V = -0.72 \pm 0.08$, providing an important consistency check of this approach.

As anticipated, the $K_L \to \pi^0\gamma\gamma$ amplitude is also interesting since it produces a CP-conserving contribution to $K_L \to \pi^0\ell^+\ell^-$ [41]. For $\ell = e$ the leading $O(p^4)$ contribution is helicity suppressed and only the $O(p^6)$ amplitude with the two photons in $J = 2$ leads to a nonvanishing $B(K_L \to \pi^0e^+e^-)_{\mathrm{CPC}}$ [34]. Given the recent experimental result [42], this should not exceed 4×10^{-12} [41]. Moreover, we stress that the Dalitz plot distribution of CP-conserving and CP-violating contributions to $K_L \to \pi^0e^+e^-$ are substantially different: in the first case the e^+e^- pair is in a state of $J = 1$, whereas in the latter it has $J = 2$. Thus in principle it is possible to extract the interesting $B(K_L \to \pi^0e^+e^-)_{\mathrm{CPV}}$ from a Dalitz plot analysis of the decay. On the other hand, the CP-conserving contribution is enhanced and more difficult to subtract in the case of $K_L \to \pi^0\mu^+\mu^-$, where the helicity suppression of the leading $O(p^4)$ contribution (photons in $J = 0$) is much less effective (see Heiliger and Sehgal in [41]).

33.5 Conclusions

Rare K decays provide a unique opportunity to perform high-precision tests of CP violation and flavor mixing, both within and beyond the SM.

A special role is undoubtedly played by $K \to \pi\nu\bar{\nu}$ decays. In some NP scenarios sizable enhancements to the branching ratios of these modes are possible and, if detected, these would provide the first evidence for physics beyond the SM. Nevertheless, even in the absence of such enhancements, precise measurements of $K \to \pi\nu\bar{\nu}$ widths will lead to unique information about the flavor structure of any extension of the SM.

Among decays into a $\ell^+\ell^-$ pair, the most interesting one from the short-distance point of view is probably $K_L \to \pi^0 e^+e^-$. However, in order to extract precise information from this mode an experimental determination (or a stringent upper bound) on $B(K_S \to \pi^0 e^+e^-)$ is also necessary.

Acknowledgments

This work is partially supported by the EEC-TMR Program, Contract N. CT98-0169.

References

[1] See, *e.g.*, R. Peccei, this volume, chapter 2; W. Molzon, *ibid.*, chapter 35; T. Rizzo, hep-ph/9809526.

[2] Z. Gagyi-Palffy, A. Pilaftsis, and K. Schilcher, Nucl. Phys. B **513**, 517 (1998).

[3] N. Cabibbo, Phys. Rev. Lett. **10**, 531 (1963); M. Kobayashi and T. Maskawa, Prog. Theor. Phys. **49**, 652 (1973).

[4] L. Wolfenstein, Phys. Rev. Lett. **51**, 1945 (1983).

[5] G. Buchalla and A.J. Buras, Nucl. Phys. B **398**, 285 (1993); Nucl. Phys. B **400**, 225 (1993); Nucl. Phys. B **412**, 106 (1994).

[6] M. Misiak and J. Urban, Phys. Lett. B **451**, 161 (1999).

[7] G. Buchalla and A.J. Buras, Nucl. Phys. B **548**, 309 (1999); G. Buchalla, this volume, chapter 54.

[8] M. Lu and M. Wise, Phys. Lett. B **324**, 461 (1994).

[9] L. Littenberg, Phys. Rev. D **39**, 3322 (1989).

[10] G. Buchalla and A.J. Buras, Phys. Rev. D **54**, 6782 (1996).

[11] G. Buchalla and G. Isidori, Phys. Lett. B **440**, 170 (1998); see also D. Rein and L.M. Sehgal, Phys. Rev. D **39**, 3325 (1989).

[12] W.J. Marciano and Z. Parsa, Phys. Rev. D **53**, R1 (1996).

[13] S. Adler *et al.* (E787 Collab.), Phys. Rev. Lett. **79**, 2204 (1997); G. Redlinger, this volume, chapter 34.

[14] G. Burdman, Phys. Rev. D **59**, 035001 (1999).

[15] G.-C. Cho, Eur. Phys. J. C **5**, 525 (1998); T. Goto, Y. Okada, and Y. Shimizu, Phys. Rev. D **58**, 094006 (1998).

[16] G. Colangelo and G. Isidori, JHEP **09**, 009 (1998).

[17] T. Hattori, T. Hausike, and S. Wakaizumi, hep-ph/9804412.

[18] Y. Grossman and Y. Nir, Phys. Lett. B **398**, 163 (1997).

[19] G. Perez, hep-ph/9907205.

[20] U. Nierste, poster session, Kaon99 Conference, Chicago, June 1999.

[21] See, *e.g.*, L. Hall, this volume, chapter 4.

[22] Y. Nir and M. Worah, Phys. Lett. B **243**, 326 (1998); A.J. Buras, A. Romanino, and L. Silvestrini, Nucl. Phys. **520**, 3 (1998).

[23] A.J. Buras *et al.*, Nucl. Phys. B **566**, 3 (2000).

[24] A.J. Buras and L. Silvestrini, Nucl. Phys. B **546**, 299 (1999).

[25] L.J. Hall, V.A. Kostelecky, and S. Rabi, Nucl. Phys. **267**, 415 (1986).

[26] Y. Nir and D. Silverman, Phys. Rev. D **42**, 1477 (1990).

[27] Y.-Y. Keum, U. Nierste, and A.I. Sanda, hep-ph/9903230.

[28] See, *e.g.*, S. Bertolini, this volume, chapter 12; A.J. Buras, *ibid.*, chapter 5; M. Ciuchini *et al.*, *ibid.*, chapter 28; M. Ciuchini *et al.*, *ibid.*, chapter 31; T. Hambye and P.H. Soldan, *ibid.*, chapter 11.

[29] C. Caso *et al.*, Eur. Phys. J. C **3**, 1 (1998).

[30] G. D'Ambrosio, G. Ecker, G. Isidori, and J. Portolés, JHEP **08**, 004 (1998).

[31] G. Ecker, A. Pich, and E. de Rafael, Nucl. Phys. B **291**, 692 (1987).

[32] R. Appel *et al.* (E865 Collab.), hep-ph/9907045; J.A. Thompson *et al.*, hep-ph/9904026; M.E. Zeller, this volume, chapter 36.

[33] A.J. Buras *et al.*, Nucl. Phys. B **423**, 349 (1994).

[34] L.M. Sehgal, Phys. Rev. D **38**, 808 (1988).

[35] G. Valencia, Nucl. Phys. B **517**, 339 (1998); G. Dumm and A. Pich, Phys. Rev. Lett. **80**, 4633 (1998).

[36] D. Ambrose *et al.* (E871 Collab.), Phys. Rev. Lett. **81**, 4309 (1998).

[37] G. D'Ambrosio, G. Isidori, and J. Portolés, Phys. Lett. B **423**, 385 (1998).

[38] G. D'Ambrosio and G. Isidori, Int. J. Mod. Phys. A **13**, 1 (1998).

[39] G. D'Ambrosio and D. Espriu, Phys. Lett. B **175**, 237 (1986); J.L. Goity, Z. Phys. C **34**, 341 (1987).

[40] G. Ecker, A. Pich, and E. de Rafael, Phys. Lett. B **189**, 363 (1987); L. Cappiello and G. D'Ambrosio, Nuovo Cimento A **99**, 155 (1988).

[41] G. Ecker, A. Pich, and E. de Rafael, Phys. Lett. B **237**, 481 (1990); L. Cappiello, G. D'Ambrosio, and M. Miragliuolo, Phys. Lett. B **298**, 423 (1993); P. Heiliger and L.M. Sehgal, Phys. Rev. D **47**, 4920 (1993); A.G. Cohen, G.

Ecker, and A. Pich, Phys. Lett. B **304**, 347 (1993); J.F. Donoghue and F. Gabbiani, Phys. Rev. D **51**, 2187 (1995); G. D'Ambrosio and J. Portolés, Nucl. Phys. B **492**, 417 (1997).

[42] A. Alavi-Harati *et al.* (KTeV Collaboration), hep-ph/9902209; J. Whitmore, this volume, chapter 39.

34

New Result on $K^+ \to \pi^+\nu\bar{\nu}$ from BNL E787

G. Redlinger[1]

Abstract

E787 at BNL has reported evidence for the rare decay $K^+ \to \pi^+\nu\bar{\nu}$, based on the observation of one candidate event. In this paper, we present the result of analyzing a new dataset of comparable sensitivity to the published result.

34.1 Introduction

In the standard model (SM), quark mixing in weak decays and CP violation are related [1]. Detailed exploration of this relationship between two longstanding mysteries of particle physics is one of the major themes of experimental work in the field over the next decade(s). The rare decay $K^+ \to \pi^+\nu\bar{\nu}$ has drawn interest in this effort as a result of the theoretically clean relationship between the branching ratio and the poorly measured quark mixing parameter $|V_{td}|$. The intrinsic theoretical uncertainty in the branching ratio (arising mainly from QCD corrections to the charm contribution to the process) is estimated [2] to be about 7%, for a given set of SM input parameters: m_t, $|V_{cb}|$, and $|V_{td}|$. Stated differently, once the branching ratio for $K^+ \to \pi^+\nu\bar{\nu}$ is known, $|V_{td}|$ can be determined to $\sim 5\%$ (given perfect knowledge of m_t and $|V_{cb}|$). In the SM, the branching ratio is expected to be [3] $B(K^+ \to \pi^+\nu\bar{\nu}) = (0.82 \pm 0.32) \times 10^{-10}$, using current data on m_t, m_c, V_{cb}, $|V_{ub}/V_{cb}|$, ϵ_K, and $B_d - \overline{B_d}$ and $B_s - \overline{B_s}$ mixing.

The E787 collaboration at BNL presented evidence for $K^+ \to \pi^+\nu\bar{\nu}$ based on the observation of one clean event from data collected in the 1995 run of the AGS [4]. The expected level of background in the signal region was 0.08 ± 0.03 events. However, the event also satisfied the most

[1]Representing the E787 Collaboration (BNL, KEK, Osaka, Princeton, TRIUMF, Alberta).

demanding criteria designed in advance for candidate evaluation; this put the event in a region where the expected background was 0.008 ± 0.005 events. If the event is due to $K^+ \to \pi^+ \nu\bar{\nu}$, this would imply a branching ratio $B(K^+ \to \pi^+ \nu\bar{\nu}) = 4.2^{+9.7}_{-3.5} \times 10^{-10}$, which is consistent with the SM range, although the central value is higher by a factor of about 4. More recently, the surprisingly large value of ϵ'/ϵ reported by the KTeV [5] and NA48 [6] groups has excited more speculation that new physics may be at work in the K system.

Amongst the vast literature on non-SM physics with the signature $K^+ \to \pi^+ +$"nothing," recent attention has focused on implications for electroweak symmetry breaking. R-parity-conserving supersymmetry (SUSY) has been considered in many papers [7]; conclusions vary depending on assumptions, but a branching ratio 2 to 3 times above the SM level seems to be possible without going to exotic models. In R-parity violating SUSY [8], $K^+ \to \pi^+ \nu\bar{\nu}$ yields the best constraints on couplings relating the first two generations. Effects from topcolor models [9] can also be significant.

The 2-body decay $K^+ \to \pi^+ X^0$ where X^0 is a long-lived noninteracting object is indistinguishable experimentally from $K^+ \to \pi^+ \nu\bar{\nu}$ except for the 2-body kinematics. An interesting possibility is where X^0 is a Nambu-Goldstone boson from breaking some global symmetry, a well-known example of which is the familon [10] from the breaking of global family symmetry. The branching ratio is expected to be $B(K^+ \to \pi^+ f) = 2.7 \times 10^{13}\ \mathrm{GeV}^2/\mathrm{F}^2$ where F is the scale at which the symmetry is broken. Cosmological considerations lead to an upper bound [11] $\mathrm{F} < 10^{12}$ GeV, implying $B(K^+ \to \pi^+ f) >\sim 10^{-11}$. A new class of axion models has also recently been considered [12].

One theme of this conference was the glorious role that kaons have played over the last fifty years in elucidating the mysteries of particle physics. Along these lines, it may be worth remarking that in the late 1960s it was the absence of the process $K^+ \to \pi^+ \nu\bar{\nu}$ (among others) at the then expected rate that led to the "new" physics (GIM mechanism, charm, etc.) of the time and the eventual establishment of the SM. It would be poetic indeed if now a higher rate than expected were to be confirmed, leading the way beyond the SM.[2]

The remainder of this paper is organized as follows. We first remind the reader of the $K^+ \to \pi^+ \nu\bar{\nu}$ detection strategy employed in E787. We then present the result from the combined analysis of the 1995–97 dataset. We conclude with a few remarks on the data collected in 1998 and on a proposed future experiment (BNL E949) based on a modest upgrade to E787.

---

[2]See the amusing exchange between B. Winstein and G. Kane in the Sept. 1998 issue of *Physics Today* on a fictitious look back from the year 2011 to the history of particle physics from the late 1990s onward.

34.2 Detection Strategy

34.2.1 The E787 Detector

The signature for $K^+ \to \pi^+\nu\bar{\nu}$ is a K^+ decay to a π^+ of momentum $P < 227$ MeV/c and no other observable product. Definitive observation of this signal requires suppression of all backgrounds to well below the sensitivity for the signal. Furthermore, reliable estimates of the residual background levels are needed.

Major sources of background include the two-body decays $K^+ \to \mu^+\nu_\mu$ ($K_{\mu 2}$) with a 64% branching ratio and $P = 236$ MeV/c, and $K^+ \to \pi^+\pi^0$ ($K_{\pi 2}$) with a 21% branching ratio and $P = 205$ MeV/c. The charged particle spectrum of the major K^+ decay modes is shown in fig. 34.1 together with the spectrum from $K^+ \to \pi^+\nu\bar{\nu}$. The $K^+ \to \pi^+\nu\bar{\nu}$ signal can be observed in the region away from these two kinematic peaks. The search described here concentrates in the region between the $K_{\pi 2}$ and $K_{\mu 2}$ peaks.

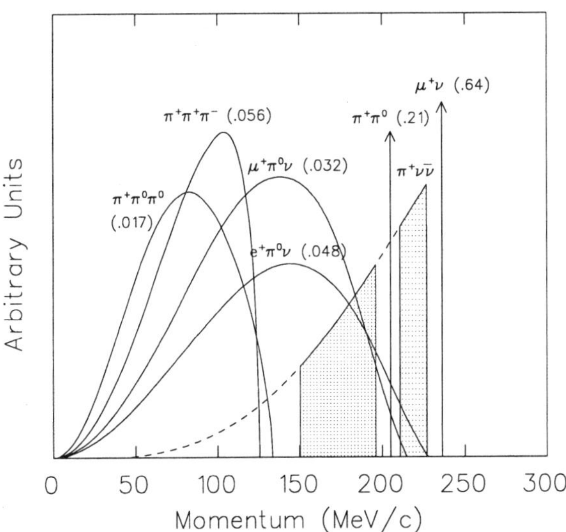

Figure 34.1: Charged-particle momentum spectrum in the K^+ rest frame for the major K^+ decay modes and for $K^+ \to \pi^+\nu\bar{\nu}$.

The only other important background sources are scattering of pions in the beam (either from kaon decay/interaction or pions from the K^+ production target) and K^+ charge exchange (CEX) reactions resulting in decays $K_L^0 \to \pi^+ l^- \bar{\nu}$, where $l = e$ or μ.

The detector has been described in detail elsewhere [13]; here we sketch the main ideas. To achieve the large kinematic suppression of the monochromatic peaks $K_{\pi 2}$ and $K_{\mu 2}$, we work in the kaon rest frame, slowing down a 790 MeV/c kaon beam in a BeO degrader and stopping it in a finely segmented, fully active, scintillating-fiber target. This drives the geometry of the detector, which is cylindrical with end caps (EC) covering the polar regions, much like a colliding-beam detector. Since vetoing of extra energy in the event is a key to the detection strategy, the detector is almost fully active.

The fiducial volume is defined by the range stack (RS) scintillators in the barrel region, covering about 50% of the solid angle. The pions from K^+ decay are stopped in the RS scintillator. This allows redundant measurements of track kinematics, namely kinetic energy and range. It also allows for a powerful technique for π/μ separation that requires the positive identification of the $\pi^+ \to \mu^+ \to e^+$decay sequence. This is achieved with flash-ADC-based 500-MHz transient digitizers (TD) [14], which digitize the phototube outputs. In addition, dE/dx separation of π from μ can be utilized. Further kinematic rejection is achieved with a momentum measurement in a 1T solenoidal field, using a low-mass drift chamber [15] with a resolution $\Delta P/P \sim 0.9\%$ at $P = 205$ MeV/c. Efficient rejection of modes with photons is achieved with nearly 4π photon veto coverage, typically about 15 radiation lengths thick, employing a variety of technologies: lead-scintillator sandwich in the barrel region (BV), undoped CsI in the EC [16], lead-glass, lead-scintillator sandwich, and lead-scintillating-fiber detectors in the beam region. Signals are digitized with 500-MHz CCD TDs [17] based on Ga-As technology. Photon veto time windows are typically a few ns, with energy thresholds ranging from 0.2 to about 3 MeV. For the incoming beam, K/π separation is achieved with a Cerenkov counter with a lucite radiator. Finely segmented tracking and good timing of the beam is achieved with MWPCs (1.27 mm wire spacing), scintillator hodoscopes, and a scintillating fiber kaon stopping target (5 mm square fiber), the latter two also digitized at 500 MHz.

At typical rates, we took about 5 MHz of incoming kaons, of which about 20% stopped in the detector. These were analyzed by a 2-level trigger. The first level provided a rejection of about 800 by requiring a kaon stopping in the target, a decay 1.5 ns later, no photons in the BV, EC, or RS, a range longer than $K^+ \to \pi^+\pi^+\pi^-$ and shorter than $K_{\mu 2}$. The second level required a $\pi^+ \to \mu^+$ decay in the counter where the charged particle stopped and no energy deposited in the counter radially outward from the stopping counter; this provided a rejection of about 20.

The DAQ [18] is Fastbus-based, with front-end readout into SLAC scanner processors (SSP). The data is transferred via the cable segment to VME processors and then to a Silicon Graphics (SGI) machine. The data transfer capability is currently > 25 Mbytes/sec. The deadtime, dominated by the readout of the front-end modules into the SSPs is currently about 17% per MHz of stopped kaons (down from 28% in 1995).

34.2.2 Offline Analysis

To elude rejection, $K_{\mu 2}$ and $K_{\pi 2}$ events have to be reconstructed incorrectly in range, energy, and momentum. In addition, any event with a muon has to have its track misidentified as a pion. The most effective weapon here is the TD analysis, requiring observation of the $\pi^+ \to \mu^+ \to e^+$ decay sequence; this provides a muon rejection factor of about 10^5. Events with photons, such as $K_{\pi 2}$ decays, are efficiently eliminated by the photon veto; the rejection factor for events with $\pi^0 s$ is around 10^6. A scattered beam pion can survive the analysis only by misidentification as a K^+ and if the track is mismeasured as delayed, or if the track is missed entirely by the beam counters after a valid K^+ stopped in the target. CEX background events can survive only if the K_L^0 is produced at low enough energy to remain in the target for at least 2 ns, if there is no visible gap between the beam track and the observed π^+ track, and if the additional charged lepton goes unobserved.

The data are analyzed with the goal of reducing the total expected background to significantly less than one event in the final sample. The same dataset is used for background studies and for the signal search so that any hardware failures or time-dependent effects are naturally accounted for in the background estimates. The offline analysis is performed "blind" in the sense that the signal region is always hidden (by inverting one or more cuts) while cuts are developed and background levels estimated. Without this procedure, it would be difficult to avoid the systematic error whereby a signal event might be eliminated with minimal acceptance loss by a slight tweaking of one of the more than 50 cuts in the analysis. Taking this one step further, the background estimation is also performed "blind" by developing the cuts on 1/3 of the data and then testing the background rejection and measuring the final remaining background with the remaining 2/3 of the data, without any further changes to the cuts. This is necessary because the background estimates are typically made with a handful of remaining events; without this protection against low statistics bias, it would be easy to eliminate a significant fraction of the remaining background events by a slight tweaking of several cuts, thereby drastically changing the estimated level of remaining background.

To develop the cuts, we take advantage of redundant, independent constraints available on each source of background to establish two independent sets of cuts. One set of cuts is relaxed or inverted to enhance the background (by up to 3 orders of magnitude) so that the other group of cuts can be evaluated to determine its power for rejection. For example, $K_{\pi 2}$ is studied by measuring the rejection of the kinematic cuts on a sample of events failing the photon veto, while the photon veto rejection is measured on a sample of events that are kinematically consistent with $K_{\pi 2}$. $K_{\mu 2}$ (including $K^+ \to \mu^+\nu_\mu\gamma$) is studied by separately measuring the rejections of the TD particle identification cuts and the kinematic cuts. The background from beam pion scattering is evaluated by separately measuring the rejections of the beam counter and timing cuts. For the CEX background, events with K^+ charge exchange in the stopping target were collected with a special trigger, triggering on the two pion decay of the K_S^0; these events are then used as input to Monte Carlo studies, replacing the K_S^0 with a

Monte Carlo K_L^0. Small correlations in the separate groups of cuts are investigated for each background source and corrected for if they exist.

Before looking in the final signal region, likelihood functions for each background type are constructed, using the method described above. For example, the $K_{\pi 2}$ kinematic likelihood function is measured with a sample of events failing the photon veto, while the $K_{\pi 2}$ photon veto likelihood is measured with a sample of events kinematically consistent with $K_{\pi 2}$. These likelihood functions are used to predict the shape of the background distributions outside the final signal region. The prediction is not guaranteed to be satisfied because the final measurement involves the simultaneous application of all likelihood functions; the success of the predictions therefore tests the independence of the individual likelihood functions. Finally, the background likelihood functions are used to assess the background likelihood of any candidate events.

34.3 Post-1995 Developments

34.3.1 Online Improvements

To increase the sensitivity per hour, the basic strategy in the post-1995 running of E787 has been to run with a lower K^+ momentum and reducing the amount of BeO degrader material. This decreases the probability that the incoming kaon is lost in the BeO degrader (from multiple Coulomb scattering and interactions), reducing the accidental rate in the detector and increasing the fraction of kaons reaching the stopping target. To maintain the overall kaon flux, the idea was to simultaneously increase the proton flux on the kaon production target; unfortunately, in practice, the proton flux remained roughly constant from year to year. In addition, improvements were made to the trigger, reducing the deadtime fraction from 28% to 17% per MHz of stopped kaons. The efficiency of the 2nd-level trigger ($\pi^+ \rightarrow \mu^+$) was increased by a factor of 1.27 over 1995. A further gain of a factor of 1.05 was achieved by extending the time range recorded by the TDs to look for the electron from $\mu^+ \rightarrow e^+$ decay.

The total exposure in the 1996 and 1997 runs amounted to about 1.5 times the 1995 exposure. The increase in sensitivity/hour was offset by a decrease in running time: 17 weeks in 1996 followed by 9 weeks in 1997, compared to 24 weeks in 1995.

34.3.2 Offline Analysis

The goal of the offline analysis was to increase the rejection to maintain the signal-to-noise ratio as the overall sensitivity grows, while maintaining (or possibly even increasing) the acceptance at the same time.

Better range resolution in both the Gaussian core and in the tails was achieved with tracking improvements in the range stack and target. For the $K_{\mu 2}$ background, a more sophisticated dE/dx analysis was developed for the range stack, reducing the number of $K_{\mu 2}$ events downshifting in range and energy due to nuclear interactions. Tracking improvements in the drift chamber and better tracking quality cuts reduced the momentum tail. On the TD side, a 17% gain in acceptance was realized for the same level of rejection as the published analysis. This was accomplished by a better understanding of the different sources of TD background in the $K_{\mu 2}$ range tail compared to $K^+ \to \mu^+ \nu_\mu \gamma$ and $K^+ \to \mu^+ \pi^0 \nu_\mu$, a better algorithm for finding the electron from $\mu^+ \to e^+$ decay, and improvements in the construction of pion likelihood functions. The beam background was lowered by improved cuts against kaons decaying in flight in the stopping target, improved use of the CCD information in the target, and better cuts against two kaons entering the detector closely spaced in time. The CEX background was lowered by the above-mentioned improvements in the target tracking and by a likelihood analysis utilizing information in the target, and beam hodoscopes.

The estimated background levels from the analysis of the 1995–97 dataset are shown in table 34.1. The statistical and systematic uncertainties on these background estimates are still being evaluated.

	1995	1996–97	Total
$K_{\pi 2}$	0.015	0.006	0.021
$K_{\mu 2}$	0.008	0.020	0.028
1-beam	0.0026	0.0015	0.004
2-beam	0.0021	0.0047	0.007
CEX	0.0045	0.0051	0.010
Total			0.07

Table 34.1: Estimated levels of background in the analysis of the 1995–97 dataset.

As a crosscheck of the background estimates and of the shapes of the background distributions near the signal region, estimates were made of the number of background events expected to appear when the cuts were relaxed in predetermined ways so as to allow orders-of-magnitude-higher levels of background. Good agreement was observed and we proceeded to finally look in the signal region.

Fig. 34.2 shows the range versus energy for the events surviving all other analysis cuts in the 1995–97 dataset. The rectangular box is a rough guide to the signal region; the actual box position on the $K_{\pi 2}$ side depends on the characteristics of each event. Only events with measured momentum in the accepted region (approximately 211 MeV/c to 229 MeV/c) are plotted. The single event in the signal region is the same event that was seen in the published analysis of the 1995 data, *i.e.*, no new signal candidates were seen.

The acceptance was determined as in the published analysis, using mainly calibration data taken simultaneously with the physics data. Monte

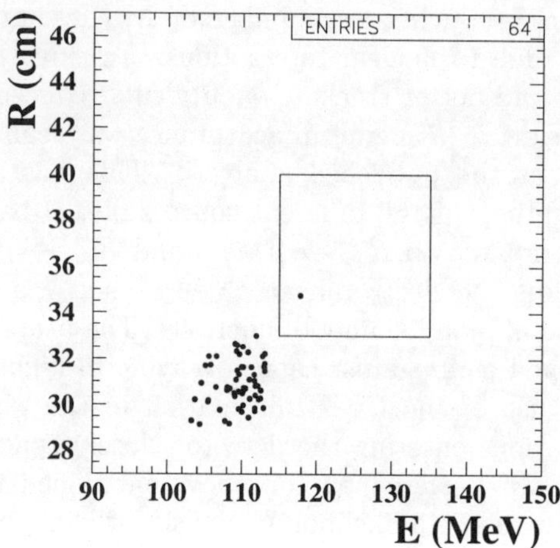

Figure 34.2: Range (R) versus energy (E) distribution for the $K^+ \to \pi^+ \nu\bar{\nu}$ data set (1995–97) with the final cuts applied. The rectangular box is a rough guide to the signal region; the actual box position on the $K_{\pi 2}$ side depends on the characteristics of each event.

Carlo was used to obtain the acceptance factors only for solid angle, $K^+ \to \pi^+ \nu\bar{\nu}$ phase space and losses from π^+ nuclear interactions and decays in flight. Final numbers for the acceptance are still being worked on; it is expected that the 1995–97 sensitivity will be about 2.5 times that of 1995 alone.

34.4 Conclusion

The E787 Collaboration has published evidence for the rare decay $K^+ \to \pi^+ \nu\bar{\nu}$ based on one clean event seen in the 1995 data sample. Analysis of the 1996–97 data, combined with a reanalysis of the 1995 data, resulted in the same single event surviving; no new signal candidates were seen. The sensitivity of the 1995–97 dataset is expected to be about 2.5 times that of 1995 alone.

The final run of E787 was completed in 1998 and analysis of that data is ongoing. The sensitivity of the 1998 run is expected to be about 2 times that of 1995–97, implying a total single-event sensitivity (1995–1998) reaching about 0.8×10^{-10}.

A new experiment, based largely on E787 with only modest upgrades and further improvement in the duty factor (and proton intensity), aims to reach a sensitivity of $(0.08 - 0.15) \times 10^{-10}$ with two years of AGS running in the RHIC era [19]. This proposal (E949) has been designated as the "highest-priority" experiment by the Brookhaven PAC for the AGS high-energy physics program during the RHIC era.

References

[1] M. Kobayashi and T. Maskawa, Prog. Theor. Phys. **49**, 652 (1973).

[2] A.J. Buras and R. Fleischer, in *Heavy Flavors II*, ed. A.J. Buras and M. Lindner (World Scientific, Singapore, 1997). See also hep-ph/9704376.

[3] G. Buchalla and A.J. Buras, Nucl. Phys. B **548**, 309 (1999).

[4] S. Adler *et al.*, Phys. Rev. Lett. **79**, 2204 (1997).

[5] Y.B. Hsiung, this volume, chapter 6. Also A. Alavi-Harati *et al.*, Phys. Rev. Lett. **83**, 22 (1999).

[6] M.S. Sozzi, this volume, chapter 8.

[7] Among many examples, Y. Nir and M. Worah, Phys. Lett. B **423**, 319 (1998); A.J. Buras, A. Romanino, and L. Silvestrini, Nucl. Phys. B **520**, 3 (1998); G. Colangelo and G. Isidori, J. High Energy Phys. **9809**, 9 (1998).

[8] K. Agashe and M. Graesser, Phys. Rev. D **54**, 4445 (1996).

[9] G. Buchalla, G. Burdman, C.T. Hill, and D. Kominis, Phys. Rev. D **53**, 5185 (1996).

[10] F. Wilczek, Phys. Rev. Lett. **49**, 1529 (1982).

[11] J. Preskill, M.B. Wise and F. Wilczek, Phys. Lett. B **120**, 127 (1983); L.F. Abbott and P. Sikivie, Phys. Lett. B **120**, 133 (1983); M. Dine and W. Fischler, Phys. Lett. B **120**, 137 (1983).

[12] M. Hindmarsh and P. Moulatsiotis, Phys. Rev. D **59**, 055015 (1999).

[13] M.S. Atiya *et al.*, Nucl. Inst. and Meth. A **321**, 129 (1992) describes the original E787 detector. References [14–18] describe most of the upgrades for the current detector.

[14] M.S. Atiya *et al.*, Nucl. Inst. and Meth. A **279**, 180 (1989).

[15] E.W. Blackmore *et al.*, Nucl. Inst. and Meth. A **404**, 295 (1998).

[16] I.-H. Chiang *et al.*, IEEE Trans. Nucl. Sci. NS **42**, 394 (1995). T.K. Komatsubara *et al.*, Nucl. Inst. and Meth. A **404**, 315 (1998). M. Kobayashi *et al.*, Nucl. Inst. and Meth. A **337**, 355 (1994).

[17] D.A. Bryman *et al.*, Nucl. Inst. and Meth. A **396**, 394 (1997).

[18] M. Burke *et al.*, IEEE Trans. Nucl. Sci. NS **41**, 131 (1994).

[19] See http://www.phy.bnl.gov/e949/e949.html

35

Search for Lepton Flavor Violation in $K_L^0 \to \mu^\pm e^\mp$ Decay

W. Molzon

Abstract

I discuss the most sensitive experiment to date to search for the muon and electron lepton number–violating decay $K_L^0 \to \mu^\pm e^\mp$. No events were detected and a limit $B(K_L^0 \to \mu^\pm e^\mp) < 4.7 \times 10^{-12}$ was set. I also discuss intrinsic limitations from rates and backgrounds that will likely restrict new experiments to about one order of magnitude improvement in sensitivity.

Despite the successes of the standard model of particle physics, a fundamental understanding of the *family structure* of the quarks and leptons remains elusive. We know that there exist 3 families of light quarks and leptons and that the quark mass eigenstates are not weak interaction eigenstates, and we have measured the parameters of the quark mixing matrix with some precision. We now also have evidence that neutrinos have mass and also mix. However, no known gauge symmetry requires conserved quantum numbers associated with each family of leptons. The observed approximate conservation of these quantum numbers is in some sense accidental; while nothing prevents lepton flavor violation (LFV), the only standard model mechanism for it is neutrino mixing. Processes like $\mu^+ \to e^+\gamma$ or $K_L^0 \to \mu^\pm e^\mp$ will occur at one loop level through ν mixing. However, this amplitude is suppressed by $\Delta M_\nu^2/M_W^2$ and the rate for this mechanism is well below what one could conceivably measure.

In contrast to the standard model, most scenarios for new physics do allow LFV. In some models (with leptoquarks, for example) tree-level processes are suppressed only by the propagator of a heavy-gauge particle that mediates the interaction and by possibly small mixing angles at the vertices. In other cases (supersymmetry, for example), LFV processes could occur through loop diagrams, suppressed by a GIM-like mechanism. LFV rates have been calculated in several models including some that give rates as large as current experimental limits: horizontal gauge interactions [1], left-right symmetry [2,3], technicolor [4], compositeness [5], and supersymmetry [6].

Experiments to search directly for LFV have been performed for many years, all with null results. Some of the best limits come from searches for $K_L^0 \to \mu^\pm e^\mp$ [7, 8], $K_L^0 \to \pi^0 \mu^\pm e^\mp$ [9], $K^+ \to \pi^+ \mu^+ e^-$ [10], $\mu^+ \to e^+ \gamma$ [11, 12], $\mu^+ \to e^+ e^+ e^-$ [13], and $\mu^- N \to e^- N$ [14]. A new upper limit on $K^+ \to \pi^+ \mu^+ e^-$ has also been reported at this conference [15]. The sensitivity of these processes to mechanisms that allow LFV varies. In general, kaon decay experiments are most sensitive for models that relate lepton and quark family numbers; an example is leptoquarks, which carry both quark and lepton number.

In the remainder of this paper, I discuss the most stringent test of LFV in the kaon sector, a search for $K_L^0 \to \mu^\pm e^\mp$. This mode is closely related to $K^+ \to \pi^+ \mu^+ e^-$. They differ in the Lorentz structure of the underlying physics; the K_L^0 mode is pseudoscalar or axial vector, while the K^+ mode is scalar or vector. For a LFV V-A interaction, the $K^+ \to \pi^+ \mu^+ e^-$ branching fraction would be smaller by a factor of more than 10 by virtue of the larger K^+ total decay rate and the phase space suppression of the 3-body decay mode.

The experimental difficulties of a search at the 10^{-12} level consist of getting the required kaon flux, building an apparatus with sufficient acceptance and rate-handling capability, and rejecting processes that could mimic a signal. Representative parameters of a beam line and experiment that has this sensitivity are a beam with $2 \times 10^8 K_L^0$ per accelerator pulse, 8% decay probability, 1.5% acceptance (including efficiency of event selection criteria), and 3×10^6 pulses (3000 hours). The performance is relatively insensitive to kaon energy, with the exception that low (< 10 GeV) energy allows the use of threshold Cerenkov counters in the electron identification, simplifying both initial event selection and subsequent analysis.

The principal background in this experiment arises from the decay $K_L^0 \to \pi e \nu$, with $\pi \to \mu \nu$ decay. Rejecting such events requires excellent kinematic measurements and particle identification. At the data rates required, coincidences of two kaon decays, each giving a lepton, are also a potential source of background. The detector is required to run at high rates (up to 1 MHz per tracking detector element) and have low mass in order to minimize scattering, the limiting contribution to kinematic resolution.

The E871 experiment was done at Brookhaven National Laboratory; the beam-line and apparatus have been described previously [16–18]. E871 used a K_L^0 beam produced using a 24 GeV proton beam from the Alternating Gradient Synchrotron (AGS). The AGS delivered $\sim 1.5 \times 10^{13}$ protons per 1.5 s spill each 3.6 s, producing a 65 μstr beam of $2 \times 10^8 K_L^0$ per pulse in the momentum interval $2 < p_K < 8$ GeV/c. Photons in the beam were attenuated using lead foils in a sweeping magnet immediately following the target. The beam was produced at an angle of 3.75° to minimize the neutron flux; the n/K ratio was about 10. This is the most intense neutral kaon beam ever made.

Kaons decayed in an 11 m long evacuated tank terminated by a thin

Kevlar and Mylar window and were detected in a magnetic spectrometer and system of particle identification counters. The apparatus is shown in fig. 35.1. The spectrometer differed from that of earlier experiments in the use of a beam-stop [18] in the first of two analyzing magnets. This reduced rates in downstream detectors and allowed them to be essentially continuous across the projected beam-line. Features of the spectrometer that

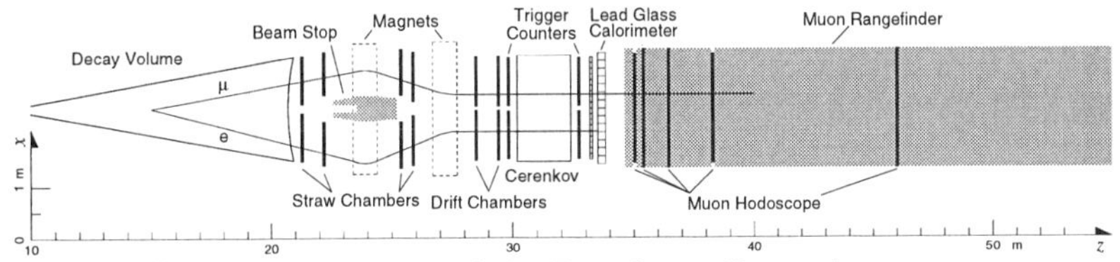

Figure 35.1: Plan view of the E871 beam line and apparatus.

were important in achieving the design goals were the use of two sequential analyzing magnets to measure redundantly the momenta of charged decay particles and the use of small-diameter straw chambers operated with a gas with high drift velocity in the upstream part of the apparatus, where rates were highest. The magnets were operated with integral B fields corresponding to 440 and 240 MeV/c momentum kicks of opposite sign, such that particles from 2-body decays of K_L^0s were approximately parallel to the beam-line following the second magnet.

The spectrometer was followed by scintillation hodoscopes (TSC), used to select events with two charged particles and define the event time. A segmented threshold Cerenkov detector (CER) and lead glass calorimeter (PBG) were used to identify electrons. Muons were identified by scintillation hodoscopes (MHO) and a range finder (MRG) downstream of an iron absorber. The amount of material between successive planes of the MRG increased with depth and corresponded to 5% increments of muon range.

Online event selection was done in multiple stages and was typical of modern, high-rate experiments. The level implemented in electronics modules required two tracks with approximately parallel trajectories after the second analyzing magnet and spatially correlated signals in lepton identification counters and resulted in a rate of ~7 kHz. Additional selection used online track reconstruction and kinematic analysis, implemented in a set of Unix processors.

The goal of the analysis was to reduce the expected background to 0.1 events or less while maintaining good efficiency for detecting $K_L^0 \to \mu^\pm e^\mp$ decays; it has been described previously [17]. To ensure that selection criteria were free of bias from knowledge of potential signal events, they were chosen by studying $K_L^0 \to \mu^\pm e^\mp$ candidates with $M_{\mu e} > 485$ MeV/c^2 but excluding a region that would contain signal events (see fig. 35.3).

To minimize accidental backgrounds, the particle times (measured in the TSCs and drift chambers) were required to be consistent within measurement uncertainty and the particle trajectories were required to project to a common vertex within the neutral beam and vacuum decay region. Events were selected on the basis of the quality of the kinematic fit in order to reject those with tracking mismeasurement or with pion decay or large angle scattering within the spectrometer. An essential feature of this background suppression was the redundant measurement of particle momenta and the requirement that the two measurements agree. Events were required to have appropriate signals in the particle identification detectors. Redundant identification methods for both muons and electrons were essential to eliminate backgrounds in which a pion is mistaken as an electron and an electron is mistaken as a muon, a rare process, but one in which the resulting value of $M_{\mu e}$ can exceed M_K.

Events satisfying the above selection criteria (but outside the exclusion region) were studied to determine their origin. Events in which the particle momenta are correctly measured are kinematically constrained to have $M_{\mu e}$ < 489.3 MeV/c^2. The Gaussian resolution in $M_{\mu e}$ inferred from the measured $K_L^0 \to \pi^+\pi^-$ mass resolution was 1.38 MeV/c^2; hence, background arising from Gaussian resolution is negligible. Kinematic distributions of events with $M_{\mu e} < 490$ MeV/c^2 were well reproduced in shape and normalization by a Monte Carlo simulation.

Larger errors in $M_{\mu e}$ were shown by Monte Carlo simulation to arise when electrons scattered at large angles (Mott scattering) in the vacuum window or in the first tracking detector (0.12% and 0.23% radiation lengths, respectively) and the pion decayed to a muon before the spectrometer. In particular, when both the scattering and decay occurred in the plane defined by $\vec{p}_e \times \vec{p}_\pi$ (the K_L^0 decay plane) events were not rejected by vertex quality criteria and could have values of $M_{\mu e}$ near M_K and small momentum transverse to the incident kaon direction (p_T). The kinematic distributions of events with $M_{\mu e} > 493$ MeV/c^2 were well reproduced by a Monte Carlo simulation of this process. The charged particle momentum asymmetry and the component of p_T in the decay plane (denoted by $p_T^{\|}$) are shown in fig. 35.2 together with the Monte Carlo predictions. The sign of $p_T^{\|}$ is taken to be positive if it lies on the electron side of the spectrometer. Upstream scatter events are characterized by large momentum asymmetry and large $p_T^{\|}$. The requirements that $(p_\mu - p_e)/(p_\mu + p_e) < 0.5$ and that $p_T^{\|}$ be small (typically 7 MeV/c, depending on electron energy) reduced this background significantly. The simulation showed that events in this region passing the selection criteria are dominated by large upstream scatters. Reducing this source of background by an additional factor of 10 would require tighter selection criteria that would result in a 50% acceptance loss.

Without additional selection criteria, accidental coincidences of two semileptonic decays were calculated to contribute significant background. Monte Carlo simulations showed that at least one of the pion trajectories is typically fully contained in the spectrometer. Events with three or more fully reconstructed tracks in the spectrometer were rejected, reducing this background by an order of magnitude. Fig. 35.2 shows the remaining ac-

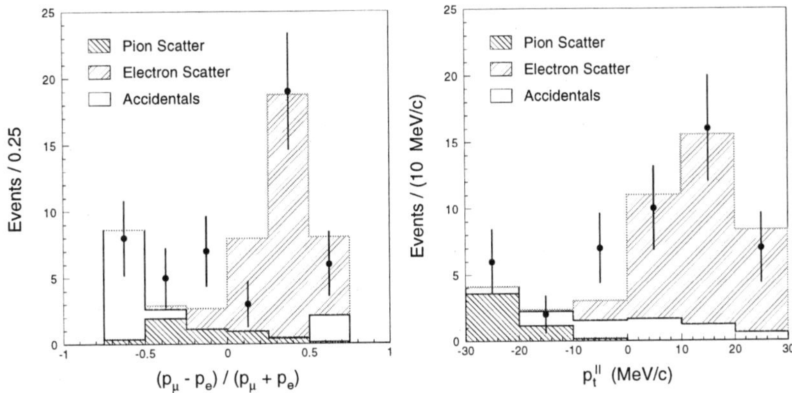

Figure 35.2: Data and Monte Carlo distributions of (a) momentum asymmetry and (b) $p_T^{||}$. Events shown satisfy all selection criteria except those on the quantities displayed above and have $M_{\mu e} > 493$ MeV/c^2 and $100 < p_T^2 < 900$ (MeV/c)^2.

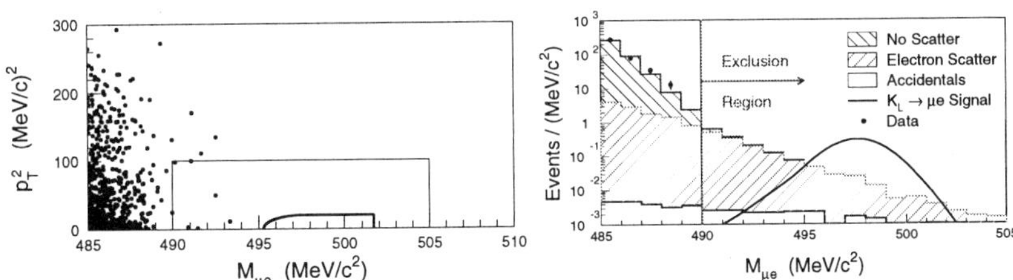

Figure 35.3: On the left is a scatter plot of p_T^2 versus $M_{\mu e}$. The exclusion region for the blind analysis is indicated by the box. The signal region is indicated by the smaller contour. The plot on the right shows the expected distributions for the signal and backgrounds, where the signal shape is shown for a branching fraction of 2×10^{-12}.

cidental events are at large negative momentum asymmetries and removed by the cut on that quantity.

Having understood the sources of background, the selection criteria, including choice of the signal region, were optimized by simultaneously varying values of relevant selection parameters to maximize the sensitivity to signal while suppressing the expected contribution from the dominant source of background to 0.1 event. After all selection criteria (including the choice of the signal region) were determined, all data (including those in the exclusion region) were reanalyzed. Fig. 35.3 shows the final distribution in p_T^2 versus $M_{\mu e}$. There are no events in the signal region and the number of events in the exclusion region is consistent with the Monte Carlo prediction.

The $K_L^0 \to \mu^\pm e^\mp$ sensitivity was determined from the number of $K_L^0 \to \pi^+\pi^-$ decays in the minimum bias sample. These events were required to satisfy an appropriate subset of the final selection criteria discussed above, and were required to have no PBG signals consistent with those of an electron. A fit in the p_T^2 versus $M_{\pi\pi}$ plane was done to subtract residual $K_L^0 \to \pi\mu\nu$ background and to determine the number of $K_L^0 \to \pi^+\pi^-$ events. Small differences in geometric acceptance and cut efficiencies were determined by Monte Carlo simulation.

In the absence of any signal events, the 90% confidence level upper limit on $B(K_L^0 \to \mu^\pm e^\mp)$ was calculated as 2.3 divided by the effective sensitivity of the experiment. The latter was derived from the number of detected $K_L^0 \to \pi^+\pi^-$ decays detected in a prescaled sample of data (with no lepton identification requirements in the trigger) divided by the known $K_L^0 \to \pi^+\pi^-$ branching fraction. The result was corrected for appropriate factors for ratios of acceptances and efficiencies for the two modes. The resulting limit is $B(K_L^0 \to \mu^\pm e^\mp) < 4.7 \times 10^{-12}$.

The extent to which a background-free experiment with better sensitivity could be done can be inferred from the studies done by E871 and the understanding of background sources that was achieved. As an example, I consider an experiment with a goal of detecting 1 event at 10^{-13} branching fraction, a factor of 20 improvement with respect to E871. Since the background is dominated by scattering in the vacuum window and first chamber, that could be reduced by reducing the mass of the chambers and putting the first (or all) straw chambers in vacuum. A total reduction in material by a factor of two could be envisaged. Using scintillating fibers would exacerbate the problem; even two layers of 500 μm fiber is equivalent to the E871 upstream scattering material. The biggest gain in background rejection is achieved by selecting events with $p_\mu < p_e$, which improves signal to noise (S/N) by a factor of ~ 10 and by requiring $p_t^{||} > 0$, which improves S/N by a factor of ~ 2.5. Each of these results in an acceptance loss of about a factor of 2.

Assuming a total S/N improvement of 10, a 50% reduction in scattering probability, and a 50% loss of acceptance due to tighter cuts, an increase in rate $\times$ acceptance $\times$ running time of 40 would be required to reach the sensitivity goal. Assuming a factor of two increased acceptance (difficult to achieve) and a running time twice longer, the K_L flux would need to increase by a factor of 10. This would impose severe constraints on the tracking system (already the single ware rates are nearly 1 MHz). The increased flux (and the increased acceptance) would result in increased losses from the requirement that no extra particle track be detected. Additional rejection could be achieved with improved timing (in the TSCs for example) and improved vertex resolution (dominated by scattering in the vacuum window and first drift chamber). An alternate approach would be to increase significantly the acceptance. However, the loss of acceptance is largely geometrical (including the effect of having no detectors in the beam, the implementation of a redundant momentum measurement, and the requirement of parallel tracks after the second magnet) and improvements are not easy, even at substantially higher cost. Clearly, a detailed design study and simulation are needed to optimize such an experiment, but the difficulties are clear from the E871 experience.

In summary, the most intense neutral kaon beam has been used to set the best limit on a LFV kaon decay: $B(K_L^0 \to \mu^\pm e^\mp) < 4.7 \times 10^{-12}$. This represents the best limit on any decay mode of a hadron. The lower limit on the mass of an intermediate particle with electroweak coupling strength and maximal mixing at quark and lepton vertices is ~150 TeV/c^2, better than can be achieved by any current or proposed K decay experiment. The sources of background that limit possible improvements on this search have been identified and modeled precisely.

Acknowledgments

I am indebted to my collaborators from BNL E871 collaboration who were responsible for constructing the apparatus and collecting and analyzing the data. In particular, S. Wojcicki proposed the use of a beam stop in the spectrometer, an idea that worked very well. R. Lee did much of the analysis of the $K_L^0 \to \mu^\pm e^\mp$ data and did the calculation of the level of different sources of background.

References

[1] R.N. Cahn and H. Harari, Nucl. Phys. B **176**, 135 (1980).

[2] P. Langacker, S.U. Sankar, and K. Schilcher, Phys. Rev. D **38**, 2841 (1988).

[3] Z. Gagyi-Palffy, A. Pilaftsis, and K. Schilcher, Nucl. Phys. B **513**, 517 (1998).

[4] S. Dimopoulos and J. Ellis, Nucl. Phys. B **182**, 505 (1981).

[5] J. Pati and H. Stremnitzer, Phys. Lett. B **172**, 441 (1986).

[6] B. Mukhopadhyaya and A. Raychaudhuri, Phys. Rev. D **42**, 3215 (1990).

[7] K. Arisaka *et al.*, Phys. Rev. Lett. **70**, 1049 (1993).

[8] T. Akagi *et al.*, Phys. Rev. D **51**, 2061 (1995).

[9] K. Arisaka *et al.*, Phys. Lett. B **432**, 230 (1998).

[10] A. Lee *et al.*, Phys. Rev. Lett. **64**, 165 (1990).

[11] R. Bolton *et al.*, Phys. Rev. D **38**, 2077 (1988).

[12] M. L. Brooks *et al.*, hep-ex/9905013, submitted to Phys. Rev. Lett. (1999).

[13] U. Bellgardt *et al.*, Nucl. Phys. B **299**, 1 (1988).

[14] C. Dohmen *et al.*, Phys. Lett. B **317**, 631 (1993)

[15] M.E. Zeller, this volume, chapter 36.

[16] D. Ambrose *et al.*, Phys. Rev. Lett. **81**, 4309 (1998).

[17] D. Ambrose *et al.*, Phys. Rev. Lett. **81**, 5734 (1998).

[18] J. Belz *et al.*, Nucl. Instrum. Methods, Sect. A, **428**, 239 (1999).

[19] A correlated cut between $p_T^{\|}$ and p_e was made: $p_T^{\|} < (3p_e - 2.5) \times 10^{-3}$, where $p_T^{\|}$ and p_e are in GeV/c.

[20] C. Caso *et al.*, Eur. Phys. J. C **3**, 1 (1998).

36

A Review of Recent Results from BNL E865: Rare K^+ Decays in Flight

Michael E. Zeller[1]

Abstract

Experiment E865 studying rare K^+ decay modes has been completed at Brookhaven. A review of some of its results is presented, including a new limit on the decay $K^+ \to \pi^+\mu^+e^-$, and branching ratio measurements for the decays $K^+ \to \pi^+e^+e^-$ and $K^+ \to \pi^+\mu^+\mu^-$.

36.1 Introduction

Brookhaven experiment E865, "An Improved Search for the Decay $K^+ \to \pi^+\mu^+e^-$" is now completed after taking data during the AGS running cycles from 1995 to 1998. Conceived as a search for the muon number–violating decay $K^+ \to \pi^+\mu^+e^-$, $K_{\pi\mu e}$, it involved a large acceptance spectrometer system with high rate and very good particle identification capabilities. As such it also acquired significantly more data for several rare decay modes of the K^+ meson than had been previously published. In this report new results from these data, as well as an overview of the experimental apparatus, are described.

36.2 Apparatus

The apparatus, shown in fig. 36.1, resided downstream from a 5 m long vacuum decay volume. A dipole magnet at the exit of the decay region separated particles by charge (negative to beam left, positive to beam right) and reduced the charged particle background originating upstream of the

[1]Representing the BNL E865 Collaboration: D. Lazarus, H. Ma, P. Rehak (Brookhaven National Laboratory); G. S. Atoyan, V. V. Issakov, A. A. Poblaguev (INR, Moscow); J. Egger, W. D. Herold, H. Kaspar (Paul Scherrer Institute); W. Menzel, H. Weyer (Basel); B. Bassalleck, S. Eilerts, H. Fischer, J. Lowe (New Mexico); R. Appel, D. N. Brown, N. Cheung, C. Felder, M. Gach, D. E. Kraus, P. Lichard, A. Sher, J. A. Thompson (Pittsburgh); D. Bergman, S. Dhawan, H. Do, J. Lozano, W. Majid, M. Zeller (Yale); S. Pislak, P. Truöl (Zurich).

detector. Pairs of proportional chamber packages (P1-P2 and P3-P4) were arrayed on either side of a second dipole magnet to form the momentum-analyzing spectrometer system. Trigger hodoscopes were located between the upstream packages (D Hod), and immediately downstream of the second pair (A Hod).

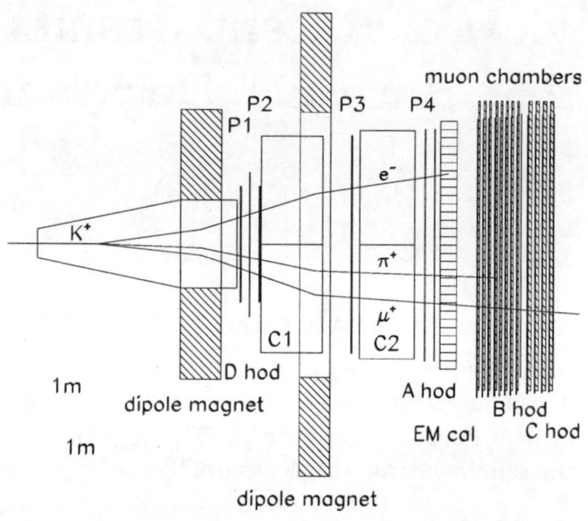

Figure 36.1: Plan view of the E865 apparatus.

Gas filled Cerenkov counters (C1, C2) at atmospheric pressure, and divided by thin membranes into left and right sides, sat fore and aft of the second dipole magnet. Since the left side of the apparatus detected electrons from the $K_{\pi\mu e}$ decay, and since a major source of potential background was misidentification of the π^- from $\pi^+\pi^+\pi^- K$ decays (termed τ decays), this side of the Cerenkov counters was filled with hydrogen to lower the probability of identifying a pion as an electron. On the right side the major potential background was from $K^+ \to \pi^+\pi^0$; $\pi^0 \to e^+e^-\gamma$ events (Dalitz decays, K_{Dal}). Thus the Cerenkov counters on this side were filled with a heavier gas (CO_2 or methane) to efficiently veto positrons.

Downstream of the last proportional chamber was an electromagnetic calorimeter followed by a muon detector consisting of 12 walls of proportional tubes with x and y readout separated by steel plates. Embedded in this muon stack was a scintillation counter array (B Hod), and behind the stack was another similar array (C Hod). The B array was used to identify muons at the trigger level.

With this apparatus we could identify electrons as those particles giving light in the appropriate Cerenkov counters and having pulse height in the calorimeter consistent with the momentum measured in the spectrometer. Muons were seen as particles giving no light in the Cerenkov counters, minimum ionizing pulse height in the calorimeter, and a range in the muon stack consistent with their momentum. Pions had no light in the Cerenkov counters and a range shorter than that expected for minimum ionizing particles.

36.3 New results

36.3.1 $K^+ \to \pi^+\mu^+e^-$

In spite of the successes of the standard model of particle physics, there are several issues that have yet to be resolved before we can say we have a true theory of elementary particles. Recognizing the inadequacies of the theory, the theoretical community has developed "extensions to the standard model", several of which address the nature of the Higgs mechanism. Some of those extensions, such as supersymmetry and technicolor, permit muon number–violating decays at levels that (when the theories were formulated) might be observable with current technology. Motivation for the study of rare decays of the K^+ meson also comes from a curiosity about the exactness of the apparent, but not theoretically motivated, "law" of muon number conservation.

In the mid and late 1980s experiment BNL E777 was performed to search for the $K_{\pi\mu e}$ decay. Prior to that experiment the upper limit on the branching ratio for that decay mode was about 4×10^{-9} [1]. E777 reduced that limit to 2×10^{-10} (90% CL) [2]. In the 1995 running period E865 achieved the same limit as E777 thus lowering the limit with the accumulated data to 1×10^{-10} [3].

The 1996 data sample was four times that of the 1995. In its analysis we have employed a likelihood method and used the resulting likelihood as an evaluator of the quality of possible events as being $K_{\pi\mu e}$ signal. In this method various distributions, *e.g.*, vertex quality, invariant mass, particle ID detector response, consistency that the reconstructed kaon momentum vector originated at the production target, etc., were normalized to unit area and used as probability density functions. The value of the probability of a particular event coming from the population implied by each function was calculated from each distribution and the logs of these probabilities were added to form the log of the likelihood for that event.

The plot with which we choose to display our results is one of log-likelihood vs. the invariant mass of the three detected particles. In order to ensure that both the log-likelihood and mass distributions were correctly simulated for the $K_{\pi\mu e}$ mode (where we have no large sample from which we can construct data distributions) we simulated such distributions for τ decays and compared these with data. The agreement was satisfactory.

Fig. 36.2 displays two-dimensional scatterplots of log-likelihood vs. mass for the $K_{\pi\mu c}$ data sample and for the simulation of this plot assuming $K_{\pi\mu e}$ decay events; the accepted mass region is between the horizontal lines shown, *i.e.*, at three standard deviations in $M_{\pi\mu e}$. Also on these plots are the 20% and 10% log-likelihood demarcation lines, *i.e.*, values of log-

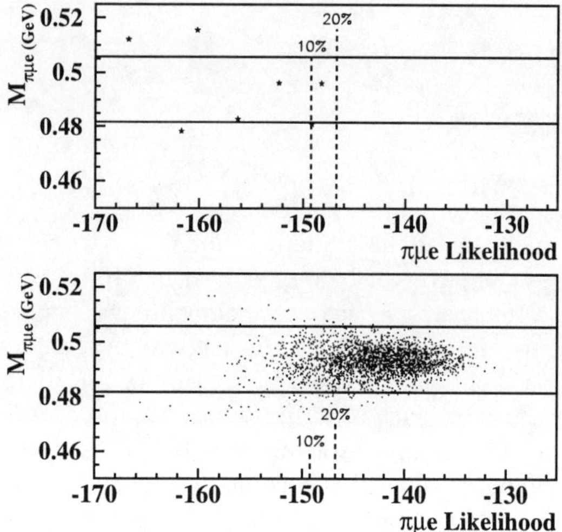

Figure 36.2: Scatter plot of E865 1996 $K_{\pi\mu e}$ data (top) and Monte Carlo (bottom). The abscissa is the log-likelihood of the reconstructed events under the $K_{\pi\mu e}$ hypothesis, the ordinate is the invariant mass of the detected particles. The horizontal lines demark the three STD mass region.

likelihood for which the probability of having worse likelihood for the $K_{\pi\mu e}$ hypotheses are 20% and 10%, respectively.

From various data distributions we could also construct log-likelihood distributions for possible background modes that might be interpreted as $K_{\pi\mu e}$. Sources of such events are other kaon decay modes where an error in kinematic measurement and/or particle identity had been made, and "accidentals" where three particles, not coming from a single decay, occur accidently in time and satisfy all cut conditions. In the region of $K_{\pi\mu e}$ events the latter mode was predicted to yield the most background events, ~ 0.25 with log-likelihood less than the 20% point.

With the known log-likelihood and mass distributions for $K_{\pi\mu e}$ and background events we employ the likelihood method of [4] to evaluate the plot in fig. 36.2. This evaluation results in the maximum likelihood occurring at 0.5 events and a 90% confidence limit at 2.0 events. After including all efficiencies and acceptances, and normalizing to the τ mode, we place a preliminary 90% CL upper limit on the $K_{\pi\mu e}$ branching ratio of 4.0×10^{-11} for the 1996 data sample.

Combining the latter results with those of E777 (2.0×10^{-10}) and the 1995 sample of E865 (2.1×10^{-10}) yields a preliminary upper limit for $K_{\pi\mu e}$ of 2.9×10^{-11} (90% CL). The 1998 data sample contains 2 times that of 1996, so if the final sample is background free, we have the statistical sensitivity of $\sim 10^{-11}$.

36.3.2 $K^+ \to \pi^+ e^+ e^-$

The decay $K^+ \to \pi^+ e^+ e^-$, $K_{\pi ee}$, proceeds via a flavor-changing neutral current and is highly suppressed by the GIM mechanism. The decay rate was first calculated [5] assuming a short-distance $s \to d\gamma$ transition at the quark level. Later it was realized that the long-distance effects dominate the decay mechanism [6]. Several recent calculations, which study the rate and invariant electron-positron mass (M_{ee}) distributions, were performed within the framework of chiral QCD perturbation theory (ChPT) [7–9], an approach that has been quite successful at modeling many decay modes of the light mesons.

The first few events for this decay were observed at CERN [10]. Two subsequent experiments at the Brookhaven AGS, E777 [6] and E851 [11], observed 500 and 800 events, and measured branching ratios of 2.75 ± 0.26 and 2.81 ± 0.20 ($\times 10^{-7}$), respectively. The new data from E865, with their greater statistics and larger kinematic reach, allow a more detailed study of the properties of this decay [12].

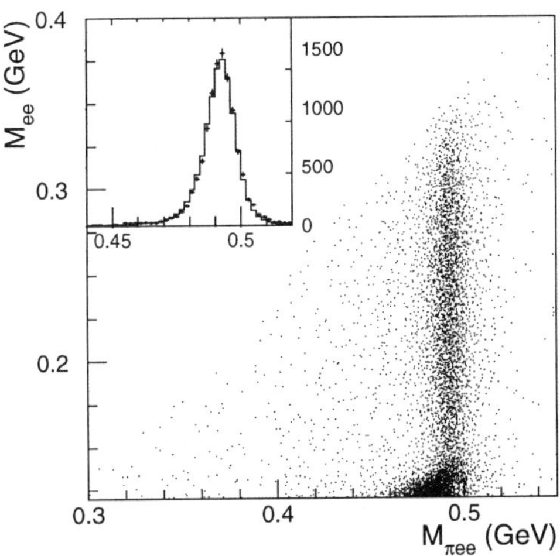

Figure 36.3: Scatter plot M_{ee} versus $M_{\pi ee}$ for $K_{\pi ee}$ candidates. Insert: $M_{\pi ee}$ mass for $K_{\pi ee}$ candidates with $M_{ee} > 0.15$ GeV. The histogram shows the Monte Carlo simulation.

The new data are displayed in fig. 36.3 [13]. Cutting at $M_{ee} > 0.15$ GeV leaves a nearly pure collection of $K_{\pi ee}$ events as seen in the insert of the figure. The onset of K_{Dal} events can be seen for $M_{ee} < m_{\pi^0}$ and $M_{\pi ee} < m_K$. The final signal sample contains 10300 $K_{\pi ee}$ candidates including 1.2 percent background events.

The essential distributions necessary for the interpretation of the data are shown in fig. 36.4. Since the decay is supposed to proceed through one photon exchange, *i.e.*, by a vector interaction (V), one expects an

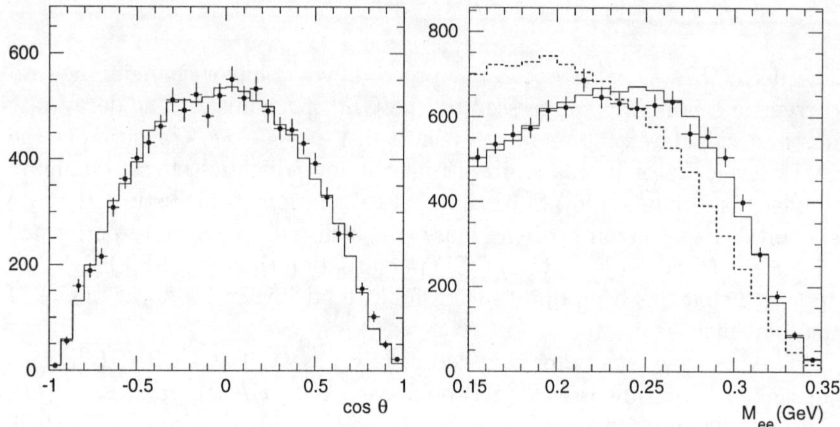

Figure 36.4: Angular (left) and invariant mass (right) distributions for $K_{\pi ee}$ events (data points) compared to Monte Carlo simulations (histogram) assuming a pure vector interaction. A linear form factor parameterization with $\delta = 2.14$ is used. The dashed histogram (right) corresponds to a constant form factor ($\delta = 0$).

angular distribution proportional to $\sin^2\theta$, where θ is the angle between the positron and pion momentum vectors in the center of mass of the e^+e^- pair. The presence of other decay mechanisms, however, may produce small admixtures of scalar (S) or tensor (T) terms that would distort this. Fitting a two-dimensional distribution, whose projections are shown in fig. 36.4, we find good agreement with a vector interaction only. At 90% CL at most 2% of the branching ratio could result from either scalar or tensor interactions.

The mass distribution for $K^+ \to \pi^+e^+e^-$ from a vector interaction can be described by $d\Gamma/dz = C(z)|f_V(z)|^2$, where $z = M_{ee}^2/m_K^2$, $C(z)$ is a function involving known constants and kinematic factors, and the form factor $f_V(z)$ can be determined by fitting the observed spectrum in the experimentally accessible range $0.1 < z < 0.51$. We have used two different parameterizations of the form factor, one model independent [eq. (36.1)] and the other derived from ChPT [9] [eq. (36.2)]:

$$f_V(z) = f_0(1 + \delta z + \delta' z^2), \tag{36.1}$$

$$f_V(z) = a_+ + b_+z + w^{\pi\pi}(z), \tag{36.2}$$

where $f_0, \delta, \delta', a_+$, and b_+ are free parameters, and $w^{\pi\pi}$ is the contribution from a pion loop graph.

The best fit for the formulation of eq. (36.1) yields $f_0 = 0.533 \pm 0.012$, $\delta = 0.97 \pm 0.44$, and $\delta' = 1.99 \pm 0.67$ with $\chi^2/n_{dof} = 16.6/17$, and that for the formulation of eq. 36.2 yields $a_+ = -0.587 \pm 0.010$ and $b_+ = -0.655 \pm 0.044$ with $\chi^2/n_{dof} = 13.3/18$. Also, a fit to eq. 36.1 without δ' has $\chi^2/n_{dof} = 22.9/18$, which is significantly reduced by including the quadratic term in z. Such a term is present in $w^{\pi\pi}$ in the ChPT approach. The resulting branching ratio is $(2.94 \pm 0.05(\text{stat.}) \pm 0.13(\text{syst.}) \pm 0.05(\text{model})) \times 10^{-7}$.

In the first, lowest nontrivial order ChPT calculation [7], $\mathcal{O}(p^4)$, the pion loop term was the dominant source of z dependence of the form factor. The large value of b_+ in the above results shows that this is a poor approximation. The $\mathcal{O}(p^4)$ ChPT formulation also related the branching ratio to the form factor in a way that is inconsistent with our results by several standard deviations, thus further indicating the need to go to next to leading order, $\mathcal{O}(p^6)$. Significant z dependence from terms other than $w^{\pi\pi}$ is expected at that order [8,9], and the amplitude linear in z [eq. (36.2)] is thought to represent all contributions other than the pion loop term [9].

36.3.3 $K^+ \to \pi^+\mu^+\mu^-$

In the standard model the decay $K^+ \to \pi^+\mu^+\mu^-$, $K_{\pi\mu\mu}$, proceeds through the same mechanism as $K_{\pi ee}$. As such, knowing the branching ratio and form factor structure of the latter allows one to make model-free predictions about those quantities for the former. However, because the branching ratio is quite small $\mathcal{O}(10^{-7})$, and because the final state leptons are significantly more massive that the electron and positron of $K_{\pi ee}$, $K_{\pi\mu\mu}$ provides a testing ground for physics beyond the standard model. In fact, the published value of the $K_{\pi\mu\mu}$ branching ratio, $5.0 \pm 1.0 \times 10^{-8}$ [14], is well below the value of $8.3 \pm 0.2 \times 10^{-8}$ expected from $K_{\pi ee}$, possibly indicating the presence of new physics.

E865 accumulated about 400 events during its 1997 run. The data acquisition and analyses were similar to those for other modes, including a log-likelihood evaluation of candidate events as having come from the $K_{\pi\mu\mu}$ mode. Distributions of the invariant mass of the three detected particles, $M_{\pi\mu\mu}$, and of $M_{\mu\mu}$ are shown in fig. 36.5. The large background at low $M_{\pi\mu\mu}$ is due to τ events where a π^+ and π^- decay into muons in their flight through the apparatus (with joint probability of $\sim 2\%$). A parameterization of the number of these events as a function of $M_{\pi\mu\mu}$ was made, as displayed in the figure, and the ratio of signal to background was found to be unaffected by details of the various cuts employed in the analysis.

As with $K_{\pi ee}$ the $\cos\theta$ distribution (where θ is the angle between the π^+ and positively charged lepton—in this case the μ^+) is consistent with the decay proceeding through a vector interaction; and the form factor parameters describing the the $M_{\mu\mu}$ spectrum are consistent with those of $K_{\pi ee}$. Finally, after normalizing to τ decays we determine a preliminary value of the $K_{\pi\mu\mu}$ branching ratio to be $(9.23 \pm 0.6_{\text{stat.}} \pm 0.6_{\text{sys.}}) \times 10^{-8}$, yielding a ratio to the above measured $K_{\pi ee}$ branching ratio, $B(\pi\mu\mu)/B(\pi ee)$, of $0.33 + 0.07$. This ratio is consistent with theoretical expectations assuming standard model physics only. Thus, we conclude that that the mechanisms for these two semirare decay modes are consistent with being the same—as theoretically expected.

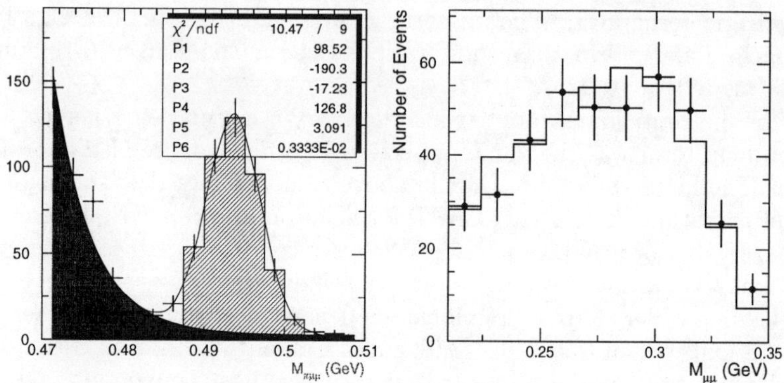

Figure 36.5: $M_{\mu\mu}$ distribution (left)—points: data; histogram: fit with $\delta = 2.14$ (that of $K_{\pi ee}$). Invariant mass distribution for $K_{\pi\mu\mu}$ events (right)—points: data; dark shaded area: fit to background; light shaded area: signal.

36.3.4 Other Decay Modes

E865 has acquired many times more data for several other K^+ decay modes than have been collected in the past. Two of those modes are $K^+ \to \mu^+\nu_\mu e^+e^-$ and $K^+ \to e^+\nu_e e^+e^-$. With roughly one hundred times the previous sample of data than that previously acquired, we have the opportunity to study the three form factors for these modes (F_V, F_A, and R) in some detail. $\mid F_A + F_V \mid$ and $\mid F_A - F_V \mid$ have recently been measured via the closely related radiative $K_{\mu 2}$ and K_{e2} modes [15], but only through the modes with virtual photons can each of the terms be explicitly determined. Our preliminary results show that the sign of each of the three factors is positive.

The K_{e4} mode has been of great interest for the last three decades because it allows one to study the low-energy $\pi\pi$ interaction in a regime with no other strongly interacting particles present, as well as being a fertile ground for testing models of weak hadronic currents, *e.g.*, ChPT. E865 has collected 300,000 events with a fairly flat acceptance over the decay phase space. This is 10 times more than the previously observed number of events [16].

Improving on the uncertainties in the published values of the $\pi\pi$ scattering length and form factors requires a very detailed understanding of the detection apparatus acceptance and response. At the writing of this report we are pursuing such an understanding. Our first results show agreement with [16] but with the expected reduction in statistical uncertainties.

36.4 Conclusion

Begun as a search for the decay $K^+ \to \pi^+\mu^+e^-$, E865 has been successful in establishing a low limit on the branching ratio for that mode and making significantly improved measurements of other semirare decays. Specifically, $B(K^+ \to \pi^+\mu^+e^-) < 2.9 \times 10^{-11}$ (90% CL, preliminary), $B(K^+ \to \pi^+e^+e^-) = (2.94 \pm 0.05_{\text{stat.}} \pm 0.13_{\text{sys.}} \pm 0.05_{\text{model}}) \times 10^{-7}$, and $B(K^+ \to \pi^+\mu^+\mu^-) = (9.23 \pm 0.6_{\text{stat.}} \pm 0.6_{\text{sys.}}) \times 10^{-8}$. While it has not found the "new physics" it sought, it has set a significant limit and provided grist for the mill of those who are trying to understand hadronic interactions.

References

[1] A.M. Daimant-Berger *et al.*, Phys. Lett. B **62**, 485 (1976).

[2] A.M. Lee *et al.*, Phys. Rev. Lett. **64**, 165 (1990).

[3] D.R. Bergman, *A Search for the Decay $K^+ \to \pi^+\mu^+e^-$*, PhD thesis, Yale University (1997); S. Pislak, *Experiment E865 at BNL; A Search for the Decay $K^+ \to \pi^+\mu^+e^-$*, PhD thesis, University of Zürich (1997).

[4] G.J. Feldman and R.D. Cousins, Phys. Rev. D **57**, 3873 (1998).

[5] M.K. Gaillard, and B.W. Lee, Phys. Rev. D **10**, 897 (1974).

[6] C. Alliegro *et al.*, Phys. Rev. Lett. **68**, 278 (1992), and references therein.

[7] G. Ecker, A. Pich, and E. de Rafael, Nucl. Phys. B **291**, 692 (1987).

[8] J. Donoghue, and F. Gabbiani, Phys. Rev. D **51**, 2187 (1995).

[9] G. D'Ambrosio *et al.*, JHEP **8**, 4 (1998).

[10] P. Bloch *et al.*, Phys. Lett. B **56**, 201 (1975).

[11] A.L. Deshpande, *A Study of the Decay of $K^+ \to \pi^+e^+e^-$, and a Measurement of the Decay $\pi^0 \to e^+e^-$*, PhD thesis, Yale University (1995).

[12] S. Eilerts, *A High-precision Study of the Rare Decay $K^+ \to \pi e^+e^-$*, PhD thesis, University of New Mexico (1998).

[13] R. Appel *et al.*, *A New Measurement of the Properties of the Rare Decay $K^+ \to \pi^+e^+e^-$*, submitted to Phys. Rev. Lett. (1999).

[14] S. Adler *et al.*, Phys. Rev. Lett. **79**, 4756 (1997).

[15] L. Littenberg, private communication (1999).

[16] L. Rosselet *et al.*, Phys. Rev. D **15**, 574 (1977).

37

I: Chiral Perturbation Theory for Kaons; II: The $\Delta I = 1/2$-Rule in the Chiral Limit

Johan Bijnens

Abstract

I: Chiral perturbation theory is introduced and its applications to semileptonic and nonleptonic kaon decays are discussed.

II: The method of large N_c is used to calculate $K \to \pi\pi$ nonleptonic matrix elements, in particular the matching procedure between long- and short-distance evolution that takes all scheme dependence correctly into account is discussed. Numerical results reproduce the $\Delta I = 1/2$ rule without the introduction of any free parameters.

37.1 Introduction

Chiral perturbation theory (CHPT) is a very large subject now so I will only discuss it briefly and then review the present status of its use in semileptonic and nonleptonic kaon decays. It has had several major successes in rare decays that are discussed in the contribution by Isidori [1]. The application to $K_L^0 \to \pi^+\pi^-e^+e^-$ is treated by Savage [2]. As described in the second part and in several other talks [3,4], it is also very relevant in calculations of the nonleptonic matrix elements. In sec. 37.2 I very briefly describe the underlying principles. The next section reviews the application to kaon semileptonic decays; this is one of the main playgrounds for CHPT and the area of some major successes. The use in kaon decays to pions is then discussed in sec. 37.4. We treat the use of CHPT in simplifying matrix element calculations in sec. 37.4.1, predictions for $K \to 3\pi$ in sec. 37.4.2, and chiral limit cancellations in B_6 in sec. 37.4.3.

Sec. 37.5 constitutes part II of this talk. Here I describe how the large-N_c method can take into account the scheme dependence of short-distance operators and first results [5].

37.2 Chiral Perturbation Theory

CHPT grew out of current algebra where systematically going beyond lowest order was difficult. The use of effective Lagrangians to reproduce current algebra results was well known and Weinberg showed how to use them for higher orders [6]. This method was improved and systematized by Gasser and Leutwyler in classic papers [7] and proving that CHPT is indeed the low-energy limit for QCD using only general assumptions was done by Leutwyler [8]. Recent lectures are [9].

The assumptions underlying CHPT are:

- Global chiral symmetry and its spontaneous breaking to the vector subgroup: $SU(3)_L \times SU(3)_R \to SU(3)_V$.
- The Goldstone bosons from this spontaneous breakdown are the only relevant degrees of freedom, *i.e.*, the only possible singularities.
- Analyticity, causality, cluster expansion, and special relativity [8].

The result is then a systematic expansion in meson masses, quark masses, momenta, and external fields. The external field method allows one to find the minimal set of parameters consistent with chiral symmetry, and the rest is basically only unitarity. With current algebra and dispersive methods it is in principle also possible to obtain the same results but the method of effective field theories is much simpler.

So for any application of CHPT two questions should be answered:

1. Does the expansion in momenta and quark masses converge?
2. If higher orders are important then:
   - Can we determine all the needed parameters from the data?
   - Can we estimate them if not directly obtainable?

37.3 Semileptonic Decays

The application of CHPT to semileptonic decays has been reviewed in [10] and in [11]. Since then first results at order p^6 have appeared. The situation order by order is:

$\mathcal{L}_2$: 2 parameters: F_0, B_0 (+quark masses).

$\mathcal{L}_4$: 10+2 parameters [7]; 7 are relevant; 3 more appear in the meson masses. In addition we also have the Wess-Zumino term and one-loop contributions.

$\mathcal{L}_6$: 90+4 parameters [12, 13]. In addition there are two-loop diagrams and one-loop diagrams with $\mathcal{L}_4$ vertices.

	F_π/F_0	F_K/F_π	$(F_K/F_\pi)^{(6)}$
p^2	1	1	—
p^4	1.07	1.22	—
p^6 set A	0.96 (1.08)	1.27 (1.30)	0.05 (0.08)
p^6 set A μ=0.9 GeV	0.96 (1.10)	1.30 (1.34)	0.08 (0.12)
p^6 set B	0.90 (1.02)	1.25 (1.28)	0.035 (0.06)

Table 37.1: Results for the ratios of F_π, F_K, and the decay constant in the chiral limit. The size of the p^6 only is in the last column.

37.3.1 General Situation

$\mathbf{p^2}$ Current algebra: 1960s

$\mathbf{p^4}$ One-loop: 1980s, early 1990s

$\mathbf{p^6}$ (see [14])
- Estimates using dispersive and/or models: done
- Double-log contributions: mostly done [15].
- Two-flavor full calculations: done.
- Three-flavor full calculations: few done, several in progress.

$\mathbf{e^2p^2}$ In progress.

Experiment Progress from DAΦNE, NA48, BNL, KTeV, ...

37.3.2 K_{l2}

These decays are used to determine F_K and test lepton universality by comparing $K \to \mu\nu$ and $K \to e\nu$. F_π is similarly determined from $\pi \to \mu\nu$. The theory is now known to NNLO (next-to-next-to-leading order) fully in CHPT [16] (for F_π also [17]) The results are shown in table 37.1 when the contributions from the p^6 Lagrangian are set to zero, i.e., the corresponding coefficients C_i^r are assumed to vanish at the scale indicated. The numbers in brackets are the extended double log approximation of [15]. The inputs are $10^3 L_{i=4,10} = (-0.3, 1.4, -0.2, 0.9, 6.9, 5.5)$; $\mu = 0.77$ GeV unless otherwise indicated and for set A $10^3 L_{i=1,3} = (0.4, 1.35, -3.5)$ while for set B $10^3 L_{i=4,10} = (-0.3, 1.4, -0.2, 0.9, 6.9, 5.5)$. We see that the variation with the p^4 input is sizable and that the extended double log approximation gives a reasonable first estimate for the correction.

37.3.3 $K_{l2\gamma}$

In this decay there are two form factors. The axial form factor is known to p^4 [18,19] and a similar calculation for $\pi \to e\nu\gamma$ [20] shows a 25% correction and a small dependence on the lepton invariant mass W^2. The vector form-factor is known to p^6 [21] and has a 10 to 20% correction in the relevant

	p^2	p^4	Ext. double log	Experiment
f_+	1	-0.0023	$-0.005 \to 0.004$	input for V_{us}
λ_+	0	0.031	$-0.006 \to -0.0044$	0.029 ± 0.002
λ_0	0	0.017	$0.003 \to 0.009$	$0.025 \pm 0.006\ K^0_{\mu 3}$
				$0.006 \pm 0.007\ K^+_{\mu 3}$
λ'_+	0	small	$0.0002 \to 0.0003$	
λ'_0	0	small	$0.0001 \to 0.0002$	

Table 37.2: CHPT and experimental results for K_{l3} decays.

phase space. The main interest in these decays is that it allows one to test the anomaly and its sign as well as the $V - A$ structure of the weak interactions.

37.3.4 K_{l2ll}

In these decays there are three vector and one axial form factor. The vector ones are known to p^4 [19] and the axial one to p^6 [21]. Especially the decays with $e\nu_e$ in the final state are strongly enhanced over bremsstrahlung. Since my last review [11] there is a new limit from BNL E787 [22] of $B(K^+ \to e^+\nu\mu^+\mu^-) \leq 5 \cdot 10^{-7}$. All data are in good agreement with CHPT.

37.3.5 K_{l3}

These decays, $K^{+,0} \to \pi^{0,-}\ell^+\nu$, are our main source of knowledge of the CKM element V_{us}. It is therefore important to have as precise predictions as possible. The form-factors

$$\langle \pi(p')|V_\mu^{4-i5}|K(p)\rangle = \frac{1}{\sqrt{2}}\left[(p+p')_\mu f_+(t) + (p-p')_\mu f_-(t)\right] \qquad (37.1)$$

are usually parametrized by $f_+(t) \approx f_+(0)\left[1 + \lambda_+ t/m_\pi^2\right]$ and $f_0(t) \equiv f_+(t) + t f_-(t)/(m_K^2 - m_\pi^2) \approx f_+(0)\left[1 + \lambda_0 t/m_\pi^2\right]$.

The CHPT calculation at order p^4 fits these parametrizations well [23]. The agreement with data is quite good except for the scalar slope, where there is disagreement between different experiments. The extended double log calculation [15] has small quadratic slopes, λ'_+ and λ'_0, and small corrections to the linear slopes. This, as shown in table 37.2, is good news for improving the precision of V_{us}. f_+ is shown for $K^0 \to \pi^- e^+\nu$, where isospin breaking is smallest.

37.3.6 $K_{l3\gamma}$

These decays have been calculated in CHPT to p^4 in [19]. There are 10 form factors and after a complicated interplay between all the various terms the final corrections to tree level are small even though individual form

	$F(0)$	$\lambda F(0)$	$G(0)$
p^2	3.74	—	3.74
p^4	set B; fit to L_1, L_2, L_3		
dispersive	set A; fit to L_1, L_2, L_3		
p^6 set A	0.86	0.38	−0.15
p^6 set A $\mu = 0.9\,\text{GeV}$	1.13	0.51	−0.20
p^6 set B	0.75	0.17	−0.04
experiment	5.59 ± 0.14	0.45 ± 0.11	4.77 ± 0.27

Table 37.3: CHPT, dispersive, partial p^6 [15], and experimental results for K_{l4}.

factors have large corrections. For example, first adding tree level, then p^4 tree level, and finally p^4 loop level contributions changes $B(K^+_{e3\gamma})$ with $E_\gamma \geq 30$ MeV and $\theta_{\ell\gamma} \geq 20^o$ from 2.8 10^{-4} via 3.2 10^{-4} to 3.0 10^{-4}. Notice that $F_K/F_\pi = 1.22$ so agreement with tree level at the 10% level is a good test of CHPT at order p^4.

Recent new results of $B(K^0_{e3\gamma}) = (3.61 \pm 0.14 \pm 0.21)\ 10^{-3}$ [24] (NA31) and $B(K^0_{\mu3\gamma}) = (0.56 \pm 0.05 \pm 0.05)\ 10^{-3}$ [25] (NA48) are in good agreement with the theory results [19] of $(3.6 \to 4.0 \to 3.8)\ 10^{-3}$ and $(0.52 \to 0.59 \to 0.56)\ 10^{-3}$ respectively. The three numbers correspond to the contributions included as above.

37.3.7 K_{l4}

In these decays, $K \to \pi\pi\ell\nu$, there are four form-factors, F, G, H, R as defined in [10,26]. The R form factor can only be measured in $K_{\mu4}$ decays and is known to p^4 [26]. F and G were calculated to p^4 in [27] and improved using dispersion relations in [26]. The main data come from [28] ($K \to \pi^+\pi^-e^+\nu$) and [29] ($K_L \to \pi^\pm\pi^0 e^\mp\nu$). The form factors were parametrized as $X = X(0)(1 + \lambda(s_{\pi\pi}/(4m_\pi^2) - 1))$ with the same slope for $X = F, G, H$. $H(0) = -2.7 \pm 0.7$ is a test of the anomaly in both sign and magnitude; see [21] and references therein. The other numbers are the main input for L_1^r, L_2^r, and L_3^r. In table 37.3 I show the tree-level results; which expression, the p^4 or dispersive improved, to determine $L^r_{1,2,3}$ of set A and B given in sec. 37.3.2; and the extended double log estimate of p^6 [15]. The results of the latter show similar patterns to the dispersive improvement. The full p^6 calculation is in progress and if the results are as indicated by the extended double log approximation a refitting of p^4 constants will be necessary. This is important since in these decays and in pionium decays the $\pi\pi$ phase shifts will be measured accurately and their main theory uncertainty is the values of these constants. A useful parametrization to determine these phases from K_{l4} can be found in [30] as well as further relevant references.

37.4 Nonleptonic Decays

For rare decays see [1]; here only $K \to 0, \pi, \pi\pi, \pi\pi\pi$ are discussed. The lowest-order Lagrangian contains three terms with parameters G_8, G_{27}, G'_8 in the notation of [31]. The term with G'_8, the weak mass term, contributes to processes with photons at lowest order and otherwise at NLO [31]. The NLO lagrangian contains about 30 parameters for the octet, denoted by E_i, and twenty-seven, denoted by D_i; octet and twenty-seven refer to the $SU(3)_L$ representation of the terms [32,33].

37.4.1 $K \to \pi$; Relation between $K \to 0$ and $K \to \pi\pi$

As shown in [31] the method of [34] can be extended to p^4 using well-defined off-shell Green functions of pseudoscalar currents. Except for one E_i and one D_i all the necessary ones can be obtained from $K \to \pi$ transitions.[1] To $K \to \pi\pi$ at order p^4 7 E_i and 6 D_i contribute in addition to the three couplings of lowest order. Of these 16 constants we can determine 14 from the much simpler K to π and vacuum transitions. This allows thus a more stringent test of various models than possible from on-shell $K \to \pi\pi$ alone. Models like factorization, etc., will probably be needed in the foreseeable future to go to $K \to 3\pi$ and various rare decays.

37.4.2 CHPT for $K \to \pi\pi$ and $K \to \pi\pi\pi$

These decays were calculated to p^4 [35], relations between them were clarified in [36], and some p^6 estimates to them were performed in [37].

The main problem is to find experimental relations after all parameters are counted. To order p^2 we have 2(1) and to p^4 7(3). The number in brackets refers to the $\Delta I = 1/2$ observables only. As observables (after using isospin) we have 2(1) $K \to \pi\pi$ rates and 2(1)(+1) $K \to \pi\pi\pi$ rates. We have 3(1)(+3) linear and 5(1)(+5) quadratic slopes. The (+i) indicates the phases, in principle also measurable and predicted but not counted here. 12 observables and 7 parameters leave five relations to be tested. The fits and results are shown in table 37.4 where we have also indicated which quantities are related. See [36,38] for definitions and references. The new CPLEAR [39] data improve the precision slightly. $K \to \pi\pi$ rates are always input. It is important to test these relations directly; the agreement at present is satisfactory but errors are large.

CP violation in $K \to 3\pi$ will be very difficult to detect. The strong phases needed to interfere with are very small; see [37] and references therein. For example, $\delta_2 - \delta_1$ in $K_L \to \pi^+\pi^-\pi^0$ is predicted to be -0.083 and the experimental result is only -0.33 ± 0.29. Asymmetries are expected to be of order 10^{-6} so we can only expect to improve limits in the near future.

---

[1] Using $K \to 0$ allows one to obtain two more constants than given in [31].

variable	p^2	p^4	experiment
α_1	74	input(1)	91.71 ± 0.32
β_1	-16.5	input(2)	-25.68 ± 0.27
ζ_1	–	$-0.47 \pm 0.18(1)$	-0.47 ± 0.15
ξ_1	–	$-1.58 \pm 0.19(2)$	-1.51 ± 0.30
α_3	-4.1	input(3)	-7.36 ± 0.47
β_3	-1.0	input(4)	-2.42 ± 0.41
γ_3	1.8	input(5)	2.26 ± 0.23
ξ_3	–	$0.092 \pm 0.030(4)$	-0.12 ± 0.17
ξ_3'	–	$-0.033 \pm 0.077(5)$	-0.21 ± 0.51
ζ_3	–	$-0.011 \pm 0.006(3)$	-0.21 ± 0.08

Table 37.4: The predictions and experimental results for the various $K \to 3\pi$ quantities. Numbers in brackets refer to the related quantities.

37.4.3 B_6 in the Chiral Limit

In the usual definitions of B_i factors in nonleptonic decays

$$B_6 \equiv \frac{\langle \text{out}|Q_6|\text{in}\rangle}{\langle \text{out}|Q_6|\text{in}\rangle_{\text{factorized}}} \tag{37.2}$$

the denominator needs to be well defined. This is *not* true for B_6 in the chiral limit. The factorizable denominator contains the scalar radius, which is infinite in the full chiral limit. This can be seen in the CHPT calculation [5].

$$G_8\Bigg|_{Q_6\text{fact}} = -\frac{80C_6(\mu)B_0^2(\mu)}{3F_0^2}\left[L_5^r(\nu) - \frac{3}{256\pi^2}\{2\ln\frac{m_L}{\nu} + 1\}\right]. \tag{37.3}$$

Here ν is the CHPT scale and m_L the meson mass; we can see that $G_{8fact} \longrightarrow \infty$ for $m_L \to 0$.

The nonfactorizable part has precisely the same divergence so that in the sum it cancels. Thus when calculating B_6 care must be taken to calculate factorizable and nonfactorizable quantities consistently so this cancellation that is required by chiral symmetry takes place and does not inflate final results.

37.5 The X-Boson Method and the $\Delta I - 1/2$ Rule in the Chiral Limit

In this section I briefly describe how in the context of the large-N_c method [3,4,40] after the improvements of the momentum routing [41] the scheme dependence [42,43] also can be described. Other relevant references to the problem of nonleptonic matrix elements are [44].

The basic underlying idea is that we have more experience in hadronizing currents. We therefore replace the effect of the local operators of

$H_W(\mu) = \sum_i C_i(\mu)Q_i(\mu)$ at a scale μ by the exchange of a series of colorless X bosons at a low scale μ. Let me illustrate the procedure in a simpler case of only one operator and neglecting penguin contributions. In the more general case all coefficients become matrices.

$$C_1(\mu)(\bar{s}_L\gamma_\mu d_L)(\bar{u}_L\gamma^\mu u_L) \Longleftrightarrow X_\mu\left[g_1(\bar{s}_L\gamma^\mu d_L) + g_2(\bar{u}_L\gamma^\mu u_L)\right]. \tag{37.4}$$

Color indices inside brackets are summed over. To determine g_1, g_2 as a function of C_1 we set matrix elements of C_1Q_1 equal to the equivalent ones of X boson exchange. This must be done at a μ such that perturbative QCD methods can still be used and thus we can use external states of quarks and gluons. To lowest order this is simple. The tree-level diagram from

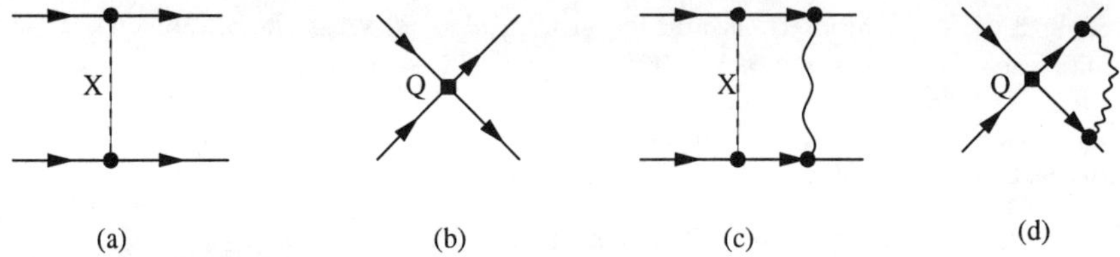

Figure 37.1: The diagrams needed for the identification of the local operator Q with X boson exchange in the case of only one operator and no penguin diagrams. The wiggly line denotes gluons, the square the operator Q and the dashed line the X exchange. The external lines are quarks.

fig. 37.1(a) is set equal to that of fig. 37.1(b) leading to $C_1 = g_1g_2/M_X^2$ .At NLO diagrams like fig. 37.1(c) and 37.1(d) contribute as well leading to

$$C_1\left(1 + \alpha_S(\mu)r_1\right) = \frac{g_1g_2}{M_X^2}\left(1 + \alpha_S(\mu)a_1 + \alpha_S(\mu)b_1\log\frac{M_X^2}{\mu^2}\right). \tag{37.5}$$

The left-hand-side (lhs) is scheme-independent. The right-hand-side can be calculated in a very different renormalization scheme from the lhs. The infrared dependence of r_1 is present in precisely the same way in a_1 such that g_1 and g_2 are scheme-independent and independent of the precise infrared definition of the external state in fig. 37.1.

One step remains, to calculate the matrix element of X boson exchange between meson external states. The integral over X boson momenta we split in two

$$\int_0^\infty dp_X\frac{1}{p_X^2 - M_X^2} \Longrightarrow \int_0^{\mu_1} dp_X\frac{1}{p_X^2 - M_X^2} + \int_{\mu_1}^\infty dp_X\frac{1}{p_X^2 - M_X^2}. \tag{37.6}$$

The second term involves a high momentum that needs to flow back through quarks or gluons and leads through diagrams like the one of fig. 37.1(c) to a four-quark operator with a coefficient

$$\frac{g_1g_2}{M_X^2}\left(\alpha_S(\mu_1)a_2 + \alpha_S(\mu_1)b_1\log\frac{M_X^2}{\mu^2}\right). \tag{37.7}$$

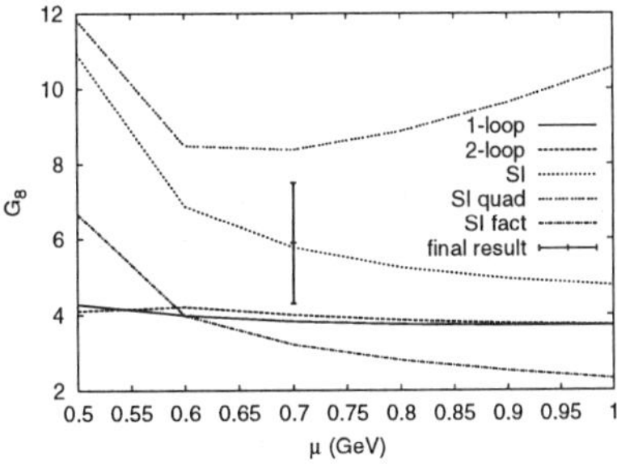

Figure 37.2: G_8 as a function of μ using the ENJL model and Wilson coefficients at one-loop, at 2-loop with and without the r_1 (SI). The factorization (SI fact) and the approach of [3] (SI quad) are shown for SI also.

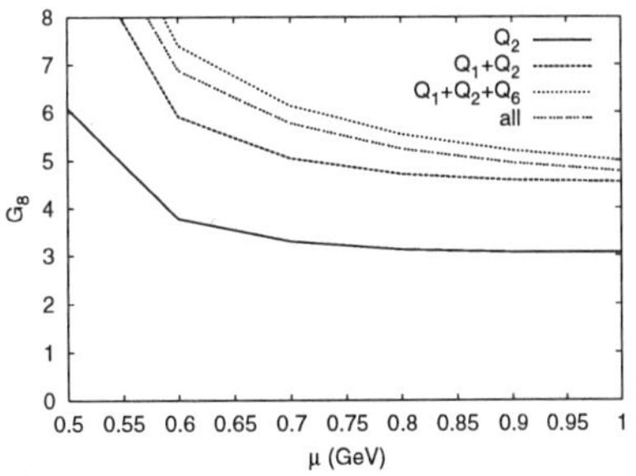

Figure 37.3: The composition of G_8 as a function of μ. Shown are Q_2, Q_1+Q_2, $Q_1+Q_2+Q_6$, and all 6 Q_i. The coefficients r_1 are included in the Wilson coefficients.

The four-quark operator needs to be evaluated only in leading order in $1/N_c$. The first term in (37.6) we have to evaluate in a low-energy model with as much QCD input as possible. The μ_1 dependence cancels between the two terms in (37.6) if the low-energy model is good enough. The coefficients r_1, a_1, and a_2 give the correction to the factorization used in previous $1/N_c$ calculations.

It should be stressed that in the end all dependence on M_X cancels out. The X boson is a purely technical device to correctly identify the four-quark operators in terms of well-defined products of nonlocal currents.

37.5.1 Numerical Results

We now use the X boson method with r_1 as given in [42] and $a_1 = a_2 = 0$, the calculation of the latter is in progress, and $\mu = \mu_1$. For B_K we can extrapolate to the pole both for the real case ($\hat{B}_K$) and in the chiral limit ($\hat{B}_K^\chi$). For $K \to \pi\pi$ we can get at the values of the octet (G_8), weak mass term (G_8'), and 27-plet (G_{27}) coupling. We obtain $\hat{B}_K^\chi = 0.25$–0.4;

$$\hat{B}_K = 0.69 \pm 0.10; \quad G_8 = 4.3\text{–}7.5; \quad G_{27} = 0.25\text{–}0.40 \text{ and } G_8' = 0.8\text{–}1.1. \tag{37.8}$$

The experimental values are $G_8 \approx 6.2$ and $G_{27} \approx 0.48$ [5, 35].

In fig. 37.2 the μ dependence of G_8 is shown and in fig. 37.3 the contribution from the various different operators. If we look inside the numbers we see that B_6, as defined with only the large N_c term in the factorizable part, is about 2 to 2.2 for μ from 0.6 to 1.0 GeV.

37.6 Conclusions

CHPT is doing fine in kaon decays, especially in the semileptonic sector, where several calculations at p^6 are now in progress. In the nonleptonic sector it provides several relations in $K \to 3\pi$ decays. Testing these is an important part since it tells us how well p^4 works in this sector. CHPT can also help in simplifying and identifying potentially dangerous parts in the calculations of nonleptonic matrix elements.

The large-N_c method allows one to include the scheme dependence appearing in short-distance operators and then when all long-distance constraints from CHPT and some other input are used, encouraging results are obtained for $K \to \pi\pi$ decays in the chiral limit.

References

[1] G. Isidori, this volume, chapter 33.

[2] M. Savage, this volume, chapter 16.

[3] T. Hambye and P.H. Soldan, this volume, chapter 11; T. Hambye *et al.*, Phys. Rev. D **58**, 014017 (1998), hep-ph/9802300; T. Hambye *et al.*, hep-ph/9902334.

[4] E. de Rafael, this volume, chapter 30; W.A. Bardeen, this volume, chapter 14.

[5] J. Bijnens and J. Prades, JHEP **01**, 023 (1999), hep-ph/9811472.

[6] S. Weinberg, Physica A **96**, 327 (1979).

[7] J. Gasser and H. Leutwyler, Ann. Phys. **158**, 142 (1984); Nucl. Phys. B **250**, 465 (1985).

[8] H. Leutwyler, Ann. Phys. **235**, 165 (1994), hep-ph/9311274.

[9] A. Pich, hep-ph/9806303; G. Ecker, hep-ph/9805500.

[10] J. Bijnens, G. Colangelo, G. Ecker, and J. Gasser, hep-ph/9411311.

[11] J. Bijnens, hep-ph/9607304.

[12] J. Bijnens *et al.*, JHEP **02**, 020 (1999), hep-ph/9902437.

[13] H. Fearing and S. Scherer, Phys. Rev. D **53**, 315 (1996), hep-ph/9408346.

[14] J. Bijnens, G. Colangelo and G. Ecker, hep-ph/9907333, Ann. Phys. (N.Y.) **280**, 100 (2000).

[15] J. Bijnens *et al.*, Phys. Lett. B **441**, 437 (1998), hep-ph/9808421.

[16] G. Amoros *et al.*, hep-ph/9907264.

[17] E. Golowich and J. Kambor, Phys. Rev. D **58**, 036004 (1998), hep-ph/9710214.

[18] J.F. Donoghue and B.R. Holstein, Phys. Rev. D **40**, 3700 (1989).

[19] J. Bijnens *et al.*, Nucl. Phys. B **396**, 81 (1993), hep-ph/9209261.

[20] J. Bijnens and P. Talavera, Nucl. Phys. B **489**, 387 (1997), hep-ph/9610269.

[21] L. Ametller *et al.*, Phys. Lett. B **303**, 140 (1993), hep-ph/9302219.

[22] S. Adler *et al.*, Phys. Rev. D **58**, 012003 (1998), hep-ex/9802011.

[23] H. Leutwyler and M. Roos, Z. Phys. C **25**, 91 (1984); J. Gasser and H. Leutwyler, Nucl. Phys. B **250**, 517 (1985).

[24] F. Leber *et al.*, Phys. Lett. B **369**, 69 (1996).

[25] M. Bender *et al.*, Phys. Lett. B **418**, 411 (1998).

[26] J. Bijnens *et al.*, Nucl. Phys. B **427**, 427 (1994), hep-ph/9403390.

[27] J. Bijnens, Nucl. Phys. B **337**, 635 (1990); C. Riggenbach *et al.*, Phys. Rev. D **43**, 127 (1991).

[28] L. Rosselet *et al.*, Phys. Rev. D **15**, 574 (1977).

[29] G. Makoff *et al.*, Phys. Rev. Lett. **70**, 1591 (1993).

[30] G. Amoros and J. Bijnens, hep-ph/9902463, accepted in J. Phys. G.

[31] J. Bijnens *et al.*, Nucl. Phys. B **521**, 305 (1998), hep-ph/9801326.

[32] J. Kambor, J. Missimer, and D. Wyler, Nucl. Phys. B **346**, 17 (1990).

[33] G. Ecker, J. Kambor, and D. Wyler, Nucl. Phys. B **394**, 101 (1993).

[34] C. Bernard *et al.*, Phys. Rev. D **32**, 2343 (1985).

[35] J. Kambor, J. Missimer, and D. Wyler, Phys. Lett. B **261**, 496 (1991).

[36] J. Kambor *et al.*, Phys. Rev. Lett. **68**, 1818 (1992).

[37] G. D'Ambrosio and G. Isidori, Int. J. Mod. Phys. A **13**, 1 (1998), hep-ph/9611284.

[38] L. Maiani and N. Paver, in *The Second DAΦNE Physics Handbook*, ed. L. Maiani *et al.* (SIS, publications of the National Labratory of Frascati, Frascati, 1995), 1:239-64.

[39] A. Angelopoulos *et al.*, Eur. Phys. J. C **5**, 389 (1998).

[40] W. A. Bardeen *et al.*, Nucl. Phys. B **293**, 787 (1987); Phys. Lett. B **192**, 138 (1987).

[41] W. A. Bardeen *et al.*, Phys. Rev. Lett. **62**, 1343 (1989); J. Bijnens *et al.*, Phys. Lett. B **257**, 191 (1991).

[42] A. J. Buras *et al.*, Nucl. Phys. B **400**, 37 (1993), hep-ph/9211304; Nucl. Phys. B **370**, 69 (1992).

[43] M. Ciuchini *et al.*, Nucl. Phys. B **415**, 403 (1994), hep-ph/9304257; M. Ciuchini *et al.*, Z. Phys. C **68**, 239 (1995), hep-ph/9501265.

[44] A. Buras, this volume, chapter 5; S. Bertolini, this volume, chapter 12; M. Ciuchini *et al.*, this volume, chapter 28; J. Bijnens, hep-ph/9907307.

38

NA48 Rare Decay Results

L. Köpke[1]

Abstract

New results by the NA48 Collaboration on rare K_L decays into the $e^+e^-\gamma$, $e^+e^-\gamma\gamma$, and $e^+e^-e^+e^-$ final states are discussed. Preliminary results on the $\Xi^0 \to \Lambda\gamma$ and $\Xi^0 \to \Sigma^0\gamma$ branching ratios and the precise determination of the Ξ^0 mass are presented.

38.1 Introduction

The main purpose of the NA48 experiment is the determination of the direct CP-violation parameter $\mathrm{Re}(\epsilon'/\epsilon)$ from simultaneous measurements of the $K_{L,S} \to \pi^+\pi^-, \pi^0\pi^0$ decay modes. At the same time, however, other rare kaon decay modes as well as hyperon decays are being recorded. In this talk I will concentrate on new results in the area of $K_L \to \gamma^*\gamma^*$ decays, where one or both photons are virtual and convert into an e^+e^- pair and on results on Ξ^0 radiative decays and the Ξ^0 mass.

38.2 Detector and Trigger Issues

The detector and trigger have been described in detail elsewhere [1, 2]. For the decays discussed here, good momentum and vertex resolution of charged tracks in a magnetic spectrometer with 4 drift chambers of 8 planes each and an excellent energy and space resolution ($<$ 1 mm) for photons and electrons in a liquid-krypton calorimeter are particularily important. Equally important is a low probability for photon conversions in the detector ($\lesssim$ 1%) and correspondingly a small probability for electron bremsstrahlung.

The triggers set up for the direct-CP analysis are also useful for rare decays. However, they are restricted to a decay range populated by K_S decays corresponding to roughly 1/3 of the decay range that would in principle be available for K_L decays. Other triggers, like the one selecting

[1]Representing the NA48 Collaboration.

Dalitz-like decays, have a wider range or no decay range restriction at all, such as the 4-track trigger.

Various rare decay topologies are preselected by a subset of 43 filter programs called in the level-3 online reconstruction and filtering program.

Most analyses presented here are based on the data taken in 1997, which represent 20% of the available data and $< 10\%$ of the data expected by the end of 1999.

38.3 The $K_L \to \gamma^{(*)}\gamma^{(*)}$ Family

Variants of the $K_L \to \gamma\gamma$ decay, with one or two photons being virtual and converting into e^+e^- or $\mu^+\mu^-$ pairs, can be studied very well with the NA48 detector. The radiative corrections to these processes are also important, in particular if a magnetic spectrometer is used.

The probability of internal $\gamma^* \to e^+e^-$ conversion has been calculated by Bergström, Masso, and Singer [4] to be about 0.9%; the $\gamma^* \to \mu^+\mu^-$ conversion probability is 10 times smaller. The exact numbers depend on the formfactors chosen. In this model (BMS), long-range contributions are assumed to proceed via pseudoscalar and vector final states. The former is enhanced by penguin diagrams. Assuming the $\Delta I = 1/2$ rule, a value $|\alpha_{K^*}| = 0.2 - 0.3$ is obtained.[2]

38.3.1 $K_L \to e^+e^-\gamma$

The $e^+e^-\gamma$ final state has been recently studied by NA48 [2]. The total number of events selected is 6864, with very small background, mainly due to $K_S \to \pi^0\pi^0_D$ (7.2 ± 1.2 events) and $K_L \to \pi e\nu$ with an additional "accidental" photon (2.9 ± 0.6 events). The correlation plot between the squared and scaled e^+e^- invariant mass and the $e^+e^-\gamma$ mass is shown in fig. 38.1.

The branching ratio is determined to be

$$B(K_L \to e^+e^-\gamma) = [1.06 \pm 0.02_{\text{stat}} \pm 0.02_{\text{sys}} \pm 0.04_{\text{br}}] \times 10^{-5},$$

using as normalization a set of 4060 $K_L \to \pi^0\pi^0_D$ decays. The major correction (3.9%) applied accounts for $K_S - K_L$ interference in the normalization channel; the major contribution to the error is due to uncertainties in the $\pi^0 \to e^+e^-\gamma$ and $K_L \to \pi^0\pi^0$ branching ratios (3rd error term). The ratio $\Gamma(K_L \to e^+e^-\gamma)/\Gamma(K_L \to \gamma\gamma) = [17.9 \pm 0.4_{\text{stat}} \pm 0.4_{\text{sys}} \pm 0.3_{\text{br}}] \times 10^{-3}$ is slightly higher than the BMS-model prediction of 16.8×10^{-3}.

Of considerable interest is the determination of the form factor in the decay, which, in the framework of the BMS model, has been parametrized with the parameter α_{K^*}. Fig. 38.2 shows a comparison of the squared and scaled e^+e^- mass with the prediction of the BMS model for $\alpha_{K^*} = -0.36$ as

[2] The model has been criticized by G. D'Ambrosio and J. Portoles because it does not reproduce the experimental results on the $K_L \to \pi^0\gamma\gamma$ decay [3].

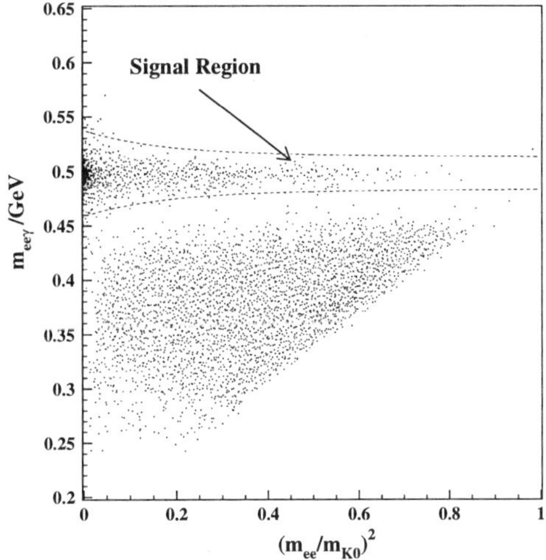

Figure 38.1: Scatter plot of the $e^+e^-\gamma$ mass versus m_{ee}^2/m_K^2.

well as with a pure QED model. A form factor is clearly needed to describe the distribution at intermediate and large $m(e^+e^-)$, which—according to the BMS model—is achieved by a vector-vector contribution given by α_{K^*}. Ten sets of Monte Carlo samples have been produced including radiative corrections [5] for different values of α_{K^*}. The χ^2 between data and Monte Carlo is minimized for

$$\alpha_{K^*} = -0.36 \pm 0.06_{\text{stat}} \pm 0.02_{\text{sys}},$$

which compares well with earlier measurements in this channel and is consistent with the BMS model.

38.3.2 $K_L \to \gamma e^+e^-\gamma$

For the understanding of kaon decays with photons and leptons in the final state, the precise knowlodge of radiative corrections is important. The decay $K_L \to e^+e^-\gamma\gamma$ in itself is of considerable interest, as it constitutes the major background to the very rare decay $K_L \to e^+e^-\pi^0$, which is expected to be dominated by direct CP violation.

Fig. 38.3 shows the $K_L \to e^+e^-\gamma\gamma$ mass plot of the data taken in 1997 and 1998. In the signal region between 482-514 MeV, 503 events are found, (7.3 ± 2.7) of which are estimated to be due to $K_L \to \pi^0\pi^0_D$ and (14.2 ± 1.2) events are due to $e^+e^-\gamma$ *plus* a photon from external electron bremsstrahlung in the detector material. The preliminary branching ratio

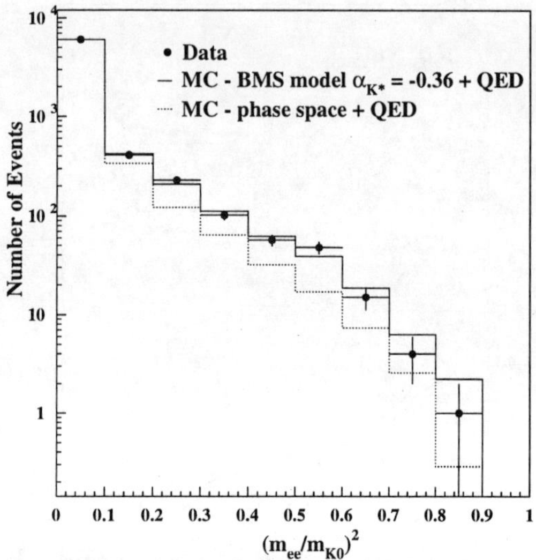

Figure 38.2: Comparison of m_{ee}^2/m_K^2 distribution in data (points with error bars) and Monte Carlo sample generated using $\alpha_{K^*} = -0.36$ (solid line) and form factor $f(x) = 1$ (dotted line).

for a minimum photon energy of 5 MeV in the K_L center-of-mass system is calculated to be

$$B(K_L \to e^+e^-\gamma\gamma) = [5.82 \pm 0.27_{\text{stat}} \pm 0.45_{\text{sys}} \pm 0.16_{\text{norm}}] \times 10^{-7},$$

using $K_L \to e^+e^-\gamma$ events for normalization. The result is in good agreement with earlier measurements. (See the contribution of J. Whitmore, this volume, chapter 39.)

38.3.3 $K_L \to e^+e^-e^+e^-$

This decay is due to double *internal* photon conversion of $K_L \to \gamma^*\gamma^*$ in e^+e^- final states. It suffers from substantial background from $K_L \to \gamma\gamma$ and $K_L \to e^+e^-\gamma$ decays with subsequent *external* conversion in the material of the detector. To reject these events, a minimal track distance of 2 cm at the spectrometer drift chambers before the magnet is required, corresponding to a cut on the minimal virtuality of the intermediate γ^*. Requiring a small transverse momentum w.r.t. the beam direction of $p_T^2 < 0.0005\,\text{GeV}^2/\text{c}^2$, 84 $K_L \to e^+e^-e^+e^-$ candidate events are selected with negligible background (see fig. 38.4).

In analogy to the classic determination of the π^0 parity, the angle between the two e^+e^- planes contains information about the CP parity of the initial state. For the CP eigenstates belonging to the eigenvalues $\pm$ 1,

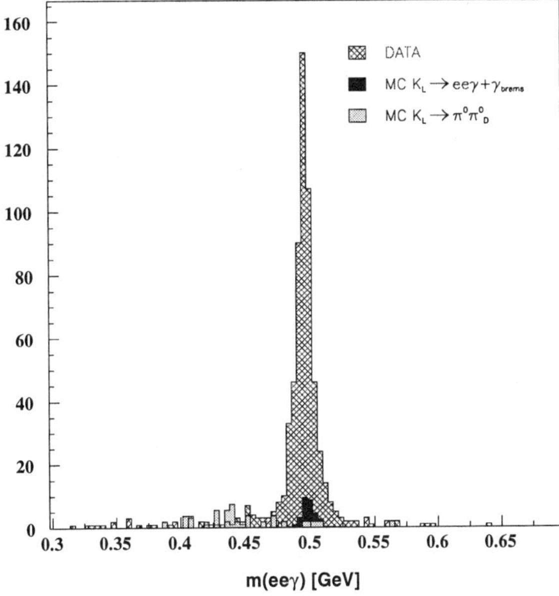

Figure 38.3: Invariant mass of $e^+e^-\gamma\gamma$ system.

the distribution is given by $dN/d\phi = 1 - \alpha^{\pm}\cos 2\phi$, with $\alpha^{\pm}$ depending on the kinematics of the event. Fits to the angular distribution give a 60% probability for CP(K_L) = −1 and a 1.5% probability for CP(K_L) = +1. In order to obtain a 10^{-3} precision measurement of the indirect CP-violation parameter ϵ, a three orders of magnitude larger sample of events would be required.

38.4 Hyperon Decays

In order to measure the direct CP-violation parameter ϵ'/ϵ, two beams are used simultaneously in the NA48 experiment. The "K_L" beam, with the target 240 m away from the detector, contains only a minor fraction of short-lived decays. The "K_S" beam, however, has a target very close to the decay region; it is also a source of hyperons. The intensity of this beam has been reduced by a factor of 40000 in order to register similar amounts of CP-allowed K_S decays and CP-violating K_L decays to the $\pi\pi$ final states. NA48 is contemplating increasing the intensity of the K_S beam for dedicated studies of K_S and hyperon decays in the future. NA48 has already published a study of Λ transverse polarization [6]. Here I will discuss a preliminary investigation of Ξ^0 radiative decays and the Ξ^0 mass.

Radiative Ξ^0 decays are difficult to calculate due to their nonleptonic weak interaction nature; not surprisingly, theoretical predictions show a wide variation. NA48 has analysed a sample of 31 $\Xi^0 \to \Lambda\gamma$ decays, almost

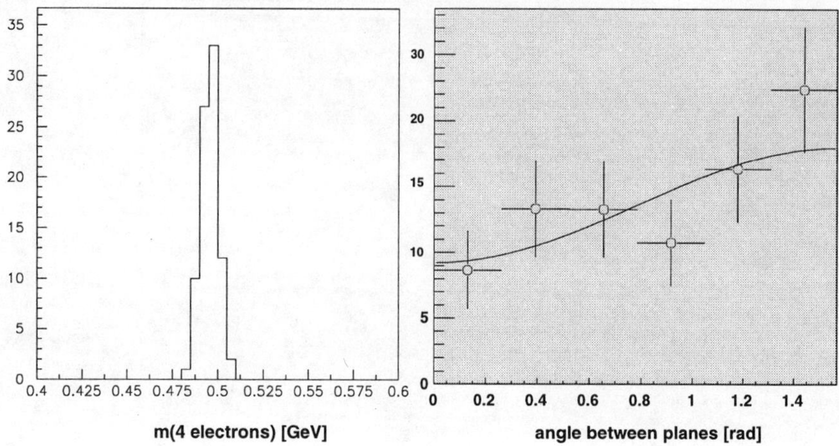

Figure 38.4: Left: m($e^+e^-e^+e^-$), right: distribution of angle ϕ between planes.

free of background, yielding a preliminary value for the branching ratio of

$$B(\Xi^0 \to \Lambda\gamma) = [1.90 \pm 0.34_{\rm stat} \pm 0.19_{\rm sys}] \times 10^{-3}.$$

Twenty candidates for the $\Xi^0 \to \Sigma^0\gamma$ decay have also been observed, with a 15% background mainly due to $\Xi^0 \to \Lambda\pi^0$ decays. A preliminary value for the branching ratio of

$$B(\Xi^0 \to \Sigma^0\gamma) = [3.14 \pm 0.76_{\rm stat} \pm 0.32_{\rm sys}] \times 10^{-3}$$

has been determined. In both cases a sample of 6669 $\Xi^0 \to \Lambda\pi^0$ decays was used for normalization. The KTeV results on both these decays are given in the contribution of N. Solomey, this volume, chapter 44.

While these results are statistically less precise than earlier determinations [7,8], the Ξ^0 mass measurement is systematically limited and can be drastically improved given the precise tracking and small uncertainty of the momentum scale in the NA48 detector. Using $\Xi^0 \to \Lambda\pi^0$ decays, the vertex is calculated from the π^0 mass constraint; the nominal π^0 and Λ masses are used as additional constraints. The resulting Ξ^0-mass resolution is about 1 MeV; see fig. 38.5. The main systematic error stems from an assumed error in the vertex position, which has been conservatively estimated to be ±30 cm. While the vertex is known to a much better precision in decays involving only electrons and photons, wide hadronic showers tend to overlap with photon showers increasing the systematic uncertainty. Still, the preliminary mass value of

$$m(\Xi^0) = 1314.83 \pm 0.06_{\rm stat} \pm 0.20_{\rm sys}{\rm MeV}$$

constitutes a factor 4 improvement w.r.t. previous measurements.

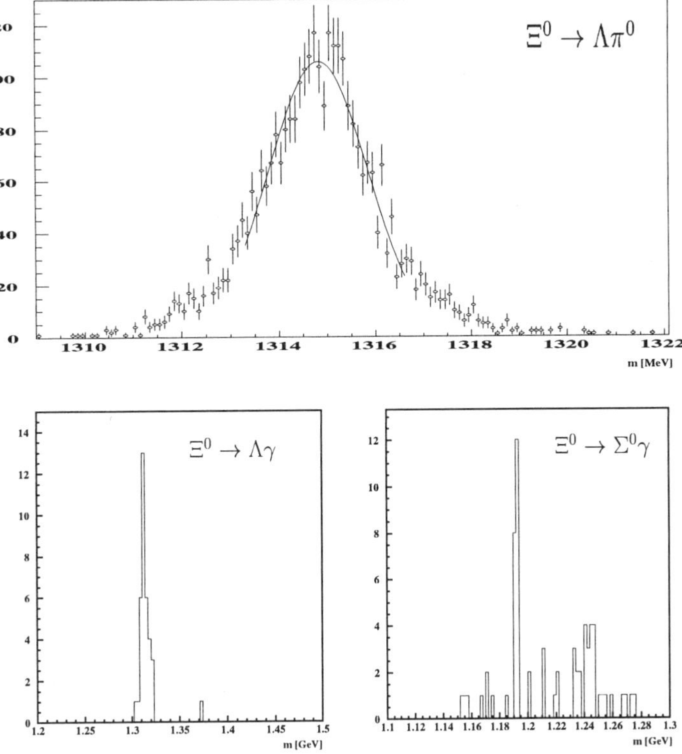

Figure 38.5: $\Lambda\pi^0$ mass distribution (top). Radiative hyperon decays. Evidence for $\Xi^0 \to \Lambda\gamma$ (bottom left) and $\Xi^0 \to \Sigma^0\gamma \to \Lambda\gamma\gamma$ (bottom right). In both cases the $\Lambda\gamma$ invariant mass distribution is shown.

38.5 Summary and Outlook

New analyses on $K_L \to e^+e^-\gamma$, $e^+e^-\gamma\gamma$, and $e^+e^-e^+e^-$ have been presented as well as preliminary results on radiative Ξ^0 decays and the Ξ^0 mass.

For the year 2000, NA48 is planning to continue data taking for the direct-CP violation measurement as well as to record a large sample of K_S decays in a dedicated run. Proposals for rare decay running after the year 2000 are being discussed in the collaboration.

References

[1] "The Beam and Detector for the Precision CP-Violation Experiment NA48," to be submitted to Nucl. Inst. Meth.

[2] NA48 Collaboration, V. Fanti *et al.*, CERN-EP/99-53 and Phys. Lett. B **458**, 503 (1999).

[3] G. D'Ambrosio and J. Portoles, Nucl. Phys. B **492**, 417 (1997).

[4] L. Bergström *et al.*, Phys. Lett B **131**, 229 (1983).

[5] H.B. Greenlee, Phys. Rev. D **42**, 3724 (1990). See also K.O. Mikaelian and J. Smith, Phys. Rev. D **5**, 1763 (1972) and **5**, 2890 (1972), and L. Roberts and J. Smith, Phys. Rev. D **33**, 3457 (1986).

[6] NA48 Collaboration, V. Fanti et al., Eur. Phys. J. C **6**, 265 (1999).

[7] C. James *et al.*, Phys. Rev. Lett. **64**, 2 (1990).

[8] S. Teige *et al.*, Phys. Rev. Lett., **63**, 12 (1990).

39

New Rare Decay Results from KTeV

J. Whitmore[1]

Abstract

Recent rare decay results from the KTeV fixed target experiment at Fermilab are shown. Results of searches for the CP violating decay modes $K_L^0 \to \pi^0 e^+ e^-$, $K_L^0 \to \pi^0 \mu^+ \mu^-$, and $K_L^0 \to \pi^0 \nu\overline{\nu}$ are presented. In addition, new branching ratio measurements of $K_L^0 \to \pi^0 \gamma\gamma$, $K_L^0 \to e^+ e^- \gamma\gamma$, and the first observation of the decay $K_L^0 \to \mu^+ \mu^- \gamma\gamma$ are discussed.

39.1 Introduction

KTeV (Kaons at the TeVatron) is a fixed-target experiment at Fermilab designed to study the properties of the neutral-kaon system. The experimental program has two main experimental goals: to make a precision measurement of the direct CP violation parameter ϵ'/ϵ via decays of neutral kaons into two pions [1] and to search for the direct CP-violating rare decay $K_L^0 \to \pi^0 e^+ e^-$ with a factor of 50 improvement in sensitivity over the previous measurement. In this article, we present results on the family of CP-violating rare decays $K_L^0 \to \pi^0 \ell\overline{\ell}$. We also discuss a new branching ratio measurement of $K_L^0 \to e^+ e^- \gamma\gamma$ and a first observation of the decay $K_L^0 \to \mu^+ \mu^- \gamma\gamma$.

The rare decays presented in this article contain leptons and photons in the final state; the salient features for detecting these particles in KTeV will be described below. The plan view of the KTeV detector in the E799-II configuration is shown in fig. 39.1. Neutral beams are produced by striking a BeO target with an 800 GeV proton beam. The neutral beam passes through a series of collimators, a Pb absorber that converts photons, and a series of sweeping magnets that bend the charged particles out of the beamline. Two nearly parallel neutral kaon beams, composed mainly of kaons and neutrons (~1:1) with a small contamination of hyperons, then enter the 65m vacuum decay region where approximately 5% of the kaons

[1] For the KTeV Collaboration.

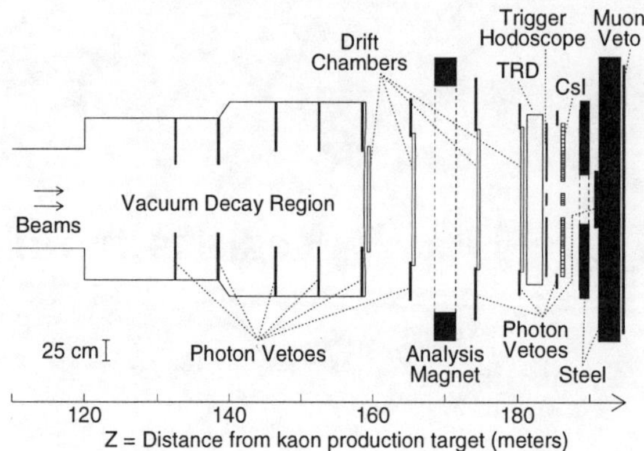

Figure 39.1: The plan view of the KTeV detector in the E799-II configuration.

decay. The range of K_L^0 momenta is 20–200 GeV/c. The charged spectometer consists of 4 planar drift chambers, two located upstream and two downstream of an analysis magnet that imparts a 205 MeV/c transverse momentum kick to the charged particles. A pure CsI calorimeter provides electron identification and photon energy measurements. Additional electron identification information comes from a set of 8 transition radiation detectors (TRD). Muons are identified with two banks of orthogonal scintillation counters located downstream of 3m of filter steel. Photon veto detectors line the vacuum decay region and the perimeters of the drift chambers and CsI and form the defining aperture of the fiducial region. All analyses described below with the exception of the decays $K_L^0 \to \pi^0 \nu\overline{\nu}$ ($\pi^0 \to 2\gamma$) and $K_L^0 \to \pi^0\gamma\gamma$ are based on the 1997 E799-II data set, which contained 2.7×10^{11} kaon decays. The $K_L^0 \to \pi^0 \nu\overline{\nu}$ ($\pi^0 \to 2\gamma$) analysis used a special 1-day run that corresponded to $6.8 \times 10^7 K_L$ decays. The data for the $K_L^0 \to \pi^0\gamma\gamma$ measurement were collected during the 1996–97 E832 running period and include 5.9×10^{10} kaon decays.

Theoretical predictions [2, 3] for the direct and indirect CP-violating components as well as the CP-conserving part of the $K_L^0 \to \pi^0 \ell\overline{\ell}$ branching ratios are listed in table 39.1.

Mode	Direct CPV	Indirect CPV	CP Conserving
$B(K_L^0 \to \pi^0 \nu\overline{\nu})$	3×10^{-11}	$\sim 10^{-15}$	$\sim 2 \times 10^{-15}$
$B(K_L^0 \to \pi^0 e^+ e^-)$	5×10^{-12}	$1 - 5 \times 10^{-12}$	$1 - 2 \times 10^{-12}$
$B(K_L^0 \to \pi^0 \mu^+ \mu^-)$	1×10^{-12}		$0.5 - 10 \times 10^{-12}$

Table 39.1: Theoretical predictions for the direct and indirect CP-violating components and CP-conserving contributions in $K_L^0 \to \pi^0 \ell\overline{\ell}$ decays.

39.2 $K_L^0 \to \pi^0 \nu \overline{\nu}$

The theoretical calculation of the branching ratio for $K_L^0 \to \pi^0 \nu \overline{\nu}$ is dominated by short-distance diagrams, allowing theoretical determination with few uncertainties. The direct CP-violating component to the branching ratio is predicted to be

$$B(K_L^0 \to \pi^0 \nu \overline{\nu})_{DIR} = (\tfrac{m_t}{m_W})^{2.2} A^4 \eta^2 \sim 3 \times 10^{-11} \text{ [4]},$$

where m_t is the top mass, m_W is the W mass, and A and η are variables from the Wolfenstein parametrization of the CKM matrix, in which η governs CP violation and represents the height of the unitarity triangle.

While the $K_L^0 \to \pi^0 \nu \overline{\nu}$ mode is compelling because it is predominantly direct CP violating, it is experimentally challenging. The more prevalent decay mode with $\pi^0 \to 2\gamma$ has significant backgrounds because the kinematics make it difficult to reconstruct the event. If the Dalitz decay of the π^0 is used to tag the event, the decay vertex helps constrain the event but there is a factor of ~ 100 reduction in acceptance due to the branching ratio and the detector geometry. KTeV has set branching ratio limits using both π^0 decay modes. The measurement using $\pi^0 \to 2\gamma$ is based on a dedicated special run in which a single kaon beam was used and events with high transverse momentum were selected. After all cuts, the single event sensitivity (SES) is 4.0×10^{-7}. One event is found in the signal region, which is consistent with the expectation from neutron interactions in material in the detector. Assuming the event is signal, we set an upper limit of

$$B(K_L^0 \to \pi^0 \nu \overline{\nu}, \text{with } \pi^0 \to 2\gamma \text{ tag}) < 1.6 \times 10^{-6} \text{ (90\% CL) [5]}.$$

This represents a first-time measurement using this decay chain. Using the full KTeV data set with the Dalitz decay tag, the SES for $K_L^0 \to \pi^0 \nu \overline{\nu}$ is 2.6×10^{-7} after all cuts and after correcting for the Dalitz decay branching ratio. A plot of the transverse momentum of the data along with overlays from signal Monte Carlo and Monte Carlo from the most prevalent backgrounds ($\Lambda \to n\pi^0$ and $\Xi \to \Lambda\pi^0$) is shown in fig. 39.2; there are no events in the signal region. The preliminary upper limit set using the Dalitz decay of the π^0 is

$$B(K_L^0 \to \pi^0 \nu \overline{\nu}, \text{with } \pi^0 \to e^+e^-\gamma \text{ tag}) < 5.9 \times 10^{-7} \text{ (90\% CL) [6]},$$

a factor of 100 improvement over the previous measurement [7]. Assuming experimental inputs for the top and W masses and the Wolfenstein parameter A, the branching ratio limit implies $\eta < 52$.

39.3 $K_L^0 \to \pi^0 e^+ e^-$ and $K_L^0 \to e^+ e^- \gamma\gamma$

Although the decay $K_L^0 \to \pi^0 e^+ e^-$ is a fundamentally easier mode to select because the final state can be fully reconstructed, it is not a pure direct CP-violating decay and it has a serious background from $K_L^0 \to e^+ e^- \gamma\gamma$.

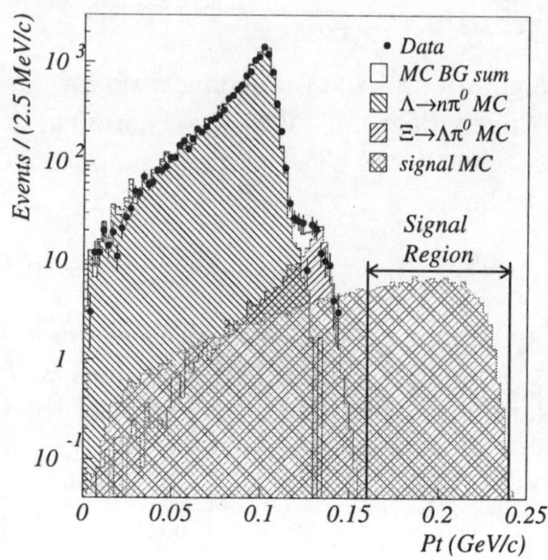

Figure 39.2: P_T distribution for $K_L^0 \to \pi^0\nu\overline{\nu}$ ($\pi^0 \to e^+e^-\gamma$) with all cuts except the P_T cut. Signal and background Monte Carlo are overlayed on data. There are no events in the signal region.

The indirect CP-violating and CP-conserving amplitudes (see table 39.1) are comparable in size to the direct CP-violating contribution and must be understood before the direct CP violation component can be determined.

The CP-conserving amplitude can be estimated from theory and from a measurement of the effective vector coupling, a_v, from $K_L^0 \to \pi^0\gamma\gamma$ [8]. KTeV accumulated a sample of $K_L^0 \to \pi^0\gamma\gamma$ events during E832 running. After all analysis cuts are applied, a significant excess in the low-mass region of the 2γ invariant mass distribution from $K_L^0 \to \pi^0\gamma\gamma$ decays is apparent and indicates contributions in the chiral perturbation model from vector meson exchange calculations [9]. The normalization mode for this measurement is $K_L^0 \to \pi^0\pi^0$. The branching ratio for $K_L^0 \to \pi^0\gamma\gamma$ is

$$B(K_L^0 \to \pi^0\gamma\gamma) = (1.68 \pm 0.07(\text{stat.}) \pm 0.08(\text{syst.})) \times 10^{-6},$$

representing a factor of 3 improvement over previous results [7]. From the shape of the 2γ invariant mass distribution, the effective vector coupling is determined to be $-0.72 \pm 0.05 \pm 0.06$ and indicates a contribution from the CP-conserved amplitude of the $B(K_L^0 \to \pi^0e^+e^-)$ of $1 - 2 \times 10^{-12}$.

The background limiting mode to the decay $K_L^0 \to \pi^0e^+e^-$ comes from the radiative Dalitz decay, $K_L^0 \to e^+e^-\gamma\gamma$, which KTeV also has measured. The previous measurement of the branching ratio for $K_L^0 \to e^+e^-\gamma\gamma$ was based on 58 events and yielded $(6.5\pm1.3)\times10^{-6}$ [7]. The major backgrounds to $K_L^0 \to e^+e^-\gamma\gamma$ come from $2\pi^0$ decays in which one π^0 Dalitz decays and $K_L^0 \to e^+e^-\gamma$ with an accidental photon. The new measurement has 1988 events in the signal region with a background estimate of 76.6 ± 3.3 events. With a photon energy cutoff of 5 GeV, the preliminary measurement of the branching ratio [10] is

$$B(K_L^0 \to e^+e^-\gamma\gamma, E_\gamma \geq 5MeV) = (6.31 \pm 0.14(\text{stat.}) \pm 0.42(\text{syst.})) \times 10^{-7},$$

roughly an order of magnitude improvement over the previous measurement.

In addition to $K_L^0 \to e^+e^-\gamma\gamma$, there are backgrounds from radiative $K_L^0 \to \pi^\pm e^\mp \overline{\nu}$ decay with an accidental photon, $K_L^0 \to \pi^\pm e^\mp \overline{\nu}$ decay with an accidental π^0, $3\pi^0$ ($2\pi^0$) decay in which two(one) pion(s) Dalitz decays, and $K_L^0 \to \pi^+\pi^-\pi^0$ decay with $\pi^\pm$ misidentified as $e^\pm$. The modes that contain a Dalitz decay are easily rejected by cutting on the invariant mass of the e^+e^- pair. Modes in which charged pions mimic electrons are reduced through the excellent π/e separation in the calorimeter and TRDs. Event selection criteria include requiring that the e^+e^- come from a good vertex and that the photons form a π^0. The event also is required to have small transverse momentum squared (P_t^2) with respect to the kaon direction.

The remaining background after all cuts have been applied comes from $K_L^0 \to e^+e^-\gamma\gamma$, which is reduced with cuts on kinematic variables (see fig. 39.3). One such variable is $\cos(\Theta_\pi)$, the direction of the photon with respect to the direction of the π^0 (defined by a momentum vector opposite the e^+e^- pair) in the pion rest frame. This variable is flat for $K_L^0 \to \pi^0 e^+e^-$ decays since the π^0 is a spinless particle and the photons emerge back to back, while it is peaked near 0 for $K_L^0 \to e^+e^-\gamma\gamma$ since the non-bremsstrahlung photon tends to go off in the direction opposite the electron-positron pair. A second variable that is used to distinguish signal from background is Θ_{MIN}, the angle between the photon and the nearest e^+ or e^-. This variable is flat for the signal mode and peaked near 0 for the bremsstrahlung photon in $K_L^0 \to e^+e^-\gamma\gamma$. A lego plot of $|\cos(\Theta_\pi)|$ versus Θ_{MIN} from signal and background Monte Carlo is also shown in fig. 39.3. The plot on the left in fig. 39.4 shows the 2γ invariant mass versus the $e^+e^-\gamma\gamma$ invariant mass with all cuts applied except those specific to suppressing $K_L^0 \to e^+e^-\gamma\gamma$. The long box is populated by $K_L^0 \to \pi^0\pi^0\pi^0$ event fragments in which one π^0 decays to 2γ while the other two π^0s Dalitz-decayed. Events in the low M($e^+e^-\gamma\gamma$) and low M(2γ) region are from $K_L^0 \to \pi^\pm e^\mp \overline{\nu}$ decays with accidental γs in which the π is misidentified. The band near the signal region is $K_L^0 \to e^+e^-\gamma\gamma$. The small box hides the signal region. A Monte Carlo background estimate predicts 1.51 ± 0.44 events in the signal region from $K_L^0 \to e^+e^-\gamma\gamma$. After all cuts, the SES is measured to be 1.26×10^{-10}. The right side of fig. 39.4 reveals the signal box after all cuts have been applied. There are 2 events observed in the signal region, which is consistent with background. We quote a preliminary measurement of the upper limit [11] of

$$B(K_L^0 \to \pi^0 e^+e^-) < 5.64 \times 10^{-10} (90\% \text{ CL}) \text{ [10]},$$

roughly an order of magnitude improvement over the previous limit. If we assume that the only contribution to the branching ratio comes from direct CP violation and we take the experimental inputs of m_t, m_W, and the Wolfenstein parameter A, we find $\eta < 5$.

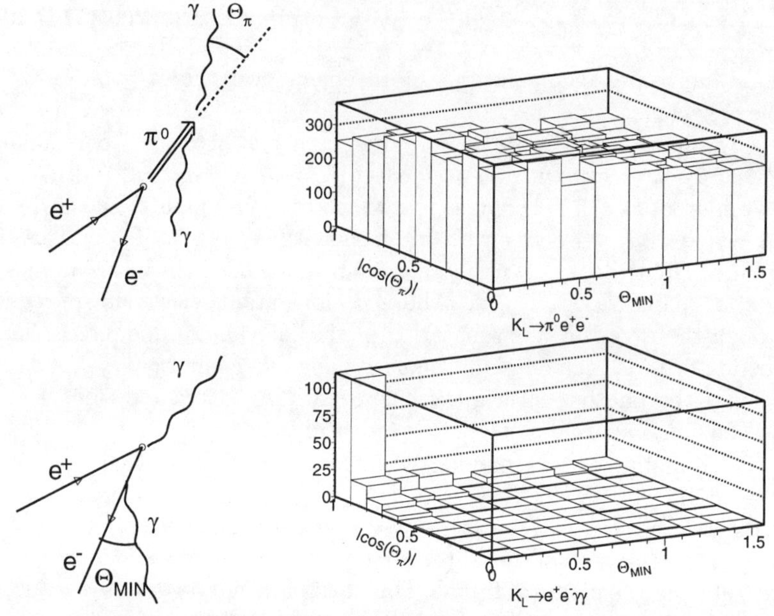

Figure 39.3: Diagrams defining the kinematic variables $|\cos(\Theta_\pi)|$ vs. $\Theta_{\rm MIN}$ and Monte Carlo lego plots of these variables for $K_L^0 \to \pi^0 e^+e^-$ (upper) and $K_L^0 \to e^+e^-\gamma\gamma$ (lower).

39.4 $K_L^0 \to \pi^0\mu^+\mu^-$ and $K_L^0 \to \mu^+\mu^-\gamma\gamma$

Like the electron mode, the decay $K_L^0 \to \pi^0\mu^+\mu^-$ is a probe of direct CP violation, although the direct CP-violating component is small relative to the CP-conserving component. In addition to the challenges of measuring a very small branching ratio, there is potential background from $K_L^0 \to \mu^+\mu^-\gamma\gamma$, which has a branching ratio calculated to be $(9.1 \pm 0.8) \times 10^{-9}$.

The event selection requires that 2 photons form a π^0 and that a good vertex is reconstructed from two oppositely charged tracks that leave minimum ionizing deposits (MIP) in the calorimeter and fire the muon counters. Backgrounds come from pion punch-through from $K_L^0 \to \pi^+\pi^-\pi^0$, pion decay-in-flight from $K_L^0 \to \pi^+\pi^-\pi^0$, and $K_L^0 \to \pi^\pm\mu^\mp\overline{\nu}$ decay with 2 accidental photons. The most significant single background comes from $K_L^0 \to \mu^+\mu^-\gamma\gamma$ and is estimated to be 0.343 ± 0.048 events in the signal region. The angular cuts that were used to reduce the background from $K_L^0 \to e^+e^-\gamma\gamma$ decays in the $K_L^0 \to \pi^0e^+e^-$ analysis are less effective for the muon mode, since the photon and muon directions are less correlated and thus are not used in this analysis. The SES for this mode corresponds to a branching ratio of 7×10^{-10}. The total background contribution is expected to be 1.02 ± 0.18 events. The normalization mode for this mea-

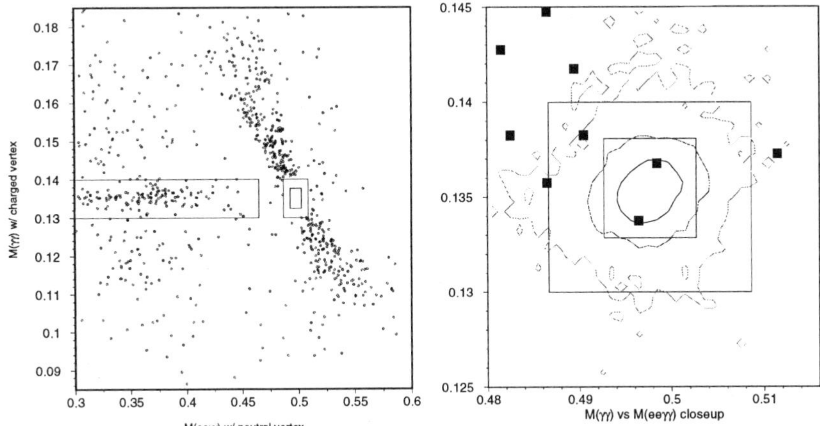

Figure 39.4: The left plot is the reconstructed $M(2\gamma)$ versus $M(e^+e^-\gamma\gamma)$ invariant masses with all cuts applied except kinematic cuts suppressing $e^+e^-\gamma\gamma$. The small box hides the signal region. The plot on the right shows the signal region box after all cuts have been applied. The curves are 68.27%, 94.45%, and 99.73% contours of the signal Monte Carlo. There are 2 events in the signal region.

surement is $K_L^0 \to \pi^+\pi^-\pi^0$. Fig. 39.5 shows the invariant mass spectrum.

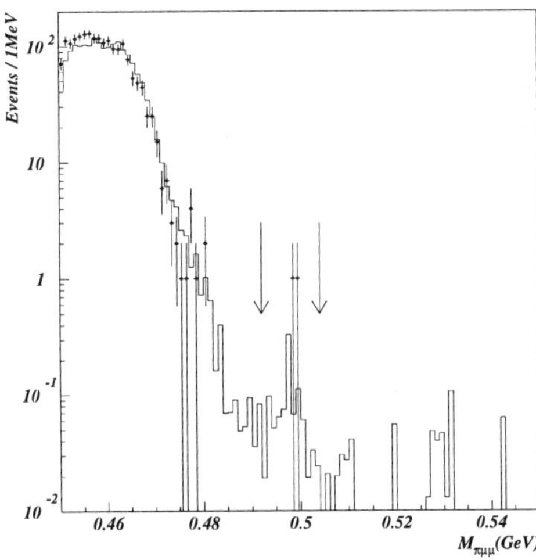

Figure 39.5: The $M(\pi^0\mu^+\mu^-)$ invariant mass distribution. Signal and background Monte Carlo (histogram) are overlaid on data (dots). The signal region is indicated by arrows.

There are 2 events in the signal region, which is consistent with background. We set a preliminary upper limit [11] on the branching ratio of

$$B(K_L^0 \to \pi^0\mu^+\mu^-) < 3.4 \times 10^{-10} \text{ (90\% CL) [12]},$$

which is an order of magnitude improvement over the previous measurement.

The decay $K_L^0 \to \mu^+\mu^-\gamma\gamma$ is of interest because it is expected to be a dangerous background to $K_L^0 \to \pi^0\mu^+\mu^-$ and it has never before been observed. The QED prediction for the branching ratio is $(9.1 \pm 0.78) \times 10^{-9}$. Candidate events are required to have two photons that do not reconstruct as a π^0, and a good vertex formed from two tracks that MIP in the calorimeter and fire the muon counters. Backgrounds come from $K_L^0 \to \pi^\pm\mu^\mp\overline{\nu}$ decay with two accidental photons and $K_L^0 \to \mu^+\mu^-\gamma$ decay with one accidental photon. Since accidental photons tend to be of low energy, these backgrounds can be reduced by a cut on the minimum photon energy. The total contribution from backgrounds is estimated to be 0.155 ± 0.081 events. The normalization mode is $K_L^0 \to \pi^+\pi^-\pi^0$. A plot of the invariant mass is shown in fig. 39.6. Four events are within the signal

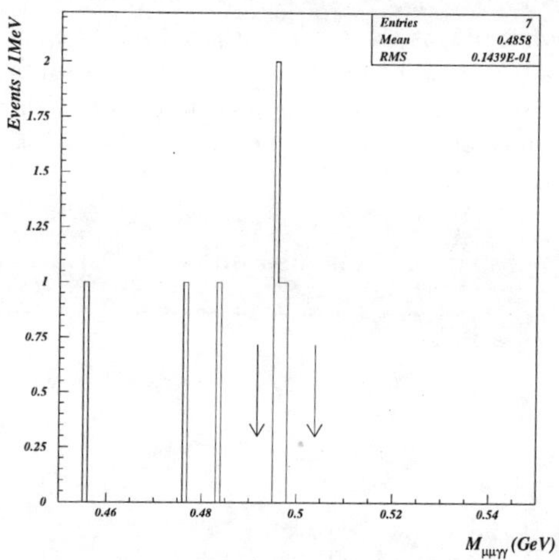

Figure 39.6: The $M(\mu^+\mu^-\gamma\gamma)$ invariant mass distribution in data. The signal region is indicated by arrows.

region, significantly above predicted background levels. The branching ratio is presented in two ways. In order to compare this with the QED prediction, the branching ratio is calculated with a $1\,\mathrm{MeV/c^2}$ cutoff in the 2γ invariant mass during Monte Carlo generation and is found to equal

$$B(K_L^0 \to \mu^+\mu^-\gamma\gamma, M_{\gamma\gamma} > 1\,\mathrm{MeV/c^2}) = (10.4^{+7.5}_{-5.9}(\mathrm{stat.}) \pm 1.0(\mathrm{syst.})) \times 10^{-9},$$

and is consistent with theoretical prediction. Finally, the preliminary measurement of the branching ratio [13] using a 10 MeV infrared cutoff for photon energies in the kaon rest frame is

$$B(K_L^0 \to \mu^+\mu^-\gamma\gamma, E_\gamma^* > 10\,\mathrm{MeV}) = 1.42^{+1.02}_{-0.81}(\mathrm{stat.}) \pm 0.14(\mathrm{syst.})) \times 10^{-9}.$$

This constitutes a first observation of this decay mode.

39.5 Summary and Future Outlook

KTeV has improved its branching ratio measurements of $K_L^0 \to e^+e^-\gamma\gamma$ and set new upper limits for the branching ratios of the direct CP-violating decays $K_L^0 \to \pi^0\nu\overline{\nu}$, $K_L^0 \to \pi^0 e^+e^-$, and $K_L^0 \to \pi^0\mu^+\mu^-$. All of these results represent approximately an order of magnitude improvement over previous measurements. KTeV also has made the first measurement of the branching ratio $K_L^0 \to \mu^+\mu^-\gamma\gamma$. Table 39.2 contains a summary of these new results.

Mode	Branching Ratio or 90% C.L. Upper Limit
$K_L^0 \to \pi^0 e^+e^-$	$< 5.64 \times 10^{-10}$
$K_L^0 \to \pi^0\mu^+\mu^-$	$< 3.4 \times 10^{-10}$
$K_L^0 \to \pi^0\nu\overline{\nu}$ with $\pi^0 \to 2\gamma$ tag	$< 1.6 \times 10^{-6}$
$K_L^0 \to \pi^0\nu\overline{\nu}$ with $\pi^0 \to e^+e^-\gamma$ tag	$< 5.9 \times 10^{-7}$
$K_L^0 \to \pi^0\gamma\gamma$	$(1.68 \pm 0.07 \pm 0.08) \times 10^{-6}$
$K_L^0 \to e^+e^-\gamma\gamma$ with $E_\gamma \geq 5$ MeV	$(6.31 \pm 0.14 \pm 0.42) \times 10^{-7}$
$K_L^0 \to \mu^+\mu^-\gamma\gamma$ with $M_{\gamma\gamma} > 1$ MeV/c^2	$(10.4^{+7.5}_{-5.9} \pm 1.0) \times 10^{-9}$
$K_L^0 \to \mu^+\mu^-\gamma\gamma$ with $E_\gamma^* > 10$ MeV	$(1.42^{+1.02}_{-0.81} \pm 0.14) \times 10^{-9}$

Table 39.2: Summary of KTeV rare decay results. The first error in the branching ratio is statistical and the second is systematic.

KTeV's rare decay program has been approved for an additional 10-week run starting in October 1999, with the goal of collecting three times the statistics of the 1997 run. With the gain in statistics from the 1999 run and a single-event sensitivity on the order of 2×10^{-11} for the charged lepton modes, KTeV will start to probe the branching ratio regions predicted when including possible enhancements from SUSY contributions [14]. In addition to near-term improvements on the $K_L^0 \to \pi^0\ell\overline{\ell}$ modes, there is a rare decay program that has been proposed at Fermilab that will use a neutral-kaon beam from the Main Injector. The main goal of the experiment, KAMI, is to collect roughly a hundred $K_L^0 \to \pi^0\nu\overline{\nu}$ events [15].

References

[1] A. Alavi-Harati *et al.*, Phys. Rev. Lett. **83**, 22, (1999); Y. B. Hsiung, this volume, chapter 6.

[2] G. Buchalla and A.J. Buras, CERN-TH-98-369, hep-ph/9901288; G. Buchalla and A.J. Buras, Phys. Rev. D **54**, 6782 (1996).

[3] J.F. Donoghue and F. Gabbiani, Phys. Rev. D **51**, 2187 (1995).

[4] B. Winstein and L. Wolfenstein, Rev. Mod. Phys. **65**, 1113 (1993).

[5] J. Adams *et al.*, Phys. Lett. B **447**, 240 (1999).

[6] K. Hanagaki, *Searches for the Decay* $K_L^0 \to \pi^0 \nu\overline{\nu}$, PhD thesis, University of Osaka, Japan, August 1998); A. Alavi-Harati *et al.*, hep-ex/9907014.

[7] C. Caso *et al.*, Eur. Jour. Phys. C **3**, 1 (1998).

[8] A. Alavi-Harati *et al.*, Phys. Rev. Lett. **83**, 917 (1999).

[9] G. D'Ambrosio and J. Portoles, Nucl. Phys. B **492**, 417 (1997).

[10] T. Yamanaka, "KTeV Rare Decay Results," presented at 34th Recontres de Moriond on Electroweak Interactions and Unified Theories, Les Arcs, France, 13–20 March, 1999.

[11] G. Feldman and R. Cousins, Phys. Rev. D **57**, 3873 (1998).

[12] L. Bellantoni, "KTeV Rare Decay Searches for Direct CP Violation," presented at Fermilab Joint Experimental-Theoretical Seminar, Batavia, IL, 30 April, 1999.

[13] B. Quinn, "Recent Rare K_L^0 Decay Results from KTeV," presented at PANIC99, 15th International Conference on Particles and Nuclei, Uppsala, Sweden, 10–16 June, 1999.

[14] A. Buras and L. Silvestrini, Nucl. Phys. B **546**, 299 (1999); L. Silvestrini, hep-ph/9906202; G. Colangelo and G. Isidori, J. High Energy Phys. **9809:009** (1998); G. Isidori, this volume, chapter 33.

[15] P. Cooper, this volume, chapter 55.

40

Measurement of $K^+ \to \pi^+\pi^0\gamma$ from BNL E787

Takeshi K. Komatsubara[1]

Abstract

We have performed a new measurement of the $K^+ \to \pi^+\pi^0\gamma$ decay with the E787 apparatus and have observed 20k signal events. The best fit to the decay spectrum gives a branching ratio for direct emission of $(4.72 \pm 0.77) \times 10^{-6}$ in the π^+ kinetic energy region of 55 to 90 MeV, with 15% systematic uncertainty, and requires no interference with inner bremsstrahlung.

40.1 Introduction

The primary motivation for studying the radiative decay $K^+ \to \pi^+\pi^0\gamma$ is to understand the direct emission (DE) component of the decay, where the γ is emitted directly due to the low-energy hadronic interaction of the mesons [1]. The inner bremsstrahlung (IB) component, which is a QED radiative correction to $K^+ \to \pi^+\pi^0$ decay, dominates the process, even though it is suppressed by the $\Delta I = 1/2$ rule. The DE component consists of the magnetic and electric transitions. The magnetic transition is known to be due to the chiral anomaly in chiral perturbation theory [2]. The electric transition can interfere with the IB component since their parities are the same. This interference term is designated by INT.

The variable $\mathsf{W}^2 \equiv (\mathsf{p}\cdot\mathsf{q})/\mathsf{m}^2_{\mathsf{K}+} \times (\mathsf{p}_+\cdot\mathsf{q})/\mathsf{m}^2_{\pi+}$ is defined in terms of the energy-momentum four-vectors of the K^+ (p), π^+ (p_+) and γ (q). In the theoretical W spectrum of the decay normalized to the IB spectrum [1], the INT component is proportional to W^2 and the DE component is proportional to W^4. To observe the DE and INT components of the decay mode in the data, a deviation of the measured spectrum from the IB spectrum at large W must be identified.

Table 40.1 lists the results of previous measurements [3–5]. In these experiments the branching ratio was measured with decays-in-flight and in

[1]Representing the E787 collaboration from BNL, KEK, Osaka, Princeton, TRIUMF and Alberta. E-mail: takeshi.komatsubara@kek.jp.

Experiment		Kaon	Events	Br(DE) $\times 10^{-5}$
BNL '72	[3]	$K^{\pm}$	2100	$1.56 \pm 0.35 \pm 0.5$
CERN '76	[4]	$K^{\pm}$	2461	2.3 ± 3.2
IHEP '87	[5]	K^{-}	140	$2.05 \pm 0.46^{+0.39}_{-0.23}$
PDG '98	[6]		-	1.8 ± 0.4

Table 40.1: Previous measurements of the $K^+ \to \pi^+\pi^0\gamma$ decay. Br(DE) is the branching ratio for DE in the region 55 MeV $< T_{\pi^+} <$ 90 MeV.

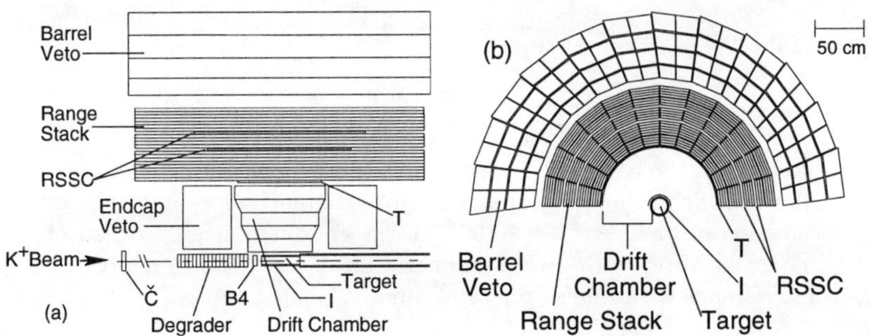

Figure 40.1: Schematic (a) side and (b) end views showing the upper half of the E787 detector. Č: beam Čerenkov counter; B4: energy loss counter; I and T: trigger scintillators; RSSC: straw tube tracking chambers.

the π^+ kinetic energy region 55 MeV $< T_{\pi^+} <$ 90 MeV. The theoretical prediction for the IB branching ratio, 2.61×10^{-4} [1], has been verified. The current Particle Data Group average for the DE component of the branching ratio is $(1.8\pm0.4)\times10^{-5}$ [6]. In the 1980s, chiral QCD theories [7] claimed to be able to account for DE branching ratios in the region of 10^{-5}. But there currently exists no definite prediction from chiral perturbation theory.

40.2 E787 Detector and Trigger

The new measurement reported here used the E787 apparatus [8,9] at the Alternating Gradient Synchrotron (AGS) of Brookhaven National Laboratory (BNL). In the 1995 data set, with which we observed the first evidence for $K^+ \to \pi^+\nu\bar{\nu}$ [10], E787 had a physics trigger called "3gamma" to detect the final state of the $K^+ \to \pi^+\pi^0(\to \gamma\gamma)\gamma$ decay and took data with it simultaneously with the $\pi^+\nu\bar{\nu}$ trigger.

In the E787 detector (fig. 40.1), which is a solenoidal spectrometer with a 1.0 Tesla field directed along the beam line, charged tracks emanating from stopped K^+ decays were measured. Kaons of 790 MeV/c were delivered to the experiment at a rate of 7×10^6 per 1.6-s spill of the AGS, and were detected by Čerenkov and energy loss (dE/dx) counters. Slowed by a BeO degrader, kaons stopped in the scintillating-fiber target at the center of the detector. The charged decay products passed through the drift chamber, lost energy by ionization loss and stopped in the range stack made of plastic scintillators and straw chambers. The output pulse-shapes of the range stack counters were recorded and the π^+ track was reconstructed with good timing accuracy. The hermetic calorimeters in the barrel and end cap detected γs and extra particles from the K^+ decays.

The barrel calorimeter, designed to veto any extra particles in the $K^+ \to \pi^+\nu\bar{\nu}$ analysis, detected the three γs from the $K^+ \to \pi^+\pi^0\gamma$ decay. It consisted of 48 azimuthal sectors and 4 radial layers, made of lead/scintillator 14 radiation lengths in depth, and covered a solid angle of about 3π sr. 30% of the energy was deposited and visible in the scintillators. The Z position (along the beam line) of the barrel hits was measured with the ADC and TDC information from phototubes on both ends of each 2-m long module.

In the 3gamma trigger a stopped kaon decay, a charged track in the range stack, and no extra particle in the range stack and end cap were required. The online delayed-coincidence (> 1.5 ns) of the timing between the stopping K^+ (via the Čerenkov counter) and the outgoing π^+ (via the I-counter surrounding the target) rejected pions scattered into the detector or kaons that decayed in flight. For the $K^+ \to \pi^+\pi^0\gamma$ decay a shorter range particle was required in the range stack, one whose kinetic energy exceeded the maximum allowed from $K_{\pi 3}$ decay and was smaller than that from $K_{\pi 2}$ decay. In addition, ≥ 3 online barrel clusters were required. The trigger was prescaled by 5 and a total of 1.1×10^7 events (5 events per spill) was recorded.

40.3 Signal Selection

The energy and direction of the three γs from the $K^+ \to \pi^+\pi^0\gamma$ decay were determined from the offline clustering in the barrel calorimeter and the decay vertex position in the target. The π^+ momentum vector and the kinetic energy were measured with the target, drift chamber, and range stack. These thirteen observables were available for the kinematical fit in this analysis. The range-momentum relation of the track in the range stack was sufficient to reject μ^+ backgrounds, so that the pulse-shape information of the decay chain $\pi^+ \to \mu^+ \to e^+$ in the stopping scintillator was not used. Any coincident signals other than the three γs and the π^+ caused the event to be vetoed.

Since the E787 detector measures stopped K^+ decays, a kinematical fit was applied to the observables with the following six constraints: energy-momentum conservation, π^+ energy and momentum consistent with m_{π^+}, and invariant mass of the two γs from π^0 equal to m_{π^0}. In order to minimize feed-down of IB events into the DE region, the following method was used to

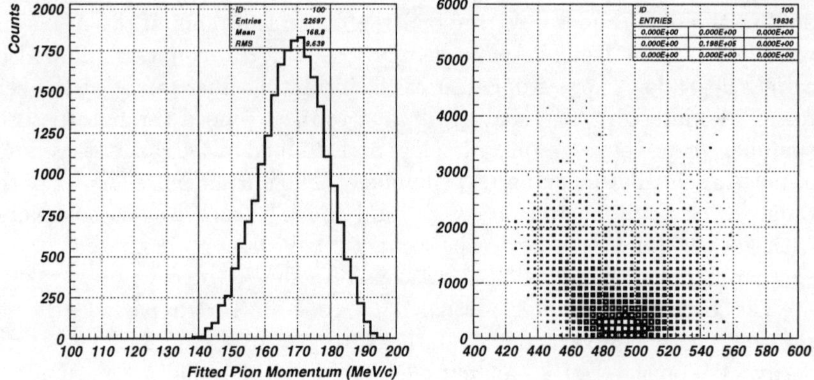

Figure 40.2: Left: Fitted π^+ momentum after the χ^2 probability cut; right: missing-momentum squared vs. total energy of the signal events.

select the two γs to be assigned to the π^0. The square of the matrix element of the IB decay was taken as a weight: the combination that maximized the product of the χ^2 probability of the kinematical fit and the IB weight was chosen. For the signal selection we required the χ^2 probability to be more than 10% and the fitted π^+ momentum (fig. 40.2, left) to be between 140 and 180 MeV/c; the latter condition was imposed to remove the $K_{\pi 3}$ and $K_{\pi 2}$ backgrounds as well as $K^+ \to \pi^+\pi^0\gamma$ decays outside the region 55 MeV $< T_{\pi^+} <$ 90 MeV. Fig. 40.2, right, shows missing-momentum squared versus total energy of the $K^+ \to \pi^+\pi^0\gamma$ events, 19836 in total, after applying all selection cuts. The signal events are clustered at m_{K^+} in the total energy and at small values of missing-momentum squared.

40.4 Spectrum Analysis

The challenge of this analysis consists in the fact that 98% of the signal events in fig. 40.2 are from the IB component of the $K^+ \to \pi^+\pi^0\gamma$ events. To address this, we performed the W spectrum analysis described earlier. In the stopped K^+ decays, W^2 was constructed from the opening angle between π^+ and γ ($\theta_{\pi^+\gamma}$), π^+ energy (E_{π^+}), momentum (P_{π^+}), and γ energy (E_γ) as $\mathsf{W}^2 = \mathsf{E}_\gamma^2 \times (\mathsf{E}_{\pi^+} - \mathsf{P}_{\pi^+} \times \cos\theta_{\pi^+\gamma})/(\mathsf{m}_{\mathsf{K}^+} \times \mathsf{m}_{\pi^+}^2)$ and is directly related to the observables in the E787 detector.

A Monte Carlo simulation of the IB, DE, and INT spectra of the $K^+ \to \pi^+\pi^0\gamma$ decay in the E787 detector was performed by generating the IB, DE, and INT decay events, respectively, with the standard matrix elements in [1] and analyzing them with the codes for real data. Thus, the effect of "incorrect pairing" of the two γs from π^0 on the Monte Carlo spectra is taken into account. The probability of incorrect pairing with the algorithm of this analysis is estimated by the Monte Carlo to be 2.6% for IB, 20% for DE, and 9% for INT.

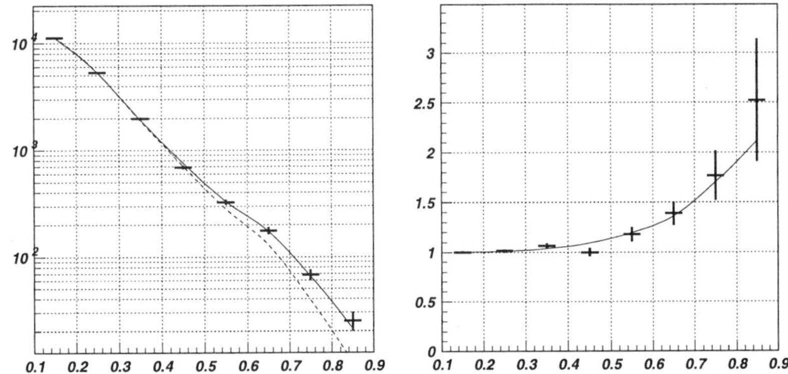

Figure 40.3: Left: W spectrum of the signal events and best fits to IB+DE [solid curve] and IB alone [dashed curve]; right: W spectrum normalized to the IB spectrum.

Fig. 40.3, left, shows the W spectrum of the signal events. The solid curve is the best fit to the sum of IB and DE; the χ^2 of the fit was 7.85 for 6 degrees of freedom. The dashed curve is the best fit to IB alone. Normalized to the IB spectrum the deviation of the measured spectrum at large W is apparent (fig. 40.3, right). The DE component was measured to be $(1.85 \pm 0.30)\%$ of the IB component with the spectrum and, normalized to the theoretical prediction for the IB branching ratio, the branching ratio for DE is

$$(4.72 \pm 0.77) \times 10^{-6}$$

in the region 55 MeV $< T_{\pi^+} <$ 90 MeV.

Several checks to verify the result were made. The best fit to the sum of IB, DE, and INT gave the INT component to be $(-0.4 \pm 1.6)\%$ of the IB component, and we concluded that no interference term is required for the fit to this spectrum. A spectrum analysis was performed on the $\cos\theta_{\pi^+\gamma}$ distribution, and a consistent result was obtained. To study systematic uncertainties, we enhanced possible problems in the analysis such as barrel calorimeter miscalibration, π^+ momentum scale error, incorrect pairing, π^+-nuclear interaction, and so on by loosening the corresponding cuts and examining the consequent changes in the result. All the changes were within 15%, no more than the statistical fluctuation due to the limited sample, and for the present we have assigned a 15% systematic uncertainty to the measurement. Background contamination in the signal sample was studied with a similar technique and was confirmed to be negligible.

The result from E787 is smaller than the results of previous measurements in table 40.1. The DE decay rate in the region $E_\gamma > 20$ MeV is calculated to be 808 s^{-1} after correction for phase-space acceptance and is close to the rate of $K^0_L \to \pi^+\pi^-\gamma$ decay, 617 s^{-1} (corresponding to the DE branching ratio $(3.19 \pm 0.16) \times 10^{-5}$ [11]).

40.5 Summary and Prospects

We have performed a new measurement of the $K^+ \to \pi^+\pi^0\gamma$ decay with more than eight times higher statistics than previous experiments and good kinematical constraints, using the hermetic E787 detector and stopped K^+ decays. In 1998, new E787 data with a modified 3gamma trigger were taken to collect the $K^+ \to \pi^+\pi^0\gamma$ decays with lower π^+ energy threshold. We aim to study the direct emission and interference components as well as to search for a new decay mode $K^+ \to \pi^0\mu^+\nu\gamma$ at the 10^{-5} level.

Acknowledgments

We gratefully acknowledge the dedicated effort of the technical staff supporting this experiment and of the Brookhaven AGS Department. This research was supported in part by the U.S. Department of Energy under Contracts No. DE-AC02-98CH10886, W-7405-ENG-36, and grant DE-FG02-91ER40671, by the Ministry of Education, Science, Sports, and Culture of Japan through the Japan-U.S. Cooperative Research Program in High Energy Physics and under the Grant-in-Aids for Scientific Research, for Encouragement of Young Scientists, and for JSPS Fellows, and by the Natural Sciences and Engineering Research Council and the National Research Council of Canada.

References

[1] G. D'Ambrosio, M. Miragliuolo, and P. Santorelli, "Radiative Non-leptonic Kaon Decays," in *The DAΦNE Physics Handbook*, vol. 1, ed. L. Maiani, G. Pancheri, and N. Paver (Servizio Documentazione dei Laboratori Nazionali di Frascati, Frascati, 1992), and references therein; G. D'Ambrosio and G. Isidori, Z. Physik C **65**, 649 (1995).

[2] J. Bijnens, this volume, chapter 37.

[3] R.J. Abrams *et al.*, Phys. Rev. Lett. **29**, 1118 (1972).

[4] K.M. Smith *et al.*, Nucl. Phys. B **109**, 173 (1976).

[5] V.N. Bolotov *et al.*, Sov. J. Nucl. Phys. **45**, 1023 (1987).

[6] C. Caso *et al.*, Particle Data Group, Eur. Phys. J. C **3**, 1 (1998).

[7] See table II of [5] and references therein.

[8] G. Redlinger, this volume, chapter 34.

[9] M.S. Atiya *et al.*, Nucl. Instr. Meth. A **321**, 129 (1992); M.S. Atiya *et al.*, Nucl. Instr. Meth. A **279**, 180 (1989); I-H. Chiang *et al.*, IEEE Trans. Nucl. Sci. NS **42**, 394 (1995); D.A. Bryman *et al.*, Nucl. Instr. Meth. A **396**, 394 (1997); E.W. Blackmore *et al.*, Nucl. Instr. Meth. A **404**, 295 (1998); T.K. Komatsubara *et al.*, Nucl. Instr. Meth. A **404**, 315 (1998).

[10] S. Adler *et al.*, Phys. Rev. Lett. **79**, 2204 (1997).

[11] E.J. Ramberg *et al.*, Phys. Rev. Lett. **70**, 2525 (1993).

41

Brookhaven E871: $K_L^0 \to \mu^+\mu^-$ and $K_L^0 \to e^+e^-$

David A. Ambrose

Abstract

This paper describes final results from the rare K_L^0 decay experiment 871, performed at the Brookhaven AGS, which measured branching ratios for the GIM- and helicity-suppressed decay modes modes $K_L^0 \to \mu^+\mu^-$ and $K_L^0 \to e^+e^-$. The experiment observed over 6200 $\mu^+\mu^-$ candidates, a factor of six improvement on all previous measurements combined, giving a branching fraction of $B(K_L^0 \to \mu^+\mu^-) = (7.18 \pm 0.17) \times 10^{-9}$, which reduces the uncertainty in this decay mode by a factor of three. E871 also detected four candidate e^+e^- events, the first observation of this rare decay, corresponding to $B(K_L^0 \to e^+e^-) = 8.7^{+5.7}_{-4.1} \times 10^{-12}$, which represents the smallest branching fraction ever measured in particle physics.

41.1 Introduction

The decay $K_L^0 \to \mu^+\mu^-$ has historical importance because its low observed rate provided motivation for the Glashow-Iliopoulos-Maiani (GIM) mechanism [1], which invoked a fourth "charmed" quark whose presence cancels flavor changing neutral currents at tree level, and strongly suppresses second order loop diagrams involving the up quark (fig. 41.1). The remaining decay rate can be expressed as the sum of real (dispersive) and imaginary (absorptive) parts of the amplitude, where $\mathrm{Re}\,\mathcal{A}$ contains contributions from the short-distance electroweak and long-distance electromagnetic diagrams

$$B(K_L^0 \to \mu^+\mu^-) = |\,\mathrm{Re}\,\mathcal{A}|^2 + |\,\mathrm{Im}\,\mathcal{A}|^2 \quad \longrightarrow \quad \mathrm{Re}\,\mathcal{A} = \mathcal{A}_{SD} + \mathcal{A}_{LD}. \quad (41.1)$$

Previous measurements show that the observed rate is almost completely saturated by an absorptive process containing a real, two-photon intermediate state, whose contribution can be calculated from QED [2] and the measured $K_L^0 \to \gamma\gamma$ branching fraction [3]

$$B(K_L^0 \to \mu^+\mu^-)_{\gamma\gamma} = \frac{1}{2}\alpha^2 \left(\frac{m_\mu}{m_K}\right)^2 \left(\ln \frac{1+\beta}{1-\beta}\right)^2 \times B(K_L^0 \to \gamma\gamma). \quad (41.2)$$

This value, called the *unitarity bound*, equals $(7.07 \pm 0.18) \times 10^{-9}$ and establishes the minimum rate we should expect for $K_L^0 \to \mu^+\mu^-$.

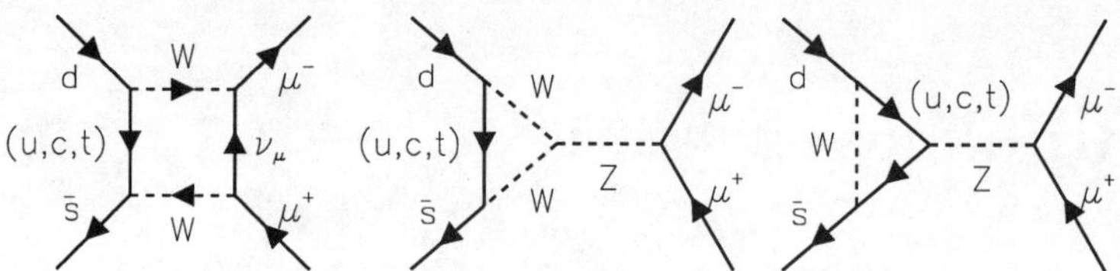

Figure 41.1: Second-order electroweak diagrams for $K_L^0 \to \mu^+\mu^-$.

The residual rate due to short-distance physics is dominated by the heavy top quark, and so is approximately proportional to $|\,\mathrm{Re}(V_{ts}^* V_{td})|^2$, or $(1-\rho)^2$ in the Wolfenstein parameterization [4] of the CKM matrix. From the next-to-leading order standard model equation [5]

$$|\mathcal{A}_{SD}|^2 = 0.9 \times 10^{-9}\,(1.2-\bar{\rho})^2 \left[\frac{\bar{m}_t(m_t)}{170\,\mathrm{GeV}}\right]^{3.1} \left[\frac{|V_{cb}|}{0.040}\right]^4, \tag{41.3}$$

where $\bar{\rho} = \rho(1-\lambda^2/2)$, we can use the measured $\mu^+\mu^-$ branching fraction to place a lower limit on ρ [6]

$$\bar{\rho} > 1.2 - (3.426 \pm 0.548) \times 10^4 \Big[|\,\mathrm{Re}\,\mathcal{A}_{exp}| + |\,\mathrm{Re}\,\mathcal{A}_{LD}|\Big] \tag{41.4}$$

provided that the long-distance dispersive amplitude can be calculated, which currently has significant theoretical uncertainty [7,8].

The decay $K_L^0 \to e^+e^-$ shares the physics of GIM suppression in the electroweak interactions, as with the $\mu^+\mu^-$ mode, but is further suppressed by helicity constraints. The short-distance helicity suppression can be compared with the K^+ decay

$$\frac{B(K_L^0 \to ee)_{SD}}{B(K_L^0 \to \mu\mu)_{SD}} \approx \frac{B(K^+ \to e\nu)}{B(K^+ \to \mu\nu)} \approx \left(\frac{m_e}{m_\mu}\right)^2 \approx 2.4 \times 10^{-5} \tag{41.5}$$

and is approximately equal to the ratio of lepton masses squared. The absorptive contribution to the decay rate is also helicity suppressed, but is enhanced by over an order of magnitude due to the logarithmic singularity in the limit of zero lepton mass [9]

$$\frac{B(K_L^0 \to \gamma\gamma \to ee)}{B(K_L^0 \to \gamma\gamma \to \mu\mu)} = \left(\frac{m_e}{m_\mu}\right)^2 \frac{\beta_e}{\beta_\mu} \frac{\left(\ln \frac{1+\beta_e}{1-\beta_e}\right)^2}{\left(\ln \frac{1+\beta_\mu}{1-\beta_\mu}\right)^2}$$

$$\Longrightarrow \beta_\ell = \sqrt{1 - 4m_\ell^2/m_K^2}. \tag{41.6}$$

The *unitarity bound* for $K_L^0 \to e^+e^-$ becomes $\sim 3 \times 10^{-12}$, so that short-distance contributions are negligible. The long-distance dispersive term is similarly enhanced over the short-distance effects and is less sensitive to the theoretical uncertainties of $\mu^+\mu^-$. Recent calculations of the long-distance amplitude in the framework of chiral perturbation theory [7, 8] show that this dispersive term dominates the total rate, for a predicted value of $B(K_L^0 \to e^+e^-) = (9.0 \pm 0.5) \times 10^{-12}$.

The major sources of background for $K_L^0 \to \mu^+\mu^-$ are the semileptonic decays $K_L^0 \to \pi\mu\nu$ and $K_L^0 \to \pi e\nu$ (referred to as $K_{\mu 3}$ and K_{e3}, respectively), where the pion decays and the electron is misidentified to mimic a $\mu^+\mu^-$ signal. These three-body backgrounds are suppressed by cutting on the net transverse momentum (p_T) of the decay with respect to the K_L^0 direction, which should equal zero for two-body decays and non-zero for semileptonics. At zero neutrino momentum, the two-body invariant mass of $K_{\mu 3}$ events reconstructed as $\mu^+\mu^-$ approaches a kinematic limit of 489 MeV/c^2, about 8 MeV/c^2 below the K_L^0 mass, so that cuts on mass and p_T remove the vast majority of $K_{\mu 3}$ background. K_{e3} events may produce a two-body mass above the K_L^0 mass, so that good muon identification and electron rejection are needed to remove this background. For $K_L^0 \to e^+e^-$, background from K_{e3} (where the pion is misidentified as an electron) is less important, since the kinematic limit on two-body e^+e^- mass is over 20 MeV/c^2 below the K_L^0 mass. However, this decay mode is susceptible to physics background sources from $K_L^0 \to e^+e^-\gamma$ ($BR \sim 9 \times 10^{-6}$) and $K_L^0 \to e^+e^-e^+e^-$ ($BR \sim 4 \times 10^{-8}$), which must be considered.

41.2 Apparatus and Analysis

Brookhaven National Laboratory Experiment 871 recorded data in 1995 and 1996 at the B5 beamline of the Alternating Gradient Synchrotron (AGS). The experiment improved upon BNL E791 [10], which measured roughly 700 $\mu^+\mu^-$ events for a branching fraction $B(K_L^0 \to \mu^+\mu^-) = (6.86 \pm 0.37) \times 10^{-9}$, while observing no candidate e^+e^- events and setting an upper limit of $B(K_L^0 \to e^+e^-) < 4.1 \times 10^{-11}$ [11]. The experimental apparatus has been described previously for the e^+e^- and $\mu^\pm e^\mp$ analyses [12, 13] and elsewhere in these proceedings [14]. Briefly, a neutral-kaon beam was produced with 24 GeV/c protons incident on a platinum target, and defined by a series of sweeping magnets and collimators, entering a vacuum decay tank about 11 m in length. Decay products exiting the tank proceeded through a double-magnet spectrometer containing 6 pairs of drift chambers, with the neutral beam terminating in a beam stop inside the first magnet [15]. The magnets had opposite polarity, giving an overall p_T kick inwards of $\sim$ 200 MeV/c, tuned to the dilepton decay momenta to allow particles to emerge with approximately parallel trajectories.

Two banks of narrow-width scintillator (TSC) hodoscopes then provided a level 0 parallel, two-body trigger. A segmented threshold Čerenkov counter using hydrogen gas triggered electrons, while a muon hodoscope

(MHO) placed downstream of an 18 in iron filter triggered muons. The segmentations of the Čerenkov and muon hodoscopes were aligned with roads in the trigger scintillators to form the parallel, level 1 dilepton triggers. Prescaled level 0, or minimum bias, triggers allowed for the simultaneous measurement of $\pi^+\pi^-$ events for normalization. A software trigger (level 3) used drift chamber and TSC information to form two-body events with a vertex within the decay volume and placed loose cuts on reconstructed invariant dilepton mass and p_T. Offline, E871 used electron identification from a lead glass (PBG) calorimeter and muon identification from a muon range finder (MRG), the latter consisting of proportional counters and four additional MHO modules interspersed between iron, marble, and aluminum filters to provide a 5% range measurement.

Data were processed with a pair of independent and complementary fitting algorithms, which used drift chamber hits and a 3-dimensional magnetic field map to determine the kinematic quantities of the decay, including reconstructed mass, p_T, vertex position, and track 3-momenta. The mass resolutions of the fitters ranged from approximately 1.1 MeV/c^2 for $\pi^+\pi^-$ to 1.2 and 1.4 MeV/c^2 for $\mu^+\mu^-$ and e^+e^- decays, respectively, so that the reconstructed K_L^0 dilepton mass lies at least 6 standard deviations above the semileptonic background thresholds. Cuts were placed on the returned χ^2 values of the fitters, along with the vertex position to ensure that decays originated within the profile of the neutral beam.

Tracks exiting the spectrometer were then projected through the downstream TSC and particle identification systems, and associated with detector hits. Electrons were identified from well-timed signals in the Čerenkov detector, and by comparing energy deposition in the PBG array with track momentum (where $E/p \sim 1$ for electrons). Muons were identified by requiring the range of track-associated hits in the MHO and MRG counters to be at least 80% of the predicted range based on momentum, in order to suppress $K_{\mu 3}$ pion decay. The projected tracks from the fitters were also required to coincide spatially with in-time TSC hits satisfying parallelism.

41.3 Branching Fraction Results

Fig. 41.2 shows the reconstructed $\pi^+\pi^-$ invariant mass and p_T^2 from the normalization sample (solid line). Since the kinematic threshold for semileptonic events reconstructed as $\pi^+\pi^-$ falls above the K_L^0 mass, an electron veto from the lead glass array is used to remove K_{e3} events, and the remaining $K_{\mu 3}$ events are simulated with Monte Carlo (dashed line in fig. 41.2) and normalized to side bands of the data to perform the background subtraction. The branching ratio for dilepton decays with respect to $K_L^0 \to \pi^+\pi^-$ is determined from the numbers of candidate $\ell^+\ell^-$ and $\pi^+\pi^-$ events according to

$$\frac{B(K_L^0 \to \ell^+\ell^-)}{B(K_L^0 \to \pi^+\pi^-)} = \frac{1}{P}\frac{N_{\ell\ell}}{N_{\pi\pi}}\frac{A_{\pi\pi}}{A_{\ell\ell}}\frac{1}{\varepsilon_{\ell\ell}^{L1}}\frac{1}{\varepsilon_{\ell\ell}^{L3}}\frac{1}{\varepsilon_{\ell\ell}^{PID}}\cdots \tag{41.7}$$

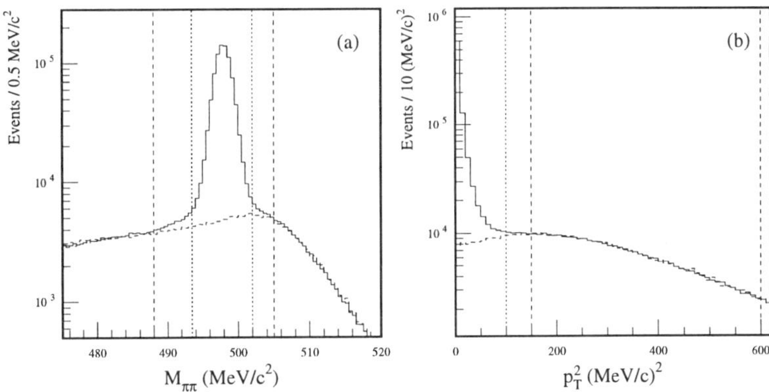

Figure 41.2: Reconstructed $\pi^+\pi^-$ invariant mass (a) and p_T^2 (b) from normalization sample (solid line), with Monte Carlo $K_{\mu 3}$ distributions (dashed line). Vertical lines denote signal (dotted) and normalization (dashed) regions.

P denotes the total prescale applied to the normalization sample. The Monte Carlo acceptances A correct for the mode dependence of the detector geometry, kinematic cuts, and trigger parallelism. The various terms ε account for efficiencies of the hardware and software triggers and particle identification, which are not simulated in the Monte Carlo.

The strategy for the $K_L^0 \to \mu^+\mu^-$ analysis involved relatively loose kinematic and particle identification cuts to reduce the background level to roughly 1% (where any errors on the subtraction method become negligible) while minimizing potential systematic errors introduced through tighter cuts. The acceptance and efficiency terms in the branching ratio were determined to within $\sim$ 0.1-0.2%, so that systematic errors would not limit the result. To avoid potential bias in these calculations through knowledge of previous measurements, an additional, random prescale factor was applied to the normalization sample, which was revealed only after all cuts were fixed and branching ratio terms finalized.

Fig. 41.3 plots the reconstructed $\mu^+\mu^-$ mass and p_T^2 after all cuts were applied. The $\mu^+\mu^-$ events appear well separated from the tail of the $K_{\mu 3}$ background distribution, with a small amount of background extending under the signal peak. The majority of background events were due to "pile-up" of two or more K_L^0 decays, and was reduced by checking for extra tracks within the TSC modules. To subtract this background, a linear fit was performed on the p_T^2 axis within a high-p_T sideband, and extrapolated into the signal region, resulting in over 6200 candidate $\mu^+\mu^-$ events with a relative error of roughly 1.3%.

The resulting branching ratio for $\mu^+\mu^-$ with respect to $\pi^+\pi^-$ becomes $(3.474 \pm 0.054) \times 10^{-6}$, corresponding to a branching fraction (using the measured $\pi^+\pi^-$ value [3]) of $B(K_L^0 \to \mu^+\mu^-) = (7.18 \pm 0.17) \times 10^{-9}$. This

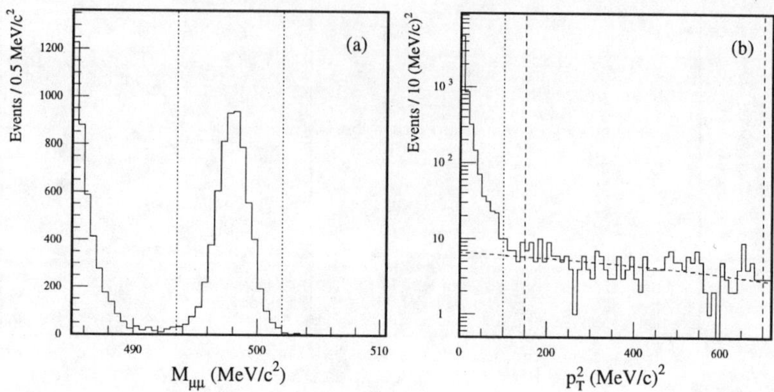

Figure 41.3: Reconstructed $\mu^+\mu^-$ invariant mass (a) and p_T^2 (b), with background subtraction from a linear fit in p_T^2. Vertical lines denote signal (dotted) and fitting (dashed) regions.

result agrees with the current world average, with a factor of three improvement in precision. By subtracting the *unitarity bound* from this improved branching fraction measurement, the dispersive contribution to the $\mu^+\mu^-$ decay rate equals $(0.11 \pm 0.25) \times 10^{-9}$, or $|\,\mathrm{Re}\,A_{exp}|^2 < 0.48 \times 10^{-9}$ (90% CL). From a recent calculation [6] of the long-distance dispersive amplitude $|\,\mathrm{Re}\,A_{LD}| < 2.9 \times 10^{-5}$ (90% CL), eq. (41.4) implies a lower bound on the Wolfenstein parameter $\rho > -0.40$ (90% CL), which is consistent with constraints from other methods [5].

The analysis of $K_L^0 \to e^+e^-$ applied a more aggressive set of cuts than those used for $\mu^+\mu^-$ in order to reduce the possibility of background events appearing in the signal region. In particular, tighter cuts on fitter χ^2 values and timing information from the TSC, Čerenkov detector, and drift chambers were applied. To prevent biasing these cuts through knowledge of potential signal events, an extended area surrounding the signal box was hidden from view until all cuts were finalized and the estimated level of background was well below one event. Data outside this exclusion region were then studied to optimize the analysis cuts.

Due to the lower reconstructed e^+e^- mass threshold for misidentified K_{e3} events, the e^+e^- data sample contained far less background compared with $\mu^+\mu^-$. In the limit of zero photon energy, however, the decay $K_L^0 \to e^+e^-\gamma$ may produce events with a reconstructed e^+e^- mass near the signal region. Since the E871 apparatus was not optimized for photon detection, $ee\gamma$ events were not actively suppressed, but an elliptical signal region was defined so that the predicted background level from this source fell below 0.1 events. Potential backgrounds from $K_L^0 \to e^+e^-e^+e^-$, though, were reduced by searching for partial tracks in the upstream drift chambers that projected to the decay vertex, leading to a predicted total background from both sources of ~ 0.2 events.

The reconstructed e^+e^- mass versus p_T^2 after the full set of analysis cuts is shown in fig. 41.4, along with the elliptical signal region within the (hatched) exclusion area, which reveals four candidate e^+e^- events, the first observation of this rare decay mode. The inset plots the predicted physics backgrounds as a function of mass, compared with the data. The probability of this 0.2 event background producing four events in the signal region equals 6×10^{-5}. The actual number of signal events was estimated from a maximum likelihood fit to the data (using as input Monte Carlo e^+e^-, $ee\gamma$, and $4e$ distributions), and was found to be $4.20^{+2.69}_{-1.94}$, resulting in a measured branching fraction for $K_L^0 \to e^+e^-$ of [12]

$$B(K_L^0 \to e^+e^-) = 8.7^{+5.7}_{-4.1} \times 10^{-12}. \tag{41.8}$$

This value is consistent with theoretical predictions [7, 8] and represents the smallest branching fraction ever measured in particle physics.

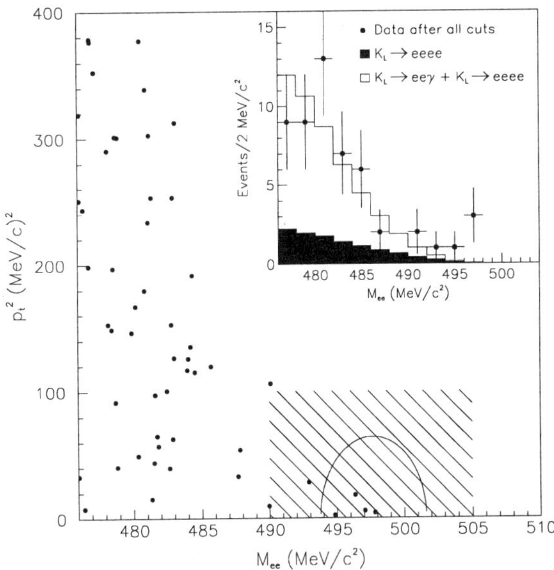

Figure 41.4: Reconstructed e^+e^- invariant mass versus p_T^2 after cuts. Inset shows comparison of data with predicted physics backgrounds by mass.

References

[1] S.L. Glashow, J. Iliopoulos, and L. Maiani, Phys. Rev. D **2**, 1285 (1970).

[2] L.M. Sehgal, Phys. Rev. **183**, 1511 (1969).

[3] Particle Data Group, Eur. Phys. J. C **3**, 1 (1998).

[4] L. Wolfenstein, Phys. Rev. Lett. **51**, 1945 (1983).

[5] A.J. Buras and R. Fleischer, hep-ph/9704376.

[6] G.D. D'Ambrosio, G. Isidori, and J. Portolés, Phys. Lett. B **423**, 385, (1998).

[7] G. Valencia, Nucl. Phys. B **517**, 339 (1998).

[8] D.G. Dumm and A. Pich, Phys. Rev. Lett. **80**, 4633 (1998).

[9] J.L. Ritchie and S.G. Wojcicki, Rev. Mod. Phys. **65**, 1149 (1993).

[10] A.P. Heinson *et al.* (BNL E791), Phys. Rev. D **51**, 985 (1995).

[11] K. Arisaka *et al.* (BNL E791), Phys. Rev. Lett. **71**, 3910 (1993).

[12] D. Ambrose *et al.* (BNL E871), Phys. Rev. Lett. **81**, 4309 (1998).

[13] D. Ambrose *et al.* (BNL E871), Phys. Rev. Lett. **81**, 5734 (1998).

[14] W.R. Molzon, this volume, chapter 35.

[15] J. Belz *et al.*, Nucl. Instrum. Methods, Sect. A, **428**, 239 (1999).

Part VI

Hyperon Physics

Strange particles include both kaons (mesons) and *hyperons* (baryons). Denoting a nonstrange quark of the type u or d by q, and a strange quark by s, the kaons are $s\bar{q}$ or $q\bar{s}$, while the hyperons are sqq, ssq, or sss. They include the Λ, $\Sigma^{+,0,-}$, $\Xi^{0,-}$, and Ω^-.

Theoretical interest in hyperon weak decays has focused on a variety of interesting problems, including linear relations between amplitudes for various two-body nonleptonic decays, the baffling pattern of weak radiative decays, and most recently the prospects for observing CP violation in such decays as $\Xi^- \to \Lambda\pi^-$ and $\Lambda \to p\pi^-$, as detailed in Pakvasa's contribution. An experiment to study these phenomena has been performed at Fermilab and will take more data this year. White describes these efforts.

Experiments studying neutral kaons also produce neutral hyperons. The KTeV Collaboration at Fermilab has a subgroup investigating various properties of the Ξ^0s and Λs that accompany the K_Ls in its beam. This work is described here by Solomey. A corresponding effort in the NA48 Collaboration at CERN is described in part V by Köpke.

42

CP Violation in Hyperon Decays

Sandip Pakvasa

Abstract

The theory and phenomenology of CP violation in hyperon decays is summarized.

42.1 Introduction

The CPT theorem was proved in 1955 [1] and soon thereafter Lüders and Zumino [2] deduced from it the equality of masses and lifetimes between particles and antiparticles. In 1958 Okubo [3] observed that CP violation allows hyperons and antihyperons to have different branching ratios into conjugate channels even though their total rates must be equal by CPT. Somewhat later, this paper inspired Sakharov [4] to his famous work on cosmological baryon-antibaryon asymmetry. Pais [5] extended Okubo's proposal to asymmetry parameters in Λ and $\bar{\Lambda}$ decays. The subject was revived in the 1980s and a number of calculations were made [6, 7]. Only now, over 40 years after Okubo's paper, are these proposals about to be tested in the laboratory.

The reason for the current interest is the need to find CP violation in places other than just $K_L - K_S$ complex. Only a number of different observations of CP violation in different channels will help us pin down the source and nature of CP violation in or beyond the standard model. From this point of view hyperon decay is one more weapon in our arsenal in addition to the K system, the B system, the D system, etc.

42.2 Phenomenology of Hyperon Decays

I summarize here the salient features of the phenomenology of nonleptonic hyperon decays [8]. Leaving out Ω^- decays, there are seven decay modes: $\Lambda \to N\pi$, $\Sigma^\pm \to N\pi$, and $\Xi \to \Lambda\pi$. The effective matrix element can be written as

$$i\ \bar{u}_{\bar{p}}(a + b\gamma_5)u_\Lambda\ \phi \tag{42.1}$$

for the mode $\Lambda \to p + \pi^-$, where a and b are complex in general. The corresponding element for $\bar{\Lambda} \to \bar{p} + \pi^+$ is then

$$i\,\bar{v}_{\bar{p}}(-a^* + b^*\gamma_5)v_{\bar{\Lambda}}\phi^+. \tag{42.2}$$

It is convenient to express the observables in terms of S and P and write the matrix element as

$$S + P\,\sigma \cdot \hat{\mathbf{q}}, \tag{42.3}$$

where $\mathbf{q}$ is the proton momentum in the Λ rest frame and

$$\begin{aligned} S &= a\sqrt{\frac{\{(m_\Lambda + m_p)^2 - m_\pi^2\}}{16\pi\, m_\Lambda^2}}, \\ P &= b\sqrt{\frac{\{(m_\Lambda - m_p)^2 - m_\pi^2\}}{16\pi\, m_\Lambda^2}}. \end{aligned} \tag{42.4}$$

In the Λ rest-frame, the decay distribution is given by

$$\begin{aligned} \frac{d\Gamma}{d\Omega} &= \frac{\Gamma}{8\pi}\{[1 + \alpha\langle\sigma_\Lambda\rangle \cdot \hat{\sigma}] \\ &+ \langle\sigma_p\rangle \cdot [(\alpha + \langle\sigma_\Lambda\rangle \cdot \hat{\mathbf{q}})\hat{\mathbf{q}} + \beta\langle\sigma_\Lambda\rangle \times \hat{\mathbf{q}} \\ &+ \gamma(\hat{\mathbf{q}} \times (\langle\sigma_\Lambda\rangle \times \hat{\mathbf{q}})]\}, \end{aligned} \tag{42.5}$$

where Γ is the decay rate and is given by

$$\Gamma = 2 \mid \mathbf{q} \mid \{\mid S \mid^2 + \mid P \mid^2\} \tag{42.6}$$

and α, β, and γ are given by

$$\begin{aligned} \alpha &= \frac{2\mathrm{Re}\,(S^*P)}{\{\mid S \mid^2 + \mid P \mid^2\}}, \\ \beta &= \frac{2\mathrm{Im}\,(SP^*)}{\{\mid S \mid^2 + \mid P \mid^2\}}, \\ \gamma &= \frac{\{\mid S \mid^2 - \mid P \mid^2\}}{\{\mid S \mid^2 + \mid P \mid^2\}}. \end{aligned} \tag{42.7}$$

For a polarized Λ, the up-down asymmetry of the final proton is given by α (α is also the longitudinal polarization of the proton for an unpolarized Λ). β and γ are components of the transverse proton polarization [9].

The observed properties of the hyperon decays can be summarized as follows: (i) the $\Delta I = 1/2$ dominance, *i.e.*, the $\Delta I = 3/2$ amplitudes are about 5% of the $\Delta I = 1/2$ amplitudes; (ii) the asymmetry parameter α is large for Λ decays, Ξ decays, and $\Sigma^+ \to p\pi^0$ and is near zero for $\Sigma^\pm \to n\pi^\pm$; and (iii) the Sugawara-Lee triangle sum rule $\sqrt{3}A(\Sigma^+ \to p\pi^0) - A(\Lambda \to p\pi^-) = 2A(\Xi \to \Lambda\pi^-)$ is satisfied to a level of 5% in both s and p wave amplitudes.

42.3 CP-Violating Observables

Let a particle P decay into several final states f_1, f_2, etc. The amplitude for $P \to f_1$ is in general

$$A = A_1 e^{i\delta_1} \quad + A_2\, e^{i\delta_2}, \tag{42.8}$$

where 1 and 2 are strong interaction eigenstates and δ_i are corresponding final state phases. Then the amplitude for $\bar{P} \to \bar{f}_1$ is

$$\bar{A} = A_1^* e^{i\delta_1} \quad + A_2^*\, e^{i\delta_2}. \tag{42.9}$$

If $\mid A_1 \mid \gg \mid A_2 \mid$, then the rate asymmetry $\Delta(= (\Gamma - \bar{\Gamma})/(\Gamma + \bar{\Gamma}))$ is given by

$$\Delta \approx -2 \mid A_2/A_1 \mid \sin(\phi_1 - \phi_2)\sin(\delta_1 - \delta_2), \tag{42.10}$$

where $A_i = \mid A_i \mid e^{i\phi_i}$. Hence, to get a nonzero rate asymmetry, one must have (i) at least two channels in the final state, (ii) CP-violating weak phases differing in the two channels, and (iii) *unequal* final state scattering phase shifts in the two channels [6]. A similar calculation of the asymmetry of α [10] shows that for a single isospin channel dominance,

$$A = (\alpha + \bar{\alpha})/(\alpha - \bar{\alpha}) = 2\tan(\delta_s - \delta_p)\sin(\phi_s - \phi_p). \tag{42.11}$$

In this case the two channels are orbital angular momentum 0 and 1; and even a single isospin mode such as $\Xi^- \to \Lambda\pi^-$ can exhibit a nonzero A. In B decays an example of a single isospin mode exhibiting CP-violating rate asymmetry is $B \to \pi\pi$, i.e. In this case the two eigenchannels with different weak CP phases and different final state phases are $B \to D\overline{D}\pi \to \pi\pi$ and $B \to \pi\pi \to \pi\pi$ [11].

To define the complete set of CP-violating observables, consider the example of the decay modes $\Lambda \to p\pi^-$ and $\bar{\Lambda} \to \bar{p}\pi^+$. The amplitudes are

$$\begin{aligned} S &= -\sqrt{\frac{2}{3}} S_1 e^{i(\delta_1 + \phi_1^s)} + \frac{1}{\sqrt{3}} S_3 e^{i(\delta_3 + \phi_3^s)}, \\ P &= -\sqrt{\frac{2}{3}} P_1 e^{i(\delta_{11} + \phi_1^p)} + \frac{1}{\sqrt{3}} P_3 e^{i(\delta_3 + \phi_3^p)}, \end{aligned} \tag{42.12}$$

where S_i, P_i are real, i refers to the final state isospin ($i = 2I$) and ϕ_i are the CPV phases. With the knowledge that S_3/S_1, $P_3/P_1 \gg 1$ one can write [12, 13]

$$\begin{aligned} \Delta_\Lambda &= \frac{(\Gamma - \overline{\Gamma})}{(\Gamma + \overline{\Gamma})} \cong \sqrt{2}\,(S_3/S_1)\sin(\delta_3 - \delta_1)\sin(\phi_3^s - \phi_1^s), \\ A_\Lambda &= \frac{(\alpha + \overline{\alpha})}{(\alpha - \overline{\alpha})} \cong -\tan(\delta_{11} - \delta_1)\sin(\phi_1^p - \phi_1^s), \\ B_\Lambda &= \frac{(\beta + \overline{\beta})}{(\beta - \overline{\beta})} \cong \cot(\delta_{11} - \delta_1)\sin(\phi_1^p - \phi_1^s). \end{aligned} \tag{42.13}$$

For $N\pi$ final states, the phase shifts at $E_{c.m.} = m_\Lambda$ are known and are $\delta_1 = 6^0$, $\delta_3 = -3.8^0$, $\delta_{11} = 1.1^0$, and $\delta_{31} = -0.7^0$. The CPV phases ϕ_i have to be provided by theory.

Similar expressions can be written for other hyperon decays. For example, for $\Lambda \to n\pi^0$, Δ is $-2\Delta_\Lambda$ and A and B are identical to A_Λ and B_Λ. For $\Xi^- \to \Lambda\pi^-$ (nd $\Xi^0 \to \Lambda\pi^0$) the asymmetries are [13]

$$\begin{aligned} \Delta_\Xi &= 0, \\ A_\Xi &= -\tan(\delta_{21} - \delta_2)\sin(\phi^p - \phi^s), \\ B_\Xi &= \cot(\delta_{21} - \delta_2)\sin(\phi^p - \phi^s), \end{aligned} \tag{42.14}$$

where δ_{21} and δ_2 are the p and s wave $\Lambda\pi$ phase shifts at m_Ξ respectively. Somewhat more complicated expressions can be and have been written for Σ decays [13]. There are no experimental proposals at the moment for measuring these.

42.4 Calculating CP Phases

In the standard model description of the nonleptonic hyperon decays, the effective $\Delta S = 1$ Hamiltonian is

$$H_{\text{eff}} = \frac{G_F}{\sqrt{2}} U_{ud}^* U_{us} \sum_{i=1}^{12} c_i(\mu) O_i(\mu) \tag{42.15}$$

after the short-distance QCD corrections (LLO + NLLO), where $c_i = z_i + y_i\tau$ ($\tau = -U_{td} U_{ts}^*/U_{ud} U_{us}$), and $\mu \sim$ O(1 GeV). For CP violation, the most important operator is

$$O_6 = \bar{d}\, \lambda_i\gamma_\mu(1+\gamma_5)s\bar{q}\lambda_i\gamma_\mu(1-\gamma_5)q, \tag{42.16}$$

and $y_6 \approx -0.1$ at $\mu \sim 1\,\text{GeV}$. To estimate the CP phases in eq. (42.12), one adopts the following procedure. The real parts (in the approximation that the imaginary parts are very small) are known from the data on rates and asymmetries. The real parts of the amplitudes have also been evaluated in the standard model with reasonable success with some use of chiral perturbation theory (current algebra and soft pion theorems) and a variety of choices for the baryonic wave functions. The MIT bag model wave function is one such choice that gives conservative results. The same procedure is adopted for calculating the imaginary parts using O_6. The major uncertainty is in the hadronic matrix elements and the fact that the simultaneous fit of s and p waves leaves a factor of 2 ambiguity [14]. In the standard model, with the Kobayashi-Maskawa (K-M) phase convention there is no CPV in $\Delta I = 3/2$ amplitudes; and for Λ decays $\phi_3 = 0$. There is a small electroweak penguin contribution to ϕ_3, which is neglected. The rate asymmetry is dominated by the s wave amplitudes, and the asymmetry A_Λ is dominated by the $\Delta I = 1/2$ amplitudes. Evaluating the matrix elements in the standard way and with the current knowledge of the K-M matrix elements one finds for the decays [13,15] $\Lambda \to p\pi^-$ and $\Xi^- \to \Lambda\pi^-$

$$\phi^s_\Lambda - \phi^p_\Lambda \cong 3.5 \times 10^{-4}; \quad \phi^s_\Xi - \phi^p_\Xi \cong -1.4 \times 10^{-4}. \tag{42.17}$$

With the $N\pi$ phase shifts known to be

$$\delta_s - \delta_p \cong 7^0 \tag{42.18}$$

one finds for the asymmetry A_Λ in the standard model a value of about -4×10^{-5}. For the $\Xi \to \Lambda\pi^-$ decay mode the phase shifts are not known experimentally and have to be determined theoretically. There are calculations from 1965 [16] that gave large values for $\delta_s - \delta_p \sim -20^0$; however, all recent calculations based on chiral perturbation theory, heavy baryon approximation, etc. agree that $\delta_s - \delta_p$ lies between 1^0 and 3^0 [17]. In this case the asymmetry A_Ξ is expected to be $\sim -(0.2 \text{ to } 0.7) \times 10^{-5}$. In table 42.1, the standard model results for the expected asymmetries in the standard model are given. Using very crude back-of-the-envelope estimates, similar results are obtained. What is needed is some attention to these matrix elements from the lattice QCD community.

An experimental measurement of the phase shifts $\delta_s - \delta_p$ in the $\Lambda\pi$ system will put the predictions for A_Ξ on a firmer basis. There is an old proposal due to Pais and Treiman [18] to measure $\Lambda\pi$ phase shifts in $\Xi \to \Lambda\pi e\nu$, but this does not seem practical in the near future. Another technique, more feasible, is to measure β and α to high precision in Ξ and $\bar{\Xi}$ decays. Then the combination

$$(\beta - \bar{\beta})/(\alpha - \bar{\alpha}) = \tan(\delta_s - \delta_p) \tag{42.19}$$

can be used to extract $\delta_s - \delta_p$. To the extent CP phases are negligible one can also use the approximate relation

$$\beta/\alpha \approx \tan(\delta_s - \delta_p). \tag{42.20}$$

42.5 Beyond Standard Model

Can new physics scenarios in which the source of CP violation is not the K-M matrix yield large enhancements of these asymmetries? We consider some classes of models where these asymmetries can be estimated more or less reliably [12, 13].

First there is the class of models that are effectively superweak [19]. Examples are models in which the K-M matrix is real and the observed CP violation is due to exchange of heavier particles; heavy scalars with FCNC, heavy quarks, etc. In all such models direct CP violation is negligible and unobservable and so all asymmetries in hyperon decays are essentially zero. Furthermore, they need to be modified to accommodate the fact that direct CP violation has now been seen in kaon decays (the fact that ϵ'/ϵ is not zero). In the three Higgs doublet model with flavor conservation imposed, the charged Higgs exchange tend to give large effects in direct CP violation as well as a large neutron electric dipole moment [20].

There are two generic classes of left-right symmetric models: the manifest left-right symmetric model (i) without $W_L - W_R$ mixing [21] and (ii) with $W_L - W_R$ mixing [22]. In (i) U^L_{KM} = real and U^R_{KM} complex with arbitrary phases but angles given by U^L_{KM}. Then one gets the "isoconjugate" version in which

$$H_{\text{eff}} = \frac{G_F\, U_{us}}{\sqrt{2}} \left[J^\dagger_{\mu L}\, J_{\mu L} + \; \eta e^{i\beta} J^\dagger_{\mu R}\, J_{\mu R} \right], \qquad (42.21)$$

where $\eta = m^2_{WL}/m^2_{WR}$ and β is the relevant CPV phase. Then $H_{p.c.}$ and $H_{p.v.}$ have overall phases $(1 + i\eta\beta)$ and $(1 - i\eta\beta)$ respectively. To account for the observed CPV in K decay, $\eta\beta$ has to be of order 4.5×10^{-5}. In this model, $\epsilon'/\epsilon = 0$ and there are no rate asymmetries in hyperon decays; but the asymmetries A and B are not zero and, *e.g.*, A goes as $2\eta\beta \sin(\delta_s - \delta_p)$. In the class of models where $W_L - W_R$ mixing is allowed, the asymmetries can be enhanced, and also ϵ'/ϵ is not zero in general.

Models where the gluon dipole operator is enhanced beyond the standard model value are especially interesting:

$$\bar{d}\lambda^\alpha (a + b\gamma_5)\sigma_{\mu\nu} s\; G^\alpha_{\mu\nu}. \qquad (42.22)$$

In the standard model, the coefficient of this operator is too small to be interesting. The parameter a (actually its imaginary part) is constrained by the known value of ϵ and contributes only to hyperon decays. The parameter b can contribute to both ϵ' as well as the CP-violating asymmetries in hyperon decays. The current range of ϵ' as given by the experimental values we heard here [23] allows b to contribute to A_Λ at a level of 5×10^{-4} [24]. Such enhancement of this operator takes place naturally in models where CP violation occurs due to the exchange of charged Higgs scalars (such as the Weinberg model) and can also occur in several scenarios based on supersymmetry [25].

42.6 Experiments

There have been several proposals to detect hyperon decay asymmetries in $\bar{p}p \to \bar{\Lambda}\Lambda$, $\bar{p}p \to \bar{\Xi}\Xi$, and in $e^+e^- \to J/\psi \to \Lambda\bar{\Lambda}$, but none of these were approved [26]. The only approved and ongoing experiment is E871 at Fermilab. In this experiment Ξ^- and $\overline{\Xi}^+$ are produced and the angular distribution of $\Xi^- \to \Lambda\pi^- \to p\pi^-\pi^-$ and $\overline{\Xi}^+$ compared. This experiment effectively measures $A_\Lambda + A_\Xi$ and is described in detail by Sharon White [27]. To summarize the implications for the measurement of $A_\Lambda + A_\Xi$ by E871: the standard model expectation is about -4×10^{-5} with a factor of two uncertainty; if new physics should contribute, it could be as large as 7×10^{-4}. A measurement by E871 at the 10^{-4} level, therefore, will already be a strong discriminant.

	SM	2-Higgs	FCNC Superweak	L-R-S (1)	L-R-S (2)
Δ_Λ	10^{-6}	10^{-5}	0	0	0
A_Λ	-4×10^{-5}	-2×10^{-5}	0	-10^5	6×10^{-4}
B_Λ	10^{-4}	2×10^{-3}	0	7×10^{-4}	-
Δ_Ξ	0	0	0	0	0
A_Ξ	-4×10^{-6}	-3×10^{-4}	0	2×10^{-5}	10^{-4}
B_Ξ	10^{-3}	4×10^{-3}	0	3×10^{-4}	-

Table 42.1: Expectations for Hyperon CPV Asymetries.

42.7 ϵ'/ϵ and Hyperon Decay Asymmetries

It might seem that now that ϵ'/ϵ has been measured and direct CP violation in $\Delta S = 1$ channel been observed, a study of CP violation in hyperon decays is unnecessary and no new information will be obtained. Why is it worthwhile measuring another $\Delta S = 1$ process like hyperon decay? The point is that there are important differences and the two are not at all identical. First, there are important differences in the matrix elements. Hyperon matrix elements do not have the kind of large cancellations that plague the kaon matrix elements. The hadronic uncertainties are present for both, but are different. Next, a very important difference is the fact that the $K \to \pi\pi$ decay (and hence ϵ') is only sensitive to CP violation in the parity-violating amplitude and cannot yield any information on parity-conserving amplitudes. Hyperon decays, by contrast, are sensitive to both. Thus, ϵ'/ϵ and hyperon decay CP asymmetries are different and complimentary. The hyperon decay measurements are as important and significant as ϵ'/ϵ.

42.8 Conclusion

The searches for direct CPV are being pursued in many channels. $K \to 2\pi, \Lambda \to N\pi$, B decays and D decays. Any observation of a signal would be the first outside of $K^0 - \overline{K}^0$ system and would be complimentary to the measurement of ϵ'/ϵ. This will constitute one more step in our bid to confirm or demolish the standard Kobayashi-Maskawa description of CP violation.

Hyperon decays offer a rich variety of CP violating observables, each with different sensitivity to various sources of CP violation. For example, Δ_Λ is mostly sensitive to parity-violating amplitudes, Δ_{Σ^+} is sensitive only to parity-conserving amplitudes, A is sensitive to both, etc. The number of experimental proposals is rather small so far. The one ongoing experiment Fermilab E871 can probe A to a level of 10^{-4}, which is already in an interesting range. In addition to more experiments, this subject sorely needs more attention devoted to calculating the matrix elements more reliably.

Acknowledgments

I am grateful to my collaborators Alakabha Datta, Xiao-Gang He, Pat O'Donnell, German Valencia, and John Donoghue and to members of the E871 collaboration for many discussions. The hospitality of Jon Rosner, Bruce Winstein, and colleagues was memorable and the atmosphere of the conference was most stimulating. This work is supported in part by USDOE under Grant #DE- FG-03-94ER40833.

References

[1] G. Lüders, Ann. Phys. **2**, 1 (1957); G. Lüders, Dan. Mat. Fys. Medd. **28**, No. 5 (1954); J.S. Bell, Proc. Roy. Soc. A **231**, 79 (1955); J. Schwinger, Phys. Rev. **91**, 713 (1953); W. Pauli, *Niels Bohr and the Development of Physics* (McGraw-Hill, New York, and Pergamon Press, London, 1955).

[2] G. Lüders and B. Zumino, Phys. Rev. **106**, 385 (1957).

[3] S. Okubo, Phys. Rev. **109**, 984 (1958).

[4] A.D. Sakharov, Zh. EK. Teor. Fiz. **5**, 32 (1967) (English translation, JETP Letters **5**, 24 (1967)).

[5] A. Pais, Phys. Rev. Lett. **3**, 242 (1959).

[6] T. Brown, S.F. Tuan, and S. Pakvasa, Phys. Rev. Lett. **51**, 1238 (1983).

[7] L-L. Chau and H.Y. Cheng, Phys. Lett. B **131**, 202 (1983); D. Chang and L. Wolfenstein, *Intense Medium Energy Sources of Strangeness*, ed. T. Goldman, H.E. Haber, and H.F.-W. Sadrozinski, AIP Conf. Proc. No. 102 (AIP, New York, 1983), p. 73; C. Kounnas, A.B. Lahanas, and P. Pavlopoulos, Phys. Lett. B **127**, 381 (1983).

[8] S.P. Rosen and S. Pakvasa, *Advances in Particle Physics*, ed. R.E. Marshak and R.L. Cool (Wiley-Interscience, NY, 1968), p. 473; S. Pakvasa and S.P. Rosen, *The Past Decade in Particle Theory*, ed. E.C.G. Sudarshan and Y. Ne'eman (Gordon and Breach, 1973), p. 437.

[9] T.D. Lee and C.N. Yang, Phys. Rev. **108**, 1645 (1957); R. Gatto, Nucl. Phys. **5**, 183 (1958).

[10] O.E. Overseth and S. Pakvasa, Phys. Rev. **184**, 1663 (1969).

[11] L. Wolfenstein, Phys. Rev. D **43**, 151 (1990).

[12] J.F. Donoghue and S. Pakvasa, Phys. Rev. Lett. **55**, 162 (1985).

[13] J.F. Donoghue, X-G. He, and S. Pakvasa, Phys. Rev. D **39**, 833 (1986).

[14] J.F. Donoghue, E. Golowich, and B. Holstein, Phys. Rep. **131**, 319 (1986); *Dynamics of the Standard Model* (Cambridge Univ. Press, 1992).

[15] M.J. Iqbal and G. Miller, Phys. Rev. D **41**, 2817 (1990); X-G. He, H. Steger, and G. Valencia, Phys. Lett. B **272**, 411 (1991); N.G. Despande, X-G. He, and S. Pakvasa, Phys. Lett. B **326**, 307 (1994); X-G. He and G. Valencia, Phys. Rev. D **52**, 5257 (1995).

[16] B. Martin, Phys. Rev. **138**, 1136 (1965); R. Nath and A. Kumar, Nuov. Cim. **36**, 669 (1965).

[17] M. Lu, M. Wise, and M. Savage, Phys. Lett. B **337**, 133 (1994); A. Datta and S. Pakvasa, Phys. Lett. B **344**, 340 (1995); A. Kamal, Phys. Rev. D **58**, 07750 (1998); A. Datta, P.J. O'Donnell, and S. Pakvasa, hep-ph/9806374.

[18] A. Pais and S.B. Treiman, Phys. Rev. **178**, 2365 (1969).

[19] Some examples can be found in the following references: R.N. Mohapatra, J.C. Pati, and L. Wolfenstein, Phys. Rev. D **11**, 3319 (1975); T. Brown, N. Despande, S. Pakvasa, and H. Sugawara, Phys. Lett. B **141**, 95 (1984); J.M. Soares and L. Wolfenstein, Phys. Rev. D **46**, 256 (1992); P. Frampton and S.L. Glashow, Phys. Rev. D **55**, 1691 (1997); H. Georgi and S.L. Glashow, hep-ph/9807399.

[20] S. Weinberg, Phys. Rev. Lett. **37**, 657 (1976).

[21] R.N. Mohapatra and J.C. Pati, Phys. Rev. D **11**, 566 (1975).

[22] D. Chang, X-G. He, and S. Pakvasa, Phys. Rev. Lett. **74**, 3927 (1995).

[23] Y.B. Hsiung, this volume, chapter 6; M.S. Sozzi, this volume, chapter 8.

[24] X-G He, S. Pakvasa, H. Murayama, and G. Valencia (in preparation).

[25] A. Masiero and H. Murayama, Phys. Rev. Lett. **83**, 907 (1999); K.S. Babu, B. Datta, and R. Mohapatra, hep-ph/9904366.

[26] N. Hamman *et al.*, CERN/SPSPLC 92-19, SPSLC/M49; S.Y. Hsueh and P. Rapidis, FNAL Proposal.

[27] S. White, this volume, chapter 43.

43

Status Report from the HyperCP Experiment at Fermilab

Sharon L. White[1]

Abstract

HyperCP (E871), a Fermilab experiment searching for direct CP violation in Ξ and Λ decays, collected over one billion Ξ^- and $\overline{\Xi}^+$ decays in 1997. A sensitivity of $\approx 2 \times 10^{-4}$ in $A_{\Xi\Lambda} = (\alpha_\Xi\alpha_\Lambda - \alpha_{\overline{\Xi}}\alpha_{\overline{\Lambda}})/(\alpha_\Xi\alpha_\Lambda + \alpha_{\overline{\Xi}}\alpha_{\overline{\Lambda}})$ is expected. A brief review of CP violation in hyperon decays is given, along with a description of the spectrometer, status of the analysis, and future prospects.

43.1 Introduction

CP violation was first observed in 1964 in the neutral kaon system [1]. Since that time, understanding of this phenomenon has progressed slowly. Until recently, only indirect CP violation had been observed through the mixing of particle and antiparticle states. The standard model predicts that direct CP violation should occur as well and this was firmly established by KTeV [2] and NA48 [3] recently.

It is important to look for CP-violating effects outside of the kaon system. Numerous searches for CP violation in B meson decays are planned or under way. Another regime where CP asymmetries are predicted is nonleptonic weak decays of hyperons. The HyperCP [4] experiment at Fermilab seeks to observe direct CP violation by comparing the decay process $\Xi^- \to \Lambda\pi^-, \Lambda \to p\pi^-$ with its antiprocess $\overline{\Xi}^+ \to \overline{\Lambda}\pi^+, \overline{\Lambda} \to \overline{p}\pi^+$. HyperCP is the first dedicated experiment to perform such a search.

---

[1]Representing the HyperCP collaboration: R. A. Burnstein, A. Chakravorty, A. Chan, Y. C. Chen, W. S. Choong, K. Clark, M. Crisler, E. C. Dukes, C. Durandet, J. Felix, G. Gidal, H. R. Gustafson, C. Ho, T. Holmstrom, M. Huang, C. James, C. M. Jenkins, T. D. Jones, D. M. Kaplan, L. M. Lederman, N. Leros, M. J. Longo, F. Lopez, G. Lopez, L. Lu, W. Luebke, K.-B. Luk, K. S. Nelson, H. K. Park, J. P. Perroud, D. Rajaram, H. A. Rubin, J. Sheng, M. Sosa, P. K. Teng, B. Turko, J. Volk, C. G. White, S. L. White, C. Yu, Z. Yu, and P. Zyla.

43.2 Theory

Several excellent references discuss the phenomenology of CP violation in hyperon decays; see for example [5]. Nonleptonic two-body weak decays of spin-1/2 hyperons are parity violating, and the final states are an admixture of parity-violating S wave states and parity-conserving P wave states. The decay distribution of a spin-1/2 daughter baryon from a hyperon decay, in the rest frame of the parent, spin averaged over final states, is

$$\frac{dN}{d\Omega} = \frac{1}{4\pi}(1 + \alpha_p \vec{P}_p \cdot \hat{p}_d), \tag{43.1}$$

where $\vec{P}_p$ is the parent hyperon polarization, and $\hat{p}_d$ is the daughter baryon momentum unit vector. The polarization of the daughter baryon is related to the polarization of the parent by

$$\vec{P}_d = \frac{(\alpha_p + \vec{P}_p \cdot \hat{p}_d)\hat{p}_d + \beta_p(\vec{P}_p \times \hat{p}_d) + \gamma_p(\hat{p}_d \times (\vec{P}_p \times \hat{p}_d))}{(1 + \alpha_p \vec{P}_p \cdot \hat{p}_d)}. \tag{43.2}$$

The Lee-Yang variables α, β, γ are defined in terms of the S and P wave amplitudes as

$$\alpha = \frac{2\,\mathrm{Re}(S * P)}{|S|^2 + |P|^2}, \quad \beta = \frac{2\,\mathrm{Im}(S * P)}{|S|^2 + |P|^2}, \quad \gamma = \frac{|S|^2 - |P|^2}{|S|^2 + |P|^2}, \tag{43.3}$$

where $\alpha^2 + \beta^2 + \gamma^2 = 1$. If the parent is unpolarized, the daughter particle is in a helicity state with polarization $\vec{P}_d = \alpha \hat{p}_d$. Some asymmetry observables have been defined that are sensitive to CP violation:

$$A = \frac{\alpha + \overline{\alpha}}{\alpha - \overline{\alpha}}, \quad B = \frac{\beta + \overline{\beta}}{\beta - \overline{\beta}}, \quad \Delta = \frac{\Gamma - \overline{\Gamma}}{\Gamma + \overline{\Gamma}}. \tag{43.4}$$

Table 43.1 shows some theoretical predictions for A for $\Lambda \to p\pi$ and charged $\Xi \to \Lambda\pi$ decays. Table 43.2 shows the current experimental limits for A_Λ. A_Ξ has not been observed directly.

Model	$A_\Xi[10^{-4}]$	$A_\Lambda[10^{-4}]$	Reference
CKM	≈ -0.04	$\approx 0.4 - 0.5$	[6]
Weinberg	≈ -3.2	≈ -0.25	[5]
Left-Right(isoconjugate)	≈ 0.25	≈ -0.11	[7]
Left-Right(with mixing)	≤ 1	≤ 6	[8]

Table 43.1: Some predictions for A_Ξ and A_Λ.

Mode	Limit on A_Λ	Experiment	Ref.
$pp \to \Lambda X, \overline{p}p \to \overline{\Lambda} X$	0.02 ± 0.14	R608	[9]
$e^+e^- \to J/\psi \to \Lambda\overline{\Lambda}$	0.01 ± 0.10	DM2	[10]
$p\overline{p} \to \Lambda\overline{\Lambda}$	0.013 ± 0.022	PS185	[11]

Table 43.2: Current experimental limits.

43.3 Experimental Technique

HyperCP will measure the sum of the asymmetry observables A_Ξ and A_Λ by comparing the angular distribution of the protons and antiprotons from the decay chains $\Xi^- \to \Lambda\pi^-, \Lambda \to p\pi^-$ and $\overline{\Xi}^+ \to \overline{\Lambda}\pi^+, \overline{\Lambda} \to \overline{p}\pi^+$. The decay distributions of the proton and antiproton are given by eq. (43.1), where the parent particle is Λ or $\overline{\Lambda}$ and the daughter is p or $\overline{p}$. HyperCP produces Ξ^- and $\overline{\Xi}^+$ hyperons with zero polarization by targeting at zero degrees; thus the Λ or $\overline{\Lambda}$ from their decays are found in pure helicity states with polarization $\vec{P}_\Lambda = \alpha_\Xi \hat{p}_\Lambda$. The data are analyzed in the Λ helicity frame, in which the Λ direction in the Ξ rest frame defines the polar axis, as shown in fig. 43.1.

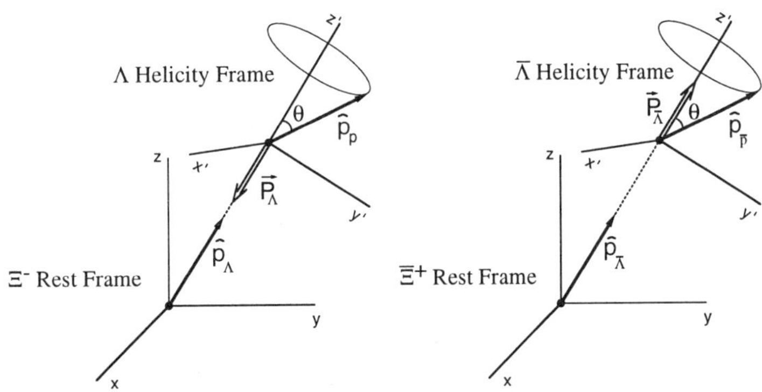

Figure 43.1: The Λ and $\overline{\Lambda}$ helicity frames.

The decay distributions of the p and $\overline{p}$ in this frame are

$$\frac{dN}{d\Omega} = \frac{1}{4\pi}(1 + \alpha_\Lambda \alpha_{\Xi^-} \cos\theta), \quad \frac{dN}{d\Omega} = \frac{1}{4\pi}(1 + \alpha_{\overline{\Lambda}} \alpha_{\overline{\Xi}^+} \cos\theta). \tag{43.5}$$

If CP symmetry is good in both Ξ and Λ decays, then $\alpha_{\Xi^-} = \alpha_{\overline{\Xi}^+}$ and $\alpha_\Lambda = -\alpha_{\overline{\Lambda}}$, and the decay distributions of the protons and antiprotons will be identical, with the slopes of the $\cos\theta$ distributions given by $\alpha_\Lambda\alpha_\Xi$. HyperCP is thus sensitive to CP violation in both Ξ and Λ decays. The asymmetry observable measured is

$$A_{\Xi\Lambda} = \frac{\alpha_\Xi \alpha_\Lambda - \alpha_{\overline{\Xi}} \alpha_{\overline{\Lambda}}}{\alpha_\Xi \alpha_\Lambda + \alpha_{\overline{\Xi}} \alpha_{\overline{\Lambda}}} \cong A_\Xi + A_\Lambda, \tag{43.6}$$

where

$$A_\Xi = \frac{\alpha_\Xi + \alpha_{\overline{\Xi}}}{\alpha_\Xi - \alpha_{\overline{\Xi}}}, \qquad A_\Lambda = \frac{\alpha_\Lambda + \alpha_{\overline{\Lambda}}}{\alpha_\Lambda - \alpha_{\overline{\Lambda}}}. \tag{43.7}$$

43.4 HyperCP Spectrometer

The primary design considerations for the HyperCP spectrometer (fig. 43.2) were that it be capable of taking data at a high rate, in order to collect the necessary large statistics, and also that it be fairly simple, in order to minimize systematic effects.

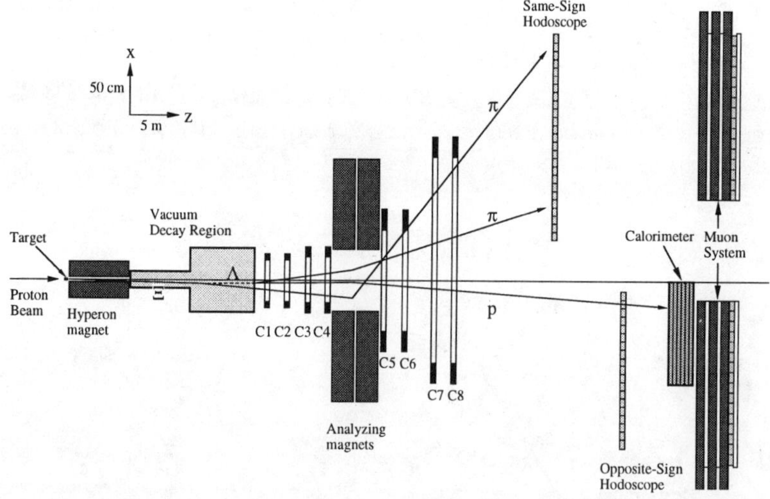

Figure 43.2: Plan view of the HyperCP spectrometer.

An 800 GeV proton beam collides with a 2 mm × 2 mm copper target. Two different target lengths are used, 6 cm to produce Ξ^-'s and 2.2 cm to produce $\overline{\Xi}^+$'s, in order to equalize the flux of particles in the spectrometer and reduce possible systematic effects related to different rates in the spectrometer for the two modes. The resulting secondary beam proceeds through a curved channel within a 1.6 T dipole magnet that selects particles with an average momentum of $\approx$ 157 GeV/c. Upon exiting this channel, the secondary beam passes through a 13 m long evacuated decay pipe before entering the spectrometer. Four high-rate multiwire proportional chambers used for tracking are followed by two BM109 dipole analyzing magnets. The BM109 magnets separate positive and negative particles in the horizontal plane and are followed by four more high-rate MWPCs. The trigger elements are downstream of the wire chambers and consist of two banks of scintillation hodoscopes and a hadronic calorimeter. A muon system consisting of proportional tubes and hodoscopes interspersed with shielding allows the study of rare decays involving muons. Helium bags are located between the detector elements and inside the analyzing magnets, to reduce multiple scattering and secondary interactions.

The trigger is fairly simple, requiring at least one hit in the same-sign hodoscope, at least one hit in the opposite-sign hodoscope, and a minimum energy deposit in the calorimeter. The calorimeter requirement makes the trigger muon blind and suppresses triggers due to secondary interactions.

When changing between Ξ^- and $\overline{\Xi}^+$ modes, the polarity of the hyperon magnet and the analysis magnets are changed. This means that the proton or antiproton is always on the calorimeter side and the two pions are always on the "same sign" side. This helps reduce possible systematic effects by having the decays occupy the same region of the spectrometer for both Ξ^- and $\overline{\Xi}^+$ decays. The polarities of the magnets are changed every few hours to avoid possible temporal systematic effects.

43.5 Summary of 1997 Run

HyperCP took data between April and September of 1997. During that time $\approx 75 \times 10^9$ events were recorded on $\approx 10,000$ 8 mm 5 Gb tapes. This should yield roughly 1.2×10^9 reconstructed Ξ^- decays and 0.3×10^9 reconstructed $\overline{\Xi}^+$ decays, giving a statistical sensitivity for $A_{\Xi\Lambda}$ of $\approx 2 \times 10^{-4}$. Analysis of the data is under way.

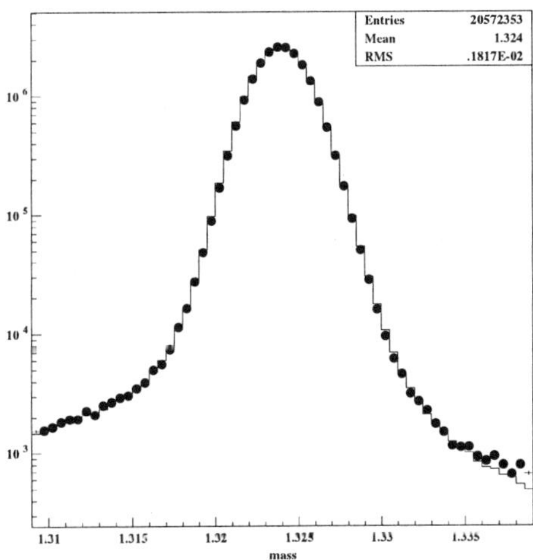

Figure 43.3: The $\Lambda\pi$ invariant mass for Ξ^- (solid line) and $\overline{\Xi}^+$ (dots) events.

A result for the asymmetry parameter is not available at this time, but fig. 43.3 shows a plot of the Ξ^- and $\overline{\Xi}^+$ reconstructed masses for a fraction of our data sample. The agreement is already quite good without any corrections.

In an experiment of this type, the most difficult part of the analysis is careful control of systematic biases. An effect that adds or subtracts unequally from the proton or antiproton $\cos\theta$ distribution might cause a false asymmetry if not corrected for. The major classes of biases are acceptance differences, differing Ξ^- and $\overline{\Xi}^+$ polarizations, differing interaction cross sections with the material in the spectrometer for p and $\overline{p}$, and π^+ and π^-, and differing backgrounds under the Λ and $\overline{\Lambda}$ and the Ξ^- and $\overline{\Xi}^+$ mass peaks. We attempted to minimize systematic biases whenever possible in the design and operation of the experiment. In addition, the analysis method itself minimizes potential biases, because the Λ helicity frame axis is not fixed with respect to the lab coordinates, but changes from event to event. Any localized acceptance variation in the spectrometer maps into a different part of the $\cos\theta$ distribution for each event. To illustrate this, in fig. 43.4 we show the proton $\cos\theta$ distributions from two Ξ^- data sets taken at production angles of ± 2.5 mrad in the horizontal plane. The two data sets have opposite polarizations and occupy different acceptance regions in the spectrometer.

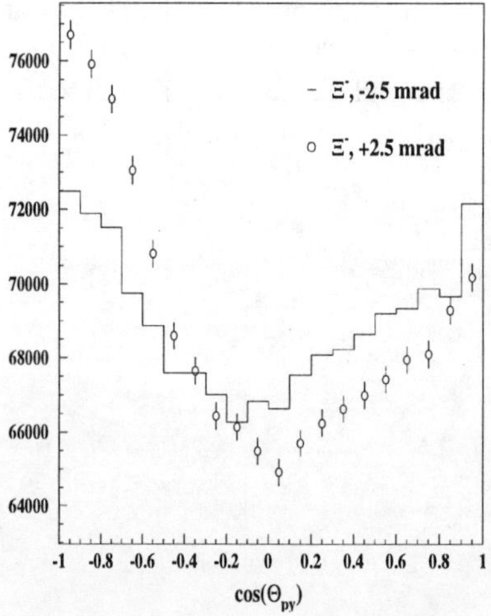

Figure 43.4: The p $\cos\theta$ distributions along the y axis in the Λ rest frame.

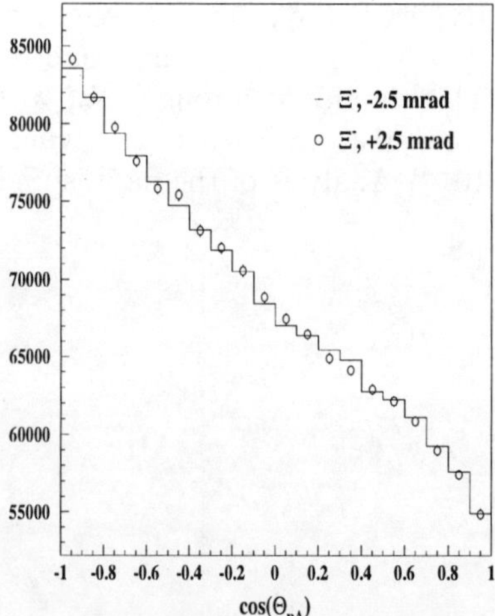

Figure 43.5: The p $\cos\theta$ distributions along the polar axis in the Λ helicity frame.

Fig. 43.4 shows the $\cos\theta$ distributions using a polar axis parallel to the laboratory y axis. Fig. 43.5 shows the same data, this time analyzed in the Λ helicity frame as we do when measuring $A_{\Xi\Lambda}$. The differences in the two distributions are greatly reduced in this frame.

43.6 Other Physics

The large data sample necessary for measurement of the asymmetry parameter puts HyperCP in a good position to study several other physics topics. We can look for lepton number nonconservation in kaon and hyperon decays such as $\Xi^- \to p\mu^-\mu^-$, $\Sigma^- \to p\mu^-\mu^-$, and $K^+ \to \pi^-\mu^+\mu^+$. We can also look for evidence of flavor-changing neutral currents in such decays as $\Sigma^+ \to p\mu^+\mu^-$ and $K^\pm \to \pi^\pm\mu^+\mu^-$, and we have observed signals for both the K^+ and K^- modes. The analysis is still in the preliminary stages. This is the first observation of the K^- mode.

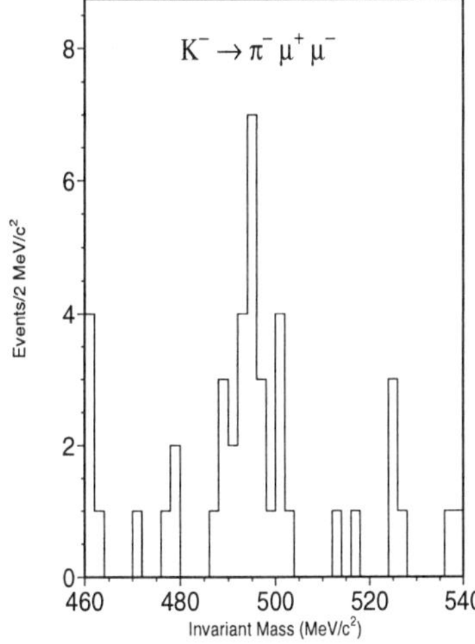

Figure 43.6: $\pi^+\mu^+\mu^-$ invariant mass.

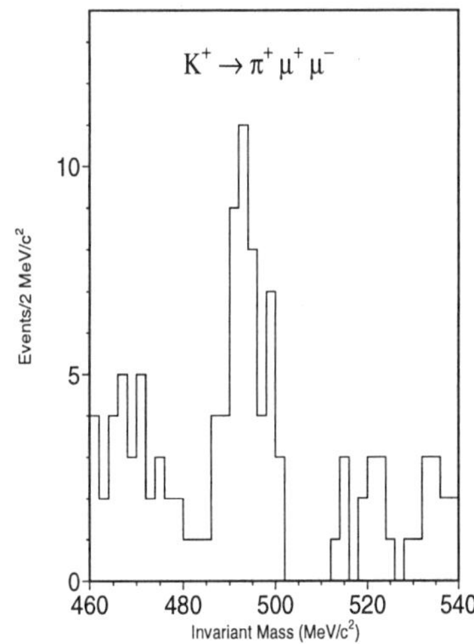

Figure 43.7: $\pi^-\mu^+\mu^-$ invariant mass.

Figs. 43.6 and 43.7 show the invariant mass plots for $K^+ \to \pi^+\mu^+\mu^-$ and $K^- \to \pi^-\mu^+\mu^-$ for a subset of our data sample.

We can also look for CP violation in $K^\pm \to 3\pi$ decays. In the $K^\pm$ rest frame, variables X and Y can be defined by

$$X = \frac{S_2 - S_1}{m_\pi{}^2}, Y = \frac{S_3 - S_0}{m_\pi{}^2}, \tag{43.8}$$

where $S_i = (p_K - p_i)^2, i = 1, 2, 3$, and $S_0 = \frac{1}{3}(m_K{}^2 + m_1{}^2 + m_2{}^2 + m_3{}^2)$. The differential decay rate is proportional to the invariant matrix element, which can be parameterized as $|M|^2 \propto 1 + gY + \cdots$. If $g_{K^-} \neq g_{K^+}$, then CP symmetry is broken. We measure the slope asymmetry $\delta g = (g - \overline{g})/(g + \overline{g})$. Plotting Y vs. X in a Dalitz plot, the number of events in the X and Y bins are counted and used to form an asymmetry

$$A = \frac{N^+(X,Y) - N^-(X,Y)}{N^+(X,Y) + N^-(X,Y)} = \frac{\epsilon^+(X,Y)(1+gY) - \epsilon^-(X,Y)(1+\overline{g}Y)}{\epsilon^+(X,Y)(1+gY) + \epsilon^-(X,Y)(1+\overline{g}Y)}, \tag{43.9}$$

where $N^{+(-)}(X,Y)$ = number of positive (negative) events in X and Y bin, and $\epsilon^\pm(X,Y)$ = acceptance. We expect to have about 109×10^6 K^+ events and 48×10^6 K^- events from our 1997 data sample, yielding a sensitivity on the order of 10^{-3} for A.

HyperCP also has a sample of several million Ω and $\overline{\Omega}$ decays, which will allow us to measure the decay parameter α_Ω. Using the decay chain $\Omega \to \Lambda K, \Lambda \to p\pi$, we measure $\alpha_\Omega \alpha_\Lambda$ from the slope of the angular distribution of the proton in the Λ helicity frame.

43.7 Outlook

HyperCP ran again in 1999, with the run ending in January 2000. Substantial improvements were made to the data acquisition system, along with various minor improvements to the spectrometer. We accumulated approximately twice the statistics of the 1997 run, corresponding to a statistical sensitivity for $A_{\Xi\Lambda}$ of $\approx 1.4 \times 10^{-4}$.

Acknowledgments

I thank the conference organizers for an excellent conference, my collaborators on HyperCP, and the Fermilab staff. This work was supported in part by the U.S. Department of Energy.

References

[1] J.H. Christenson *et. al.*, Phys. Rev. Lett. **13**, 138 (1964).

[2] Y.B. Hsiung, this volume, chapter 6.

[3] M.S. Sozzi, this volume, chapter 8.

[4] J. Antos *et al.*, Fermilab Proposal P-871 (revised version), 26 March, 1994.

[5] J.F. Donoghue and S. Pakvasa, Phys. Rev. Lett. **55**, 162 (1985); J.F. Donoghue, X.-G. He, and S. Pakvasa, Phys. Rev. D **34**, 833 (1986).

[6] S. Pakvasa, this volume, chapter 42.

[7] S. Pakvasa, hep-ph/9808472, 1998, presented at Workshop on CP Violation, Adelaide, 3–8 July, 1998.

[8] D. Chang, X.-G. He, and S. Pakvasa, Phys. Rev. Lett. **74**, 3927 (1995).

[9] P. Chauvat *et al.*, Phys. Lett. B **163**, 273 (1985).

[10] M.H. Tixier *et al.*, Phys. Lett. B **212**, 523 (1988).

[11] P.D. Barnes *et al.*, Phys. Rev. C **54**, 1877 (1996); Nucl. Phys. A **526**, 575 (1991); Phys. Lett. B **199**, 147 (1987).

44

Recent Results in Weak Hyperon Decays

Nickolas Solomey[1]

Abstract

Weak hyperon decays are of interest for improving our understanding of the electroweak theory and may even be of help in expanding this theory to include the role of the strong nuclear force [1]. The current experimental understanding of weak hyperon decays is presented, followed by recent results from current experiments. The most notable recent achievements in weak hyperon decays come from the KTeV experiment at Fermilab, which was used to observe Ξ^0 hyperon beta and muonic semileptonic decays and weak radiative decays, to perform polarization studies, and to establish a new limit for $\Delta S = 2$ processes. More important is that a high statistics study of these decays can permit a detailed understanding of hyperon form factors from beta decays and asymmetry parameters in radiative decays, both of which can lead to possible new physics beyond the standard model.

44.1 Semileptonic Weak Hyperon Decays

The standard model of electroweak interactions, SU(2)×U(1), is not capable of fully describing hyperon beta decays because it fails to include the strong force. However, beta decays are still capable of being described by the older $V - A$ theory, which includes more than just simple electroweak interactions [2]. The $V - A$ theory was abandoned because it could not handle neutral currents (*e.g.*, decays of hyperons in which the baryons before and after the decay have the same charge), and at high energies it violates probability conservation. However, none of these limitations applies to hyperon beta decays.

In this section a simple theoretical overview is presented of semileptonic hyperon decays and their relevance for further study. Then the experimental status before 1997 will be summarized, and a description of the recent results from the KTeV experimental hyperon program will end the paper.

[1]For the KTeV collaboration.

44.1.1 Theoretical Review

The transition amplitude in the $V - A$ theory for hyperon beta decays $A \to Be^-\bar{\nu}$ is

$$M = \frac{G}{\sqrt{2}} \langle B|J^\lambda|A\rangle \bar{u}_e \gamma_\lambda (1+\gamma_5) u_\nu, \tag{44.1}$$

where G is the weak coupling constant. The $V - A$ hadronic current can be written as

$$\begin{aligned}\langle B|J^\lambda|A\rangle &= Ci\bar{u}(B)\Bigg[f_1(q^2)\gamma^\lambda + f_2(q^2)\frac{\sigma^{\lambda\upsilon}\gamma_\upsilon}{M_A} + f_3(q^2)\frac{q^\lambda}{M_A} \\ &\quad + \left[g_1(q^2)\gamma^\lambda + g_2(q^2)\frac{\sigma^{\lambda\upsilon}\gamma_\upsilon}{M_A}\right. \\ &\quad \left. + g_3(q^2)\frac{q^\lambda}{M_A}\right]\gamma_5\Bigg]u(A), \end{aligned} \tag{44.2}$$

where C is the CKM matrix element and q is the momentum transfer. The leptonic component of the decay permits the determination of the form factors that appear in the hadronic matrix element, with the subsequent possibility to determine how quarks bind together in hadrons without assumptions about the lack of interplay between electroweak and strong force as in some of the more current models. There are 3 vector form factors: f_1 (vector), f_2 (weak magnetism), and f_3 (an induced scalar); plus 3 axial-vector form factors: g_1 (axial vector), g_2 (weak electricity), and g_3 (an induced pseudoscalar). By time reversal invariance we know all 6 form factors are real, but even if time reversal is not a perfect symmetry of nature it is at most the order of magnitude of CP violation (1 part in 1000), which is sufficiently small for our purposes. We do not let the fact that the formalism we are using is the older $V - A$ theory instead of the SU(2)×U(1) model of electroweak interactions, because a field theory of quark and lepton interactions must lie behind it, even if we do not know what that theory looks like. Whatever the true theory of a unified strong and electroweak theory is, it simplifies to the $V - A$ or SU(2)×U(1) theory in the proper limit. Hence the justification for our experimental enquiry.

When the mass of the lepton is small compared to that of the accompanying baryon, as in the case of the electron, the g_3 and f_3 form factors are essentially zero. Then only 4 form factors are relevant: f_1, f_2, g_1, and g_2. To make the standard model of electroweak interactions, SU(2)×U(1), match this older $V - A$ theory, g_2 has to be set to zero. This is done because a nonzero g_2 cannot fit into the formalism of the electroweak theory. Hence, observation of a nonzero g_2 would be direct evidence of physics beyond the standard model, while showing g_2 to be equal to zero would confirm the validity of the assumption made in SU(2)×U(1), and that experimental evidence showing how the strong force is incorporable into the standard model's theory must come from elsewhere. The Ξ^0 muon decay is very similar to its beta decay, yet this decay has some additional benefits not

Decay	B.R.	# events	Q [MeV]
$n \to pe^-\bar{\nu}$	100%	large	0.8
$\Lambda \to pe^-\bar{\nu}$	8.32×10^{-4}	20 K	177
$\Sigma^- \to ne^-\bar{\nu}$	1.02×10^{-3}	4.1 K	257
$\Sigma^- \to \Lambda e^-\bar{\nu}$	5.73×10^{-5}	1.8 K	81
$\Sigma^+ \to \Lambda e^+\nu$	2.0×10^{-5}	21	73
$\Xi^- \to \Lambda e^-\bar{\nu}$	5.63×10^{-4}	2868	205
$\Xi^- \to \Sigma^0 e^-\bar{\nu}$	8.7×10^{-5}	154	128

Table 44.1: Previously seen hyperon beta decays, their branching ratio, the number of events used, and the energy released in the decay.

possible with normal beta decay [3]. Here the muon mass is substantial, which has the effect that the g_3 form factor is no longer negligibly small. However, because the Ξ^0 muon decay has only 20 MeV of energy released in its decay, compared to 120 MeV with Ξ^0 beta decay, it is expected to be suppressed by 113 owing to reduced phase space, making it difficult to collect a large sample of muon decays.

44.1.2 Experimental Review

All of the previously seen hyperon beta decays are listed in table 44.1 [4], along with the number of events seen to determine a branching ratio and the energy released. Form factors have been measured for only a few of the observed hyperon beta decays where statistics were sufficiently large. However, most form factor measurements have concentrated only on g_1, f_1, and f_2. Attempts to determine g_2 have been in two forms, those that assume $g_2 = 0$ and investigate whether the observed g_1, f_1, and f_2 are consistent with expectations, and those that tried to measure g_2 independently.

What evidence for or against a nonzero g_2 form factor exists? In unpolarized Λ beta decay there is perfect agreement between the world data [5] and the standard model for g_1/f_1 and f_2/f_1. However, in polarized Λ beta decay data the outgoing asymmetries of the proton, electron, and reconstructed neutrino direction can be compared with the standard model predictions. Here we find that the neutrino asymmetry is anomalous in that it is predicted to be $A_\nu = 0.974$, while the best measurement is $A_\nu = 0.821 \pm 0.060$, a 2.6 sigma effect [6]. A more recent high-statistics experiment, E715 at Fermilab [7], using Σ^- beta decay, finds that g_2/f_1 fits best to 0.6, with g_1/f_1 of -0.2. The g_1/f_1 is very close to its predicted value but g_2/f_1 is displaced from zero by 1.5 sigma. This hyperon beta decay has a neutron in its final state, limiting the energy and momentum reconstruction accuracy, so it too should be used cautiously. Furthermore another experiment, WA-2 at CERN [8], using Ξ^- beta decay found no need for a

Decay	B.R.	# events	Q [MeV]
$\Sigma^- \to n\mu^-\bar{\nu}$	4.5×10^{-4}	174	152
$\Lambda \to p\mu^-\bar{\nu}$	5.17×10^{-4}	28	72
$\Xi^- \to \Lambda\mu^-\bar{\nu}$	3.5×10^{-4}	1	100

Table 44.2: Previously seen hyperon muonic semileptonic decays, their branching ratio, the number of events seen, and the energy released in the decay.

nonzero g_2, but low statistics limited their ability to resolve this question. Therefore, these results taken together warrant further experimental study.

All previous experiments have had low statistics or difficult decay products to analyze. The perfect experiment would use a hyperon with essentially no background, easy to accurately reconstruct by avoiding neutrons in the final state, and either 100% polarized or with 100% analyzing power of its polarization. While constructing the KTeV experiment it was realized that a never-before-observed beta decay of the Ξ^0 had amazing properties [9]. First, KTeV would have a large sample of Ξ^0 beta decays, and, second, the excellent CsI electromagnetic calorimeter would permit an accurate reconstruction of the Σ^+ decay into $p\pi^0$, which has a 98% analyzing power [10]. Since the Ξ^0 are produced with about 10% polarization, this was a built-in check to the analysis.

Since the Σ^+ decay contains all of the polarization information, working in the center of mass reference frame of the Σ^+ and aligning all of the events such that the proton points along a major axis is convenient. Then the asymmetries of the decay are simply found by plotting the angular distributions of the electron, initial Ξ^0, and reconstructed neutrino that is found from the missing transverse momentum (p_t). With good momentum and energy resolution these asymmetries can be accurately determined. A small correction for the fact that the Σ^+ has only 98% and not 100% polarization-analyzing power must be taken out, and corrections for momentum-transfer dependence and radiative effects must be applied.

There is also the possibility of extending this analysis to decays with the heavier lepton of the muon type. Presently very few experiments have observed these muonic decays, and those that have are of very low statistics (see table 44.2 [4]), not permitting any form factor analysis.

44.1.3 Two Types of Ξ^0 Beta Decays from KTeV

Since the KTeV experiment has only long-lived neutral particles in the decay volume, the easiest means of searching for Ξ^0 beta decay candidates is to find a proton and a π^0 that reconstruct into a Σ^+, under the assumption that there is no other known source of Σ^+ in the decay volume except for Ξ^0 beta decay. One problem with this decay is the lack of a charged particle vertex to localize the downstream position well enough to permit an accurate particle identification by reconstructed mass. However, by making the assumption that the two highest-energy photons that shower in the CsI calorimeter come from the π^0 decay, and forcing the reconstructed mass to be that of the π^0, an accurate location of the vertex is possible. When a data run was analyzed with these simple criteria, plus requiring that the other charged track be negative and identified by the CsI calorimeter as an electron, a Σ^+ mass peak is observed. To improve the ability to use the Ξ^0 beta decays as a precise asymmetry and form factor measurement, it is desirable to reduce the background to the level of a few percent. This is done by cutting further on the topology; the backgrounds are $\Xi^0 \to \pi^0\Lambda(\to pe^-\bar{\nu})$ and $K^0 \to \pi^+\pi^0e^-\bar{\nu}$ (π^+ misidentified as a proton). Since none of these decays has the $p\pi^0$ vertex downstream of the e^- vertex this selection is an added benefit, and it is further possible to add particle identification from the TRD system [11, 12] for the negative track which will reduce the background to the 1% level. Requiring the reconstructed electron vertex to be 1 to 20 meters upstream of the reconstructed $p\pi^0$ vertex eliminates 75% of the background while keeping nearly 90% of the signal. Furthermore, requiring the ratio of the proton momentum to electron momentum to be above 3.5 provides a good means to eliminate the $p\pi^-$ from a Λ decay, coming from another background: $\Xi^0 \to \Lambda\pi^0$. After making these major analysis cuts and some additional minor cuts, such as requiring the reconstructed vertex to be inside the beam line, and only accepting vertices that reconstruct to the upstream half of the decay volume, a much cleaner signal peak is obtained; see fig. 44.1 (left). By changing the negative particle identification to be that of a muon, we obtain five candidate events in a 90% selection box, shown in fig. 44.1 (right), for this muonic semileptonic beta decay.

In this analysis some assumptions were made, such as forcing the two photons to have the π^0 mass in order to obtain a vertex. Thus, how do we know these events are truly Ξ^0 beta decays and not some K^0 interaction that could produce Σ^+ particles? In addition to the event topology there are two additional proofs. First, the invariant mass of the Σ^+e^- pair looks like that expected from Ξ^0 beta decays, and second, the lifetime determined from the observed distribution of events in the decay volume matches the expected lifetime distribution of Ξ^0 particles; see fig. 44.2. For the Ξ^0 muon decay, a reliance on the Monte Carlo simulation of all possible background and interactions was used to support the conjecture that the five events are from Ξ^0.

From only six weeks of data taking we observe 153 Ξ^0 beta decays and 30,175 normal mode decays ($\Xi^0 \to \Lambda\pi^0$). Using Monte Carlo simulation of the KTeV detector, trigger, and analysis code to determine the acceptance for these decay modes, a preliminary branching ratio of $[2.7 \pm 0.2(\text{stat.}) \pm 0.3(\text{sys.})] \times 10^{-4}$ [13] was obtained. This is in agreement with expectations of 2.6×10^{-4}. Systematic errors are high because this analysis used the online calibration constants; eventually this is expected to improve by a factor of five or more. It is estimated that a thousand Ξ^0 beta decays

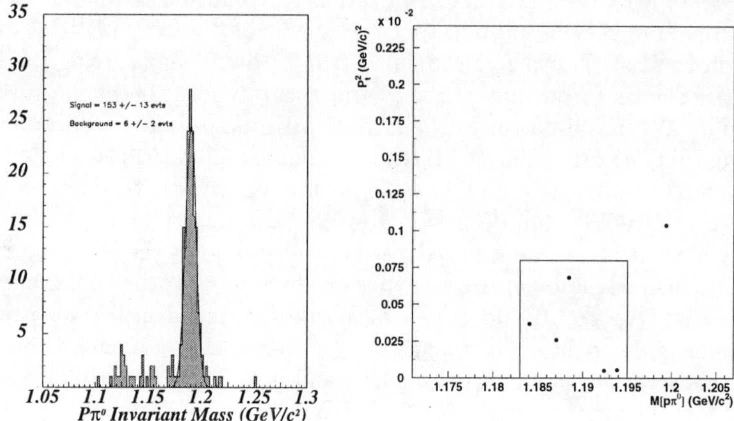

Figure 44.1: The left figure is the invariant mass distribution of the $p\pi^0$ for $\Xi^0 \to \Sigma^+ e^- \bar{\nu}$. There are 153 ± 13 Σ^+ candidate events from Ξ^0 beta decay with an estimated background of 6 ± 2 events. (This represents a small sample of the total data collected.) The right figure is the invariant mass distribution versus p_t^2 for the reconstructed $\Xi^0 \to \Sigma^+ \mu^- \bar{\nu}$, using the full data sample available from the 1997 KTeV run.

or more can be used to analyze asymmetries and form factors. The Ξ^0 muon decay with the full data sample has a preliminary branching ratio of $[2.8^{+1.5}_{-1.2}(\text{stat.}) \pm 0.8(\text{sys.})] \times 10^{-6}$ with Poisson statistical error for the 68% CL, and has a Monte Carlo estimate of no background. This should be compared to the theoretical expectation of 2.3×10^{-6}.

Under some restricted assumptions a Ξ^0 form factor analysis is available. It used only 70% of the data from the KTeV 1997 run, which is ~900 events. If the form factor f_2 has the charged vector current expected value of 2.56, and the second-class-current contribution g_2 vanishes, then $f_1 = 1.10 \pm 0.35$ and $g_1 = 1.20 \pm 0.13$, which is consistent with SU(3) predictions [14]. This result also gives an improved new preliminary branching ratio of $[2.54 \pm 0.11(\text{stat.}) \pm 0.16(\text{sys.})] \times 10^{-4}$. Form factor analysis is still in progress and the goal is to have a final independent determination with the whole sample shortly; however, more data were collected in 1999 that tripled this sample.

44.2 Weak Radiative Hyperon Decays

The radiative decays of hyperons are of interest because they too have the special feature of being a weakly dominated process where the strong nuclear force plays a very important and badly understood rule. Apart from their branching ratios, the only other variable that can be measured in these decays is the outgoing gamma ray direction relative to the initial hyperon polarization.

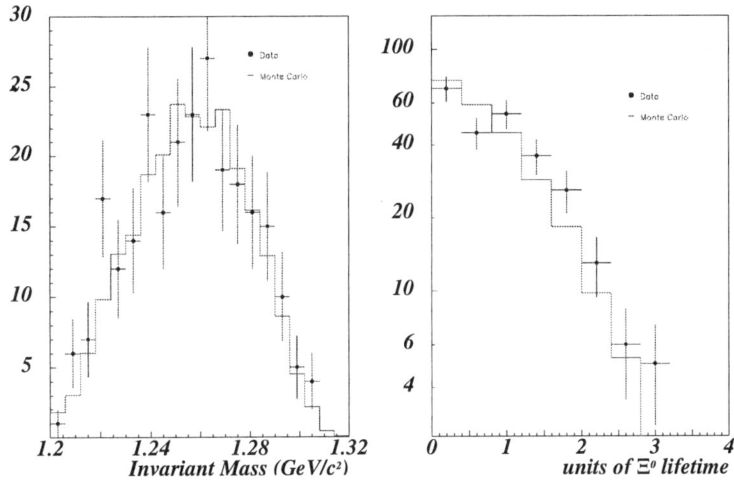

Figure 44.2: Using the candidate Ξ^0 beta decays from the peak of fig. 44.1, on the left is the invariant mass of the (Σ^+e^-), and on the right the distribution of the events in units of Ξ^0 lifetimes. Data is plotted as points, and the line is Monte Carlo simulation.

44.2.1 Theoretical and Experimental Review

Because strangeness-changing neutral currents are known not to exist from detailed neutrino interaction studies, this implies that the seemingly simple weak radiative hyperon decays are actually very complex. However, they have relatively high branching ratios on the order of 10^{-3}, but there is no theoretical framework that can predict with any accuracy their branching ratio, let alone the emitted photon direction relative to the initial hyperon polarization. In the SU(3) theory of indifference between d and s quark interchange and CP invariance it was shown that the asymmetry in the emitted photon direction should be zero [15]. Recent theoretical work suggests that an intermediate virtual state in the decay can create an asymmetry [16]. To fully understand these predictions it is necessary to greatly improve the experimental observations before any progress can truly be made.

All of the previously seen hyperon radiative decays are listed in table 44.3 [4], along with the number of events used in the branching ratio measurement and the released energy. The only reliable asymmetry measured is in the decay $\Sigma^+ \to p\gamma$. It was found to be -0.76 ± 0.08 [17], and does not agree with expectations. A previous experiment that observed $\Xi^0 \to \Sigma^0\gamma$ saw an asymmetry of 0.20 ± 0.32 [18], but the analysis neglected a 33% depolarization due to the subsequent Σ^0 decay. The resulting distortion of experimental acceptance that this may have created makes the measurement unreliable. Until a way can be found to correct for this effect

Decay	B.R.	# events	Q [MeV]
$\Lambda \to n\gamma$	1.75×10^{-3}	2100	176
$\Sigma^+ \to p\gamma$	1.23×10^{-3}	32 K	250
$\Sigma^0 \to \Lambda\gamma$	100%	large	77
$\Xi^- \to \Sigma^-\gamma$	1.27×10^{-4}	220	124
$\Xi^0 \to \Lambda\gamma$	1.06×10^{-3}	116	199
$\Xi^0 \to \Sigma^0\gamma$	3.5×10^{-3}	85	122

Table 44.3: Previously seen hyperon radiative decays, their branching ratio, the number of events used in this measurement, and the energy released in the decay.

this measurement should be disregarded. Therefore there is great interest in a detailed radiative hyperon decay experiment with high statistics to improve the measurement of the asymmetry as well as the branching ratio.

44.2.2 Two Ξ^0 Radiative Decays from KTeV

As with the Ξ^0 beta decays, the radiative decays of the Ξ^0 do not have a charged particle vertex. The decay proceeds through two stages: first, the radiative the decay $\Xi^0 \to \Sigma^0\gamma$, followed by another radiative decay $\Sigma^0 \to \Lambda\gamma$, where Λ is reconstructed through its charged particle decay: $\Lambda \to p\pi^-$. Although the Λ can be accurately reconstructed with the tracking chambers, some assumptions are needed to identify the intermediate states Σ^0. In this analysis the source of two photons was assumed to be the Ξ^0 radiative decay and its mass was constrained; the highest-energy photon was assumed to be associated with the Ξ^0 radiative decay, which was correct 80% of the time. If a two-dimensional plot of the reconstructed $\Lambda\gamma$ mass and a $\gamma\gamma$ mass is made (see fig. 44.3 left), a horizontal band is visible at the Σ^0 mass, but the problem background is the dominant decay $\Xi^0 \to \Lambda\pi^0$ (where $\pi^0 \to \gamma\gamma$), a dense vertical band that was removed. Here several other problems also arise in finding a reconstruction technique, but these backgrounds are minimal thanks to the tight energy cuts that can be imposed with KTeV's exceptional electromagnetic calorimeter, which has better than 1% energy resolution down to 5 GeV/c in momentum.

The total number of cleanly identified events from this reconstruction for $\Sigma^0\gamma$ radiative decays was 4048 (see fig. 44.3 right) above a background of 804 events, which permitted a new branching ratio measurement of $3.34 \pm 0.05(\text{stat.}) \pm 0.11(\text{sys.})] \times 10^{-3}$. However, more importantly, these events were used to measure the decay asymmetry, and a preliminary number of -0.65 ± 0.13 is in circulation.

Another radiative decay, $\Xi^0 \to \Lambda\gamma$, has also been observed by KTeV. It is slightly harder to reconstruct because of accidental γ in the high-rate environment causing a higher background of events. The latest improved branching ratio for this decay is $[0.94 \pm 0.03] \times 10^{-3}$, with 1100 reconstructed

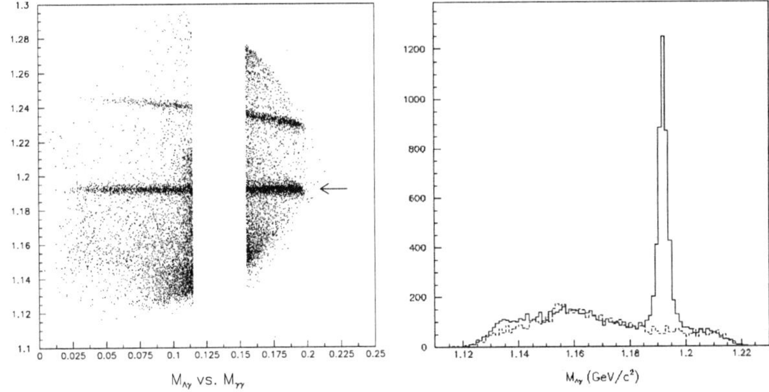

Figure 44.3: Reconstructed radiative decay chain for $\Xi^0 \to \Sigma^0\gamma$ (where $\Sigma^0 \to \Lambda\gamma$ and subsequently $\Lambda \to p\pi^-$ 64% of the time).

events above a background. No asymmetry measurement is available yet for this decay mode.

44.3 Other Weak Hyperon Physics

There are many other hyperon physics topics that need theoretical explanation and further experimental investigation. The most perplexing phenomena to explain are the experimental observations of hyperon polarization in hadronic production [19], where hyperons produced at a nonzero angle are polarized (those produced at zero production angle necessarily have zero polarization) while antihyperons have no such polarization effect. Other topics such as hyperon charge radii, hyperon mass, and second-order weak hyperon decays also are of interest.

If two polarization states, here called $d_\uparrow$ and $d_\downarrow$, are created and the distributions as a function of angle (θ) for one of the outgoing particles in the rest frame of the decaying particle for the two body decay are plotted relative to the initial polarization direction, then the polarization (P) and a constant asymmetry parameter (α)

$$D = \frac{d_\uparrow - d_\downarrow}{d_\uparrow + d_\downarrow} - 1 = 2\alpha P \cos\theta \tag{44.3}$$

can be used to fit the data. From fitting this expression, corrected for any acceptance difference in the two data samples, a polarization can be obtained. The surprise was that hyperons produced from unpolarized protons at 400 GeV/c were observed to be highly polarized when produced in the nonforward direction (with no polarization, of course, for forward-produced hyperons). These experimental observations cannot be answered by QCD, with the exception that SU(6) theories in which the valence and sea quarks are polarized in the collision process have been proposed, but not confirmed. Another interesting prediction about polarization that has not been answered with experiments is that at large p_t QCD implies that the observed polarization should change and reverse or go back towards zero polarization. However, experiments with large targeting angles have not been performed to see this effect since it is not expected until p_t of 1–3 GeV/c. There is the possibility of new results with polarization from the HyperCP experiment that has the ability to look at Ξ^- and Ω hyperons, the KTeV experiment with Λ and Ξ^0 hyperons, and the SELEX experiment with $\Sigma^\pm$ hyperons. All of these and their antiparticles are also of interest to study. Fig. 44.4 is a recent preliminary result from Ξ^0 decays with the KTeV experiment using 800 GeV/c protons at a 4.8 mrad targeting angle [20].

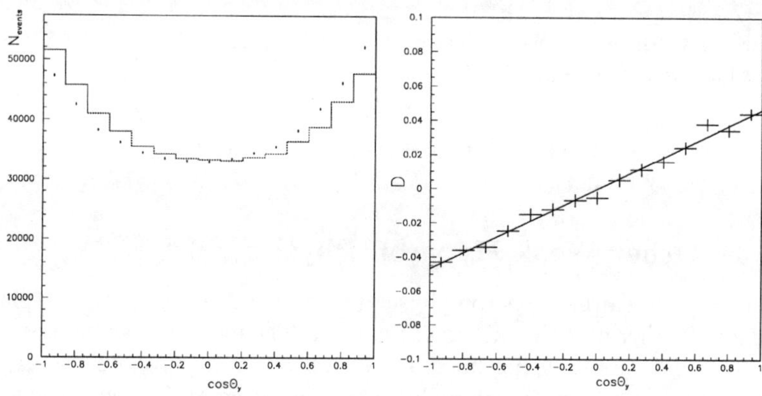

Figure 44.4: Ξ^0 polarization for only one direction. On the left are the two distributions $d_\uparrow$ (dots) and $d_\downarrow$ (histogram) and on the right is the ratio D; the two other views have no observed polarization, as expected when the two opposite polarization samples are controlled better than 1%.

The charge radius of the proton is observed to be ~0.8 fm^2, while that of the neutron is consistent with zero. For charged mesons such as the $\pi^\pm$ or $K^\pm$ it is about half that of the proton, or ~0.4 fm^2; for neutral mesons such as the K^0, it is observed to act like that of the neutron. Color confinement into states with only three quarks to form baryons, and quark-antiquark pairs to form mesons, can be tested by measuring the charge radius of hyperons; when one of the normal quarks in the proton or neutron is replaced with a strange quark color confinement requires them to behave identically. If they do not, this could be a signal that color confinement is not perfect, an important result for understanding the strong nuclear force. The SELEX experiment at Fermilab has recently found the charge radius of the the Σ^- to be $0.6 \pm 0.1\,\text{fm}^2$ [21], while WA89 has measured this same charge radius to be $0.9 \pm 0.4\,\text{fm}^2$ [22]. Both of these measurements are consistent within error bars with the proton charge radius. Similar measurements with the neutral hyperons (Σ^0 and Λ) are also of interest to compare with the neutron and K^0 charge radii measurements.

The Coleman-Glashow mass relation [23] under perfect SU(3) symmetry requires

$$\begin{aligned}(M_{\Xi^-} - M_{\Xi^0}) &= (M_p - M_n) + (M_{\Sigma^-} - M_{\Sigma^+}) \\ 6.420 \pm 0.614\,\mathrm{MeV/c^2} &= 6.773 \pm 0.077\,\mathrm{MeV/c^2} \qquad (44.4)\end{aligned}$$

with standard PDG mass and errors [4]. The major source of uncertainty in testing this relation is the mass of the Ξ^0 particle, which has an error of 0.6 MeV/c^2. Although this mass relationship is not expected to hold better than 3% as a result of known SU(3) symmetry breaking [24], it could be a test of further physics if it is seen to be broken at the 10% level, which only requires improving the mass measurement of one crucial hyperon. Recent results from the NA48 collaboration [25] include a preliminary new Ξ^0 mass with better errors, but they are not ready to release a final number.

Currently the only known $\Delta S = 2$ process is the K^0 transition into $\overline{K}^0$; the small mass difference between the long- and short-lived K^0 provides the experimental evidence that $\Delta S = 2$ transitions are permitted only in second-order weak decays. In the standard model $\Xi^0 \rightarrow p\pi^-$ decays can occur at the level of 10^{-13} or smaller. However, a superweak type of interaction [26] with this small K^0 mass difference due to an odd-parity transition puts this Ξ^0 branching ratio to be higher than 10^{-6}. A previous experiment has looked for this decay [27], placing a limit of 3.6×10^{-5}. A recent analysis from the KTeV data sample puts a new limit on this decay at less than 1.7×10^{-5} [28]. This result is an improvement over the previously published result, extending the excluded range, and is free of background. As such it constrains the theoretically allowed decay, almost completely ruling out this type of odd-parity $\Delta S = 2$ transition.

44.4 Conclusions

The hyperon program in KTeV has observed many types of Ξ^0 decays. A high statistics sample for beta and radiative decays is in hand and currently under study for measuring form factors and decay asymmetries. Furthermore, rare decays like the muonic decay of the Ξ^0 have been observed with enough data to measure its branching ratio, but more data will be needed from the 1999 run to do an asymmetry study. The successful observation of the radiative decay $\Xi^0 \rightarrow \Lambda\gamma$ has shown the potential for greatly improving knowledge about its asymmetry. These results are summarized in table 44.4. Further exciting hyperon decay topics show the wealth of physics being covered.

References

[1] J.F. Donoghue *et al.*, Physics Reports **131**, 321 (1986).

Quantity	Value
$B(\Xi^0 \to \Sigma^+ e^- \bar{\nu}_e)$	$[2.7 \pm 0.2\text{ (stat.)} \pm 0.3\text{ (sys.)}] \times 10^{-4}$
$B(\Xi^0 \to \Sigma^+ e^- \bar{\nu}_e)$ (a)	$[2.54 \pm 0.11\text{ (stat.)} \pm 0.16\text{ (sys.)}] \times 10^{-4}$
$B(\Xi^0 \to \Sigma^+ \mu^- \bar{\nu}_\mu)$	$[2.8^{+1.5}_{-1.2}\text{ (stat.)} \pm 0.8\text{ (sys.)}] \times 10^{-6}$
$B(\Xi^0 \to \Sigma^0 \gamma)$	$[3.34 \pm 0.05\text{ (stat.)} \pm 0.11\text{ (sys.)}] \times 10^{-3}$
$A(\Xi^0 \to \Sigma^0 \gamma)$	-0.65 ± 0.13
$B(\Xi^0 \to \Lambda \gamma)$	$[0.94 \pm 0.03] \times 10^{-3}$

(a) Branching ratio assuming $f_2 = 2.56$, $g_2 = 0$ (see text).

Table 44.4: Summary of preliminary KTeV results on Ξ^0 branching ratios (B) and decay asymmetry (A).

[2] A. Garcia and P. Kielanowski, *The Beta Decay of Hyperons* (Springer-Verlag, Berlin, 1985).

[3] V. Linke, Nucl. Phys. B **12**, 669 (1969).

[4] C. Caso *et al.*, Eur. Phys. J. C **3**, 1 (1998).

[5] M. Aguilar-Benitez et al., Phys. Lett. B **239**, 1 (1990).

[6] J. Lindquist *et al.*, Phys. Rev. Lett. **27**, 612 (1971); Phys. Rev. D **16**, 2104 (1977).

[7] S.Y. Hsueh *et al.*, Phys. Rev. D **38**, 2056 (1988).

[8] M. Bourquin *et al.*, Z. Phys. C **21**, 1 (1983); Z. Phys. C **21**, 37 (1983).

[9] N. Solomey *et al.*, "A Proposal for Hyperon Physics at KTeV," Univ. of Chicago note EFI 93-25, April 1993.

[10] The author first proposed the advantage in this decay of using the Σ^+ decay analyzing power as an alternative to the more traditional method of two opposite-polarization samples.

[11] K. Arisaka *et al.*, KTeV design report, Fermilab report FN-580, June 1992.

[12] N. Solomey, Nucl. Instrum. and Methods A **419**, 637 (1998).

[13] A. Affolder *et al.*, Phys. Rev. Lett. **82**, 3751 (1999).

[14] P. G. Ratcliffe, Phys. Rev. D **59**, 014038 (1999).

[15] Y. Hara, Phys. Rev. Lett. **12**, 378 (1964).

[16] B. Borasoy and B. Holstein, Phys. Rev. **D59**, 054019 (1999).

[17] M. Foucher *et al.*, Phys. Rev. Lett. **68**, 3004 (1992).

[18] S. Teige *et al.*, Phys. Rev. Lett. **63**, 2717 (1989).

[19] L. Pondrom, Physics Reports **122**, 57 (1985).

[20] T. Rooker, *Ξ^0 and $\overline{\Xi}^0$ Polarization Analysis at KTeV*, master's thesis in Physics, University of Wisconsin, Aug. 1999.

[21] I. Eschrich, Hyperon Physics Results from SELEX, AIP Conf. Proc. **459**, 303 (1999).

[22] M. Adamevich *et al.* (WA89 collaboration), Submitted to Eur. Phys. J. C.

[23] S. Coleman and S.L. Glashow, Phys. Rev. Lett. **6**, 428 (1961).

[24] E. Jenkins, Phys. Rev. D **54**, 4515 (1996); Phys. Rev. D **55**, 10 (1997); Ann. Rev. Nucl. Part. Sci. **48**, 81 (1998).

[25] L. Köpke, this volume, chapter 38.

[26] L.B. Okun, Sov. J. Nucl. Phys. **1**, 806 (1965).

[27] C. Geweniger *et al.*, Phys. Lett. B **57**, 193 (1975).

[28] T. Kreutz, *Limiting the Branching Ratio of* $\Xi^0 \to p\pi^-$, bachelor's thesis in Physics, The University of Chicago, June 1998.

Part VII

Charm: CP Violation and Mixing

Although the main focus of the present volume is the physics of kaons, many of the same issues arise for particles containing heavier quarks. In the present part some of these are discussed for charmed quarks, while the part that follows is devoted to b (bottom, or beauty) quarks.

The lightest charmed particle is the D^0 meson, consisting of a charmed quark c and a light antiquark $\bar{u}$. The $D^0 = c\bar{u}$ can mix with its antiparticle, $\bar{D}^0 = u\bar{c}$, in the same way that K^0 and $\bar{K}^0$ can mix with one another. However, while the mixing of K^0 and $\bar{K}^0$ is expected to be so complete as to give mass eigenstates that are essentially the sum and difference of these particles, the situation is very different for charm. Whereas K^0 and $\bar{K}^0$ share a dominant decay mode ($\pi\pi$), the major decay modes of the D^0 and $\bar{D}^0$ have opposite strangeness and thus are distinct from one another. Moreover, the few states that they do share appear to cancel one another to a significant degree in their mixing effects. Thus, mixing between D^0 and $\bar{D}^0$ is expected to be extremely small. (Its actual magnitude is a matter of some debate.) Detection of a large mixing amplitude would be evidence for new physics.

Similarly, CP violation in D decays is expected to be small in the standard model. Again, observation of large CP-violating effects in charmed particle decays could be evidence for new physics.

The present part is devoted to some studies of CP violation and mixing in charmed particle decays, performed in hadron beams (Purohit) and photon beams (Pedrini) at Fermilab and in electron-positron collisions at the Cornell Electron Storage Ring, or CESR (Nelson).

45

Results from Fermilab Charm Experiment E791

Milind V. Purohit

Abstract

Recent results from Fermilab experiment E791 are summarized. Topics covered include mixing and CP violation in charm decays, searches for FCNC, LFV, and LNV decays, and some examples of other results such as a measurement of the D_s^+ lifetime and results of a resonant analysis and polarization of Λ_c^+ decays.

45.1 Experiment E791 at Fermilab

Fermilab E791 is a fixed-target pion-production charm collaboration representing over 17 institutions. Over 20 billion events were collected during 1991–92 and analyses of these data are now almost complete. The experiment used a segmented target with five thin foils (each about 1.2 mm thick) with about 1.5 cm center-to-center separation. With this target configuration and with 23 planes of silicon detectors we were able to suppress the large backgrounds due to combinatorics and secondary interactions. Complementing this vertex detector is a complete 2-magnet spectrometer with 35 planes of drift chambers and with Cherenkov detectors and calorimeters for particle identification and energy measurement. Analyses of these data have led to over 20 publications and should be complete in a couple of years.

45.2 $D^0 - \overline{D}^0$ Mixing

As in the kaon sector, the neutral charm mesons D^0 and $\overline{D}^0$ are produced in strong interactions and are strong eigenstates. Diagrams such as the box diagram lead to mixing of the two particles and one can define the CP eigenstates $D_{1,2}^0$ by

$$D_{1,2}^0 = \frac{D^0 \pm \overline{D}^0}{\sqrt{2}}. \tag{45.1}$$

If there is CP violation in the mixing the mass eigenstates $D^0_{H,L}$ may differ slightly from $D^0_{1,2}$. Defining Δm and $\Delta\Gamma$ as the mass and decay rate differences between the two neutral D mass eigenstates we can define

$$x \equiv \frac{\Delta m}{\Gamma} \qquad \text{and} \qquad y \equiv \frac{\Delta\Gamma}{2\Gamma}. \tag{45.2}$$

Integrated over time, the ratio of the rate for wrong-sign decays such as $D^0 \to K^+\pi^-$ to that for right sign decays such as $D^0 \to K^-\pi^+$ is given by

$$r_{\rm mix} \equiv \frac{\Gamma(D^0 \to \overline{D}^0 \to \overline{f})}{\Gamma(D^0 \to f)} = \frac{x^2+y^2}{2}. \tag{45.3}$$

In the standard model, the quantities x and y are of the order of 10^{-5} yielding an extremely small mixing rate $\sim 10^{-10}$. Long-distance contributions are not expected to raise this rate beyond 10^{-7} or so [1,2]. This implies great opportunities to observe new physics in the charm sector.

45.2.1 The E791 $D^0 - \overline{D}^0$ Mixing Limit

We used our data to study $D^0 - \overline{D}^0$ mixing using both the hadronic final states $K^\mp\pi^\pm$ and $K^\mp\pi^\pm\pi^+\pi^-$ as well as the semileptonic final states $K^\mp l^\pm \nu_l$ of the D^0 (where l stands for e or μ) to look for $D^0 - \overline{D}^0$ mixing. Using Ds from $D^{*+} \to D^0\pi^+$ decays the flavor of the Ds is tagged at birth by the charge of the pion. Their flavor at decay is tagged by the sign of the decay kaon. Ds that appear to have the same flavor at birth and decay are called "right-sign" decays, while the others, the "wrong-sign" decays, are mixing / Double Cabibbo Suppressed (DCS) decay candidates and we measure their rate relative to the "right-sign" decays. The time dependence of mixed and DCS events is quite different. Mixed events (in the first approximation) have a time dependence $\sim t^2 \exp(-t/\tau)$ while DCS events have a $\exp(-t/\tau)$ time dependence. We use these different time dependences to separate the two components in the hadronic decays; the semileptonic "wrong-sign" decays do not have any DCS background.

The hadronic mixing data were fit in several different ways. The most "likely" fit in the standard model allows for CP violation only in the interference terms, while the most general fit allows for CP violation in the mixing, interference, and DCS terms. The other fits were a "no-interference" fit (done for comparison to E691) and a "no-mixing" fit (for comparison to CLEO). Results of these four fits and from the semileptonic decay fit are shown in table 45.1. The semileptonic data could be analyzed without complications from DCS contributions which compensates for the poorer statistics (see fig. 45.1). It is fair to say that the two methods achieved about equal sensitivity and the present limit on $r_{\rm mix}$ is 0.5% at the 90% CL.

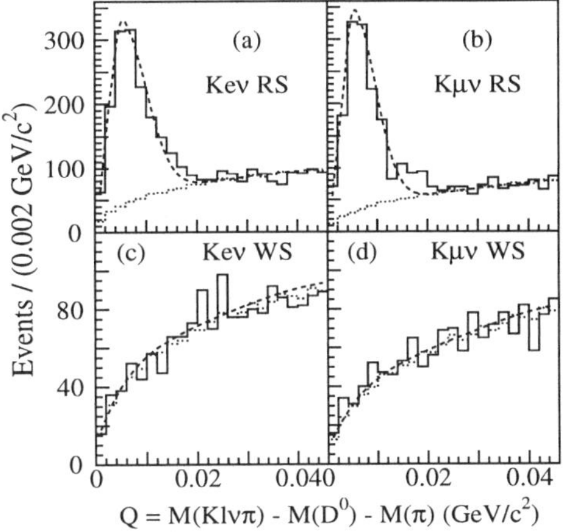

Figure 45.1: Q value projections of the E791 $D^0 \to K\mu\nu$ and $D^0 \to Ke\nu$ right-sign and wrong-sign data with fit projections.

45.3 $\Delta\Gamma$ from a Lifetime Difference Measurement

If one ascribes the entire limit of 0.5% from our mixing studies to a wrong sign rate due to $\Delta\Gamma/\Gamma$, one obtains a limit of

$$|\Delta\Gamma| < 0.48 \text{ ps}^{-1}. \tag{45.4}$$

Ted Liu [7] pointed out that a direct limit from measuring lifetimes may actually be better. In the presence of mixing, even assuming CP invariance, the decay $D^0 \to K^-\pi^+$ is not purely exponential in time.

$$\Gamma_{K\pi}(t) = A_{K\pi}e^{-\Gamma t}\cosh(\frac{\Delta\Gamma t}{2}), \tag{45.5}$$

where $\Gamma \equiv (\Gamma_1 + \Gamma_2)/2$ and $\Delta\Gamma \equiv \Gamma_1 - \Gamma_2$.

Of course, decays to CP eigenstates such as $D^0 \to K^-K^+$ remain purely exponential:

$$\Gamma_{KK}(t) = A_{KK}e^{-\Gamma_1 t}. \tag{45.6}$$

Therefore, we searched for a difference in D^0 lifetimes in the two decay modes $K^-\pi^+$ and K^-K^+. We find that

$$\Delta\Gamma = 0.04 \pm 0.14 \pm 0.05 \text{ ps}^{-1} \tag{45.7}$$

or, expressed as a limit,

$$-0.20 < \Delta\Gamma < 0.28 \text{ ps}^{-1} \quad \text{at the 90\% CL.} \tag{45.8}$$

Fit type	E791 & Other Results
CP violation only in interference term	$r_{\text{mix}} = (0.39^{+0.36}_{-0.32} \pm 0.16)\%$ $r_{\text{mix}} = (0.11^{+0.30}_{-0.27})\%$ (E791 [3]) (Semileptonic decays)
Most general, no CP assumptions	$r_{\text{mix}}(\overline{D}^0 \to D^0) = (0.18^{+0.43}_{-0.39} \pm 0.17)\%$ $r_{\text{mix}}(D^0 \to \overline{D}^0) = (0.70^{+0.58}_{-0.53} \pm 0.18)\%$
No CP violation, no interference	$r_{\text{mix}} = (0.21^{+0.09}_{-0.09} \pm 0.02)\%$ $r_{\text{mix}} = (0.05 \pm 0.20)\%$ (E691 [4])
No mixing	$r_{dcs}(K\pi) = (0.68^{+0.34}_{-0.33} \pm 0.07)\%$ $r_{dcs}(K\pi\pi\pi) = (0.25^{+0.36}_{-0.34} \pm 0.03)\%$ $r_{ws}(K\pi) = (0.77 \pm 0.25 \pm 0.25)\%$ (CLEO [5])

Table 45.1: A summary of values from our four hadronic decay fits. The top line describes the most likely case for extensions to the standard model that produce large mixing. The second line describes our most general fit, providing the most conservative results. The third line matches the assumptions of previous experiments, which are included for completeness. The bottom line describes the standard model case. Results from other experiments are also listed for comparison. Unless mentioned, the reference for the results is [6].

45.4 Measurement of the D_s^+ Lifetime

Among the recent measurements made by E791 has been the lifetime of the D_s^+. Many experiments have measured this lifetime in the past. What is unique about this new measurement is that it is significantly larger than previous measurements thereby showing for the first time that the lifetime of the D_s^+ is substantially different from the D^0. E791 finds that

$$\tau_{D_s^+} = 0.518 \pm 0.014 \pm 0.007, \tag{45.9}$$

which yields for the ratio of lifetimes

$$\frac{\tau_{D_s^+}}{\tau_{D^0}} = 1.25 \pm 0.04 \qquad \text{(E791)} \tag{45.10}$$

compared to 1.13 ± 0.04 for the same ratio from the PDG 1998 [8] world average of lifetimes.

45.5 CP Asymmetries

CP violation in charm decays in the standard model is most likely to occur in Singly Cabibbo Suppressed (SCS) decays of charm mesons where penguin diagrams provide a second amplitude for the decays. Again, the predicted asymmetries are much smaller than current experimental limits and there is a large window for new physics [9, 10]. The search for CP violation was carried out in several modes and the limits on the CP asymmetry in particle-antiparticle decay rates are found to be at the few percent level [11, 12].

45.6 FCNC, LFV, and LNV Decays

E791 has searched for flavor changing neutral currents (FCNC), lepton flavor violating (LFV), and lepton number violating (LNV) charm decays. The first are expected to have small branching ratios ($\lesssim 10^{-8}$) while the last two are forbidden in the standard model. A total of 24 modes were searched and typical limits are up to an order of magnitude better than present PDG limits [8, 13].

45.7 $\Lambda_c^+ \to pK^-\pi^+$ Resonant Decay Analysis

Using 886±43 $\Lambda_c^+ \to pK^-\pi^+$ decays we did a 5-dimensional analysis to search for resonances in this decay. We found the fit fractions listed in table 45.2. As a by-product of this analysis we measured the polarization of the Λ_c^+ in pion production. Our errors are large, but preliminary indications are that the polarization gets substantially large and negative as $p_T(\Lambda_c^+)$ increases.

Mode	Fit Fraction (%)
$p\overline{K}^{*0}(890)$	19.5±2.6±1.8
$\Delta^{++}(1232)K^-$	18.0±2.9±2.9
$\Lambda(1520)\pi^+$	7.7±1.8±1.1
Nonresonant	54.8±5.5±3.5

Table 45.2: The decay fractions for $\Lambda_c^+ \to pK^-\pi^+$ with statistical and systematic errors from the final fit.

45.8 Upcoming Results

E791 has published approximately 20 papers so far, on a variety of topics including rare decays, production asymmetries, the Σ_c mass splitting, the $\rho l\nu$ branching ratio, $D-\pi$ correlations, CP violation searches, $D^0-\overline{D}^0$ mixing, DCS decays, 4-body branching ratios, form factors, pentaquark limits, results on form factors, a limit on $\Delta\Gamma/\Gamma$, charm pairs, and the D_s^+ lifetime. We will soon submit for publication new results on hyperon asymmetries, $D^+ \to K^+K^-K^+$ decays, Dalitz analyses, diffractive dijets, production of D^0 and D^{*+}, and the Λ_c^+ resonant substructure.

Acknowledgments

I would like to thank members of my collaboration (Fermilab E791) for sharing their results. This work was supported by a grant from the U.S. Department of Energy. Special thanks go to organizers of this conference including Profs. Rosner and Winstein for bringing us together and discussing precision results in kaon physics.

References

[1] L. Wolfenstein, Phys. Rev. Lett. **75**, 2460 (1995).

[2] T. Ohl, G. Ricciardi, and E.H. Simmons, Nucl. Phys. B **403**, 605 (1993).

[3] E.M. Aitala *et al.*, PRL **77**, 2384 (1996).

[4] J.C. Anjos *et al.*, PRL **60**, 1239 (1988).

[5] D. Cinabro *et al.*, PRL **72**, 1406 (1994).

[6] E.M. Aitala *et al.*, PRD **57**, 13 (1998).

[7] T. Liu *et al.*, Harvard University Report HUTP-94/E021, hep-ph/9408330 (July 1994).

[8] Particle Data Group, EPJC **3**, 1 (1998).

[9] M. Golden and B. Grinstein, PLB **222**, 501 (1989).

[10] F. Buccella *et al.*, PLB **302**, 319 (1993).

[11] E.M. Aitala *et al.*, PLB **421**, 405 (1998).

[12] E.M. Aitala *et al.*, PLB **403**, 377 (1997).

[13] E.M. Aitala *et al.*, Fermilab Pub-99-183-E, hep-ex/9906045 (June 1999).

46

CP Violation, $D^0 - \overline{D}^0$ Mixing, and Rare and Forbidden Decays in FOCUS

D. Pedrini[1]

Abstract

An overview of Fermilab experiment FOCUS is presented. During the 1996–97 fixed target run at Fermilab we collected more than 6.3 billion events. From this sample we reconstructed more than 1 million charmed particles. Preliminary results on CP violation, $D^0 - \overline{D}^0$ mixing, and rare and forbidden D decay modes are reported.

46.1 Introduction

The standard model contains two key mysteries: the origin of the masses and the existence of multiple fermion generations. While the first mystery could be resolved by LHC experiments looking for the Higgs boson, the

[1]On behalf of the FOCUS (E831) collaboration. Coauthors: J. Link, M. Reyes, P. M. Yager (UC Davis); J. Anjos, I. Bediaga, C. Gobel, J. Magnin, I. M. Pepe, A. C. Reis, A. Sánchez-Hernández, F. R. A. Simão (CPBF, Rio de Janeiro); S. Carrillo, E. Casimiro, H. Mendez, M. Sheaff, C. Uribe, F. Vasquez (CINVESTAV, Mexico City); L. Cinquini, J. P. Cumalat, J. E. Ramirez, B. O'Reilly, E. W. Vaandering (CU Boulder); J. N. Butler, H. W. K. Cheung, I. Gaines, P. H. Garbincius, L. A. Garren, A. Gourlay, P. H. Kasper, A. E. Kreymer, R. Kutschke (Fermilab); S. Bianco, F. L. Fabbri, S. Sarwar, A. Zallo (INFN Frascati); C. Cawlfield, E. Gottschalk, K. S. Park, A. Rahimi, J. Wiss (UI Champaign); R. Gardner (IU Bloomington); Y. S. Chung, J. S. Kang, B. R. Ko, J. W. Kwak, K. B. Lee, S. S. Myung, H. Park (Korea University, Seoul); G. Alimonti, M. Boschini, B. Caccianiga, A. Calandrino, P. D'Angelo, M. DiCorato, P. Dini, M. Giammarchi, P. Inzani, F. Leveraro, S. Malvezzi, D. Menasce, M. Mezzadri, L. Milazzo, L. Moroni, F. Prelz, A. Sala, S. Sala (INFN and Milano); T. F. Davenport III (UNC Asheville); V. Arena, G. Boca, G. Bonomi, G. Gianini, G. Liguori, M. Merlo, D. Pantea, S. P. Ratti, C. Riccardi, L. Viola, P. Vitulo (INFN and Pavia); A. M. Lopez, L. Mendez, A. Mirles, E. Montiel, D. Olaya, C. Rivera, W. Johns, Y. Zhang (Mayagucz, Puerto Rico); N. Copty, M. V. Purohit, J. R. Wilson (USC Columbia); K. Cho, T. Handler (UT Knoxville); D. Engh, M. Hosack, M. Nehring, M. Sales, P. D. Sheldon, M. Webster (Vanderbilt); K. Stenson (Wisconsin, Madison); Y. Kwon (Yonsei University, Korea).

second mystery appears to originate at higher mass scales and therefore can be studied only indirectly.

CP violation in charm decays, $D^0 - \overline{D}^0$ mixing, and rare and forbidden decay modes can be used to investigate the physics at these new scales.

Why *charm*? Because in the charm sector the standard model predictions for these processes are extremely small. As a consequence, any observation of a signal should be a clear indication of physics beyond the standard model. In addition *charm* is the unique probe of the *up-type* quark sector.

FOCUS means **Pho**toproduction of **C**harm with an **U**pgraded **S**pectrometer, with a lexical license. The word "upgrade" refers to the upgrade of the E687 (the predecessor experiment) spectrometer [1].

The charmed particles were produced by the interaction of high energy photons (obtained by means of bremsstrahlung of electron and positron beams) with a beryllium oxide target. The mean energy of the photon beam was about 200 GeV. The data were collected at Fermilab during the 1996–97 fixed target run. More than 6.3 billion triggers were collected. The initial goal was to increase the statistics of E687 by a factor of 10, but we actually exceeded this goal by reconstructing more than 1 million charmed particles.

46.2 CP Violation and $D^0 - \overline{D}^0$ Mixing

It is well known that CP-violating effects occur in a decay process only if the decay amplitude is the sum of two different parts, whose phases are made of a weak (CKM) and a strong (FSI) contribution [2]. The weak contributions to the phases change sign when going to the CP-conjugate process, while the strong ones do not.

In Cabibbo first-forbidden D decays, the penguin terms in the effective Hamiltonian provide the different phases of the two weak amplitudes. The expected asymmetries are around 10^{-3} [2].

Experimentally one looks at the Cabibbo-suppressed decay modes that have the largest branching fractions, taking into account also the detection efficiency. For this reason the selected decay modes are $D^+ \to K^- K^+ \pi^+$ and $D^0 \to K^- K^+$ (throughout this paper the charge conjugate state is implied). To tag the neutral D as either a D^0 or a $\overline{D}^0$ we use the sign of the bachelor pion in the $D^{*\pm}$ decay.

Before searching for a CP asymmetry we must account for the different D and $\overline{D}$ production rates in photoproduction[1]. This is done using the Cabibbo-favored modes $D^0 \to K^- \pi^+$ and $D^+ \to K^- \pi^+ \pi^+$. There is the additional benefit that most of the corrections due to inefficiencies cancel out, therefore reducing the systematic uncertainties. An implicit

[1] The main reason for this difference is the associated production of a charmed baryon with a $\overline{D}$.

Decay mode	Preliminary FOCUS 59% of the data	Best measurement
$D^+ \to K^-K^+\pi^+$	-0.004 ± 0.017	-0.014 ± 0.029 [3]
$D^0 \to K^-K^+$	0.003 ± 0.039	$-0.010 \pm 0.049 \pm 0.012$ [4]

Table 46.1: CP asymmetry in D decays.

assumption is that there is no measurable CP violation in the Cabibbo-favored decays.

The CP asymmetry can be then written as

$$A_{CP} = \frac{\eta(D) - \eta(\overline{D})}{\eta(D) + \eta(\overline{D})},$$

where η is (considering for example the decay mode $D^0 \to K^-K^+$)

$$\eta(D) = \frac{N(D^0 \to K^-K^+)}{N(D^0 \to K^-\pi^+)}$$

and $N(D^0 \to K^-K^+)$ is the efficiency-corrected number of candidate decays.

In fig. 46.1 the uncorrected invariant mass for the decay modes $D^+ \to K^-K^+\pi^+$(a), $D^- \to K^+K^-\pi^-$(b), $D^0 \to K^-K^+$(c), and $\overline{D}^0 \to K^+K^-$(d) are shown. These plots are obtained from 59% of our total sample. From this analysis we can compute preliminary results on the CP asymmetry. These preliminary results are summarized and compared to the current best measurements in the table 46.1:

In the case of the D^0 the asymmetry is not a direct CP asymmetry–related parameter, but rather a measurement of combined direct and indirect CP asymmetries [5]. $D^0 - \overline{D}^0$ mixing can cause indirect CP violation. In the standard model the expected value for the mixing parameter r_{mix} is in the range $10^{-8} - 10^{-10}$ [6].

Experimentally r_{mix} is given by

$$r_{\text{mix}} = \frac{\Gamma(D^0 \to \overline{D}^0 \to \overline{f})}{\Gamma(D^0 \to f)} = \frac{WS}{RS};$$

that is the ratio of the D^0 that converts in $\overline{D}^0$ and decays as $\overline{D}^0$ (wrong sign) to the unmixed decay mode of D^0 (right sign).

The method consists of two steps: first to tag the flavor of the produced neutral D by means of D^* and then to identify the final D using the reconstructed decay mode. To study the mixing, one would select the copious (Cabibbo-favored) hadronic modes. However, the semileptonic modes give a cleaner signature because there is no *contamination* from doubly

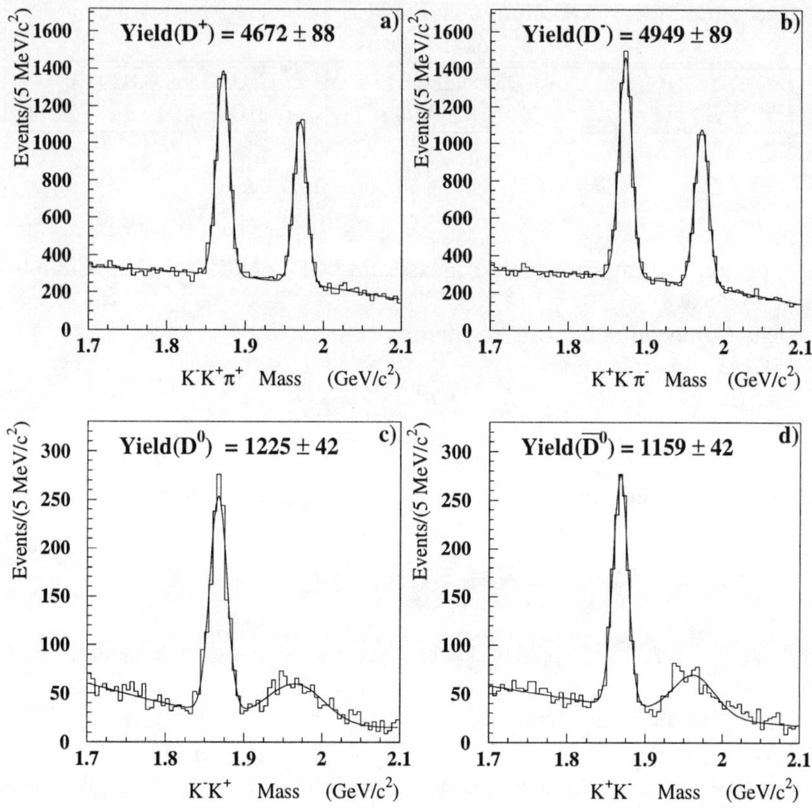

Figure 46.1: Invariant mass, from 59% of our total sample, for: (a) $D^+ \to K^-K^+\pi^+$, (b) $D^- \to K^+K^-\pi^-$, (c) $D^0 \to K^-K^+$, and (d) $\overline{D}^0 \to K^+K^-$

Cabibbo-suppressed decays (DCSD). For example the time distribution of the decay mode $D^0 \to K^+\pi^-$ can be written as [7]

$$I(t) \sim e^{-\Gamma t}\left[r_{\rm DCSD} + r_{\rm mix} \times \frac{\Gamma^2 t^2}{2} + \sqrt{2 r_{\rm mix} r_{\rm DCSD}} \cos\phi \times \Gamma t\right],$$

where the terms in the brackets are the DCSD contribution, the mixing term, and the interference between mixing and DCSD. This relation implies that it is necessary to have a very good decay time resolution in order to disentangle the different contributions.

FOCUS intends to study $D^0 - \overline{D}^0$ mixing using both the hadronic and the semileptonic decays; however, in this paper we present a preliminary study (from 30% of our sample) for the decay mode $D^0 \to K^-\mu^+\nu_\mu$. In fig. 46.2 the $D^* - D^0$ mass difference distributions for the right sign (solid histogram) and for the background (shaded histogram) are shown. We did not look at the wrong-sign signal region (blind analysis) to prevent any bias in selecting analysis cuts.

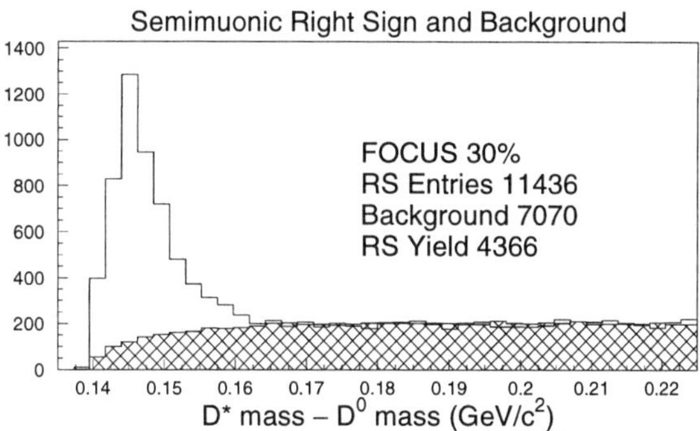

Figure 46.2: Mass difference for RS (solid) and background (shaded).

The main source of background is from random soft pions. We obtain the background shape from data by replacing the pions in the primary vertex from one event with pions from a different event. The background shape is normalized using the events outside the signal region. The best 90% CL upper limit for the semileptonic r_{mix} has been obtained by E791 [8]: $r_{\text{mix}} < 0.50\%$. From the previous analysis we estimate our sensitivity[2] for the whole sample to be 0.18%. Including the electron mode, $D^0 \to K^- e^+ \nu_e$, this estimate drops to 0.13%.

46.3 Rare and Forbidden *D* Decay Modes

In the charm sector the rare and forbidden decay modes can be split mainly into three categories:

1) flavor-changing neutral current (FCNC) such as $D^0 \to \ell^+\ell^-$ and $D^+ \to h^+\ell^+\ell^-$

2) lepton family number–violating (LFNV) such as $D^+ \to h^+\ell_1^+\ell_2^-$

3) lepton number–violating (LNV) such as $D^+ \to h^-\ell_1^+\ell_{1,2}^+$,

where h stands for π, K and ℓ for e, μ.

The first decay modes (FCNC) are *rare*, where rare usually means a process that proceeds via an internal quark loop in the standard model [9]. The FCNC decay mode $D^0 \to \ell^+\ell^-$ can proceed via a W box diagram, and the theoretical estimates [10] for the branching fraction are of the order of $\sim 10^{-19}$. The predictions for the other FCNC decay modes, $D^+ \to h^+\ell^+\ell^-$, are considerably larger. These decay modes can proceed via several penguin diagrams [9], and the branching fraction estimates for these modes are of

[2]That is, the limit we would get if the number of WS (or rare) events in the signal region minus the expected number of background events is zero.

Decay mode	FOCUS expected sensitivity 45% of the data	E791 90% C.L. lim.	PDG 98 90% C.L. lim.
$D^+ \to K^+\mu^+\mu^-$	8.1×10^{-6}	4.4×10^{-5}	9.7×10^{-5}
$D^+ \to K^-\mu^+\mu^+$	12.1×10^{-6}		1.2×10^{-4}
$D^+ \to \pi^+\mu^+\mu^-$	7.8×10^{-6}	1.5×10^{-5}	1.8×10^{-5}
$D^+ \to \pi^-\mu^+\mu^+$	7.1×10^{-6}	1.7×10^{-5}	8.7×10^{-5}
$D^+ \to \mu^-\mu^+\mu^+$	4.4×10^{-6}		

Table 46.2: Limits for rare and forbidden D decay modes.

the order of $\sim 10^{-9}$ [10, 11]. In addition to these short-distance diagrams, there are contributions from long-distance effects.

The other two decay modes (LFNV and LNV) are strictly *forbidden* in the standard model.

To determine the branching fraction we use the branching ratio method; that is we compute the branching ratio relative to a normalization mode. Using the same cuts for both the normalization and the rare mode (where possible) many systematics cancel in the relative efficiency ratio.

We performed a preliminary blind analysis for the following five rare modes. The expected background in the signal region is inferred from the sidebands. Our expected sensitivity[2] for 45% of the data is compared to the 90% CL limit of the Particle Data Group [12] and to very recent results from E791 [13] in table 46.2:

Based on these estimates an analysis of the total sample should substantially improve the current limits.

46.4 Conclusions

A reasonable estimate is that with more than 1 million charm particles reconstructed, FOCUS should improve all the current limits on CP violation in D decays, $D^0 - \overline{D}^0$ mixing, and rare and forbidden D decays.

References

[1] E687 Collaboration, P.L. Frabetti *et al.*, Nucl. Instr. Meth. A **320**, 519 (1992).

[2] F. Buccella *et al.*, Phys. Rev. D **51**, 3478 (1995).

[3] E791 Collaboration, E.M. Aitala *et al.*, Phys. Lett. B **403**, 377 (1997).

[4] E791 Collaboration, E.M. Aitala *et al.*, Phys. Lett. B **421**, 405 (1998).

[5] W.F. Palmer and Y.L. Wu, Phys. Lett. B **350**, 245 (1995).

[6] G. Burdman, CHARM2000, Fermilab, hep-ph/9407378 (June 1994).

[7] T. Liu, FCNC 97, Santa Monica, hep-ph/9706477 (February 1997).

[8] E791 Collaboration, E.M. Aitala *et al.*, Phys. Rev. Lett. **77**, 2384 (1996).

[9] A.J. Schwartz, Mod. Phys. Lett. A **8**, 967 (1993).

[10] J.L. Hewett, LISHEP 95 Session C: Heavy Flavour Physics (1995) 171, hep-ph/9505246.

[11] S. Pakvasa, Chin. J. Phys. **32**, 1163 (1994) (hep-ph/9408270).

[12] Particle Data Group, C. Caso *et al.*, Eur. Phys. J. C **3**, 1 (1998).

[13] E791 Collaboration, E.M. Aitala *et al.*, hep-ex/9906045 (30 Jun 1999). See also M. Purohit, this volume, chapter 45.

47

Search for $D^0 - \overline{D}^0$ Mixing

Harry N. Nelson

Abstract

We report on a search for $D^0 - \overline{D}^0$ mixing made by a study of the "wrong-sign" process $D^0 \rightarrow K^+\pi^-$. The data come from an integrated luminosity of e^+e^- collisions at $\sqrt{s} \approx 10\,\text{GeV}$ consisting of 9.0 fb^{-1}, recorded with the CLEO-II.V detector. We measure the time-integrated rate of the wrong-sign process $D^0 \rightarrow K^+\pi^-$, relative to that of the Cabibbo-favored process $\overline{D}^0 \rightarrow K^+\pi^-$, to be $R_{\text{ws}} = (0.34 \pm 0.07 \pm 0.06)\%$. We study that rate as a function of the decay time of the D^0, to distinguish the rate of direct doubly-Cabibbo-suppressed decay from $D^0 - \overline{D}^0$ mixing. The amplitudes that describe $D^0 - \overline{D}^0$ mixing, x' and y', are consistent with zero. The one-dimensional limits, at the 95% CL, that we determine are $(1/2)x'^2 < 0.05\%$, and $-5.9\% < y' < 0.3\%$. All results are preliminary.

Studies of the evolution of a K^0 or B^0_d into the respective antiparticle, a $\overline{K}^0$ or $\overline{B}^0_d$, have guided the form and content of the standard model, and permitted useful estimates of the masses of the charm and top quark masses prior to direct observation of those quarks. In this paper, we present the results of a search for the evolution of the D^0 into the $\overline{D}^0$, where the principal motivation is to glimpse new physics outside the standard model prior to direct observation of that physics at the high-energy frontier.

A D^0 can evolve into a $\overline{D}^0$ through on-shell intermediate states, such as K^+K^- with $m_{K^+K^-} = m_{D^0}$, or through off-shell intermediate states, such those that might be present due to new physics. We denote the amplitude through the former (latter) states by $-iy$ (x), in units of $\Gamma_{D^0}/2$ [1], and we neglect all types of CP violation.

For comparison, in the $K^0 - \overline{K}^0$ system the analogous y and $|x|$ are both near unity [2, 3]. The prediction that an initial K^0 will decay with two lifetimes, as described by the nonzero y, was famously made by Gell-Mann and Pais [4]. In the $B^0_d - \overline{B}^0_d$ system theory firmly predicts that y is negligible, and experiments have not yet sought y out. For the $D^0 - \overline{D}^0$ system, y is likely to receive significant contributions from the standard model [5]. It may even be that y dominates the $D^0 \rightarrow \overline{D}^0$ amplitude.

It has been x, for both the $K^0 - \overline{K}^0$ and $B^0_d - \overline{B}^0_d$ systems, that has provided information about the charm and top quark masses. A report of the first measurement in 1961 of $|x|$ for the $K^0 - \overline{K}^0$ system noted, "We cannot compare our experimental value for $|x|$ with any theoretical calculation" [6]. By 1974, theory had caught up and exploited $|x|$ for the $K^0 - \overline{K}^0$ system to predict the charm quark mass [7], just before that quark was discovered. The tiny, CP-violating $\mathrm{Im}(x)$ for the K^0 is sensitive to the value of the top quark mass. The large value for the B^0_d of $|x|$ (now 0.73 ± 0.03 [2]) indicated that the top quark is very massive [8], in distinction to contemporaneous data from the high-energy frontier [9].

Many predictions for x in the $D^0 \to \overline{D}^0$ amplitude have been made [10]. The standard model contributions are suppressed down to at least $|x| \approx \tan^2 \theta_C \approx 0.05$ because D^0 decay is Cabibbo-favored; the GIM [11] cancellation could suppress $|x|$ down to $10^{-6} - 0.01$. Many nonstandard models, particularly those that address patterns of quark flavor, predict $|x| \approx 0.01$ or greater. Contributions to x at this level can result from the presence of new particles with masses as high as 100 TeV [12]. Decisive signatures of such particles might include $|y| \ll |x|$, or CP-violating interference between a substantial imaginary component of x and either y, or a direct decay amplitude. In order to accurately assess the origin of a $D^0 - \overline{D}^0$ mixing signal, the effects described by y must be distinguished from those that are described by x.

We report here on a study of the process $D^0 \to K^+\pi^-$. We use the charge of the "slow" pion, π^+_s, from the decay $D^{*+} \to D^0\pi^+_s$ to deduce production of the D^0, and then we seek the rare "wrong-sign" $K^+\pi^-$ final state (WS), in addition to the more frequent "right-sign" final state, $K^-\pi^+$ (RS). The wrong-sign process, $D^0 \to K^+\pi^-$, can proceed either through direct doubly-Cabibbo-suppressed decay (DCSD), or through mixing followed by the Cabibbo-favored decay (CFD), $D^0 \to \overline{D}^0 \to K^+\pi^-$. Both processes contribute to the time integrated wrong-sign rate, R_{ws}:

$$R_{\mathrm{ws}} = \frac{\Gamma(D^0 \to K^+\pi^-)}{\Gamma(\overline{D}^0 \to K^+\pi^-)}.$$

To disentangle the two processes that could contribute to $D^0 \to K^+\pi^-$, we study the distribution of wrong-sign final states as a function of the proper decay time, t, of the D^0. The proper decay time is in units of the mean D^0 lifetime, $\tau_{D^0} = 415 \pm 4\,\mathrm{fs}$ [2]. The differential wrong-sign rate, relative to $\Gamma(\overline{D}^0 \to K^+\pi^-)$, is [13, 14]

$$r_{\mathrm{ws}}(t) \equiv [R_D + \sqrt{R_D}y't + \frac{1}{4}(x'^2 + y'^2)t^2]e^{-t}, \tag{47.1}$$

where the modified mixing amplitudes x' and y' are given by

$$\begin{aligned} y' &\equiv y\cos\delta - x\sin\delta, \\ x' &\equiv x\cos\delta + y\sin\delta, \end{aligned}$$

and δ is a possible strong phase between DCSD and CFD amplitudes:

$$-\sqrt{R_D}e^{-i\delta} \equiv \frac{\langle K^+\pi^-|T|D^0\rangle}{\langle K^+\pi^-|T|\overline{D}^0\rangle}.$$

The symbol R_D represents the DCSD rate, relative to the CFD rate. There are plausible theoretical arguments that δ is small [15,16]. If both D^0 and $\overline{D}^0$ exclusively populate the $I = 1/2$ amplitude of $K^+\pi^-$, then δ would be zero. For the CFD, $\overline{D}^0 \to K^+\pi^-$, the $I = 3/2$ amplitude is indeed disfavored, with $|A_{3/2}/A_{1/2}| = 0.27 \pm 0.02$ [17]. A nonzero δ could develop if $D^0 \to K^+\pi^-$ populates the $I = 3/2$ amplitude differently than the CFD does. Crudely, one might guess $|\delta| <\sim |A_{3/2}/A_{1/2}| \sim 15°$. The size of δ could be settled by measurements of the DCSD contributions to $D^+ \to K^+\pi^0$, $D^+ \to K_L\pi^+$, and $D^0 \to K_L\pi^0$, which are now feasible with our data set.

For decays to wrong-sign final states other than $K^+\pi^-$, such as $K^+\pi^-\pi^0$ or $K^+\pi^-\pi^+\pi^-$, there will be distinct strong phases. Moreover, the broad resonances that mediate those multibody hadronic decays, such as the K^*, ρ, etc., modulate those phases as a function of position on the Dalitz plot. Thus, multibody hadronic wrong-sign decays might afford an opportunity to distinguish x and y from x' and y'.

An important aspect of eq. (47.1) for the two-body decay $D^0 \to K^+\pi^-$ is that the dependence on y' and x' is distinguishable due to the interference with the direct decay amplitude, which induces the term linear in t. Such behavior is complementary to the differential decay rate to CP eigenstates such as $D^0 \to K^+K^-$, which is sensitive to y alone, or that of $D^0 \to K^+\ell^-\overline{\nu}_\ell$, which is sensitive to $r_{\rm mix} \equiv (x^2+y^2)/2 = (x'^2+y'^2)/2$ alone.

Our data were accumulated from January 1996 to February 1999 from an integrated luminosity of e^+e^- collisions at $\sqrt{s} \approx 10\,\text{GeV}$ consisting of 9.0 fb^{-1}, provided by the Cornell Electron Storage Rings (CESR). The data were taken with the CLEO-II multipurpose detector [18], upgraded in 1995 when a silicon vertex detector (SVX) was installed [19] and the drift chamber gas was changed from argon-ethane to helium-propane. The upgraded configuration is named CLEO-II.V, where the "V" is short for Vertex.

We reconstruct candidates for the decay sequences $D^{*+} \to \pi_s^+ D^0$, followed by either $D^0 \to K^+\pi^-$ (wrong-sign) or $D^0 \to K^-\pi^+$ (right-sign). The sign of the slow charged pion tags the charm state at $t = 0$ as either D^0 or $\overline{D}^0$. The broad features of the reconstruction are similar to those employed in the recent CLEO-II.V measurement of the D meson lifetimes [20]. The following points are important in understanding how we have improved our sensitivity to $D^0 \to K^+\pi^-$, relative to earlier CLEO work [21,22]:

1. The SVX allows substantially more precise measurement of charged particle trajectories in the dimension parallel to the colliding beam axis. When combined with improvements in track-fitting, the CLEO-II.V resolution for $Q = M_{\pi_s K\pi} - M$, where M is the mass of the

$K^+\pi^-$ system, is $\sigma_Q = 190 \pm 6\,\text{KeV}$, compared with the earlier value of $\sigma_Q = 750\,\text{KeV}$ [21].

2. The use of helium-propane, in addition to improvements in track-fitting, have reduced the CLEO-II.V resolution for M to $\sigma_M = 6.4 \pm 0.1\,\text{MeV}$, compared with the earlier value of $\sigma_M = 11\,\text{MeV}$ [21].

3. Improved mass resolution, as well as rejections based on the momentum asymmetry of the two charged tracks, allow clean separation of the signal $D^0 \to K^+\pi^-$ from $D^0 \to K^+K^-$, $D^0 \to \pi^+\pi^-$, and $D^0 \to K^-\pi^+$, and from multibody decays of the D^0, at a cost of about 35% of the signal acceptance. We use the modest π^+/K^+ separation provided by measurement of dE/dx in the CLEO-II.V drift chamber primarily for systematic studies. Addition of a new device with perfect π^+/K^+ identification and acceptance might enable the recovery of the 35% acceptance loss.

Multiple scattering on the field wires, which constitute 70% of the material, as measured in radiation lengths, in the CLEO-II.V drift chamber, appears to dominate the current σ_Q and σ_M. Should the track-fitting be altered to treat scattering on the field wires as discrete and localized, both σ_Q and σ_M might improve by as much as a factor of 2.

Our signal for the wrong-sign process $D^0 \to K^+\pi^-$ is shown in fig. 47.1. We determine the background levels by performing a fit to the plane of $0 < Q < 20\,\text{MeV}$ versus $1.76 < M < 1.97\,\text{GeV}$, which has an area about 150 times larger than our signal region. Event samples generated by the Monte Carlo method and fully simulated in our detector, corresponding to $90\,\text{fb}^{-1}$ of integrated luminosity, are used to estimate the background shapes in the Q–M plane. The shapes are allowed to float in a fit to the data; the results of the fit are displayed in fig. 47.1. The excess of events in the signal region is prominent.

We describe the signal shape with the right-sign data that is within $7\,\sigma$ of the nominal CFD value in the Q–M plane. The results of the fit to the wrong-sign data are summarized in table 47.1.

No acceptance corrections are needed to compute $R_{\text{ws}} = (0.34 \pm 0.07)\%$ from table 47.1. The dominant systematic errors all stem from the potentially inaccurate modeling of the initial and acceptance-corrected shapes of the background contributions in the Q-M plane. We assess these systematic errors by substantial variation of the fit regions, dE/dx criteria, and kinematic criteria; the total systematic error we assess is 0.06%.

Our complete result for R_{ws} is summarized in table 47.2. As we describe later, our data are consistent with an absence of $D^0 - \overline{D}{}^0$ mixing, so our best estimate for the relative DCSD rate, R_D, is that it equals R_{ws}.

There are two directly comparable measurements of R_{ws}: one is from CLEO-II [21], $R_{\text{ws}} = (0.77 \pm 0.25 \pm 0.25)\%$, which used a data set independent of that used here; the second is from ALEPH [23], $R_{\text{ws}} =$

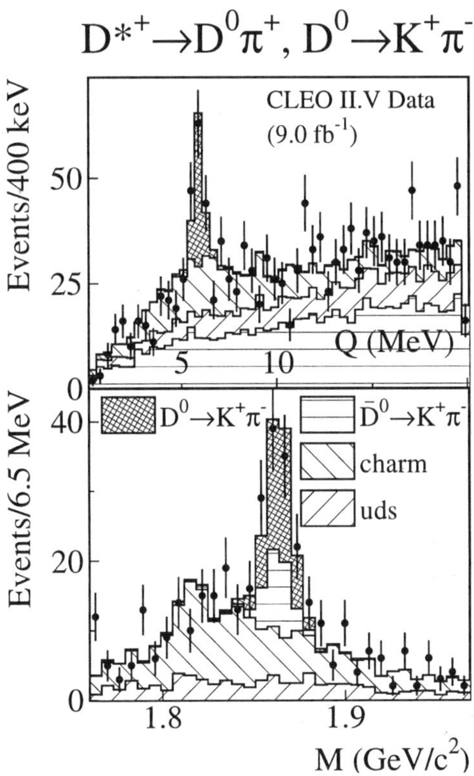

Figure 47.1: Signal for the wrong-sign process $D^0 \to K^+\pi^-$. For the top plot, M is within 14 MeV of the nominal CFD value, and for the bottom plot, Q is within 500 KeV of the nominal CFD value. The data are the full circles with error bars, the fit to the signal is cross-hatched, and the fits to the backgrounds are singly hatched. The results of the fit are summarized in table 47.1.

$(1.84 \pm 0.59 \pm 0.34)\%$; comparison of our result and these are marginally consistent with $\chi^2 = 6.0$ for 2 DoF, for a CL of 5.0%.

We have split our sample into candidates for $D^0 \to K^+\pi^-$ and $\overline{D}^0 \to K^-\pi^+$. There is no evidence for a CP-violating time-integrated asymmetry. From table 47.1, it is straightforward to estimate the 1σ statistical error on the CP-violating time-integrated asymmetry as $\sqrt{107}/54.8 \approx 20\%$.

Given the absence of a significant time-integrated CP asymmetry, we undertake a study of the decay time dependence wrong-sign rate based upon eq. (47.1). We reconstruct the proper D^0 decay time primarily from the vertical displacement of the $K^+\pi^-$ vertex from the e^+e^- collision "ribbon," which is infinitesimal in its vertical extent. We require a well-reconstructed vertex in 3-dimensions, which causes a loss of about 12% of the candidates described in table 47.1. Our resolution, in units of the mean D^0 life, is about 1/2. Study of the plentiful right-sign sample

Component	# Events
$D^0 \to K^+\pi^-$ (WS Signal)	54.8 ± 10.8
random $\pi^\pm + D^0/\overline{D}^0$	24.3 ± 1.8
$c\overline{c}$	12.3 ± 0.8
uds	8.6 ± 0.4
$D^0 \to$ Pseudoscalar-Vector	7.0 ± 0.4
$\overline{D}^0 \to K^+\pi^-$ (RS Normalization)	16126 ± 126

Table 47.1: Event yields in a signal region of $2.4\,\sigma$ centered on the nominal Q and M values. The total number of candidates is 107. The bottom row describes the normalization sample.

Quantity	Result
R_{ws}	$(0.34 \pm 0.07 \pm 0.06)\%$
$R_{\text{ws}}/\tan^4\theta_C$	$(1.28 \pm 0.25 \pm 0.21)$
$\mathcal{B}(D^0 \to K^+\pi^-)$	$(1.31 \pm 0.26 \pm 0.22 \pm 0.03) \times 10^{-4}$

Table 47.2: Result for R_{ws}. For the branching ratio $\mathcal{B}(D^0 \to K^+\pi^-)$ we take the absolute branching ratio $\mathcal{B}(\overline{D}^0 \to K^+\pi^-) = (3.85 \pm 0.09)\%$. The third error results from the uncertainty in this absolute branching ratio. As discussed in the text, our best estimate is that $R_D = R_{\text{ws}}$.

allows us to fix our detailed resolution function and shows that biases in the reconstruction of the proper decay time contribute negligibly to the wrong-sign results.

The distribution of proper decay times, t, for wrong-sign candidates that are within $2.4\,\sigma$ of the nominal CFD value in the $Q-M$ plane is shown in fig. 47.2. Maximum-likelihood fits are made to those data. The backgrounds are described by levels and shapes deduced from the fit to the Q–M plane, and from study of the simulated data sample. The wrong-sign signal is described by eq. (47.1), folded with our resolution function.

For the benchmark fit to the wrong-sign data, x' and y' are constrained to be zero. This fit has a confidence level of 84%, indicating a good fit.

The mixing amplitudes x' and y' are then allowed to freely vary, and the best fit values are shown in fig. 47.2 and in table 47.3. The fit improves slightly when x' and y' are allowed to float to the values that maximize the likelihood. However, the small value of the likelihood change $\sqrt{-2\Delta\ln\mathcal{L}} = 1.8\,\sigma$ (including systematic errors) does not permit us to eliminate the possibility that the improvement is due to a statistical fluctuation.

Therefore, our principal results concerning mixing are the one-dimensional intervals, which correspond to a 95% confidence level, that are given in the second column of table 47.3.

Additionally, we evaluate a contour in the two-dimensional plane of y' versus x' which, at 95% confidence level, contains the true value of x' and

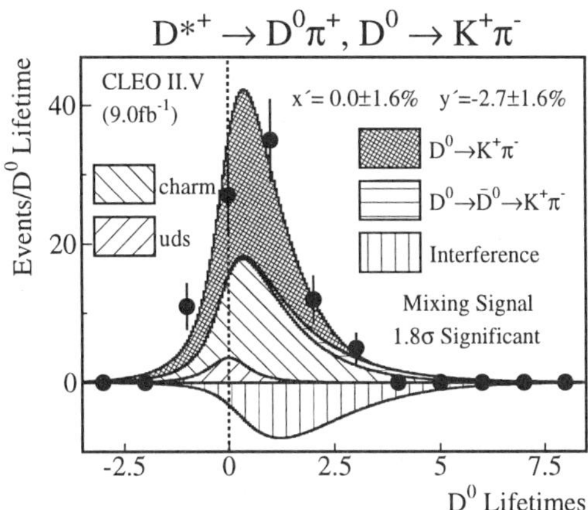

Figure 47.2: Distribution in t for the $D^0 \to K^+\pi^-$ candidates. The data are shown as the full circles with error bars. All other information comes from the fit to the data. The smooth curves show the various specific contributions as labeled. The cross-hatched region is the net contribution from $D^0 \to K^+\pi^-$, after incorporation of the (destructive) interference and mixing.

y'. To do so, we determine the contour around our best fit values where the $-\ln \mathcal{L}$ has increased by 3.0 units. All other fit variables, including the DCSD rate and background contributions, are allowed to float to distinct, best fit values at each point on the contour. The interior of the contour is shown, as the small, dark, cross-hatched region near the origin of fig. 47.3. On the axes of x' and y', this contour falls slightly outside the one-dimensional intervals listed in table 47.3, as expected.

Parameter	Best Fit	95% CL
R_D	$(0.50^{+0.11}_{-0.12} \pm 0.08)\%$	$0.22\% < R_D < 0.77\%$
y'	$(-2.7^{+1.5}_{-1.6} \pm 0.2)\%$	$-5.9\% < y' < 0.3\%$
x'	$(0 \pm 1.6 \pm 0.2)\%$	$\lvert x'\rvert < 3.2\%$
$(1/2)x'^2$		$< 0.05\%$

Table 47.3: Results of the fit, where x' and y' are free to float, to the distribution of $D^0 \to K^+\pi^-$ candidates in t. The data and the fit components are shown in fig. 47.2.

We have evaluated the allowed regions of other experiments [24–27] at 95% CL, and shown those regions in fig. 47.3.

All results described here are preliminary.

If we assume that δ is small, which is plausible [15, 16], then $x' \approx x$ and we can indicate the impact of our work in limiting predictions of $D^0 - \overline{D}^0$ mixing from extensions to the standard model. Eighteen of the predictions recently tabulated [10] have some inconsistency with our limit. Among

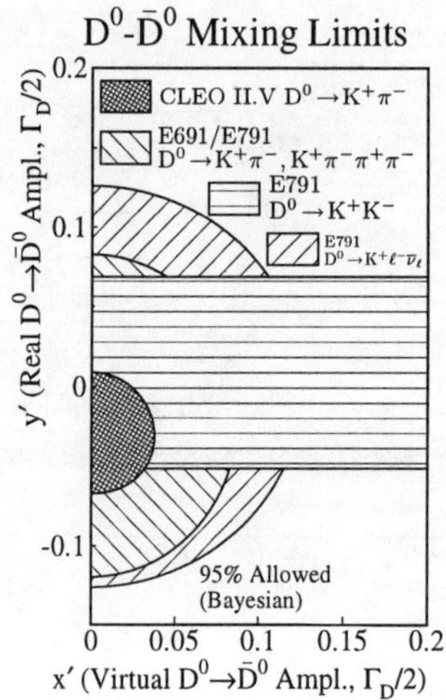

Figure 47.3: Limits in the y' vs. x' plane. Our experiment limits, at 95% C.L., the true values of x' and y' to occupy the cross-hatched region near the origin. Also shown are the similar zones from other recent experiments. We assume $\delta = 0$ to place the recent work of E791 that utilized $D^0 \to K^+K^-$; a non-zero δ would rotate the E791 confidence region clockwise about the origin by an angle of δ.

those predictions, some authors have made common assumptions, however.

Because our data is consistent with an absence of $D^0 - \overline{D}^0$ mixing, our best information on the DCSD rate, R_D, is that it equals R_{ws}, as summarized in table 47.2.

We will soon complete studies where we allow various types of CP violation to modify eq. (47.1). Also, the D^0 decay modes K^+K^-, $\pi^+\pi^-$, $K_S\phi$, $K_S\pi^+\pi^-$, $K^+\pi^-\pi^0$, $K^+\pi^-\pi^+\pi^-$, and $K^+\ell^-\overline{\nu}_l$ are under active investigation, using the CLEO-II.V data. The $(K_S\pi^+)$ resonance bands in $D^0 \to K_S\pi^+\pi^-$ permit investigation of $D^0 \to \overline{D}^0$, with sensitivity heightened by coherent interference that is modulated by resonant phases.

The entire CLEO-II.V data set, suitably exploited, could be used to observe $D^0 - \overline{D}^0$ mixing if either $|x|$ or $|y|$ exceed approximately 1%.

Acknowledgments

I gratefully acknowledge the efforts of the CLEO Collaboration, the CESR staff, and the staff members of CLEO institutions. This work is David Asner's Ph.D. dissertation topic, and he made the principal contributions to this study, and Tony Hill has made important contributions. Rolly Morrison and Mike Witherell provided intellectual guidance.

I thank the hosts of Kaon 99, particularly Y. Wah and B. Winstein, for running a hospitable and stimulating conference. I thank J. Rosner for help with this manuscript.

This work was supported by the Department of Energy under contract DE-AC03-76SF00098.

References

[1] T.D. Lee, R. Oehme, and C.N. Yang, Phys. Rev. **106**, 340 (1957); A. Pais and S.B. Treiman, Phys. Rev. D **12**, 2744 (1975).

[2] C. Caso, *et al.* European Physical Journal C **3**, 1 (1998).

[3] Specifically, from the previous reference, for the K^0 system $x = -0.947 \pm 0.003$, and $y = 0.99655 \pm 0.00003$. In our convention $|y|$ is bounded by 1, and signed by the CP eigenvalue of the dominant real intermediate state(s), which is 2π (to which we assign CP=+1) for $K^0 \to \overline{K}^0$. For $D^0 \to \overline{D}^0$ there are real intermediate states that are not CP eigenstates, such as $K\pi$, but the convention is straightforward to generalize. Then, we define x as proportional to the mass difference between the eigenstate that decays preferentially to CP=+1 final states and the other eigenstate, that decays preferentially to CP=−1 states. The study reported here is sensitive (when $\delta = 0$) to the magnitude $|x|$ for the $D^0 \to \overline{D}^0$ amplitude and to the magnitude, and sign, of y for the $D^0 \to \overline{D}^0$ amplitude.

[4] M. Gell-Mann and A. Pais, Phys. Rev. **97**, 1387 (1955).

[5] E. Golowich and A.A. Petrov, Phys. Lett. B **427**, 172 (1998).

[6] R.H. Good *et al.*, Phys. Rev. **124**, 1223 (1961).

[7] M.K. Gaillard, B.W. Lee, and J. Rosner, Rev. Mod. Phys. **47**, 277 (1975).

[8] J.L. Rosner, in proceedings of *Hadron 87 (2nd Int. Conf. on Hadron Spectroscopy)* Tsukuba, Japan, 16–18 April 1987, ed. Y. Oyanagi, K. Takamatsu, and T. Tsuru (KEK, 1987), p. 395.

[9] G. Arnison *et al.*, Phys. Lett. B **147**, 493 (1984).

[10] H.N. Nelson, hep-ex/9908021, submitted to the 1999 Lepton-Photon Symposium (unpublished).

[11] S.L. Glashow, J. Iliopolous, and L. Maiani, Phys. Rev. D **2**, 1285 (1970); R.L. Kingsley, S.B. Treiman, F. Wilczek, and A. Zee, Phys. Rev. D **11**, 1919 (1975).

[12] M. Leurer, Y. Nir, and N. Seiberg, Nucl. Phys. B **420**, 468 (1994).

[13] S.B. Treiman and R.G. Sachs, Phys. Rev. **103**, 1545 (1956).

[14] G. Blaylock, A. Seiden, and Y. Nir, Phys. Lett. B **355**, 555 (1995).

[15] L. Wolfenstein, Phys. Rev. Lett. **75**, 2460 (1995).

[16] T.E. Browder and S. Pakvasa, Phys. Lett. B **383**, 475 (1996).

[17] M.S. Witherell, in *Proceedings of the International Symposium on Lepton Photon Interactions at High Energies*, Ithaca, 1993, ed. P.S. Drell and D.L. Rubin (Amer. Inst. Physics, 1994), p. 198.

[18] Y. Kubota, Nucl. Instrum. Meth. A **320**, 66 (1992).

[19] T.S. Hill, Nucl. Instrum. Meth. A **418**, 32 (1998).

[20] G. Bonvicini *et al.*, Phys. Rev. Lett. **82**, 4586 (1999).

[21] D. Cinabro *et al.*, Phys. Rev. Lett. **72**, 1406 (1994).

[22] More detailed descriptions of the $D^0 \to K^+\pi^-$ study can be found in M. Artuso *et al.*, hep-ex/9908040, submitted to the 1999 Lepton-Photon Symposium (unpublished).

[23] R. Barate *et al.*, Phys. Lett. B **436**, 211 (1998).

[24] E.M. Aitala *et al.*, Phys. Rev. D **57**, 13 (1998).

[25] J.C. Anjos *et al.*, Phys. Rev. Lett. **60**, 1239 (1988).

[26] E.M. Aitala *et al.*, Phys. Rev. Lett. **77**, 2384 (1996).

[27] E.M. Aitala *et al.*, Phys. Rev. Lett. **83**, 32 (1999).

Part VIII

The Physics of B Mesons

In the standard model of electroweak interactions, it is the b quark and its partner the t quark that are responsible, through the Kobayashi-Maskawa theory, for the observed CP violation in the kaon system. Thus, any discussion of kaon physics should necessarily contain a treatment of the physics of particles containing the b quark. This physics turns out to be very rich, and is being addressed by a number of experiments worldwide.

B mesons are much closer in their properties to K mesons than are either to D mesons. Neutral B mesons mix readily with their antiparticles, for example.

CP-violating effects are expected to be quite remarkable in the decays of B mesons. Whereas CP violation in kaon decays is characterized by the small parameter ϵ (and now the even smaller parameter ϵ'), the anticipated CP-violating asymmetries in certain B decays can reach tens of percent in such processes as $B^0 \to J/\psi K_S$.

The present part is devoted to the physics of B mesons with particular emphasis on its relation to kaon decays. Gronau surveys the prospects for CP-violation studies, while Ligeti reviews the accuracy with which one can determine crucial elements of the Cabibbo-Kobayashi-Maskawa (CKM) matrix. Experimental surveys of present and anticipated results are then given by representatives of the CLEO Collaboration at Cornell (Briere), the CDF Collaboration at Fermilab (Kroll), and new "B factories" at KEK in Japan (Kinoshita) and SLAC (Varnes).

48

CP Violation and B Physics

Michael Gronau

Abstract

This is a quick review of CP nonconservation in B physics. Several methods are described for testing the Kobayashi-Maskawa single-phase origin of CP violation in B decays, pointing out some limitations due to hadronic uncertainties. A few characteristic signatures of new physics in B decay asymmetries are listed.

48.1 The CKM Matrix

In the standard model of electroweak interactions CP violation is due to a nonzero complex phase [1] in the Cabibbo-Kobayashi-Maskawa (CKM) matrix V, describing the weak couplings of the charged gauge boson to quarks. The unitary matrix V, given by three mixing angles $\theta_{ij}(i < j = 1,2,3)$ and a phase γ, can be approximated by $(s_{ij} \equiv \sin\theta_{ij})$ [2,3]

$$V \approx \begin{pmatrix} 1 - \frac{1}{2}s_{12}^2 & s_{12} & s_{13}e^{-i\gamma} \\ -s_{12} & 1 - \frac{1}{2}s_{12}^2 & s_{23} \\ s_{12}s_{23} - s_{13}e^{i\gamma} & -s_{23} & 1 \end{pmatrix}. \tag{48.1}$$

Within this approximation, the only complex elements are V_{ub}, with phase $-\gamma$ and V_{td}, the phase of which is denoted $-\beta$.

The measured values of the three mixing angles and phase are [3]

$$s_{12} = 0.220 \pm 0.002, \quad s_{23} = 0.040 \pm 0.003, \quad s_{13} = 0.003 \pm 0.001, \tag{48.2}$$

$$35^0 \leq \gamma = \mathrm{Arg}(V_{ub}^*) \leq 145^0. \tag{48.3}$$

First evidence for a nonzero phase γ came 35 years ago with the measurement of ϵ, parameterizing CP violation in $K^0 - \bar{K}^0$ mixing. The second evidence was obtained recently through the measurement of $\mathrm{Re}(\epsilon'/\epsilon)$ [4,5] discussed extensively at this meeting.

Unitarity of V implies a set of 6 triangle relations. The db triangle,

$$V_{ud}V_{ub}^* + V_{cd}V_{cb}^* + V_{td}V_{tb}^* = 0, \tag{48.4}$$

is unique in having three comparable sides, which were measured in $b \to u\ell\nu$, $b \to c\ell\nu$, and $\Delta M_{d,s}$, respectively. Whereas V_{cb} was measured quite precisely, V_{ub} and V_{td} are rather poorly known at present. The three large angles of the triangle lie in the ranges $35° \leq \alpha \leq 120°$, $10° \leq \beta \leq 35°$, and eq. (48.3). As we will show in the next sections, certain B decay asymmetries can constrain these angles considerably beyond present limits.

For comparison with K physics, note that because of the extremely small t quark side of the ds unitarity triangle

$$V_{ud}V_{us}^* + V_{cd}V_{cs}^* + V_{td}V_{ts}^* = 0, \tag{48.5}$$

this triangle has an angle of order 10^{-3}, which accounts for the smallness of CP violation in K decays. The area of this triangle, which is equal to the area of the db triangle [6], can be determined by fixing its tiny height through the rate of $K_L \to \pi^0\nu\bar{\nu}$. This demonstrates the complementarity of K and B physics in verifying or falsifying the assumption that CP violation originates solely in the single phase of the CKM matrix.

As we will show, the advantage of B decays in testing the KM hypothesis is the large variety of decay modes. This permits a detailed study of the phase structure of the CKM matrix through various interference phenomena that can measure the two phases γ and β. New physics can affect this interference in several ways to be discussed below.

48.2 CP Violation in $B^0 - \bar{B}^0$ Mixing

The wrong-sign lepton asymmetry

$$A_{sl} \equiv \frac{\Gamma(\bar{B}^0 \to X\ell^+\nu) - \Gamma(B^0 \to X\ell^-\bar{\nu})}{\Gamma(\bar{B}^0 \to X\ell^+\nu) + \Gamma(B^0 \to X\ell^-\bar{\nu})}, \tag{48.6}$$

measures CP violation in $B^0 - \bar{B}^0$ mixing. Top-quark dominance of $B^0 - \bar{B}^0$ mixing implies that this asymmetry is of order 10^{-3} or smaller [7]:

$$\begin{aligned} A_{sl} = 4\mathrm{Re}\epsilon_B = \mathrm{Im}\left(\frac{\Gamma_{12}}{M_{12}}\right) &= \frac{|\Gamma_{12}|}{|M_{12}|}\mathrm{Arg}\left(\frac{\Gamma_{12}}{M_{12}}\right) \\ &\simeq \left(\frac{m_b^2}{m_t^2}\right)\left(\frac{m_c^2}{m_b^2}\right) \leq \mathcal{O}(10^{-3}). \end{aligned} \tag{48.7}$$

Present limits are at the level of 5% [8].

Writing the neutral B mass eigenstates as

$$|B_L> = p|B^0> \; + \; q|\bar{B}^0>, \quad |B_H> = p|B^0> \; - \; q|\bar{B}^0>, \tag{48.8}$$

one has $2\mathrm{Re}\epsilon_{\mathrm{B}} \approx 1 - |q/p| \leq \mathcal{O}(10^{-3})$. Thus, to a very high accuracy, the mixing amplitude is a pure phase

$$\frac{q}{p} = e^{2i\mathrm{Arg}(V_{td})} = e^{-2i\beta}. \tag{48.9}$$

48.3 The Asymmetry in $B^0(t) \to \psi K_S$

When an initially produced B^0 state oscillates in time via the mixing amplitude that carries a phase $e^{-2i\beta}$,

$$|B^0(t)> \;=\; |B^0> \cos(\Delta m t/2) + |\bar{B}^0> ie^{-2i\beta}\sin(\Delta m t/2), \qquad (48.10)$$

the B^0 and $\bar{B}^0$ components decay with equal amplitudes to ψK_S. The interference creates a time-dependent CP asymmetry between this process and the corresponding process starting with a $\bar{B}^0$ [9]

$$A(t) = \frac{\Gamma(B^0(t)\to\psi K_S) - \Gamma(\bar{B}^0(t)\to\psi K_S)}{\Gamma(B^0(t)\to\psi K_S) + \Gamma(\bar{B}^0(t)\to\psi K_S)} = -\sin(2\beta)\sin(\Delta m t). \qquad (48.11)$$

The simplicity of this result, relating a measured asymmetry to an angle of the unitarity triangle, follows from having a single weak phase in the decay amplitude that is dominated by $b \to c\bar{c}s$. This single-phase approximation holds to better than 1% [10] and provides a clean measurement of β.

A recent measurement by the CDF Collaboration at the Tevatron [11], $\sin(2\beta) = 0.79 \pm 0.39 \pm 0.16$, has not yet produced a significant nonzero result. It is already encouraging, however, to note that this result prefers positive values and is not in conflict with present limits, $0.4 \leq \sin 2\beta \leq 0.8$.

48.4 Penguin Pollution in $B^0 \to \pi^+\pi^-$

By applying the above argument to $B^0 \to \pi^+\pi^-$, in which the decay amplitude has the phase γ, one would expect the asymmetry in this process to measure $\sin 2(\beta+\gamma) = -\sin(2\alpha)$. However, this process involves a second amplitude due to penguin operators that carry a different weak phase than the dominant current-current (tree) amplitude [10, 12]. This leads to a more general form of the time-dependent asymmetry, which includes a new term due to direct CP violation in the decay [10]

$$A(t) = a_{\rm dir}\cos(\Delta m t) + \sqrt{1 - a^2_{\rm dir}}\,\sin 2(\alpha+\theta)\sin(\Delta m t). \qquad (48.12)$$

Both $a_{\rm dir}$ and θ, the correction to α in the second term, are given roughly by the ratio of penguin to tree amplitudes, $a_{\rm dir} \sim 2(\text{penguin/tree})\sin\delta$, $\theta \sim (\text{penguin/tree})\cos\delta$, where δ is an unknown strong phase. A crude estimate of the penguin-to-tree ratio, based on CKM and QCD factors, is 0.1. Recently, flavor SU(3) was applied [13] to relate $B \to \pi\pi$ to $B \to K\pi$ data, finding this ratio to be in the range 0.3 ± 0.1. Precise knowledge of this ratio could provide very useful information about α [10, 14].

One way of eliminating the penguin effect is by measuring also the time-integrated rates of $B^0 \to \pi^0\pi^0$, $B^+ \to \pi^+\pi^0$ and their charge conjugates [15]. The three $B \to \pi\pi$ amplitudes obey an isospin triangle relation,

$$A(B^0 \to \pi^+\pi^-)/\sqrt{2} + A(B^0 \to \pi^0\pi^0) = A(B^+ \to \pi^+\pi^0). \qquad (48.13)$$

A similar relation holds for the charge-conjugate processes. One uses the different isospin properties of the penguin ($\Delta I = 1/2$) and tree ($\Delta I = 1/2, 3/2$) contributions and the well-defined weak phase (γ) of the tree amplitude. This enables one to determine the correction to $\sin 2\alpha$ in the second term of eq.(48.12) by constructing the two isospin triangles.

Electroweak penguin contributions could spoil this method [16] since they involve $\Delta I = 3/2$ components. This implies that the amplitudes of $B^+ \to \pi^+\pi^0$ and its charge conjugate differ in phase, which introduces a correction at the level of a few percent in the isospin analysis. It was shown recently [17] that this small correction can be taken into account analytically in the isospin analysis, since the dominant electroweak contributions are related by isospin to the tree amplitude. Other very small corrections can come from isospin breaking in strong interactions [18].

The major difficulty of measuring α without knowing the ratio penguin/tree is experimental rather than theoretical. The first signal for $B^0 \to \pi^+\pi^-$ reported recently [19, 20], $\mathrm{BR}(B^0 \to \pi^+\pi^-) = [0.47^{+0.18}_{-0.15} \pm 0.06) \times 10^{-5}$, is somewhat weaker than expected. Worse than that, the branching ratio into two neutral pions is expected to be at most an order of magnitude smaller. This estimate is based on color suppression, a feature already observed in CKM-favored $B \to \bar{D}\pi$ decays. Here it was found [2] that $\mathrm{BR}(B^0 \to \bar{D}^0\pi^0)/\mathrm{BR}(B^0 \to D^-\pi^+) < 0.04$. If the same color suppression holds in $B \to \pi\pi$, then $\mathrm{BR}(B^0 \to \pi^0\pi^0) < 3 \times 10^{-7}$, which would be too small to be measured with a useful precision. Constructive interference between a color-suppressed current-current amplitude and a penguin amplitude can increase the $\pi^0\pi^0$ rate somewhat. Limits on this rather rare mode can be used to bound the uncertainty in determining $\sin(2\alpha)$ from $B^0 \to \pi^+\pi^-$ [21]

$$\sin(\delta\alpha) \leq \sqrt{\frac{\mathcal{B}(B \to \pi^0\pi^0)}{\mathcal{B}(B^\pm \to \pi^\pm\pi^0)}}. \tag{48.14}$$

Other ways of treating the penguin problem were discussed in [22].

48.5 *B* Decays to Three Pions

The angle α can also be studied in the processes $B \to \pi\rho$ [23], which have already been seen with branching ratios larger than those of $B \to \pi\pi$ [24], $\mathrm{BR}(B^0 \to \pi^\pm\rho^\mp) = (3.5^{+1.1}_{-1.0} \pm 0.5) \times 10^{-5}$, $\mathrm{BR}(B^\pm \to \pi^\pm\rho^0) = (1.5 \pm 0.5 \pm 0.4) \times 10^{-5}$. An effective study of α, which can eliminate uncertainties due to penguin corrections, requires

- a separation between B^0 and $\bar{B}^0$ decays;
- time-dependent rate asymmetry measurements in $B \to \pi^\pm\rho^\mp$;
- measuring the rates of processes involving neutral pions, including the color-suppressed $B^0 \to \pi^0\rho^0$.

This will not be an easy task.

48.6 γ from $B \to K\pi$ and Other Processes

While discussing $B^\pm$ decays to three charged pions, we note that these decays are of high interest for a different reason [25]. When two of the pions form a mass around the charmonium $\chi_{c0}(3415)$ state, a very large CP asymmetry is expected between B^+ and B^- decays. In this case the direct decay amplitude into three pions ($b \to uu\bar{d}$) interferes with a comparable amplitude into $\chi_{c0}\pi^\pm$ ($b \to c\bar{c}d$) followed by $\chi_{c0} \to \pi^+\pi^-$. The large asymmetry (proportional to $\sin\gamma$), of order several tens of percent, follows from the 90° strong phase obtained when the two-pion invariant mass approaches the charmonium mass.

A method for determining the angle γ through $B^\pm \to DK^\pm$ decays [26], which in principle is completely free of hadronic uncertainties, faces severe experimental difficulties. It requires measuring separately decays to states involving D^0 and $\bar{D}^0$. Tagging the flavor of a neutral D by the charge of the decay lepton suffers from a very large background from B decay leptons, while tagging by hadronic modes involves interference with doubly Cabibbo-suppressed D decays. A few variants of this method were suggested [27]; however, due to low statistics, it seems unlikely that these variants can be performed effectively in near-future facilities.

Much attention was drawn recently to studies of γ in $B \to K\pi$, motivated by measurements of charge-averaged $B \to K\pi$ decay branching ratios [19, 20]

$$\begin{aligned}
\mathrm{BR}(B^\pm \to K\pi^\pm) &= (1.82^{+0.46}_{-0.40} \pm 0.16) \times 10^{-5}, \\
\mathrm{BR}(B^\pm \to K^\pm\pi^0) &= (1.21^{+0.30+0.21}_{-0.28-0.14}) \times 10^{-5}, \\
\mathrm{BR}(B^0 \to K^\pm\pi^\mp) &= (1.88^{+0.28}_{-0.26} \pm 0.13) \times 10^{-5}, \\
\mathrm{BR}(B^0 \to K^0\pi^0) &= (1.48^{+0.59+0.24}_{-0.51-0.33}) \times 10^{-5}.
\end{aligned} \tag{48.15}$$

The first suggestion to constrain γ from $B \to K\pi$ was made in [28], where electroweak penguin contributions were neglected. The importance of electroweak penguin terms was noted in [29], which was followed by several ideas about controlling these effects [30]. In the present discussion we will focus briefly on very recent work along these lines [17, 31–33], simplifying the discussion as much as possible.

Decomposing the $B^+ \to K\pi$ amplitudes into contributions from penguin (P), color-favored tree (T), and color-suppressed tree (C) terms [34],

$$A(B^+ \to K^0\pi^+) = P, \quad A(B^+ \to K^+\pi^0) = -(P + T + C)/\sqrt{2}, \tag{48.16}$$

P has a weak phase π, while T and C each carry the phase γ. Some information about the relative magnitudes of these terms can be gained by using SU(3) and comparing these amplitudes to those of $B \to \pi\pi$ [13]. This implies

$$r \equiv \frac{T + C}{P} = 0.24 \pm 0.06. \tag{48.17}$$

Defining the ratio of charge-averaged rates [31]

$$R_*^{-1} = \frac{2\mathcal{B}(B^\pm \to K^\pm\pi^0)}{\mathcal{B}(B^\pm \to K\pi^\pm)}, \tag{48.18}$$

one has

$$R_*^{-1} = 1 - 2r\cos\delta\cos\gamma + r^2, \tag{48.19}$$

where δ is the penguin-tree strong phase-difference. Any deviation of this ratio from one would be a clear signal of interference between $T + C$ and P in $B^+ \to K^+\pi^0$ and could be used to constrain γ.

So far, electroweak penguin contributions have been neglected. These terms can be included in the above ratio by relating them through flavor SU(3) to the corresponding tree amplitudes. This is possible since the two types of operators have the same (V-A)(V-A) structure and differ only by SU(3). Hence, in the SU(3) limit, the dominant electroweak penguin term and the tree amplitude have the same strong phase, and the ratio of their magnitudes is given simply by a ratio of the corresponding Wilson coefficients multiplied by CKM factors [17,31]

$$\delta_{EW} \equiv \frac{|\mathrm{EWP}(B^+ \to K^0\pi^+) + \sqrt{2}\mathrm{EWP}(B^+ \to K^+\pi^0)|}{|T + C|} \tag{48.20}$$

$$= -\frac{3}{2}\frac{c_9 + c_{10}}{c_1 + c_2}\frac{|V_{tb}^*V_{ts}|}{|V_{ub}^*V_{us}|} = 0.6 \pm 0.2, \tag{48.21}$$

where the error comes from $|V_{ub}|$. Consequently, one finds instead of (48.19)

$$R_*^{-1} = 1 - 2r\cos\delta(\cos\gamma - \delta_{EW}) + \mathcal{O}(r^2), \tag{48.22}$$

implying

$$|\cos\gamma - \delta_{EW}| \geq \frac{|1 - R_*^{-1}|}{2r}. \tag{48.23}$$

If $R_*^{-1} \neq 1$, this constraint can be used to exclude a region around $\gamma = 50°$. The present value of R_*^{-1} is consistent with one. Experimental errors must be substantially reduced before drawing any conclusions.

The above constraint is based only on charge-averaged rates. Further information on γ can be obtained by measuring separately B^+ and B^- decay rates. The $B^+ \to K\pi$ rates obey a triangle relation with $B^+ \to \pi^+\pi^0$ [17,28,31]

$$\sqrt{2}A(B^+ \to K^+\pi^0) + A(B^+ \to K^0\pi^+) = \tilde{r}_u A(B^+ \to \pi^+\pi^0)\left(1 - \delta_{EW}e^{-i\gamma}\right), \tag{48.24}$$

where $\tilde{r}_u = (f_K/f_\pi)\tan\theta_c \simeq 0.28$ contains explicit SU(3) breaking. This relation and its charge conjugate permit a determination of γ that does not rely on $R_*^{-1} \neq 1$.

This analysis involves uncertainties due to errors in r and δ_{EW}, which are expected to be reduced to the level of 10%. Additional uncertainties follow from SU(3) breaking in eq. (48.20) and from rescattering effects in $B^+ \to K^0\pi^+$, which introduce a term with phase γ in this process. The latter effects can be bounded by the U-spin related rate of $B^+ \to K^+\bar{K}^0$ [35]. Present limits on rescattering corrections are at a level of 20% and can be reduced to 10% in future high statistics experiments. Such rescattering corrections introduce an error of about 10° in determining γ [32]. Summing up all the theoretical uncertainties, and neglecting experimental errors, it is unlikely that this method will determine γ to better than 20°. Nevertheless, this would be a substantial improvement over the present bounds eq. (48.3).

We conclude this section with a simple observation [36] which enables an early detection of a CP asymmetry in $B \to K\pi$. Using $A(B^0 \to K^+\pi^-) = -P - T$, the hierarchy among amplitudes [34], $|P| \gg |T| \gg |C|$, implies $\text{Asym}(B^\pm \to K^\pm\pi^0) \approx \text{Asym}(B \to K^\pm\pi^\mp)$. This may be used to gain statistics by measuring the combined asymmetry in these two modes. The magnitude of the asymmetry depends on an unknown final state strong phase. Very recently a 90% confidence level upper limit was reported $\text{Asym}(B \to K^\pm\pi^\mp) < 0.35$ [19,37].

48.7 Signals of New Physics

The purpose of future B physics is to overconstrain the unitarity triangle. $|V_{ub}|$ can at best be determined to 10% [38] and $|V_{td}|$ relies on future measurements of the higher-order $B_s^0 - \bar{B}_s^0$ mixing [11] and $K^+ \to \pi^+\nu\bar{\nu}$ [39]. Constraining the angles α, β , and γ by CP asymmetries is complementary to these CP-conserving measurements. The asymmetry measurements involve discrete ambiguities in the angles, which ought to be resolved [40].

Hopefully, these studies will not only sharpen our knowledge of the CKM parameters but will eventually show some inconsistencies. In this case, the first purpose of B physics will be to identify the source of the inconsistencies in a model-independent way. Let us discuss this scenario briefly by considering a few general possibilities.

Physics beyond the standard model can modify CKM phenomenology and predictions for CP asymmetries by introducing additional contributions in three types of amplitudes:

- $B^0 - \bar{B}^0$ and $B_s^0 - \bar{B}_s^0$ mixing amplitudes;
- penguin decay amplitudes;
- tree decay amplitudes.

The first case is the most likely possibility, demonstrated by a large variety of models [41]. New mixing terms, which can be large and which often also affect the rates of electroweak penguin decays, modify in a universal way the interpretation of asymmetries in terms of phases of $B^0 - \bar{B}^0$ and $B_s^0 - \bar{B}_s^0$ mixing amplitudes. These contributions can be identified either by measuring asymmetries that lie outside the allowed range, or by comparison with mixing-unrelated constraints. On the other hand, new contributions in decay amplitudes [42] are usually small, may vary from one process to another, and can be detected be comparing asymmetries in different processes. Processes in which the KM hypothesis implies extremely small asymmetries are particularly sensitive to new amplitudes.

To conclude this brief discussion, let us list a few examples of signals for new physics.

- $A_{sl} \geq \mathcal{O}(10^{-2})$;
- sizable asymmetries in $b \to s\gamma$ or $B_s \to \psi\phi$;
- "forbidden" values of angles, $|\sin 2\beta - 0.6| > 0.2, \quad \sin\gamma < 0.6$;
- different asymmetries in $B^0(t) \to \psi K_S,\ \phi K_S,\ \eta' K_S$;
- contradictory constraints on γ from $B \to K\pi,\ B \to DK,\ B_s \to D_s K$;
- rate enhancement beyond standard model predictions for electroweak penguin decays, $B \to X_{d,s}\ell^+\ell^-,\ B^0/B_s \to \ell^+\ell^-$.

48.8 Conclusion

The CP asymmetry in $B \to \psi K_S$ is related cleanly to the weak-phase β and can be used experimentally to measure $\sin 2\beta$. In other cases, such as in $B^0 \to \pi^+\pi^-$, which measures $\sin 2\alpha$ and $B \to DK$ which determines $\sin\gamma$, the relations between the asymmetries, supplemented by certain rates, and the corresponding weak phases are free of significant theoretical uncertainties. However, the applications of these methods are expected to suffer from experimental difficulties due to the small rates of color-suppressed processes.

While one expects qualitatively that color supression is affected by final-state interactions, these long-distance phenomena are not understood quantitatively. The case of $B \to K\pi$ demonstrates the need for a better undersanding of these features and the need for a reliable treatment of SU(3) breaking. That is, whereas the short-distance effects of QCD in weak hadronic B decays are well understood [43], we are in great need of a theoretical framework for studying long-distance effects. An interesting suggestion in this direction was made very recently in [44].

We discussed mainly the very immediate B decay modes, for which CP asymmetries can provide new information on CKM parameters. Asymmetries should be searched in *all B decay processes*, including those which are plagued by theoretical uncertainties due to unknown final state interactions, and those where the KM framework predicts negligibly small asymmetries. After all, our understanding of the origin of CP violation is rather limited and surprises may be right around the corner.

Acknowledgments

I thank the SLAC Theory Group for its very kind hospitality. I am grateful to Gad Eilam, David London, Dan Pirjol, Jon Rosner, and Daniel Wyler for collaborations on topics discussed here. This work was supported in part by the United States–Israel Binational Science Foundation under research grant agreement 94-00253/3, and by the Department of Energy under contract number DE-AC03-76SF00515.

References

[1] M. Kobayashi and T. Maskawa, Prog. Theor. Phys. **49**, 652 (1973).

[2] C. Caso *et al.*, Eur. Phys. J. C **3**, 1 (1998).

[3] R. Peccei, this volume, chapter 2, also discusses the Wolfenstein parameterization in terms of λ, A, ρ, and η.

[4] Y.B. Hsiung, this volume, chapter 6.

[5] M.S. Sozzi, this volume, chapter 8.

[6] C. Jarlskog, Phys. Rev. Lett. **55**, 1039 (1985).

[7] I.I. Bigi *et al.*, in *CP Violation*, ed. C. Jarlskog (World Scientific, Singapore, 1992).

[8] OPAL Collaboration, K. Ackerstaff *et al.*, Z. Phys. C **76**, 401 (1997).

[9] A.B. Carter and A.I. Sanda, Phys. Rev. Lett. **45**, 952 (1980); Phys. Rev. D **23**, 1567 (1981); I.I. Bigi and A.I. Sanda, Nucl. Phys. B **193**, 85 (1981).

[10] M. Gronau, Phys. Rev. Lett. **63**, 1451 (1989).

[11] I.J. Kroll, this volume, chapter 51.

[12] D. London and R.D. Peccei, Phys. Lett. B **223**, 257 (1989); B. Grinstein, Phys. Lett. B **229**, 280 (1989).

[13] A. Dighe, M. Gronau, and J.L. Rosner, Phys. Rev. Lett. **79**, 4333 (1997).

[14] F. DeJongh and P. Sphicas, Phys. Rev. D **53**, 4930 (1996); P.S. Marrocchesi and N. Paver, Int. J. Mod. Phys. A **13**, 251 (1998).

[15] M. Gronau and D. London, Phys. Rev. Lett. **65**, 3381 (1990).

[16] N.G. Deshpande and X.G. He, Phys. Rev. Lett. **74**, 26, 4099(E) (1995).

[17] M. Gronau, D. Pirjol, and T.M. Yan, Phys. Rev. D **60**, 034021 (1999).

[18] S. Gardner, Phys. Rev. D **59**, 077502 (1999).

[19] R. Poling, Rapporteur talk at the 19th International Lepton Photon Symposium, Stanford, CA, 9–14 August, 1999.

[20] CLEO Collaboration, Y. Kwon *et al.*, hep-ex/9908029.

[21] Y. Grossman and H.R. Quinn, Phys. Rev. D **58**, 017504 (1998).

[22] J. Charles, Phys. Rev. D **59**, 054007 (1999); D. Pirjol, Phys. Rev. D **60**, 54020 (1999); R. Fleischer, Phys. Lett. B **459**, 306 (1999).

[23] H.J. Lipkin, Y. Nir, H.R. Quinn, and A. Snyder, Phys. Rev. D **44**, 1454 (1991); M. Gronau, Phys. Lett. B **265**, 389 (1991); H.R. Quinn and A. Snyder, Phys. Rev. D **48**, 2139 (1993).

[24] CLEO Collaboration, M. Bishai *et al.*, hep-ex/9908018.

[25] G. Eilam, M. Gronau, and R.R. Mendel, Phys. Rev. Lett. **74**, 4984 (1995); N.G. Deshpande *et al.*, Phys. Rev. D **52**, 5354 (1995); I. Bediaga, R.E. Blanco, C. Gobel, and R. Mendez-Galain, Phys. Rev. Lett. **81**, 4067 (1998); B. Bajc *et al.*, Phys. Lett. B **447**, 313 (1999).

[26] M. Gronau and D. Wyler, Phys. Lett. B **265**, 172 (1991).

[27] D. Atwood, I. Dunietz, and A. Soni, Phys. Rev. Lett. **78**, 3257 (1997); M. Gronau, Phys. Rev. D **58**, 037301 (1998); M. Gronau and J.L. Rosner, Phys. Lett. B **439**, 171 (1998).

[28] M. Gronau, J.L. Rosner, and D. London, Phys. Rev. Lett. **73**, 21 (1994).

[29] N.G. Deshpande and X.G. He, Phys. Rev. Lett. **74**, 26 (1995). Electroweak penguin effects in other B decays were studied earlier by R. Fleischer, Phys. Lett. B **321**, 259 (1994).

[30] R. Fleischer, Phys. Lett. B **365**, 399 (1996); A.J. Buras and R. Fleischer, Phys. Lett. B **365**, 390 (1996); M. Gronau and J.L. Rosner, Phys. Rev. Lett. **76**, 1200 (1996); A.S. Dighe, M. Gronau, and J.L. Rosner, Phys. Rev. D **54**, 3309 (1996); R. Fleischer and T. Mannel, Phys. Rev. D **57**, 2752 (1998).

[31] M. Neubert and J.L. Rosner, Phys. Lett. B **441**, 403 (1998); Phys. Rev. Lett. **81**, 5076 (1998); M. Neubert, JHEP **9902**, 014 (1999).

[32] M. Gronau and D. Pirjol, Phys. Rev. D **61**, 013005 (2000).

[33] A.J. Buras and R. Fleischer, Eur. Phys. J. C **11**, 93 (1999).

[34] M. Gronau, O. Hernández, D. London, and J.L. Rosner, Phys. Rev. D **50**, 4529 (1994); Phys. Rev. D **52**, 6374 (1995).

[35] A. Falk, A.L. Kagan, Y. Nir, and A.A. Petrov, Phys. Rev. D **57**, 4290 (1998); M. Gronau and J.L. Rosner, Phys. Rev. D **57**, 6843 (1998); **58**, 113005 (1998); R. Fleischer, Phys. Lett. B **435**, 221 (1998); Eur. Phys. J. C **6**, 451 (1999).

[36] M. Gronau and J.L. Rosner, Phys. Rev. D **59**, 113002 (1999).

[37] CLEO Collaboration, T.E. Coan *et al.*, hep-ex/9908029.

[38] Z. Ligeti, this volume, chapter 49.

[39] G. Redlinger, this volume, chapter 34.

[40] Y. Grossman and H.R. Quinn, Phys. Rev. D **56**, 7529 (1997); L. Wolfenstein, Phys. Rev. D **57**, 6857 (1998).

[41] C.O. Dib, D. London, and Y. Nir, Int. J. Mod. Phys. A **6**, 1253 (1991); M. Gronau and D. London, Phys. Rev. D **55**, 2845 (1997).

[42] Y. Grossman and M.P. Worah, Phys. Lett. B **395**, 241 (1997); D. London and A. Soni, Phys. Lett. B **407**, 61 (1997).

[43] G. Buchalla, A.J. Buras, and M.E. Lautenbacher, Rev. Mod. Phys. **68** 1125 (1996).

[44] M. Beneke, G. Buchalla, M. Neubert, and C.T. Sachrajda, Phys. Rev. Lett. **83**, 1914 (1999).

49

$|V_{cb}|$ and $|V_{ub}|$ from B Decays: Recent Progress and Limitations

Zoltan Ligeti

Abstract

The determination of $|V_{cb}|$ and $|V_{ub}|$ from semileptonic B decay is reviewed with a critical discussion of the theoretical uncertainties. Future prospects and limitations are also discussed.

49.1 Introduction

The purpose of K and B physics in the near future is testing the Cabibbo-Kobayashi-Maskawa (CKM) picture of quark mixing and CP violation. The goal is to overconstrain the unitarity triangle by directly measuring the sides and (some) angles in several decay modes. If the value of $\sin 2\beta$, the CP asymmetry in $B \to J/\psi\, K_S$, is near the CDF central value [1], then searching for new physics will require a combination of precision measurements. This talk concentrates on $|V_{cb}|$ and $|V_{ub}|$; the latter is particularly important since it largely controls the experimentally allowed range for $\sin 2\beta$ in the standard model.

49.2 Exclusive Decays

In mesons composed of a heavy quark and a light antiquark (plus gluons and $q\bar{q}$ pairs), the energy scale of strong processes is small compared to the heavy quark mass. The heavy quark acts as a static pointlike color source with fixed four-velocity, since the soft gluons responsible for confinement cannot resolve structures much smaller than $\Lambda_{\rm QCD}$, such as the heavy quark's Compton wavelength. Thus the configuration of the light degrees of freedom becomes insensitive to the spin and flavor (mass) of the heavy quark, resulting in a $SU(2n)$ spin-flavor symmetry [2] (n is the number of heavy quark flavors). Heavy quark symmetry (HQS) helps us to understand the spectroscopy and decays of heavy hadrons from first principles.

The predictions of HQS are particularly restrictive for $\bar{B} \to D^{(*)}\ell\bar{\nu}$ decays. In the infinite mass limit all form factors are proportional to a universal Isgur-Wise function, $\xi(v\cdot v')$, satisfying $\xi(1) = 1$ [2]. The symmetry-breaking corrections can be organized in a simultaneous expansion in α_s and $\Lambda_{\rm QCD}/m_Q$ ($Q = c, b$). The $\bar{B} \to D^{(*)}\ell\bar{\nu}$ decay rates are given by

$$\begin{aligned}
\frac{d\Gamma(\bar{B}\to D^*\ell\bar{\nu})}{dw} &= \frac{G_F^2 m_B^5}{48\pi^3}\, r_*^3\, (1-r_*)^2\, \sqrt{w^2-1}\, (w+1)^2 \\
&\quad \times \left[1 + \frac{4w}{1+w}\, \frac{1-2wr_*+r_*^2}{(1-r_*)^2}\right] |V_{cb}|^2\, \mathcal{F}_{D^*}^2(w)\,, \\
\frac{d\Gamma(\bar{B}\to D\ell\bar{\nu})}{dw} &= \frac{G_F^2 m_B^5}{48\pi^3}\, r^3\, (1+r)^2\, (w^2-1)^{3/2}\, |V_{cb}|^2\, \mathcal{F}_D^2(w),
\end{aligned} \tag{49.1}$$

where $w = v \cdot v'$ and $r_{(*)} = m_{D^{(*)}}/m_B$. $\mathcal{F}_{D^{(*)}}(w)$ is equal to the Isgur-Wise function in the $m_Q \to \infty$ limit, and in particular $\mathcal{F}_{D^{(*)}}(1) = 1$, allowing for a model-independent determination of $|V_{cb}|$. Including symmetry-breaking corrections one finds

$$\begin{aligned}
\mathcal{F}_{D^*}(1) &= 1 + c_A(\alpha_s) + \frac{0}{m_Q} + \frac{(\ldots)}{m_Q^2} + \ldots\,, \\
\mathcal{F}_D(1) &= 1 + c_V(\alpha_s) + \frac{(\ldots)}{m_Q} + \frac{(\ldots)}{m_Q^2} + \ldots\,.
\end{aligned} \tag{49.2}$$

The perturbative corrections, $c_A = -0.04$ and $c_V = 0.02$, have been computed to order α_s^2 [3], and the unknown higher-order corrections should affect $|V_{cb}|$ at below the 1% level. The vanishing of the order $1/m_Q$ corrections to $\mathcal{F}_{D^*}(1)$ is known as Luke's theorem [4]. The terms indicated by $(\ldots)$ are only known using phenomenological models at present. Thus the determination of $|V_{cb}|$ from $\bar{B} \to D^*\ell\bar{\nu}$ is theoretically more reliable than that from $\bar{B} \to D\ell\bar{\nu}$ (unless using lattice QCD for $\mathcal{F}_{D^{(*)}}(1)$—see below), although for example QCD sum rules predict that the order $1/m_Q$ correction to $\mathcal{F}_D(1)$ is small [5]. Due to the extra $w^2 - 1$ suppression near zero recoil, $\bar{B} \to D\ell\bar{\nu}$ is also harder experimentally.

The main uncertainty in this determination of $|V_{cb}|$ comes from the estimate of nonperturbative corrections at zero recoil. In the case of $\bar{B} \to D^*\ell\bar{\nu}$, model calculations [6] and sum rule estimates [7] suggest about -5%. Assigning a 100% uncertainty to this estimate, I will use

$$\mathcal{F}_{D^*}(1) = 0.91 \pm 0.05\,, \qquad \mathcal{F}_D(1) = 1.02 \pm 0.08\,. \tag{49.3}$$

The most promising way to reduce these uncertainties may be calculating directly the deviation of the form factor from unity, $\mathcal{F}_{D^{(*)}}(1) - 1$, in lattice QCD from certain double ratios of correlation functions [8]. Recent quenched calculations give $\mathcal{F}_D(1) = 1.06 \pm 0.02$ and $\mathcal{F}_{D^*}(1) = 0.935 \pm 0.03$ [8], in agreement with eq. (49.3) but with smaller errors.

Another uncertainty comes from extrapolating the experimentally measured quantity, $|V_{cb}|\,\mathcal{F}_{D^{(*)}}(w)$, to zero recoil. Recent theoretical developments largely reduce this uncertainty by establishing a model-independent relationship between the slope and curvature of $\mathcal{F}_{D^{(*)}}(w)$ [9]. This may also become less of an experimental problem at asymmetric B factories, where the efficiency may fall less rapidly near zero recoil.

Eq. (49.3) and the experimental average, $|V_{cb}|\mathcal{F}_{D^*}(1) = 0.0347 \pm 0.0015$ [10], obtained using the constraints on the shape of $\mathcal{F}_{D^*}(w)$ yield

$$|V_{cb}| = (38.1 \pm 1.7_{\rm exp} \pm 2.0_{\rm th}) \times 10^{-3}\,. \tag{49.4}$$

The value obtained from $\bar{B} \to D\ell\bar{\nu}$ is consistent with this, but the experimental uncertainties are significantly larger.

For the determination of $|V_{ub}|$ from exclusive heavy to light decays, heavy quark symmetry is less predictive. It neither reduces the number of form factors parameterizing these decays nor determines the value of any form factor. Still, there are model-independent relations between B and D decay form factors, *e.g.*, the form factors that occur in $D \to K^*\bar{\ell}\nu$ can be related to those in $\bar{B} \to \rho\ell\bar{\nu}$ using heavy quark and chiral symmetry [11]. These relations apply for the same value of $v \cdot v'$ in the two processes, *i.e.*, from the measured $D \to K^*\bar{\ell}\nu$ form factors one can predict the $\bar{B} \to \rho\ell\bar{\nu}$ rate in the large q^2 region [12]. Such a prediction has first-order heavy quark and chiral symmetry breaking corrections, each of which can be 15–20%. Lattice QCD also works best for large q^2, but the existing calculations are still all quenched. Light cone sum rules [13] are claimed to yield predictions for the form factors with small model dependence in the small q^2 region. Recently CLEO made the first attempt at concentrating at the large q^2 region to reduce the model dependence, and obtained [14]

$$|V_{ub}| = (3.25 \pm 0.14^{+0.21}_{-0.29} \pm 0.55) \times 10^{-3}\,. \tag{49.5}$$

A determination of $|V_{ub}|$ from $\bar{B} \to \pi\ell\bar{\nu}$ is more complicated because very near zero recoil "pole contributions" [15] spoil the simple scaling of the form factors with the heavy quark mass. Still, in the future some combination of the soft pion limit, model-independent bounds based on dispersion relations and analyticity [16], and lattice results may provide a determination of $|V_{ub}|$ from this decay with small errors.

If experimental data on the $D \to \rho\bar{\ell}\nu$ and $\bar{B} \to K^*\ell\bar{\ell}$ form factors become available in the future, then $|V_{ub}|$ can be extracted with ~10% theoretical uncertainty [12] using a "Grinstein-type double ratio" [17], which deviates from unity only due to corrections that violate both heavy quark and chiral symmetries. Such a determination is possible even if only the q^2 spectrum in $D \to \rho\bar{\ell}\nu$ and the integrated $\bar{B} \to K^*\ell\bar{\ell}$ rate in the large q^2 region are measured [18].

49.3 Inclusive Decays

Inclusive B decay rates can be computed model independently in a series in $\Lambda_{\rm QCD}/m_b$ and $\alpha_s(m_b)$, using an operator product expansion (OPE) [19–21]. The $m_b \to \infty$ limit is given by b quark decay, and for most quantities of interest it is known including the dominant part of the order α_s^2 corrections. Observables that do not depend on the four-momentum of the hadronic final state (*e.g.*, total decay rate and lepton spectra) receive no correction at order $\Lambda_{\rm QCD}/m_b$ when written in terms of m_b, whereas differential rates with respect to hadronic variables (*e.g.*, hadronic energy and invariant mass spectra) also depend on $\bar\Lambda/m_b$, where $\bar\Lambda$ is the $m_B - m_b$ mass difference in the $m_b \to \infty$ limit. At order $\Lambda_{\rm QCD}^2/m_b^2$, the corrections are parameterized by two hadronic matrix elements, usually denoted by λ_1 and λ_2. The value $\lambda_2 \simeq 0.12\,{\rm GeV}^2$ is known from the $B^* - B$ mass splitting. Corrections to the $m_b \to \infty$ limit are expected to be under control in parts of the $b \to q$ phase space, where several hadronic final states are allowed (but not required) to contribute with invariant masses satisfying $m_{X_q}^2 \gtrsim m_q^2 +$ (few times)$\Lambda_{\rm QCD} m_b$.

The major uncertainty in the predictions for such "sufficiently inclusive" observables is from the values of the quark masses and λ_1, or equivalently, the values of $\bar\Lambda$ and λ_1. These quantities can be extracted, for example, from heavy meson decay spectra. A theoretical subtlety is related to the fact that $\bar\Lambda$ (or the heavy quark pole mass) cannot be defined unambiguously beyond perturbation theory [22], and its value extracted from data using theoretical expressions valid to different orders in the α_s may vary by order $\Lambda_{\rm QCD}$. These ambiguities cancel [23] when one relates consistently physical observables to one another. One way to make this cancellation manifest is by using short-distance quark mass definitions, but recent determinations of such b quark masses still have about 50–100 MeV uncertainties [24].

The shape of the lepton energy [25–27] or hadronic invariant mass [27–29] spectra in $\bar B \to X_c \ell \bar\nu$ decay can be used to determine $\bar\Lambda$ and λ_1. Last year the CLEO Collaboration measured the first two moments of the hadronic invariant mass-squared distribution. Each of these measurements gives an allowed band in the $\bar\Lambda - \lambda_1$ plane, and their intersection gives [30]

$$\bar\Lambda = (0.33 \pm 0.08)\,{\rm GeV}, \qquad \lambda_1 = -(0.13 \pm 0.06)\,{\rm GeV}^2. \tag{49.6}$$

This result agrees well with the one obtained from an analysis of the lepton energy spectrum in ref. [25]. CLEO also considered moments of the lepton spectrum, however, without any restriction on the lepton energy, yielding unlikely central values of $\bar\Lambda$ and λ_1. Since this analysis uses a model dependent extrapolation to $E_\ell < 0.6\,{\rm GeV}$, I consider the result in eq. (49.6) more reliable [31]. The unknown order $\Lambda_{\rm QCD}^3/m_b^3$ terms not included in eq. (49.6) introduce a sizable uncertainty [27, 29], which could be significantly reduced when more precise data on the photon energy spectrum in $\bar B \to X_s \gamma$ becomes available [32, 33].

The significance of eq. (49.6) is that, taken at face value, it gives $|V_{cb}| = 0.0415$ from the $\bar{B} \to X_c \ell \bar{\nu}$ width with only 3% uncertainty. The theoretical uncertainty hardest to quantify in the inclusive determination of $|V_{cb}|$ is the size of quark-hadron duality violation [34]. Studying the shapes of these $\bar{B} \to X_c \ell \bar{\nu}$ decay distributions may be the best way to constrain this experimentally, since it is unlikely that duality violation would not show up in a comparison of moments of different spectra. Thus, testing our understanding of these spectra is important to assess the reliability of the inclusive determination of $|V_{cb}|$, and especially that of $|V_{ub}|$ (see below).

A new approach to replace the b quark mass in theoretical predictions with the $\Upsilon(1S)$ mass was proposed recently [35]. The crucial point of this "upsilon expansion" is that for theoretical consistency one must combine different orders in the α_s perturbation series in the expression for B decay rates and m_Υ in terms of m_b. As the simplest example, consider schematically the $\bar{B} \to X_u \ell \bar{\nu}$ rate, neglecting nonperturbative corrections,

$$\Gamma(\bar{B} \to X_u \ell \bar{\nu}) = \frac{G_F^2 |V_{ub}|^2}{192\pi^3}\, m_b^5 \left[1 - (\ldots)\frac{\alpha_s}{\pi}\,\epsilon - (\ldots)\frac{\alpha_s^2}{\pi^2}\,\epsilon^2 - \ldots\right]. \quad (49.7)$$

The coefficients denoted by $(\ldots)$ are known, and the parameter $\epsilon \equiv 1$ denotes the order in the upsilon expansion. In comparison, the expansion of the $\Upsilon(1S)$ mass in terms of m_b has a different structure,

$$m_\Upsilon = 2m_b \left[1 - (\ldots)\frac{\alpha_s^2}{\pi^2}\,\epsilon - (\ldots)\frac{\alpha_s^3}{\pi^3}\,\epsilon^2 - \ldots\right]. \quad (49.8)$$

In this expansion one must assign to each term one less power of ϵ than the power of α_s [35]. At the scale $\mu = m_b$ both of these series appear badly behaved, but substituting eq. (49.8) into eq. (49.7) and collecting terms of a given order in ϵ gives [35]

$$\Gamma(\bar{B} \to X_u \ell \bar{\nu}) = \frac{G_F^2 |V_{ub}|^2}{192\pi^3} \left(\frac{m_\Upsilon}{2}\right)^5 \left[1 - 0.115\epsilon - 0.035\epsilon^2 - \ldots\right]. \quad (49.9)$$

The perturbation series, $1 - 0.115\epsilon - 0.035\epsilon^2$, is far better behaved than the series in eq. (49.7) in terms of the b quark pole mass, $1 - 0.17\epsilon - 0.13\epsilon^2$, or the series expressed in terms of the $\overline{\text{MS}}$ mass, $1 + 0.30\epsilon + 0.19\epsilon^2$. The uncertainty in the decay rate using eq. (49.9) is much smaller than that in eq. (49.7), both because the perturbation series is better behaved, and because m_Υ is better known (and better defined) than m_b. The relation between $|V_{ub}|$ and the $\bar{B} \to X_u \ell \nu$ rate is [35]

$$|V_{ub}| = (3.06 \pm 0.08 \pm 0.08) \times 10^{-3} \left(\frac{\mathcal{B}(\bar{B} \to X_u \ell \bar{\nu})}{0.001}\, \frac{1.6\,\text{ps}}{\tau_B}\right)^{1/2}. \quad (49.10)$$

The upsilon expansion also improves the behavior of the perturbation series for the $\bar{B} \to X_c \ell \bar{\nu}$ rate, and yields

$$|V_{cb}| = (41.9 \pm 0.8 \pm 0.5 \pm 0.7) \times 10^{-3} \left(\frac{\mathcal{B}(\bar{B} \to X_c \ell \bar{\nu})}{0.105}\, \frac{1.6\,\text{ps}}{\tau_B}\right)^{1/2}. \quad (49.11)$$

These results agree with other estimates [36] within the uncertainties. The first error in eqs. (49.10) and (49.11) come from assigning an uncertainty equal to the size of the ϵ^2 term, the second is from assuming a 100 MeV uncertainty in eq. (49.8), and the third error in eq. (49.11) is from a 0.25 GeV^2 error in λ_1. The most important uncertainty is the size of nonperturbative contributions to m_Υ other than those which can be absorbed into m_b, for which we used 100 MeV. By dimensional analysis it is of order $\Lambda_{\rm QCD}^4/(m_b\alpha_s)^3$, however, quantitative estimates vary in a large range. It is preferable to constrain such effects from data [32, 37].

For the determination of $|V_{ub}|$, eq. (49.10) is of little use by itself, since $\mathcal{B}(\bar{B} \to X_u \ell \bar{\nu})$ cannot be measured without significant cuts on the phase space. The traditional method for extracting $|V_{ub}|$ involves a study of the electron energy spectrum in the endpoint region $m_B/2 > E_\ell > (m_B^2 - m_D^2)/2m_B$ (in the B rest frame), which must arise from $b \to u$ transition. Since the width of this region is only 300 MeV (of order $\Lambda_{\rm QCD}$), an infinite set of terms in the OPE may be important, and at the present time it is not known how to make a model-independent prediction for the spectrum in this region. Another possibility for extracting $|V_{ub}|$ is based on reconstructing the neutrino momentum. The idea is to infer the invariant mass-squared of the hadronic final state, $s_H = (p_B - p_\ell - p_{\bar{\nu}})^2$. Semileptonic B decays satisfying $s_H < m_D^2$ must come from $b \to u$ transition [38–40].

Both the invariant mass region $s_H < m_D^2$ and the electron endpoint region $E_\ell > (m_B^2 - m_D^2)/2m_B$ receive contributions from hadronic final states with invariant masses between m_π and m_D. However, for the electron endpoint region the contribution of states with masses nearer to m_D is strongly suppressed kinematically. This region may be dominated by the π and the ρ, and includes only of order 10% of the total $\bar{B} \to X_u \ell \bar{\nu}$ rate. The situation is very different for the low-invariant-mass region, $s_H < m_D^2$, where all such states contribute without any preferential weighting towards the lowest-mass ones. In this case the π and the ρ exclusive modes comprise a smaller fraction, and only of order 10% of the $\bar{B} \to X_u \ell \bar{\nu}$ rate is excluded from the $s_H < m_D$ region. Consequently, it is much more likely that the first few terms in the OPE provide an accurate description of the decay rate in the region $s_H < m_D^2$ than in the region $E_\ell > (m_B^2 - m_D^2)/2m_B$.

Since m_D^2 is not much larger than $\Lambda_{\rm QCD} m_b$, one needs to model the nonperturbative effects in both cases. However, assigning a 100% uncertainty to these estimates affects the extracted value of $|V_{ub}|$ much less from the $s_H < m_D^2$ than from the $E_\ell > (m_B^2 - m_D^2)/2m_B$ region. Such estimates suggest that the theoretical uncertainty in $|V_{ub}|$ determined from the hadronic invariant mass spectrum in the region $s_H < m_D^2$ is about $\sim$10%. If experimental constraints force one to consider a significantly smaller region, then the uncertainties increase rapidly. The first analyses of LEP data utilizing this idea were performed recently [41], but it is not transparent how they weigh the Dalitz plot, which affects crucially the theoretical uncertainties.

The inclusive nonleptonic decay rate to "wrong sign" charm ($\bar{B} \to X_{u\bar{c}s}$) may also give a determination of $|V_{ub}|$ with modest theoretical uncertainties [42], if such a measurement is experimentally feasible.

49.4 Conclusions

The present status of $|V_{cb}|$ and $|V_{ub}|$ is approximately

$$|V_{cb}| = 0.040 \pm 0.002, \qquad |V_{ub}/V_{cb}| \simeq 0.090 \pm 0.025. \tag{49.12}$$

The central value and error of $|V_{cb}|$ come from first principles, and the uncertainty in both its exclusive and inclusive determination is of order $1/m_Q^2$. On the other hand, the above error on $|V_{ub}|$ is somewhat ad hoc, since it is still estimated relying on phenomenological models.

Within the next 3–5 years, in my opinion, an optimistic scenario is roughly as follows. The theoretical error of $|V_{cb}|$ might be reduced to 2–3%. This requires better agreement between the inclusive and exclusive determinations, since in the exclusive determination the nonperturbative corrections to $\mathcal{F}_{D^{(*)}}(1)$ are at the 5% level and model dependent, while in the inclusive determination it is hard to constrain the model independently of the size of quark-hadron duality violation. It will give confidence in lattice calculations of $\mathcal{F}_{D^*}(1)$ and $\mathcal{F}_D(1)$ if they give the same value of $|V_{cb}|$, and the deviations of the form factor ratios conventionally denoted by $R_{1,2}(w)$ from unity can also be predicted precisely. Quark-hadron duality violation in the inclusive determination of $|V_{cb}|$ can be constrained by comparing the measured shapes of $\bar{B} \to X_c \ell \bar{\nu}$ decay spectra in different variables (*e.g.*, lepton energy, hadronic invariant mass, etc.).

At the same time, the theoretical error of $|V_{ub}|$ might be reduced to about 10%. Again, a better agreement between the inclusive and exclusive determinations is needed. At this level only unquenched lattice calculations will be trusted, and they ought to give consistent values of $|V_{ub}|$ from $\bar{B} \to \pi \ell \bar{\nu}$ and $\bar{B} \to \rho \ell \bar{\nu}$. From exclusive decays a double ratio method discussed in sec. 49.2 may give $|V_{ub}|$ with ~10% error. In inclusive $\bar{B} \to X_u \ell \bar{\nu}$ decay, the hadron invariant mass spectrum should be measured up to a cut as close to m_D as possible. It would be reassuring as a check if varying this cut in some range leaves $|V_{ub}|$ unaffected.

Acknowledgments

I would like to thank Jon Rosner and Bruce Winstein for inviting me and for organizing a very interesting and stimulating workshop. I also thank Adam Falk and Andreas Kronfeld for comments on the manuscript. Fermilab is operated by Universities Research Association, Inc., under DOE contract DE-AC02-76CH03000.

References

[1] K. Pitts (CDF Collaboration), Fermilab Joint Experimental Theoretical Physics Seminar, 5 February, 1999.

[2] N. Isgur and M.B. Wise, Phys. Lett. B **232**, 113 (1989); Phys. Lett. B **237**, 527 (1990).

[3] A. Czarnecki, Phys. Rev. Lett. **76**, 4124 (1996); A. Czarnecki and K. Melnikov, Nucl. Phys. B **505**, 65 (1997).

[4] M.E. Luke, Phys. Lett. B **252**, 447 (1990).

[5] Z. Ligeti, Y. Nir, and M. Neubert, Phys. Rev. D **49**, 1302 (1994).

[6] A.F. Falk and M. Neubert, Phys. Rev. D **47**, 2965 (1993); T. Mannel, Phys. Rev. D **50**, 428 (1994).

[7] I. Bigi, M. Shifman, N.G. Uraltsev, and A. Vainshtein, Phys. Rev. D **52**, 196 (1995); A. Kapustin, Z. Ligeti, M.B. Wise, and B. Grinstein, Phys. Lett. B **375**, 327 (1996).

[8] S. Hashimoto *et al.*, FERMILAB-PUB-99-001-T [hep-ph/9906376]; S. Hashimoto, to appear in the proceedings of Lattice '99, Pisa, Italy; J. Simone, to appear in the proceedings of Lattice '99, Pisa, Italy.

[9] C.G. Boyd, B. Grinstein, and R.F. Lebed, Phys. Lett. B **353**, 306 (1995); Nucl. Phys. B **461**, 493 (1996); Phys. Rev. D **56**, 6895 (1997); I. Caprini, L. Lellouch, and M. Neubert, Nucl. Phys. B **530**, 153 (1998); C.-W. Chiang and A.K. Leibovich, hep-ph/9906420.

[10] S. Stone, hep-ph/9904350.

[11] N. Isgur and M.B. Wise, Phys. Rev. D **42**, 2388 (1990).

[12] Z. Ligeti and M.B. Wise, Phys. Rev. D **53**, 4937 (1996).

[13] P. Ball and V.M. Braun, Phys. Rev. D **58**, 094016 (1998); P. Ball, JHEP **09**, 005 (1998).

[14] B.H. Behrens *et al.*, CLEO Collaboration, hep-ex/9905056.

[15] N. Isgur and M.B. Wise, Phys. Rev. D **41**, 151 (1990); M.B. Wise, Phys. Rev. D **45**, 2188 (1992); G. Burdman and J.F. Donoghue, Phys. Lett. B **280**, 287 (1992); L. Wolfenstein, Phys. Lett. B **291**, 177 (1992); T.-M. Yan *et al.*, Phys. Rev. D **46**, 1148 (1992); G. Burdman, Z. Ligeti, M. Neubert, and Y. Nir, Phys. Rev. D **49**, 2331 (1994).

[16] C.G. Boyd, B. Grinstein, and R.F. Lebed, Phys. Rev. Lett. **74**, 4603 (1995).

[17] B. Grinstein, Phys. Rev. Lett. **71**, 3067 (1993).

[18] Z. Ligeti, I.W. Stewart, and M.B. Wise, Phys. Lett. B **420**, 359 (1998).

[19] J. Chay, H. Georgi, and B. Grinstein, Phys. Lett. B **247**, 399 (1990); M. Voloshin and M. Shifman, Sov. J. Nucl. Phys. **41**, 120 (1985).

[20] I.I. Bigi, N.G. Uraltsev, and A.I. Vainshtein, Phys. Lett. B **293**, 430 (1992) [(E) Phys. Lett. B **297**, 477 (1993)]; I.I. Bigi, M. Shifman, N.G. Uraltsev, and A. Vainshtein, Phys. Rev. Lett. **71**, 496 (1993).

[21] A.V. Manohar and M.B. Wise, Phys. Rev. D **49**, 1310 (1994); B. Blok, L. Koyrakh, M. Shifman, and A.I. Vainshtein, Phys. Rev. D **49**, 3356 (1994); T. Mannel, Nucl. Phys. B **413**, 396 (1994).

[22] I.I. Bigi, M.A. Shifman, N.G. Uraltsev, and A.I. Vainshtein, Phys. Rev. D **50**, 2234 (1994); M. Beneke and V.M. Braun, Nucl. Phys. B **426**, 301 (1994).

[23] M. Beneke, V.M. Braun, and V.I. Zakharov, Phys. Rev. Lett. **73**, 3058 (1994); M. Luke, A.V. Manohar, and M.J. Savage, Phys. Rev. D **51**, 4924 (1995); M. Neubert and C.T. Sachrajda, Nucl. Phys. B **438**, 235 (1995).

[24] A.H. Hoang, hep-ph/9905550; M. Beneke and A. Signer, hep-ph/9906475; K. Melnikov and A. Yelkhovsky, Phys. Rev. D **59**, 114009 (1999); M. Jamin and A. Pich, Nucl. Phys. Proc. Suppl. **74**, 300 (1999).

[25] M. Gremm, A. Kapustin, Z. Ligeti, and M.B. Wise, Phys. Rev. Lett. **77**, 20 (1996); M. Gremm and I. Stewart, Phys. Rev. D **55**, 1226 (1997).

[26] M.B. Voloshin, Phys. Rev. D **51**, 4934 (1995).

[27] M. Gremm and A. Kapustin, Phys. Rev. D **55**, 6924 (1997).

[28] A.F. Falk, M. Luke, and M.J. Savage, Phys. Rev. D **53**, 2491 (1996); Phys. Rev. D **53**, 6316 (1996).

[29] A.F. Falk and M. Luke, Phys. Rev. D **57**, 424 (1998).

[30] J. Bartelt *et al.*, CLEO Collaboration, CLEO CONF 98-21.

[31] Z. Ligeti, FERMILAB-CONF-99-058-T [hep-ph/9904460].

[32] Z. Ligeti, M. Luke, A.V. Manohar, and M.B. Wise, Phys. Rev. D **60**, 034019 (1999) [hep-ph/9903305]; A. Kapustin and Z. Ligeti, Phys. Lett. B **355**, 318 (1995).

[33] A.F. Falk, M. Luke, and M.J. Savage, Phys. Rev. D **49**, 3367 (1994); C. Bauer, Phys. Rev. D **57**, 5611 (1998); A.L. Kagan and M. Neubert, Eur. Phys. J. C **7**, 5 (1999).

[34] N. Isgur, Phys. Lett. B **448**, 111 (1999); B. Chibisov, R.D. Dikeman, M. Shifman, and N. Uraltsev, Int. J. Mod. Phys. A **12**, 2075 (1997).

[35] A.H. Hoang, Z. Ligeti, and A.V. Manohar, Phys. Rev. Lett. **82**, 277 (1999); Phys. Rev. D **59**, 074017 (1999) [hep-ph/9811239].

[36] I. Bigi, M. Shifman, and N. Uraltsev, Annu. Rev. Nucl. Part. Sci. **47**, 591 (1997); I.I. Bigi, hep-ph/9907270.

[37] A.H. Hoang, CERN-TH-99-152 [hep-ph/9905550].

[38] A.F. Falk, Z. Ligeti, and M.B. Wise, Phys. Lett. B **406**, 225 (1997).

[39] R.D. Dikeman and N.G. Uraltsev, Nucl. Phys. B **509**, 378 (1998); I. Bigi, R.D. Dikeman, and N. Uraltsev, Eur. Phys. J. C **4**, 453 (1998).

[40] V. Barger *et al.*, Phys. Lett. B **251**, 629 (1990); J. Dai, Phys. Lett. B **333**, 212 (1994);

[41] R. Barate *et al.*, ALEPH Collaboration, CERN EP/98-067; DELPHI Collaboration, ICHEP98 Conference paper 241; M. Acciarri *et al.*, L3 Collaboration, Phys. Lett. B **436**, 174 (1998).

[42] A.F. Falk and A.A. Petrov, hep-ph/9903518; J. Chay, A.F. Falk, M. Luke, and A.A. Petrov, hep-ph/9907363; M. Beneke, G. Buchalla, and I. Dunietz, Phys. Lett. B **393**, 132 (1997).

50

Recent CLEO Results on Rare B Decays

Roy A. Briere

Abstract

Some recent CLEO results in B physics are reviewed. Included are measurements of the rate and CP asymmetry for $b \to s\gamma$ and rates for rare charmless $b \to u$ decays. We follow with a brief survey of other B decays of interest for CP-violation studies results, and conclude with a look at future directions.

50.1 Introduction

CLEO has been investigating B physics for 20 years. Overconstraining the CKM sector of the standard model is of great current interest; b quark decays can access much of this matrix. While semileptonic decays measure CKM element magnitudes such as $|V_{cb}|$ and $|V_{ub}|$, other modes are also important. Electroweak processes allow unfettered glimpses of loop diagrams, while rare hadronic processes can allow CP-violating phases to show up as interference effects, both versus time and in rate asymmetries.

The data discussed here include 5 fb^{-1} taken with the CLEO II detector from 1989–1995, and also portions of the 9 fb^{-1} taken with CLEO II.5 from 1995–1999. CLEO II was distinguished by the addition of a CsI calorimeter, while CLEO II.5 further includes a silicon vertex detector, the first to be operated at the $\Upsilon(4S)$ resonance. More details on the CLEO II detector may be found elsewhere [1, 2].

The CLEO detector operates at the Cornell Electron Storage Ring, which achieved a peak luminosity of nearly $8 \times 10^{32}\text{cm}^{-2}\text{s}^{-1}$, the highest at any collider, allowing CLEO to log 600 pb^{-1} of data in its best month. Standard running comprises two energies: on the $\Upsilon(4S)$ resonance, just above $B\bar{B}$ threshold, and in the continuum region 60 MeV below it. Two-thirds of our luminosity is taken on the resonance; with this ratio, 1.4 fb^{-1} total data yields 10^6 $B\bar{B}$ pairs. The *total* luminosity of the data used for each of our results is quoted. Note that all results not yet published are preliminary and hence subject to minor changes.

50.2 Kinematics of Full Reconstruction

With our symmetric beams, the energy of a B is given by $E_B = E_{beam}$, which is well-known. The average B meson momentum is small, $|P_B| \simeq 325$ MeV/c. Summing over all B decay daughters, we define $\Delta E \equiv \sum E_i - E_{beam}$, expressing energy conservation. This is sensitive to both missing particles and particle (*e.g.*, K-π) misidentification. The resolution is about 25 MeV for $B \to K\pi$; however, it is much improved, to 8 MeV, for final states like $B \to D^{*+}D^{*-}$ with large rest-mass. The "beam-constrained" mass is defined as $M_B \equiv \sqrt{E_{beam}^2 - |\sum \vec{P}_i|^2}$, expressing momentum conservation. Using E_{beam} instead of $\sum E_i$ improves the mass resolution by a factor of 10, to 2.5 MeV, for two-body modes like $B \to K\pi$.

These variables are used extensively for hadronic decays where the decay products are all accessible; they are also used for some semileptonic modes, where the neutrino is inferred from global four-momentum balance [3].

50.3 $b \to s\gamma$

Since tree-level $b \to s$ transitions are absent in the standard model, new physics can compete with the leading electroweak loop diagram which is suppressed. The inclusive rate for $b \to s\gamma$, first measured by CLEO [4], is an important constraint on many extended models, for example, the charged Higgs in minimal SUSY.

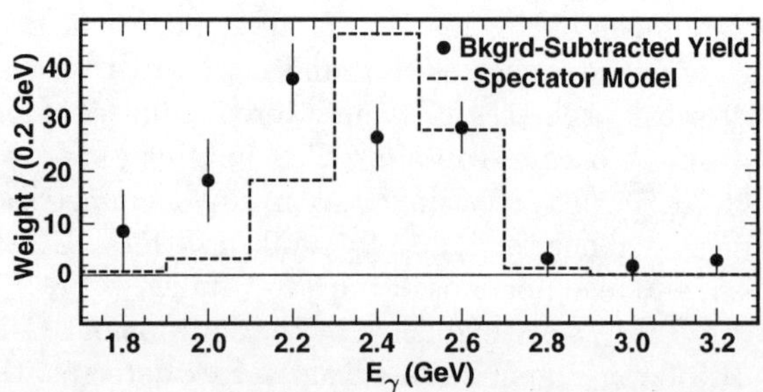

Figure 50.1: The final background-subtracted E_γ spectrum from the $b \to s\gamma$ analysis, showing a clear excess. The data is compared to a simple hadronization model shown as the dashed histogram.

Experimentally, the key handle for an inclusive measurement is the high-energy photon. The largest backgrounds are continuum $q\bar{q}$ processes and initial-state radiation; stiff photons are rare in generic B decays.

CLEO's analysis combines two techniques, each yielding a weight characterizing the signal vs. background quality of the events. Combining weights for events common to both analyses yields the greatest statistical power.

Decay Mode	Branching Ratio ($\times 10^{-5}$)
$B \to K^*\ell^+\ell^-$	< 0.68 90% CL
$B \to K\ell^+\ell^-$	< 0.70 90% CL
$b \to s\ e^+e^-$	< 5.7 90% CL
$b \to s\ \mu^+\mu^-$	< 5.8 90% CL
$b \to s\ e^\pm\mu^\mp$	< 2.2 90% CL

Table 50.1: Upper limits on electroweak penguin decay rates.

One method is a shape analysis that looks only for the photon. A sophisticated combination of variables, describing event shape and energy flow, is used to suppress background. The other technique utilizes a modified full reconstruction analysis. A stiff photon, a kaon ($K^\pm$ or K_S), and from 0–4 pions (where at most one is a π^0) are combined into a B candidate. The event weight is a function of ΔE, M_B, and one shape variable.

The measured E_γ spectrum is detailed in fig. 50.1, showing a clear excess in the energy range expected from a quasi–two body γX_s final state. Our new preliminary result based on 5 fb^{-1} of data is [5]

$$\mathcal{B}(b \to s\gamma) = (3.15 \pm 0.35(\text{stat}) \pm 0.32(\text{syst}) \pm 0.26(\text{model})) \times 10^{-4}. \quad (50.1)$$

This is consistent with theory, $\mathcal{B}(b \to s\gamma) = (3.28 \pm 0.33) \times 10^{-4}$ [6]. For further discussion and some charged Higgs limit curves, see ref. [7].

A CP-violating rate asymmetry, A, may be constructed if the separate b and $\bar{b}$ rates are known. Effects might show up in A with little change in the rate, depending on the phases of the new physics and of strong rescattering. The required B flavor tag (B or K charge) comes from the reconstruction analysis, formerly used only to suppress background to the E_γ spectrum. Monte Carlo events are used to correct for mistags.

We find that $A = (0.16 \pm 0.14 \pm 0.05) \times (1.00 \pm 0.04)$, where errors are statistical, additive systematic, and multiplicative systematic. Expressed as a limit, $-0.09 < A < 0.42$, 95% CL.

Results on related electroweak penguin modes from the same data sample are beginning to approach standard model predictions; these are summarized in table 50.1 [8, 9].

50.4 Rare Charmless *B* Decays

Let us now turn to the rare charmless B decays, first discussing some common aspects of the analyses.

To maintain high efficiency, we use likelihood fits to extract yields. Quantities fit include ΔE, M_B, resonance masses and helicities, particle ID information, and shape variables for suppressing continuum backgrounds. Plots of results, however, necessarily include additional requirements beyond the loose cuts used to select events for the fitter.

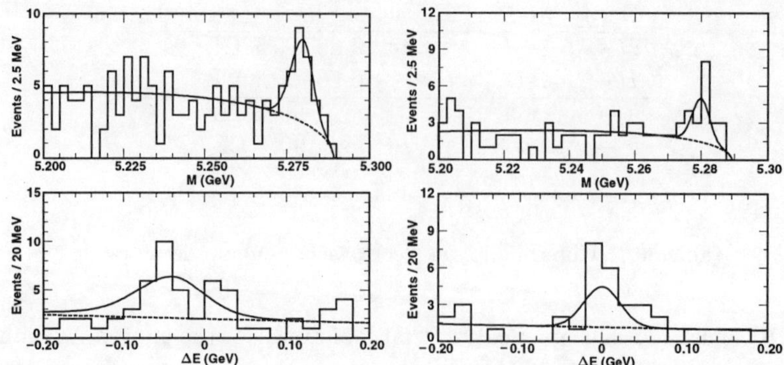

Figure 50.2: Fit projections for $K^+\pi^0$ (left column) and π^+K_S (right column) modes.

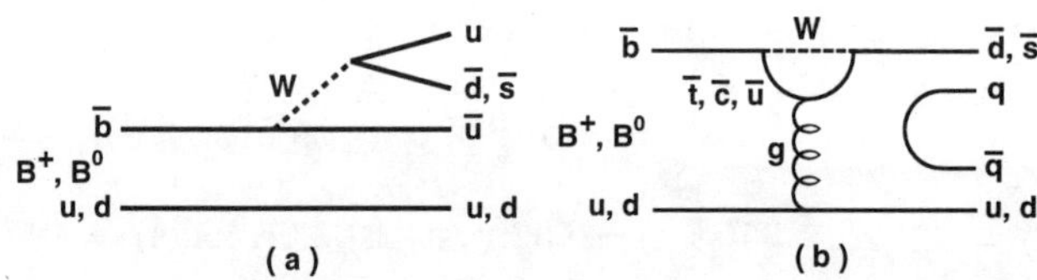

Figure 50.3: Spectator (a) and penguin (b) diagrams in charmless B decays.

In CLEO II.5, the K-π separation at 2.6 GeV/c from dE/dx is about 2σ and an additional, independent 1.7σ is obtained from the mass dependence inherent in calculating ΔE. Thus, simultaneous fits are used; for example, the rates for $B^+ \to \eta' K^+$ and $B^+ \to \eta' \pi^+$ are extracted from a single fit.

CLEO recently published three major papers, all based on 5 fb^{-1} of CLEO II data. These publications covered $K\pi$ and related final states [10], final states with an η or η' [11], and final states containing an ω or ϕ [12]. We now summarize the recent, unpublished updates to these results, based on 8.5 fb^{-1} of data.

Decay Mode	Branching Ratio ($\times 10^{-5}$)
$B^0 \to \pi^+\pi^-$	< 0.84 90% CL
$B^+ \to \pi^+\pi^0$	< 1.6 90% CL
$B^0 \to K^+\pi^-$	$1.4 \pm 0.3 \pm 0.2$
$B^+ \to K^+\pi^0$	$1.5 \pm 0.4 \pm 0.3$
$B^+ \to K^0\pi^+$	$1.4 \pm 0.5 \pm 0.2$
$B^0 \to K^+K^-$	< 0.24 90% CL
$B^+ \to K^+\bar{K}^0$	< 0.93 90% CL

Table 50.2: Summary of $B \to K\bar{K}, K\pi$, and $\pi\pi$ rates; note that $K^+\pi^0$ is new first observation.

Decay Mode	Branching Ratio ($\times 10^{-5}$)
$B^+ \to \eta' K^+$	$7.4^{+0.8}_{-1.3} \pm 1.0$
$B^0 \to \eta' K^0$	$5.9^{+1.8}_{-1.6} \pm 0.9$
$B^+ \to \eta' \pi^+$	< 1.2 90% CL

Table 50.3: New $B \to \eta'$ rate results.

Decay Mode	Branching Ratio ($\times 10^{-5}$)
$B^+ \to \pi^+ \rho^0$	$1.5 \pm 0.5 \pm 0.4$
$B^0 \to \pi^\mp \rho^\pm$	$3.5^{+1.1}_{-1.0} \pm 0.5$
$B^0 \to \pi^\pm K^{*0}$	< 2.7 (90% CL).
$B^0 \to \pi^\pm K^{*\mp}$	$2.2^{+0.8}_{-0.6} \pm ^{+0.4}_{-0.5}$

Table 50.4: New $B \to PV$ decay rates.

Results for the $B \to KK, K\pi$ and $\pi\pi$ modes are given in table 50.2 [13]. Some typical fit projections are shown in fig. 50.2. Since $K^\pm \pi^\pm$ is significant, and not $\pi^\pm \pi^\mp$, penguin diagrams appear dominant; this can be seen from the CKM factors in these decay diagrams; see fig. 50.3. Other new results include updated rates for $B \to \eta' K$, shown in table 50.3 [14]; these remain surprisingly large.

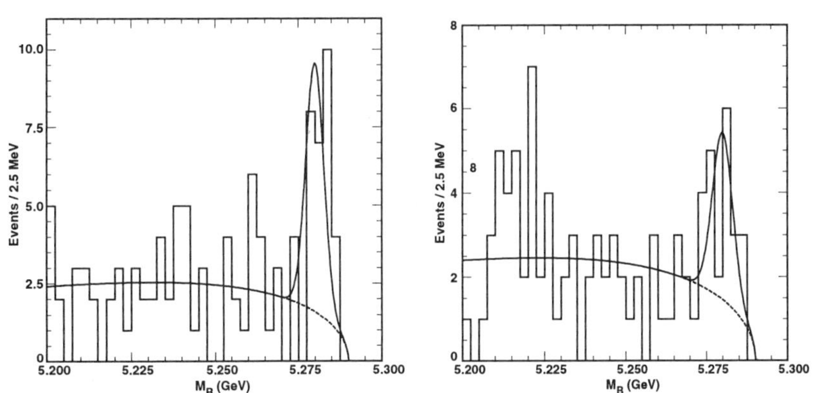

Figure 50.4: Beam-constrained mass plots for $B \to \pi\rho$ decays.

CLEO recently observed several new PV (pseudoscalar vector) modes [15]. We use $\rho^{0,+} \to \pi^+\pi^{-,0}$ and $K^{*0,+} \to K + \pi^{-,0}$ decays, and fit $\pi^+\rho$ and $K^+\rho$ simultaneously. The $K^{*+}\pi^-$ result is further supplemented using the $K^{*+} \to K_S \pi^+$ decay. Results are summarized in table 50.4; we find the $\pi\rho$ modes (see fig. 50.4) to be comparable to the πK^*, unlike the $B \to PP$ and $K\pi$ vs. $\pi\pi$.

Decay Mode	Branching Ratio ($\times 10^{-4}$)
$B^- \to \psi(2S)K^-$	$(7.1 \pm 0.9 \pm 0.7)$
$B^- \to \psi(2S)K^{*-}$	$(11.1 \pm 3.2 \pm 1.3)$
$B^0 \to \psi(2S)K^0$	$(5.2 \pm 1.4 \pm 0.5)$
$B^0 \to \psi(2S)K^{*0}$	$(8.4 \pm 1.8 \pm 1.0)$

Table 50.5: Rates for $\psi' K^{(*)}$ modes.

50.5 Other Hadronic B Decays

There are also many $b \to c$ decays useful for CP violation studies.

The first DK mode, subject of many rate triangles, was seen at the expected rate, $\mathcal{B}(B^- \to D^0 K^-) = (2.57 \pm 0.65 \pm 0.32) \times 10^{-4}$ [16]. The first $B^0 \to D^{(*)}\bar{D}^{(*)}$ mode, which may complement $\psi K^{(*)}$ for $\sim 2\beta$ measurement, has also been observed: $\mathcal{B}(B^0 \to D^{*+}D^{*-}) = (6.2^{+4.0}_{-2.9} \pm 1.0) \times 10^{-4}$ [17]. We are currently working on related modes.

CLEO is also active in modes containing charmonium. New branching ratios results for $B \to \psi(2S)K^{(*)}$ modes, using 8.7 fb^{-1} of data, are summarized in table 50.5. The CP eigenstate $B^0 \to \psi(2S)K^0$ is seen for the first time. First observations of $B \to \psi\pi^0$ and $B \to \chi_{c1}K_S$, performed with 9.5 fb^{-1} of data, are displayed in fig. 50.5. The branching ratios are included in table 50.5.

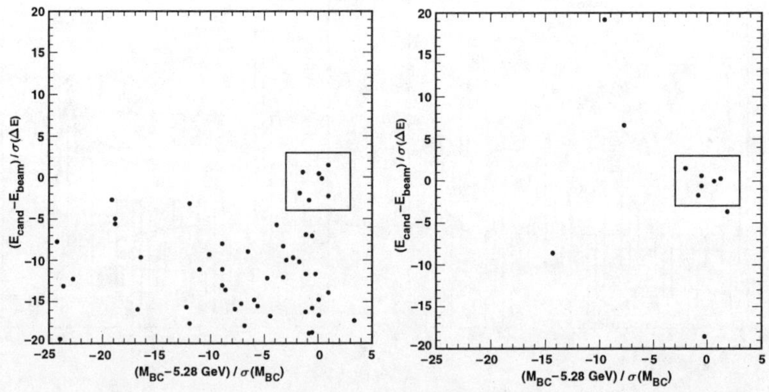

Figure 50.5: $M_B - \Delta E$ plots for $B \to \psi\pi^0$ and $B \to \chi_{c1}K_S$.

Statistics for various CP eigenstate decay modes suitable for the standard $\sin 2\beta$ analysis are collected in table 50.6. There are clean modes to supplement the usual $B^0 \to J/\psi K^0$; $K_S \to \pi^+\pi^-$. The CP +1 eigenstate is also available, with relative statistics sufficient to allow a check of the expected inversion of the time-dependent decay asymmetry with CP.

Decay Mode	Events	CP	Branching Ratio
$B^0 \to J/\psi K^0$			
$K_S \to \pi^+\pi^-$	75	-1	$(0.92 \pm 0.11 \pm 0.11) \cdot 10^{-3}$
$K_S \to \pi^0\pi^0$	15	-1	$(1.21 \pm 0.31 \pm 0.25) \cdot 10^{-3}$
$B^0 \to J/\psi\pi^0$	7	$+1$	$(3.4^{+1.7}_{-1.5} \pm 0.4) \cdot 10^{-5}$
$B^0 \to \chi_{c1} K^0$	6	-1	$(4.5^{+2.8}_{-1.8} \pm 0.9) \cdot 10^{-4}$
$B^0 \to \psi(2S) K^0$	15	-1	$(5.2 \pm 1.4 \pm 0.5) \cdot 10^{-4}$

Table 50.6: Summary of relative signal sizes in exclusive B decays to CP eigenstates with charmonia.

50.6 Conclusion and Outlook

CLEO remains the major source of information on rare B decays; many results from our full 14 fb^{-1} of data are in press. We plan to include searches for CP-violating rate asymmetries in $K\pi$ and $\eta' K$. Other items in progress in the CKM matrix arena include updated measurements of the magnitudes of $|V_{ub}|$ and $|V_{cb}|$.

CLEO is also a major force in charm physics (including $D - \bar{D}$ mixing [18]), charm baryon spectroscopy, and tau, Upsilon, and two-photon physics.

The CLEO III upgrade includes a ring-imaging Cherenkov detector for improved particle identification. In addition, a new drift chamber, silicon vertex detector, beam pipe, and interaction region will be installed. The concurrent CESR machine upgrade includes superconducting RF cavities and new superconducting focusing quadrupoles. Peak luminosities should approach $2 \times 10^{33}\text{cm}^{-2}\text{s}^{-1}$, yielding 15 fb^{-1} of data per year. This will allow CLEO to contribute to B physics for many more years.

References

[1] Y. Kubota *et al.*, Nucl. Instrum. Methods Phys. Res., Sec. A **320**, 66 (1992).

[2] T.S. Hill, Nucl. Instrum. Methods Phys. Res., Sec. A **418**, 32 (1998).

[3] J. Alexander *et al.*, Phys. Rev. Lett. **77**, 5000 (1996).

[4] M.S. Alam *et al.*, Phys. Rev. Lett. **74**, 2885 (1995).

[5] S. Glenn *et al.*, CLEO CONF 98-17; ICHEP 1011.

[6] K. Chetyrkin, M. Misiak, and M. Munz, Phys. Lett. B **400**, 206 (1997), E: *ibid.*, **425**, 414 (1998).

[7] F. Borzumati and C. Greub, Phys. Rev. D **59**, 057501 (1999), and references therein.

[8] S. Glenn *et al.*, Phys. Rev. Lett. **80**, 2289 (1998).

[9] R. Godang *et al.*, CLEO CONF 98-22.

[10] R. Godang *et al.*, Phys. Rev. Lett. **80**, 3456 (1998).

[11] B.H. Behrens *et al.*, Phys. Rev. Lett. **80**, 3710 (1998).

[12] T. Bergfeld *et al.*, Phys. Rev. Lett. **81**, 272 (1998).

[13] M. Artuso *et al.*, CLEO CONF 98-20; final results are reported in D. Cronin-Hennessy *et al.*, CLNS 99/1650 (to appear in Phys. Rev. Lett.).

[14] B.H. Behrens *et al.*, CLEO CONF 98-09; final results are reported in S.J. Richichi it et al., CLNS 99/1649 (to appear in Phys. Rev. Lett.).

[15] Y. Gao and F. Wurthwein, hep-ex/9904008; final results are reported in C.P. Jessop *et al.*, CLNS 99/1652 (submitted to Phys. Rev. Lett.).

[16] M. Athanas *et al.*, Phys. Rev. Lett. **80**, 5493 (1998).

[17] M. Artuso *et al.*, Phys. Rev. Lett. **82**, 3020 (1999); final results are reported in E. Lipeles *et al.*, CLNS 00/1663 (to appear in Phys. Rev. D).

[18] H.N. Nelson, this volume, chapter 47.

51

CP Violation in B Decays at the Tevatron

I. Joseph Kroll

Abstract

From 1992 to 1996, the CDF and D0 detectors each collected data samples exceeding 100 pb^{-1} of $p\bar{p}$ collisions at $\sqrt{s} = 1.8$ TeV at the Fermilab Tevatron. These data sets led to a large number of precision measurements of the properties of B hadrons including lifetimes, masses, neutral B meson flavor oscillations, and relative branching fractions, and to the discovery of the B_c meson. Perhaps the most exciting result was the first look at the CP violation parameter $\sin(2\beta)$ using the world's largest sample of fully reconstructed $B^0/\bar{B}^0 \to J/\psi K_s^0$ decays. A summary of this result is presented here. In the year 2000, the Tevatron will recommence $p\bar{p}$ collisions with an over order of magnitude expected increase in integrated luminosity (1 fb^{-1} per year). The CDF and D0 detectors will have undergone substantial upgrades, particularly in the tracking detectors and the triggers. With these enhancements, the Tevatron B physics program will include precision measurements of $\sin(2\beta)$ and B_s^0 flavor oscillations, as well as studies of rare B decays that are sensitive to new physics. The studies of B_s^0 mesons will be particularly interesting as this physics will be unique to the Tevatron during the first half of the next decade.

51.1 Introduction

In this paper, we review results on CP violation in B decays from the Fermilab Tevatron and discuss prospects in the near future. The results are based on the analysis of 110 pb^{-1} of $p\bar{p}$ collisions at $\sqrt{s} = 1.8$ TeV, collected from August 1992 to February 1996, which we refer to as Tevatron Run I. At present, only the CDF Collaboration has presented results on CP violation in B decays from these data.

The Tevatron will recommence $p\bar{p}$ collisions in August 2000 at $\sqrt{s} =$ 2.0 TeV. This future data-taking period is referred to as Run II. The new crucial accelerator component, the main injector, has been commissioned successfully and will increase the rate of production of antiprotons by a factor of three above previous rates. The expected data rate is 2 fb^{-1} in the first two years of operation. This corresponds to approximately 10^{11} $b\bar{b}$ pairs per year.

The Tevatron will continue to operate beyond these first two years. Ultimately a data sample of more than 20 fb^{-1} may be collected prior to the turn-on of the LHC. At that time, Fermilab may continue operation of the Tevatron with an experiment dedicated to the study of B physics. The physics motivation and capabilities of such a detector are not discussed here, but an excellent discussion can be found in [1].

As has been discussed in numerous other talks at this conference, the study of B hadron decays plays a unique role in the test of the standard model of electroweak interactions and the study of the Cabibbo-Kobayashi-Maskawa matrix V_{CKM}. Measurements of the decays of B hadrons determine the magnitudes of five of the nine elements of V_{CKM} as well as the phase.

The unitarity of V_{CKM} leads to nine unitarity relationships, one of which is of particular interest:

$$V_{\text{ud}}V_{\text{ub}}^* + V_{\text{cd}}V_{\text{cb}}^* + V_{\text{td}}V_{\text{tb}}^* = 0. \tag{51.1}$$

This sum of three complex numbers forms a triangle in the complex plane, commonly referred to as *the* unitarity triangle. Measurements of the weak decays of B hadrons and the already known CKM matrix elements determine the magnitudes of the three sides of the unitarity triangle, and CP asymmetries in B meson decays determine the three angles. The primary goal of B physics in the next decade is to measure precisely both the sides and angles of this triangle and test consistency within the standard model.

We can use several approximations to express eq. (51.1) in a more convenient form. The elements $V_{\text{ud}} \simeq 1$ and $V_{\text{cd}} \simeq -\lambda = -\sin\theta_{\text{C}}$, where θ_{C} is the Cabibbo angle, are well measured. Although the elements V_{tb} and V_{ts} are not well measured, the theoretical expectations are that $V_{\text{tb}} \simeq 1$ and $V_{\text{ts}} \simeq -V_{\text{cb}}^*$. With these assumptions, eq. (51.1) becomes

$$\frac{V_{\text{ub}}^*}{\lambda V_{\text{cb}}^*} - 1 - \frac{V_{\text{td}}}{\lambda V_{\text{ts}}} = 0. \tag{51.2}$$

A pictorial representation of this equation as a triangle in the complex plane is shown in fig. 51.1; the vertical axis is the complex axis. In the Wolfenstein approximation [2,3], the apex of this triangle is (ρ, η).

The next experimental steps in studying this triangle will be the measurement of the angle β from the asymmetry in the decays $B^0/\bar{B}^0 \to J/\psi K_{\text{S}}^0$, and the measurement of $|V_{\text{td}}/V_{\text{ts}}|$ from the ratio of mass differences $\Delta m_d/\Delta m_s$ determined from $B^0 - \bar{B}^0$ and $B_s^0 - \bar{B}_s^0$ flavor oscillations. Since B_s^0 mesons are not produced on the $\Upsilon(4S)$ resonance, the measurement of $B_s^0 - \bar{B}_s^0$ flavor oscillations as well as the study of B_s^0 decays will be unique to experiments operating at hadron machines.

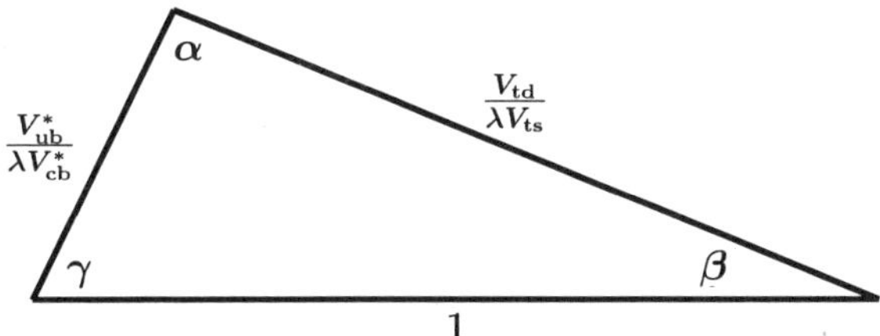

Figure 51.1: A pictorial representation of the unitarity triangle in the complex plane. The vertical axis is the complex axis.

In the B system, the measurements of CP violation that are cleanly related (*i.e.*, without large theoretical uncertainties) to angles in the unitarity triangle are from asymmetries in the decays of neutral B mesons to CP eigenstates. The most popular mode is $B^0/\bar{B}^0 \to J/\psi K^0_S$. It is lucky that at hadron colliders the leptonic decays $J/\psi \to \mu^+\mu^-$ and $J/\psi \to e^+e^-$ form the basis of a practical trigger for this decay mode and the means to reconstruct the decay mode with excellent signal to noise.

CP violation is observed as an asymmetry $\mathcal{A}_{CP}$ in decay rates:

$$\mathcal{A}_{CP} = \frac{\frac{dN}{dt}(\bar{B}^0 \to J/\psi\, K^0_S) - \frac{dN}{dt}(B^0 \to J/\psi\, K^0_S)}{\frac{dN}{dt}(\bar{B}^0 \to J/\psi\, K^0_S) + \frac{dN}{dt}(B^0 \to J/\psi\, K^0_S)}, \tag{51.3}$$

where $\frac{dN}{dt}(\bar{B}^0 \to J/\psi\, K^0_S)$ is the rate of observed $J/\psi\, K^0_S$ given the particle produced was a $\bar{B}^0$. The asymmetry is produced by the interference of direct decays ($\bar{B}^0 \to J/\psi\, K^0_S$) and decays that occur after mixing ($\bar{B}^0 \to B^0 \to J/\psi\, K^0_S$). The asymmetry oscillates as a function of proper decay time t with a frequency Δm_d, and the amplitude of the oscillation is $\sin(2\beta)$:

$$\mathcal{A}_{CP}(t) = \sin(2\beta)\sin(\Delta m_d t). \tag{51.4}$$

Unlike experiments operating at the $\Upsilon(4S)$, the time-integrated asymmetry of eq. (51.3) (*i.e.*, replacing the rates in eq. (51.3) with the total observed numbers) does not vanish in hadron collisions:

$$\mathcal{A}_{CP} = \frac{x_d}{1+x_d^2}\sin(2\beta) \approx 0.5\sin(2\beta), \tag{51.5}$$

where $x_d = \Delta m_d/\tau(B^0) = 0.732 \pm 0.032$ [3]. The value of x_d falls fortuitously close to 1.0, the value that maximizes the coefficient in front of $\sin(2\beta)$. In contrast, the large value of x_s ($x_s > 14.0$ at 95% CL [3]) implies that even if there were large CP violation due to B^0_s mixing, the asymmetry cannot be observed using a time integrated asymmetry.

Even though the time-integrated asymmetry can be used to extract $\sin(2\beta)$, it is better to measure the asymmetry as a function of proper decay time (*i.e.*, the time-dependent asymmetry), if possible. The improvement comes from two sources. First, there is more statistical power in the time-dependent asymmetry. Decays at low lifetime exhibit a small asymmetry because there has not been adequate time for mixing to occur to create the interference leading to CP violation. Second, a substantial fraction of the combinatoric background occurs at low values of t, well below the value of t (about 2.2 lifetimes) where $\mathcal{A}_{CP}(t)$ is a maximum.

51.2 Current Results

So far at the Tevatron, only the CDF experiment has had the capability to measure CP violation in B decays. This will change in Run II, when major upgrades of both the CDF and D0 detector will make both detectors capable of unique and important measurements of B hadron decays.

The features of the Run I CDF detector [4] crucial for B physics included a four-layer silicon microstrip detector, a large-volume drift chamber, and excellent electron and muon identification. The silicon microstrip detector provided an impact parameter (d_0) resolution for charged tracks of $\sigma(d_0) = (13 + 40/p_T)\,\mu$m, where p_T is the magnitude of the component of the momentum of the track transverse to the beam line in units of GeV/c. This impact parameter resolution made the precise measurement of B hadron proper decay times t possible. The drift chamber was 1.4 m in radius and was immersed in a 1.4 T axial magnetic field. It provided excellent momentum resolution $(\delta p_T/p_T)^2 = (0.0066)^2 \oplus (0.0009 p_T)^2$ (where p_T is in units of GeV/c) and excellent track reconstruction efficiency making it possible to fully reconstruct B hadron decays with excellent mass resolution and high signal to noise. Electron (e) and muon (μ) detectors in the central rapidity region ($|y| < 1$) made it possible to detect (and trigger on) B hadrons using semileptonic decays ($b \to \ell X, \ell = e, \mu$) or using $B \to J/\psi X, J/\psi \to \mu^+\mu^-$.

The measurement of $\mathcal{A}_{CP}(t)$ of eq. (51.3) has three crucial experimental components: (1) reconstructing the decay mode $B^0/\bar{B}^0 \to J/\psi K^0_S$ with good signal to noise; (2) measuring the proper decay time t; and (3) determining whether the meson that was produced was a B^0 (*i.e.*, $\bar{b}d$) or a $\bar{B}^0$ (*i.e.*, $b\bar{d}$). This last component is known as "b flavor tagging," and it is the most challenging of the above three requirements. The CDF experiment has demonstrated that the first two are possible by extracting large-statistics, low-background signals of fully reconstructed B decays using $J/\psi \to \mu^+\mu^-$ and partially reconstructed semileptonic decays (*e.g.*, $B^0_s \to \ell^+ D^-_s X$ and charge conjugate). With these signals, CDF has made some of the most precise species-specific measurements of B hadron lifetimes [3] to date.

The performance of the b flavor tags may be quantified conveniently by their efficiency ϵ and dilution D. The efficiency is the fraction of B candidates to which the flavor tag can be applied ($0 < \epsilon < 1$). The dilution is related to the probability $\mathcal{P}$ that the tag is correct: $D = 2\mathcal{P} - 1$, so a perfect tag has $D = 1$, and a random tag has $D = 0$.

The experimentally measured amplitude of the asymmetry in eq. (51.4) is reduced by the dilution of the tag:

$$\mathcal{A}_{CP}^{\text{meas}}(t) = D\sin(2\beta)\sin(\Delta m_d t). \tag{51.6}$$

The statistical error on the true asymmetry $\mathcal{A}$ is approximately (for $D^2\mathcal{A} \ll 1$)

$$\delta\mathcal{A} \approx \sqrt{\frac{1}{\epsilon D^2 N}}, \tag{51.7}$$

where N is the total number of candidates (signal and background) before applying the flavor tag. The statistical power of the data sample scales with ϵD^2. At a hadron collider, a flavor tag with $\epsilon D^2 \geq 1\%$ is respectable.

To extract $\sin(2\beta)$ from the measured asymmetry, the value of the dilution of the flavor tag(s) must be determined quantitatively. The most reliable means of determining D is from the data themselves. Measurements of neutral B meson flavor oscillations requires b flavor tagging as well. In this case, the measured asymmetry $\mathcal{A}_{\text{mix}}$ is given by

$$\mathcal{A}_{\text{mix}}(t) = \frac{N_{\text{unmixed}}(t) - N_{\text{mixed}}(t)}{N_{\text{unmixed}}(t) + N_{\text{mixed}}(t)} = D \cdot \cos(\Delta m t), \tag{51.8}$$

where $N_{\text{unmixed}}(t)$ $[N_{\text{mixed}}(t)]$ are the number of candidates observed to have decayed with the same [opposite] b flavor as they were produced with. CDF has used measurements of $B^0 - \bar{B}^0$ flavor oscillations to determine D for three different b flavor tagging methods.

The methods of b flavor tagging fall into two categories: (1) opposite-side flavor tags (OST) and (2) same-side flavor tags (SST). The dominant production mechanisms of b quarks in hadron collisions produce $b\bar{b}$ pairs. Opposite-side flavor tags exploit this fact: to identify the production flavor of the B hadron of interest (*e.g.*, the one that eventually leads to a $J/\psi K^0_S$), we identify the flavor of the second B hadron in the event and *infer* the flavor at production of the first B. Since the Run I CDF central drift chamber was fully efficient only for the central rapidity region ($-1 < \eta < 1$, where η is pseudorapidity), full reconstruction of B decays and b flavor tagging were usually restricted to this central region. If a $J/\psi K^0_S$ candidate is detected in this region, then the second B is in this central region as well only about 50% of the time. This means that opposite-side flavor tags at CDF have a maximum efficiency of $\epsilon = 0.5$.

So far, two opposite side flavor tags have been used. (1) a lepton tag and (2) a jet-charge tag. The lepton tag is based on b semileptonic decay: $b \to \ell^- X$, but $\bar{b} \to \ell^+ X$, where $\ell = e, \mu$. The charge of the lepton identifies the flavor of the b. The low semileptonic branching fraction ($\sim 10\%$ per lepton flavor) limits the efficiency of this tag, although the flavor tag dilution is high. Of course leptons can originate from the decays of secondary decay products in B decay, *e.g.*, $b \to c \to \ell^+ X$, and this will cause the wrong production flavor to be assigned, reducing the dilution. The jet-charge tag is based on the momentum weighted charge average Q^b_{jet} of the charged particles produced in the fragmentation of a b quark and in the subsequent decay of the B hadron. On average, a b quark will produce a jet charge less than zero ($Q^{b}_{\text{jet}} < 0$), but a $\bar{b}$ quark will produce a jet charge larger than zero ($Q^{\bar{b}}_{\text{jet}} > 0$). The jet-charge tag is more efficient, but has a smaller dilution, than the lepton-tag. The jet-charge tag was especially effective in measurements of neutral B meson flavor oscillations performed at e^+e^- colliders operating on the Z^0 resonance, but it is much more challenging to apply this method in the environment of hadronic collisions. Both the lepton flavor tag and the jet-charge flavor tag were used in a precise measurement of Δm_d [5] by CDF.

The same-side flavor tag exploits the correlation between the b flavor and the charge of the particles produced by the b quark fragmentation, as illustrated in fig. 51.2.

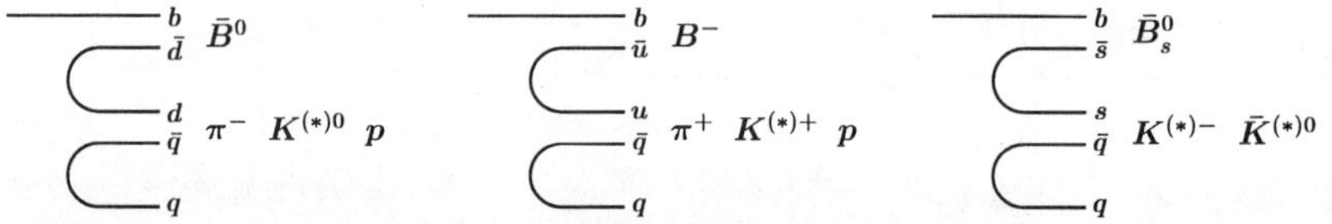

Figure 51.2: The same-side flavor tag is based on the correlation between b flavor and the charge of particles produced in b quark fragmentation.

A π^+ tags a B^0, but a π^- tags a $\bar{B}^0$. If a charged B hadron is produced, the correlation between pion charge and b flavor is the *opposite* to the correlation in the case of the B^0. This has important consequences when utilizing fully reconstructed charged B hadron decays (*e.g.*, $B^\pm \to J/\psi K^\pm$) to quantify the performance of the SST when applied to B^0. The idea of the SST was originally proposed by Gronau, Nippe, and Rosner [6]. The decays of P wave B^{**} mesons produce the same b flavor-pion charge correlations. Since the SST is based on particles produced with the B of interest, if this B is in the experimental acceptance, then it is likely that these fragmentation particles are in the acceptance as well. As a result, the efficiency of the SST is potentially larger than the efficiency of the opposite-side tags. The SST was used successfully [7] as a flavor tag in a precise measurement of Δm_d by CDF using approximately 6 000 partially reconstructed semileptonic B^0 decays in the decay modes $B^0 \to \ell^- D^{(*)+} X$. The mixing signal is shown in fig. 51.3.

This figure depicts the experimental asymmetry described by eq. (51.8). The points with error bars are the measured asymmetry plotted as a function of proper decay length ct. The upper plot is for B^+ decays and the lower two plots are for B^0 decays. The dashed curves are the expected asymmetry fit to the data. The two lower curves show the expected $\cos(\Delta m_d t)$ behavior from B^0 mixing. The intercept on the vertical axis at $ct = 0$ is the dilution D_0, where the subscript indicates this is the dilution for neutral B^0. Charged B^+ do not oscillate, so the curve in the uppermost

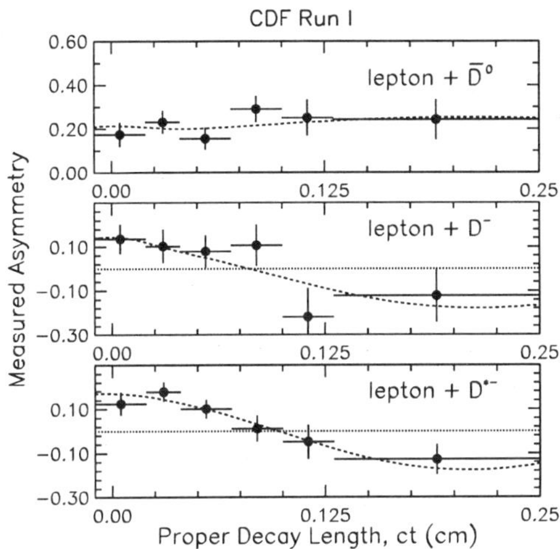

Figure 51.3: The mixing asymmetry $\mathcal{A}_{\text{mix}}$ observed in a sample of partially reconstructed B^+ and B^0 semileptonic decays using a same-side b flavor tag.

plot is a straight line (the small oscillation is due to a $\sim$ 15% contamination of B^0 in this sample). The intercept determines the dilution D_+ for the same-side tag for B^+. The charged dilution D_+ is larger than the neutral dilution D_0. Monte Carlo studies [7] confirm that this is due to charged kaons, which increase D_+, but decrease D_0, as illustrated in fig. 51.2.

Using this same-side flavor tag, and a signal of 198 ± 17 $B^0/\bar{B}^0 \to J/\psi K^0_S$ decays, CDF published [8] a first look at $\sin(2\beta)$ from the Run I data. The final state is reconstructed using the decays $J/\psi \to \mu^+\mu^-$ (this is also the basis of the trigger) and $K^0_S \to \pi^+\pi^-$. The measured asymmetry $\mathcal{A}$ as a function of proper decay length ct is shown in fig. 51.4.

The points with error bars are the experimental asymmetries; the curves are fits of these points to the expected form given in eq. (51.6). The resulting amplitude is $D\sin(2\beta) = 0.31 \pm 0.18\,(\text{stat.}) \pm 0.03\,(\text{syst.})$, where the largest contributions to the systematic error are the error on world average value of Δm_d and the ability to constrain potential detector biases that could produce false asymmetries. If the error on this measurement were small enough to establish that this amplitude is not zero, then this would be sufficient to establish CP violation in the B system. In this case, knowledge of the dilution is necessary only to determine the corresponding value of $\sin(2\beta)$. Using the value of the dilution $D = 0.166 \pm 0.018\,(\text{data}) \pm 0.013\,(\text{Monte Carlo})$ determined mainly with the B^0 mixing measurement discussed previously, the value of $\sin(2\beta)$ is $1.8 \pm 1.1\,(\text{stat.}) \pm 0.3\,(\text{syst.})$. The uncertainty on the dilution is contained in the systematic error on $\sin(2\beta)$. The nonphysical value is possible with low statistics; the corresponding limit is $-0.2 < \sin(2\beta)$ at 95% CL.

As this first result was limited by statistics, the CDF collaboration has

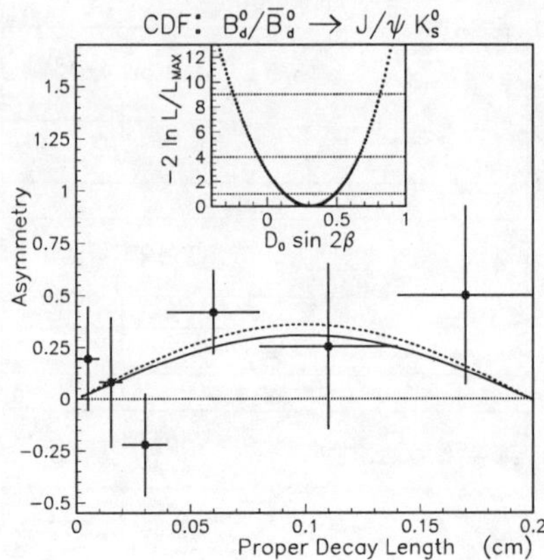

Figure 51.4: The measured asymmetry $\mathcal{A}_{cp}^{\text{meas}}$ as a function of proper decay length ct for $B^0/\bar{B}^0 \rightarrow J/\psi K^0_S$ decays.

continued to analyze the Run I data set with the aim of improving the statistical precision on $\sin(2\beta)$ as much as possible. An updated published measurement [9] supersedes the first publication. The statistical power of the first published result was improved in two ways: (1) the signal of $B^0/\bar{B}^0 \rightarrow J/\psi K^0_S$ was doubled, and (2) two opposite-side flavor tags (lepton tag and jet-charge tag) were added. The first published analysis used only candidates reconstructed in the silicon microstrip detector, to ensure a precise measurement of the B proper decay time t. Due to the large length (r.m.s of 30 cm) of the distribution of $p\bar{p}$ collisions along the beam line, only about 60% of the interactions are contained in the acceptance of the 50 cm long silicon detector. The newer analysis adds candidates that are not reconstructed in the silicon microstrip detector. The total signal of $B^0/\bar{B}^0 \rightarrow J/\psi K^0_S$ is 395 ± 31. The normalized mass[1] distributions of the candidates reconstructed both inside and outside of the acceptance of the silicon microstrip detector are shown in fig. 51.5. The mass resolution is approximately $\sigma_M \sim 10\,\text{MeV}/c^2$.

Candidates fully reconstructed in the silicon microstrip detector have a proper decay time resolution of $\sigma_{ct} \approx 60\,\mu\text{m}$, whereas the candidates not fully contained in this detector have $\sigma_{ct} \sim 300 - 900\,\mu\text{m}$.

Three methods of b flavor tagging are applied to this sample: the same-side tag described previously, a lepton tag, and a jet-charge tag. The lepton tag is identical to the tag used in a published mixing analysis [5]. The jet-charge tag is based on the jet-charge tag from the published mixing analysis [5], except that the method of associating charged particles to the jet was modified to increase the tag efficiency. Any candidate track for the

[1]By normalized mass we mean the difference of the measured mass from the world average B^0 mass divided by the estimated error on the measured mass: $[M(\mu^+\mu^-\pi^+\pi^-) - M(B^0)]/\sigma_M$.

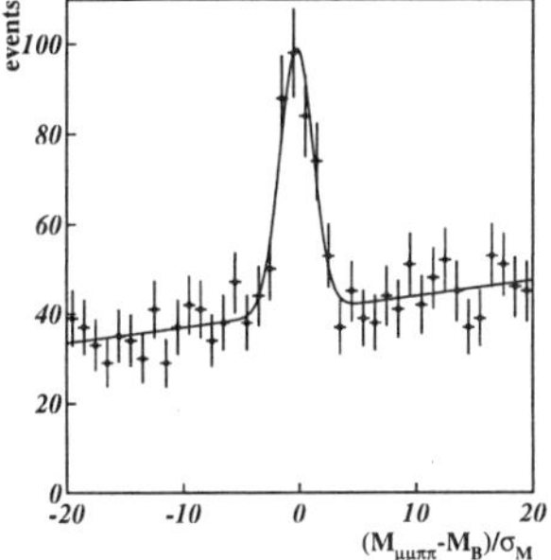

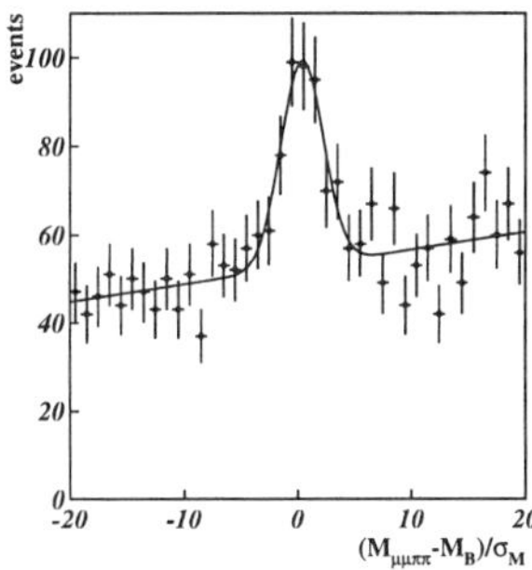

Figure 51.5: The normalized mass distribution of $B^0/\bar{B}^0 \to J/\psi K^0_S$ candidates, where $J/\psi \to \mu^+\mu^-$ and $K^0_S \to \pi^+\pi^-$. The left-hand (right-hand) plot is for candidates contained (not contained) in the silicon microstrip detector.

Flavor tag	efficiency (ϵ)	dilution (D)	ϵD^2
Lepton	$(5.6 \pm 1.0)\%$	$(62.5 \pm 14.6)\%$	$(2.2 \pm 1.0)\%$
Jet charge	$(40.2 \pm 2.2)\%$	$(23.5 \pm 6.6)\%$	$(2.2 \pm 1.3)\%$
Same side (in Si)	$\sim 70\%$	$(16.6 \pm 2.2)\%$	$(2.1 \pm 0.5)\%$
Same side (not in Si)		$(17.4 \pm 3.6)\%$	

Table 51.1: The performance of the three flavor tags as applied to the $B^0/\bar{B}^0 \to J/\psi K^0_S$ data sample. For the same-side tag, ϵ and ϵD^2 are for the total data sample (with and without silicon).

same-side tag is explicitly removed from the jet-charge determination. If a lepton tag exists, then the jet-charge tag is ignored, since the dilution of the lepton tag is much larger than the dilution of the jet-charge tag. These two requirements reduce the effect of correlations between the tags and simplify the analysis.

The performance of the two opposite-side tags is quantified using a signal of 998 ± 51 $B^\pm \to J/\psi K^\pm$ decays. This data sample has similar kinematics to the $B^0/\bar{B}^0 \to J/\psi K^0_S$ sample. As mentioned previously, the same-side tag is expected to perform differently for B^+ than for B^0, so this fully reconstructed charged B sample cannot be used to quantify the same-side tag. Instead the performance is determined as described in the published analysis [8]. The resulting flavor tag performance is summarized in table 51.1. The performances of the individual tags are comparable; the combined tagging effectiveness of all three flavor tags is $\epsilon D^2 = (6.3 \pm 1.7)\%$. The efficiency for tagging a $J/\psi K^0_S$ with at least one tag is $\sim 80\%$.

The value of $\sin(2\beta)$ is determined from the data using the method of maximum likelihood. The likelihood probability density includes the possibility of detector biases (*e.g.*, differences in reconstruction efficiency for positive and negative tracks) that may create false asymmetries. There are also terms that account for the behavior of backgrounds to the signal.

These backgrounds are dominated by two sources: (1) short-lived prompt background from prompt $p\bar{p} \to J/\psi + X$ production with a random K^0_S, and (2) long-lived backgrounds from $B \to J/\psi + X$ with a random K^0_S. A plot showing the data and the result of the fit is shown in fig. 51.6. The result is $\sin(2\beta) = 0.79^{+0.41}_{-0.44}$, where the error includes contributions from both the statistics and systematic effects (discussed below).

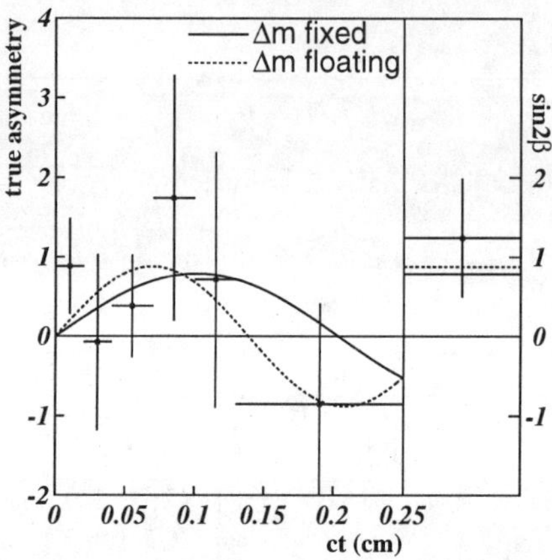

Figure 51.6: The updated determination of $\sin(2\beta)$.

The left-hand side of the figure shows the data for which the candidates are contained in the silicon microstrip detector; the points are the data corrected by the dilution so that they represent the true asymmetry (this is in contrast to fig. 51.4). The right-hand side of the figure shows the value of $\sin(2\beta)$ for the data not fully contained in the silicon microstrip detector. The two curves are the results of the fit to both the left-hand and right-hand data combined. The amplitude of the solid curve is the value of $\sin(2\beta)$ quoted above. The dashed curve is the same fit except that the value of Δm_d is included as a fit parameter. The result of this fit is $\sin(2\beta) = 0.88^{+0.44}_{-0.41}$ and $\Delta m_d = 0.68 \pm 0.17\,\text{ps}^{-1}$. For comparison, a time-integrated determination of $\sin(2\beta)$ from these same data yields $\sin(2\beta) = 0.71 \pm 0.63$; the resulting error is about 50% larger, which gives an indication of the statistical improvement made possible by making a time-dependent determination of $\sin(2\beta)$.

The statistics of the $J/\psi K^0_S$ sample contribute 0.39 to the total error. The main systematic contribution to the error is 0.16, which comes from the uncertainty in the dilutions of the flavor tags. This uncertainty is due to the limited statistics of the data used to calibrate the performance of the tags. As more data are accumulated in Run II, the sizes of both the $J/\psi K^0_S$ sample and the calibration samples will increase so that the statistics of the signal will continue to dominate the error on $\sin(2\beta)$.

The variation of the negative logarithm of the likelihood with $\sin(2\beta)$

follows a parabola near the minimum, so a confidence limit on $\sin(2\beta)$ may be derived in a straightforward manner. The Bayesian limit is $0 < \sin(2\beta) < 1$ at 95% CL. If the true value of $\sin(2\beta)$ were zero, then the probability of observing $\sin(2\beta) > 0.79$ is 3.6%. This result is the first compelling evidence that there is CP violation in B hadron decays.

51.3 Future Expectations

At present, the CDF and D0 experiments are undergoing upgrades to improve the performance of the detectors for the upcoming data-taking period that should begin in the year 2000. The new accelerator component, the main injector, will increase the production rate of antiprotons by approximately a factor of three over Run I rates. This increased rate will allow a substantial increase in instantaneous luminosity. The number of proton and antiproton bunches will be increased from six each in Run I, with bunches colliding every 3.5 μs, to thirty-six each in Run II, with bunches colliding every 396 ns. Eventually the machine will incorporate about 100 bunches each with collisions every 132 ns. The term "upgrade" is an understatement as the changes to the detectors are so substantial that much of the original components will be replaced. For example, both CDF and D0 are replacing their entire charged-particle tracking systems, and D0 is adding a 2 T superconducting solenoid. The electronics for components that were not upgraded have to be replaced to accommodate the shorter time between bunch crossings. Below we highlight the changes that are crucial for B physics.

At CDF the important changes [10] are to the tracking system, the trigger, and the particle identification capabilities. There will be a new eight-layer silicon system that extends from a radius of $r = 1.4$ cm from the beam line to $r = 28$ cm. This system will include $r - \phi$ and $r - z$ information to allow reconstruction in three dimensions. It will cover out to a pseudorapidity of $|\eta| < 2$, which will double the acceptance of the Run I detector. A new central drift chamber has been constructed. The drift cell size has been reduced by a factor of four, and ultimately with use of a fast gas, the drift time will be reduced by a factor of eight from Run I. The goal is to maintain the excellent reconstruction efficiency and momentum resolution of the Run I drift chamber. The trigger system is pipelined to allow "deadtimeless" operation. In particular, fast tracking is now possible at level one, and information from the silicon detectors will be available at level two, making it possible to trigger on the presence of tracks originating from the decay of long-lived B hadrons. This should allow triggering on all hadronic decays of B hadrons such as $B_s^0 \to \pi^+ D_s^-$, with $D_s^- \to \phi\pi^-$, which is important for B_s^0 mixing, and $B^0/\bar{B}^0 \to \pi^+\pi^-$, which is important for CP violation. Finally, a time-of-flight detector will make kaon identification possible in a momentum range that is especially interesting for b flavor identification.

The D0 detector is undergoing an even more radical change [11] than

CDF. A superconducting solenoid (2 T) has been installed to allow momentum measurements. The main tracking device will be an eight-layer scintillating fiber detector with full coverage out to $|\eta| < 1.7$. A silicon microstrip tracker will be installed. It will consist of six barrel detectors, each with four layers, with both $r - \phi$ and $r - z$ read out, and sixteen disks extending out to $|z| < 1.2$ m along the beam line. With this new spectrometer and tracking system, D0 will tag B decays using displaced vertices. Finally, with improvements to the muon system and trigger, as well as the existing liquid argon calorimeter, D0 should be competitive with CDF for B physics in Run II.

CDF and D0 will address many important questions in B physics in Run II. Many of the relevant measurements have already been investigated using the Run I data. Among the more important goals are (1) the precise measurement of $|V_{\rm td}/V_{\rm ts}|$ from $B_s^0 - \bar{B}_s^0$ flavor oscillations (†) (or from a measurement of $\Delta\Gamma_s/\Gamma_s$) or from radiative decays, *e.g.*, the rate of $B_s^0 \to K^{*0}\gamma$ compared to $B_s^0 \to \phi\gamma$ (†); (2) the observation of CP violation in $B^0/\bar{B}^0 \to J/\psi K_{\rm S}^0$ and the precise measurement of $\sin(2\beta)$; (3) the observation of CP violation in $B^0/\bar{B}^0 \to \pi^+\pi^-$ and a precise measurement of the CP asymmetry, which is related to $\sin(2\alpha)$; (4) the search for large CP violation in $B_s^0/\bar{B}_s^0 \to J/\psi\,\phi$ (†) (a large asymmetry would be an unambiguous signal of physics beyond the standard model); (5) the observation of decay modes related to angle γ: $B_s^0 \to D_s^{\pm}K^{\mp}$ (†) and $B^+ \to \bar{D}^0K^+$; (6) the observation of rare decays such as $B^+ \to \mu^+\mu^-K^+$, $B^0 \to \mu^+\mu^-K^{*0}$, and $B_s^0 \to \mu^+\mu^-\phi$ (†); (7) the study of the B_c^+ meson (†) and b baryons (†) . As long as the B factories remain on the $\Upsilon(4S)$ resonance, the topics marked by a dagger (†) will be unique to hadron colliders.

Many studies of these future topics have been performed. Here we summarize expectations for two more important measurements: (1) $\sin(2\beta)$ and (2) Δm_s. A more complete discussion can be found in [12]. These projections (*e.g.*, flavor tag performance) are based on Run I data from CDF as much as possible and assume an integrated luminosity of 2 fb^{-1} collected during the first two years of operation. Although the expectations are specific to CDF, similar sensitivity should be possible with the D0 detector.

For the measurement of $\sin(2\beta)$, the signal size should increase to 10 000 $B^0/\bar{B}^0 \to J/\psi K_{\rm S}^0$ decays, where $J/\psi \to \mu^+\mu^-$ and $K_{\rm S}^0 \to \pi^+\pi^-$. The b flavor tags will be calibrated with samples of 40 000 $B^+ \to J/\psi K^+$ (and charge conjugate) decays and 20 000 $B^0 \to J/\psi K^{*0}$ (and charge conjugate) decays. The expected combined flavor tag effectiveness of the same-side, lepton, and jet-charge tags discussed earlier is $\epsilon D^2 = 6.7\%$. The resulting estimate of the error on $\sin(2\beta)$ is $\delta(\sin(2\beta)) = 0.084$. This uncertainty includes the systematic uncertainties in the dilutions due to the statistics of the calibration samples. The time-of-flight detector will make it possible to use a new opposite-side flavor tag based on kaons: the decays of B hadrons containing $\bar{b}$ (b) quarks usually produce K^+ (K^-). With this additional flavor tag, the total flavor tag effectiveness could increase to $\epsilon D^2 = 9.1\%$ [13].

Based on current experimental measurements and theoretical predictions, it is possible to predict the value of $\sin(2\beta)$. For example, S. Mele [14] predicts $\sin(2\beta) = 0.75 \pm 0.09$ (not including the latest measurement of $\sin(2\beta)$ from CDF). This current projection rivals the expected precision of the measurement in Run II, therefore, to really test our predictions and consistency within the standard model, even more precise measurements of $\sin(2\beta)$ will be necessary. This motivates the need for future experiments such as BTeV [1] at the Tevatron.

The other crucial measurement, which is attainable only at hadron colliders, is the measurement of Δm_s. The upgraded CDF trigger will make it possible to trigger on fully reconstructed hadronic B decays. A large sample of 20 000 fully reconstructed B_s^0 decays in the modes $B_s^0 \to D_s^-\pi^+, D_s^-\pi^+\pi^-\pi^+$, with $D_s^- \to \phi\pi^-, \bar{K}^{*0}K^-$ is expected. The time-of-flight detector will double the expected b flavor tagging efficiency to $\epsilon D^2 = 11.3\%$ by exploiting kaon identification in the same-side tag of B_s^0 mesons (see fig. 51.2). The CDF silicon system will have excellent impact parameter resolution, and the expected proper lifetime resolution is 45 fs. With all these improvements combined, a better than five-sigma measurement of Δm_s is possible for $\Delta m_s < 40\,\text{ps}^{-1}$ [13].

51.4 Summary

In Run I many important measurements of B hadron decay properties were made at the Tevatron, despite severe trigger restrictions. Many of these measurements are similar or better in precision to measurements from e^+e^- colliders; some are unique to the Tevatron. In particular, a first indication of CP violation in the B system was found yielding $\sin(2\beta) = 0.79^{+0.41}_{-0.44}$.

In Run II, improved detectors will increase the scope of B physics at the Tevatron. The precision on $\sin(2\beta)$ should be competitive with (and complementary to) measurements at the B factories. Studies of CP violation in B_s^0 decays and the measurement of Δm_s will be unique to hadron colliders. The Tevatron will play a crucial, unique role in our test of the CKM matrix.

Acknowledgments

I would like to thank the organizers of this conference for an enjoyable informative, and well-organized meeting. This work is supported by DOE grant number DE-FGO2-95ER40893.

References

[1] The BTeV Proposal (May 2000); see www-btev.fnal.gov/btev.html.

[2] L. Wolfenstein, Phys. Rev. Lett. **51**, 1945 (1983).

[3] C. Caso *et al.*, Eur. Jour. Phys. C **3**, 1 (1998).

[4] CDF Collaboration, F. Abe *et al.*, Nucl. Instrum. Methods A **271**, 387 (1988); CDF Collaboration, F. Abe *et al.*, Phys. Rev. D **50** (1994) 2966; D. Amidei *et al.*, Nucl. Instr. Methods A **350**, 73, (1994); P. Azzi *et al.*, Nucl. Instr. Methods A **360**, 137 (1995).

[5] CDF Collaboration, F. Abe *et al.*, Phys. Rev. D **60**, 072003 (1999).

[6] M. Gronau, A. Nippe, and J. Rosner, Phys. Rev. D **47**, 1988 (1993); M. Gronau and J. Rosner, Phys. Rev. D **49**, 254 (1994).

[7] CDF Collaboration, F. Abe *et al.*, Phys. Rev. Lett. **80**, 2057 (1998); Phys. Rev. D **59**, 032001 (1999).

[8] CDF Collaboration, F. Abe *et al.*, Phys. Rev. Lett. **81**, 5513 (1998).

[9] CDF Collaboration, T. Affolder *et al.*, Phys. Rev. D **61**, 072005 (2000).

[10] CDF II Collaboration, The CDF II Detector Technical Design Report; see also www-cdf.fnal.gov/upgrades/upgrades.html.

[11] See www-d0.fnal.gov/hardware/upgrade/upgrade.html.

[12] M. Paulini, Int. J. Mod. Phys. A **14**, 2791 (1999).

[13] *Proposal for Enhancement of the CDF II Detector: An Inner Silicon Layer and Time of Flight Detector* (P-909), and *Update to Proposal P-909: Physics Performance of the CDF II with an Inner Silicon Layer and a Time of Flight Detector*. Both of these documents may be obtained at www-cdf.fnal.gov/upgrades/upgrades.html.

[14] S. Mele, Phys. Rev. D **59**, 113011 (1999).

52

Status of Belle

K. Kinoshita

The Belle detector at the KEKB storage ring (KEK laboratory, Japan) is designed for the study of CP violation in B meson decays produced via asymmetric energy e^+e^- collisions at the $\Upsilon(4S)$ resonance. The detector was installed in the ring in the spring of 1999. Reported here are results from commissioning and first results from early running, which demonstrate the performance of the detector and prospects for running over the next year. Due to space limitations, only a fraction of the plots shown at Kaon99 are included here, although most are described.

The collaboration currently consists of around 300 physicists from 52 institutions, representing 10 nations and one region. The Belle detector was completed in late 1998 and rolled into its final position in early May 1999. Its main components, from the beam line outward, are the silicon vertex detector (SVD), central drift chamber (CDC), aerogel Čerenkov counter (ACC), time-of-flight counter (TOF), CsI-based electromagnetic calorimeter (ECL), 1.5T superconducting solenoid, K_L/muon counter (KLM), and extreme forward calorimeter (EFC). The reader is referred to the Belle TDR [1] for details.

The KEKB commissioning began prior to the installation of Belle in the experimental area, during the period 12/1/98–4/19/99. During this period the Belle collaboration operated a detector known as BEAST (**B**ackground **E**xorcism for **A ST**able Belle experiment), intended to provide information in real time to characterize the radiation environment, to enable KEKB to maximize luminosity while maintaining an acceptable radiation level for Belle, and to accomplish this without risk to the detector. The BEAST consisted of a number of actual Belle detector modules as well as electronics and other components that served as radiation monitors. These included 12 pairs of PIN photodiodes, 12 MOSFETs, 53 drift tubes, a partial array of BGO (bismuth germanate) from the EFC, an array of CsI blocks, and two spare SVD ladders.

By the end of commissioning, KEKB had achieved a luminosity of $6 \times 10^{30}\mathrm{cm}^{-2}\mathrm{s}^{-1}$ concurrently with acceptable background levels. Information collected by BEAST had been used to identify and correct several sources of

	e^-	e^+
Beam energy (GeV)	8.0	3.5
Energy spread (MeV)	5.7	2.3
Current (mA)	1100 (514)	2600 (532)
# bunches	5000 (800)	5000 (1024)
$\sigma_x(\mu\mathrm{m})$	77 (170)	77 (170)
$\sigma_y(\mu\mathrm{m})$	1.9 (5.7)	1.9 (5.7)
$\sigma_z(\mathrm{mm})$	4.0 (5.6)	4.0 (5.6)

Table 52.1: Selected parameters of KEKB, design and, in parentheses, achieved, during commissioning. σ_x, σ_y, and σ_z are the beam size at the collision point in the horizontal, vertical, and beam directions, respectively.

instability at the injection stage. The remaining unresolved limitations to luminosity at the time were an orbit drift on a 10-minute time scale, a 14 Hz vertical vibration, instability of beams in the low energy ring at currents above 350 mA, short collision lifetime, and a disagreement between the expected and measured chromaticities. Selected KEKB parameters, both design and achieved, are shown in table 52.1. It is anticipated that these will be solved during the summer run, and that steady progress will be made toward the design luminosity of $1 \times 10^{34}\mathrm{cm}^{-2}\mathrm{s}^{-1}$. There have in fact already been some substantial improvements in the first few weeks of running with the detector, notably to the vertical beam size, which is now near design, as well to the collision lifetime and some instabilities.

A major goal of commissioning was to characterize the principal sources of the radiation delivered to Belle, the known possibilities being synchrotron radiation, or beam particles undergoing bremsstrahlung or Coulomb interactions with the residual gas in the vacuum chamber. (Because KEKB is configured to collide beams at a finite crossing angle, there is no beam loss or background from parasitic collisions.) It is important to quantify each source because each has different effects and requires different solutions. Intense synchrotron radiation can cause damage and high occupancies, particularly in the SVD. Beam-gas collisions can create triggerable events either directly or by perturbing particles into unstable orbits and shortening beam lifetimes. The principal concerns for Belle are the trigger rate, radiation damage to the detector and electronics, and occupancy. Background studies have indicated that the radiation levels increase quadratically with current, and it is believed that this points to beam particles rather than synchrotron radiation as the primary source. By comparing data with simulations, several locations for additional collimation were identified. With the conditions as optimized at the end of commissioning, it was determined that the radiation dose to SVD will be 330 krad/yr running with $\mathcal{L} = 2 \times 10^{33}\mathrm{cm}^{-2}\mathrm{s}^{-1}$. As SVD is able to survive up to 200 krad, we can conclude that it will survive the first year, when luminosity is not expected to reach this level. After a year, it is anticipated that SVD will be replaced.

Given the current conditions, the luminosity is limited more by occupancy in Belle than by radiation dose. As the experiment relies on online sparsification to manage event size, high occupancy results in unacceptably large event size, reducing live time and slowing offline reconstruction. This experience has prompted actions to reduce the detector's sensitivity to backgrounds, including the reduction of shaping times in SVD and ECL from 2.5 μs to 1.0 μs. Information gathered by BEAST was also used to develop the initial triggering strategy and prompted a decision to integrate background measurements, using parts from BEAST, into the Belle data acquisition system.

The process of rolling the Belle detector into its final position began on 1 May, 1999. KEKB resumed operation on 24 May with the detector in place. The first data-taking run occurred on 2 June with the full detector in place and nearly all in operation. Due to high radiation levels, the inner parts of the CDC were operated with reduced high voltage. The first run collected approximately 10 hadronic events, several hundred Bhabhas, and one μ pair. Steady progress was made over the next ten days until 11 June, when a vacuum break forced a temporary shutdown of around two weeks.

Up to this shutdown, the detector had logged 527.2 nb^{-1} of integrated luminosity and collected 1330 hadronic events, including ~250 $\Upsilon(4S) \to B\bar{B}$ events. Offline event reconstruction was completed within several days after acquisition. Events were classified after first boosting all detected particles to the center of mass under the hypotheses that all charged tracks are pions and all neutral calorimeter clusters not matched to tracks are photons. To be classified as hadronic, an event was required to have at least two well-reconstructed tracks, a total visible center-of-mass energy at least 0.20 times the known energy (E_{cm}), calorimetric energy between 2% and 90% times E_{cm}, and a component along the beam line of total event momentum that is less than 50% times E_{cm}. For events with only two tracks, there was an additional requirement that the total momentum must be less than 90% times E_{cm}.

We use Bhabha and hadronic events to evaluate the performance of various pieces of the detector. All plots shown here, unless specifically indicated otherwise, were obtained from these events. In asymmetric e^+e^- collisions the two tracks of a Bhabha event are not back-to-back. They are, however, if viewed projected in the plane perpendicular to the beam line ($r - \phi$), and the distance in this projection between the points of closest approach of the two tracks, the "miss distance," is due to errors in track position extrapolated to the event vertex. This error is dominated by the resolution of silicon tracking. The precision at this initial stage is found to be better in the horizontal ($\sigma = 33\mu$m) than in the vertical ($\sigma = 43\mu$m) direction, due to the anisotropy of the cosmic rays with which the relative alignment of SVD modules has been calibrated.

The CDC has been operating with reduced high voltage due to high backgrounds in this initial running period. Nonetheless, we have been able to measure tracks and make ionization (dE/dx) measurements with mod-

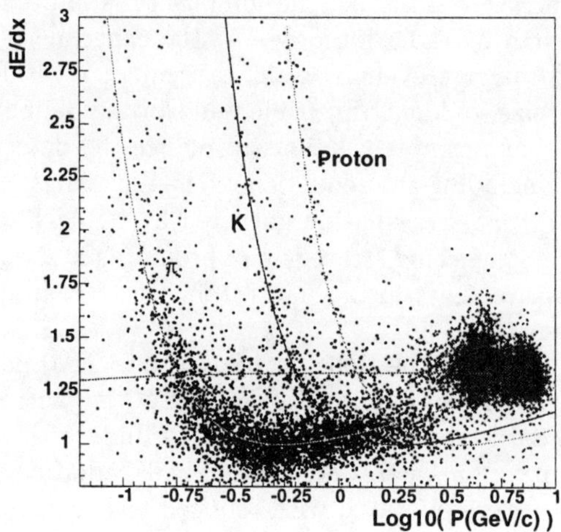

Figure 52.1: Specific ionization $(dE/dx)/(dE/dx)_{min}$ plotted against momentum measured in the central tracking system, for charged tracks found in hadronic and Bhabha events selected from data.

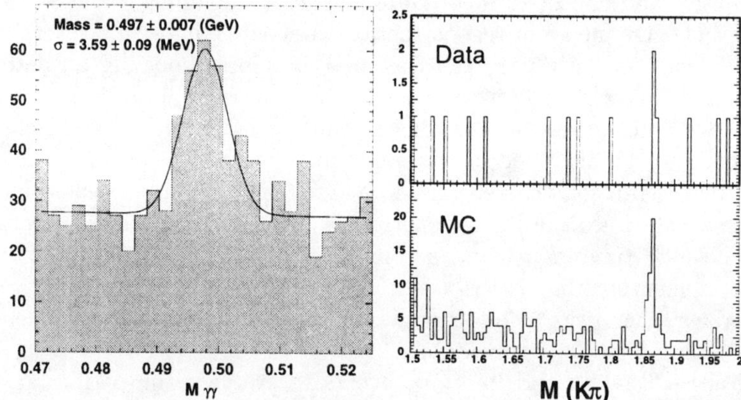

Figure 52.2: Invariant mass distributions, in GeV/c^2. Left: $\pi - \pi$ invariant mass of track pairs originating at a single vertex separated by at least 5 mm from the event vertex shows a clear K_s peak. Right: $K^{\mp}\pi^{\pm}$ invariant mass of oppositely charged track pairs shows an accumulation of D^0 candidates in data (top) at the expected mass.

erately good efficiency in the events collected thus far. A plot of dE/dx vs. momentum (fig. 52.1) shows clearly separated bands of pions, kaons, and protons, as well as a cluster of electrons. Fig. 52.2 (left) shows the distribution in $\pi - \pi$ invariant mass of K_S candidates required to have a minimum decay length of 5 mm and fitted to a Gaussian plus flat background shape. The mean mass and width are in good agreement with the expectation based on Monte Carlo studies. Reconstruction of the decay

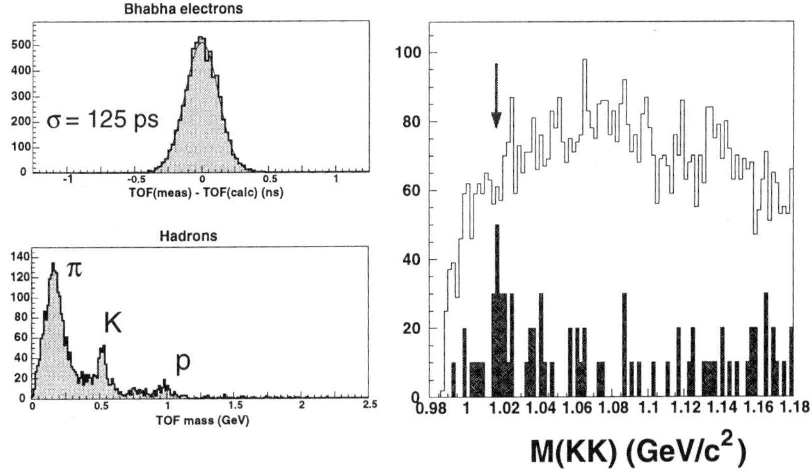

Figure 52.3: Left: Performance of time-of-flight. Top: Difference between measured TOF and expected TOF for Bhabha electrons. Bottom: Mass calculated from momentum and time-of-flight, for charged tracks with momentum less than 1.2 GeV/c reconstructed in hadronic events. Right: $K-K$ invariant mass for oppositely charged pairs of tracks in hadronic events, without (unshaded histogram) and with (shaded) particle identification requirements. The vertical scale on the shaded plot has been magnified by a factor of 10.

$D^0 \to K^-\pi^+$ is a test of tracking at higher average momentum. The distribution in invariant mass obtained by assigning the K^- and π^+ masses to oppositely charged tracks (fig. 52.2 (right)) shows three candidates in the expected interval, in good agreement with the expectation from ~1000 hadronic events.

The response of the ACC aerogel counter, expressed in number of photoelectrons collected, has been compared with Monte Carlo simulations for Bhabha electrons, which are above Cerenkov threshold, and for protons below threshold identified by dE/dx and TOF criteria. These initial studies show very good agreement between data and simulations.

The performance of TOF can be seen directly with Bhabha electrons by observing the difference between expected and measured times (fig. 52.3 (left, top)). We find a resolution of 125 ps. This detector is intended to identify hadrons with momentum below 1.2 GeV/c, and its capability may be seen by plotting the mass derived from the measurements of momentum and TOF for tracks in this momentum range (fig. 52.3 (left, bottom)). The resolution can be expected to approach the design value of 100 ps as tracking and calibrations are refined and more data are accumulated.

The information from dE/dx, TOF, and ACC are combined to achieve good charged hadron identification over nearly all of the relevant momentum range, < 4.0 GeV/c. Fig. 52.3 (right) shows that a clean signal for

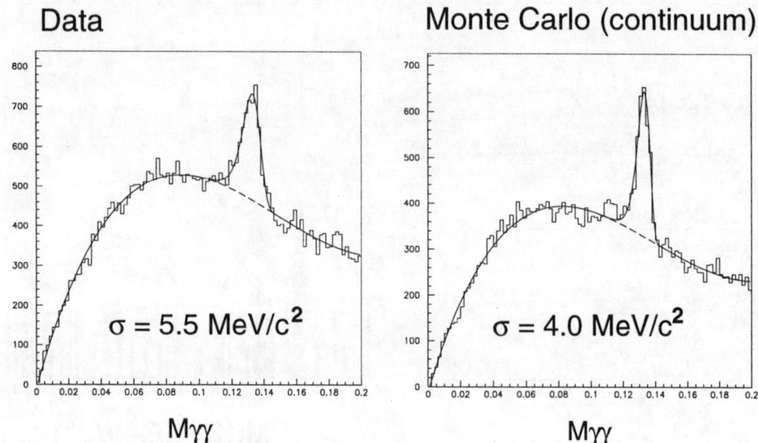

Figure 52.4: The $\gamma - \gamma$ invariant mass distribution for calorimeter clusters not associated with tracks, data (left) and Monte Carlo simulation (right). Photon candidates are required to have a minimum energy of 20 MeV.

$\phi \to K^+K^-$ can be extracted from an overwhelming background of pions by the requiring that candidate tracks be identified as kaons.

The ECL resolution can be checked directly by looking at the total calorimetric energy of Bhabha events, where a fit gives a resolution of 2.34%, in good agreement with the expectation of 2.24%. The $\gamma - \gamma$ invariant mass distribution for calorimeter clusters not associated with tracks shows a peak at the π^0 mass with a width of (5.5 ± 0.4) MeV/c^2, not far from the design expectation of 4.0 MeV/c^2 (fig. 52.4).

K_L mesons may be observed by looking at tracks in the KLM detector that are not matched to charged tracks in the CDC. According to simulations, a substantial source of missing momentum in Υ(4S) events is neutral hadrons, the other major sources being detection inefficiency and neutrinos. Plots of angular differences between KLM clusters and event missing momentum reveal broad but well-defined peaks with signal/background greater than 1 and good agreement with Monte Carlo simulations.

The beam energies at KEKB are known approximately but must be calibrated by locating the peak of the hadronic cross section at the Υ(4S) resonance. Although Belle has thus far collected data at only a single center-of-mass energy, it is possible to know the proportion of Υ(4S) events in the sample and thus its proximity (but not direction) to the resonance by looking at the distribution in R_2, which differs substantially between resonant and continuum events. R_2 is the ratio of the second and zeroth Fox-Wolfram moments [2] and is given by

$$R_2 = \frac{\Sigma_{i,j}|p_i||p_j|(3\cos^2\theta_{ij} - 1)}{\Sigma_{i,j}2|p_i||p_j|},$$

where p_i is the momentum of track i and θ_{ij} is the angle between the momentum vectors of tracks i and j. For "jetty" events R_2 tends toward 1 while for the more isotropic $B\bar{B}$ events the average R_2 is closer to zero. The distribution in data is shown in fig. 52.5 with fits to expected distributions of resonant plus nonresonant events. The fitted fraction of resonant events in this sample is $(18.7 \pm 1.9)\%$. Rather hard event selection criteria were used in this sample, and the fraction of resonant events at the peak of the $\Upsilon(4S)$ is expected to be significantly higher. We conclude that the center-of-mass energy for this run is roughly 10 MeV off the resonance.

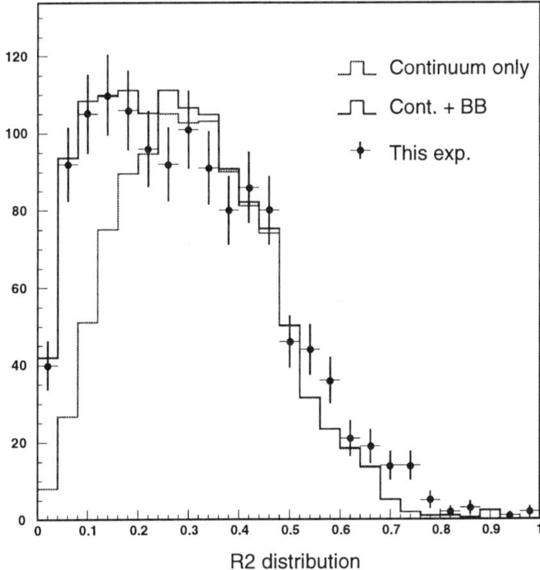

Figure 52.5: Distribution of R_2 for hadronic events in data, fitted to a sum of expected distributions for continuum and $B\bar{B}$ events. Fitted shape is shown as a heavy line, the continuum part is shown as a light line, and data are shown as symbols with error bars.

To summarize, the KEKB ring is operating with luminosities approaching $10^{31}\text{cm}^{-2}\text{s}^{-1}$, and the Belle detector has collected over 1000 hadronic events in a period of 10 days. The hadronic events have been processed through the offline analysis and show that all of the detector systems are working as well as can be expected at this stage. We can expect both KEKB and Belle to improve steadily over the coming weeks and months. Operations will resume as soon as the vacuum chamber repair is complete and continue through early August. After that, running is scheduled in the fall, October through December, with a 2-week pause over the new year, then continuing through the end of July. The target luminosity by the next year is $5 \times 10^{33} \text{cm}^{-2}\text{s}^{-1}$ with integrated luminosity of a few fb^{-1}.

References

[1] Belle TDR, KEK Proceedings 95-1, available at http://bsunsrv1.kek.jp/bdocs/tdr.html.

[2] G. Fox and S. Wolfram, Phys. Rev. Lett **41**, 1581 (1978).

53

Progress and Results from the SLAC B Factory

Erich W. Varnes[1]

Abstract

The B Factory at SLAC, designed to study CP violation in the B meson system, has recently begun to record data. The design and commissioning of the PEP-II accelerator and BaBar detector, which comprise the B Factory, are reviewed and early results are presented.

53.1 Physics Goals

The B Factory at the Stanford Linear Accelerator Center is designed to produce about 3×10^7 $b\bar{b}$ events per year. With this sample an extensive program of b quark, c quark, τ lepton, and two-photon physics will be carried out. The primary goal, however, is to investigate CP nonconservation in the B_d meson system. This effect was first observed in the K meson system over thirty years ago [1] and has been studied extensively in that system. The standard model can accommodate CP violation through the presence of a physical phase in the Cabibbo-Kobayashi-Maskawa (CKM) matrix (which relates the quark weak eigenstates to the mass eigenstates for three quark generations) [2]. If this is truly the sole origin of CP violation, several predictions are made for the B_d system. Most significantly, it is expected that there are B decay modes that have large CP asymmetries, a well-understood relationship between the symmetry and CKM matrix parameters, and branching ratios of order 10^{-5}. These decay modes are clearly accessible at the B Factory. By directly measuring the CP asymmetries in such modes (the best among them is $B \to J/\psi K_S$) one can overconstrain the values of the CKM matrix parameters. Any inconsistency between measurements would be a signal that at least some portion of CP violation is caused by physics beyond the standard model. A complete review of the planned B Factory physics program may be found in ref. [3].

[1] Representing the BaBar Collaboration.

53.2 Experimental Design

The B Factory produces b quarks via the reaction $e^+e^- \to \Upsilon(4S) \to b\bar{b}$. With the e^+e^- center-of-mass energy set to the $\Upsilon(4S)$ mass the first reaction has a cross section of about one nb, above a continuum cross section of 3 nb. The $\Upsilon(4s)$ decays to $b\bar{b}$ with a branching ratio greater than 95%. In half the cases, the $b\bar{b}$ state will produce a pair of neutral B mesons. As these mesons propagate they oscillate between the B and $\overline{B}$ states. After one of the Bs decays, the other will continue to oscillate. By collecting events in which one of the Bs decays to a CP eigenstate f, and the other in a mode for which the b flavor can be tagged, one can measure the rates $\Gamma(B^0(t) \to f)$ and $\Gamma(\overline{B^0}(t) \to f)$. In these relations, $t = 0$ is the decay time of the flavor-tagging B, and $B^0(t)$ denotes the state that was known to be a B^0 at $t = 0$. CP violation then appears as a nonzero value for the quantity

$$a_f(t) \equiv \frac{\Gamma(B^0(t) \to f) - \Gamma(\overline{B^0}(t) \to f)}{\Gamma(B^0(t) \to f) + \Gamma(\overline{B^0}(t) \to f)}. \tag{53.1}$$

Measuring the time dependence in the above relation is crucial because $a_f(t)$ is proportional to $\sin(\Delta M t)$ (ΔM is the mass difference between the two B_d mass eigenstates) and therefore vanishes when integrated over all times. Since the Bs are nearly at rest in the $\Upsilon(4S)$ frame, measuring the decay time difference is not possible if the e^+e^- center-of-mass frame coincides with the laboratory frame. The solution is to collide beams with asymmetric energies, thereby producing an $\Upsilon(4S)$ that is boosted in the laboratory frame. This allows the difference in decay times of the B mesons to be measured by measuring the distance between their decay vertices along the boost direction.

53.2.1 The PEP-II Accelerator

For BaBar, the PEP-II accelerator provides the asymmetric beams. PEP-II consists of two accelerator rings located in the same tunnel. The high-energy ring (HER) accelerates electrons to 9 GeV while the low-energy ring (LER) accelerates positrons to 3 GeV. At the interaction region the LER bends to intersect the HER with head-on collisions.

Since the B decays to CP eigenstates have low branching ratios, the experiment relies on PEP-II providing high luminosity. The design is $3 \times 10^{33}\mathrm{cm}^{-2}\mathrm{s}^{-1}$, roughly four times higher than the current record for a storage ring. Achieving this luminosity while maintaining a background level low enough for BaBar to record meaningful data is a considerable challenge. Commissioning of the accelerator began in 1998; a summary of some of PEP-II's design parameters, and the values achieved to date, is given in table 53.1.

	Design	Achieved
No. of bunches	1658	262
Luminosity ($cm^{-2}s^{-1}$)	3×10^{33}	5.5×10^{32}
Low-energy ring (LER) current (mA)	2140	750
High-energy ring (HER) current (mA)	750	240
LER lifetime (min)	240 @ 2A	120 @ 750 mA
HER lifetime (min)	240 @ 2A	350 @ 240 mA

Table 53.1: Some of the parameters of the PEP-II accelerator, with their design values and those achieved since the roll-in of BaBar

53.2.2 The BaBar Detector

The BaBar detector is a state-of-the-art device featuring precision vertex measurement, accurate tracking, particle identification, high-resolution calorimetry, and muon and neutral hadron detection [4]. All of these features are necessary to carry out the intended physics program, and they are implemented by a set of subsystems, each of which performs one or more of the above tasks. This section gives a brief description of each BaBar subsystem.

The system nearest the beam line is a silicon microstrip tracker. This tracker consists of five layers of double-sided silicon wafers, providing high-precision track reconstruction in three dimensions. The resolution for each hit is 10 μm in ϕ and 12 μm in z. With this hit resolution, the z position of a track when extrapolated to the beam line is 60 μm. This ability to precisely measure track vertices along the beam direction is crucial in reconstructing the difference in decay times of the two B mesons in the event. The silicon wafers and the on-board electronics are radiation hard up to 2 Mrad, allowing them to survive the expected ten-year lifetime of the experiment.

Next is a drift chamber, which uses a helium-based gas and aluminum field wires to reduce mass and thus multiple coulomb scattering. The chamber has forty layers of wires, which are grouped into ten "superlayers." In four of the superlayers the wires are strung axially, and in the remainder a small stereo angle is used to allow measurment of the z coordinate. While the primary role of the chamber is to reconstruct charged particle tracks and measure their momenta, it also measures dE/dx to aid in particle identification (especially for particles with energies below the Cerenkov emission threshold). The design resolutions are 140 μm in position and 7% in charge, and both of these have been exceeded in cosmic ray tests.

The primary particle identification device is a detector of internally reflected Cerenkov light (DIRC). To collect the Cerenkov light, a cylinder of thin quartz bars is mounted outside of the drift chamber. The light propagates by internal reflection to the end of the bar, at which point it enters a stand-off region (filled with water to more closely match the index of refraction of quartz) in which the rings expand before their positions are recorded by phototubes. The end of the bar opposite the stand-off region is mirrored to reflect the light back to the instrumented end. BaBar's physics goals require that πs and Ks be identified with 95% confidence up to momenta of 4 GeV/c, which requires a Cerenkov angle resolution of 2 mr. To acheive this resolution, both the purity and surface quality of the quartz bars must be extremely good, and these tolerances have proven difficult for industry to reach. For this reason, only 5 of the 12 sets of quartz bars are installed at present; the balance will be put in place when they are available this fall. Aside from these missing radiator bars, the entire DIRC system (including all phototubes and front-end electronics) has been installed and commissioned.

Photon and electron energies are measured using a cesium iodide calorimeter consisting of 7000 crystals arranged in a barrel and forward end cap. The crystals are arranged in a slightly nonprojective geometry to minimize the loss of energy from particles that traverse the interface between two crystals. The mechanical support and readout phototubes are mounted on the rear of the crystals to minimize the upstream material. This design achieves a resolution of $\sigma(E)/E = 1\%/\sqrt[4]{E(\mathrm{GeV})} \oplus 1.2\%$, which is important for the reconstruction of B mesons in decay modes that include one or more π^0s.

All of the above systems reside in the bore of a superconducting solenoid, which produces a magnetic field of 1.5 T in the tracking region. The magnetic flux is returned through a yoke comprised of 20 layers of iron, with the gaps between the layers instrumented with resistive plate chambers. The primary goal of this instrumented flux return is to identify muons and neutral hadrons. Although there is no capability to measure neutral hadrons' energies, detection of their presence is important in identifying decay modes such as $B \to J/\psi K_L^0$.

Data from all of the detector subsystems is digitized in the front-end electronics, with the resulting signals transported via optical fiber to the electronics house. Custom-made VME boards receive the data and perform initial processing and reduction on it using embedded CPUs. The data is then transmitted to a farm of thirty-two UNIX nodes in which monitoring histograms are filled and a software trigger algorithm is performed. Following reconstruction, the data is stored to an object-oriented database for analysis. The system is designed to handle hardware triggers at 2 kHz, and to record and reconstruct events at 100 Hz. With this capacity BaBar can record the full rate of inelastic e^+e^- collisions at the design luminosity.

53.3 Results from First Data

BaBar was rolled onto the PEP-II beamline in late April, and the first collisions were observed on 26 May, 1999. A total of roughly 100 pb^{-1} of data has been recorded as of this writing. With this initial sample the focus has been on calibrating and aligning the detector. One of the important steps in this process is to develop efficient selection criteria for Bhabha, dimuon, and multihadron events, since all of these are important for the calibration of various detector components.

While all analyses are in their early stages at this point, the detector performance has been demonstrated by comparing the angular distribution of Bhabha electrons with the QED prediction, and finding clear $\pi^0 \to \gamma\gamma$ and $K_s^0 \to \pi^+\pi^-$ peaks.

The most significant analysis performed to date was the joint effort of BaBar and PEP-II to locate the $\Upsilon(4S)$ resonance. The procedure started with setting the accelerator energies to the best guess for the resonance value, and then scanning steps in ten-MeV steps on each side of this guess. About 0.5 pb^{-1} was recorded at each point to ensure a significant sample of multihadron events. The analysis consisted of determining the ratio of multihadron to Bhabha events at each energy point. Bhabhas were selected by requiring two high-momentum tracks (greater than 2 (1) GeV/c for the e^- (e^+) track) that are back-to-back in theta when boosted to the center-of-mass frame. In addition, two matching calorimeter clusters with energies greater than 1 GeV were required, and events with more than four charged tracks were rejected as a further veto against beam-gas events. Multihadron candidates were required to have more than two tracks, with at least one exceeding 0.5 GeV/c. To supress beam-gas events, the total charge for the event was required to be less than four, the maximum momentum of any track in the center-of-mass frame less than 4.5 GeV/c. In addition, the square of the invariant mass for all tracks in the event was required to be greater than 5 GeV^2. To enhance the contribution due to B meson events, the ration of Fox-Wolfram moments H_2/H_0 was required to be less than 0.2.

The preliminary result is shown in fig. 53.1, which is based on an analysis of one-third of the data collected. The $\Upsilon(4S)$ peak is clearly visible, and is located very near PEP-II's original guess.

53.4 Future Plans

BaBar continued to take data for the remainder of 1999, with a shutdown as the remaining DIRC quartz bars were installed. The goal was to reach a peak luminosity of 10^{33} cm^{-2} s^{-1} and to integrate 3-5 fb^{-1} of data durign this period. Continued running and development in 2000 are expected to bring the B Factory to its design luminosity.

Although the commencement of data taking and the rapid improvement of PEP-II luminosity indicate that BaBar and PEP-II are well on their way to these goals, significant challenges remain. In particular, increasing the luminosity while keeping the background radiation low will require greater understanding of the sources of background, and perhaps the addition of more collimators and shielding in the accelerator. Also, the offline reconstruction throughput needs to be improved significantly to process and store the planned 100 Hz event rate.

In light of the progress already made, it seems reasonable to assume

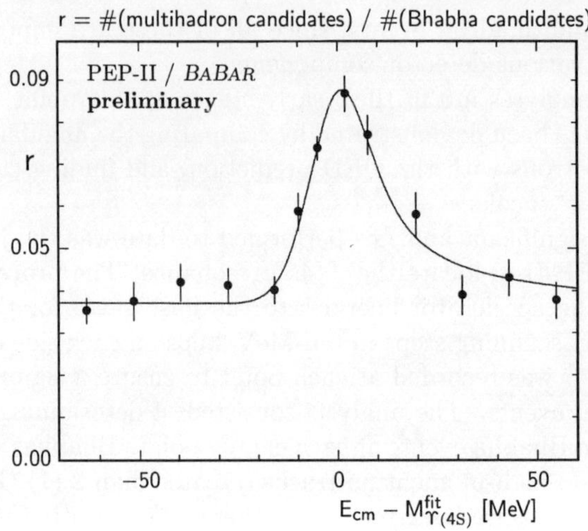

Figure 53.1: The result of the PEP-II energy scan, with the $\Upsilon(4S)$ peak clearly visible.

that these remaining issues, though difficult, will also be resolved. If so, BaBar will begin to shed light on CP violation in the B meson system in the next year, and we will be on our way to learning if there is more to this phenomenon than the standard model explanation.

References

[1] J.H. Christenson *et al.*, Phys. Rev. Lett. **13**, 138 (1964).

[2] M. Kobayashi and T. Maskawa, Prog. Theor. Phys. **49**, 652 (1973).

[3] P.F. Harrison and H.R. Quinn, eds., SLAC Report No. SLAC-R-504 (1998).

[4] BaBar Collaboration, G.S Abrams *et al.*, SLAC Report No. SLAC-R-95-457 (1995).

Part IX

Future Opportunities in K Physics

For nearly the past decade, the prime goals in kaon physics have been the establishment of direct CP violation in neutral K decays (ϵ'/ϵ), and the observation of the $K^+ \to \pi^+\nu\bar{\nu}$ decays. As we see from the earlier parts of this volume, these goals are being met.

Major new experiments at Fermilab (KTeV) and CERN (NA48) were constructed over the past five years to study ϵ'/ϵ. KTeV has reported a nearly 7 standard deviation direct CP-violating effect, and NA48 has reported a new consistent measurement as well. These results are consistent with evidence first reported by NA31 a decade ago. Both have substantially more data, and KLOE at Frascati is just under way, so a truly precision measurement of ϵ'/ϵ can now be foreseen.

BNL 787 has one golden $K^+ \to \pi^+\nu\bar{\nu}$ event, in a recently enlarged sample of K^+ decays. Their upgrade should allow about 10 events to be collected, which will be a great success and allow a clean determination of the important parameter V_{td}.

Now that the main goals are being met, is it worth continuing any of the K programs? The contributions in this part address this issue and treat certain physics issues that can still be best addressed in precision studies of rare K decays.

While searches for lepton flavor violation and CPT symmetry violations can be significantly and uniquely extended in rare K decays, it is generally agreed that the most important topic is the precision measurement of both the K^+ and K^0 decay to a pion and a neutrino-antineutrino pair. If the goals of the existing proposals to make these measurements are met, the combination of the two can give the equivalent of $\sin(2\beta)$ in the neutral B decay, providing a crucial consistency check of the standard model mechanism for CP violation. The importance of these modes is stressed in the constributions of Buchalla, Gilman, and Franzini.

Fermilab, which has had a continuous series of kaon experiments since it first turned on in 1972, has constructed a new, rapid cycling proton accelerator at 120 GeV with the primary purpose of improving the luminosity in collider operations. This also allows a substantial fixed target program, and its focus and reach are described (Cooper). Although the BNL AGS, with an even longer history of K decay studies, is being turned over to RHIC, it is possible to continue a significant program in K decay physics there, with experiments that exploit the unique characteristics of that machine (Littenberg). The NA48 group at CERN has been contemplating future efforts. Interesting options there, including high-sensitivity K_S decay studies, are presented (Kalmus). In Japan, a joint KEK/JAERI project is the construction of a very intense 50 GeV proton synchrotron; this opens up an important new set of possibilities (Inagaki). And there are still many topics that will be addressed at IHEP in Russia (Landsberg).

These experiments are important and the theory behind them is "clean," but they are highly challenging. The community is convinced that backgrounds can be rejected at the required levels but full detector design needs a significant amount of research and development.

Since the Chicago Kaon 99 Conference, there have been at least two important developments toward future efforts. First, the BNL upgrade to E787 has been approved by the DOE. Second, significant R&D funding for both the proposed K^+ and $K_L^0 \to \pi^0 \nu\bar{\nu}$ experiments at Fermilab has been made available.

54

Kaon Physics: Future Opportunities

G. Buchalla

Abstract

We comment on the future potential of kaon physics, emphasizing in particular the unique possibilities provided by $K \to \pi\nu\bar{\nu}$ decays.

54.1 Introductory Remarks

The detailed study of kaons during the past 50 years has proved to be extremely fruitful for our understanding of the basic properties of matter. In fact, many of the key elements of the standard model have their roots in the investigation of kaon decays: The very existence of K mesons and their properties led to the concept of strangeness, which proved crucial for the emergence of the quark model, defining the framework for the eventual development of QCD. The violation of parity in weak decays was first suggested by kaon experiments ("θ-τ puzzle"), and is now reflected in the chiral nature of the weak gauge interactions. CP violation, discovered 35 years ago in $K \to \pi^+\pi^-$ decays, has even deeper consequences, not yet fully elucidated today, by establishing a fundamental difference between matter and antimatter and assigning a special role to the replication of fermion generations. Finally, the careful analysis of rare transitions, such as $K_L \to \mu^+\mu^-$ or $K - \bar{K}$ mixing, provided the basis for the GIM mechanism and allowed the prediction of the existence and properties of charm.

These examples illustrate that investigating in detail the often rather subtle effects in low-energy phenomena can give profound insight into fundamental physics at much higher energy scales. The kaon system has been particularly rich in this respect and, as shall be argued below, still continues to provide excellent opportunities for future discoveries.

54.2 Current Highlights

The strong potential of kaon physics, even after 50 years of intense research, is amply demonstrated by many recent highlights, made possible by the continued and impressive progress in experimental techniques. These developments underline the very promising capabilities for future discoveries. It is therefore worthwhile to briefly recall a few of these highlights here.

First of all, indirect CP violation described by the parameter ε is firmly established and very precisely measured. The effect has been observed in the decays $K_L \to \pi\pi$, $K_L \to \pi l \nu$, $K_L \to \pi^+\pi^-\gamma$, $K_L \to \pi^+\pi^- e^+ e^-$, where the last mode is a recent addition to this list. Within some theoretical uncertainties for ε, the experimental result agrees with the standard model description. In view of the characteristic interplay of several parameters of the flavor sector (in particular m_t, V_{cb}, V_{ub}) with their specific values, this agreement is certainly nontrivial. Moreover, ε, and more generally K–$\bar{K}$ mixing, yield important constraints both within the standard model and for its extensions (in particular for the flavor sector in supersymmetry). Within the standard model the ε constraint implies a substantial CKM phase. This is the basis for the expectation of large CP violation in $B \to J/\Psi K_S$ decays and will allow a crucial test of the CKM model at the B physics facilities.

Until very recently the various manifestations of ε have been the only well established signals of CP violation in the laboratory. One of the big highlights of this year is the new determination of direct CP violation in $K \to \pi\pi$ decay, parametrized through ε'/ε, by experiments at Fermilab (KTeV) and CERN (NA48). They establish a nonzero effect, $\varepsilon'/\varepsilon = (21.2 \pm 4.6) \cdot 10^{-4}$, confirming earlier evidence from CERN experiment NA31. More than 30 years after the discovery of CP violation, this is the first time that a fundamentally new aspect of this phenomenon has been unambiguously demonstrated. Unfortunately, a detailed quantitative interpretation of this result is still plagued by considerable hadronic uncertainties, despite continuing progress also on the theoretical side (see contribution to this volume by Buras (chapter 5)).

Further recent results that have deservedly attracted considerable interest are the first direct observation of T violation in $K \leftrightarrow \bar{K}$ transitions by the CPLEAR collaboration, and the measurement of a large ($\sim 14\%$) T-odd, CP-violating asymmetry in the final-state angular distribution of $K_L \to \pi^+\pi^- e^+ e^-$ (KTeV, NA48). The latter effect had been predicted by Sehgal and Wanninger (see contributions by Sehgal (chapter 15) and Savage (chapter 16)). Both phenomena serve to demonstrate new and intriguing facets of the well-known indirect CP violation effect in neutral kaon decays.

An impressive illustration of the potential of kaon physics to explore extremely rare processes is the measurement by Brookhaven experiment E781 of $B(K_L \to e^+e^-) = 8.7^{+5.7}_{-4.1} \cdot 10^{-12}$, the smallest decay branching ratio yet observed in particle physics. Another, particularly important, example of such highly suppressed decay modes is the transition $K^+ \to \pi^+\nu\bar{\nu}$. The observation of a single clean event by Brookhaven experiment E787 in 1997 translates into a branching ratio of $B(K^+ \to \pi^+\nu\bar{\nu}) = 4.2^{+9.7}_{-3.5} \cdot 10^{-10}$. The analysis of additional data (see contribution by Redlinger(chapter 34)) reveals no new candidates and is expected to lower the central value by a factor of 2–3. The particular relevance of these exciting results is that they open the way to precision flavor physics, as will be further discussed below.

54.3 Opportunities

The study of kaons offers a very broad field of investigation, providing insight into a large variety of issues in fundamental physics (see contribution by Isidori (chapter 33) for more details on rare decays). These range from high-sensitivity searches of lepton-flavor-violating processes ($K_L \to \mu e$, $K \to \pi\mu e$), over tests of the chiral perturbation theory framework ($K^+ \to \pi^+ l^+ l^-$, $K_L \to \pi^0\gamma\gamma$, $K_S \to \gamma\gamma$, $K_S \to \pi^0 e^+ e^-$, ...) to flavor-changing neutral currents (FCNC) and tests of discrete symmetries as CP or T ($K_L \to \pi^0 e^+ e^-$, $K_L \to \mu^+\mu^-$, $K^+ \to \pi^0\mu^+\nu$, ...). In the following we focus on the rare decays $K^+ \to \pi^+\nu\bar{\nu}$ and $K_L \to \pi^0\nu\bar{\nu}$, which are particularly promising. (See also the contribution of Isidori.) In these modes the FCNC transition $s \to d$ is probed by a neutrino current, which couples only to heavy gauge bosons (W, Z). Correspondingly, the GIM pattern of the $\bar{s} \to \bar{d}\nu\bar{\nu}$ amplitude has, roughly speaking, the form $\lambda_i m_i^2$, summed over $i = u, c, t$ ($\lambda_i = V_{is}^* V_{id}$). The powerlike (rather than logarithmic) mass dependence strongly enhances the short-distance contributions, coming from the heavy flavors c and t. The transition thus proceeds through an effectively local, semileptonic $(\bar{s}d)_{V-A}(\bar{\nu}\nu)_{V-A}$ interaction, where the only hadronic matrix element required, $\langle\pi|(\bar{s}d)_V|K\rangle$, can be obtained from $K^+ \to \pi^0 l^+\nu$ decay. $K \to \pi\nu\bar{\nu}$ is therefore theoretically exceptionally well under control. While $K^+ \to \pi^+\nu\bar{\nu}$ receives both top and charm contributions, $K_L \to \pi^0\nu\bar{\nu}$ probes direct CP violation [1] and is dominated entirely by the top sector. The $K \to \pi\nu\bar{\nu}$ modes have been studied in great detail over the years to quantify the degree of theoretical precision. Important effects come from short-distance QCD corrections. These were computed at leading order in [2]. The complete next-to-leading-order calculations [3] reduce the theoretical uncertainty in these decays to $\sim 5\%$ for $K^+ \to \pi^+\nu\bar{\nu}$ and $\sim 1\%$ for $K_L \to \pi^0\nu\bar{\nu}$. This picture is essentially unchanged when further effects are considered, including isospin breaking in the relation of $K \to \pi\nu\bar{\nu}$ to $K^+ \to \pi^0 l^+\nu$ [4], long-distance contributions [5, 6], the CP-conserving effect in $K_L \to \pi^0\nu\bar{\nu}$ in the standard model [5, 7] and two-loop electroweak corrections for large m_t [8]. The current standard model predictions for the branching ratios are $B(K^+ \to \pi^+\nu\bar{\nu}) = (0.8 \pm 0.3) \times 10^{-10}$ and $B(K_L \to \pi^0\nu\bar{\nu}) = (2.8 \pm 1.1) \times 10^{-11}$.

54.4 $K \to \pi\nu\bar{\nu}$ – Phenomenology

The study of $K \to \pi\nu\bar{\nu}$ can give crucial information for testing the CKM picture of flavor mixing. This information is complementary to the results expected from B physics and is much needed to provide the overdetermination of the unitarity triangle necessary for a real test. Let us briefly illustrate some specific opportunities. $K_L \to \pi^0\nu\bar{\nu}$ is probably the best measure of the Jarlskog parameter $J_{CP} \sim \text{Im}\lambda_t$, the invariant measure of CP violation in the standard model [9]. For example a 10% measurement of $B(K_L \to \pi^0\nu\bar{\nu}) = (3.0 \pm 0.3) \times 10^{-11}$ would directly give $\text{Im}\lambda_t = (1.37 \pm 0.07) \times 10^{-4}$, a remarkably precise result. Combining 10% measurements of both $K_L \to \pi^0\nu\bar{\nu}$ and $K^+ \to \pi^+\nu\bar{\nu}$ determines the unitarity triangle parameter $\sin 2\beta$ with an uncertainty of about ± 0.07, comparable to the precision obtainable for the same quantity from CP violation in $B \to J/\Psi K_S$ before the LHC era. A measurement of $B(K^+ \to \pi^+\nu\bar{\nu})$ to 10% accuracy can be expected to determine $|V_{td}|$ with similar precision. As a final example, using only information from the ratio of $B_d - \bar{B}_d$ to $B_s - \bar{B}_s$ mixing, $\Delta M_d/\Delta M_s$, one can derive a stringent and clean upper bound

$$B(K^+ \to \pi^+\nu\bar{\nu}) < 0.4 \cdot 10^{-10} \left[P_{charm} + A^2 X(m_t) \frac{r_{sd}}{\lambda} \sqrt{\frac{\Delta M_d}{\Delta M_s}} \right]^2 . \quad (54.1)$$

Note that the ε constraint and V_{ub} with their theoretical uncertainties are not needed here. Using $V_{cb} \equiv A\lambda^2 < 0.043$, $r_{sd} < 1.4$ (describing SU(3) breaking in the ratio of B_d to B_s mixing matrix elements) and $\sqrt{\Delta M_d/\Delta M_s} < 0.2$, gives the bound $B(K^+ \to \pi^+\nu\bar{\nu}) < 1.67 \cdot 10^{-10}$, which can be confronted with future measurements of $K^+ \to \pi^+\nu\bar{\nu}$ decay.

In conclusion, flavor physics is one of the major topics in particle phenomenology. The continued study of kaons offers a rich perspective in this regard and includes opportunities, such as $K \to \pi\nu\bar{\nu}$, that have the potential to make decisive contributions to our understanding of fundamental physics. Every effort should be made to exploit these unique possibilities.

References

[1] L. Littenberg, Phys. Rev. D **39**, 3322 (1989).

[2] C. Dib, I. Dunietz, and F.J. Gilman, Mod. Phys. Lett. A **6**, 3573 (1991).

[3] G. Buchalla and A.J. Buras, Nucl. Phys. B **400**, 225 (1993); **412**, 106 (1994); **548**, 309 (1999); M. Misiak and J. Urban, Phys. Lett. B **451**, 161 (1999).

[4] W. Marciano and Z. Parsa, Phys. Rev. D **53**, R1 (1996).

[5] D. Rein and L.M. Sehgal, Phys. Rev. D **39**, 3325 (1989).

[6] J.S. Hagelin and L.S. Littenberg, Prog. Part. Nucl. Phys. **23**, 1 (1989); M. Lu and M. Wise, Phys. Lett. B **324**, 461 (1994).

[7] G. Buchalla and G. Isidori, Phys. Lett. B **440**, 170 (1998).

[8] G. Buchalla and A.J. Buras, Phys. Rev. D **57**, 216 (1998).

[9] G. Buchalla and A.J. Buras, Phys. Rev. D **54**, 6782 (1996).

55

A Future Kaon Physics Program at Fermilab

Peter S. Cooper

Abstract

The program of fixed target kaon experiments proposed for the new Fermilab Main Injector are described. Three experiments are being considered: CP/T, to make precision studies in the K^0_S system; CKM, to measure the magnitude of V_{td} by measuring the branching ratio of $K^+ \to \pi^+ \nu\overline{\nu}$; and KAMI, to measure the imaginary part of V_{td} by measuring the branching ratio of $K_L \to \pi^0 \nu\overline{\nu}$. The goals, plans, and prospects for this program are discussed.

The Fermilab Main Injector is a new rapid cycling 120 GeV proton accelerator that has recently been commissioned. It will provide opportunities for a rich and powerful kaon physics program. Three fixed target experiments are being considered. I will briefly describe the physics goals and experimental methods of each of these experiments. I will conclude with the current status, plans, and prospects for these experiments and the program as a whole.

CP/T is an experiment to study CP violation and search for CPT violation in the K^0_L–K^0_S system. It will measure $\eta_{+-0}, \eta_{000}, \eta_{+-\gamma}$, and η_{+-} with high precision and search for CPT violation to the level of the Planck scale. CP/T is a Fermilab, IHEP Protvino, Rutgers, TRIUMF, and Wisconsin collaboration lead by Gordon Thomson. They submitted a full proposal in April 1999.

CP/T would have a beam of pure K^0_S produced by K^+ charge exchange on a target using a superconducting RF separated beam at $\sim$ 22 GeV/c. This separated beam line would be shared with the CKM experiment. R&D is presently under way to develop the 3.9 GHz superconducting cavity technology required. The beam line would yield up to 200 MHz of separated K^+ with a purity approaching 90%. The CP/T spectrometer would be a conventional neutral-kaon decay spectrometer with charged and neutral particle detection.

CKM[1] is a proposal to measure the ultrarare kaon decay $K^+ \to \pi^+ \nu\overline{\nu}$

[1] http://www.fnal.gov/projects/ckm/Welcome.html.

to a branching fraction precision of 10%. This measurement determines the magnitude of the CKM matrix element V_{td} with a statistical precision of 5% and an overall precision, including theoretical uncertainties, of 10%. This experiment is a collaboration of Fermilab, IHEP Protvino, the University of San Luis Potosi, and the University of Texas. I am the spokesman of this outfit. We submitted a full proposal to Fermilab in April 1998, and our R&D proposal was approved November 1998.

The standard model prediction for the $K^+ \to \pi^+ \nu\overline{\nu}$ decay mode is $[0.9 \pm 0.3] \times 10^{-10}$. The theoretical uncertainty in this prediction is well quantified and dominated by the lack of knowledge of the mass of the charmed quark. The primary physics goal of CKM to the observed $\sim$100 $K^+ \to \pi^+ \nu\overline{\nu}$ events with a background of less than 10%. With the present design this will require 2 years of data taking. Several other searches and measurements in K^+ decay are also planned.

CKM is a decay-in-flight experiment. The apparatus is a long instrumented beam line with momentum and velocity spectrometers for both the incident K^+ and the decay daughter π^+, a vacuum decay volume, and full-coverage photon and muon vetos. Momenta are measured in conventional magnetic spectrometers with silicon strip detectors for the K^+ and straw tubes for the π^+. Vector velocities are measured with phototube ring-imaging Cherenkov counters. All four spectrometers have approximately matched resolutions of $\sim 1\%$ in terms of momentum. The time resolution of the RICHs are better than 1 nsec. The photon vetos need to reject π^0 with an inefficiency of better than 2×10^{-7}. All the detector technologies used have been established in previous experiments. The CKM beam is the same SCRF separated beam used by CP/T. For CKM operation the K^+ flux is 30 MHz with a K^+ decay rate in the sensitive volume of 6 MHz. A debunched proton beam from the main injector will be required.

KAMI[2] submitted an expression of interest to Fermilab in 1997 and also was approved as an R&D project in November 1998. This experiment is a direct descendent of KTeV sharing most of the same collaboration. It uses the same beam line and detector as KTeV with much-enhanced photon veto and tracking systems. Their major physics goal is a measurement of the branching ratio of the pure direct CP-violating decay $K_L \to \pi^0 \nu\overline{\nu}$ to $\sim 10\%$ precision to measure η_{CKM} to $\sim 5\%$. The present standard model prediction for this branching ratio is $[2.8 \pm 1.7] \times 10^{-11}$ with very small theoretical uncertainty. They have a broad physics program including $K_L \to \pi^0 e^+ e^-$, $K_L \to \pi^0 \mu^+ \mu^-$, $K_L \to \pi^0 \mu^+ e^-$, $K_L \to \pi^+ \pi^- e^+ e^-$, and other ultrarare K_L^0 decay modes at the $\sim 10^{-13}$ level.

Their apparatus stresses open geometry, low rates, and high-precision photon detection. Achieving the required photon detection inefficiencies is the major challenge to the experiment. For example, they require photon inefficiencies below 1×10^{-6} for photons above 3 GeV. Inefficiency tests on prototype vetos have already begun (at INS, Tokyo). Design specifications

[2] http://fnphyx-www.fnal.gov/experiments/ktev/kami/kami.html.

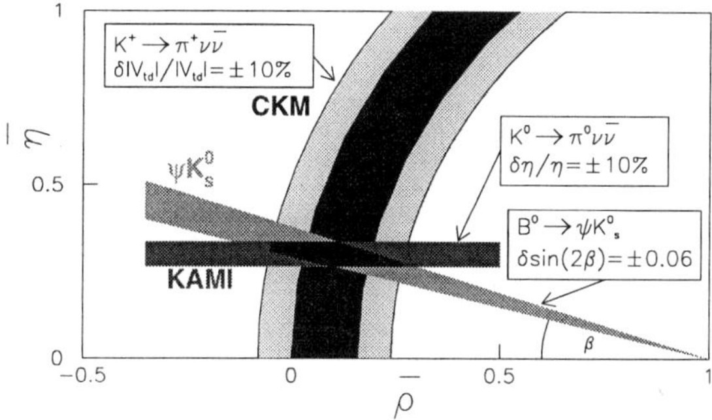

Figure 55.1: The Unitarity Triangle with illustrative results.

are already exceeded for photons in coincidence with emitted neutrons.

At the time of the conference these three experiments were being discussed by the Fermilab PAC at their annual Aspen retreat. The R&D proposals approved in 1998 were unfunded. This kaon physics program had, thus far, received encouragement but no commitment from Fermilab. In the subsequent weeks the message which came down from the mountains was largely very positive. "[T]he two kaon experiments CKM and KAMI, if proven technically feasible, would form the core of the 120 GeV fixed-target program." The bad news was that CP/T was viewed to have insufficient physics interest and would be given "no further consideration." Many of us in the kaon community would disagree with this latter judgment and we surely regret this loss. For CKM and KAMI the challenge now is to complete their R&D projects in time for consideration of full experimental proposals in 2001. Assuming approval, these experiments would be built beginning in 2002–3.

The prospects appear good for a vigorous and challenging kaon physics program at Fermilab that will begin to produce new results in the middle of the next decade. Fig. 55.1 shows the physics reach of CKM, KAMI, and that expected from B experiments. Things will get interesting if we find that these three curves don't, in fact, meet at one point!

56

Future Kaon Initiatives at BNL

L. Littenberg

Abstract

Although the Brookhaven AGS will become an injector to RHIC, it will still be available for external proton beam experiments. I discuss a number of new K decay experiments that have been proposed for this facility

56.1 Introduction

This month, the Brookhaven AGS begins its new career as an injector to the RHIC heavy ion collider. Although this will be its primary function, once RHIC is running routinely the AGS will still be available more than 20 hours per day for fixed-target proton experiments. Several such experiments have been proposed, among which are four very topical K decay experiments.

56.2 E927

E927 is designed to determine $|V_{us}|$ through a measurement of $\Gamma(K^+ \to \pi^0 e^+ \nu_e)$ [1]. The present determination of $|V_{us}|$ is based on K_{e3} measurements more than 20 years old. It is usually assigned an error of about 1%, but this degree of precision is achieved by combining data from experiments with quite different systematics [2]. E927 seeks to measure $\Gamma(K_{e3}^+)$ to $\pm 0.7\%$, which in principle would allow a determination of $|V_{us}|$ to $\sim 0.35\%$. Aside from the fundamental importance of this parameter, Bill Marciano [3] has noted that there is $\sim 2\sigma$ violation of CKM unitarity when the present best values of $|V_{us}|$ and $|V_{ud}|$ are used

Unlike previous measurements of this type, E927 is a stopping experiment. A 700 MeV/c separated beam containing $\sim 30,000$ K^+ per spill is incident on a scintillating fiber stopping target, situated in the central cavity of the crystal ball detector. Augmenting the latter will be end cap crystals to complete the solid angle coverage, a cylindrical drift chamber to detect charged particles, and a plexiglass Cerenkov counter outside the drift chamber. The technique of the experiment is to detect and identify the large majority of the K^+ decays to obtain the K^+_{e3} branching ratio, and use the well-known K^+ lifetime to convert it to a partial rate.

56.3 E923

Although the standard model (SM) predicts a negligible level of T-violating μ^+ polarization in $K^+ \to \pi^0\mu^+\nu$ decay, some BSM models predict polarizations as large as 10^{-3} [4]. In his talk at this conference, Roberto Peccei [5] emphasized the desirability of a measurement at the 10^{-4} level. AGS-E923 proposes to do just that, reaching a one-σ sensitivity of 0.00013 [6]. It is a lineal descendent of the last round of AGS experiments [7] on this subject, which reached a level of ~ 0.005. The major improvements with respect to the older experiments are the use of a separated (2 GeV/c) kaon beam, and a more granular polarimeter with a higher analyzing power. The proposed apparatus is shown in fig. 56.1. Roughly 20 million K^+ per AGS cycle would be incident on the apparatus. A relatively large acceptance is achieved by allowing the muons to pass through the (shashlyk) calorimeter on their way to the polarimeter. Gammas from the π^0 are detected in the calorimeter. The polarimeter stopping material is graphite, which allows an analyzing power of ~ 0.36. As in the predecessor experiments the polarization is measured by comparing the clockwise-going vs counter-clockwise-going decay electrons from the muons stopped in the polarimeter wedges. The experiment is also sensitive to the T-violating μ^+ polarization in $K^+ \to \gamma\mu^+\nu$ decay. Great efforts have been devoted to a design in which systematic errors can be controlled to the 10^{-4} level.

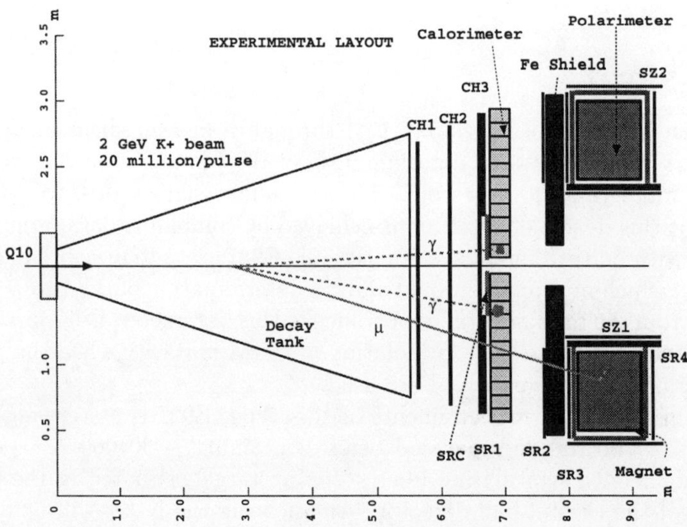

Figure 56.1: E923 apparatus.

56.4 E949

E949 is an upgrade to E787 designed to measure $K^+ \to \pi^+\nu\bar{\nu}$ at the 10^{-11}/event level [8]. E787 is expected to reach a single-event sensitivity of $\sim 8 \times 10^{-11}$/event for this process when all its data is analyzed. This is roughly the present SM predicted level [9]. There are very strong reasons for continuing this study. In the SM, measuring $B(K^+ \to \pi^+\nu\bar{\nu})$ is the theoretically cleanest known way of measuring $|V_{td}|$. If the branching ratio exceeds $\sim 1.2 \times 10^{-10}$, it is a clear indication of new physics [10]. Thus one either gets a good measurement at the SM level or overthrows the SM!

Projections for E949 are firmly grounded in the experience of E787. The most conspicuous upgrade to the E787 apparatus is an extra layer of photon veto counter. This is calculated to improve to π^0 veto efficiency by a factor 2–3. It is not essential for measuring the branching ratio in the kinematic region exploited so far—the background level for this region achieved in the analysis of the 1995–97 data of E787 is already low enough for a measurement at the SM level—but it will allow the experiment to be sensitive to softer π^+, potentially doubling its acceptance. Other substantial improvement factors are made possible because E949 will have the AGS to itself. For example, the much higher proton intensity, 65×10^{12} protons/pulse (*cf.* 13×10^{12} for E787 in 1995), can be used to extend the spill and reduce the beam momentum, each of which will improve the sensitivity without increasing instantaneous rates. Additional vetoing, electronic, and data-acquisition improvements will also be made.

The proposal for E949 is presently in the hands of the DOE. Since slow beam running of the AGS is no longer routinely funded by the agency, each experiment must be individually approved.

56.5 E926

There is now wide agreement that the most compelling experiment in the kaon system, and one of the most important in all of flavor physics, is a measurement of the rate of $K_L \to \pi^0\nu\bar{\nu}$. However, it is obviously a considerable experimental challenge: To reach the necessary sensitivity one must improve on the present state of the art by a factor of more than 10^5. AGS-E926 takes up this challenge, proposing to measure $B(K_L \to \pi^0\nu\bar{\nu})$ to the few $\times 10^{-13}$/event level [11]. This experiment is the poster child for AGS-2000, fully exploiting the unique features of the AGS: the intense flux at medium energies, the great flexibility of operation and the deep expertise in rare K decay techniques.

The E926 apparatus is shown in fig. 56.2. The AGS proton beam is microbunched on extraction so that the protons arrive in ~ 200 ps bunches every 40–50 ns. The neutral channel is taken off at an extremely large angle ($\sim 40°$) so that a very low-energy K_L beam results ($\langle p_K \rangle \sim 0.7$ GeV/c). This allows the momentum of the K_L to be determined to a few percent by timing against the microbunches. Other advantages of the wide-angle beam are a neutron spectrum predominantly below π^0 production threshold and the small number of surviving hyperons, both of which suppress potential backgrounds. Photon directions are measured by a live preradiator and the measurement of their energy and arrival time is completed by a shashlyk calorimeter. Thus all possible physical observables are measured, which is crucial for background rejection. The π^0 vertex (and thus the K_L vertex) is directly measured, rather than inferred, and it is possible to work in the K_L cm system, in which many backgrounds can be readily identified and eliminated. The kinematic handle on backgrounds relieves the otherwise onerous burden on photon vetoing; one need do no better than E787 has done. In addition, the existence of independent redundant methods of rejecting backgrounds makes it possible for E926 to actually measure their level.

A three-year run would yield about 50 events at the SM level, with S/B $\sim$ 5 : 1. This would make possible a direct and unambiguous 15% measurement of CKM η.

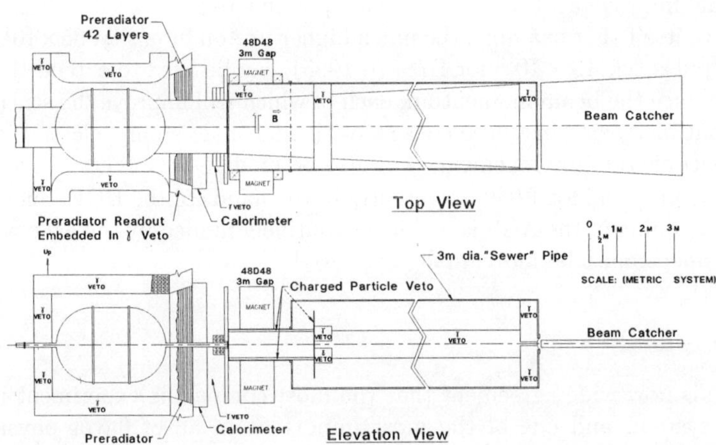

Figure 56.2: E926 apparatus.

56.6 Conclusion

Running parasitically to RHIC, an extremely cost effective program of kaon decay experiments could be carried out at the AGS, including some of the currently most compelling measurements. The experiments are designed and the collaborations are ready to start. All that is needed is the political will.

Acknowledgments

This work was supported by the Department of Energy under Contract DE-AC02-98CH10886.

References

[1] T. Kycia *et al.*, AGS Proposal 927, *Measurement of the K^+_{e3} Decay Rate and Spectrum*, September 1996.

[2] H. Leutwyler and M. Roos, Z. Physik C **25**, 91 (1984).

[3] W. Marciano, this volume, chapter 62.

[4] S. Weinberg, Phys. Rev. Lett. **37**, 657 (1976); M. Leurer, Phys. Rev. Lett. **62**, 1967 (1989); P. Castoldi, J.M. Frère, and G.L. Kane, Phys. Rev. D **39**, 2633 (1989); G. Bélanger and C.Q. Geng, Phys. Rev. D **44**, 2789 (1991); C.Q. Geng and S.K. Lee, Phys. Rev. D **51**, 99 (1995); M. Kobayashi, T.T. Lin, and Y. Okada, Prog. of Theor. Phys. **95**, 361 (1996); G.H. Wu and J.N. Ng, Phys. Rev. D **55**, 2806 (1997); M. Fabbrichesi and F. Vissani, Phys. Rev. D **55**, 5334 (1997).

[5] R.D. Peccei, this volume, chapter 2.

[6] M.V. Diwan *et al.*, AGS Proposal 923, *Search for T-Violating Muon Polarization in the $K\mu3$ Decay, $K^+ \to \pi^0\mu^+\nu$*, December 1996.

[7] M. Schmidt *et al.*, Phys. Rev. Lett. **43**, 556 (1979); W. Morse *et al.*, Phys. Rev. D **21**, 1750 (1980); Phys. Rev. Lett. **47**, 1032 (1981); S. Blatt *et al.*, Phys. Rev. D **27**, 1056 (1983).

[8] M. Aoki, *et al.*, AGS Proposal 949, *An Experiment to Measure the Branching Ratio $B(K^+ \to \pi^+\nu\bar{\nu})$*, October 1998.

[9] G. Buchalla and A.J. Buras, Nucl. Phys. B **548**, 309 (1999), hep-ph/9901288.

[10] A.J. Buras, hep-ph/9905437.

[11] I-H. Chiang *et al.*, AGS Proposal 926, *Measurement of $K_L \to \pi^0\nu\bar{\nu}$*, September 1996.

57

Future CERN Kaon Program

George Kalmus

57.1 Introduction

The scientific program of CERN foreseen at present is summarized in fig. 57.1.

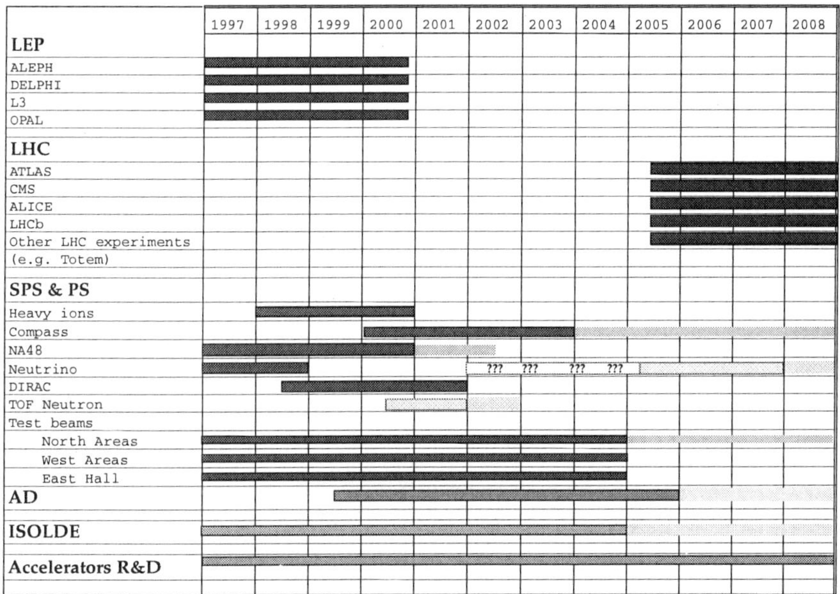

Figure 57.1: The scientific program of CERN.

It can be seen that NA48 continues through the year 2000. Beyond that, it might be extended to 2002 if a good enough scientific case can be made. NA48 is the only component of the CERN program containing a K decay element.

NA48 was designed primarily to measure ϵ'/ϵ, but it was foreseen that it would address various rare decay modes as well. Results from some of these have been presented by Wronka and Köpke at this conference.

Beam	P_p (GeV/c)	Cycle (sec/sec)	Protons/pulse on target	Total decays per year (120 days, $\epsilon = 0.5$) in accepted P and Z range	SES (Single event sensitivity) assuming 10% acceptance	Estimated instantaneous rates in detector compared to present NA48
1 K_L^0	450	2.5/14.4	1.5×10^{12}	6.7×10^{10}	1.5×10^{-10}	1
2 K_L^0	400	5.0/19.2	4×10^{12}	3.3×10^{11}	3×10^{-11}	3
3 K_L^0	400	5.0/19.2	1.2×10^{13}	1×10^{12}	10^{-11}	10
4 K_L^0	400	5.0/19.2	2.4×10^{13}	2×10^{12}	5×10^{-12}	20
5 K_S^0	450	2.5/14.4	3×10^{7}	8.7×10^{7}	1.2×10^{-7}	$\ll 1$
6 K_S^0	400	5.0/19.2	1×10^{10}	3.3×10^{10}	3×10^{-10}	< 3
7 $K^+ + K^-$	400	5.0/19.2	1×10^{12}	$(6.3 + 3.5) \times 10^{10}$	$(1.5 + 3) \times 10^{-10}$	2
8 K_L^0	400	5.0/19.2	1×10^{12}	8.3×10^{10}	1.2×10^{-10}	2

Table 57.1: Some parameters of beams that could be installed in the present NA48 beam line (For additional information see 'Notes')

About a year ago a study started within the NA48 collaboration to determine whether the experiment and its beam could be modified in such a way as to allow very rare K_L^0 decay modes to be sought. (The guidelines were that changes to the present NA48 setup should neither be too costly nor take too much time or effort.) In particular the channel $K_L^0 \to \pi^0 e^+ e^-$ was targeted. The conclusion of this study was that it appeared to be possible, but not without some effort, to modify the experimental conditions so as to be able to reach a single-event sensitivity (SES) of O(10^{-11}) per running year in a typical channel.

This is a factor of about 30 lower than the present NA48 running conditions would allow, and is entering the interesting range for $K_L^0 \to \pi^0 e^+ e^-$, where the standard model branching ratio is expected to be O(10^{-11}). Recent studies [1, 2, 3], however, point out that the branching ratio for this channel might be enhanced to levels above 10^{-10}. A detailed study of backgrounds to this channel showed that the irreducible background due to $K_L^0 \to \gamma\gamma e^+ e^-$, as was pointed out by Greenlee [4], was at a level of 10^{-10} under the NA48 experimental conditions.

There is an ongoing study looking into less rare K_L^0 decay modes as well as K_S^0 modes and hyperon decays. A separate study is also being made into the feasibility of measuring the asymmetry in the Dalitz plot of K^+/K^- decays into 3 pions. A beam capable of transporting simultaneously K^+ and K^- down the present beam line has been designed.

Table 57.1 gives some of the parameters of beams that have been studied and would fit into the existing beam line without a major rebuild.

A decision by NA48 on which direction to follow after this year's running will be taken this autumn and presented to the SPS committee for approval.

There have also been some thoughts on the possibility of attacking the most interesting channels: $K_L^0 \to \pi^0 \nu\bar{\nu}$ and $K^+ \to \pi^+ \nu\bar{\nu}$. However, the conclusion reached was that neither of these very challenging experiments could be seriously attempted without a completely new setup. My appraisal of the situation at CERN at present is that it would require the push from a sizeable community with considerable resources to launch such an ambitious program.

57.2 Notes on Beams

1) Beam 1 and beam 5 are the two beams currently running simultaneously for NA48. The protons used for beam 5 are channeled to the K_S^0 target by means of a bent crystal.

2) Beam 2 is an upgrade of the present K_L^0 beam, in which the K_S^0 beam option has been removed, the V and H acceptances have been increased, and the proton intensity has been increased. The proton momentum has been reduced from 450 to 400 GeV/c to obtain a higher duty cycle. This would require modest modifications.

3) Beams 3 and 4 are similar to beam 2 but would require considerable modifications to handle the higher proton flux on target.

4) Beam 6 is a stand-alone K_S^0 beam, the bent crystal being replaced by a magnet.

5) Beam 7 is a 60 ± 6 GeV/c K^+ and K^- unseparated beam. The 2 charged beams travel down the decay pipe coincident in both time and space.

6) Beam 8 is a K_L^0 beam that can be run simultaneously with beam 7.

References

[1] G. Colangelo and G. Isidori, JHEP **09**, 009 (1998).

[2] A.J. Buras and L. Silvestrini, Nucl. Phys. B **546**, 299 (1999).

[3] L. Silvestrini, TUM-HEP-350/99 (1999).

[4] H.B. Greenlee, 1990, Phys. Rev. D **42**, 3724 (1990).

58

Future Plan for Kaon Physics at KEK

Takao Inagaki

Abstract

A joint project between KEK and JAERI is going to start to construct a 50-GeV proton synchrotron (PS). It will change the ability of the kaon experiments that have been performed by using the present low-energy and low-intensity KEK 12-GeV PS. One of the future plans for kaon physics is an experimental study on the $K_L^0 \to \pi^0 \nu\overline{\nu}$ decay. A pilot experiment for the decay, E391a at the present 12-GeV PS, has started construction.

Since the construction of the 12-GeV PS in the late 1970s, about ten weak-interaction experiments have been performed at KEK, including two that are running (E362 and E246) and one that is being prepared (E391a), as listed in table 58.1. Most of those weak-interaction experiments are K decay experiments and they have made some contribution to the progress of particle physics, although the 12-GeV PS is much less powerful than other accelerators in energy and intensity. The concentration of resources by management is one reason; for example, these experiments, though only 10% of all those executed, have spent more than half of the budget of PS. Another is the fact that in the case of a low-energy K beam the production

E10 (KD)	Study of $K^+ \to \pi^+ \nu\overline{\nu}$ decay	1979, 1980
E49 (Λ)	Asymmetry in $\Lambda \to pe(\mu)\nu$ decay	1981
E89/104 (Kμ)	Search for heavy ν by $K^+ \to \mu^+\nu$ decay	1981, 1983
E92 (Σ)	Asymmetry in $\Sigma^+ \to p\gamma$ decay	1983
E99 (KμP)	μ^+ polarization in $K^+ \to \mu^+\nu$ decay	1982
E137 (KL)	Study of $K_L^0 \to \mu e$, $\mu\mu$ and ee decays	1988–1990
E162 (CP)	Study of $K_L^0 \to \pi^0 ee$ and $\pi\pi ee$ decays	1995–1997
E195 (Pμ)	Long. μ^+ polarization in $K^+ \to \mu^+\nu$ decay	1989
E246 (T-viol.)	Trans. μ^+ polarization in $K^+ \to \pi^0\mu^+\nu$ decay	1995–*running*
E362 (K2K)	Long baseline ν oscillation	1999–*running*
E391a (KL)	Study of $K_L^0 \to \pi^0 \nu\overline{\nu}$ decay	2001–*preparing*

Table 58.1: Weak interaction experiments at KEK.

of secondary K does not largely depend on the energy of primary protons above 12 GeV. As far as we select the experiment that takes advantage of a low energy K beam, the factor of low-energy primary proton can be overcome.

Nevertheless, most KEK experiments are statistically limited due to low intensity, as indicated in the $K_L^0 \to \mu\mu$ branching ratio measured by E137 as $(7.9 \pm 0.6(\text{stat}) \pm 0.2(\text{sys})$ [1] and the transverse μ polarization in the $K_{\mu 3}$ decay by E246 as $-0.0042 \pm 0.0049 \pm 0.0009$ [2]. So, the construction of a powerful accelerator is our constant desire, and we have now a joint project of KEK and JAERI to construct a 50-GeV PS. It will increase the proton intensity by a factor of 60. If we take a few-GeV K beam, the primary energy enhances the production by another factor of 10. Such a beam is regarded to be suitable for a $K_L^0 \to \pi^0\nu\overline{\nu}$ experiment to avoid the heavy hyperon background at high energy and the large inefficiency of calorimeters at low energy. The production cross section of few-GeV K_L^0 increases with proton energy below 50 GeV and saturates above it.

One of the prospective K decays at the 50-GeV PS is the $K_L^0 \to \pi^0\nu\overline{\nu}$ decay. The decay is a flavor-changing neutral-current process of CP violation with a pure $\Delta S = 1$ transition. The theoretical ambiguity due to QCD corrections and long-distance effects is very small for the $K_L^0 \to \pi^0\nu\overline{\nu}$ decay. Measurement of the decay branching ratio, which is predicted to be around 3×10^{-11} by the standard model, will offer clean information on Im V_{td}, which is a basic parameter of the present particle physics.

No dedicated experiment has been done for the $K_L^0 \to \pi^0\nu\overline{\nu}$ decay. The best experimental limit of 1.6×10^{-6} [3] was given by a parasitic run of KTeV-97 at Fermilab but it is still far from the standard model prediction. Our plan is unique among several plans in the world as a redundant approach by series experiments: E391a at the present 12-GeV PS [4] and an experiment using 50-GeV PS of the joint project.

E391a is a kind of pilot experiment to check experimental feasibility near the sensitivity of the SM prediction. We will learn many critically valuable things about signal and background from the E391a run. E391a is also valuable to uncover the SUSY-enhanced range for the $K_L^0 \to \pi^0\nu\overline{\nu}$ branching ratio [5].

The experiment at the 50-GeV PS is expected to observe hundreds of SM events, being projected from the natural extension of E391a. The sensitivity is somehow an ultimate goal for the present technology and background. It is consistent with the ultimate limit of detector counting rate caused by the K_L^0 decays around the detector for the running of a few years. The statistical error is equal to the errors of the branching ratio of other K_L^0 decays ($K_L^0 \to \gamma\gamma$, $\pi^0\pi^0$, and $\pi^0\pi^0\pi^0$), which will be utilized for the normalization of the number of incident K_L^0 , and it is comparable with the theoretical ambiguity of a few percent. Moreover, the top quark mass, which links to the SM calculation, has been measured in a similar error.

Although the joint project is not formally approved yet, the site selection (JAERI-Tokai) and the agreement between the two ministries, to which KEK and JAERI belong, have been achieved recently. The promotion office in KEK expects its operational start around 2005. An important message, which they asked me to pass on before this trip, is that the joint project should be encouraged by international contributions.

E391a is now under construction. Many parts of the detectors are recycles of what were used by previous PS and TRISTAN experiments. The experiment is expected to start a run in 2001, and the detectors will be gradually upgraded to meet the high-sensitivity experiment at the 50-GeV PS.

References

[1] T. Akagi *et al.*, Phys. Rev. D **51**, 2061 (1995).

[2] G.Y. Lim, this volume, chapter 21.

[3] J. Adams *et al.*, Phys. Lett. B **447**, 240 (1999).

[4] T. Inagaki *et al.*, *KEK-E391 Proposal*, KEK-Internal 96-13 (1996).

[5] G. Isidori, this volume, chapter 33.

59

Experiments in the Separated Kaon Beam of the IHEP Accelerator

L. G. Landsberg and V. F. Obraztsov

Future prospects of kaon physics at IHEP are based on a new proposal for the development of a separated kaon beam with the Karlsruhe-CERN superconductive RF cavities. These separators were built in Karlsruhe (Germany) and were used at CERN in 1978–80 for kaon and antiproton beams for the Ω spectrometer [1]. We may be able to use this sophisticated equipment due to a special CERN-IHEP agreement. The separators were delivered from CERN to IHEP in 1998 and are now in the process of adjustment and testing with the IHEP cryogenic system.

The parameters of a new kaon beam, which is being developed now at the IHEP 70 GeV accelerator on the basis of these separators, are presented in table 59.1.

The experiment on the separated beam will be carried out with a kaon facility that will include a wide-aperture magnetic spectrometer, two lead glass γ-spectrometers, muon detector, Cherenkov detectors and other equipment (it is proposed to use 3 main IHEP setups—SPHINX, ISTRA-M, and GAMS-2000—in this facility).

As is seen from table 59.1, the expected sensitivity of the IHEP kaon decay experiment for different decay modes lies in the range of $5 \cdot 10^{-12} \div 4 \cdot 10^{-11}$. This sensitivity is an order of magnitude higher than in proposed experiments at the DAΦNE ϕ factory [2]. On the other hand, our sensitivity is moderate as compared to possibilities of the next generation of separated kaon beams at the main injector of Fermilab. Thus, we have no intention to make experiments with the highest sensitivity (such as the study of $K^+ \to \pi^+\nu\nu^-$ or the search for $K^+ \to \pi^+\mu e$). But the IHEP experiment offers a good opportunity for precise study of a broad set of kaon decays with "intermediate branching ratios" $BR \sim 10^{-2} - 10^{-8}$.

Three main directions in our decay research can be identified.

A. Kaon decays as a laboratory for investigation of the strong interactions at low energy and testing chiral perturbative theory (CHPT).

Plenty of interesting processes in this field are available. For example, we hope to collect $\sim 1.5 \times 10^7$ radiative decays $K^+ \to \pi^+\pi^0\gamma$, to separate

Parameters of the beam	
Primary proton energy	70 GeV
Primary proton intensity	10^{13} p/spill
Number of spills/hour	$4 \cdot 10^2$
Secondary beam momentum	12.5 GeV/c
$\Delta p/p$	$\pm 4\%$
Horizontal acceptance	$\pm$ 10 mrad
Vertical acceptance	$\pm$ 1.9 mrad
Intensity of K^+ beam	$3 \div 5 \cdot 10^6$ K/spill
Duration of the spill	1.8 sec
π^+, p contamination	$\lesssim 50\%$
muon halo	$\sim 100\%$
Parameters of superconductive separators	
Operating frequency	2865 MHz
Wavelength λ	10.47 cm
Iris opening, ϕ	40 mm
Effective length of deflector	2.74 m
Number of cells in deflector	104
Mean field	$1.0 \div 1.2$ MV/m
Working T	$2.5 \div 2.8$ K
Distance between two separators	76.3 m
Possibilities of kaon decay experiment	
Decay path in the setup	$10 \div 12$ m (decay probability $\simeq 11 \div 13\%$)
Duration of the measurements	$(3.0 \div 3.5) \cdot 10^3$ hours $\simeq$ $(1.2 \div 1.4) \cdot 10^6$ spills
Number of kaon decays	$\sim (5 \div 6) \cdot 10^{11}$
Detection efficiency for different decay channels	$\varepsilon = 0.05 \div 0.3$
Sensitivity of the experiment	$4 \cdot 10^{-11} \div 5 \cdot 10^{-12}$

Table 59.1: Separated kaon beam of the IHEP accelerator

direct photon emission, and to study electric and magnetic form factors in this process. We are planning to detect $\sim 5 \times 10^6$ events of $K^+ \to l^+ \nu_l \pi\pi$ decays and to study $\pi^+\pi^-$ and $\pi^0\pi^0$ interactions under pure conditions. The investigation of $K^+ \to \pi^+ l^+ l^-$ and $K^+ \to \pi^+\gamma\gamma$ decays at a statistical level of (few) $\times 10^4$ events also looks very promising for testing CHPT.

B. Search for new pseudoscalar (PS), scalar (S), and tensor (T) weak interactions.

To this end the precise study of several decay channels with high statistics is planned: $K^+ \to e^+\nu_e$ ($\sim 2 \times 10^6$ events), $K^+ \to e^+\nu_e\pi^0$ ($\sim 5 \times 10^9$

events), $K^+ \to e^+\nu_e\gamma$ ($\sim 3 \times 10^6$ events). We may be able to observe new interactions based on the search for their interference with the "normal" Standard Model amplitudes A_{SM}:

$$|A|^2 = |A_{SM} + A_{add}|^2 = |A_{SM}|^2 + |A_{add}|^2 + 2Re(A_{SM} \cdot A^*_{add}) \\ \simeq |A_{SM}|^2(1 \pm 2|\tfrac{A_{add}}{A_{SM}}|).$$

The main influence of a small admixture of new interaction has manifested itself in the interference term (I.T.). Thus, the measured effect is more sensitive to the mass of a new intermediate boson (as M^{-2}_{boson}) as compared to processes without I.T. (like $K^+ \to \pi^+\mu e$) where the effect is proportional to $|A_{add}|^2$ (e.g. M^{-4}_{boson}). For example, let us consider $K^+ \to e^+\nu_e$ decay, which is specially sensitive to the PS interaction because A_{SM} for this decay is helicity suppressed. For $K^+ \to e^+\nu_e$,

$$R = BR(K^+ \to e^+\nu_e)/BR(K^+ \to \mu^+\nu_\mu) \simeq R_{SM}[1 + \delta_R \pm 2\Delta_{PS}] \\ = 2.569 \times 10^{-5}[1 - 0.0378 \pm 0.0004 \pm 2\Delta_{PS}].$$

Here $\delta_R = -0.0378 \pm 0.0004$ is the radiative correction that is calculated with a high precision [3]; $\Delta_{PS} = |A_{PS}/A_{SM}|$. The last value can be measured in the new IHEP experiment with a precision $< 10^{-3}$, which corresponds to $BR[K^+ \to e^+\nu_e]_{PS} < 10^{-11}$ (the existing data for Δ_{PS} are of ± 0.045 precision). It can be shown that K_{e2} decay is much more sensitive to the PS interaction than $K^0_L \to e^+e^-$ (which is also helicity suppressed in the SM).

C. Search for direct CP violation in $K^\pm$ decays.

It is not clear yet if it is possible to explain a large value of direct CP nonconservation in $K^0 \to 2\pi$ ($\varepsilon'/\varepsilon \simeq 2 \times 10^{-3}$) in the framework of the standard model or if we must consider some new mechanisms for this violation (see, for example, [4]). Thus, the search for direct CP violation in different processes is of primary interest nowadays. We hope to reveal such effects in the experiments with "twinkling" K^+ and K^- beams by studying $K^\pm \to \pi^\pm\pi^+\pi^-(\tau^\pm)$ and $K^\pm \to \pi^\pm\pi^0\pi^0(\tau'^\pm)$ decays with statistics $\sim 8 \times 10^9\tau^\pm$ and $\sim 8 \times 10^8\tau'^\pm$ and by measuring the asymmetry in the slopes of the Dalitz plot distributions for these decays ($\delta g_\pm = g(K^+) - g(K^-)$). The statistical precision of these measurements is expected to be $\delta g_\pm/2g \simeq 1 \times 10^{-4}$ (for both τ and τ'). Certainly, the major problems in these very difficult measurements are systematic effects that are now under study. The theoretical predictions for this asymmetry in the standard model are disappointingly small ($[\delta g_\pm/2g]_{SM} \simeq$ (few) $\times 10^{-6}$. But in the work of E. Shabalin for the model with the spontaneous CP violation more promising predictions were obtained: $(\delta g_\pm/2g) \simeq$ (few) $\times 10^{-4}$ [5].

The search for CP asymmetry in other $K^\pm$ decays as well as for the T-odd triple correlation $\vec{P}_\pi(\vec{P}_l \times \vec{P}_\gamma)$ in $K^+ \to \pi^0 l^+\nu_l\gamma$ decays on the level of $\lesssim 10^{-3}$ are foreseen for IHEP experiments.

We are also intending to study strong $K^{\pm}$ interactions in separated beams with sensitivity around 3×10^2 events/nb. This program includes the study of rare decays of several mesons with strange quarks (for example, radiative decays $f_2'(1520) \to \phi + \gamma$) and the search for exotic hybrid mesons $s\bar{s}g$ and pentaquark baryons $qqss\bar{s}$.

Now the superconducting separators are in the process of being tested. If these separators are in good shape, we hope to complete the work with a new kaon beam and to begin our kaon experiments in the years 2001–2002.

References

[1] A. Citron *et al.*, NIM **155**, 93 (1978); **164**, 31 (1979).

[2] J. Lee-Franzini, *The Second DAΦNE Phys. Handbook*, ed. L. Maiani *et al.* (SIS, publications of LNF, Frascati, 1995), p. 761.

[3] M. Finkemeier, *The Second DAΦNE Phys. Handbook*, ed. L. Maiani *et al.* (SIS, publications of LNF, Frascati, 1995), p. 389.

[4] E. Shabalin, Proc. of the Workshop on K Phys., Orsay, 1996, p. 71.

[5] E. Shabalin, Preprint ITEP 98-8, Moscow, 1998.

60

The Role of Future Kaon Experiments

Paolo Franzini

Kaons were discovered in 1943 by Leprince-Ringuet and M. l'Heritier [1] who observed in a Wilson chamber a track of a particle of mass 506±91 MeV, in modern units. (See, however, the discussion by Dalitz [2].) The decays of K mesons, both charged and neutral, were first observed, also in a Wilson chamber, by Rochester and Butler in 1947 [3]. Since then, the study of kaons (and hyperons) has led to many fundamental discoveries and concepts in particle physics. Dalitz [2] reminded us of the history of the "$\theta - \tau$ puzzle," which led to the parity revolution.

The study of K mesons has been instrumental for the introduction of strangeness, the idea of quarks, and flavors. Flavor mixing was first proposed in 1963 by Cabibbo, possible only if $\Delta S = \Delta Q$ as observed in K decays. Mixing was extended to four quarks in 1970 to suppress the decay $K^0 \to \mu^+\mu^-$, not observed at that time. The mystery of the Δ I=1/2 rule in nonleptonic decays of kaons and hyperons is not yet solved.

Meanwhile CP violation was discovered in 1964 in neutral K decays [4]. With six quarks, the mixing matrix allows CP violation, without however telling us why and how much.

To date the most accurate test of CPT comes from the neutral kaon system, even if we do not quite know what sensitivity we should reach for $(\mathrm{m}(K^0) - \mathrm{m}(\bar{K}^0))/\mathrm{m}(K)$. Still the limit is around 10^{-18}–10^{-19}, depending on assumptions. Quite impressive [5].

Since 1964 we were left with the question of whether direct CP violation exists in kaon decay, with a long history of difficult experiments and calculations. Most authors came to the conclusion, a few years ago, that $\Re(\epsilon'/\epsilon)$ is very small indeed, with a few discordant voices. As a result, the recent announcements of a nonzero value for $\Re(\epsilon'/\epsilon)$ around 2×10^{-3} have immediately raised speculations on whether we have a signal of new physics. I believe that this is quite premature at this point. The calculations of $\Re(\epsilon'/\epsilon)$ are still incomplete and calculations of A_2/A_0, which are *possibly* connected those of $\Re(\epsilon'/\epsilon)$, are just beginning to get the right answer. And we still have a poor knowledge of the values of most parameters of the mixing matrix.

However, the real interesting fact is that $\Re(\epsilon'/\epsilon)$ is truly measurable today and in the near future we will know its value to 5–10%; three experiments are working on it. I hope that this fact will encourage more complete calculations and therefore allow the use of the measured value of $\Re(\epsilon'/\epsilon)$ in extracting the mixing matrix parameters, which are still rather poorly known.

CP studies in the B system, soon to come, will add much information but will not close nor overdetermine the *unitarity triangle* as initially expected or, to say the least, without calculations that are not so different from those required for $\Re(\epsilon'/\epsilon)$. It appears that all CP violation (CP̸) experiments are difficult. Lots of time is necessary to accumulate good data, to analyze them, to understand systematics.

Very interesting are the final results from Run I of CDF [6], which for the first time measures a value of $\sin 2\beta$ that is different from zero by 2 standard deviations. It's quite likely that the better results for CP̸ in the B system will come from hadron colliders. The so-called Run II at the Tevatron begins in about a year, and both CDF and DØ will take data. BTeV, if approved, will be on its way. Somewhat farther in the future we will have ATLAS, CMS, and LHCB at the LHC, who will all also be able do something with strange Bs.

$\Re(\epsilon'/\epsilon)$ will probably never be measured again, after the present round of precision experiments, which will take maybe two more years of work. Kaon physics, however, has not finished its program and will continue to remain important in the pre-LHC era. KTeV and NA48 are beginning to obtain beautiful results in rare K decays, and KTeV can still improve their limits by a factor of 10. The BNL experiments are reaching their sensitivity goals of $\mathcal{O}(10^{-11})$ for branching ratios.

We are quite ready to attack the unique opportunity of kaon physics, *i.e.*, to measure the branching ratio for $K \to \pi^0 \nu\bar{\nu}$. Too many people have discussed this point in the last few days to belabor it here. However, reaching a sensitivity of 10^{-13} appears feasible. Experiments of this kind do still require some learning as well as R&D, but the sensitivity above allows the measurement of the decay rate to an accuracy of 10% and therefore to determine η to $\mathcal{O}(5\%)$, by far the most precise measurement of the Wolfenstein parameter η in the foreseeable future.

Considering the efforts and money being invested in B physics, it would not be very wise to forget this potential contribution of kaons to the understanding of CP. A well-supported five-year program would give the accuracy required. Ultimately, in order to truly test the viability of the standard model, one has to perform as complete a set of measurements as possible, helped by precise calculations, both in the K system and in the B system. Only after comparison of the two complete and complementary results will we be able to conclude either that the standard model wins as usual or that the new physics signal is truly there.

References

[1] L. Leprince-Ringuet and M. l'Heritier, Comptes Rendus Acad. Science de Paris, séance du 13 Dec. 1944, p. 618.

[2] R.H. Dalitz, this volume, chapter 1.

[3] G.D. Rochester and C.C. Butler, Nature **160**, 855 (1947).

[4] J. Christenson, J. Cronin, V. Fitch and R. Turlay, Phys. Rev. Lett. **13**, 138 (1964).

[5] See, however, V.A. Kostelecký, this volume, chapter 23.

[6] I.J. Kroll, this volume, chapter 51.

61

Future Opportunities in K Physics

Frederick J. Gilman

61.1 Theoretical Background

The standard model has a natural place for CP violation in the Cabibbo-Kobayashi-Maskawa (CKM) matrix [1]. Flavor-changing neutral-current (FCNC) processes, while absent at tree level, are generated through loops.

We know that all three angles that characterize the CKM matrix are nonzero; there is every expectation that the single nontrivial phase should be nonzero as well. These yield both a natural scale and a special pattern for FCNC processes and for CP-violating effects in K, D, and B decays.

Thus, the question is not, as it once was, "Where does CP violation come from?" Rather, the question is "Does the single phase in the CKM matrix of the standard model provide the sole source of CP violation?" Put in a more exploratory fashion, we ask "Can we find FCNC processes and CP-violating effects from physics beyond the standard model?" New physics will generally bring in its own characteristic phases and/or have a different weighting of the same (CKM) phase to various low-energy processes.

There are many ways to test the standard model and to look for new physics; K experiments should be looked at in the more general context of FCNC and CP violation in transitions among the charge $-1/3$ quarks:

- $b \leftrightarrow d$, which in the standard model has an amplitude dominated by diagrams with a virtual top quark that brings in a factor $V_{tb}V_{td}^*$, and is typically studied in the B system;
- $b \leftrightarrow s$, which in the standard model has an amplitude dominated by diagrams with a virtual top quark that brings in a factor $V_{tb}V_{ts}^*$, and is also typically studied in the B system;
- $s \leftrightarrow d$, which in the standard model has an amplitude dominated by diagrams with virtual charm and top quarks that bring in factors of $V_{cs}V_{cd}^*$ and $V_{ts}V_{td}^*$, respectively, and is typically studied in the K system.

Each of these transitions can be looked upon as the side of a quark-flavor triangle with s, d, and b quarks at the vertices. Whether probing the standard model or new physics, we seek to understand the physics through its footprint on each of these transitions and through the consistency of determinations of the same fundamental parameter(s) in more than one process.

61.2 U.S. Plans

The report [2] of the HEPAP subpanel that I chaired to plan for the next decade of U.S. high-energy physics recognized that "intense K and muon beams offer the possibility of adding greatly to our understanding of rare quark and lepton flavor-changing transitions and of CP violation." The subpanel was charged with making a recommendation on use of the AGS at Brookhaven for high-energy physics in the RHIC era, but it considered this in the more general context of understanding the physics of flavor and pushing the "precision frontier." Aside from a general recognition of the importance of this part of the U. S. high-energy physics program, the subpanel's recommendations that specifically impact on K physics were that:

- the possibility be held open for experiments that use AGS beams at BNL after RHIC operation begins;
- a single advisory body look at which experiments be approved when they might be carried out at either BNL or Fermilab.

61.3 Some Specific Processes

Several K physics processes discussed at this meeting are of special note with regard to the remarks above. A very personal list as we look out over the next several years is:

- ϵ'/ϵ: It appears that the twenty-year roller coaster ride to establish experimentally that ϵ' is nonzero is finally over [3]! Furthermore, stepping back a bit to gain some theoretical perspective, the sign and rough magnitude are in accord with the standard model. Theory, particularly because of the difficulties in calculating hadronic matrix elements of certain operators, is not yet able to deliver a stark confrontation with experiment [4], although the indications [5] coming from some of the calculations point toward an enhancement of the matrix elements of penguin-related operators so as to make agreement of the standard model with experiment more likely. Theorists (which mostly means those doing the next-generation lattice calculations) have much hard work ahead to make a fully quantitative comparison and perhaps even to convert this into one of the better tests of the standard model.

- $K^+ \to \pi^+\nu\bar{\nu}$: Here we are nearing the moment of truth as the experimental sensitivity [6] has roughly reached the level of standard model expectations. There is every reason to pursue this to about a 10% measurement of the branching ratio, something that now seems doable in the next several years.

- $K_L \to \pi^0\nu\bar{\nu}$: The amplitude for this process, whose theoretical prediction is especially clean [7], is proportional to $\mathrm{Im}(V_{td}V_{ts}{}^*)$, which, aside from a factor of the sine of the Cabibbo angle, is just twice the area of the oft-noted unitarity triangle, *i.e.*, its magnitude is that of the invariant measure of CP violation [8], or what I used to call the "price of CP violation in the standard model." Thus, for theorists, $K_L \to \pi^0\nu\bar{\nu}$ is a "must-do." For experimentalists, the question is whether it is a "can-do." If it can be done, it should be given high priority.

- $K_L \to \pi^0\ell^+\ell^-$: As I work down my list, things are getting more difficult experimentally. I must admit to a bit of discouragement here. Twenty years ago, after Mark Wise and I carried out our investigation of ϵ'/ϵ, we did a related renormalization group analysis [9] of $K_L \to \pi^0\ell^+\ell^-$ in the six-quark standard model. We found that one would likely have to deal with roughly comparable contributions from the direct CP-violating amplitude in which we were particularly interested, the indirect CP-violating amplitude proportional to ϵ times the $K_S \to \pi^0\ell^+\ell^-$ amplitude, and the CP-conserving amplitude involving two intermediate photons. What we did not know is that unavoidable experimental backgrounds [10] appear to overwhelm the potential signal for $K_L \to \pi^0e^+e^-$ and make its observation unlikely. I would still like to see us push on $K_L \to \pi^0\mu^+\mu^-$, where the corresponding backgrounds should be much less, and to make a direct measurement of $K_S \to \pi^0\mu^+\mu^-$ in order to fix, independent of any theoretical calculation, the magnitude of the indirect CP-violating amplitude.

Acknowledgment

This work was supported in part by Department of Energy contract DE-FG02-91ER40682 .

References

[1] N. Cabibbo, Phys. Rev. Lett. **10**, 531 (1963); M. Kobayashi and T. Maskawa, Prog. Theo. Phys. **49**, 652 (1973).

[2] HEPAP Subpanel Report on Planning for the Future of U.S. High-Energy Physics, Department of Energy Report DOE/ER-0718, February 1998.

[3] Y.B. Hsiung, this volume, chapter 6; M.S. Sozzi, this volume, chapter 8.

[4] See, for example, the ϵ'/ϵ panel at this conference: W.A. Bardeen, this volume, chapter 14; S. Bertolini, this volume, chapter 12; A.J Buras, this volume, chapter 5; and M. Ciuchini *et al.*, this volume, chapter 28.

[5] See ref. [4] and T. Hambye and P.H. Soldan, this volume, chapter 11; M. Ciuchini *et al.*, this volume, chapter 31; and S. Ryan, this volume, chapter 29.

[6] G. Redlinger, this volume, chapter 34.

[7] The clean theoretical prediction and the importance of measuring this process has been repeatedly stressed in recent years by Buras and collaborators; see for example G. Buchalla and A.J. Buras, Phys. Rev. D **54**, 6782 (1996).

[8] C. Jarlskog, Phys. Rev. Lett. **55**, 1039 (1985); Z. Phys. C **29**, 491 (1985).

[9] F.J. Gilman and M.B. Wise, Phys. Rev. D **21**, 3150 (1980).

[10] J. Whitmore, this volume, chapter 39.

Part X

Summary and Outlook

62

KAON 99: Summary and Perspective

William J. Marciano

Abstract

An overview of KAON 99 with commentary is presented. Emphasis is placed on the state of CKM mixing and CP violation. The Jarlskog invariant, J_{CP}, is shown to provide a useful quantitative comparison of K and B phenomenology. The potential of future rare and "forbidden" decay experiments to probe $\mathcal{O}(3000\ \text{TeV})$ "new physics" is also described.

62.1 Conference Overview and Commentary

For more than 50 years, kaon physics has played a leading role in unveiling nature's fundamental intricacies and challenging our creative imaginations [1,2]. The concept of hadronic "flavor" has its roots in the associated production of kaons and introduction of "strangeness" as a nearly conserved quantum number. $SU(3)_F$, current algebra, and the quark model all stemmed, to a large extent, from extensive follow-up studies of that discovery.

The θ-τ puzzle in $K \to 2\pi$ and 3π (final states with different parities) provided the stimulus for Lee and Yang's parity violation conjecture. Today, we easily accommodate parity violation via the chiral nature of the standard model's $SU(2)_L \times U(1)_Y$ local gauge symmetry. However, that left-right asymmetry remains a deep fundamental mystery with potentially profound implications about the short-distance properties of space-time and origin of mass.

Early null results in rare K decay searches also led to important physics insights. The observed suppression of flavor-changing neutral currents (FCNC) in $K_L \to \mu^+\mu^-$, $K \to \pi\nu\bar{\nu}$, etc. motivated the GIM (Glashow-Iliopoulos-Maiani) mechanism and introduction of charm. Today, medium rare $\mathcal{O}(10^{-8})$, branching ratios such as $K_L \to \mu^+\mu^-$ are routinely measured with high precision and used to search for or constrain potential "new physics" effects.

The special (unique) $\Delta S = 2$ mixing features of the K^0-$\bar{K}^0$ system allowed CP violation to be unveiled in $K_L \to 2\pi$ decays. To explain that enigmatic effect, Kobayashi and Maskawa (KM) boldly proposed [3] the now-discovered third generation of quarks, t and b. Their parametrization of CP violation via angles and phases in a unitary 3×3 quark mixing matrix provided a simple but elegant solution to that outstanding puzzle. It also suggested many interesting predictions for FCNC and direct CP violation effects [4]. The recent measurement of ϵ'/ϵ in $K \to 2\pi$ and initial studies of $B \to J/\psi K_s$ lend strong support to their hypothesis.

In a sense, the KM model of CP violation trivialized that previously mysterious phenomenon. It suggested that a mere nonvanishing weak interaction phase and quark mixing, rather than some new superweak interaction, was responsible for CP violation. That beautiful solution now seems almost obvious. Also, if additional new interactions are eventually uncovered, it seems likely that they will similarly have relative phases that would provide additional sources of CP violation. That would be a welcome discovery, since electroweak baryogenesis [5] seems to require additional CP violation beyond the standard model.

Given its already rich and glorious history, what more can we hope to learn from K decays? Are kaon studies passe, or competitive with B physics and other ways to investigate CP violation?

This conference is proof of the excitement K physics continues to generate. Its copious production cross-section and relative long lifetime make the kaon very special and experimentally popular worldwide. Indeed, there are many ongoing diverse experimental programs at labs around the world along with exciting ideas and proposals for new initiatives. I give in table 62.1 a list of the kaon programs discussed at this meeting. Experiments at those facilities measure CKM (Cabibbo-Kobayashi-Maskawa) matrix elements, probe CP and possible CPT violation, search for very rare or forbidden decays, etc. In addition, they thoroughly study medium rare decays and other properties of kaons, thus providing an arena for refining theoretical skills such as chiral perturbation theory, lattice techniques, large-N_c approaches, and perturbative QCD.

Program	Speakers
KEK 12 GeV PS $\to$ 50 GeV PS	T. Inagaki, G. Y. Lim
BNL 30 GeV AGS + Booster	L. Littenberg, W. Molzon, M. Zeller
FNAL KTEV $\to$ KAMI	P. Cooper, J. Whitmore
CERN SPS	G. Kalmus, L. Köpke
CP LEAR (Completed)	P. Bloch
Frascati-DAΦNE	P. Franzini, S. Di Falco
Novosibirsk	L. Landsberg, N. Ryskulov

Table 62.1: Ongoing and future kaon physics programs reported on at this meeting

With regard to determining the CKM quark mixing matrix, $V_{\rm CKM}$, K

and B measurements both play special key roles. Their importance is well illustrated by the Wolfenstein parametrization [6]

$$V_{\mathrm{CKM}} = \begin{pmatrix} V_{ud} & V_{us} & V_{ub} \\ V_{cd} & V_{cs} & V_{cb} \\ V_{td} & V_{ts} & V_{tb} \end{pmatrix} \tag{62.1}$$

$$= \begin{pmatrix} 1-\frac{\lambda^2}{2} & \lambda & A\lambda^3(\rho - i\eta) \\ -\lambda & 1-\frac{\lambda^2}{2} & A\lambda^2 \\ A\lambda^3(1-\rho-i\eta) & -A\lambda^2 & 1 \end{pmatrix} + \mathcal{O}(\lambda^4)$$

One would like to measure λ, A, ρ, and η as precisely and with as much redundancy as possible. In that way, the unitarity conditions

$$\sum_i V_{ij}V_{ik}^* = \sum_i V_{ji}^* V_{ki} = \delta_{jk} \tag{62.2}$$

can be tested. A deviation from expectations in any mode would signal new physics. Let me discuss some important experiments.

The theoretically cleanest direct measurement of the cornerstone CKM parameter, λ, comes from K_{e3} decays ($K \to \pi e \nu$) [7]

$$\lambda = 0.2196 \pm 0.0023 \quad (K_{e3}) \tag{62.3}$$

where the theoretical and experimental uncertainties (added in quadrature) are comparable. That value is to be compared with results from Hyperon and nuclear beta decays

$$\lambda = 0.226 \pm 0.003 \qquad \text{(hyperons)}, \tag{62.4}$$

$$\lambda = 0.2265 \pm 0.0026 \qquad (\beta\text{decay}). \tag{62.5}$$

There is some inconsistency. Hopefully, ongoing efforts at BNL, FNAL, and Novosibirsk to remeasure the K_{e3} decay rates for both K^+ and K_L will help clarify the situation.

The parameter A is obtained from V_{cb} as measured in semileptonic B decays (the counterpart of K_{e3}). Currently, one finds [8]

$$A = 0.83 \pm 0.05. \tag{62.6}$$

In Ligeti's talk it was suggested that ongoing and future studies of $B \to D^* e\nu$ decays may lead to a reduction in the A uncertainty by a factor of 2 or 3 during the next 3–5 years. Such improvement would be a welcome advancement.

The ρ and η parameters are constrained by a combination of K and B measurements. For example, within the standard model, the CP-violating mixing parameter $|\epsilon| = 2.28(1) \times 10^{-3}$ provides a determination of the combination

$$A^4\lambda^{10}\eta(1-\rho+0.44) \simeq 5.6(1.1) \times 10^{-8}, \tag{62.7}$$

where the error is primarily due to the K^0-$\bar{K}^0$ matrix element uncertainty. For a given A and λ, that constraint leads to a hyperbola in the ρ, η plane. However, the $\pm 20\%$ uncertainty in (62.7) is amplified by the current $\pm 29\%$ uncertainty in $A^4\lambda^{10}$. So, the ρ, η plane is not very favorable for displaying constraints from K decays. In contrast, it reduces the uncertainties in constraints from B decays, thus presenting them in a very favorable light. K decay presentations should resist the lure of the ρ, η plane.

B physics already provides some powerful constraints on ρ and η. Most useful is the ratio $\Gamma(b \to u)/\Gamma(b \to c)$, which implies the relatively narrow band [9]

$$(\rho^2 + \eta^2)^{1/2} = 0.363 \pm 0.073. \tag{62.8}$$

Taken together with the B_d^0-$\bar{B}_d^0$ mixing constraint $|1 - \rho - i\eta| = 1.01 \pm 0.22$ and eq. (62.7), it suggests (roughly)

$$\rho \simeq 0.13, \quad \eta \simeq 0.34. \tag{62.9}$$

Other constraints from B_s^0-$\bar{B}_s^0$ mixing and $B \to J/\psi K_s$ are consistent with values in that general region. Overall, there is good support for CKM mixing and unitarity.

Given the success of the CKM model, what more can we learn from CP violation, further CKM studies, and K decays? There are compelling reasons to push those efforts further. Precision studies of CKM elements can not only further confirm the standard model, but can help explain the origin of electroweak mass and perhaps help uncover new physics. Indeed, as I will later demonstrate, CP violation and FCNC measurements are sensitive to effects originating from scales as high as 3000 TeV!

If a true deviation from the standard model in K decays or other rare reactions is uncovered, there will certainly not be a lack of interesting explanations. SUSY [10, 11], dynamical symmetry breaking, large extra dimensions, etc. can potentially provide new significant sources of CP violation and FCNC effects.

62.2 New CP Violation Results

The most exciting kaon physics announcement of 1999 was the measurement of Re ϵ'/ϵ by KTeV [12] and NA48 [13]. Taken together with earlier studies, those new results

$$\begin{aligned} \mathrm{Re}\epsilon'/\epsilon \;=\; & 23.0 \pm 6.5 \times 10^{-4} \quad \text{NA31} \\ & 7.4 \pm 5.9 \times 10^{-4} \quad \text{FNAL} \\ & 28.0 \pm 4.1 \times 10^{-4} \quad K\text{TeV} \\ & 18.5 \pm 7.3 \times 10^{-4} \quad \text{NA48} \end{aligned} \tag{62.10}$$

give an average (with PDG expanded error) [13]

$$(\mathrm{Re}\epsilon'/\epsilon)_{\mathrm{Ave}} = 21.2 \pm 4.6 \times 10^{-4}. \tag{62.11}$$

That rather solid observation of direct CP violation rules out (old) superweak models. Is it consistent with CKM expectations? Pre-1999, the main theory predictions were (labeled by their home cities) [14, 15]

$$\begin{aligned}(\mathrm{Re}\epsilon'/\epsilon)_{\mathrm{Theory}} &= 4.6 \pm 3.0 \pm 0.4 \times 10^{-4} \quad \text{(Rome)} \\ &= \left.\begin{array}{c} 3.6 \pm 3.4 \times 10^{-4} \\ (10.4 \pm 8.3) \times 10^{-4} \end{array}\right\} \quad \text{(Munich)} \\ &= 17^{+14}_{-10} \times 10^{-4} \quad \text{(Trieste)}. \end{aligned} \tag{62.12}$$

The broad range of those estimates does not allow for a definitive conclusion. The experimental result does, however, appear to be somewhat high.

Let me comment on the utility of ϵ'/ϵ to probe sources of CP violation beyond the standard model. It is quite conceivable that some part of the experimentally observed ϵ'/ϵ comes from new physics. However, the current theoretical uncertainty of at least ±100% (probably more) makes such an interpretation very premature. Nevertheless, as discussed in Isidori's talk [16], even with that large a theory error one can still obtain interesting constraints on, for example, potentially large new CP-violating $Z_\mu \bar{d}_L \gamma^\mu s_L$ interactions induced by SUSY loops in some models [17].

The ongoing experiments (KTeV and NA48) were, however, designed to reach a $\Delta\epsilon'/\epsilon$ of ± 1–2×10^{-4}, *i.e.* a ±5–10% determination of that important quantity. In addition, the KLOE experiment [18] at Frascati will provide independent confirmation with very different systematic uncertainties. It would be a shame if such elegant measurements could not be fully utilized because of theoretical shortcomings.

To significantly reduce the theoretical uncertainty in ϵ'/ϵ requires a systematic first-principles calculation of the $K \to 2\pi$ amplitudes in, for example, a lattice gauge theory approach. With today's powerful QCD teraflop computers and new theoretical methods such as domain wall fermions, much more precise calculations may, in fact, be possible. Indeed, T. Blum [15] described just such an ongoing effort at the RIKEN BNL Research Center. That collaboration aims for about ±20% theoretical uncertainty. Of course, before any new method is accepted, it must undergo close theoretical scrutiny and pass various consistency checks. For example, it should quantitatively explain the $\Delta I = 1/2$ amplitude enhancement relative to $\Delta I = 3/2$ amplitude in K decays (a factor of 22). Also, it should demonstrate control of isospin-violating effects which can feed $\Delta I = 1/2$ enhancements into the $\Delta I = 3/2$ amplitudes of ϵ'/ϵ. Perhaps most important, as emphasized by Ciuchini *et al.* [15], the lattice approach should be self-contained. Rather than patch together pieces of calculations from other prescriptions, it should be as complete as possible.

If a ±20% theoretical calculation of ϵ'/ϵ is achieved, it will provide a very interesting confrontation with experiment. It would either allow for a powerful precise determination of the standard model CP violation parameter or point to new physics. Either case justifies the effort.

Further confirmation of CKM mixing and CP violation is also starting to come from B decays. (Of course B studies offer tremendous potential for future studies.) CDF [19] has been able to observe an asymmetry in $\overset{(-)}{B} \to J/\psi K_S$. Using a time-integrated sample of 400 events, they have determined β of the unitarity triangle

$$\sin 2\beta = 0.79^{+0.41}_{-0.44}. \tag{62.13}$$

That result is in good accord with standard model expectations (see [20]). Although currently only a 2σ effect, CDF expects to reduce the error in eq. (62.13) by a factor of 5 to ± 0.084. In the longer term, B factories (now up and running), BTeV, and LHC-B hope to achieve ± 0.02 precision. The CDF result indicates that CDF and $D\emptyset$ with their significant upgrades can be expected to play major roles in future b physics.

Other probes of CP violation discussed at this meeting include: (1) measurement of T-odd asymmetries in $K_L \to \pi^+\pi^-e^+e^-$ and $p\bar{p} \to K^\pm\pi^\pm \overset{(-)}{K}{}^0$, (2) search for transverse muon polarization in $K_{\mu 3}$ decay, (3) hyperon decay asymmetries, and (4) electric dipole moments.

The measured experimental 13.6% T-odd asymmetry between the $\pi^+\pi^-$ and e^+e^- planes in $K_L \to \pi^+\pi^-e^+e^-$ observed at Fermilab (see [21]) is in good accord with the 13–14% expectation due to ϵ in the standard model (see [22] and [23]). That result was based on 1811 events at KTeV. Such a large asymmetry in K decays is quite spectacular and was to most people very surprising. Future efforts at KAMI could yield 10^5 decays in that channel and perhaps provide another probe of direct CP violation.

We were also reminded here of an earlier T-odd study from CPLEAR [24]

$$A = \frac{R(\bar{K}^0 \to \pi^-e^+\nu) - R(K^0 \to \pi^+e^-\bar{\nu})}{R(\bar{K}^0 \to \pi^-e^+\nu) + R(K^0 \to \pi^+e^-\bar{\nu})} = 6.6 \pm 1.3 \pm 1.6 \times 10^{-3}. \tag{62.14}$$

That Kabir test is in good agreement with the standard model prediction $A = 6.4 \times 10^{-3}$, again due to ϵ.

G. Y. Lim reported a recent KEK result for the muon transverse polarization in $K_{\mu 3}(K^+ \to \pi^0\mu^+\nu_\mu)$ decay [25]

$$p^T_\mu = \hat{s}_\mu \cdot (\hat{p}_\mu \times \hat{p}_\pi). \tag{62.15}$$

They have reached

$$p^T_\mu = -0.0042 \pm 0.0049 \pm 0.0009 \tag{62.16}$$

and aim for 10^{-3} sensitivity. The standard model predicts $P^T_\mu \sim 0$; so a nonzero experimental result would directly point to a new source of CP violation. The leading candidate would be a charged Higgs exchange amplitude with a relatively large CP-violating phase [26]. Such direct searches for completely new sources of CP violation are extremely important and must be pushed as far as possible. An approved BNL experiment would reach 10^{-4} sensitivity, but unfortunately, it may never get to take data because of uncertainties in future AGS running for fixed target experiments.

Larger than expected CP-violating asymmetries in hyperon decays (see [27]) could also point to new physics. An extensive hyperon decay program is being proposed at Fermilab.

Perhaps the most promising way to uncover new sources of CP violation is the study of electric dipole moments. Such effects are predicted to be nonzero, but unobservably small in the CKM framework. However, new physics of the type needed in some baryogenesis scenarios [5] , for example, could provide much larger edm signals, near the current experimental bounds. In the talk by M. Romalis [28] we heard of ambitious efforts to push the sensitivity for the neutron and electron edm's from $6.3 \times 10^{-26} \rightarrow 10^{-28}e$-cm and $4 \times 10^{-27} \rightarrow 10^{-31}e$-cm respectively. A proposal by the g_μ-2 collaboration at BNL would also greatly extend the search for a muon edm from $10^{-18} \rightarrow 10^{-24}e$-cm. All such advances should be strongly encouraged, since a positive finding would be revolutionary and may, in fact, be just waiting to be unveiled.

A general theoretical framework for discussing CPT and Lorentz invariance violation was given by A. Kostelecký [29]. K physics studies currently provide the most sensitive tests of CPT [30]. Measurements of m_{K^0}-$m_{\bar{K}^0}$ at KTeV, NA48, and CPLEAR have reached the incredible 10^{-18} GeV level and are beginning to approach the interesting m_K^2/m_{planck} sensitivity. Future measurements at Frascati (see S. Di Falco and M. Incagli [30]) will further advance the cause.

62.3 Medium Rare, Rare, and Forbidden Decays

In recent years, great progress has been made in the study of flavor changing neutral current (FCNC) decays of K mesons. Experimental studies have been accompanied by an expansion in our theoretical arsenal of tools that now includes: chiral perturbation theory, large N_c, lattice gauge theories, etc. Together, they have allowed us to test the standard model as well as to probe for and constrain possible new physics. Here, I divide rare decays into three categories: (1) medium rare, which includes roughly 10^{-5}–10^{-9} branching ratios, (2) rare decays with branching ratios $\lesssim 10^{-9}$, and (3) forbidden decays, which do not occur in the standard model. For the last of those, I will discuss only muon-number nonconservation, because it provides such a sensitive probe of new physics.

At this meeting, we heard about many measurements of medium rare decays [31]. In table 62.2, I list some of the results that were discussed. Most impressive to me is the fact that some measurements of historical importance such as $K_L \rightarrow \mu^+\mu^-$ have gone from a handful of events to precision measurements based on 5–10 thousand events [32]. Indeed, they now confront the standard model at its quantum loop level so as to constrain new physics such as SUSY or Technicolor-inspired models. In addition, the abundance of events in $K_L \to \mu^+\mu^-$, $\pi^+\pi^-e^+e^-$, $K^+ \to \pi^+\mu^+\mu^-$, etc. suggest that they may be further used to study CP violation effects in the future. Note also that those measurements have been very useful in fine tuning the parameters of chiral perturbation theory and advancing its techniques (see [33]).

Decay Mode	Branching Ratio	Comments
$K^+ \to \pi^+e^+e^-$	$2.82 \pm 0.04 \pm 0.07 \times 10^{-7}$	BNL E865 (preliminary)
$K^+ \to \pi^+\mu^+\mu^-$	$9.23 \pm 0.6 \pm 0.6 \times 10^{-8}$	
$K^+ \to \pi^+\pi^-e^+\nu_e$	$\sim 3.9 \times 10^{-5}$	(300,000 events)
$K^+ \to \pi^+\pi^0\gamma$	$4.72 \pm 0.77 \times 10^{-6}$	BNL E787
$K_L \to \pi^+\pi^-e^+e^-$	$4.4 \pm 1.3 \pm 0.5 \times 10^{-7}$	KEK E162
$K_L \to e^+e^-\gamma$	$1.06 \pm 0.02 \pm 0.02 \pm 0.04 \times 10^{-5}$	CERN NA48
$K_L \to \mu^+\mu^-$	$7.18 \pm 0.17 \times 10^{-9}$	BNL E871

Table 62.2: Examples of medium rare K decay branching ratios.

Rare K decay experiments have made spectacular progress in measuring incredibly small branching ratios or pushing bounds. D. Ambrose [34] reported the smallest branching ratio ever measured in a decay process

$$B(K_L \to e^+e^-) = 8.7^{+3.7}_{-4.1} \times 10^{-12}. \tag{62.17}$$

That result is in good agreement with the standard model prediction of 9×10^{-12}. It indicates that even such rare decays can be cleanly observed and measured with precision. That bodes well for other more interesting rare decays for which only bounds currently exist [35]

$$\begin{aligned} B(K_L \to \pi^0e^+e^-) &< 5.6 \times 10^{-10} && (a) \\ B(K_L \to \pi^0\mu^+\mu^-) &< 3.4 \times 10^{-10} && (b) \quad (62.18) \\ B(K_L \to \pi^0\nu\bar{\nu}) &< 5.9 \times 10^{-7} && (c) \end{aligned}$$

but are expected to occur at about 5×10^{-12}, 1×10^{-12}, and 3×10^{-11} respectively. Each of those decays provides a nice test of direct CP violation, if a real measurement can be achieved. The golden mode [36] $K_L \to \pi^0\nu\bar{\nu}$ is particularly attractive because it is theoretically pristine (with only about ±1–2% theoretical uncertainty). In fact, as I will subsequently describe, it has the unique potential of determining the extremely important Jarlskog [37] CP-violating parameter J_{CP} at about the ±5% level. Such a measurement is so compelling that it must be carried out if experimentally feasible (more commentary and discussion of experimental goals later).

Similar to $K_L \to \pi^0\nu\bar{\nu}$ is the rare decay $K^+ \to \pi^+\nu\bar{\nu}$ being pursued by the E787 collaboration at BNL. That group saw a single event in its 1995 run. Further analysis, as described by G. Redlinger [38], did not uncover additional candidates. The collaboration has not updated their 1 event branching ratio, but one expects that it now corresponds to about 1.5×10^{-10} with fairly large errors. About 2 times as much data remains to be analyzed, but already the experiment appears to be consistent with the standard model expectation $B(K^+ \to \pi^+ \nu\bar{\nu}) \simeq 0.9 \times 10^{-10}$.

The theoretical error [39] on $B(K^+ \to \pi^+ \nu\bar{\nu})$ due to charm mass and QCD uncertainties is only about $\pm 7\%$. So, it would be extremely useful to measure that branching ratio with a similar $\pm 10\%$ experimental error, both as a means of determining CKM mixing parameters and constraining new physics. E787 could wind up with several events when the analysis is complete. Its approved follow-up E949 at BNL has a goal of 0.8×10^{-11} sensitivity, or about 10 standard model events (about a $\pm 30\%$ determination of $B(K^+ \to \pi^+ \nu\bar{\nu})$). In the longer term, the CKM proposal at Fermilab's KAMI facility would aim for 100 events or $\pm 10\%$. As I will describe later, a $\pm 10\%$ measurement of that important branching ratio will allow new physics to be probed beyond the 1000 TeV (PeV) level!

Muon-number-violating (forbidden) decays have also been searched for with impressive sensitivities. Kaon and muon decays have achieved the bounds [32] given in table 62.3. If no events appear in the ongoing E865 analysis at BNL, the bound on $K^+ \to \pi^+ \mu e$ is expected to reach 8×10^{-12}. Searches for those forbidden K decays could probably be pushed by about another order of magnitude at future high-intensity kaon facilities. However, currently most planning activity involves forbidden muon decays such as $\mu^+ \to e^+ \gamma$ and $\mu^- N \to e^- N$ (coherent muon conversion in muonic atoms) because ideas for extending the current experimental sensitivity by 3 or 4 orders of magnitude exist. (New forbidden decay searches should generally strive for at least 2 orders of magnitude improvement.)

Decay Mode	Current Bound		Future Potential
$B(K_L \to \mu e)$	$< 4.7 \times 10^{-12}$	BNL E781	Probably could be
$B(K^+ \to \pi^+ \mu e)$	$< 4.8 \times 10^{-11}$	BNL E865	pushed to a few
$B(K_L \to \pi^0 \mu e)$	$< 3.2 \times 10^{-9}$	FNAL-KTeV	$\times 10^{-13}$
$B(\mu^+ \to e^+ \gamma)$	$< 1.2 \times 10^{-11}$	MEGA	10^{-14} PSI Proposals
$B(\mu^+ \to e^+ e^- e^+)$	$< 1 \times 10^{-12}$		—
$B(\mu^- N \to e^- N)$	$< 6 \times 10^{-13}$	SINDRUM II	5×10^{-17} MECO at BNL

Table 62.3: Current bounds on muon-number violating decays and future potential.

Coherent muon-electron conversion, $\mu^- N \to e^- N$, is a particularly powerful probe of new physics. Its discovery potential is very robust, including SUSY loops, heavy neutrino mixing, Z' bosons, Multi-Higgs models, compositeness, etc. To demonstrate its reach, consider the muon-number-nonconserving four-fermion interaction

$$\mathcal{L} = \frac{4\pi}{\Lambda^2}\eta_q \bar{e}\gamma_\alpha \mu \bar{q}\gamma^\alpha q \qquad q = u, d, \tag{62.19}$$

where Λ is a generic scale of new physics and η_q represents a model-dependent combination of couplings, mixing parameters, etc. At a sensitivity of 5×10^{-17}, the goal of the proposed MECO experiment at BNL, one is probing (approximately)

$$\Lambda \sim 3000 \text{ TeV} \sqrt{\eta_q}. \tag{62.20}$$

Few experiments are capable of exploring such short-distance scales. Of course, a discovery would be revolutionary. Given its potential, experiments such as MECO must be pushed as far and as soon as possible.

62.4 Quantitative Tests of CKM Unitarity: CP Violation

The 3×3 CKM mixing matrix, V_{CKM}, must be unitary. A convenient parametrization

$$\begin{aligned} V_{CKM} &= \begin{pmatrix} c_1 c_3 & s_1 c_3 & s_3 e^{-i\delta} \\ -s_1 c_2 - c_1 s_2 s_3 e^{i\delta} & c_1 c_2 - s_1 s_2 s_3 e^{i\delta} & s_2 c_3 \\ s_1 s_2 - c_1 c_2 s_3 e^{i\delta} & -c_1 s_2 - s_1 c_2 s_3 e^{i\delta} & c_2 c_3 \end{pmatrix} \\ c_i &= \cos\theta_i \\ s_i &= \sin\theta_i \end{aligned} \tag{62.21}$$

exhibits the features that allow the orthonormal relationships in eq. (62.2) to be satisfied.

One can test the standard model and search for new physics by making clean precision measurements of V_{CKM} elements and seeing if unitarity is satisfied. For example, 4 measurements determine θ_1, θ_2, θ_3, and δ (or Wolfenstein's λ, A, ρ, and η). A fifth measurement then tests unitarity. Alternatively, each of the individual relationships in eq. (62.2) can be tested by 3 (or more) measurements. Unitarity can be tested within K or B decays alone or in comparison with one another. Of course, in all cases theoretical uncertainties should be minimized. Also, it is useful to have as many different consistency checks as possible, since that allows many potential new physics effects to be explored.

CP violation and FCNC effects are particularly good probes of new physics, because the standard model predictions are generally so small. Which system is more sensitive, K or B decays? How does one compare the potential of K and B studies in an unbiased manner? A nice answer is provided by Cecilia Jarlskog's J_{CP} parameter. Let me describe its utility.

The six orthogonal relations in eq. (62.2) with $j \neq k$ give rise to so-called unitarity triangles. I will label the 6 distinct triangles by their (j, k)

indices. The (1,3) or (d, b) triangle

$$V_{ud}V_{ub}^* + V_{cd}V_{cb}^* + V_{td}V_{tb}^* = 0 \tag{62.22}$$

is best known because of its general use in illustrating b physics studies. B programs aim to measure the angles and sides of those triangles in as many ways as possible. A deviation from closure or single inconsistent measurement would signal new physics. In addition, if one factors out $V_{cd}V_{cb}^*$ from that relation, the remaining triangle is nicely illustrated in the ρ, η plane.

In K physics there is also a useful unitarity triangle, the (1,2) or (d, s) relation

$$V_{ud}V_{us}^* + V_{cd}V_{cs}^* + V_{td}V_{ts}^* = 0. \tag{62.23}$$

Both triangles are illustrated in fig. 62.1. The (1,2) triangle has angles near 0 and 90°, which imply very small CP-violating decay asymmetries (in contrast with B decays). Does that make it uninteresting? No. As pointed out by C. Jarlskog, the most interesting feature of any unitarity triangle is its area and that quantity is the same for all 6 triangles.

"All CKM triangles are created equal in area!"

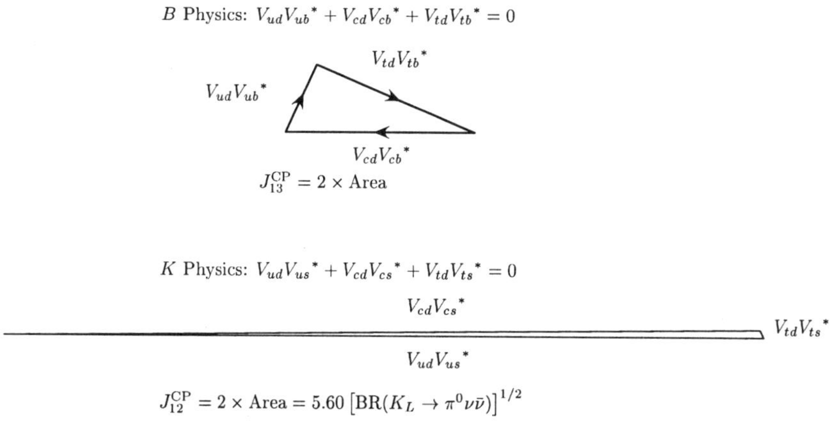

Figure 62.1: Unitarity triangles for B and K studies.

In fact, she observed that a quantity $J_{CP} = 2 \times$ the triangle area was the unique real measure of CP violation in the standard model. Unitarity requires

$$J_{CP} = J_{12} = J_{13} = J_{23} = J_{21} = J_{31} = J_{32}. \tag{62.24}$$

In terms of the parametrizations of eq. (62.21) or eq. (62.1)

$$J_{CP} = s_1 s_2 s_3 c_1 c_2 c_3^2 \sin\delta \simeq A^2\lambda^6\eta. \tag{62.25}$$

Standard model CP violation is tested by measuring J_{CP} as precisely and in as many distinct ways as possible [40]. A deviation would signal new physics.

Currently, a global fit to all K and B studies indicates [17]

$$J_{CP} = 2.7 \pm 1.1 \times 10^{-5}, \tag{62.26}$$

i.e., it is determined (very conservatively) to about ±40%. How well can the next generation of K and B studies individually determine J_{CP}? In the case of B physics, the long term prospects are that J_{13} will be measured to about ±15%. Pushing to ±5% is extremely difficult, but worth trying to achieve.

In the case of K decays, we are extremely fortunate. The decay $K_L \to \pi^0 \nu\bar{\nu}$ directly determines the height of the (1,2) triangle and the base is already well known from β decay and K_{e3} decays. One finds

$$J_{12} = J_{CP} = 5.60[B(K_L \to \pi^0 \nu\bar{\nu})]^{1/2}. \tag{62.27}$$

That result is extremely clean. Theoretical uncertainties are at the level of 1–2%. So the only real limitation is how well $B(K_L \to \pi^0 \nu\bar{\nu})$ can be measured. A proposed measurement at the ±25% level (about 16 events) would determine J_{CP} to about ±12 1/2%, which is better than long-term B physics expectations. In the longer term, a 10% measurement of $B(K_L \to \pi^0 \nu\bar{\nu})$ would give J_{CP} to ±5%. Of course, we need at least 2 measurements of similar precision J_{CP} for comparison; so it would be nice if B efforts could remain competitive.

How might determinations and comparison of J_{12} and J_{13} with high precision be utilized? As a simple illustration, consider a strangeness-changing interaction [41]

$$\mathcal{L} = \frac{4\pi}{\Lambda^2} B \bar{d}_L \gamma_\alpha s_L \bar{\nu}_i \gamma^\alpha \nu_i + h.c. \tag{62.28}$$

due to new physics at scale Λ. A ±10% measurement of $B(K_L \to \pi^0 \nu\bar{\nu})$ would probe $\Lambda \sim 3000$ TeV $(\mathrm{Im}B)^{1/2}$. Note that ±10% precision in $B(K^+ \to \pi^+ \nu\bar{\nu})$ provides similar probing power. Clearly, studies of $K^+ \to \pi^+ \nu\bar{\nu}$ and $K_L \to \pi^0 \nu\bar{\nu}$ must be pushed as far as possible.

62.5 Concluding Remarks (Future Outlook)

Direct CP violation in $K \to 2\pi$ decays has finally been unambiguously observed. Ongoing experimental efforts should eventually determine $\mathrm{Re}\epsilon'/\epsilon$ to ±5–10%. Theoretical calculations must strive to reach a similar level of precision.

B physics has come of age. Studies at CLEO will soon share the spotlight and be challenged by asymmetric *B* factories with CP violation as their primary goal. CDF and *D∅* will also be important players in the future along with LHCB, BTeV, etc.

B studies open a new exciting frontier, but they do not close the door on K or rare muon decays. The kaon system is still the best place to look for CPT violation and the new ϕ factory at Frascati will be at the forefront of that effort. The rare decays $K^+ \to \pi^+\nu\bar{\nu}$ and $K_L \to \pi^0\nu\bar{\nu}$ are exceptionally clean theoretically. Besides testing CKM mixing with great precision, they are capable of probing new physics up to about the 3000 TeV level. The muon-number-violating reaction $\mu^- N \to e^- N$ similarly probes 3000 TeV physics, but in a very different channel. Such outstanding experimental opportunities are extremely scarce. They must be seized and pushed as far as possible.

The decay $K_L \to \pi^0\nu\bar{\nu}$ is very special. It alone can determine the all-important Jarlskog parameter J_{CP} to about $\pm 5\%$ in the long term. It would then set the standard for comparing other manifestations of CP violation in K and B decays. It must be pursued with the same zeal and priority as B physics.

Other rare K decays, $K_L \to \pi^0 e^+ e^-$, $K_L \to \pi^0\mu^+\mu^-$, $K^+ \to \pi^+\mu^+\mu^-$, etc., can also contribute to our understanding of CP violation and search for new physics. Kinematic and polarization asymmetries may be particularly useful in those endeavors.

Kaon physics has had a glorious history. It continues to be exciting (*e.g.*, ϵ'/ϵ, $K^+ \to \pi^+\nu\bar{\nu}$, $K_L \to \pi^0\nu\bar{\nu}$, etc.) Are there any future big surprises or great discoveries waiting still to be uncovered in the kaon system? We will find out only if we continue to expand our efforts and follow our instinct to explore.

References

[1] R.H. Dalitz, this volume, chapter 1.

[2] L. Wolfenstein, this volume, chapter 9.

[3] M. Kobayashi and T. Maskawa, Prog. Theor. Phys. **49**, 652 (1973).

[4] F. Gilman and M. Wise, Phys. Lett. B **83**, 83 (1979).

[5] M. Worah, this volume, chapter 3.

[6] L. Wolfenstein, Phys. Rev. Lett. **51**, 1945 (1983).

[7] H. Leutwyler and M. Roos, Z. Phys. C **25**, 91 (1984).

[8] Z. Ligeti, this volume, chapter 49.

[9] W. Marciano, Nucl. Phys. B (Proc. Suppl.) **59**, 339 (1997).

[10] L. Hall, this volume, chapter 4.

[11] H. Murayama, this volume, chapter 10.

[12] Y.B. Hsiung, this volume, chapter 6.

[13] M.S. Sozzi, this volume, chapter 8.

[14] For a review see S. Bertolini, M. Fabbrichesi, and J. Eeg to be published in Reviews of Modern Physics, hep-ph/9802405.

[15] Theoretical discussions of ϵ'/ϵ were given at this meeting by S. Bertolini, this volume, chapter 12; M. Ciuchini *et al.*, this volume, chapter 28; W.A. Bardeen, this volume, chapter 14; E. de Rafael, this volume, chapter 30; A.J. Buras, this volume, chapter 5; R.D. Peccei, this volume, chapter 2; T. Blum (unpublished); T. Hambye and P.H. Soldan, this volume, chapter 11; G. Buchalla, this volume, chapter 54; and F.J. Gilman, this volume, chapter 61.

[16] G. Isidori, this volume, chapter 33.

[17] A. Buras, Lake Louise Winter Institute lectures, Feb. 1999, hep-ph/9905437.

[18] A. Antonelli, this volume, chapter 7.

[19] I.J. Kroll, this volume, chapter 51.

[20] M. Gronau, this volume, chapter 48.

[21] A. Ledovskoy, this volume, chapter 17.

[22] L.M. Sehgal, this volume, chapter 15.

[23] M.J. Savage, this volume, chapter 16.

[24] P. Bloch, this volume, chapter 20.

[25] G.Y. Lim, this volume, chapter 21.

[26] R.D. Peccei, this volume, chapter 2.

[27] Hyperon physics was discussed in S. Pakvasa, this volume, chapter 42; S.L. White, this volume, chapter 43; and N. Solomey, this volume, chapter 44.

[28] M. Romalis, this volume, chapter 22.

[29] A. Kostelecký, this volume, chapter 23.

[30] Experimental tests of CPT were discussed in S. Di Falco and M. Incagli, this volume, chapter 24; and P. Bloch, this volume, chapter 20.

[31] Medium rare decays were discussed in M.E. Zeller, this volume, chapter 36; T. Nomura, this volume, chapter 19; L. Köpke, this volume, chapter 38; J. Whitmore, this volume, chapter 39; and T.K. Komatsubara, this volume, chapter 40.

[32] W. Molzon, this volume, chapter 35.

[33] J. Bijnens, this volume, chapter 37.

[34] D.A. Ambrose, this volume, chapter 41.

[35] J. Whitmore, this volume, chapter 39.

[36] L. Littenberg, Phys. Rev. D **39**, 3322 (1989).

[37] C. Jarlskog, Phys. Rev. Lett. **55**, 1039 (1985).

[38] G. Redlinger, this volume, chapter 34.

[39] G. Buchalla and A. Buras, Nucl. Phys. B **412**, 106 (1994).

[40] A similar discussion concerning J_{CP} has been given by G. Buchalla and A. Buras, Phys. Rev. D **54**, 6782 (1996).

[41] W. Marciano and Z. Parsa, Phys. Rev. **53**, R1 (1996).

Contributors

David A. Ambrose
Department of Physics and Astronomy
University of Pennsylvania
Philadelphia, PA 19104

Antonella Antonelli
I.N.F.N. – Laboratori Nazionali di Frascati
P.O. Box 13
I-00044 Frascati (Rome), Italy

William A. Bardeen
Fermilab, MS 106
P. O. Box 500
Batavia, IL 60510

Stefano Bertolini
INFN, Sezione di Trieste
Scuola Internazionale Superiore di Studi Avanzati
via Beirut 4, I-34013 Trieste, Italy

Johan Bijnens
Department of Theoretical Physics 2
Lund University
Sölvegatan 14A, S-22362 Lund, Sweden

Philippe Bloch
CERN
CH-1211 Geneva 23, Switzerland

Roy A. Briere
Department of Physics
Harvard University
Cambridge, MA 02138

Gerhard Buchalla
Theory Division, CERN
CH-1211 Geneva 23, Switzerland

Andrzej J. Buras
Technische Universität München
Physik Department T30
D-85747 Garching, Germany

Marco Ciuchini
Dip. di Fisica, Università di Roma Tre
and I.N.F.N., Sezione di Roma III
via della Vasca Navale 84
I-00146 Rome, Italy

Peter S. Cooper
Fermi National Accelerator Laboratory, MS 122
P.O. Box 500
Batavia, IL 60510

Richard H. Dalitz
Department of Theoretical Physics
Oxford University
Oxford, U.K.

Eduardo de Rafael
CPT, CNRS–Luminy
Case 907 F-13288
Marseille Cedex 9, France

Stefano Di Falco
I.N.F.N., Sezione di Pisa
via Livornese 1291
S. Piero a Grado (PI)
I-56010 Pisa, Italy

Enrico Franco
Dipartimento di Fisica
Università di Roma "La Sapienza"
and I.N.F.N., Sezione di Roma
P.le A. Moro 2
I-00185 Rome, Italy

Paolo Franzini
Dipartimento di Fisica
Università di Roma "La Sapienza"
P.le A. Moro 2
I-00185 Rome, Italy

Frederick J. Gilman
Department of Physics
Carnegie Mellon University
Pittsburgh, PA 15213

Leonardo Giusti
Department of Physics
Boston University
Boston, MA 02215

Michael Gronau
Department of Physics
Technion—Israel Institute of Technology
Haifa 32000, Israel

Rajan Gupta
Theoretical Division, MS-B285
Los Alamos National Laboratory
Los Alamos, NM 87545

Lawrence Hall
Department of Physics
University of California
Berkeley, CA 94720
and
Theory Group
Lawrence Berkeley Laboratory
Berkeley, CA 94720

Thomas Hambye
I.N.F.N.—Laboratori Nazionali di Frascati
P. O. Box 13
I-00044 Frascati (Rome), Italy

Yee Bob Hsiung
Fermi National Accelerator Laboratory, MS 122
P. O. Box 500
Batavia, IL 60510

Takao Inagaki
Institute of Particle and Nuclear Studies
High Energy Accelerator Research Organization, KEK
1-1 Oho, Tsukuba, Ibaraki 305-0801, Japan

Marco Incagli
I.N.F.N., Sezione di Pisa
via Livornese 1291
S. Piero a Grado (PI)
I-56010 Pisa, Italy

Gino Isidori
I.N.F.N.—Laboratori Nazionali di Frascati
P. O. Box 13
I-00044 Frascati (Rome), Italy

George Kalmus
Rutherford Appleton Laboratory
Chilton, Didcot, Oxon. OX11 0QX, England

Kay Kinoshita
Department of Physics
University of Cincinnati
P. O. Box 0011
Cincinnati, OH 45221-0011

Takeshi K. Komatsubara
KEK, Tanashi Branch
Tokyo 188-8501, Japan

Lutz Köpke
Institut für Physik
Johannes-Gutenberg Universität
D-55099 Mainz, Germany

V. Alan Kostelecký
Physics Department
Indiana University
Bloomington, IN 47405

I. Joseph Kroll
Department of Physics and Astronomy
University of Pennsylvania
Philadelphia, PA 19104

Leonid G. Landsberg
Institute for High Energy Physics
Protvino, 142280, Russia

Alexander Ledovskoy
Department of Physics
University of Virginia
Charlottesville, VA 22901

Zoltan Ligeti
Theory Group, Fermilab
P. O. Box 500
Batavia, IL 60510

Gei Youb Lim
Institute of Particle and Nuclear Studies
High Energy Accelerator Research Organization, KEK
1-1 Oho, Tsukuba, Ibaraki 305-0801, Japan

Laurence Littenberg
Brookhaven National Laboratory
Upton, NY 11973

Vittorio Lubicz
Dip. di Fisica, Università di Roma Tre
and I.N.F.N., Sezione di Roma III
via della Vasca Navale 84
I-00146 Rome, Italy

William J. Marciano
Brookhaven National Laboratory
Upton, NY 11973

Guido Martinelli
Dipartimento di Fisica
Università di Roma "La Sapienza"
and I.N.F.N., Sezione di Roma
P.le A. Moro 2
I-00185 Rome, Italy

William Molzon
Department of Physics
University of California
Irvine, CA 92697-4575

Hitoshi Murayama
Department of Physics
University of California
Berkeley, CA 94720
and
Theory Group
Lawrence Berkeley Laboratory
Berkeley, CA 94720

Harry N. Nelson
Physics Department
University of California
Santa Barbara, CA 93106-9530

Tadashi Nomura
Department of Physics
Kyoto University
Kyoto 606-8502, Japan

Vladimir F. Obraztsov
Institute for High Energy Physics
Protvino, 142280, Russia

Sandip Pakvasa
Department of Physics and Astronomy
University of Hawaii
Honolulu, HI 96822

Roberto D. Peccei
Department of Physics and Astronomy
University of California at Los Angeles
Los Angeles, CA 90095-1547

Daniele Pedrini
I.N.F.N., Sezione di Milano
Via Celoria 16
I-20133 Milano, Italy

Alexey A. Petrov
Department of Physics and Astronomy
Johns Hopkins University
Baltimore, MD 21218

Milind V. Purohit
Department of Physics and Astronomy
University of South Carolina
Columbia, SC 29208

George Redlinger
Brookhaven National Laboratory
Upton, NY 11973

Michael Romalis
Department of Physics
University of Washington
Seattle, WA 98195-1560

Sinéad Ryan
Theory Group, Fermilab
P.O. Box 500
Batavia, IL 60510

Nikolay M. Ryskulov
Budker Institute of Nuclear Physics
Novosibirsk, 630090, Russia

Martin J. Savage
Department of Physics
Box 351560
University of Washington
Seattle, WA 98195-1560

Lalit M. Sehgal
Institute of Theoretical Physics
RWTH Aachen
D-52056 Aachen, Germany

Luca Silvestrini
Technische Universität München
Physik Department T30
D-85747 Garching, Germany

Peter H. Soldan
Institut für Physik
Universität Dortmund
D-44221 Dortmund, Germany

Evgeny P. Solodov
CMD-2 Collaboration
Budker Institute of Nuclear Physics
Novosibirsk, 630090, Russia

Nickolas Solomey
Enrico Fermi Institute
University of Chicago
Chicago, IL 60637

Marco S. Sozzi
Scuola Normale Superiore
Piazza dei Cavalieri 7
I-56126 Pisa, Italy

Erich W. Varnes
Stanford Linear Accelerator Center
P.O. Box 4349
Stanford, CA 94309-4349

Sharon L. White
Department of Physics
Illinois Institute of Technology
Chicago, IL 60616

Juliana Whitmore
Fermi National Accelerator Laboratory
Batavia, IL 60510

Lincoln Wolfenstein
Department of Physics
Carnegie Mellon University
Pittsburgh, PA 15213

Mihir P. Worah
Department of Physics
University of California
Berkeley, CA 94720